Second Edition

Plant Propagation Concepts and Laboratory Exercises

Second Edition

Plant Propagation Concepts and Laboratory Exercises

Edited by **Caula A. Beyl and Robert N. Trigiano**

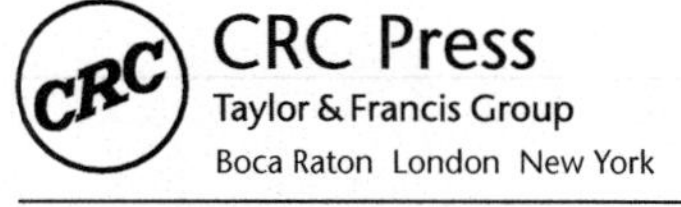

CRC Press is an imprint of the
Taylor & Francis Group, an **informa** business

CRC Press
Taylor & Francis Group
6000 Broken Sound Parkway NW, Suite 300
Boca Raton, FL 33487-2742

Printed on acid-free paper
Version Date: 20140317

International Standard Book Number-13: 978-1-4665-0387-8 (Paperback)

Library of Congress Cataloging-in-Publication Data

Plant propagation concepts and laboratory exercises / editors: Caula A. Beyl, Robert N. Trigiano. -- Second edition.
pages cm
Includes bibliographical references and index.
ISBN 978-1-4665-0387-8 (pbk. : alk. paper)
1. Plant propagation. 2. Plant propagation--Laboratory manuals. I. Beyl, Caula A. II. Trigiano, R. N. (Robert Nicholas), 1953-

SB119.P56 2014
631.5'3--dc23 2014007204

Visit the Taylor & Francis Web site at
http://www.taylorandfrancis.com

and the CRC Press Web site at
http://www.crcpress.com

Contents

Acknowledgments

Foremost, just as in the first edition (2008), we wish to recognize the extraordinary efforts and talents of all the contributing authors—their creativity, support, advice, understanding, and patience throughout the conception and long-delayed development of the second edition of this book was phenomenal—and the Institute of Agriculture at the University of Tennessee for granting us the time and the financial support necessary to complete the book. We also thank our colleagues and students who suggested changes to the chapters and the format; which have improved the content and presentation of materials in the book. We also extend very special thanks to Randy Brehm, senior editor and good friend at CRC Press, who worked tirelessly to see this project through to completion and never lost confidence in us and our contributors in finishing the task. Lastly, RNT sincerely thanks PKT, RLB, CGS, ANT, DJG, and BHO for their advice, insight, friendship, love, and constant support and encouragement—RNT could have never done it without you and you have all made his life much richer. CAB thanks CAC for his talent, JPS for the careful reading, and her family for their unflinching support.

Editors

Caula A. Beyl, dean of the College of Agricultural Sciences and Natural Resources, received a bachelor's degree in biology with majors in botany and zoology from Florida Atlantic University, Boca Raton, Florida, in 1974, and a master's degree in horticulture from Purdue University in West Lafayette, Indiana, in 1977. In 1979, she obtained a PhD from Purdue University in the area of stress physiology. She began her academic career as a postdoctoral researcher at Alabama A&M University, Huntsville, Alabama, and after a year, she joined the faculty in horticulture in the Department of Natural Resources and Environmental Studies. She was promoted to associate professor in 1987 and professor in 1992. Beyl provided leadership as a director of the Center for Plant Science within the Department of Plant and Soil Science from 1988 to 2002. In more than 33 years as a researcher, teacher, and administrator, she has served as a principal investigator or co-investigator on 42 funded research projects in various areas of horticulture, stress physiology, and space biology. She has been a major advisor to 8 doctoral candidates and 28 master's students, half of whom were minority students. Her research and academic work has resulted in 40 refereed research publications; 16 book chapters; and 117 abstracts, presentations, or proceedings, 19 of which were on institutional research topics. Beyl is a member of the American Society for Horticultural Science and the honor society of Gamma Sigma Delta. She served as the editor for the *Plant Growth Regulation Quarterly* from 2001 to 2005 and the associate editor in the area of environmental stress physiology for the *Journal of the American Society for Horticultural Science* from 2000 to 2002. In 1995, she received the School of Agriculture Outstanding Researcher award at Alabama A&M University and, in 1998, the Abbott award for outstanding research paper from the Plant Growth Regulation Society. In 2005, she won the Alabama A&M University Researcher of the Month award. As an undergraduate and graduate educator, Beyl has taught 14 different courses including agricultural leadership and a graduate-level scientific writing course, and was honored for outstanding teaching with the Alabama A&M University Outstanding Teacher award in 1998 and the School of Agriculture Outstanding Teacher award in 1998 and 1991. In 2003, she was the recipient of the Distinguished Alumna award from the horticulture program, Purdue University.

In 2002, she assumed the role of director of the Office of Institutional Planning, Research, and Evaluation. As such, she revitalized the office and helped revise and write new strategic and effectiveness documents for the university, and was essential in guiding Alabama A&M University through its reaffirmation of accreditation process, including the development of a Quality Enhancement Plan. The planning documents that she helped develop were honored by the Southern Association of Institutional Researchers with the Outstanding Planning Document award in 2004. She has served as a consultant to Bethune Cookman College, Bennett College for Women, and Southern University at Baton Rouge for SACS, QEP, and learning outcome development. She also served as the Quality Enhancement Plan evaluator for SACS during its on-site visit to the University of Central Florida. She has presented on a variety of retention, institutional planning, learning outcomes, and QEP development topics to NACDRAO, SAIR, SEF, and SACS, among others. She served as an Alabama Baldrige quality award examiner in 2006. In 2007, she was appointed interim dean of graduate studies at Alabama A&M University. Since joining the University of Tennessee in June of 2007 as the dean of the College of Agricultural Sciences and Natural Resources, she has been active in implementing strategies to focus on retention, recruitment, and diversity issues as well as in integrating technological paradigms into the classroom to enhance learning of the millennial generation. In 2008, she received the Outstanding College of Agriculture Alumna award from Purdue University, and she was inducted into the Alumni Hall of Fame of Florida Atlantic University. In the same year, she was elected fellow in the American Society for Horticultural Science and, in 2009, was honored with their Outstanding Graduate Educator award. As the dean of the College of Agricultural Sciences and Natural Resources, her time is spent developing global engagement for the college, teaching, guiding the college through its strategic planning process, and writing.

Robert N. Trigiano received a bachelor's degree with an emphasis in biology and chemistry from Juniata College, Huntingdon, Pennsylvania, in 1975, and a master's degree in biology (mycology) from Pennsylvania State University in 1977. He was an associate research agronomist working with mushroom culture and plant pathology for Green Giant Co., Le Sueur, Minnesota, until 1979, and then a mushroom grower for Rol-Land Farms, Ltd., Blenheim, Ontario, Canada, in 1979 and 1980. He completed a PhD in botany and plant pathology (co-majors) at North Carolina State University at Raleigh in 1983. After concluding postdoctoral work in the Plant and Soil Science Department at the University of Tennessee, he was appointed an assistant professor in the Department of Ornamental Horticulture and Landscape Design at the same university in 1987 and promoted to associate professor in 1991 and to professor in 1997. He served as interim head of the department from 1999 to 2001. He then joined the Department of Entomology and Plant Pathology at the University of Tennessee in 2002 and was interim head from 2012 to 2013.

Trigiano is a member of the American Phytopathological Society and the American Society for Horticultural Science (ASHS), and the honorary societies of Gamma Sigma Delta, Sigma Xi, and Phi Kappa Phi. He received the T. J. Whatley Distinguished Young Scientist award (Institute of Agriculture, University of Tennessee) and the Gamma Sigma Delta Research Individual and Team Award of Merit at the University of Tennessee. He is the recipient of the publication awards for the most outstanding educational and ornamental papers in ASHS and the Southern region ASHS L. M. Ware distinguished research award. In 2006, he was elected fellow of ASHS. Trigiano was awarded the B. Otto and Kathleen Wheeley Award of Excellence in Technology Transfer and the founder and manager of Creative Agricultural Technologies, LLC, by the University of Tennessee Research Foundation in 2007. He has been an editor for *Plant Disease*; ASHS journals; *Plant Cell, Tissue, and Organ Culture*; and *Plant Cell Reports* and is currently the co-editor-in-chief of *Critical Reviews in Plant Sciences and associate editor for Horticultural Research*. Additionally, he has co-edited ten books, all published by CRC Press (Boca Raton, FL), including *Plant Tissue Culture Concepts and Laboratory Exercises* (two editions: 1996 and 2000), *Plant Pathology Concepts and Laboratory Exercises* (two editions: 2004 and 2008), and *Plant Tissue Culture Development and Biotechnology* (2011).

Trigiano has been the recipient of research grants from the U.S. Department of Agriculture and Forest Service and the Horticultural Research Institute, and from private industries and foundations. He has published more than 200 research papers, book chapters, patents, and popular press articles. He teaches a graduate course in scientific writing, has presented numerous workshops on English scientific writing in Germany, the People's Republic of China, and Brazil, and founded Scientific Writing Experts, an educational business. His current research interests include molecular markers for breeding ornamental plants, population studies of pathogens and native plants, diseases of ornamental plants, somatic embryogenesis, and micropropagation of ornamental species.

Contributors

Jeffrey A. Adkins
Department of Agriculture
Stephen F. Austin State University
Nacogdoches, Texas

Caula A. Beyl
Department of Plant Sciences
University of Tennessee
Knoxville, Tennessee

David W. Burger
Department of Plant Sciences
University of California
Davis, California

Zong-Ming Cheng
Department of Plant Sciences
University of Tennessee
Knoxville, Tennessee

Michael E. Compton
School of Agriculture
University of Wisconsin–Platteville
Platteville, Wisconsin

Richard A. Criley
Department of Tropical Plant and Soil Sciences
University of Hawaii
Honolulu, Hawaii

Adam Dale
Department of Plant Science
McGill University
Ste-Anne-de-Bellevue, Quebec, Canada

Danielle J. Donnelly
Department of Plant Science
McGill University
Ste-Anne-de-Bellevue, Quebec, Canada

Christopher Eisenschenk
Saliwanchik, Lloyd & Saliwanchik
Gainesville, Florida

Jennifer A. Franklin
Department of Forestry, Wildlife, and Fisheries
The University of Tennessee
Knoxville, Tennessee

Amy Fulcher
Department of Plant Sciences
University of Tennessee
Knoxville, Tennessee

Jeffrey H. Gilman
Department of Horticultural Science
University of Minnesota
St. Paul, Minnesota

Dennis J. Gray
Central Florida Research Education Center
University of Florida
Apoka, Florida

John L. Griffis, Jr.
Division of Ecological Studies
Florida Gulf Coast University
Fort Myers, Florida

Michael E. Kane
Environmental Horticulture Department
University of Florida
Gainesville, Florida

Philip J. Kauth
Environmental Horticulture Department
University of Florida
Gainesville, Florida

Sherry L. Kitto
Department of Plant and Soil Science
University of Delaware
Newark, Delaware

Gerald L. Klingaman
Operation Director
Botanical Garden of the Ozarks
Fayetteville, Arkansas

William E. Klingeman
Department of Plant Sciences
University of Tennessee
Knoxville, Tennessee

Chieri Kubota
School of Plant Sciences
The University of Arizona
Tucson, Arizona

Malcolm M. Manners
Horticulture Department
Florida Southern College
Lakeland, Florida

Neil S. Mattson
Department of Horticulture
Cornell University
Ithaca, New York

Brian Maynard
Department of Plant Sciences and Entomology
University of Rhode Island
Kingston, Rhode Island

Garry V. McDonald
Department of Horticulture
University of Arkansas
Fayetteville, Arkansas

Mark P. McQuilken
Department of Horticulture and Landscape
Scotland's Rural College–Riverside Campus
Ayrshire, Scotland

William B. Miller
Department of Horticulture
Cornell University
Ithaca, New York

Kenneth W. Mudge
Department of Horticulture
Cornell University
Ithaca, New York

Margaret A. Norton
Department of Natural Resources and Environmental Sciences
University of Illinois
Urbana–Champaign, Illinois

Leopold M. Nyochembeng
Department of Biology and Environmental Sciences
Alabama A&M University
Normal, Alabama

Lori D. Osburn
Department of Plant Sciences
University of Tennessee
Knoxville, Tennessee

J. Kim Pittcock
Dean of Agriculture
Arkansas State University
Jonesboro, Arkansas

John E. Preece
Research Leader—USDA-ARS
The National Clonal Germplasm Repository
Davis, California

Paul E. Read
Department of Agronomy and Horticulture
University of Nebraska–Lincoln
Lincoln, Nebraska

Timothy A. Rinehart
USDA-ARS MSA
Southern Horticultural Laboratory
Poplarville, Mississippi

John M. Ruter
Department of Horticulture
University of Georgia
Athens, Georgia

Holly L. Scoggins
Department of Horticulture
Virginia Polytechnic Institute and State University
Blacksburg, Virginia

Govind C. Sharma
Department of Biology and Environmental Sciences
Alabama A&M University
Normal, Alabama

Manan Sharma
Environmental Microbial and Food Safety Laboratory
USDA-ARS, BA, BARC, EMFSL
Beltsville, Maryland

Robert M. Skirvin
Department of Natural Resources and Environmental Sciences
University of Illinois
Urbana–Champaign, Illinois

Haley S. Smith
Department of Entomology and Plant Pathology
University of Tennessee
Knoxville, Tennessee

Wolfgang Spethmann
Leibniz University of Hannover
Division of Tree Nursery Science
Institute of Floriculture and Woody Plant Science
Hannover, Germany

James J. Stapleton
Kearney Agricultural Center
University of California
Parlier, California

Scott L. Stewart
Environmental Horticulture Department
University of Florida
Gainesville, Florida

Robert N. Trigiano
Department of Entomology and Plant Pathology
University of Tennessee
Knoxville, Tennessee

Robert R. Tripepi
Plant, Soil, and Entomological Sciences
University of Idaho
Moscow, Idaho

Phillip A. Wadl
Department of Entomology and Plant Pathology
University of Tennessee
Knoxville, Tennessee

Alan S. Windham
Department of Entomology and Plant Pathology
University of Tennessee
Knoxville, Tennessee

Part I

Introduction

1 Introduction to Plant Propagation

Caula A. Beyl and Robert N. Trigiano

Concept Box 1.1

- Early Greeks, Romans, and Chinese knew and used various techniques of plant propagation, including rooting cuttings, air layering, and grafting.
- Knowledge of plant growth hormones and the role of the endogenous growth regulator, auxin, in promoting the initiation of roots was a major milestone in plant propagation development.
- Various methods are used to propagate plants, but ultimately, the methodology used depends on whether a plant is desired to be genetically identical to the original.
- Sexual processes result in propagation by seed, particularly important to the vegetable, bedding plant, and nursery industries.
- Asexual methods include division; cuttage using various parts of the plant, including stems, leaves, and roots; budding; and grafting.
- There are many challenges to successful plant propagation, including lack of knowledge, recalcitrance associated with the phase change from juvenility to maturity, and various pathogens that thrive in the propagation environment.
- Future approaches to plant propagation may include genetic and physiological studies of rooting, the manipulation through tissue culture to stabilize the juvenile phenotype, adjusting conditions in which stock plants are held to optimize the rooting of the propagules taken from them, and the use of special properties of certain bacteria such as *Agrobacterium rhizogenes* to induce roots, in this case, hairy roots.

INTRODUCTION

Man's dependence on plants as the most significant source of food depends predominantly on the ability not only to cultivate plants and utilize them, but also to propagate them. This knowledge of plant propagation, the multiplication or making of more plants, is both a fascinating art and an exciting science. Merely walking through the produce department of a grocery store and examining the array of fruits, vegetables, and flowers available should engender an appreciation of all the various techniques of plant propagation used. If the plants themselves could reveal how they were propagated, an impressive array of techniques would be described from the relatively simple methods of seed germination and grafting to the more cutting-edge techniques of tissue culture, which often includes molecular genetics.

HISTORY OF PLANT PROPAGATION

The origins of plant propagation are hard to document, but it is reasonable to believe that plant propagation co-developed with agriculture approximately 10,000 years ago. The earliest propagation may have been an inadvertent sowing of seeds gathered during collection and harvesting activities, which evolved into deliberate agriculture. Descriptions of early horticulture in Egypt, Babylon, China, and other countries suggest that the culture of ornamental and food crops was fairly well understood and that they could be propagated easily. The Old Testament contains the following passage from Ezekiel 17:22: "…I myself will take a shoot from the very top of a cedar and plant it. I will break off a tender sprig from its topmost shoots and plant it on a high and lofty mountain…it will produce branches and bear fruit and become a splendid cedar…" This indicates that the concept of taking cuttings was well known at that time. Babylon and Assyria were known for terraced gardens and parks. Such deliberate and extensive cultivation of ornamentals required knowledge of how they could be propagated. The Greek philosopher Theophrastus (371–287 B.C.), a student of Aristotle, made observations on the suckering of olive, pear, and pomegranate.

FIGURE 1.1 Detached scion grafting, part of a larger mosaic calendar featuring different agricultural activities throughout the year found in St. Roman-en-gal, Vienne, France, from the third century C.E. (Mudge, K. W., J. Janick., S. Scofield, and E. E. Goldschmidt: A history of grafting. *Hort Rev.* 2009. 35. 437–493. Copyright Wiley-VCH Verlag GmbH & Co. KGaA. Reproduced with permission.)

With respect to grafting, he even wrote in *De Causis Plantarum*, "It is also reasonable that grafts should best take hold when scion and stock have the same bark, for the change is smallest between trees of the same kind…" Roman writings contain references to budding and grafting, and Roman mosaics depict grafting (Figure 1.1). Cato, a Roman statesman in the second century, described cuttings or scions of apple grafted onto sturdy rootstocks. Graftage and topworking were illustrated in medieval treatises (Figure 1.2).

The Wardian case, an interesting invention by Dr. Nathaniel Ward in the early 1800s, enabled not only the germination of fern spores and orchids, but also the transportation of newly germinated and delicate plants across long distances. This enabled the plants to survive and arrive at their destinations in good condition. Wardian cases not only became quite popular for growing orchids, but also were instrumental in helping establish the tea plantations of Assam and rubber trees in Ceylon. Today, replicas of Wardian cases are still used by garden enthusiasts for propagating and protecting sensitive plants (Figure 1.3).

FIGURE 1.2 16th-century plate depicting the steps to be used for topworking trees successfully.

FIGURE 1.3 Modern replica of the traditional Wardian case commonly used for the propagation of ferns and orchids and the culture of plants needing protection.

Much of modern plant propagation is unchanged from practices in use before the 1900s. The following types of cuttings that we use today were all used at that time: softwood, semi-hardwood, hardwood, herbaceous, leaf, and root. Propagators were aware that many cuttings root at the nodes and were maximizing cutting production by using single-node (leaf-bud) cuttings. Growers were propagating plants using bulbs, rhizomes, and stolons, with techniques such as separation and division. Both layering and grafting were used commercially before the 1900s, and methods used then remain largely unchanged today.

Transpiration of cuttings was controlled using high-humidity environments, such as glass-covered rooting chambers, some of which had supplemental bottom heat provided by lamps or recirculating water systems. Disease (likely caused by *Botrytis* and other pathogens) was a problem in these rooting chambers. Mist (Figure 1.4a) and fog (Figure 1.4b) for propagation were not developed until the 20th century.

The major discoveries and advancements in plant propagation in the 20th century are the developments of intermittent mist and fog systems; the discovery of plant growth substances; sanitation and disease control; the knowledge of juvenility and chimeras; micropropagation; and the use of micropropagation techniques to escape pathogens, such as fungi, viruses, mycoplasms, and bacteria.

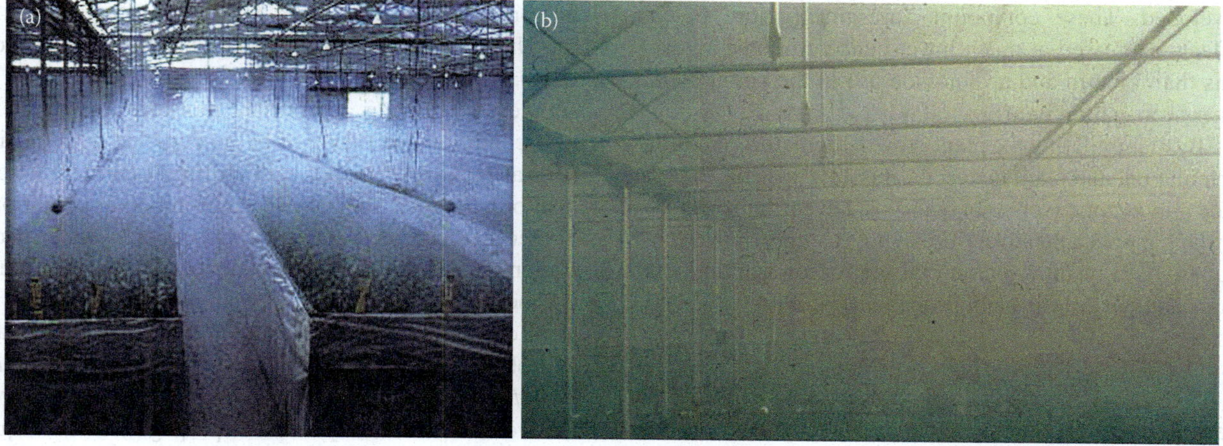

FIGURE 1.4 Misting in greenhouse. (Courtesy of John Ruter, University of Georgia.)

HEAT (STEAM) DISINFESTATION OF MEDIA

Many losses of propagated materials were incurred because of the presence of weeds and pathogens in the soil before the 20th century. In the early 1900s, growers were beginning to use steam to disinfest soil prior to planting. By the late 1930s to the early 1940s, people began disinfesting (pasteurizing) greenhouse and propagation media because the plants grew better than when the soils were sterilized by overheating. By the 1950s, people were aware that nitrifying and other beneficial microbes were killed by steam sterilization, and that has led to modern steaming methods aimed at killing insects, pathogens, and weed seeds while maintaining beneficial microbes and the physical structure of the media.

DEVELOPMENT OF MIST AND FOG PROPAGATION SYSTEMS

Prior to the 20th century, transpiration from cuttings was reduced using high-humidity chambers; shading; cutting leaves in half and, otherwise, reducing leaf area; keeping the soil moist; and frequently sprinkling cuttings with water. By the 1940s and 1950s, intermittent mist and fog were being used for plant propagation. Commercial adoption of fog lagged behind mist because of disease problems. By the 1970s, ventilation combined with fog reduced the incidence of disease, and therefore, the use of fog in plant propagation has become more commonplace at commercial nurseries. Fog and intermittent mist are now indispensable in the cutting propagation of plants.

AUXINS

One of the most important discoveries in the history of plant propagation were auxins and their role in root initiation. It was known for some time prior to the 1930s that the presence of leaves on cuttings was critical for adventitious rooting. In fact, it was known that concentrated "juice" from leaves could stimulate the rooting of cuttings. In the early 1930s, the auxin indole-3-acetic acid was isolated and shown to stimulate root formation. This hormone is produced in the growing points of shoots, especially in the young, expanding leaves. By the late 1930s, indole-3-butyric acid, naphthaleneacetic acid, and their commercial formulations "Auxan" and "Rootone" were available and being tested for rooting cuttings of many different species of plants. The use of auxin has increased the percentage of cuttings that root and the number and distribution of roots on cuttings. The quick commercial adoption of auxins for rooting speaks to the great need for root-inducing substances.

MICROPROPAGATION

Early in the 20th century, pioneers in plant tissue culture first developed *in vitro* techniques for the germination of orchid seeds. Later, a medium was developed, which supported the growth of roots floating on its surface. However, *in vitro* culture was far from being commercially viable because of the lack of a plant growth regulator that would stimulate cell division. By the early 1950s, researchers had shown that water extracts from the vascular tissue of tobacco, malt extract, liquid *Cocos nucifera* L. (coconut) endosperm, and an extract from solid coconut endosperm all induced cell division. Additionally, autoclaved DNA from herring sperm and from calf thymus stimulated cell division in tobacco wound callus tissue cultured *in vitro*. This activity was not present if the DNA was not autoclaved.

Cytokinins are a class of growth regulators that induce cell (*cyto*) division (*kinin*). The first cytokinin, kinetin, was isolated, purified, and crystallized from autoclaved animal DNA and was shown to stimulate cell division and shoot multiplication. Since then, benzyladenine, thidiazuron, and many others have been developed. These compounds are much more potent (effects expressed at much lower concentrations) cytokinins than kinetin and are in wide use in the commercial micropropagation industry today.

By manipulating the plant growth regulators, medium formulation, and plant materials, it is now possible to produce non-zygotic (somatic) embryos, adventitious shoots, axillary shoots, and adventitious roots using *in vitro* techniques. As a result, commercial micropropagation is an important part of the plant propagation industry.

ESCAPING PATHOGENS

Plant diseases, especially those caused by viruses, are especially problematic because plants with viruses cannot be "cured." During the late 1940s, there was an outbreak of spotted wilt disease in *Dahlia pinnata* (Dahlia), which is caused by a virus. Tip cuttings rooted from infected plants showed no spotted wilt symptoms. By the early 1950s, scientists were excising very small 250-μm apical meristematic domes from dahlia plants with symptoms of dahlia mosaic virus and were placing them *in vitro* where they elongated. When the shoots were 1 to 2 cm long, they were grafted onto young virus-free plants, where they grew normally without dahlia mosaic symptoms.

By the mid- to late-1950s, scientists were combining thermotherapy with tissue culture. Virus-infected plants were grown at 25 to 40°C prior to excising meristematic domes and placing them *in vitro*. A percentage of the resulting plants were free from certain viruses by grafting onto indicator plants or using immunoassay techniques, such as enzyme-linked immunosorbent assay, which was developed in the 1970s for the detection of viruses. Therefore, plant pathologists have helped plant propagators in the 20th century by developing methods for steaming and disinfestation of media, benches, and planting containers. In turn, plant propagators have helped plant pathologists by developing techniques, including rooting of cuttings, grafting, and tissue culture, which can be used to escape viruses. Escaping viruses is an important part of the fruit, nut, and vegetable industries of the 21st century.

ROLE AND IMPORTANCE OF PLANT PROPAGATION TODAY

Everything that is dependent on plants in any way is also dependent on the ability of plants to be propagated. Large-scale agriculture of crops, such as for grains and certain vegetables, depends on successful seed germination. Looking at it another way, you could say that seeds are the first necessary step in the world's food chain, and this dependency has resulted in a huge industry. The estimated value of the global domestic seed market was $45 billion per year for 2012, $12 billion of which was for the United States alone (International Seed Federation, 2013). The requirements to germinate seeds can be as simple as requiring appropriate moisture and temperature conditions or they can be as complex as requiring several months of moist, cold chilling treatment (stratification), or other specialized environmental conditions. Sometimes, the seed coat with its barrier to water penetration must be broken or weakened (scarification). Large numbers of seeds can be planted mechanically, but in some cases, they must be planted by hand to avoid damage and obtain a better stand such as for buckeye (*Aesculus parviflora* Walt.; Figure 1.5). Nursery industries all over the world use a variety of plant propagation techniques from T-budding dogwoods in Tennessee (Figure 1.6) and June-budding peaches in Georgia to stool-bedding apple rootstocks in England, a process taking a year or more. Nurserymen propagating ornamental trees, shrubs, and grasses or specializing in fruit trees must know the best way to propagate each species and have a thorough knowledge of the best "window of opportunity" that works for each species.

The floral and nursery industries are recognized as the two fastest growing segments of horticulture in recent years. The growth and demand for bedding plants has probably been the most significant factor in this growth over the last 20 years. Establishment of bedding plants is dependent on successful and uniform seed germination.

FIGURE 1.5 Seeds of Ohio buckeye (*Aesculus parvifolia*) being planted by hand at Shadow Nursery, Winchester, Tennessee, to prevent damage and increase the number that will germinate successfully.

FIGURE 1.6 Fall budding dogwoods in the field at Shadow Nursery, Winchester, Tennessee. The bud stick from which the bud of the desired cultivar was taken can be seen in the propagator's right hand. Notice that the budding rubber used to wrap the T-bud is wrapped from below the insertion point upward.

Another area of horticulture dependent on successful seed germination and production of transplants is the vegetable production industry. The Chinese described grafting of vegetable crops as early as the fifth century, and today, both Japan and Korea have developed robots and grafting machines to produce large numbers of grafted tomatoes, eggplants, and peppers.

Even the forest industry is very interested in plant propagation techniques being used to improve the performance of trees, from the establishment of seed orchards containing superior tree selections to the use of cuttings of superior clones. Some approaches have focused on non-zygotic embryogenesis or creating an embryo identical to the original from the somatic portion of the plant. Clonal approaches are being studied all over the world for species, such as eucalyptus and pine, and molecular genetics techniques are being explored to boost the genes for auxin production in aspen to increase how rapidly the transgenic tree will grow.

The craft and science of plant propagation is particularly important for the preservation of rare and endangered plants. Both traditional and specialized techniques are used to propagate species recognized as either rare or endangered so that they may be preserved and, in some cases, re-introduced into new and old habitats. Sometimes, these techniques involve *in vitro* seed germination, and in other cases, various tissues may be used in tissue culture to create a large number of plants very rapidly for species that respond well to the *in vitro* environment. Plant propagators working for arboreta and botanical gardens must have a thorough knowledge of plant propagation to be able to germinate seeds sent to them from all parts of the world or to increase numbers of rare or highly valued specimen plants to share with other arboreta and botanical gardens.

HOW THIS BOOK IS ORGANIZED

This book is based on a successful model of organization used in *Plant Pathology Concepts and Laboratory Exercises*, Second Edition (Trigiano, Windham, and Windham, 2008), and *Plant Tissue Culture, Development, and Biotechnology* (Trigiano and Gray, 2010), which combines concept chapters with accompanying laboratory exercises to provide additional in-depth information and hands-on tasks that illustrate the principles found in the concept chapters. This not only aids those using the book for teaching, but also provides those using the book as a reference with examples of techniques that can be used as models for other species. The various chapters are contributed by horticultural scientists with expertise in the disciplines of plant propagation, breeding, pathology, tissue culture, seed technology, among others, and who have had extensive experience teaching plant propagation.

We have reorganized and revised this edition of the propagation book according to the many suggestions by our authors and instructors and, more importantly, students who have used the book. The most obvious change is that we have fashioned a beginning section that collects basic botanical information for those students who are unfamiliar with botany or just need a quick "refresher" on the subjects. Other chapters throughout the book still focus on traditional horticultural practices much like those found in the first edition. We have also added a few more laboratory exercises that we hope will better illustrate the concepts in the main chapters.

The concept chapters always begin with a list of some of the more important ideas, the concept box, contained in the section. These bulleted lists are intended to be a type of "executive summary" for the chapter as well as to alert students to the major points of the topics in the chapter. The laboratory exercises are meant to illustrate the ideas and procedures from the concept chapters. They are organized in a standard format throughout the book, including a brief introduction or introductory remarks to the topic. Teachers and students are provided a list of materials necessary to complete the experiments, followed by step-by-step procedures that detail the exact methodology for that exercise. This book is unique in propagation teaching aids as it provides, in general terms and descriptions, what the anticipated results of the laboratory exercises should be. The last section of the laboratory exercises lists a set of questions that are intended to stimulate discussion and thought about the experiments. The laboratory exercises included in the book have been used repeatedly by our instructors and proven to be very reliable for classroom use. They provide a broad exposure of the concepts presented in the chapters to students of hands-on application.

We realize that instructors arrange topics for their propagation courses in a manner specific to meet the needs of their classes. The organization of the chapters in this book reflects that which we have been told is "typical" of many courses; however, most chapters have been written to be independent or a "free standing unit of information" regardless of their presentation position in the book. Therefore, this edition offers a collection of informational chapters and laboratory exercises that may be used in any sequence that provides a good fit for the course without loss of clarity and usefulness.

After the Introduction to the book in Part I, Chapter 1, Part II, deals with "Botanical Concepts for Plant Propagation," which should provide students with a solid foundation of basic plant science subdisciplines. Chapter 2 serves as a basic primer for anatomy and morphology, which is necessary to understand the cells, tissues, and organs of plants. We have chosen to retain the black-and-white photomicrograph format in this chapter for clarity. The next three chapters (3, 4, and 5) delve into the very interesting field of plant physiology, including a general chapter on basic plant physiology and how it impacts the propagation systems used (Chapter 3), followed by the very specific content area of plant growth regulators (Chapter 4). What horticultural propagation text would be complete without a chapter on sexual reproduction and breeding? Chapter 5 not only provides clear, easy-to-understand diagrams and explanations of mitosis and meiosis, but also addresses the processes of sexual reproduction in angiosperms and some genetic and breeding concepts. Once again, we have retained the black-and-white instead of colored photomicrographs in this chapter for clarity. Chapter 6 deals with juvenility. The change of phase of a plant from juvenile to physiologically mature impacts not only flowering and fruiting, but also propagation success. Cuttings of some species are very difficult to impossible to root once they have become mature, and propagators have used a number of different techniques to "restore" juvenility, such as serial propagation, severe heading back, and *in vitro* culture. This section ends with a chapter (7) on chimeras. Chimeras are plants composed of layers or sectors that have a different genotype, so depending on where new shoots arise to become new plants, they may have a different phenotype than the parent plant. The genotype is the heritable genetic blueprint for the plant, and the phenotype is the manifestation of the plant that results from putting that genetic blueprint to work—its structure, function, and behavior. The concept of chimeras is often difficult for students to understand, but our authors have written very clear, understandable explanations, and the chapter is beautifully illustrated.

In Part III, the book leads readers through a progression, beginning with propagation structures of different types and how these can be managed for more effective propagation in Chapter 8. Chapter 9 explores the rationale for the design of mist or fog propagation systems for the production of cuttings or for germinating seeds. Intermittent mist/fog systems are one of the most significant innovations in asexual propagation, enabling the successful propagation of many species by providing a high-humidity environment conducive to rooting. Chapter 9 compares the rooting performance of cuttings in a propagation tent with those rooted under mist. This chapter allows the study of the various control systems that determine misting frequency and how to determine whether the mist is being distributed uniformly across the bench. A complete book on plant propagation cannot neglect containers and media used for seed germination, cutting propagation, and acclimatization of cuttings. Chapter 10 details the types of media that take the place of soil and their characteristics since they provide many advantages over soil, including lightness and ease of handling as well as freedom from pathogens, pests, weed seed, or other contaminants. Chapter 10 also deals with the factors that must be considered before an appropriate medium can be chosen and gives procedures for determining bulk density and pore space, important characteristics of media.

Part IV, "Plant Propagation Diseases, Insects, and the Importance of Sanitation," contains five chapters. Sanitation can determine the success or failure of seed or cutting propagation and can make the difference whether a commercial propagation establishment is economically viable. Part IV deals with the major threats to successful plant propagation offered by plant diseases. In Chapter 11, the characteristics of various common pathogens are presented along with information on how they are typically spread. Chapter 12 includes an exercise that illustrates the impact of *Botrytis* and other propagation pathogens. It also contains laboratory exercises on the chemical control of gray mold, the effect of ventilation on its development, and the biological control of *Pythium*. These concepts are reinforced by laboratory exercises on how to disinfest soil and planting media in Chapter 13. Chapter 14 provides important information on the concept of producing specific pathogen-free plants and crop certification programs in Canada. Integrated pest management, which has almost become ubiquitous in literature on greenhouse and nursery management, is detailed in Chapter 15, with a focus on its application to propagation systems.

Part V contains only one chapter, but it is a very important chapter for anyone who conducts plant propagation research, whether an undergraduate or a graduate student, or a professional in the discipline. Chapter 16 is concerned with how to evaluate propagation experiments, and collect, handle, and analyze data. This issue is ignored completely or mentioned only briefly in most books on plant propagation but is an important aspect of horticultural education.

Parts VI to XII explore the various types of propagation, starting with the use of stem cuttings in Part VI. In Part VI, there are four chapters: a concept presentation followed by three novel laboratory exercises that explore the general concepts for successful vegetative propagation. Chapter 17 introduces the concept of cloning plants using stem cuttings. Clones are exact copies of the original plant and are used extensively in horticulture. This chapter is followed by a laboratory exercise (Chapter 18) that explores the adventitious rooting of woody plants. We introduce in this edition of the book a chapter (19) illustrating some cutting-edge woody propagation techniques using long cuttings developed in Germany in the last decade. The last chapter in the section (20) investigates the rooting of cuttings from tropical plants.

In Part VII, propagation by leaf and root cuttings is explored. For many, the idea that you can make a new plant genetically identical to the first by taking a cutting, inserting it into a rooting medium, keeping it moist and humid, and then watching new roots form is tremendously exciting. Home gardeners share cuttings this way quite frequently. The types of cuttings and what kinds of conditions help them root successfully are explored in Chapter 21.

Many people are familiar with the use of stem cuttings, but leaves and roots can also serve as very effective propagules. Many plants can be successfully rooted from just the leaf blade, and others, from the leaf and the petiole (Chapter 22). Many home horticulturists do this with leaf and petiole cuttings of African violets rooted in water. For other plants, particularly those that sucker readily, roots can serve as a source of cuttings or can be forced to produce shoots that, in turn, can be used as cuttings. There are many examples among woody trees of new trees developing from the roots of older ones until dense strands are formed (aspen and sumac). Chapter 23 uses sumac, nandina, mahonia, and crape myrtle to demonstrate propagation by root cuttings.

Part VIII contains a single chapter (24) that describes the time-honored technique of layering to produce new clonal plants on their own roots. Layering involves techniques that encourage roots to form before the cutting is severed from the parent plant. Layering techniques probably were adopted through close observation of what occurs in nature with some shrubs or trees. Low-lying branches that come into prolonged contact with the ground may form roots at that point. The technique, although slow, is still valuable for some species that are extremely difficult to root with cutting propagation.

Techniques of grafting and budding in Part IX require skill as well as science. Depictions of early knowledge and application of grafting has occurred in historical depictions from the 16th century (Figure 1.5). Grafting and budding both involve combining two plants: one serving as the rootstock or bottom portion of the plant and the other serving as the top portion of the plant (Chapter 25). With grafting, the scion or a piece of stem containing several buds is cut so that it can be joined with the rootstock. With budding, the scion piece consists of only one bud. These techniques are extremely important in the propagation of many fruit tree species, including apple and peach. Chapter 26 offers a variety of very easy-to-complete laboratory exercises on grafting techniques using rose, coleus, and boxwood, and Chapter 27 demonstrates the art of grafting vegetables.

Part X deals with bulbs and plants with special structures. Chapter 28 describes the use of underground storage structures to propagate plants asexually. We take advantage of the underground storage organ, such as bulbs, corms, etc., and use various techniques to encourage the bulbs to form more propagules. The laboratory exercise in Chapter 29 includes experiments with the propagation of lily, a nontunicate bulb, and hyacinth, a tunicate bulb. Tunicate bulbs like onion are protected by a dry papery sheath—a modified leaf around continuous concentric layers. It also includes an experiment with Irish potato (a stem tuber) and dahlia (a tuberous root). The chapter ends with a general observational activity of many bulbous and tuber crops in the field.

In the next section (Part XI), the chapters describe the exciting world of micropropagation, where small pieces of tissue or even single cells can be grown in a sterile environment on media containing all the nutrients, minerals, vitamins, growth regulators, etc. that the tissue (or explant) needs to proliferate more cells (Chapter 30). The mass of cells that grows from that original explant is called "callus." Depending on what growth regulators are used, explants can be tweaked to form shoots and roots or, via another route, non-zygotic embryos. Because a sterile (axenic) environment is needed and explants have such precise requirements for what is contained in the tissue culture medium, micropropagation laboratories have certain requirements in common (Chapter 31). Many of the operations manipulating the explants and placing them on sterile autoclaved medium are conducted in sterile transfer hoods. In Chapter 32, the procedures for the micropropagation of mint are described from disinfesting the shoots for use as explants and preparing the medium to placing them in the culture vessels. Chapter 33 describes the micropropagation of tropical plants, and Chapter 34 describes the special procedures used with woody plants.

The most fundamental classification of plant propagation systems is not by the technique used, but whether they are the result of sexual or asexual process. The previous parts of the book depicting the types of plant propagation have all dealt with asexual or clonal propagation, in which the propagule is an identical genetic copy of the original plant. Part XII of the book deals with seed propagation. Seeds, unless they are apomictic, are the result of the sexual reproduction of plants. Apomixis is a special case of seed propagation that is asexual. This occurs with some species of citrus and Kentucky bluegrass, among other species. The embryo results either from cells of the ovule or from an egg with 2n chromosomes, which can develop without fertilization. Details of sexual reproduction in plants are featured in Chapter 5, which includes examples of higher plants, ferns, and mosses. Understanding not only the processes involved in developing a seed, but also how to produce and germinate a seed are important. Chapter 5 also explores plant breeding, a process that is essential for the transfer of traits between cultivars and the development of new cultivars with enhanced traits. Chapter 35 deals with seed production, analysis, and processing, and provides a laboratory exercise to emphasize the topic. Seeds have the potential to germinate into seedlings, which then develop into plants. Some seeds only require appropriate temperature and moist conditions to germinate readily. Others require much more complex conditions before they will germinate. Scarification treatments are used to weaken or remove hard seed coats, which are impediments to germination. Some seeds also require a period of cold moist stratification. In Chapter 36, the laboratory exercises acquaint students with various techniques used for scarification and the effectiveness of scarification coupled with stratification. Finally, some seeds are allowed to imbibe or take up water, which allows the initial events in germination to occur, but they are prevented from germinating fully by imbibing them in a germination solution containing solutes or an osmoticum, which keeps germination from proceeding beyond imbibition. This is called "priming," and its effects are explored in one of the laboratory exercises. When the seeds are dried and then imbibed again, they germinate much faster for having had the head start. The last exercise in that chapter deals with the stimulation of germination by the chemical constituents of smoke. Chapter 37 contains information on producing seedlings and bedding plants, without which garden centers in spring would be much less colorful and gardens would not get an early start.

The last section (Part XIII) contains chapters that are best described as "in conclusion: special topics." Chapter 38 takes a light-hearted, often humorous, and sometimes hilarious look at some propagation myths—those "truths" handed down to you by your grandmother. This chapter may also be described as "myth busters." Intellectual property protection for plants (Chapter 39) has become an increasing important concern for the horticultural industry in the last several decades. Horticulture is a big business, and tremendous resources are invested in developing new plants. These investments are now typically protected by legal processes, which may include national and international jurisdictions. This chapter provides an overview of plant protection instruments, which may take the form of trademarks, copyrights, patents, and plant breeder's rights. It also outlines the process for the awarding of exclusive rights to plants and briefly addresses the remedies or the enforcement of these rights. Many of the policies and rules governing patents changed in 2013, so this is a MUST read. The last chapter (40) speculates about the future directions of plant propagation research and some of the most interesting recent discoveries impacting the discipline that have come from the study of molecular genetics.

Lastly, we have included a DVD along with the book, which contains most of the figures appearing in the book. The figures are presented in color, with captions and/or notes in PowerPoint format, which are amenable to classroom or individual study. We have intentionally "mounted" the figures on a blank background so that they may be easily incorporated into your own presentations. Supplemental photographs have been added to some of the chapters to further illustrate important concepts.

FUTURE OF PLANT PROPAGATION

Many different approaches are being pursued to develop new methods of plant propagation, and these include a better understanding of the genetic and physiological bases of rooting; the manipulation through tissue culture to stabilize the juvenile phenotype; adjusting conditions in which stock plants are held to optimize the rooting of the propagules taken from them; and the use of special properties of certain bacteria such as *Agrobacterium rhizogenes* to induce roots, in this case, hairy roots. Another approach may employ genetic engineering techniques to identify genes that contribute to the ease of rooting and insert those genes into more recalcitrant species.

Just examining the scientific literature devoted to plant propagation and its various techniques such as tissue culture can be overwhelming. For someone wanting the latest information about plant propagation and the physiology associated with the induction of rooting, a membership in the American Society for Horticultural Science (ASHS) would be an excellent approach. ASHS provides outlets for peer-reviewed research in its three journals, which range from predominantly applied research to somewhat basic research. Another excellent source of current information about the latest techniques and how they have been applied to various species that are difficult to propagate is the International Plant Propagator's Society. Since 1951, this society has been sharing information about plant propagation. It has eight regions, including three in the United States—the Eastern, Southern, and Western regions. Attending its meetings is a delight for anyone who wants to focus specifically on propagating plants.

As you learn more about plant propagation in the following chapters, you will appreciate the variety of techniques used to propagate plants and the diversity of the kinds of propagules used. We hope that this knowledge will prepare you to be successful in your future plant propagation endeavors. We look forward to receiving any comments or suggestions that you may have to improve either the content or the presentations in the book and the DVD.

LITERATURE CITED AND SUGGESTED READING

International Seed Federation. 2013. Estimated value of the domestic seed market in selected countries for the year 2012. (http://www.worldseed.org/cms/medias/file/ResourceCenter/SeedStatistics/Domestic_Market_Value_2012.pdf). Accessed July 3, 2013.

Janick, J. 2002. Ancient Egyptian agriculture and the origins of horticulture. *Acta Hort* 582:23–39.

Janick, J. 2005. The origins of fruits, fruit growing, and fruit breeding. *Plant Breed Rev* 25:255–320.

Janick, J. 2006. History of horticulture lecture series. (http://newcrop.hort.purdue.edu/newcrop/history/default.html). Accessed August 27, 2006.

Mudge, K. W., J. Janick., S. Scofield, and E. E. Goldschmidt. 2009. A history of grafting. *Hort Rev* 35:437–493.

Theophrastus (ca. 300 B.C.). The modes of propagation in woody and herbaceous plants: Propagation in another tree—Grafting. In J. Janick, ed., *Classic Papers in Horticultural Science*. Prentice Hall: American Society for Horticultural Science, pp. 6–13.

Trigiano, R. N. and D. J. Gray. 2010. *Plant Tissue Culture, Development, and Biotechnology*. Boca Raton, Florida: CRC Press LLC, 583 pp.

Trigiano, R. N., M. T. Windham, and A. S. Windham. 2008. *Plant Pathology Concepts and Laboratory Exercises*, 2nd Edition. Boca Raton, Florida: CRC Press, LLC, 558 pp.

Part II

Botanical Basics

2 A Brief Introduction to Plant Anatomy and Morphology

Robert N. Trigiano, Jennifer A. Franklin, and Dennis J. Gray

Concept Box 2.1

- Plant cells can be classified into the following basic types: meristematic and their immediate derivatives—parenchyma, collenchyma, and sclerenchyma.
- Simple tissues consist of one cell type, whereas complex tissues contain more than one of the basic cell types.
- The four plant organs are roots, stems, leaves, and flowers.
- Generally, the organization of monocots and dicots are similar, but they each have distinctive arrangements of cells/tissues in their organs.
- The four units of a flower are sepals, petals, anthers, and pistils.
- The study of plant anatomy is useful for understanding plant development, and determining the origin of cells and tissues *in situ* and *in vitro*.

This chapter, "A Brief Introduction to Plant Anatomy and Morphology," originally appeared in *Plant Tissue Culture, Development, and Biotechnology* (Trigiano and Gray, 2010), but is very appropriate for beginning propagation students. We included this chapter because many horticultural students lack basic botanical knowledge about how plants are assembled. We have also elected to retain the figures as in the original chapter as structures are exceeding clear in black and white. This chapter, therefore, explores some of the internal organization (cells, tissues, and organs) or anatomy of vascular plants. For simplicity, we have organized and illustrated the material by first looking at cell types and then comparing and contrasting the anatomy of the tissues and the organs of the monocotyledonous (monocot) and dicotyledonous (dicot) angiosperms, gymnosperms, and pteridophytes (ferns).

For the purposes of this book, we will consider the following four plant organs: roots, stems, leaves, and flowers (reproductive structures). Note that some authors omit the flower as an organ. It is impossible to discuss adequately all the details of the anatomy and development of these organs in this short chapter. Therefore, most treatments of cell types, tissue, and organs are described in broad, widespread terms, and students are cautioned that many exceptions to our generalizations can be found. The relationship of anatomy to the common forms and shapes, or morphology, of these organs will also be touched upon. Readers with greater interests in more exhaustive details of anatomy and development are directed to some botany and anatomy textbooks cited at the end of this chapter. Most of the material in this chapter is derived from Esau (1960) and Fahn (1990). Readers should also appreciate that while plant anatomy and morphology typically are studied through the use of static materials, such as histological sections, it is important to view these in the context that they represent growing, changing, three-dimensional organisms. In this way, an understanding of plant development can be better achieved.

Plants, like other complex organisms, are constructed of cells, the basic unit of life. However, unlike animal cells, plant cells are surrounded by a wall composed of structural polymers, which may include pectin, cellulose, lignin, and hemicellulose. The wall may be relatively thin and flexible as in many parenchyma cells or rather thick as in collenchyma and sclerenchyma cell types (see later discussion). In fact, the structure of the plant cell wall imparts, to some degree, the function of the cell. Cells may only have a primary cell wall, which is more or less defined as the wall material deposited while the cell is increasing in size and having cellulose microfibrils that are laid down randomly or in more or less parallel orientations (Esau, 1960). The primary wall usually contains cellulose, hemicellulose, and pectic compounds referred to as the "middle lamella." The wall is flexible and stretches as the cell grows, with the orientation of cellulose fibers determining the direction of growth and eventual cell shape. The middle lamella acts as a "cement" between adjacent cells. Secondary walls found in some cells are deposited to the inside of the primary wall and the middle lamella after the primary wall has been completed and can be very thick. Lignin may then be deposited into primary and secondary cell walls, making the cell rigid. Secondary walls containing cellulose and hemicellulose and lignin may or may not be present (Esau, 1960).

CELL TYPES

Let us consider the basic cell types of plants before examining the internal arrangement of cells into tissues and, in turn, tissues into organs. For the purposes of this chapter, plant cells can be simply classified into the following types and their variations: (1) meristematic, (2) parenchyma, (3) collenchyma, and (4) sclerenchyma. Note that most references consider the meristematic cells to be a form of parenchyma cells.

1. Meristematic cells have the following characteristics:
   - Very thin-walled cells that undergo mitosis to increase the length (apical meristem) or thickness (lateral meristem) of the organ. The meristematic initials reproduce themselves as well as form new cells, termed "derivatives," which increase the body of the plant. These derivative cells usually continue to divide several times before any significant differentiation into other cell types occurs. The initials and their derivative cells constitute the apical meristem (Esau, 1960), which can be found at the shoot (Figure 2.4a) and root (Figure 2.1a, b) tips. The stem apical meristem of monocots and dicots may be divided further into the tunica and the corpus. The tunica (coat) is one to several cell layers thick, divides only by anticlinal divisions to increase the surface area of the tip, and surrounds the corpus (body), which consists of a number of cells that divide in different planes to increase the volume of the meristem. This two-part arrangement of the meristem is absent in roots. The apical meristem of ferns and most other seedless nonvascular plants consists of a single large, triangular cell from which the surrounding apical meristem cells are derived, and the tunica/corpus organization is not present. In gymnosperms, there is no clear outer layer of tunica, but zones can be discerned within the apical meristem. An upper, lens-shaped region of large cells called the "central mother cell zone" and the surrounding

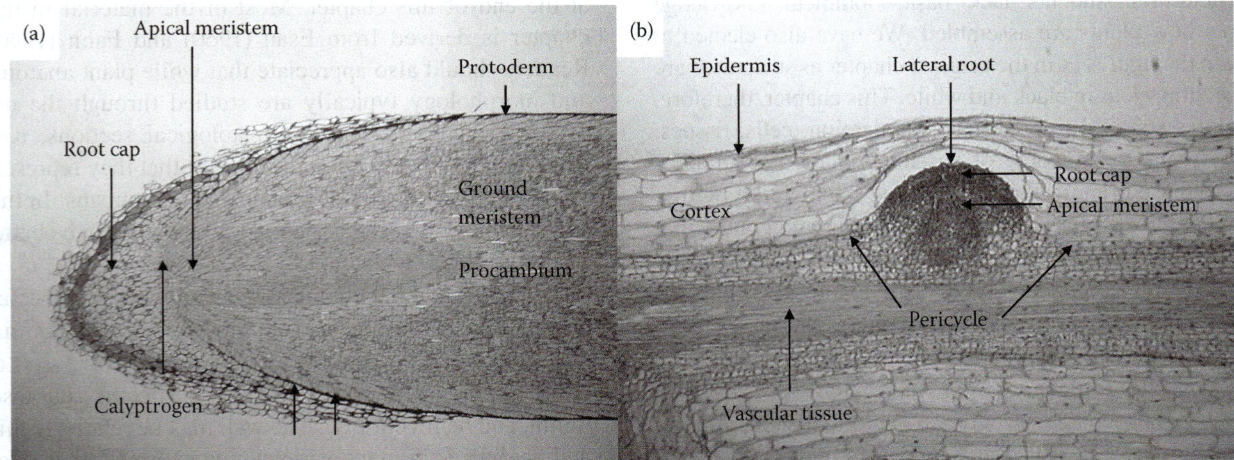

FIGURE 2.1 Root apical meristem and lateral root origin. (a) Near-median longitudinal section through a young corn (*Zea mays*) root. The calyptrogen, in this case, gives rise to the cells of the root cap. In other roots, including those of many dicots, the root cap is derived from the apical meristem. The three meristematic areas are the protoderm (gives rise to the epidermis), the ground meristem (cortex), and the procambium (vascular tissue). Arrows indicate mucilaginous wall substance. Slide courtesy of Carolina Biological Supply Co., Burlington, North Carolina. (b) A young lateral (secondary) bean (*Phaseolus vulgaris*) root that originated from the pericycle. Note that, as the lateral root develops, it pushes through and crushes the cortical and epidermal tissues of the primary root. The architecture of lateral roots is similar to that of primary roots.

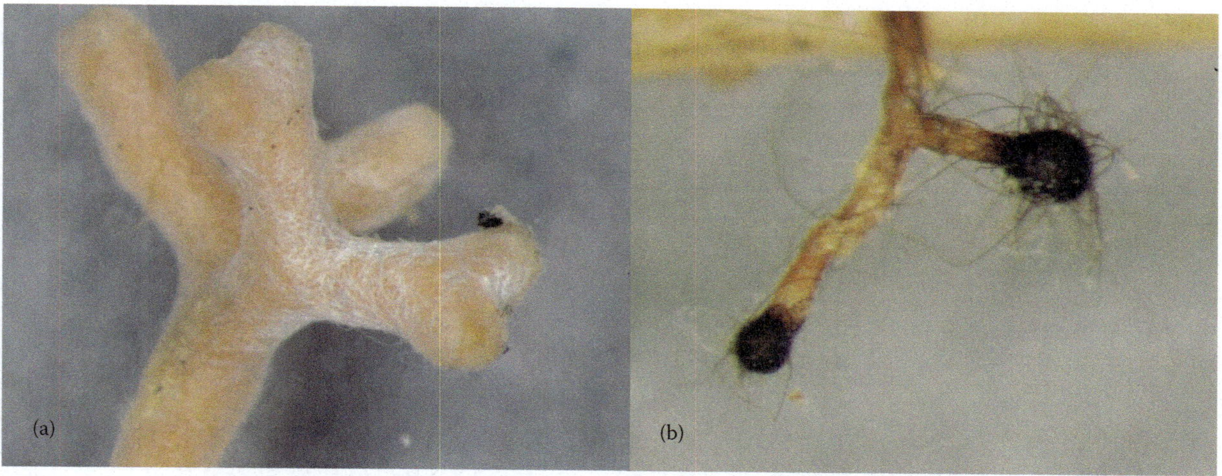

FIGURE 2.2 (a) Roots of *Pinus* colonized by light-colored *Scleroderma* mycorrhizal hyphae, showing a cluster of the short, highly branched roots characteristic of many ectomycorrhizal associations. (b) Roots of *Castanea dentata* colonized at their tips by the dark hyphae of the ectomycorrhizae *Cenococcum*. (Photographs courtesy of Jenise Bauman.)

initials derived from it divide infrequently. Cells on the periphery of this zone are active in cell division, and below the apex, the arrangement is similar to that of monocots and dicots. The shoot apical meristem can be divided into a peripheral zone, which gives rise to leaves, buds, and flowers (lateral organs), and a rib zone, which produces the stem tissues. Several cell layers distal to the apical meristem and in the rib zone, the procambium (vascular), ground meristem (cortex), and protoderm (epidermis) of the stem are differentiated and give rise to the primary tissues of the plant body.

- Primary growth of the plant is brought about by the activities of apical meristems and subsequent divisions and the differentiation of the derivative cells into the tissues and the organs of the plant. All tissues originating from a primary meristem are termed "primary tissues," that is, the primary xylem and the epidermis. Most ferns, monocots, and some dicots complete their growth and development via primary growth only.
- Secondary growth exhibited by many dicots and gymnosperms is achieved through specialized lateral meristems. The vascular cambium (Figures 2.3b and 2.5a), located between the primary phloem and the primary xylem, produces cells that differentiate into additional vascular tissue, that is, the secondary xylem and the phloem, which increases the girth of stems and roots. Another lateral meristem, the phellogen or cork cambium, is found near the exterior of stems and roots and arises in the primary cortex. It produces phellem and phelloderm cells that replace the epidermis and the cortex, respectively, which is lost or crushed due to the expanding diameter of the root or the stem. Collectively, this new tissue is called the "periderm."

2. Parenchyma cells generally have the following characteristics:
   - They are typically nearly isodiametic (about as long as they are wide); however, cells may vary in shape, being elongated or even lobed.
   - The primary cell wall of this cell type is relatively thin and composed mainly of cellulose and hemicellulose, with a layer of pectic substances, the middle lamella, on the exterior of the primary wall. Note that some parenchyma cells, especially in vascular tissue, may develop a secondary wall or become sclerified with lignin (see sclerenchyma cells).
   - Parenchyma cells always have a nuclei and functioning protoplasts (cytoplasm).
   - These cells are generally considered to be relatively undifferentiated (compared to sclerenchyma) and capable of resuming meristematic activities by dedifferentiation. Indeed, this cell type and tissue is involved in the development of adventitious roots and shoots, wound healing, and other activities. Note that some parenchyma cells can be very differentiated and specialized in their function.
   - This cell type can be found through the body of the plant in primary and secondary tissues.

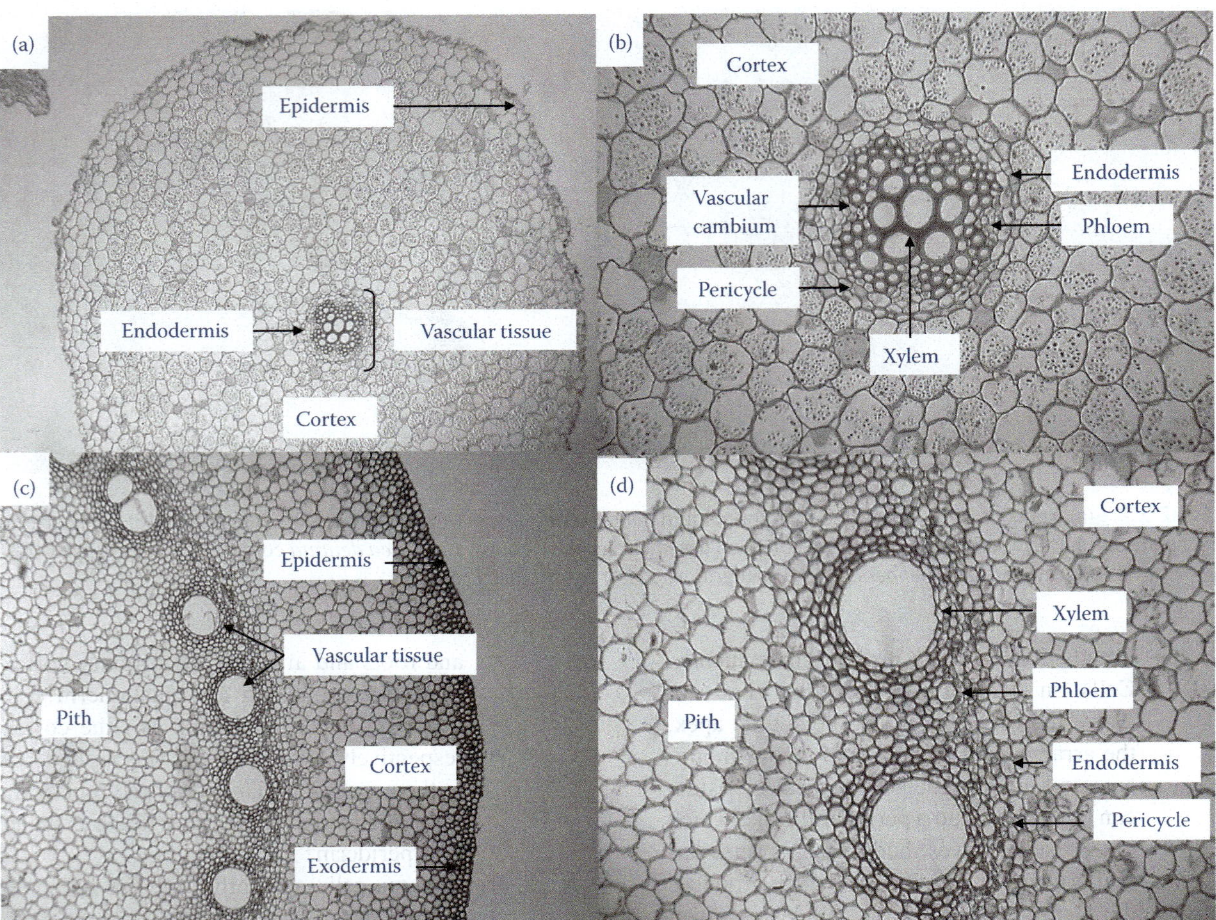

FIGURE 2.3 Dicot and monocot root structure. (a) Cross section of a buttercup (*Ranunculus* species), a dicot root. Note the arrangement of tissues and the lack of pith in the center. (b) Higher magnification of the central area shown in Figure 2.3a. The "stele" includes all vascular tissue (xylem and phloem), the vascular cambium (discernible only by relative position in the young root), and the pericycle. It does not include the endodermis. Note the large xylem vessels in the center and the thickened walls (Casparian strips) of the endodermal cells. (c) Cross section of a corn (*Zea mays*), a monocot root. Note the central core of pith and very large xylem vessels. (d) Enlargement of the vascular area of the cross section shown in Figure 2.3c. (All slides provided courtesy of Carolina Biological Supply Co., Burlington, North Carolina.)

3. Collenchyma cells have the following characteristics:
   - They are typically more elongated than parenchyma cells and are specialized to function as a mechanical support for the plant.
   - Collenchyma cells have soft, pliable, unevenly thickened primary walls composed mostly of cellulose, with some pectin, but never lignin.
   - Collenchyma cells are similar to parenchyma cells in having a nuclei and a living protoplasm and are capable of dedifferentiation and meristematic activity.
   - This type of cell is generally found in young stems (Figure 2.5d) and leaf petioles and functions as flexible supporting tissues.
4. Sclerenchyma cells have the following characteristics:
   - Sclerenchyma cells can be long and thin (fibers; Figure 2.5a) or isodiametic to elongated (sclerids) and are involved in mechanical support and water conduction. This cell type is widely distributed in the four major plant organs. Specific cells types include fibers, vessel elements, tracheids, and sclerids (various forms including astrosclerids [star-shaped], stone cells, etc.).
   - The hallmark of sclerenchyma cells is the deposition of a secondary wall on the interior of the primary wall. Proceeding from the outside of the cell inward, the wall layers encountered in these cells are the middle lamella, the primary wall, and, lastly, the secondary wall. By definition, the secondary wall is laid down after the growth of the

cell ceases. The wall is composed primarily of cellulose arranged in parallel fibers and usually lacks pectic components. If a cell becomes lignified, deposition of lignin starts at the middle lamella and proceeds inward.

- Although many sclerenchyma cells are dead (lack a protoplasm) at functional maturity, some types of cells may retain a living protoplasm. However, the protoplasm of these cells appears to be physiologically nonfunctional or inactive (Esau, 1960).
- Sclerenchyma cells are highly differentiated and are usually considered not to be capable of dedifferentiation and resuming meristematic activity.

Cells are organized into simple tissues (one cell type) and complex tissues (more than one cell type). For example, young ground and pith tissue found in stems is a simple tissue made up of parenchyma cells (Esau, 1960), as is the mesophyll tissue in a leaf. Another example would be the collenchyma tissue found in the four corners of a mint stem (Figure 2.5d) or the "strings" in a celery stalk (petiole). An excellent example of a complex tissue is the secondary vascular tissues found in many dicots. This tissue contains representatives of both parenchyma and sclerenchyma cell types. Tissues, in turn, are organized into the four primary organs—roots, stems, leaves, and reproductive structures. The remainder of this chapter will be devoted to illustrating the general arrangement of cells and tissues within these organs and a brief account of seed anatomy. Anatomical studies of the tissue culture regeneration of organs are included where applicable.

ROOTS

Roots serve as the primary water- and mineral-absorbing organ of plants. They also act to anchor the plant in the soil and may also function as storage organs and in vegetative (asexual) reproduction. Roots vary greatly in diameter, density of root hairs, and degree of lateral branching. Dicots and gymnosperms typically have a persistent taproot and may exhibit secondary growth, whereas with many monocots, the taproot is ephemeral and replaced with a fibrous root system consisting of many adventitious roots. The root morphology of ferns is fibrous and adventitious, similar to that of monocots, but the internal anatomy is similar with that of dicots.

While the general morphology of the root system is genetically determined, both overall appearance and internal anatomy can be modified by the growth environment. Primary roots typically range from 0.04 to 1 mm in diameter, with monocots often having roots that are smaller in diameter than those of dicots and gymnosperms. Very fine root-like structures produced by the fern gametophyte are greatly elongated single cells, rather than true roots, and are referred to as "rhizoids." Roots can contain a wide variety of pigments, and the color of roots ranges from white to brightly colored to nearly black. Young roots generally contain no pigments and, thus, appear white. Exposure to light may result in a pink pigmentation. However, in some species, the above-ground portions of roots contain chlorophyll, so are green in color. While older roots often appear to be darker in color, pigmentation is not an accurate indicator of maturity or internal anatomy.

The anatomy of roots is extremely variable, but general models may be developed for both monocots and dicots. The anatomy of ferns and gymnosperms is similar to that of dicots. The apical meristem of most roots (Figure 2.1a) appears less conspicuous than shoot meristems (Figure 2.4a), which are arranged as a tunica and a corpus. Ferns have a single large, triangular, meristematic apical cell rather than the multicellular meristem found in seed plants. In many plants, the root apical meristem gives rise to the cells of both the root cap and the primary meristems; however, in some grasses, a group of cells, the calyptrogen, produces the cells of the root cap (Figure 2.1a). In many instances, a quiescent ("quiet") zone is located in the apical meristem. This region exhibits low mitotic activity, but cell division typically resumes distal to this zone. The most noticeable differentiation of cells is in the vascular tissue—the primary phloem differentiates first, followed by the primary xylem. Cells continue to mature by elongation and enlargement. Root hairs, extensions of epidermal cells that increase the surface area to absorb water and minerals, are typically first seen behind the zone of mitotic activity and mark the maturation of the first xylem elements. Root hair length and density is genetically determined and fairly consistent within a species, but is highly variable between species.

Lateral or secondary roots greatly increase the absorptive area of the root system and are usually initiated from the pericycle, a layer or layers of cells in the vascular tissue or stele, at some distance behind the apical meristem (Figure 2.1b). Several adjacent pericycle cells divide to form a root primordium, and continued divisions force the developing root through and crush the endodermis, cortex, and epidermis of the primary root. Vascular tissue within the lateral root is connected to similar elements in the primary or parent root by differentiation of pericycle cells (Esau, 1960). The number of lateral branches is influenced by water, nutrient availability, and biotic interactions. Genetic and hormonal control of lateral branch production is fairly well understood in dicots, whereas less is known concerning the control of adventitious root proliferation in monocots (Osmont *et al.*, 2007). Associations with ectomycorrhizal fungi often result in clusters of short lateral roots that are covered with the fungal sheath. These are often visibly distinct, being dark in color, and occurring in clusters or in pairs

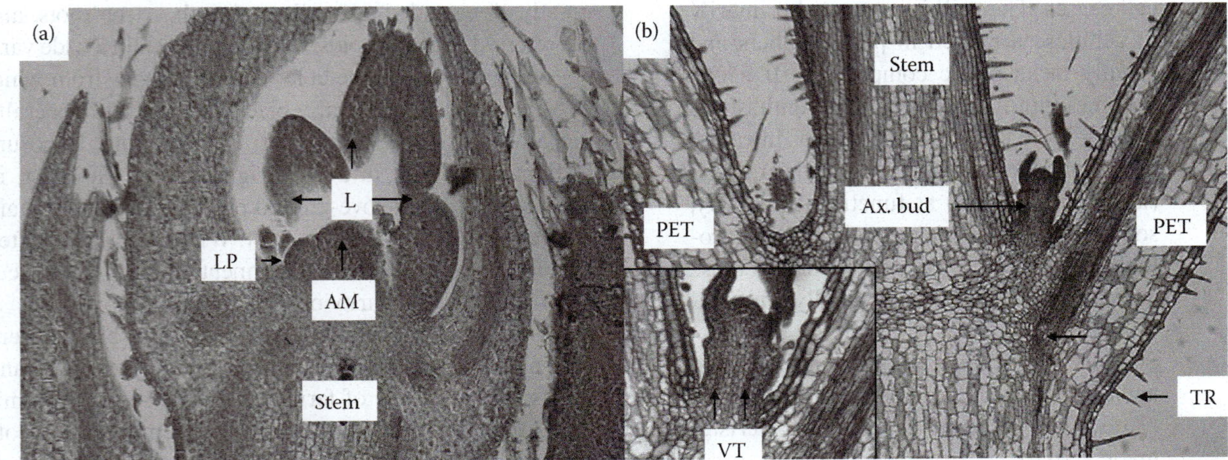

FIGURE 2.4 Apical and lateral (axillary) shoot meristems. (a) Longitudinal section through the stem tip of bean (*Phaseolus vulgaris*). The apical meristem (AM) is surrounded by a number of very young leaf primordia (LP) and more developed leaves (L). (b) Longitudinal section through a node of catnip (*Nepeta cataria*). The axillary bud (Ax. bud) is located in the axil formed by the leaf petiole (PET) and the stem. Note the vascular tissue (VT) extending from the petiole and connecting to the vascular tissue of the stem (arrow). The apical meristem (AM) is dome shaped, and the vascular tissue (VT) has differentiated (insert). TR = trichome.

(Figure 2.2a, b). The pattern of branching is termed "root architecture." Lateral roots growing directly from the first, or primary, root are termed "first-order lateral roots"; laterals growing from those are termed "second-order lateral roots," and so on.

Most of the same cells and tissues are present in monocots and dicots; however, the arrangement of these tissues is somewhat different. A summary of the main differences between the two divisions of angiosperms is presented in Table 2.1. The most conspicuous difference between the dicot and the monocot root anatomy is the arrangement of the primary vascular tissue. Most dicot roots lack central pith tissue, but instead, the core of the root is occupied by large metaxylem vessels with relatively large-diameter lumens (Figure 2.3a, b). Vascular tissue is contained in a single central area called the "stele," with the primary xylem arranged in arches or arms and the primary phloem located between the arms. Ferns generally have only two arms, termed "diarch development." Gymnosperms and dicots often have three to four arms, triarch or tetrarch development, although some species have as many as nine. A vascular cambium is located between the primary xylem and the phloem in dicot roots. In contrast, many primary monocot roots

TABLE 2.1
A Summary Guide to the Morphological and Anatomical Traits of Monocots and Dicots[a]

Organ	Monocot	Dicot
Root	• Usually fibrous • Pith present • Lacks vascular and cork cambia or secondary growth	• Usually a taproot • Lacks pith • Vascular and cork cambia and secondary growth present • Primary xylem typically arranged in "arches"
Stem	• Lacks pith, but vascular bundles embedded in pith-like fundamental tissue • Vascular and cork cambia typically lacking; no secondary growth, but may have primary thickening meristems	• Pith present • Vascular and cork cambia typically present; secondary growth evident manifested as rings of vascular tissue
Leaf	• Typically blade-like with parallel venation • Leaf mesophyll generally undifferentiated into distinct layers	• Variously shaped with net venation • Mesophyll may be differentiated into spongy and palisade parenchyma layers
Flower and seed	• Typically three-merous (flower parts in three or multiples of three) • Embryo has one cotyledon	• Typically four or five-merous (flower parts in four or five or multiples of four or five) • Embryo has two cotyledons

[a] Although this table provides very broad characterizations and contrasts of monocot and dicot plants, the reader is cautioned that there are many exceptions to these generalizations.

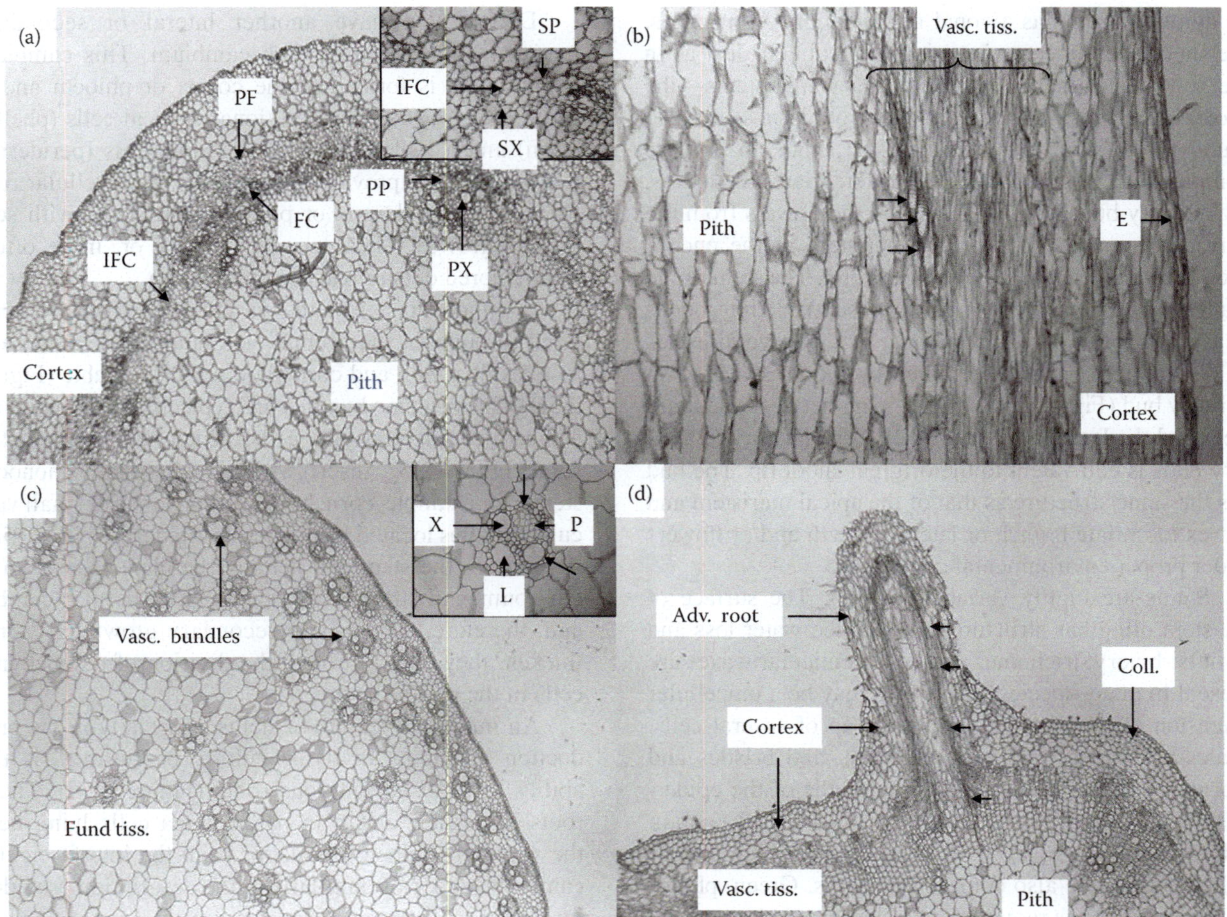

FIGURE 2.5 Anatomy of dicot and monocot stems. (a) Cross section of a bean (*Phaeseolus vulgaris*) stem. Note the prominent pith and the cap of phloem fibers (PF). Secondary xylem (SX) and secondary phloem (SP) tissues, which originated from the interfascicular vascular cambium (IFC) and fascicular cambium (FC), are also common in dicot stems (insert). PP = primary phloem; PX = primary xylem. (b) Longitudinal section of bean illustrates the arrangement of tissues in the stem. Arrows indicate xylem vessels with helical secondary wall patterns. E = epidermis. (c) Cross section of an older corn (*Zea mays*) stem. Notice that vascular bundles (Vasc. bundles) are scattered throughout the stem and "embedded" in fundamental tissue (Fund. tiss.). The vascular bundles contain only primary xylem (X) and phloem (P) tissues (vascular cambium absent) with a prominent lacuna (L) or air space (insert). Arrows indicate sclerenchyma cells of the sheath. (Slide courtesy of Carolina Biological Supply Co., Burlington, North Carolina.) (d) Cross section through the stem of catnip (*Nepeta cataria*) and a longitudinal section through an adventitious root (Adv. root). The root originated near the primary phloem, and vascular tissue (arrows) has differentiated and connected to the vascular tissue (Vasc. tiss.) in the stem. Coll = collenchyma tissue.

have pith in the center of the primary root, surrounded by many arms of vascular tissue (Figure 2.3c, d). Dicot roots generally exhibit secondary growth via the vascular cambia located between the primary xylem and the phloem and by the cork cambia, which form in the cortex (see discussion in stems). Monocots and ferns lack lateral cambia or meristems.

Some species have roots that are highly specialized for a specific function, and these may have an internal structure that is somewhat different than the other portions of the root system. Dicots may have roots specialized for storage; these have a large diameter due to a proliferation of parenchyma cells within the secondary xylem and phloem, and may have several concentric layers of vascular cambia producing secondary vascular tissue. Aerial roots, or aerial portions of a root, generally lack root hairs, have smaller vessels than subterranean roots, and may be green due to the presence of chlorophyll in the cortex. Some roots, both above and below the ground, have a cortex that appears "spongy" to facilitate the diffusion of gases. Specialized sections of root may be produced to house associated organisms such as mycorrhizal fungi and certain types of bacteria. Specialized anatomy can also be found in parasitic roots.

STEMS

As noted before, the shoot meristem (Figure 2.4a) of monocots and dicots is multicellular and organized into two parts—the tunica and the corpus—while that of gymnosperms has a zonal organization. As in roots, the shoot apex of ferns is a single apical cell. Just as in the root apical meristem, the primary meristems—the ground, the procambium, and the protoderm—are also found in the shoot tip. However, the shoot tip is more complex than the root apex as it differentiates the leaves, the axillary buds (Figure 2.4a), and the flowers from the peripheral zone. The primary meristem at the end of each growing branch is the terminal bud. The stem of the plant is divided into nodes (leaves present) and internodes (leaves absent). There may be one, two, or several leaves at each node. Leaves are defined by the presence of an axillary bud (Figure 2.4b, insert) lying between the main stem and the leaf petiole (the axil). The anatomy of axillary buds is equivalent to the original shoot tip. The bud has the same structure as that of the apical meristem and serves to initiate branch or lateral growth and/or flowers under proper environmental conditions.

Stems are highly variable in form. The surface of the stem often has structures that reduce water loss and limit herbivory; trichomes and heavy cuticular waxes are present in many species. Trichomes may be a unicellular extension of the epidermis or made up of several cells. Scales are papery and one cell thick, and bristles and thorns are rigid, multicellular outgrowths of the epidermis. Many young stems and some older stems contain chlorophyll, so are green. Anthocyanins, which appear red or purple, are also common in stems. Green, photosynthesizing stems require a means of gas exchange, and this is provided by the stomata in very young stems.

The arrangement of tissues in stems is variable and very different than that in roots. Just as in root anatomy, dicot and monocot stems are generally different from one another, and these differences are summarized in Table 2.1. Dicots typically have primary vascular bundles (fasciculars) arranged in a ring around the central pith (Figure 2.5a, b). The primary xylem is located toward the pith, whereas the primary phloem is located toward and contiguous with the cortex. Note that, in some instances, the primary phloem may lie on either side of the primary xylem. This arrangement is common in ferns, which are similar in structure to dicots, although some primitive ferns lack pith. The fascicular may also have a cap of very conspicuous phloem fibers (Figure 2.5a). A portion of the vascular cambium (intrafascicular cambium) is located between the primary xylem and the phloem, and another portion (interfascicular cambium), between the vascular bundles. The activities of the vascular cambium produce secondary vascular tissues, which, in due course, surround the pith with a continuous ring of vascular tissue. The secondary vascular tissues eventually crush and obliterate the primary tissues, especially in perennials and woody plants. It is this continuous ring of vascular tissue that easily differentiates the pith from the cortex in dicots.

Dicots also have another lateral or secondary meristem—phellogen or cork cambium. This cambium or meristem is formed in the cortex or phloem and is responsible for forming additional cortical cells (phelloderm) and a replacement for the epidermis (periderm). Gas exchange is provided by lenticels, multicellular ruptures of the epidermis or periderm, in stems with secondary growth. These appear as dark or, more often, light-colored dots or patches on the stem.

In monocot stems, the vascular bundles are scattered throughout the ground or fundamental tissue, and as a result, a pith and cortex are not discernable (Figure 2.5c). The vascular bundles are typically surrounded by a sheath of sclerenchyma cells, which helps support the stem (Figure 2.5c, insert). Additionally, many monocot stems, for example, corn, have an abundance of small vascular bundles located just under the epidermis, imparting stiffness to the stem and helping the plant to withstand environmental stresses. Monocots lack vascular cambia and, therefore, do not have secondary growth. If stems thicken, they do so through the division of parenchyma cells in the ground tissue.

An important feature of stems, especially in the production of shoots in tissue culture or cuttings, is the ability to form adventitious roots (Figure 2.5d). These roots have their origin in parenchyma cells lying near the vascular (phloem) tissue, or from the interfascicular cambium, and grow similarly to the description provided for lateral roots. In many species, adventitious roots form most readily at nodes.

There are many modified stems among different plant taxa. Rhizomes are common in monocots and are the primary stem form of ferns. These grow horizontally underground, generally have short internodes, and produce both roots and leaves. Stolons are similar to rhizomes but are more often found in dicots, and generally have long internodes, grow horizontally above the ground, and produce a new individual at each node. Tubers, corms, and bulbs are below-ground stems modified for starch storage. The term "tuber" refers to a swollen below-ground structure. This can be a root, or a modified stem as in a potato, which can produce new shoots from axillary buds at, sometimes inconspicuous, internodes.

LEAVES

Leaves are the primary photosynthesizing organs of vascular plants. Most angiosperm leaves are relatively thin and flat (large surface area compared to volume) and adapted for capturing light and facilitating gas/water exchange with the atmosphere. Fern leaves are similar, but are called "fronds." The leaves of most gymnosperms are elongated, may be round, oval, or triangular in cross section, and are commonly referred to as "needles." Other gymnosperms have small scalelike leaves that grow closely oppressed to the stem. There are, of course, many exceptions to the above statements, and leaves can exhibit extensive, often unusual, modifications, depending on the environment coupled with genetic inputs and evolution.

Leaves are found at the nodes of the stem and are variously arranged (phyllotaxis)—alternate (one leaf at the node), opposite (two leaves at the node), or whorled (three or more leaves at the node). Some plants have leaves that are very difficult to recognize as leaves, whereas others are very obvious. Generally, leaves have three parts—blades or lamina, petioles, and stipules—although it is not unusual for many leaves to lack the petiole (sessile leaf) and/or the stipules. The blade may be simple with smooth or toothed margins; shallowly or deeply lobed; or compound, with the leaf divided into leaflets (Figure 2.6). There are many arrangements found in compound leaves, the variations of which are two common patterns: palmate, where the leaflets are all connected at the top of the petiole, and pinnate, with the leaflets forming two rows along the central midvein. All leaves must have an axillary bud associated with them, and the bud is located in the axis formed by the petiole and the stem. Compound leaves do not have buds at the base of the branch points or where leaflets join the common axis.

Many dicot and fern leaves exhibit net venation patterns with major and minor veins (Figure 2.6a), whereas monocots and gymnosperms typically have a parallel venation arrangement, with most veins of more or less equal size (Figure 2.6b). Note that there are many exceptions to this general rule. Many gymnosperms have a single vascular bundle, or two bundles that run side by side, down the length of the needle. Anatomically, the vascular tissue in the node will exhibit a leaf gap, parenchyma tissue in the vascular cylinder of the stem where the leaf trace(s) (vascular tissue connecting the leaf to the stem) is/are bent toward the leaf petiole (Esau, 1960).

Leaves typically do not exhibit secondary growth and, therefore, only have primary tissues. Leaf tissue can contain all the three basic cell types discussed previously. The epidermis of leaves has many special features, including a cuticle to prevent wall loss, various trichomes (hairs), guard cells and accessory cells to regulate water and gas exchange, and other cells with specialized functions (Figure 2.7a, b). Large bulliform cells in the epidermis of many grasses can "deflate," allowing the blade to fold or roll. A hypodermis beneath the epidermis of conifer needles protects inner tissue from mechanical injury. The leaf mesophyll of many dicots and ferns is differentiated into one or more palisade layers toward the adaxial (top) surface of the leaf, and the spongy mesophyll, toward the abaxial (bottom) surface. The palisade cells are elongated (shoe box–shaped) and arranged "tightly" and orthogonally (right angle) to the leaf surface. The cells of the spongy mesophyll tissue are isodiametric to slightly

FIGURE 2.6 Simple and compound leaves, with a petiole indicated by the arrow in each image. (a) Simple leaves with opposite leaf arrangement in privet (*Lingustrum japonicum*). (b) Simple leaves with an alternate leaf arrangement in redbud (*Cercis canadensis*). (c) Simple leaves of *Taxus* species, with whorled leaf arrangement. (d) Simple linear leaves of a monocot, a fescue grass (*Festuca* spp.). Note the horizontal stem, a rhizome (RH), from which the roots and the vertical stem are growing. (e) Pinnately compound leaves of *Cardamine hirsuta*. (f) Palmately compound leaf of buckeye (*Aesculus flava*).

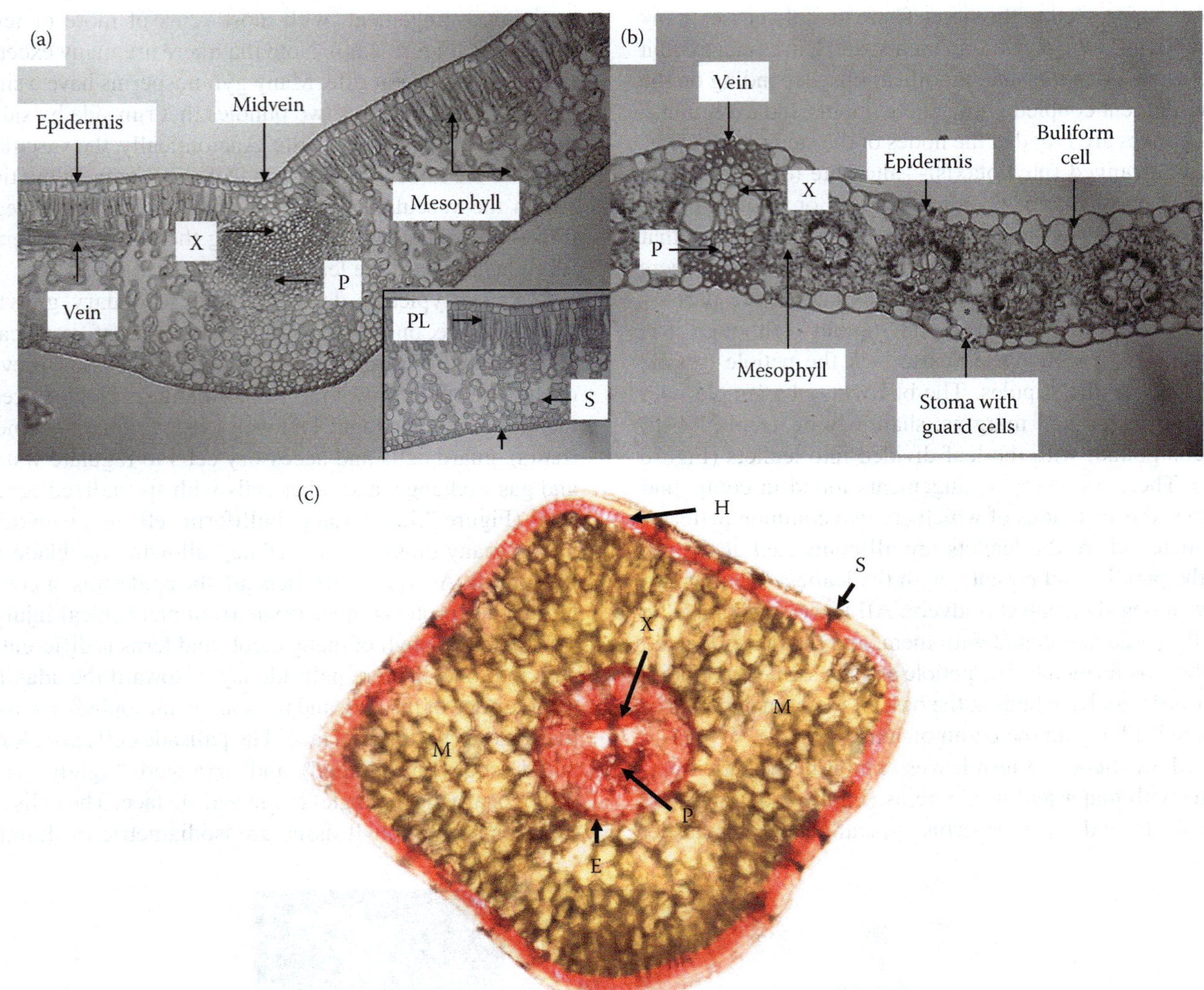

FIGURE 2.7 Leaf anatomy. (a) Cross section of a dicot leaf (privet: *Ligustrum* species). A large midvein with xylem (X) toward the top (adaxial) surface and phloem (P) oriented toward the bottom (abaxial) surface is shown. The mesophyll is differentiated into palisade (PL) and spongy layers (S), with stomata and guard cells (arrow) on the bottom surface (insert). (b) Cross section through a corn (*Zea mays*) leaf, a typical monocot. Notice that the epidermis contains specialized bulliform cells and that the mesophyll tissue is not differentiated into layers. Stomata and guard cells are associated with large substomatal cavities (*). (a and b courtesy of Carolina Biological Supply Co., Burlington, North Carolina.) (c) Anatomy of a spruce leaf. Cross section of a gymnosperm leaf (*Picea pungens*). The single vein of xylem (X) and phloem (P) are encircled by the endodermis (E). The mesophyll (M) is dense and undifferentiated. Notice that immediately below the epidermis is a layer of sclerenchyma, the hypodermis (H), which contains channels for airflow from the stomatal opening (S) to the mesophyll.

elongated and arranged "loosely," with large intercellular spaces (Figure 2.7a). Xylem tissue is usually toward the top of the leaf, and the phloem, toward the bottom of the leaf. The mesophyll of monocot leaves and primitive ferns is typically not differentiated into layers and only has spongy parenchyma (Figure 2.7b). Vascular tissue exists as bundles running through the mesophyll of the leaf. The mesophyll of gymnosperms is also undifferentiated and compact, often having highly invaginated cell walls. Large, circular resin ducts may be found within the mesophyll. The vascular tissue is surrounded by transfusion tissue, made up of parenchyma and thin-walled tracheids, which, in turn, are surrounded by a suberized endodermis, similar with that of roots (Figure 2.7c). In many plants, the vascular tissue may be surrounded by a sheath of sclerenchyma cells. Vascular tissue of grasses is enclosed by a layer of parenchyma called "bundle sheath cells," which, in C_4 plants, contain many large chloroplasts (Figure 2.7b).

REPRODUCTIVE STRUCTURES

Flowers are organs unique to angiosperms. They are incredibly diverse in morphology and range from very showy and conspicuous to very bland and almost imperceptible. They may occur singly or be arranged in different types of inflorescences (multiple flowers on a common axis). If a plant has male flowers (staminate) and female flowers (pistillate) on one plant, we call the plant "monoecious." If male flowers exist on one plant, and female flowers on another, we call the plant "dioecious" (two houses). Meiosis and alternation of generations occur in the ovule (female) and the anther (male). The male and female gametophytes are greatly reduced, consisting only of a few cells.

For the sake of simplicity, we will only consider perfect flowers (all flower parts, including both male and female structures, are present). The four basic units in a flower are the sepals, petals, stamens, and pistils (Figure 2.8a–c). The floral parts are arranged as whorls on the receptacle (Figure 2.8b). Working from the outside, inward at the base of the receptacle, the sepals, usually green, leaf-like structures, are the first whorl encountered, and these are collectively termed the "calyx." The next layer is composed of the petals, which can be green but are generally white or colored. The petals are usually larger than the sepals and are collectively termed the "corolla." The petals and sepals, when taken together, are also called the "perianth." In some cases, when the sepals and the petals are similar or almost indistinguishable, they are called "tepals" (Figure 2.8c). The stamens

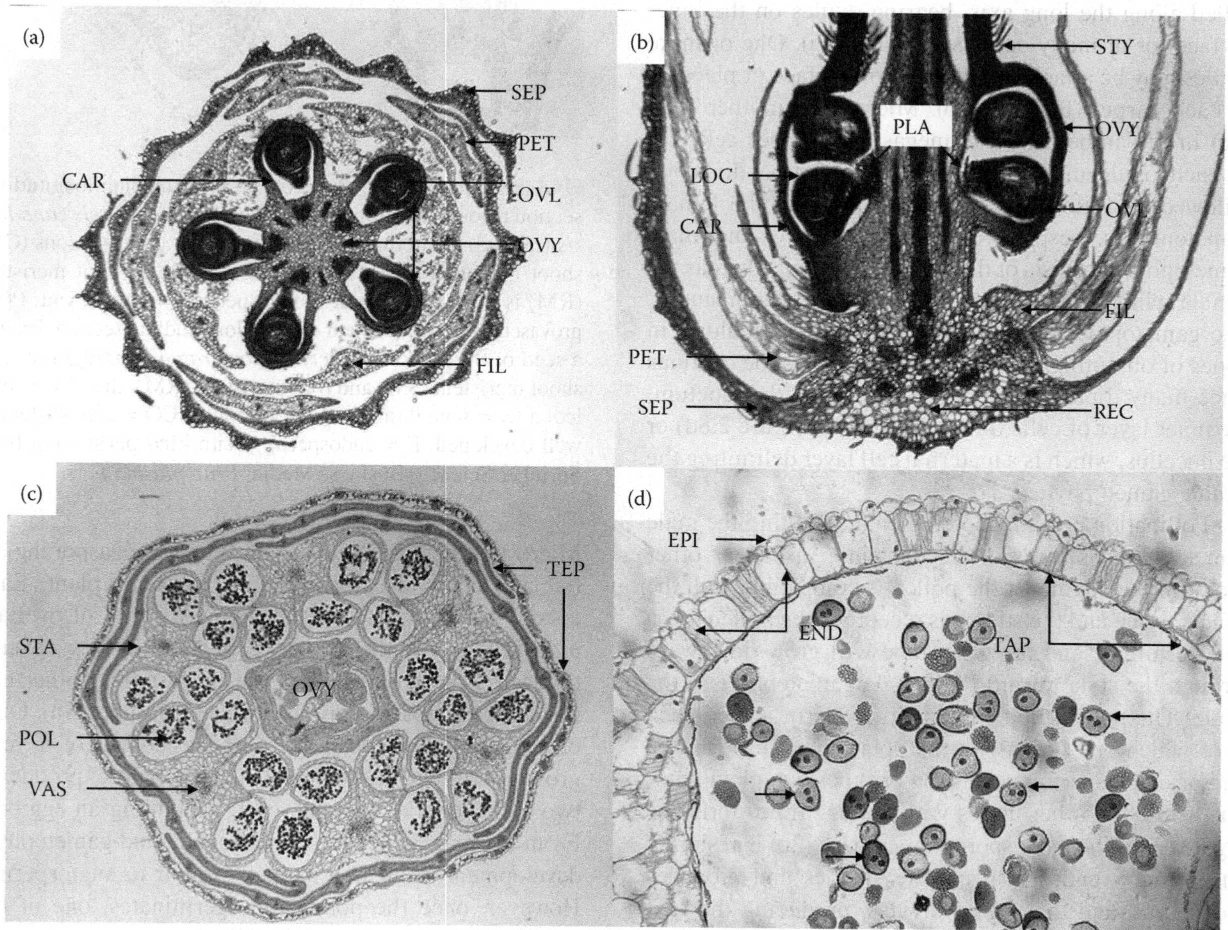

FIGURE 2.8 Flower anatomy. (a) Cross section through a *Geranium* species (Stork's bill) flower. Progressing from the outside inward, sepals (SEP), petals (PET), and filaments (FIL: the stalks of the anther) are borne. Occupying the center of the flower is the ovary (OVY), which is composed of five carpels (CAR) in which two ovules (OVL) are borne. Note that only a single ovule can be seen in each locule (space) with this section. (b) Longitudinal section through the Stork's bill flower seen in (a). In this view, two carpels (CAR) and locules (LOC) of the ovary (OVY) are visible, but each contains only two ovules (OVL) of which only one will mature placenta. Note the trichomes (hairs) on the style (STY) portion of the ovary. The sepals (SEP), petals (PET), and filaments (FIL) are inserted below the ovary on the receptacle (REC). (c) Cross section of a lily (*Lilium* species) flower. In this view, six stamens (STA) containing pollen (POL) are shown. The vascular tissue (VAS) in the filament and four lobes of each anther are evident. Sepals and petals are collectively termed "tepals" (TEP) for lily flowers. OVY = ovary. (d) Higher magnification of an anther of lily. The anther wall is composed of the epidermis (EPI), the endothecium (END), and the tapetum (TAP), which, in this case, has degenerated and is only represented by the remnants of the cells. Many pollen grains are present—some in the binucleate stage (arrows). (c and d courtesy of Carolina Biological Supply, Burlington, North Carolina.)

or the male portion of the flower are the next set of structures found in the flower. A stamen consists of the filament or stalk (Figure 2.8a, b) that terminates in the anther, which contains pollen (Figure 2.8c, d). The pistil or female structure, which is more or less flask-shaped, occupies the center of the flower and consists of a swollen base, the ovary, a smaller-diameter stalk, the style, which ends in a somewhat swollen tissue, the stigma (Figure 2.8b; stigma not shown). Dicots typically have sepals, petals, and stamens in whorls of four or five, or multiples of four or five (Figure 2.8a), whereas monocots generally have these structures in groups of three or multiples of three (Figure 2.8c). Pistils may be composed of a single carpel (a structure analogous to a leaf rolled along the long axis, bearing ovules on the inner surface) or of many carpels (Figure 2.8a). One or more ovules may be attached to a common surface or placenta in each carpel (Figure 2.8b). Microspore mother cells (2n) in the anther and one megaspore mother cell (2n) in each ovule undergo meiosis (a reduction division), followed by mitosis, to form the male and the female gametophytes, respectively. When mature, the male gametophyte consists of three cells or nuclei, whereas the female gametophyte typically has eight cells or nuclei. The gametophytes are often placed in tissue culture in hopes of obtaining haploid plants. One must be cautious since many diploid plants may arise from the tapetum, the inner layer of cells lining the anther (Figure 2.8d) or the nucellus, which is a maternal cell layer delimiting the female gametophyte.

Pollination occurs when pollen (containing the male gametophyte) is transported via wind, insect, or other vector to the stigma. If the pollen is compatible with the female tissue and the stigma is receptive, the pollen grain will germinate and the germ tube will grow downward through the style toward the female gametophyte in the ovule. The pollen tube contains two sperm nuclei (each n) of which one fuses with the polar nuclei to form the primary endosperm nucleus (3n or other polyploid number), while the other unites with the egg (n) to form the zygote (2n). Thus, the sporophytic phase (2n) is restored. The primary endosperm nucleus divides and produces endosperm, and the zygote divides, producing the suspensor and the embryo. The embryo develops into a bipolar structure, possessing both shoot and root meristems and exhibiting bilateral symmetry. The dicot zygotic embryo has two cotyledons (Figure 2.9a), and the monocot embryo has one cotyledon (scutellum; Figure 2.9b). The integuments (seed coat) covering the ovule harden, and the ovule is now a seed (Figure 2.9a). The endosperm may be completely absorbed by the embryo or may remain outside the embryo.

Gymnosperms produce seeds, but do not flower, and with the exception of several non-coniferous gymnosperms, the cone is the primary reproductive structure.

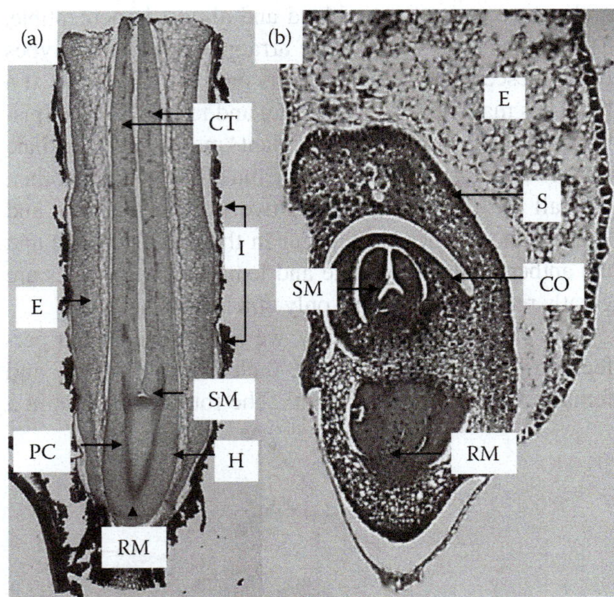

FIGURE 2.9 Zygotic embryos. (a) Near-median longitudinal section through an immature seed of the dicot *Cercis canadensis* (redbud). Note the well-developed pair of cotyledons (CT), shoot meristem (SM), and hypocotyl (H); the root meristem (RM) is inconspicuous. E = endosperm; I = integument; PC = provascular tissue. (b) Near-median longitudinal section through a seed of the monocot *Dactylis glomerata* (orchardgrass). The shoot meristem (SM) and root meristem (RM), the single cotyledon (S = scutellum), and the first leaf (CO = coleoptile) are well developed. E = endosperm. (With kind permission from Springer Science+Business Media: *Protoplasma*.)

Microsporangiate cones (male) and megasporangiate (female) cones usually occur on the same plant. Each cone consists of an axis supporting a spiral of overlapping scales. On the female cone, two ovules develop on each scale within which the megaspore mother cell undergoes meiosis to produce four megaspores (n). Only one of these develops and, over the course of a year, grows to thousands of cells in size before producing two archegonia at one end, each containing an egg cell. From the male cone, pollen production and gametophyte development occur in a manner similar to angiosperms. However, once the pollen tube germinates, one of the sperm fuses with the egg to produce the zygote (2n), and there is no endosperm produced. Here, the seed coat and the thin layer of tissue to the inside of it originate from the female parent (2n), but the tissue that surrounds the zygote is megagametophyte (n). The embryo develops as in angiosperms, but with three to six cotyledons.

In ferns, no seeds are produced, and the alternation of generations is more apparent, with the gametophyte being a free-living haploid (n) stage known as a "prothallus." The prothallus is generally 2 to 20 mm in diameter and composed entirely of parenchyma cells. These are variously shaped, but are often flat and heart-shaped, with thin rhizoids to serve the function of the roots. They may contain chlorophyll and grow on the soil surface, or be colorless and dependent on mycorrhizal associations. The prothallus produces cup-shaped archegonia (female), inside which a single egg is produced, and antheridia (male) that produce numerous motile sperm. A single prothallus may produce both structures, or they may occur on separate individuals. Water is required for the sperm to reach the egg and for fertilization to occur. The resulting zygote grows into the familiar sporophyte generation (2n). When mature, clusters of sporangia called "sori" are produced on specialized fronds, or in rows or along the margins of regular fronds. Within the sporangia, the spore mother cell undergoes two meiotic divisions to create a tetrad of spores (n) that are shed and germinate to grow into the prothallus. Numerous variations on this cycle exist.

CONCLUSIONS

We have presented a very brief account of the anatomy of angiosperms, gymnosperms, and ferns, and we hope that we have whetted your appetite to further your study of this area of plant science. The study of plant anatomy allows us to understand more fully the process of plant development from the union of the egg and the sperm to the formation of the zygotic embryo to the development of the various tissues in the embryo and the mature plant—fascinating processes. The study of anatomy also allows us to somewhat understand the origin and the development of tissues from various propagules, including rooted cuttings and grafting.

If there is a laboratory associated with your course, we suggest that the following microscope slides (one supplier is Carolina Biological Supply, Burlington, North Carolina) may be useful for students to study:

- Dicot & Monocot Roots c.s. (30-1898)
- Typical Monocot & Dicot Stems c.s. (30-2642)
- Privet Leaf (Typical Dicot) *Ligustrum* c.s. (30-3838)
- Corn Leaf (Typical Monocot) *Zea* c.s. (30-4054)
- Typical Monocot & Dicot Flower Buds c.s. (30-4258)
- Shepherd's Purse Mature Embyo *Capsella* l.s. (30-4888)

LITERATURE CITED AND SUGGESTED READING

Dickinson, W. C. 2000. *Integrative Plant Anatomy*. San Diego, CA: Academic Press, 533 pp.

Esau, K. 1960. *Anatomy of Seed Plants*. New York: John Wiley and Sons, Inc., 376 pp.

Fahn, A. 1990. *Plant Anatomy*. Fourth Edition. New York: Pergamon Press. 588 pp.

Gifford, E. M. and R. H. Wetmore. 1957. Apical meristems of vegetative shoots and strobili in certain gymnosperms. *Proceedings of the National Academy of Sciences of the United States of America* 43:571–576.

Jones, D. L. 1987. *Encyclopaedia of Ferns*. Melbourne, Australia: Lothian Publishing Company, 433 pp.

McDaniel, J. K., B. V. Conger, and E. T. Graham. 1982. A histological study of tissue proliferation, embryogenesis, and organogenesis from tissue cultures of *Dactylis glomerata* L. *Protoplasma* 110:121–128.

Osmont, K. S., R. Sibout, and C. C. Hardtkey. 2007. Hidden branches: Developments in root system architecture. *Annu Rev Plant Biol* 58:93–113.

Raven, P. H., R. F. Evert, and S. E. Eichhorn. 1999. *Biology of Plants*. Sixth Edition. New York: W. H. Freeman and Co., 944 pp.

Smucker, A. J. M. 1993. Soil environmental modifications of root dynamics and measurement. *Annu Rev Phytopathol* 31:191–213.

Trigiano, R. N. and D. J. Gray. 2010. *Plant Tissue Culture, Development, and Biotechnology*. Boca Raton, Florida: CRC Press LLC, 583 pp.

Wilson, C. L., W. E. Loomis, and T. A. Steeves. 1971. *Botany*. Fifth Edition. New York: Holt, Rinehart, and Winston, 752 pp.

3 Plant Physiology Concepts Important for Propagation Success

Caula A. Beyl and Govind C. Sharma

Concept Box 3.1

- An understanding of both the anatomy and the physiology of plants is critical for successfully propagating plants whether by sexual or asexual propagation techniques.
- Vegetative (asexual) propagation and sexual propagation differ in their ability to give rise to true-to-type progenies.
- Conceptually, each plant cell is endowed with the genetic information to potentially divide, grow, and differentiate into specialized cells and tissues, leading to a fully functional organism. This concept is referred to as "totipotency."
- Meristems are stem cells of plants; they proliferate, regulate, and direct plant growth.
- Plant propagation principles are well established, but their application is species and often individual genotype and growth stage dependent.
- Rooting of cuttings is one of the most common forms of plant propagation and success can be influenced even before the cutting is taken as a result of how the stock plant is grown and treated.
- Natural plant hormones interact and shape growth, differentiation, dedifferentiation, and many physiological events during their vegetative and sexual growth cycles. Propagators can manipulate these and even apply synthetic plant growth regulators to induce the responses that they desire.
- Some plant species are recalcitrant to propagation, and numerous combinations of genotype, media, hormones, and environment are used by plant propagators to overcome recalcitrance.
- Juvenile plants respond more readily to cutting propagation than do mature plants.
- Propagators use various techniques to "rejuvenate" stock plants and increase the likelihood of rooting cuttings. These include severe pruning, serial propagation onto juvenile seedlings, and the use of micro-propagated plants.
- Techniques such as blanching, etiolation, banding, and stool bed layering all work via the reduction of light exposure to enhance rooting.

Plant physiology is the study of how plants function, and an understanding of the basic plant physiological principles helps explain the success or the lack of success of the various techniques and practices used in plant propagation. A multitude of genetic or physiological factors influence plant response, whether the propagation occurs in the field, greenhouse, or in the laboratory. Metabolic and physiological processes in the plant are profoundly influenced by the environment of the plant, including light (intensity; day length; and quality, including light sources and wavelength differences), temperature, water relations, photoperiod, and season. A plant propagator must make decisions about the techniques to be employed, and these will be influenced by the type of plant that is being propagated (annual, biennial, or perennial) and whether a clone is desired. A clone is an offspring that is genetically identical to the parent and produced with mitotic divisions of cells and not meiosis (Chapter 5). The use of cloning goes back to primitive agriculture, where pieces of woody stems were inserted into the ground to root, thus producing new plants. Clonal propagation maintains the genetic integrity of propagules through generations; it is also utilized to bypass or reduce the juvenile period and the characteristics of the growth forms associated with it, such as thorniness, juvenile leaf morphology, and the vegetative or nonflowering state. Much more can be learned about juvenility in Chapter 6. Clones of selected seed-propagated plants allow the widespread exploitation of unique traits or characters. Through the process of budding or grafting, the traits of a single desirable tree or even a branch (with a sport or desirable mutation) can be vegetatively perpetuated. A wonderful example of this is the 'Delicious' apple, which originated as a chance seedling in Iowa and is now commonly budded or grafted onto rootstocks to retain its unique features. Foresters also use cloning to perpetuate valuable traits in trees and describe the original seedling as an "ortet" and use the term "ramet" to designate its vegetative offspring or clone.

The environment of a plant interacts with the genotype to shape the characteristics of the plant, referred to as its "phenotype." For example, clones grown in two parts of the same field, although genetically identical, may show differences in size mainly due to microclimatic or soil properties and reflect gross or subtle differences in water and mineral nutrient availability. Such differences are largely environmentally induced. Just as both genetics and environment interact to shape plants, successful propagation is also affected by both genetics and environment. Furthermore, not only intrinsic and environmental conditions may have a profound influence on the propagule after it is separated from the parent plant, but also conditions that exist(ed) for the donor or parent plant that is used as a source of the propagule or explant.

The influence of individual aspects of the environment can differ dramatically depending on how they interact and affect the processes within the plant. For example, light plays a myriad of roles with respect to propagation and for plant growth. Light drives photosynthesis, which provides energy and carbohydrates to permit the growth and regeneration of roots on cuttings. Reduction of light has a profound effect on the way plants grow, and in the extreme, the absence of light can result in the etiolation of the plant. Etiolated plants, with their characteristic lack of chlorophyll and elongated stems, have a greater tendency to form roots. Light, particularly red and far-red light, regulates the light-sensitive pigment, phytochrome, which, in some plants, determines whether or not seeds will germinate (Chapter 36) and plants will flower. Other photoreceptors are sensitive to blue light and are involved in phototropism. The length of light relative to the dark period or photoperiod is important in signaling plants what part of the year or season it is and when the time is right for budbreak, flowering, senescence, and dormancy. Some species are able to form roots on cuttings in only a very narrow seasonal window during the year.

The following sections give a brief summary of some basic metabolic processes and their regulation within the plant, and examples of how those processes affect propagation success.

BASIC PLANT PROCESSES THAT IMPACT PROPAGATION SUCCESS

Light and Carbohydrate Metabolism (Photosynthesis and Respiration)

Photosynthesis is the "light-harvesting process" in plants. In photosynthesis, carbon dioxide molecules are taken in through the stomata and reduced by the water molecules that are split using the energy of light impinging on the chlorophyll complex located in specialized cell organelles called "chloroplasts." This photosynthetic apparatus of most higher plants reduces CO_2 by using the electrons from the splitting of the water molecule and converts them into molecules of the simple sugar, glucose, while releasing oxygen to our environment (Figure 3.1). The glucose molecules act as stored energy to enable growth and differentiation. Longer-term storage of the carbon fixed during photosynthesis occurs when the monosaccharide glucose is converted to starch, a polysaccharide that consists of long chains or polymers of glucose, or it may be utilized in the synthesis of the polymer cellulose, in forming cell wall lignin, or in a plethora of compounds utilized in cellular functions. Photosynthesis and respiration are basically reciprocal processes in a plant. In respiration, the plant uses the glucose and oxygen to release stored energy, water, and carbon dioxide. Although photosynthesis requires light and takes place in the chloroplasts, respiration can occur in the light or the dark, and occurs in the cytoplasm and the mitochondria of cells irrespective of their location. Photorespiration is also unique and especially deleterious to some angiosperms. Having adequate carbohydrate reserves in the form of

Photosynthesis

$$6\,CO_2 + 6\,H_2O + \text{energy (from light)} \rightarrow C_6H_{12}O_6 + 6\,O_2$$

Aerobic respiration

$$C_6H_{12}O_6 + 6\,O_2 \rightarrow 6\,CO_2 + 6\,H_2O + \text{energy}$$

FIGURE 3.1 Equations for photosynthesis depicting the fixation of carbon dioxide, with addition of water and light to form glucose and release oxygen. Aerobic respiration is the reciprocal of that for photosynthesis.

sugars and starches is important not only for the cutting that needs these resources to generate new roots, but also for the seeds that must have sufficient stored reserves to allow the germination and establishment of the seedling until it becomes photosynthetically competent.

Temperature can affect both photosynthesis and respiration. Photosynthesis typically increases with an increase in temperature up to an optimum (usually within the range of 25–30°C) and then declines more rapidly after that. Respiration typically doubles for every 10° increment over the range of 5 to 35°C (Talling, 1961) until the temperature exceeds the point where plants respire more than what the photosynthesis can compensate or under even higher temperature conditions where cellular enzymes begin to denature. Exact temperature optima can vary with the native environment of the species, the stage of the life cycle, and even the part of the plant. Temperature may have an effect on propagation success, for example, if stock plants experience high nighttime temperatures, the increased respiration can reduce carbohydrates stores, and this will have a negative impact on the rooting ability of the cuttings taken from the stock plants. In another case, it is customary to provide bottom heat when rooting evergreen cuttings in late fall or hardwood cuttings over the winter. The bottom heat provides moderate temperatures for the root zone, promoting callusing and adventitious root formation. The air temperature around the exposed portion of the cutting remains cool and keeps top growth from occurring too soon for the newly formed root system to support it.

Light and Photoperiod Effects on Flowering

Light exerts effects on flowering and seed production, important to the plant breeder and the plant propagator. This effect is termed "photoperiodism." The name is somewhat of a misnomer because it is not so much the period of light that is important, but the length of the uninterrupted dark period. Plants that are responsive to day length are called either "long-day plants" or "short-day plants." Short-day plants, like chrysanthemums and poinsettias, flower only if the uninterrupted night period is longer than a certain critical length (Figure 3.2). Under long-day conditions, long-day plants, like clover, flower if the night period is shorter than a critical period. Plants responsive to day length are further divided into those that are obligate or facultative. Obligate short-day (like African marigold) or long-day (like dill or flax) plants must have the critical night period satisfied before they will flower, whereas facultative short-day (like dahlia) or long-day (like lily) plants will flower eventually once they reach a certain size, but the flowering response is hastened or enhanced by having the critical day length conditions satisfied.

Growers can create the right conditions for flowering for long-day plants even during naturally short-day conditions by extending the day length at the end of the day, a technique known as "day extension," or by providing a 4-h exposure to light in the middle of the dark period, known as "night interruption." Short-day conditions are typically created by pulling black cloth over

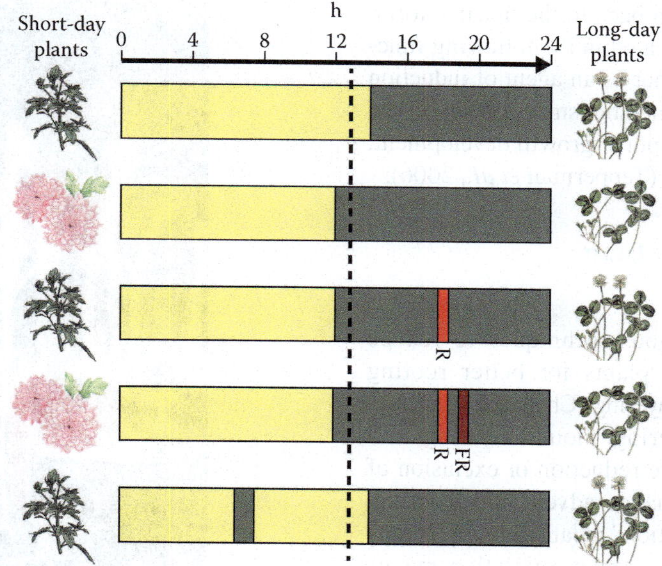

FIGURE 3.2 In the top bar, long-day (short-night) conditions promote flowering for the long-day clover and keep the chrysanthemum vegetative. In the next bar, chrysanthemum flowers are under short-day (long-night) conditions, and the clover remains vegetative. A night break causes short-day conditions to exert the same effect as long days by converting a long night into two short nights. Far-red light (FR) exposure after red light (R) exposure reverses the outcome due to its effect on phytochrome (see Figure 3.3). A dark break provided in the middle of the light period does not affect flowering, indicating that it is the length of uninterrupted night that is critical to flowering responses.

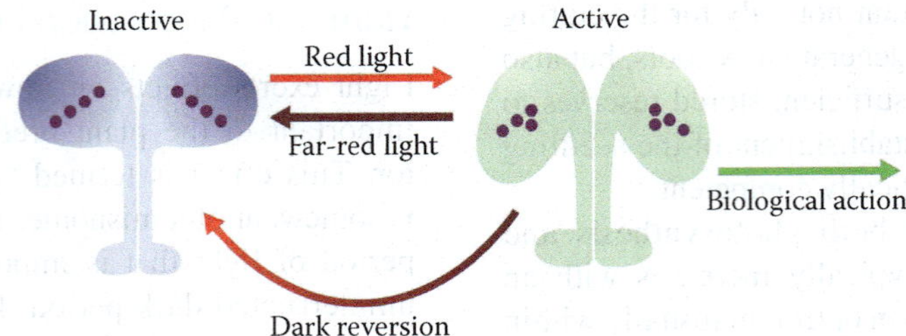

FIGURE 3.3 Phytochrome exists in two forms. The inactive Pr form converts to the active Pfr form upon exposure to red light. This form is the one that triggers biological responses that are sensitive to photoperiod. The active form, Pfr, converts back to the inactive form with exposure to far-red light or slowly through dark reversion. (The shape of the phytochrome molecule was adapted from Smith, H., *Nature* 400:710–713, 1999.)

the plants to create a longer night inductive of flowering. Providing a dark break during the long light period has no impact on flowering since it is the length of the uninterrupted dark period that regulates the flowering response (Figure 3.2).

Photoperiodism is regulated by the light-sensitive pigment "phytochrome" that exists in two forms: an inactive form Pr and an active form Pfr (Figure 3.3). The red form, Pr, is maximally sensitive to 660-nm wavelength light, and the far-red form, Pfr, is maximally sensitive to 730-nm light. With exposure to red light (as you would have in bright daylight), the Pr form is converted to Pfr, which is the form that triggers light-sensitive responses in the plant. Exposure to far-red light (as you would have predominate in shade or in the evening) causes the Pfr form to revert back to the Pr inactive form. In the dark, there is a slow reversion back to the inactive form. Phytochrome has been implicated in the following functional categories of genes either as an agent of induction or repression: transcription, metabolism, photosynthesis/chloroplast, signaling, transport, growth/development, hormones, and stress/defense (Tepperman *et al.*, 2006).

Exclusion or Reduction of Light to Promote Rooting

Propagators have used various techniques to reduce light to precondition stock plants for better rooting responses. Various types of layering (Chapter 24), including air layering, simple layering, mound layering, and stool bed layering, all involve reduction or exclusion of light to some degree to enhance adventitious rooting. Technically, the term "etiolation" means growing plants without light, but to a nurseryman, etiolation means covering stock plants so that the new shoots emerge and grow under very heavy shade or in the darkness (Maynard and Bassuk, 1988). For some species, as little as a 50% reduction of light is effective to promote adventitious rooting as in the case of three Dahlia cultivars, increasing rooting from 7% to 75% (Biran and Halevy, 1973). Reducing the irradiance of stock plants has promoted rooting in such diverse plant species such as pea (Bertram and Veierskov, 1989), juniper (Lin and Molnar, 1980), aspen and willow (Eliasson and Brunes, 1980), and apple (Christensen *et al.*, 1980).

When the shade or light-impervious cover is removed, the base of etiolated shoots is often covered with an opaque band such as an ordinary black, insulating tape or even Velcro bands (Figure 3.4). It is important that the tissues that are banded be as undifferentiated or close to the tip as possible (Gardener, 1937). Applying a wide band is more effective than using a narrow band. With *Carpinus betulus* "Fastigiata," rooting of 87% was obtained with a 7.5-cm band and much less with bands that were thinner (63% with a 5-cm band and 53% with

FIGURE 3.4 Etiolated shoots of cut back oak hybrids before being treated with growth regulator to promote roots. Note the characteristic lack of chlorophyll; elongated shoots; small leaves; and soft, succulent stem tissue. (Courtesy of Nina Bassuk.)

a 2.5-cm band) (Maynard and Bassuk, 1988). After the shoot becomes green again, the shoot is severed just below the band and then placed into the rooting medium after treatment with a growth regulator to enhance rooting (Maynard and Bassuk, 1988).

Increasing the duration of the stem banding of M.9 and MM.106 apple rootstocks of up to 21 days increased both the percentage of rooting (Figure 3.5a) and the number of roots per cutting (Figure 3.5b; Sun and Bassuk, 1991). Two different techniques that have been proven effective with oaks are based on the same principles of etiolation and severe cutting back of the stock plant (Amissah and Bassuk, 2009; Figure 3.6). Both involve cutting back the dormant tree severely and inducing budbreak, followed by etiolation. For cuttings, shoots are banded when they reach 6 to 10 cm in length, and at 28 days, cuttings are harvested, treated with indolebutyric acid (IBA), and placed under mist. After pinching, rooted shoots are ready to pot up at 58 days. A modification of the same technique can be used with dormant trees in the field. After severe cutting back, budbreak, and etiolation, shoots are treated at the base with IBA

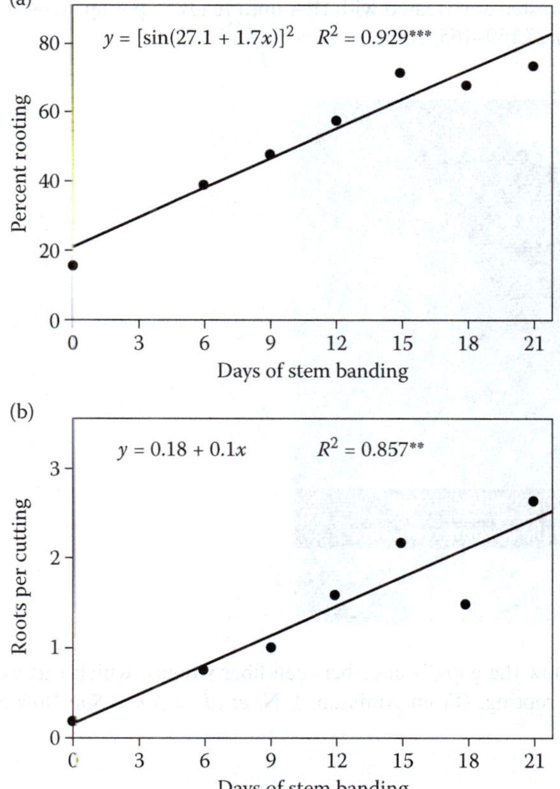

FIGURE 3.5 Percentage rooting (a) and number of roots per cutting (b) of MM. 106 apple rootstock cuttings as a function of the number of days of stem banding after being dipped for 10 sec in 500-mg/L IBA in 50% ethanol. (Redrawn from Sun, W. and N. L. Bassuk, *Hortic Science* 26:1368–1370, 1991.)

when they reach 6 to 10 cm. A bottomless pot is placed on the soil surface over the etiolated/treated shoots, and this pot is filled with a peat–perlite mixture to just under the growing tips. As the shoots grow, more mixture is added, until at 49 days, the bottomless pot can be removed and the rooted shoots can be harvested.

"Blanching" is a term used to describe placing a light-excluding barrier such as a tape around a green stem for several weeks. Reducing light, whether by etiolation or blanching, has several effects on the tissue, which help promote rooting. Tissue beneath the light barrier becomes chlorotic and softer, such as this rose stem after 1 week of being blanched with 2 in (5 cm) of Velcro (Figure 3.4). Other effects include less lignification (Reid, 1922), more undifferentiated tissue (Gardner, 1937), and greater sensitivity to exogenously applied auxin (Herman and Hess, 1963).

"Girdling" (Gardner, 1937), or "ring barking" (Delargy and Wright, 1978), that is, the removal of all or part of the phloem, including the cambium layer, below the blanched area on apple stems enhanced the positive effect on rooting, most likely due to the disruption of auxin transport through the phloem basipetally and allowing carbohydrates produced by the shoot above the girdle to accumulate. However, in the latter case, the positive effect was eliminated when the ring barking was done above the blanched zone. Girdling is also used for air layering of tropical foliage plants and as a preconditioning treatment prior to taking cuttings of species such as sycamore, sweetgum, and pine. Root-promoting growth regulators can be applied to the girdled area, or the plant growth regulator in talc can be applied to the Velcro bands, with dramatic results in some cases coupled with etiolation (Maynard and Bassuk, 1986). With the notoriously difficult-to-root (recalcitrant) *Quercus coccinea*, 48% rooting of cuttings taken from a one-year-old seedling was achieved with both Velcro banding coated with 0.8% IBA and etiolation. No rooting was achieved with light-grown cuttings (banded or not) and with etiolated cuttings with no band.

Physical barriers to developing roots have been suggested as one of the reasons for the inability to form adventitious roots in some hardwood species, and the sclerenchyma layer has been suggested by several studies as an impediment (Beakbane, 1961; Goodin, 1965; Edwards and Thomas, 1980). In a study using *Q. bicolor*, a good correlation was found between the percentage of parenchymous gaps between fibers of sclerenchyma in the stem cross section and the percentage of rooting (Figure 3.7; Amissah *et al.*, 2008), although a low gap percentage was not deemed a causal factor for the poor rooting of *Q. macrocarpa*. Roots emerged from cuttings in a linear fashion (Figure 3.8a), corresponding to the location of the origin of the root primordia associated with the phloem outside each distinct lobe of pith (Figure 3.8b).

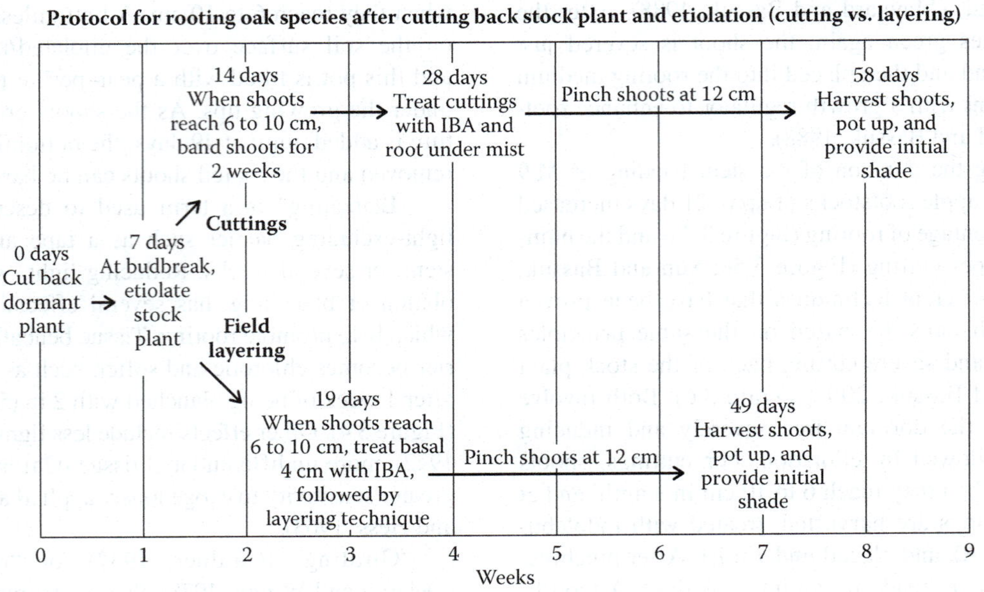

Protocol for rooting oak species after cutting back stock plant and etiolation (cutting vs. layering)

0 days
Cut back dormant plant →

7 days
At budbreak, etiolate stock plant

Cuttings

14 days
When shoots reach 6 to 10 cm, band shoots for 2 weeks →

28 days
Treat cuttings with IBA and root under mist

Pinch shoots at 12 cm →

58 days
Harvest shoots, pot up, and provide initial shade

Field layering

19 days
When shoots reach 6 to 10 cm, treat basal 4 cm with IBA, followed by layering technique

Pinch shoots at 12 cm →

49 days
Harvest shoots, pot up, and provide initial shade

0 1 2 3 4 5 6 7 8 9
Weeks

FIGURE 3.6 Protocol for rooting oaks from cut back dormant plants in the field after they had received their chilling. For field layering (bottom procedure), inverted pots were placed over the stump to allow shoots emerging from the stump to elongate and etiolate. When they had achieved sufficient size (6–10 cm in length), they were treated with IBA, and a bottomless pot was placed over them and filled with a peat–perlite mixture up to just under the shoot tips. Additional medium was added to keep the bases of the growing shoot in darkness until they were rooted and harvested. For cuttings (top procedure), etiolated shoots were banded at the base with Velcro tape after reaching the sufficient size. At 4 weeks, they were harvested and treated with IBA until ready to pot up at 58 days. (Procedure redrawn from Amissah, J. N. and N. Bassuk, *J Environ Hort* 27:159–165, 2009.)

FIGURE 3.7 Transverse section of *Quercus bicolor*. Yellow arrows show the gap distance between fiber strands, which was used to calculate the percentage gap later correlated with the percentage of rooting. (From Amissah, J. N. *et al.*, *J Amer Soc Hort Sci* 133:479–486, 2008.)

Increasing the irradiance has also been reported to increase rooting for some species (Moe and Andersen, 1988). Bertram (1992) proposed a hypothesis that helps explain apparently conflicting evidence. Essentially, as the irradiance of the stock plant increases, auxin also increases until an optimum (level) occurs for rooting, after which both sensitivity to auxin decreases and/or differentiation of cells increases. The increased sensitivity to auxin of etiolated tissue helps explain the enhanced success rate for adventitious root formation of some species under reduced or absent light (Eliasson and Brunes, 1980; Maynard and Bassuk, 1988).

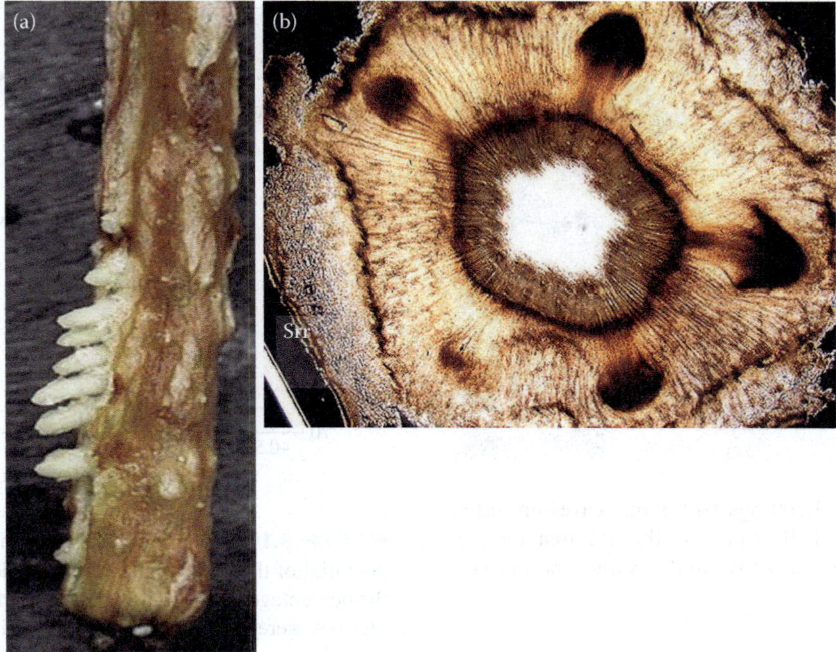

FIGURE 3.8 Root emergence from the cuttings of *Quercus bicolor* in terms of days after the application of 6000-mg/L IBA. Emerging roots (a) (From Amissah, J. N. *et al.*, *J Amer Soc Hort Sci* 133:479–486, 2008.) line up vertically due to the position of the root initiation of the root primordial radially in line with the lobes of the pith depicted at 14 to 16 days after treatment with IBA (b). (Courtesy of Nina Bassuk.)

Transpiration and Water Relations

Water relations are extremely important to consider during propagation as water plays an essential role in plant metabolism, transport, support, and cooling. Water enters the plant primarily through the roots and is transported through the xylem along with various mineral nutrient and other factors throughout the plant. Water vapor exits the plant through the stomata on the leaves and provides evaporative cooling through the process of transpiration in response to a difference in the concentration of water vapor inside the leaf relative to that outside the leaf. A waxy cuticle on the surface of the leaf also provides some resistance to water loss, but much of transpiration is regulated by the stomata. At the same time that water vapor is leaving the plant, carbon dioxide is entering the leaf as long as the stomata are open. Water in the plant cells provides positive pressure against the cell walls, which is termed "turgor." When there is adequate turgor in guard cells, stomata open and enhanced gas exchange occurs. Plants under stress from water deficit display wilting due to the loss of turgor; stomates close; and other metabolic changes occur, including the production of the phytohormone abscisic acid (ABA). It is often difficult to separate the effect of light or irradiance levels and moisture status because impinging light on a leaf raises temperature. Increasing temperature, in turn, raises the internal vapor pressure of water in leaves, causing water vapor to exit the leaf via the stomates. Some modulation of the effect of temperature is achieved because most of the dicot angiosperm species locate their stomata on the underside of the leaf, thus self-shading the stomata.

The importance of water relations to propagation success can be appreciated, particularly with the rooting of the cuttings. When a cutting is taken, the integrity of the tissues is disturbed, and unless care is taken, the cutting can wilt. One way that dehydration is avoided is by placing cuttings under a water misting or fogging system, which increases the relative humidity, thus decreasing water loss via transpiration. Propagators use many techniques to decrease transpirational water loss and optimize the water status of cuttings. Cuttings taken early in the morning are the most turgid and have less negative water potential than they will have later in the day as temperature and light intensity increase. Once cuttings are taken, they are often wrapped to retain moisture and placed on ice (chilling), thereby reducing ambient temperature and water loss. Cuttings are placed under mist systems, which reduce the difference in relative humidity between the leaf and the ambient surrounding air, a driving force for transpirational water loss. Shading also helps by reducing temperature. Propagators can also minimize the transpirational leaf area by trimming the leaf (Figure 3.9). These measures are necessary until a new set of roots has formed and can resume the role of water uptake.

A very good example of the relationship between leaf water potential and the percentage of rooting is illustrated in two experiments using cuttings of Hebe 'Caledonia,'

FIGURE 3.9 Softwood cuttings with large leaves are often trimmed to remove one half or more of the leaf area, thereby reducing transpirational water loss until a competent root system has formed again.

a compact purple-flowering shrub conducted by Loach and Whalley (1978). In their first experiment, they compared the leaf water potential of cuttings under four different regimes: conventional intermittent mist, a mesh cover underneath the intermittent mist but above the cuttings, clear polyethylene, or white polyethylene over the misting bays. In their second experiment, they used increasing levels of shade consisting of white polyethylene and additional green shade netting to achieve four levels of shade relative to ambient outdoor radiation. As water potential, measured in units of megapascals (MPa), becomes more negative, water stress increases. In Figure 3.10, past a threshold point of about −1.0 MPa, the more negative the water potential of the cutting, the greater the decrease in the percentage of rooting. Different species may have their own specific critical leaf water potentials before rooting is impacted, but in general, better rooting is accomplished by maintaining turgidity in cuttings.

Another good example that underscores the importance of water relations in propagation success occurs during tissue culture propagation. Newly rooted shoots produced *in vitro* are often incompetent with respect to stomatal closure and lack effective cuticular resistance to water loss. For example, fully turgid leaves of *Solanum laciniatum* plants cultured *in vitro* lost water at a greater rate than those from either the original parent plants or the acclimatized plants. Examination of the epidermis of these leaves revealed that half the stomata of the leaves cultured *in vitro* was still open after 30 min compared to the full closure of stomata on detached leaves from the parent plant (Conner and Conner, 1984). During micropropagation, propagators must sometimes subject rooted plantlets to a process of acclimatization to accustom them to the rigors of the *ex vitro* environment. This is done

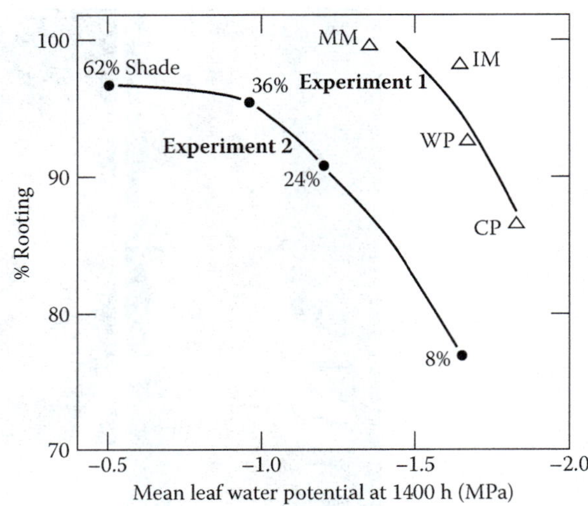

FIGURE 3.10 Two experiments comparing the leaf water potential of the cuttings of Hebe 'Caledonia' and the effect on the percentage of rooting. In the first experiment, four different regimes were compared: conventional intermittent mist (IM), a mesh cover underneath the intermittent mist but above the cuttings (MM), clear polyethylene (CP), or white polyethylene (WP) over the misting bays. In the second experiment, WP and additional green shade netting achieved four levels of shade (the actual percentage shade is noted on the graph) relative to ambient outdoor radiation. As water potential became more negative past a threshold point of about −1.0 MPa, the greater was the decrease in the percentage of rooting. (Redrawn from Loach, K. and D. N. Whalley, *Acta Hortic* 79:161–168, 1978.)

by shading and gradually increasing the venting of the container to which the plantlets have been transferred (Figure 3.11). In the case of *Prunus serotina*, incorporating paclobutrazol, a growth retardant, into the rooting medium prior to acclimatization resulted in increased

FIGURE 3.11 Propagation dome used to help acclimatize *Prunus serotina* microshoots after rooting. As plants developed, the lid of the dome was propped up for increasing amounts of time until the plants were fully acclimatized. (From Eliasson, M. K., *In vitro* acclimatization of genotypes of *Prunus serotina* using triazole growth retardant pretreatments. PhD dissertation. Alabama A&M University, Normal, Alabama, 157 p. 1992.)

survival and percentage of rooting, and reduced water loss from detached leaves (Eliasson, 1992).

Meristems, Growth, and Differentiation

Propagation success depends on the development of meristems and their subsequent differentiation. In the simplest terms, growth can be thought of as an increase in size. Growth of a multicellular organism like a plant is the result of both cell division and cell enlargement. On the other hand, development refers to growth plus cell specialization or cell differentiation, resulting in tissues with particular functions. Much more on the anatomy and the morphology of plants and how they relate to plant propagation was discussed in Chapter 2. The architects of a plant's growth and development are individual genes that give rise to enzymes, plant hormones or phytohormones, and a myriad of secondary plant products. Their impact on adventitious root formation varies depending on the phytohormone, the concentration, and the phase of rooting. In 1999, de Klerk *et al.*, proposed the following phases for the rooting of the microcuttings of apple, which help explain differential sensitivity to auxins and other compounds during adventitious rooting:

- Phase 1 (0–24 h): Dedifferentiation
  - Activation of cells accompanied by accumulation of starch grains.
  - Action of wound-related compounds and auxin.
- Phase 2 (24–96 h): Induction
  - Start of cell division (first visible at 24 h), with 30-cell meristemoids appearing at about 96 h.
  - High sensitivity to the rhizogenic activity of auxin. Cytokinins are inhibitory.
- Phase 3 (96–120 h): Outgrowth in stem
  - Typical dome-shaped primordial develops from meristemoids.
  - Auxins becomes inhibitory.
  - Decrease in sensitivity to cytokinin.

The role of phytohormones in addition to auxin and cytokinin, such as ethylene, and other compounds (wound related and phenolic compounds) cannot be investigated for their effect on adventitious root formation without an understanding of what phase of rooting is being studied.

Phytohormones and Plant Growth Regulators

Plant growth and development are under hormonal, genetic, and environmental regulations as are many other important processes in a plant, including maturation, flowering, dormancy, and senescence. Plant hormones or phytohormones are organic in nature and synthesized naturally by the plant. They affect the growth and development of the plant at low concentrations. On the other hand, plant growth regulators are synthetic compounds that can act like hormones by modifying the actions of hormones or affecting growth and development in their own unique way. Plant hormones are classified into auxins, ethylene, gibberellins, ABA, cytokinins, or brassinosteroids. Each plays a role in regulating plant growth and development, and by understanding their roles, plant propagators can be successful. Plant hormones and growth regulators that are relevant to propagation and how to use them are discussed in greater detail in Chapter 4. Phytohormones that are naturally present within the plant are termed "endogenous." If phytohormones or growth regulators are supplied to the entire "intact" plant or to an "excised" piece of the plant, they are termed "exogenous."

Auxins (Greek "auxein," which means "to increase") are the most important growth regulator for the conventional rooting of cuttings. The first auxin, discovered in 1934, was indole-3-acetic acid (IAA), and it was soon found to promote adventitious root formation especially in cuttings. Charles Darwin and his son Francis, in the year 1880, experimented with phototropism, a process where the shoot bends toward light, which is regulated by auxin. Auxins have several other physiological roles, such as inhibition of lateral buds and activation of cambial cells. The synthetic auxin, indolebutyric acid, has been the most commonly used adventitious rooting agent, followed by naphthalene acetic acid. Most woody and several herbaceous species (e.g., apples, willows, and Lombardy poplar) have preformed adventitious root primordia that remain "arrested" until stimulated by an auxin. Auxin shapes gene expression to regulate growth and development in intact plants as well as in excised stem and root cuttings. The primary response to auxin is almost exclusively through the upregulation of genes. Tissue-specific auxin responses correlate with the differential expression of Aux/IAA genes. Auxin regulates the development of xylem and the genes associated with the biosynthesis, catabolism, and the signaling pathways of other phytohormones. Polarity in plants is established by the "basipetal" (from shoot tip to root) movement of IAA, which occurs via the phloem. It is critical that polarity in both stem and root cuttings is properly maintained during rooting and shoot emergence. Generally, cuttings give rise to shoots at the "distal" end (closest to the shoot tip), whereas the roots are initiated at the "basal" end. It is common for propagators taking hardwood cuttings to make a slanting cut at the end closest to the root zone and a horizontal cut closest to the shoot tip (Figure 3.12) so that the polarity of the cutting can be maintained and the cutting can be inserted into the rooting medium right side up.

FIGURE 3.12 Hardwood cuttings of willow bundled, labeled with cultivar name, and ready to be placed in propagation medium. Note the slanting cut used for the base of the cuttings to differentiate the base from the top so that polarity could be maintained. (From http://www.willowbasketmaker.com/tag/cuttings/.)

There is a substantial body of evidence that auxins contribute to the root initiation of lateral and adventitious roots. Adventitious rooting at the base of a cutting is largely regulated by endogenous (naturally produced) or exogenous (externally applied) auxin. The adventitious and lateral root primordia form in the phloem parenchyma regions that are interspersed in and around key phloem cells (sieve tube and the companion cells). Adventitious root initiation and growth is a complex physiological process that is influenced by different endogenous and environmental factors (Haissig, 1986). The cut end often develops a necrotic zone, followed by dedifferentiation from suppressed primordia in specific tissues and dedifferentiation into callus-like tissue, a precursor of adventitious roots.

Since their discovery, the application of exogenous auxins has become a routine horticultural practice to encourage and enhance rooting. Once initiated, root growth is an energy-requiring process that needs a supply of carbohydrates. Both auxins and carbohydrates are often interrelated and coincide with increased sugar availability at the site of root primordia development (Sadhu *et al.*, 1978). Cuttings with leaves root more successfully than those without leaves, supporting the role of leaves as both a source of endogenous auxins and carbohydrates. Auxin stimulates the mobilization of carbohydrates in leaves and the upper stem, and increases the translocation of assimilates toward the rooting zone. There is also evidence for the role of nitrous oxide as a signal molecule in auxin-induced adventitious root formation. Treatment of *Lavendula dentata* and *Chamaecyparis* spp. cuttings with a nitrous oxide donor (10 μM sodium nitroprusside) induced adventitious rooting (Lanteri *et al.*, 2009).

"Cytokinins" are another class of growth regulators whose name indicates one of their most important roles, that is, regulating cell division. In 1954, Carlos Miller and Folke Skoog identified a breakdown product of Herringbone sperm DNA, which was very active in the promotion of cell division (cytokinesis), and named this compound "kinetin." Subsequent work has identified biosynthesized hormones with strong cytokinin-like properties, such as the plant hormone zeatin and zeatin riboside, its conjugated form with the pentose sugar ribose. Now, many, mostly synthetic substituted adenine compounds have been found to have cytokinin-like properties. Such hormones are in higher concentrations in young organs that are actively undergoing cell divisions, such as seeds, developing fruits, and leaves. Skoog and Miller (1957) demonstrated that undifferentiated tobacco callus cells develop differently into organs based on auxin and cytokinin ratios. The success of shoot micropropagation in the *in vitro* culture of horticultural plants, especially rapidly growing foliage and ornamentals, can be attributed to the effective use of cytokinins, especially for shoot multiplication during micropropagation. This effort has also been aided considerably by the effective use of a nutrient-rich tissue culture media (Murashige and Skoog, 1962) for faster-growing explants in culture. For slower-growing woody and near-recalcitrant tissues, media with lower macroelements have been developed and utilized. Cytokinins generally inhibit adventitious root formation in the second stage of adventitious root formation, induction (de Klerk *et al.*, 1999), but there have been some reports of cytokinins increasing adventitious root formation (Pijut *et al.*, 2011).

Other hormones and growth regulators play a somewhat lesser role in the propagation of plants. "Gibberellins" are a large group of hormones that overcome genetic dwarfism; mediate the response of rosette plants to temperature and day length; and modulate seed and bud dormancy, and, therefore, find some utility in propagation. Generally, gibberellins have no effect or an inhibitory effect on rooting, but in the case of globe artichoke, low concentrations of GA_3 have been shown to have a stimulatory effect (Morzadec and Hourmant, 1997). Gibberellin sprays on mature English ivy caused them to revert back to a more juvenile state (Robbins, 1960). Ethylene, a gaseous plant hormone, has been identified as the agent responsible for fruit ripening, a stimulator of senescence and abscission, and in germinating seedlings, ethylene regulates emergence via the classic triple response (epinasty or bending, lateral growth, and inhibition of elongation). Ethylene has had a positive effect on adventitious root formation in many plant species (Zimmerman and Hitchcock, 1933). Ethylene may impact endogenous auxin through the regulation of synthesis and transport.

ABA regulates the dormancy and germination of seeds and plant response to stress. When water stress occurs, ABA usually increases. Higher levels of ABA in the leaves of avocado were correlated with declining rooting ability with age (Raviv *et al.*, 1986). Brassinosteroids have been generally recognized as an additional class of plant growth regulators and were so named because extracts from *Brassica napus* were found to have an activity that promoted stem elongation and cell division. They also affect the mechanical properties of cell walls; play a role in the differentiation of vascular tissue; affect pollen tube elongation; and enhance the yields of crops, particularly those under stress. Cuttings of Norway spruce, treated with a brassinosteroid, had 92% rooting relative to the 50% rooting of controls (Ronsch *et al.*, 1993).

Plant growth regulators are often synthetically produced, and may mimic hormones or have their own unique properties. They also affect physiological processes at low concentrations. For example, thidiazuron (TDZ; N-phenyl-N′-1,2,3-thidiazol-5-ylurea) is a substituted phenylurea, which was developed for the mechanized harvesting of cotton balls and has now emerged as a highly effective bioregulant of morphogenesis in the tissue culture of many plant species. TDZ exhibits the unique properties of both auxins and cytokinins on the growth and differentiation of cultured explants, although it is structurally different from either auxins or purine-based cytokinins. Other TDZ effects include the modification of cell membranes, energy levels, nutrient uptake, or nutrient assimilation (Murthy *et al.*, 1998). Although not considered classical growth regulators, polyamines such as putrescine and spermidine play a role especially in selected *in vitro* tissue cultures in affecting cell growth and development, and stress response.

Phenolic compounds are found to be abundant in recalcitrant *Protea cynaroides* stem cuttings (Wu and du Toit, 2011), and the high total phenol content was associated with significantly higher rooting percentage and increased number of roots formed. Blanching reduced the time needed for rooting. An application of phenolic compounds, p-hydroxybenzoic acid, p-coumaric acid, and ferulic acid, in most cases, promoted the auxin-induced rooting of cuttings of mango, a difficult-to-root plant (Sadhu *et al.*, 1978), in chestnut hybrid clones and apples. On the other hand, in micropropagation and tissue culture systems, polyphenols released from the cut end in contact with the media interfere with propagation and culture. *In vitro* systems do permit the addition of antioxidants such as ascorbic acid or polyvinylpyrrolidone to reduce polyphenol release and inhibition. Ascorbic acid solutions of 2.5% have been used as a 1 to 2 h soak for cuttings of difficult-to-root 'Frank Houser' camellia, and this practice has increased rooting by 50% (Crawford, 2005).

Wound Responses

Some meristems are preformed, but some are induced when wounding occurs. When a propagator takes a cutting, many events unfold as a result of the wounding. Possibly, in response to electrical or hydraulic rapid response signals, early events in the wounding response are closure of stomata and decrease in water and gas exchange through the stomata. Very quickly, there is a large rise in endogenous jasmonic acid and a transient rise in ABA (Hlaváckovà and Nauš, 2007). Ethylene, associated with rooting, is also produced as a result of wounding (Yang and Hoffman, 1984). In addition to the wounding caused by severing a cutting from the parent plant, propagators may also choose to use additional techniques to wound the base of the cutting further. These may include stripping leaves and young branches off by hand, slicing one or two sides of the base, scraping or abrading the stem, making incisions into the xylem, crushing the base of the cut, or splitting the base. Wounding may facilitate better penetration and effectiveness of rooting compounds. New roots emerge from cut surfaces, and wounding may also allow new roots to emerge more easily without mechanical impediment. Roots can also arise directly from the callus tissue formed as a result of the wounding.

Juvenility and Maturity

The juvenile stage of plants typically lacks reproductive capability and roots more easily. For several monocots, an immature embryo is in the juvenile state, but a mature sexually produced embryo, just a few days later, is already a mature tissue. By the time a corn embryo matures, much of this embryo is a miniature of the future plant, and the assigned cells that will become the tassels can be identified as can the cells that will become the ear of the corn. The juvenile phase differs greatly among species and even cultivars. It may be weeks long for an annual plant, months long for a biennially flowering species, or a lengthy 40 years for certain beech *Fagus sylvatica* and bamboo *Bambusa* species. Oftentimes, the juvenile and adult stages differ in leaf shape or heterophylly, as commonly seen in the changes in leaf shape in English ivy, common bean, and some tree species like Liriodendron. Hormones play a critical role in heterophylly, which can be changed by gibberellic acid and abscisic concentrations and treatments. There are some pines that propagate readily as juveniles, but they are recalcitrant as adults. From the plant propagation perspective, the juvenile state is the most desirable for the rooting of cuttings, and if flowering and seed production is desired, the mature state is needed. Many species require a specific state of juvenility or maturity to undergo embryogenesis *in vitro*, and some differ whether the juvenile or the mature stage of a meristem can undergo organogenesis and extensive proliferation in micropropagation (Sharma and Bello, 1982). Coast redwood *Sequoia sempervirens* has the ability to retain pockets of juvenility even on ancient trees. On roots and stems of old trees, they produce burls in response to stress. These burls have extensive quiescent adventitious root and shoot primordia that can grow and give rise to shoots if the burl is separated from the tree and exposed to water and light. Some Eucalyptus species can sprout from lignotubers, swellings at the base of the trunk, and the sprouts retain juvenile characteristics. The base of the tree is the oldest chronologically but the youngest (most juvenile) with respect to ontogeny, which affects the ability to root adventitiously. Cuttings taken closer to the juvenile portion of the tree often root more easily and with greater percentages than do those taken from the top of the tree.

Adventitious rooting ability tends to decline with ontogenetic (developmental) age. Aside from the observation that it is easier to root cuttings taken from juvenile plant material, the stage of maturity or ontogeny can have a persistent effect on clonal material. The tendency for cuttings taken from a particular part of the plant with its associated degree of juvenility or maturity may reflect that state, and this is referred to as "topophysis" (Bonga, 1982). For example, a coffee plant cutting from an upright (orthotrophic) branch maintains its upright growth, whereas one from a lateral branch (plagiotrophic) will exhibit lateral growth habit. This same phenomenon can be observed with cuttings of junipers, cacao, and Norfolk Island pine (Figure 3.13).

Flowering

In general, flowering and adventitious rooting are contraindicated, so it is wise to select cutting materials from stock plants that are not flowering. If that is not an option, removing the flower buds early before development increases rooting for woody species (Johnson, 1970). Cuttings obtained from a mature jujube tree approximately 30 ft tall, which was flowering prolifically, were rooted successfully after flower buds were removed and cuttings were taken from the more juvenile zone at the base of the tree (unpublished data).

CHARACTERISTICS OF THE STOCK PLANT

Adventitious roots form on cuttings partially because they have been severed from the mother plant. Considerable research has been done on post-severance techniques to enhance rooting success, but there is also a strong appreciation that the management and conditioning of propagation stock or "mother" plants dedicated for propagation to promote subsequent success in the rooting of cuttings can be just as critical.

Age of Donor

The age of the parent plant also has a profound effect on the rooting of cuttings, and younger age favors a higher frequency of rooting. For example, yellow cypress

FIGURE 3.13 Norfolk Island pine grown from a terminal cutting exhibiting orthotropic growth (a) versus one growing from a lateral cutting exhibiting a typical plagiotrophic growth habit (b) in the phenomenon termed "topophysis." (Courtesy of Dr. Mack Thetford, University of Florida.)

Callitropsis nootkatensis (formerly *Chamaecyparis nootkatensis*) rooted cuttings (ramets) originating from pruned donor hedges (ortets) were grown for 3 years in southwestern British Columbia. Ramets were cut when ortets were 3-, 7-, 11-, and 15-years-old and compared; selected ramets were also serially propagated for one, two, or three cycles, 4 years apart. The detrimental effects of ortet aging were detected only at age 15, and ramets from them had the lowest rooting success. Serially propagated ramets from 15-year-old ortets had smaller shoots (Krakowski *et al.*, 2005). Rooting in a recalcitrant Australian woody species *Backhousia citriodora* declined with the stock plant maturity, and it was genotype dependent. The presence of axillary buds on the cutting had an inhibitory effect on rooting, and the juvenility of stock plants could be prolonged with continuous pruning (Kibbler *et al.*, 2004). For larch (*Larix laricina*), as donor trees age from 1 to 20 years old, the success in rooting cuttings declined by 50% (Greenwood *et al.*, 1989). For loblolly pine, the decline in rooting ability with age occurs much more precipitously by the second year (Greenwood and Weir, 1995).

Methods used to "rejuvenate" stock plants so that they will yield cuttings capable of rooting include techniques such as hedging, coppicing, or severe pruning, resulting in growth that more resembles juvenile tissue because it arises from the base of the tree, the most juvenile area ontogenetically. Hedging of stock plants has been successful with both lilac and hazelnut (Cameron *et al.*, 2003). Serial grafting of mature scions onto young juvenile seedlings improves the rooting of shoots that then grow from the scions, and this technique has been used commercially for Eucalyptus (Hackett, 1988). Using the shoots that emerge from root cuttings has proven to be successful for species of *Populus*, *Ulmus*, and *Malus*. *In vitro* propagation has also been proven to be a method to increase the adventitious rooting potential of plants that have been serially cultured. After about 10 subcultures, each about 1 month apart, the shoot tips of apple yielded microcuttings with increased rooting (Mullins, 1985).

Carbohydrate and Nitrogen Status

The relationship between carbohydrate status and subsequent rooting has been a controversial one, and research dealing with it has been complicated by various limitations in the experimental approaches designed to determine their role (Veierskov, 1988). For example, some approaches have examined carbohydrate contents in stock plants or cuttings, but there are many forms and examining selected ones may not give a whole picture. Manipulating stock plants or cuttings to impact carbohydrate pools may also impact other plant processes (such as endogenous auxin transport, inactivation, or destruction) that profoundly impact rooting success. Veierskov

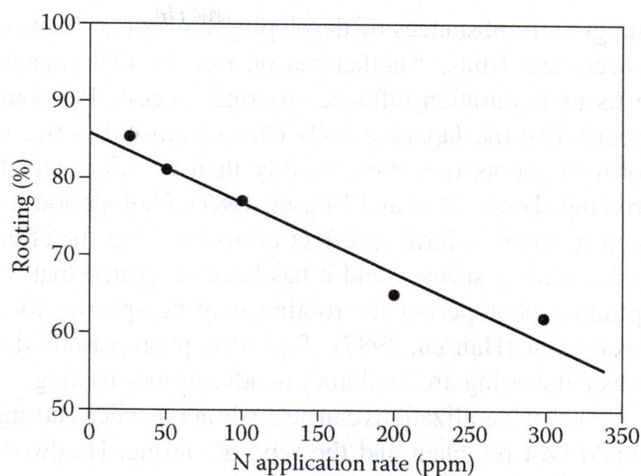

FIGURE 3.14 Rate of nitrogen fertilizer had a strong negative linear correlation with the percentage of rooting of the stem cuttings of *Ilex crenata* "Rotundifolia." (Redrawn from Rein, W. H. *et al.*, *J Environ Hort* 9:83–85, 1991.)

(1988) makes an important point that there is the possibility that "carbohydrates play different roles in the different stages." Kraus and Kraybill (1918) proposed that the ratio of carbon to nitrogen was related to rooting success and found that stems of tomato with a high carbon-to-nitrogen ratio rooted better than those with a low ratio. When stock plants of 'Rotundifolia' holly were fertilized throughout the growing season with a range of nitrogen rates, the rooting success of cuttings taken from them in the fall decreased with increasing nitrogen rates (Figure 3.14; Rein *et al.*, 1991). Two reasons for this were given: increases in shoot growth and decreased tissue maturation.

Carbohydrates have been shown to be related to rooting for some fruit species (Hartmann and Kester, 1983) and tree species like pine (Haissig, 1984). For the tropical legume tree *Albizia guachapele*, a negative interaction was found between increasing irradiance and nutrient supply (0.25%–1.25% NPK) on the stock plant growth and subsequent rooting of leafy cuttings. Rooting percentage decreased from 53.8% with low irradiance/low nutrients to 11.2% with high irradiance/high nutrients (Mesen *et al.*, 2001). Carbohydrates, irrespective of form, may not play a regulatory role in the induction of rooting; however, it stands to reason that sufficient carbohydrate reserves should be available for use in root initiation and subsequent root development.

Seasonality

Some plants have very narrow windows of opportunity in the year during which cuttings may be taken and successfully rooted. For example, *Chionanthus retusus* can only be rooted during 2 weeks out of the year (Stoutemyer, 1942). Conifers appear to root best when cuttings are taken in autumn or early spring (Moe and Andersen, 1988). Some of these may be related to the competition for carbohydrates and growth substances by developing vegetative shoots or flowers and fruits. Another reason may be that changes in tissue maturation influence rooting success. For some species like the Japanese holly (*Ilex crenata*), inactive or dormant shoots root more readily than do soft, actively growing shoots (Dirr and Heuser, 1987). Photoperiod has been reported to have an effect on rooting, but the effect differs among species, and it has been suggested that the optimum photoperiod for rooting may be specific for a given plant (Hansen, 1987). Typically, photoperiods that induce flowering are inhibitory to adventitious rooting.

Some generalizations can be made about when cuttings should best be taken and the type of cutting. Hardwood cuttings collected during the dormant season and prior to bud swelling work best for deciduous plants. The cuttings can be stored at low temperatures until the time for lining out to encourage rooting and budbreak. This approach works well with species like red twig dogwood, viburnum, and forsythia. In mid- to late summer, broad leaf evergreens like hollies, camellia, and photinia can be propagated using semi-hardwood cuttings. Shoots chosen for the cuttings will have wood that is partially mature and stems that have typically stopped active growth. In late fall to late winter, cuttings of narrow-leaf evergreens such as juniper and arborvitae can be taken and rooted successfully. They are not as desiccation tolerant as hardwood cuttings, so care must be taken that they do not dry out.

Genotype and Species Effects

Genotype has a profound influence on the ability to root adventitiously. In a study to determine recalcitrance, 100 forest tree species (representing 41 families and 78 genera) were collected in a tropical rainforest in Sarawak, Malaysia (Itoh *et al.*, 2002). Leafy cuttings of natural forest saplings were placed in a non-mist propagation system with IBA treatment. During the 6-month experiment, 66 species rooted with an overall mean rooting percentage of 37.7% (range, 0%–100%). There were differences among families of trees; species within the *Dipterocarpaceae* and *Lauraceae* had low rooting ability compared to those in the *Euphorbiaceae*, *Rubiaceae*, and *Annonaceae* families (Itoh *et al.*, 2002). In another study with cuttings from 75 half-sib families of three-year-old white spruce, both root initiation and root development phases were under strong genetic control (Gravel-Grenier *et al.*, 2011).

There are still many woody plant species that are too difficult to root in acceptable numbers. It would be great if we could manipulate a single gene to enhance rooting, but we know that the adventitious root formation of cuttings is a complex process involving many genes. It was recently reported that some 220 genes are differentially expressed during the adventitious root development in lodgepole pine (*Pinus contorta*) hypocotyl cuttings (Brinker *et al.*, 2004). Genes were upregulated (increased gene expression) or downregulated (decreased gene expression) during various stages of rooting. Genes are important because they are expressed through the production of proteins, some of which are enzymes that help drive biochemical reactions. Hence, a mature, difficult-to-root plant species has certain genes that are being turned off or on, whereas the juvenile, more easy-to-root stage of the same species differs in its gene expression, although the genome (gene composition) is the same between the two.

CONCLUSIONS

It would be wonderful if the induction of adventitious rooting were a one-size-fits-all proposition, with one approach that worked for all plants reliably. The reality is much more complex with not only the external environment of the propagule playing a role, but also the plant's intrinsic ontogeny, physiology, morphology, and genetics. This chapter has presented information on the many factors that influence adventitious rooting success not only for the propagule after separation from the parent plant, but also for the stock plant. An understanding of all these processes allows the plant propagator to optimize the chances for success.

EXERCISE

Experiment 1. Comparison of Etiolation Versus Blanching on the Ability to Root Softwood Cuttings from Shrubs and Trees

It has long been recognized by propagators that reduction of light to stock plants promotes the rooting success of cuttings taken from them. Earlier in this chapter, various techniques for etiolating plants were offered. These included heavy shade or opaque coverings, and a modified layering technique using bottomless pots placed over shoots developing from cut back dormant trees and filled with a peat–perlite mixture to exclude light from the basal portion of the stem. Blanching, or excluding light from already green tissues, can be accomplished with black electrical insulation tape and even Velcro tape applied to the stem. In this exercise, three treatments are being compared for their effect on the subsequent rooting success of cuttings: control with no light exclusion, stem blanched with Velcro band, and etiolated shoots induced from a cut back branch and grown in an "etiolation chamber" made from an 8-oz. plastic water bottle.

MATERIALS

The following materials will be needed for each student or team of students:

- Several woody trees or shrubs either in a field setting or in pots, ideally one per team of students. Many different species are amenable to this exercise, and suggested species include the

following: abelia, azalea, birch, blueberry, boxwood, camellia, crabapple, crape myrtle, elm, forsythia, gardenia, holly, maple, rhododendron, rose, althea, and willow.
- Pruning shears
- Scissors
- Labels and marking pen
- 0.8% IBA in talc
- Disposable Petri dish
- Velcro tape (1 in. or 2.5 cm wide)
- 8-oz. (250 mL) plastic water bottles
- Silver spray paint
- Silver duct tape
- Paper towel
- Cooler for transporting cuttings to the greenhouse
- Plastic flats (at least three per student or team) containing 1 peat:3 perlite (by volume) rooting medium

Procedure 3.1

Comparison of Etiolation versus Blanching on the Ability to Root Softwood Cuttings from Shrubs and Trees

Step	Instructions and Comments
1	The etiolation chamber should be prepared ahead of time so that it is ready to be used when the class meets. The 8-oz. (250 mL) water bottle should be cut longitudinally along its axis and painted with the silver spray paint to make it opaque (Figure 3.15). One etiolation chamber should be prepared for each student.
2	The species of choice should be chosen, and branches with actively growing shoots should be selected. There will be one branch chosen for the Velcro application and one branch chosen for the etiolation chamber.
3	The branch for the etiolation chamber should be cut back at about 6 to 8 cm from the tip to encourage branching and new growth. The etiolation chamber should be opened at the slit and placed over and sufficiently down the branch to be well anchored. Duct tape can be used to close the slit in the plastic and to attach the chamber to the branch at the base.
4	The Velcro tape should be cut into 1-in. (2.5 cm) pieces for application at about 4 cm down from the tip of the shoot. Matching pieces of Velcro should be pressed gently so that they adhere together, sandwiching the stem (Figure 3.16a).
5	One week after placing the etiolation chamber, the chamber should be opened gently by peeling back the duct tape placed along the slit edge. When etiolated shoots are 6 to 8 cm long, they should be harvested, wrapped in a moist paper towel, and placed in a cooler until treated with IBA.
6	At the same time, remove the Velcro from the blanched stems (Figure 3.16b) and collect the cutting by severing it at the base of the blanched area. A control cutting of the same size should be chosen from a branch that has not been blanched. Wrap cuttings in a moist paper towel and place it in a cooler until the time for IBA treatment.
7	Place a small amount of the IBA (0.8% in talc) in a disposable Petri dish and dip the base of the cuttings in the IBA. Never put the remaining IBA back into the original container. Discard the remnant safely. The cuttings are now ready to be placed under intermittent mist in the greenhouse for rooting.
8	After 4 weeks, examine the cuttings for rooting. Collect data on the percentage of cuttings rooted and the length of the longest root.

Many variations of this exercise can be performed:

- The Velcro tape may be cut at different lengths to determine the effect of the length of the blanched zone on subsequent rooting.
- The Velcro tape can be left in place for varying lengths of time. This procedure specifies 1 week; however, the Velcro tape could be left in place for 2 or 3 weeks and compared with the 1-week duration.
- In another variation, IBA in talc can be applied by coating the Velcro band before placing it on the stem.
- Shoots to be blanched can be selected from different parts of the tree or shrub to determine if the location of the blanched shoot (and its juvenility) have an effect on rooting.
- Of course, many different species can be evaluated and compared for their response to this technique.

FIGURE 3.15 Etiolation chamber made from an 8-oz. plastic water bottle coated with reflective silver spray paint to prevent overheating of etiolated shoots that emerge from the cut back stem. The bottle was slit down its length to make its placement and removal easier, and the slit was sealed with duct tape (just visible alongside the lower edge). Duct tape was also used to attach the bottle to the azalea stem.

FIGURE 3.16 (a) Velcro tape, 2 in. long, attached to an actively growing shoot of rose to blanch the stem prior to the application of IBA. The Velcro tape can be easily removed and cut into varying lengths to determine what length is the most effective. (b) The stem after the Velcro tape was removed only 1 week later, showing the visible signs of blanching.

ANTICIPATED RESULTS

Cuttings from the etiolation and blanching treatments will root more readily and more quickly than the control cuttings. For hard-to-root species, the control cuttings may not even root at all despite IBA treatment. Etiolation may be more effective than blanching for enhancing adventitious root formation.

QUESTIONS

- Compare the appearance of etiolated, blanched, and control shoots. How would you expect the morphology and sensitivity to auxins to be different, if at all?
- Can you design an alternative system for etiolating shoots in intact plants in the field, which would be more efficient? Describe your system.
- What are some other ways of applying the IBA besides in a talc formulation, which would work with this system?

LITERATURE CITED

Amissah, J. N. and N. Bassuk. 2009. Cutting back stock plants promotes adventitious rooting of stems of *Quercus bicolor* and *Quercus macrocarpa*. *J Environ Hort* 27: 159–165.

Amissah, J. N., J. P. Dominick Jr., and N. Bassuk. 2008. Adventitious root formation in stem cuttings of *Quercus bicolor* and *Quercus macrocarpa* and its relationship to stem anatomy. *J Amer Soc Hort Sci* 133:479–486.

Beakbane, B. 1961. Structure of the plant stem in relation to adventitious rooting. *Nature* 192:954–955.

Bertram, L. and B. Veierskov. 1989. A relationship between irradiation, carbohydrates, and rooting in cuttings of *Pisum sativum*. *Physiol Plant* 76:81–85.

Bertram, L. 1992. The physiological background for regulation of adventitious root formation by the irradiance to stock plants: A hypothesis. *Acta Hortic* 314:291–300.

Biran, I. and A. H. Halevy. 1973. Stock plant shading and rooting of Dahlia cuttings. *Scientia Hortic* 1:125–131.

Bonga, J. M. 1982. Vegetative propagation in relation to juvenility maturity and rejuvenation. In Bonga, J. M. and D. J. Durzan, eds., *Tissue Culture of Forest Trees*. Amsterdam, Netherlands: Elsevier, 420 pp.

Brinker, M., L. van Zyl, W. Liu, D. Craig, R. R. Sederoff, D. H. Clapham, and S. von Arnold. 2004. Microarray analyses of gene expression during adventitious root development in *Pinus contorta*. *Plant Physiol* 135:1526–1539.

Cameron, R., R. Harrison-Murray, M. Fordham, H. Judd, Y. Ford, T. Marks, and R. Edmondson. 2003. Rooting cuttings of *Syringa vulgaris* cv. Charles Joly and *Corylus avellana* cv. Aurea: The influence of stock plant pruning and shoot growth. *Trees* 17:451.

Christensen, M. V., E. N. Eriksen, and A. S. Andersen. 1980. Interaction of stock plant irradiance and auxin in the propagation of apple rootstocks by cuttings. *Sci Hortic* 12:11–17.

Conner, L. N. and A. J. Conner. 1984. Comparative water loss from leaves of *Solanum laciniatum* plants cultured *in vitro* and *in vivo*. *Plant Sci Lett* 36:241–246.

Crawford, M. 2005. Methods and techniques to improve root initiation of cuttings. *Comb Proc Intl Plant Prop Soc* 55:581–585.

de Klerk, G., W. van der Krieken, and J. C. de Jong. 1999. The formation of adventitious roots: New concepts, new possibilities. *In Vitro Cell Dev Biol (Plant)* 35:189–199.

Delargy, J. A. and C. E. Wright. 1978. Root formation in cuttings of apple (cv. Brambley's seedlings) in relation to ring barking and to etiolation. *New Phytol* 81:117–127.

Dirr, M. A. and C. W. Heuser. 1987. *The Reference Manual of Woody Plant Propagation*. Athens, Georgia: Varsity Press.

Edwards, R. A. and M. B. Thomas. 1980. Observations on physical barriers to root formation in cuttings. *Plant Prop* 26:6–8.

Eliasson, L. and L. Brunes. 1980. Light effects on root formation in aspen and willow cuttings. *Physiol Plant* 48:78–82.

Eliasson, M. K. 1992. *In vitro* acclimatization of genotypes of *Prunus serotina* using triazole growth retardant pretreatments. PhD dissertation. Alabama A&M University, Normal, Alabama, 157 pp.

Gardner, F. E. 1937. Etiolation as a method of rooting apply variety stem cuttings. *Proc Amer Soc Hort Sci* 34:323–329.

Goodin, J R. 1965. Anatomical changes associated with juvenile-to-mature growth phase transition in *Hedera. Nature* 208:504–505.

Gravel-Grenier, J., M. S. Lamhamedi, J. Beaulieu, S. Carlos, H. A. Margolis, M. Rioux, D. C. Stowe, and L. Lapointe. 2011. Utilization of family genetic variability to improve the rooting ability of white spruce cuttings. *Can J Forest Res* 41:1308–1318.

Greenwood, M. S., C. A. Hopper, and K. W. Hutchison. 1989. Maturation in larch: Part 1. Effect of age on shoot growth, foliar characteristics, and DNA methylation. *Plant Physiol* 90:406–412.

Greenwood, M. S. and R. J. Weir. 1995. Genetic variation in rooting ability of loblolly pine cuttings: Effects of auxin and family on rooting by hypocotyl cuttings. *Tree Physiol* 15:41–45.

Hacket, W. P. 1988. Donor plant maturation and adventitious root formation. In Davis, T. D., B. E. Haissig, and N. Sankhla, eds., *Adventitious Root Formation in Cuttings*. Portland, Oregon: Dioscorides Press, pp. 11–28.

Haissig, B. E. 1984. Carbohydrate accumulation and partitioning in *Pinus banksiana* seedlings and cuttings. *Physiol Plant* 61:13–19.

Haissig, B. E. 1986. Metabolic processes in adventitious rooting of cuttings. In Jackson, M. B., ed., *New Root Formation in Plants and Cuttings*. Dordrecht, Netherlands: Martinus Nijhoff Publishers, pp. 141–189.

Hansen, J. 1987. Stock plant lighting and adventitious root formation. *Hortic Science* 22:746–749.

Hartmann, H. T. and D. E. Kester. 1983. *Plant Propagation: Principles and Practices*, 4th Edition. Englewood Cliffs, New Jersey: Prentice-Hall, Inc., 727 pp.

Herman, D. E. and C. E. Hess. 1963. The effect of etiolation upon the rooting of cuttings. *Proc Intl Plant Prop Soc* 13:42–62.

Hlaváčková, V. and J. Nauš. 2007. Chemical signal as a rapid long-distance information messenger after local wounding in a plant? *Plant Signal Behav* 2:103–105.

Itoh, A., T. Yamakura, M. Kanzaki, T. Ohkubo, P. A. Palmiotto, J. V. LaFrankiee, J. J. Kendawang, and H. S. Leef. 2002. Rooting ability of cuttings relates to phylogeny, habitat preference, and growth characteristics of tropical rainforest trees. *Forest Ecol Manage* 168:275–287.

Johnson, C. R. 1970. The nature of flower bud influence on root regeneration in the *Rhododendron* shoot. PhD dissertation. Oregon State University, Corvallis, Oregon, 98 p.

Kibbler, H., M. E. Johnston, and R. R. Williams. 2004. Adventitious root formation in cuttings of *Backhousia citriodora* F. Muell: Part 1: Plant genotype, juvenility, and characteristics of cuttings. *Sci Hort* 102:133–143.

Krakowski, J., A. Benowicz, J. H. Russell, and Y. A. El-Kassaby. 2005. Effects of serial propagation, donor age, and genotype on *Chamaecyparis nootkatensis* physiology and growth traits. *Can J Forest Res* 35:623–632.

Kraus, E. J. and H. R. Kraybill. 1918. Vegetation and reproduction with special reference to tomato. *Oregon Agric Coll Exp Sta Bull* 149: 90.

Lanteri, L., G. Pagnussat, A. M. Laxalt, and L. Lamattina. 2009. Nitric oxide is downstream of auxin and required for inducing adventitious root formation in herbaceous and woody plants. In Niemi, E. K. and C. Scagel, eds., *Adventitious Root Formation of Forest Trees and Horticultural Plants: From Genes to Applications*. pp. 31–50.

Lin, W. C. and J. M. Molnar. 1980. Carbonated mist and high-intensity supplementary lighting for propagation of selected woody ornamentals. *Proc Intl Plant Prop Soc* 30:104–109.

Loach, K. and D. N. Whalley. 1978. Water and carbohydrate relationships during the rooting of cuttings. *Acta Hortic* 79:161–168.

Maynard, B. K. and N. Bassuk. 1986. Etiolation as a tool for rooting cuttings of difficult-to-root woody plants. *Proc Intl Plant Prop Soc* 36:488–495.

Maynard, B. K. and N. Bassuk. 1988. Etiolation and banding effects on adventitious root formation. In Davis, T. D., B. E. Haissig, and N. Sankhla, eds., *Adventitious Root Formation in Cuttings*. Portland, Oregon: Dioscorides Press, pp. 29–46.

Mesen, F., R. R. B. Leakey, and A. C. Newton. 2001. The influence of stock plant environment on morphology, physiology, and rooting of leafy stem cuttings of *Albizia guachpele. New Forests* 22:213–227.

Moe, R. and A. S. Andersen. 1988. Stock plant environment and subsequent adventitious rooting. In Davis, T. D., B. E. Haissig, and N. Sankhla, eds., *Adventitious Root Formation in Cuttings*. Portland, Oregon: Dioscorides Press, pp. 214–234.

Morzadec, J. M. and A. Hourmant. 1997. *In vitro* rooting of globe artichoke (cv. *Camus de Bretagne*) by GA_3. *Sci Hort* 72:59–62.

Mullins, M. G. 1985. Regulation of adventitious root formation in microcuttings. *Acta Hortic* 166:53–61.

Murashige, T. and F. Skoog. 1962. A revised medium for rapid growth and bioassay with tobacco tissue cultures. *Physiol Plant* 15:473–497.

Murthy, B. N. S., S. J. Murch, and P. K. Saxena. 1998. Thidiazuron: A potent regulator of *in vitro* plant morphogenesis. *In Vitro Cell Dev Biol (Plant)* 34:267–275.

Pijut, P. M., K. E. Woeste, and C. H. Michler. 2011. Promotion of adventitious root formation of difficult-to-root hardwood tree species. *Hortic Rev* 38:213–251.

Raviv, M., O. Reuveni, and E. E. Goldschmidt. 1986. The physiological basis for loss of rootability with age of avocado seedlings. *Tree Physiol* 3:112–115.

Reid, O. 1922. The propagation of camphor by stem cuttings. *Trans Proc Bot Soc* 28:184–188.

Rein, W. H., R. D. Wright, and D. D. Wolf. 1991. Stock plant nutrition influences the adventitious rooting of "Rotundifolia" holly stem cuttings. *J Environ Hort* 9: 83–85.

Robbins, W. J. 1960. Further observations on juvenile and adult *Hedera. Amer J Bot* 47:485–491.

Ronsch, H., G. Adam, J. Matschke, and G. Schachler. 1993. Influence of (22S, 23S)-homobrassinolide on rooting capacity and survival of adult Norway spruce cuttings. *Tree Physiol* 12:71–80.

Sadhu, M. K., S. Bose, and L. Saha. 1978. Auxin synergists in the rooting of mango cuttings. *Sci Hortic* 9:381–387.

Sharma, G. C. and L. Bello. 1982. A scanning electron microscope study of surface differentiation in triticale callus. *Crop Sci* 22:428–430.

Skoog, F. and C. Miller. 1957. Chemical regulation of growth and organ formation in plant tissues cultured *in vitro*. *Symp Soc Exp Biol* 11:118–131.

Smith, H. 1999. Phytochrome: Tripping the light fantastic. *Nature* 400:710–713.

Stoutemyer, V. T. 1942. The propagation of *Chionanthus retusus* by cuttings. *Nat Hort Mag* 21:175–178.

Sun, W. and N. L. Bassuk. 1991. Stem banding enhances rooting and subsequent growth of M.9 and MM.106 apple rootstock cuttings. *Hortic Science* 26:1368–1370.

Talling, F. 1961. Photosynthesis under natural conditions. *Ann Rev Plant Physiol* 12:133–154.

Tepperman, J. M., Y. S. Hwang, and P. H. Quall. 2006. phyA dominates in transduction of red-light signals to rapidly responding genes at the initiation of *Arabidopsis* seedling etiolation. *Plant J* 48:728–742.

Veierskov, B. 1988. Relations between carbohydrates and root formation. In Davis, T. D., B. E. Haissig, and N. Sankhla, eds. *Adventitious Root Formation in Cuttings*. Portland, Oregon: Dioscorides Press, pp. 70–78.

Wu, H. C. and E. S. du Toit. 2011. Role and significance of total phenols during rooting of *Protea cynaroides* L. cuttings. *Afr J Biotech* 10:12,542–12,546.

Yang, S. F. and N. E. Hoffman. 1984. Ethylene biosynthesis and its regulation in higher plants. *Annu Rev Plant Physiol* 35:155–189.

Zimmerman, P. W. and A. E. Hitchcock. 1933. Initiation and stimulation of adventitious roots caused by unsaturated hydrocarbon gases. *Contrib Boyce Thompson Inst* 5:351–369.

4 Plant Growth Substances Used in Propagation

Caula A. Beyl, David W. Burger, and Zong-Ming Cheng

Concept Box 4.1

- Plant growth regulators (PGRs) are organic molecules that profoundly affect different plant processes such as growth and morphogenesis in small concentrations. Naturally occurring ones are called "phytohormones" or "hormones." Synthetic PGRs are often more resistant to enzymatic breakdown and thus exhibit heightened effectiveness.
- The classical five PGRs are auxins, cytokinins, gibberellins, abscisic acid, and ethylene. Auxins are the most important for propagation by cuttings. Auxins and cytokinins are commonly used for micropropagation.
- Additional classes of PGRs have been discovered, including polyamines, jasmonates, salicyclic acid, and brassinosteroids.
- PGRs produced in one tissue or part of the plant may be transported and exert effects in other parts of the plant.
- Auxins, such as naphthalene acetic acid (NAA) and indolebutyric acid (IBA), are used in the propagation industry to induce adventitious root formation on stem cuttings. Commercially, they are available as dry formulations in talc, gel formulations, or liquid formulations.
- Synthetic auxins, such as 2,4-dichlorophenoxyacetic acid (2,4-D) and dicamba, are typically not used to promote adventitious rooting, but are used as broad-leaf herbicides. These auxins are often employed in tissue culture protocols to induce somatic embryogenesis.

Plant growth regulators (PGRs) are a group of organic substances that are capable of exerting large effects in plants in terms of gene expression, growth, and development, although they occur in relatively low concentrations, usually in nano- (10^{-9}) to micro- (10^{-6}) molar amounts. Endogenous PGRs, those that the plant biosynthesizes for itself and are naturally occurring, are called "hormones." The word "hormone" is derived from the Greek word *hormone*, which means "that which sets in motion." These plant hormones or phytohormones control processes such as cell division and enlargement, root and bud initiation, dormancy, flowering, and ripening. Horticulturists make use of the unique features of synthetic growth regulators and apply them exogenously (originating outside of the plant) to generate desired responses in the plant. When a plant is developing from a germinating seed, endogenous growth regulators or hormones are responsible for directing the development of shoots and roots via cell division and enlargement. In traditional plant propagation with cuttings, PGRs have enhanced adventitious rooting for even many hard-to-root species. PGRs have been used to stimulate germination, enhance flowering, and enlarge fruit. In tissue culture, we make use of the ability to direct growth using growth regulators to stimulate the development of non-zygotic embryos, the growth and development of callus, the proliferation of axillary shoots, and the development of adventitious roots. This has enabled the application of tissue culture techniques to fundamental studies of how cells function, the growth and development of the various tissues of the plant, commercial clonal plant propagation, germplasm storage, the development of pathogen-free plants, the use of bioreactors to produce secondary plant products, genetic transformation, and even the production of artificial seed by means of non-zygotic (somatic) embryogenesis (Figure 4.1).

Traditionally, the following five categories of hormones have been classified: auxins, gibberellins (GAs), cytokinins, abscisic acid (ABA), and ethylene. Each has unique effects, and yet they are capable of working

FIGURE 4.1 Chrysanthemum non-zygotic (somatic) embryos encapsulated in calcium alginate to produce "synthetic seeds." (Courtesy of Robert Trigiano.)

together either synergistically or antagonistically to exert other responses within the plant. In addition to these five classical plant hormone classes, polyamines, jasmonates, salicylic acid, and brassinosteroids all have been listed as plant hormones, although it is unclear whether these compounds have universal effects or act in just a few special growth and developmental events (Davies, 2004a, b). Another group of synthetic compounds, plant growth retardants, are PGRs, but not plant hormones, and have been used widely in greenhouse crop production but do not typically play a large role in plant propagation. Each of these is discussed further below with respect to the history of its discovery, general effects, and importance in the propagation of plants.

AUXINS

Hints about the existence of a growth-promoting substance were given by Darwin (1880), who worked with canary grass. He discovered that unilateral light would cause the tips of coleoptiles to bend in the direction of the light and that this would not occur if the tips were cut off or covered with an opaque cap. Went (1926) later confirmed the role of a growth-regulating substance when he allowed the excised tips of oat seedlings to diffuse into agar blocks, and these blocks, when placed on one side of the decapitated oat seedlings, would restore the ability of the seedling to bend in the direction away from the agar block. This diffusible substance was auxin. The role that auxins play is consistent with the Greek origin of the word *auxein*, which means to enlarge or to grow. In general, auxins promote cell enlargement at the cellular level and the root initiation and tissue/organ level of development. In the whole plant, auxins play a role in phototropism and gravitropism, apical dominance, root induction, and wounding responses. Commercially, auxins are used to induce parthenocarpy, to prevent fruit abscission, to stimulate root formation on cuttings and, in the case of stronger synthetic auxins, to serve as herbicides.

The most common naturally occurring auxin is indole-3-acetic acid (IAA), and it was discovered in 1935. Indolebutyric acid (IBA) also occurs naturally in some plants (Ludwig-Muller and Epstein, 1994) and is the auxin most often used to stimulate adventitious rooting. Auxins are characterized as having an aromatic ring separated from a carboxyl group that is at least two carbons long. The structures of some of the most common auxins used for plant propagation are shown in Figure 4.2. Although, IAA is the most commonly found naturally occurring auxin, its use for plant propagation is limited by its sensitivity to light and its tendency to be metabolized by the tissues of the plant or broken down by microorganisms. Synthetic auxins such as naphthalene acetic acid (NAA) and strong auxins such as 2,4-dichlorophenoxyacetic acid (2,4-D), dicamba (DIC), and picloram (PIC), used horticulturally as herbicides against dicotyledonous weeds, are commonly used in tissue culture to induce callus growth and non-zygotic embryogenesis. Endogenous auxins are synthesized from tryptophan and occur primarily in leaf primordia, young leaf tissues, and developing seeds (Bandurski *et al.*, 1995). Compounds that are precursors of IAA, such as indoleacetaldehyde, may also exhibit some auxin activity. Auxin that is synthesized in leaves is transported basipetally through cell-to-cell

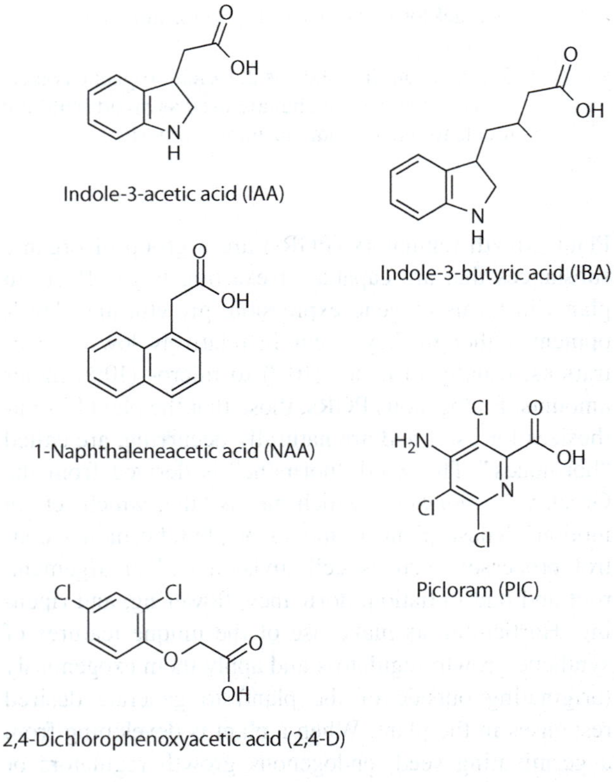

Indole-3-acetic acid (IAA)

Indole-3-butyric acid (IBA)

1-Naphthaleneacetic acid (NAA)

Picloram (PIC)

2,4-Dichlorophenoxyacetic acid (2,4-D)

FIGURE 4.2 Natural and synthetic auxins most commonly used in tissue culture, including indole-3-acetic acid (IAA), indole-3-butyric acid (IBA), 1-naphthaleneacetic acid (NAA), 2,4-dichlorophenoxyacetic acid (2,4-D), and 4-amino-3,5,6-trichloropicolinic acid or picloram (PIC).

movement into the stem. Transport in stems to roots may also involve the phloem tissues (Davies, 2004b).

Since auxin plays many roles in plant growth and development, the concentrations of these phytohormones are critical and highly regulated by the plant. One way that plants regulate the amount of IAA in tissues is by controlling its biosynthesis. Another regulatory mechanism involves the conjugation of auxins with sugars and amino acids into inactive forms. The formation of auxin conjugates permits plants to store and transport the auxins without expressing the growth regulator effects. Conjugated auxins can be activated rapidly by environmental stimuli. The third mechanism for regulating auxin levels is by the oxidation of IAA and, presumably, IBA, by the enzyme IAA oxidase (Bandurski *et al.*, 1995). Conjugates of IAA and synthetic auxins, such as 2,4-D, cannot be destroyed by the activity of IAA oxidase.

Shortly after natural auxin was discovered, auxin was synthesized and became available to plant biologists. Much of our earlier knowledge on the functions of auxin was derived from its exogenous applications to experimental plant systems. In the last 20 years, the characterization of many auxin-related mutants (Reid and Howell, 1995) and genetic engineering with auxin biosynthesis genes from *Agrobacterium tumefaciens* (Cheng *et al.*, 2005) have also contributed to our current understanding of auxin biosynthesis, mode of action, and signal transduction pathways, as well as interactions with other hormones.

Auxins are essential for numerous aspects of plant growth and development (Davies, 2004a). The most relevant role of auxins in plant propagation is that they stimulate root initiation on stem and leaf cuttings and the development of branch roots (Chapter 17). This function of auxin is essential for the cutting propagation of many plant species used in horticultural and forestry industries. Auxins have been used as a powerful agent to induce adventitious root formation in many species that are naturally difficult to root and would not root without their use (Dirr, 1998). Because of this, the horticultural industry has become more efficient in production and more flexible in producing plants all year around. Auxins are used commonly in tissue culture for inducing adventitious root formation in microcuttings.

Although auxin is required or helpful for inducing root primordia, high concentrations of auxin often inhibit root primordia growth after the induction and elongation of roots from stem cuttings or microcuttings. This has been demonstrated by the high-level expression of the *iaa*M gene from *A. tumefaciens* (Klee *et al.*, 1987; Sitbon *et al.*, 1992). In those transgenic plants, abundant root primordia exist, but few developed into normal roots.

Since IAA is broken down rapidly to inactive products by light and microorganisms, it is not widely used commercially for plant propagation. Because IBA and NAA are more resistant to degradation by microbes and plants, they have a more pronounced and longer effect than IAA and are, therefore, more useful in the horticultural industry for plant propagation. Responses of individual plant species to different auxins are not universal; therefore, trials with different auxins and various concentrations and carriers are necessary for each species before successful rooting procedures can be developed and used for propagation. Several synthetic, phenoxy-type of herbicides, such as 2,4-D (Figure 4.2) and 2,4,5-T used at low concentrations, have auxin-like effects in promoting callus formation either alone or in combination with cytokinins. However, they are rarely used for inducing adventitious roots from cuttings because of the propensity to form callus, excessive amounts of which are undesirable during rooting. Instead, these synthetic compounds are widely used for inducing non-zygotic embryos in the tissue culture of many species, including cereal crops and turfgrasses.

In addition to the promotion of adventitious root formation, auxins synthesized in apical buds and young leaf tissues suppress the growth of lateral buds, thus maintaining apical dominance. This effect of auxin can be overcome by pinching the terminal shoot tip (thus removing the primary source of auxin) or by applying exogenous cytokinin.

Auxin also stimulates cell division in the cambium and vascular tissue differentiation, including xylem and phloem (Aloni, 1995). In combination with cytokinin, it also induces cell division and differentiation of organs (Schwarz *et al.*, 2004), such as shoots and non-zygotic embryos (Gray, 2004). It also stimulates root differentiation from somatic tissues in tissue culture. Other effects induced by auxins include mediating tropistic (movements, such as bending toward or away from a stimulus) responses of shoots to light and of roots and shoots to gravity, delaying leaf senescence, stimulating parthenocarpic fruits (fruits set without the fertilization) and seedlessness, delaying fruit ripening, and promoting femaleness in dioecious flowers via the stimulation of ethylene production (Davies, 2004a).

CYTOKININS

Cytokinins are a group of plant hormones that are essential to plant development because of their role in promoting cell division. Unlike auxins whose discovery occurred through observation on tropistic effects, cytokinins were discovered for their ability to promote cell division *in vitro*. An excellent narration of the events and the persons involved in the discovery of cytokinins can be found in Amasino (2005). In Folke Skoog's laboratory in the 1950s, research was underway on looking at the chemical control of new shoot formation on excised tobacco stem pieces (he called the newly formed shoots "buddies"; Amasino, 2005). Variable results with tobacco stem pieces from the greenhouse led him to seek an alternative system using tobacco pith parenchyma cells. He found that they could be induced to enlarge with auxin, but did not undergo cell division. Carlos Miller joined Skoog's laboratory as a postdoctoral researcher and found that when he supplied yeast extract he was able to get cell division. These materials contained purines, and when he tested purine-containing herring sperm DNA, he found that aged or autoclaved herring sperm DNA stimulated cell division, whereas fresh materials did not. The active component, isolated and purified, was named "kinetin" (Figure 4.3; Miller *et al.*, 1955). The cytokinin zeatin (ZEA; Figure 4.3), or 6-(4-hydroxy-3-methylbut-2-enylamino) purine, was first purified from the immature kernels of *Zea mays* (Letham, 1963). The relationship between auxins and cytokinins in regulating the formation of either the shoots or the roots from callus was reported in a classic publication by Skoog and

trans-Zeatin (ZEA)
6-(4-hydroxy-3methylbut-2-enylamino)purine

2-iP
N^6-(2-isopentyl)adenine

Kinetin (KIN)
6-furfurylaminopurine

Benzyladenine (BA)
6-benzylaminopurine

Thidiazuron (TDZ)
N-phenyl-N'-1,2,3-thiadiazol-5-ylurea

Meta-topolin (mT)
N^6-(*meta*-hydroxybenzyl)adenine

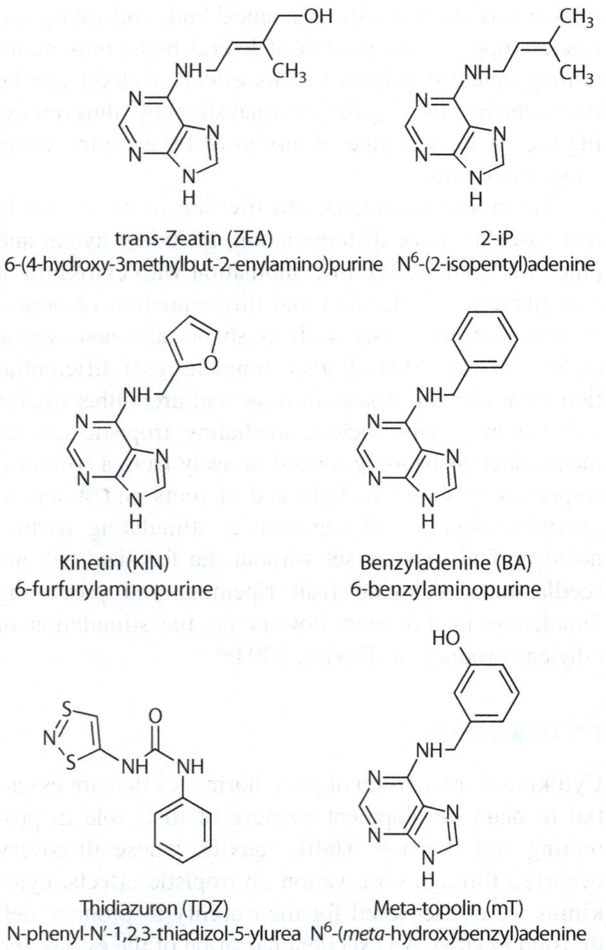

FIGURE 4.3 Cytokinins most commonly used in tissue culture, including trans-zeatin (ZEA), kinetin or 6-furfurylaminopurine (KIN), 6-benzylaminopurine or benzyladenine (BA), thidiazuron (TDZ), and a relative newcomer to the kinetin group, metatopolin (mT).

Miller (1957). If the ratio of auxin to cytokinin is high, root formation is favored, and if the ratio of cytokinin to auxin is high, shoot formation occurs (Figure 4.4). High and roughly equal concentrations of both PGRS promote callus formation.

Probably the two most commonly used cytokinins are benzyladenine (BA; Figure 4.3) and kinetin (KIN; Figure 4.3). These cytokinins have been proven to be effective in stimulating shoot proliferation and callus formation in a wide range of woody and herbaceous species. Naturally occurring cytokinins commonly used in tissue culture are isopentenyl adenine (2-iP) and ZEA, isolated from *Zea mays* kernels (Letham, 1963). Synthetic cytokinins include 6-furfurylaminopurine (KIN) and BA. The substituted phenylurea (1-phenyl-3-(1,2,3-thiadiazol-5-yl)urea), thidiazuron (TDZ), originally used as a defoliant for cotton, has proven to be a potent cytokinin exhibiting activity at concentrations of as low as 10 pM (Preece *et al.*, 1991), relative to the effective range of amino purine cytokinins (1–10 μM). Although it is different in structure from either the auxins or the more familiar cytokinins, TDZ has effects similar to both cytokinins and auxins (induction of callus, formation of shoots, somatic embryogenesis, androgenesis, and gynogenesis) and has the ability to work alone or with other PGRs to stimulate regeneration *in vitro*. Many species have proven responsive to TDZ in terms of embryogenesis, including tobacco, peanut, geranium, chickpea, neem, and St. John's wort (Murthy *et al.*, 1998).

Cytokinins are synthesized in root tips and developing seeds and transported from roots to shoots via the xylem. Cytokinins are present in relatively high

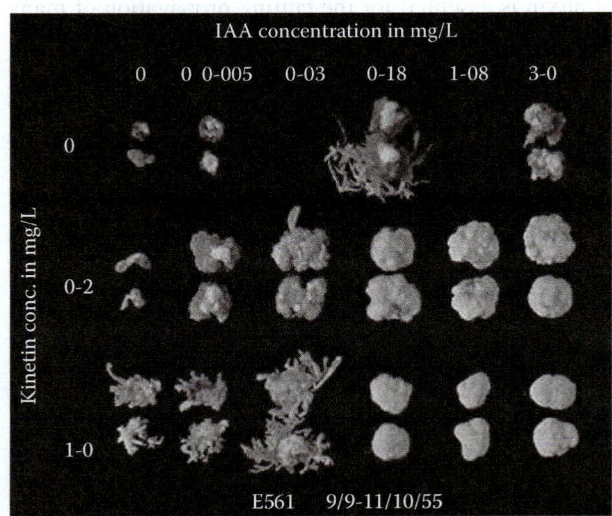

FIGURE 4.4 Classic paper of Skoog and Miller (1957) showing the interaction between various concentrations of indoleacetic acid (IAA) and kinetin (KIN) in promoting shoot formation, root formation, or callus.

concentrations (mM) in developing seeds, but decrease quickly as the seeds mature. Cytokinins also interact with gibberellic acid (GA) and ABA in controlling seed dormancy. Cytokinins also counteract the effects of ABA and have a synergistic effect with GA in breaking seed and bud dormancy, but do not seem to have a direct role in seed germination. In the whole plant, cytokinin release of lateral buds from apical dominance is controlled by auxin. Cytokinins applied after petal fall cause apple fruits to elongate with a greater length-to-diameter ratio and increase fruit set (Carlson and Crovetti, 1988). Application of exogenous cytokinins results in "bushy" or excessively "branchy" phenotypes. Another major role of cytokinins is to delay the senescence of leaves and detached flowers. This delay may be related to the effect of cytokinins in retarding the degradation of chlorophyll and cellular proteins. The anti-senescence effects of cytokinins are critical to the cut flower industry.

GIBBERELLINS

Gibberellins (GAs) are a group of tetracyclic, diterpenoid acids with a common *ent*-gibberellane ring structure (Figure 4.5). Rice seedlings infected with *Gibberella fujikoroi*, the causal fungus of "foolish rice seedling disease" or *bakanae* first caught the interest of a Japanese scientist (Kurosawa, 1926), and GA was first isolated in 1935 from the fungus (Yabuta and Sumiki, 1938). GAs are ubiquitous in both angiosperms (flowering plants) and gymnosperms (conifers, cycads, and gingko), as well as in ferns. They have also been isolated from lower plants, such as mosses and algae, and some fungal and bacterial species. All GAs are assigned numbers as GA_{1-x} irrespective of their origin. To date, there have been 136 GAs isolated, many of which are most likely nonessential to the plant. Instead, these forms are probably inactive precursors or breakdown products of active GAs (Sponsel, 1995).

(a) HO, H, CH$_3$, CO, O, H, H, COOH, OH, =CH$_2$ — GA$_3$

(b) CH$_3$ CH$_3$, CH$_3$, OH, COOH, O, CH$_3$ — ABA

(c) $CH_2{=}CH_2$ — Ethylene

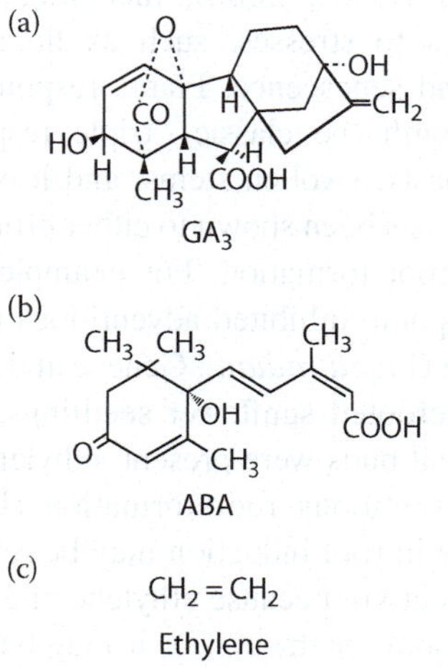

FIGURE 4.5 Chemical structures of PGRs of varied actions. (a) Gibberellic acid (GA_3). (b) Abscisic acid (ABA). (c) Ethylene. (Reprinted from Gaba, V. P., Plant growth regulators in plant tissue culture and development. In Trigiano, R. N. and D. J. Gray, eds., *Plant Development and Biotechnology*. Boca Raton, Florida, CRC Press, pp. 87–99, 2005.)

The most widely available GA is GA_3, which is a fungal secondary product. GA_1 is the primary GA responsible for stem elongation in plants (Thomas and Sun, 2004). GAs are synthesized in the young tissues of shoots, although the exact location is uncertain, and in developing seeds. It is also uncertain whether or not GA is synthesized in the roots. GAs are probably transported both in the xylem and phloem tissues (Davies, 2004b).

GAs play many roles in plant growth and development. One of the major roles is to promote stem elongation by stimulating both cell elongation and cell division, thus resulting in tall plants. One of the seven traits, dwarf-tall, that Mendel studied in peas to establish the Mendelian inheritance law, is now known to involve GA biosynthesis. The dwarf plant has the recessive gene, *le*, which encodes a defective enzyme to synthesize GA. The tall plant possesses the dominant gene, *Le*, which encodes a functioning enzyme for GA biosynthesis. This can be confirmed by the fact that the dwarf phenotype can be transformed into the tall phenotype by treatment with GA_3. Similar results have been obtained with dwarf mutants of other plant species.

GAs cause the stem elongation of plants in response to long days, resulting in the bolting of long-day plants in the spring. In a number of long-day plants that are cultivated under short-day conditions, or vernalization-requiring plants, such as *Hyoscyamus*, *Daucus*, *Crepis*, and *Silene*, flower formation does not take place without a cold stimulus. Treatment with GA_3 induces flower development even without the necessary external environmental conditions. This is attributed to the elongation of the stem axis that is a prerequisite for flower formation and bolting. GAs do not affect flower formation in short-day plants and most other long-day plants whose stem axis does not elongate immediately before flower formation. Commercially, GAs are used to increase the size of seedless grapes.

The major roles of GA in plant propagation are inducing seed germination for some seeds that normally require cold treatment (stratification) or light exposure, and releasing buds from dormancy. In seeds that require light to germinate (e.g., lettuce), GAs are able to substitute for P_{FR} (far-red light). GA also regulates the formation and secretion of hydrolytic enzymes, such as alpha-amylase, a protease, and a ribonuclease, in the aleurone layer (the part of the endosperm tissues of cereal seeds) of cereal grains during the early stages of embryo germination. These enzymes help mobilize endosperm storage compounds supplying the embryo with nutrients during germination.

ABSCISIC ACID

ABA (Figure 4.5) was originally identified by Addicott *et al.* (1968), studying abscission in cotton and one of the compounds isolated that was called "abscisin II." Wareing, studying bud dormancy, discovered the same compound and called it "dormin." The name ABA was the compromise name settled on for the phytohormone (Salisbury and Ross, 1985). In whole plants, ABA plays a role in bud and seed dormancy, stomatal control, leaf abscission, and senescence. In tissue culture systems, ABA is generally positive in effect at relatively low concentrations, but high concentrations inhibit both callus growth and the formation of buds, roots, and embryos (Gaspar *et al.*, 1996). ABA is most often used to enhance the maturation and normal growth of non-zygotic embryos. In the pursuit of embryogenesis in cereals such as wheat, enhanced embryogenesis occurred when low concentrations of ABA were included in the medium (Brown *et al.*, 1989). When germplasm preservation is desired, proliferation of shoots can be prevented by the addition of ABA (Singha and Powell, 1978). For cell cultures, ABA has been used to increase freezing tolerance (Gaspar *et al.*, 1996). ABA has also been useful to pretreat the somatic embryos of carrot to increase the survival rate of embryos encapsulated with 5% polyethylene oxide and then dried to a constant weight (Kitto and Janick, 1985).

In whole plants, ABA is involved in the regulation of many aspects of plant growth and development, including embryo maturation; seed dormancy; germination; root development; stomata function; cell division and elongation; and responses to environmental stresses such as flood, salinity, cold, drought, pathogen attack, and UV radiation (Davies, 2004a; b; Koiwai *et al.*, 2004; Gonzalez-Guzman *et al.*, 2004). However, ABA, unlike the name indicated, does not appear to control directly organ (e.g., leaf) abscission, but does promote senescence and/or stress responses, the processes directly preceding abscission. ABA has historically been considered as a growth inhibitor; however, some young, actively growing tissues have relatively high levels of PGR. In contrast, ABA-deficient mutant plants are severely stunted; therefore, the name of ABA does not match with the roles that it plays in plant growth and development.

The role of ABA in plant propagation is primarily associated with the following two aspects of growth and physiology of plants: seed and bud dormancy, and somatic embryo maturation in tissue culture. ABA appears to play a key role in preventing the "viviparous germination" of developing embryos (Figure 4.6) and inducing primary dormancy. ABA has been isolated from the seed coats of dormant peach, plum, apple, rose, and walnut. Another indication that ABA may be involved in seed dormancy is that the concentration usually decreases during seed stratification. ABA's role in preventing viviparous germination and inducing dormancy has also been observed similarly in bud dormancy/dormancy release. The application of ABA can inhibit the germination of nondormant seeds and counteract the effect of applied GA, which promotes germination.

FIGURE 4.6 Abscisic acid plays a role in preventing the premature germination of seeds (vivipary) while still in the fruit. (Courtesy of Rosie Lerner, Department of Horticulture and Landscape Architecture, Purdue University.)

ETHYLENE

Ethylene (Figure 4.5) is a colorless, odorless gas with a simple structure and is the only member of its class. The endogenous synthesis of ethylene by plants was first demonstrated in 1934 by Gane, and ethylene was proposed as a plant hormone in 1935 by Reid and Howell (1995). It is ubiquitous to all higher plants and usually associated with fruit ripening (Reid and Howell, 1995).

Although ethylene is a simple compound, as a plant hormone, it affects many plant processes (Davies, 2004a; Bleecker and Kende, 2000), including fruit ripening and responses to stresses, such as flooding, drought, wounding, and senescence. Plants respond to exposure to ethylene with the classic "triple response," which includes epinasty, swollen stems, and loss of gravitropism. Ethylene has been shown to either promote or inhibit adventitious root formation. For example, the ethylene releaser, ethephon, inhibited adventitious root formation in mung bean (*Vigna radiata*; Geneve and Heuser, 1983), whereas in derooted sunflower seedlings, when cotyledons and apical buds were present, ethylene or ethephon promoted adventitious root formation (Liu and Reid, 1992). Its role in root induction may be related to how it interacts with auxin because ethylene and auxin biosyntheses share some pathways, or it may be related to the stress environment. In flooded maize, ethylene stimulates the outgrowth of root primordia already preexisting in the shoot base (Figure 4.7) and the development of cortical air spaces (Drew *et al.*, 1979), which may help the plant survive in waterlogged soils. Ethylene is also involved in the induction of new root primordia in flooded sunflower

FIGURE 4.7 Effect of approximately 2 weeks of waterlogging on the formation of replacement adventitious roots at the soil surface by plants of maize (*Zea mays*) (a) and sunflower (*Helianthus annus*) (b). In A, the roots are outgrowths of preexisting primordia. In B, the original root system has been killed and replaced by newly initiated roots formed at the shoot base and the hypocotyl. Ethylene appears to be involved in both processes. (Courtesy of Michael Jackson; http://plantstress.com/articles/waterlogging_i/waterlog_i.htm).

(Figure 4.7). In another example, in a wetland plant, *Rumex palustris*, flooding induced ethylene production. Plants in flooded soil exhibited an increased sensitivity of the root-forming tissues to endogenous IAA, thus inducing adventitious roots (Visser *et al.*, 1996). However, since a high concentration of auxin induces ethylene synthesis, it is difficult to differentiate the roles of auxin and ethylene. Clark *et al.* (1999) conducted an elegant experiment using ethylene-insensitive (NR) tomato (*Lycopersicon esculentum*) and petunia (*Petunia* × *hybrida*) plants to determine if normal or adventitious root formation is affected by ethylene insensitivity. They found that the application of auxin (IBA) increased adventitious root formation on the stem cuttings of wild-type plants, but had little or no effect on the rooting of NR mutant plants and in NR transgenic petunia plants. The application of I-aminocyclopropane-I-carboxylic acid (ACC) (a precursor of ethylene) increased adventitious root formation on vegetative stem cuttings from NR and wild-type plants, but NR cuttings produced fewer adventitious roots than wild-type cuttings. They concluded that the promotive effect of auxin on adventitious rooting may be mediated by ethylene.

OTHER REGULATORS OF PLANT GROWTH USEFUL IN PLANT PROPAGATION

Polyamines (PAs) are organic molecules with two or more amino groups that play a role in many processes in whole plants, including cell division and morphogenesis. Polyamines in plants appear to be essential to growth and development because mutants lacking the ability to synthesize PAs are unable to grow and develop normally. Furthermore, the addition of PAs to these mutants generally restores normal growth and development. They tend to be found in higher concentrations in juvenile tissues rather than in mature ones (Rey *et al.*, 1994). PAs have a number of functions, including stabilizing membranes and nucleic acids, and controlling protein structure and enzyme activity. The compounds also have been linked to cell division, vascular differentiation, somatic embryo formation in tissue culture, root initiation, adventitious shoot formation, flower initiation, development, and the control of fruit ripening and senescence (Galston and Kaur-Sawhney, 1995). PAs can enhance rooting in stem cuttings when applied with auxin, but PAs alone seems to have a limited effect in adventitious root formation, at least not considered as essential. PAs also appear to enhance *vir* gene induction and T-DNA transfer during infection by *Agrobacterium tumefaciens* (Kumar and Rajam, 2005), which may help molecular geneticists achieve a higher frequency of transformation.

Jasmonic acid (JA) and its fragrant methyl ester, methyl jasmonate (meJA), are widely present in species of the plant kingdom. meJA is a fragrant oil that contributes to the distinctive aroma of certain fruits and flowers, such as jasmine, and is where the name jasmonic acid was derived. Since both compounds display biological activity, they are collectively called "jasmonates." Jasmonates inhibit many plant growth and developmental processes, including seed germination. They also promote senescence, abscission, fruit ripening, pigment formation, and tendril coiling. One of the important roles of jasmonates appears to be in plant defense by inducing the synthesis of proteinase inhibitors that deter insect feeding (Davies, 2004a). Jasmonates also promote potato tuber formation. The role of jasmonates in plant propagation has not been well established. It promotes adventitious root formation in mung bean derooted cuttings, as well as in potato *in vitro* shoots (Zhang and Cheng, 1996).

Brassinosteroids is a group of over 60 different steroidal compounds represented by the compound brassinolide (BR_1) that was first isolated from *Brassica napus* pollen. They are widely distributed in plants. The BR_x promotes stem elongation, ethylene synthesis, and epinasty (bending of leaves outward and downward), and inhibits root growth and development. BR_x do not seem to have a direct role in plant propagation (Arteca, 1995; Davies, 2004a). Various brassinosteroids improve the rooting efficiency and survival of Norway spruce (Ronsch *et al.*, 1993), increase the cell division of *Petunia* leaf protoplasts (Oh and Clouse, 1998), and substitute for cytokinin in *Arabidopsis* callus and suspension cells (Hu *et al.*, 2000). The brassinosteroid 24-epibrassinolide was used successfully to initiate protocorm-like bodies and the successful regeneration of *Cymbidium elegans* (Malabadi and Nataraja, 2007). Brassinosteroids act synergistically with auxins and, overall, play a role in improving the efficiency of regeneration.

Salicylic acid is found in the bark of willow (*Salix* species) trees, where the name was derived, and is only recently recognized as a plant growth hormone. Salicylic acid is a phenolic phytohormone involved in plant growth regulation and development (Raskin, 1992), but it is widely known for its role in defense mechanisms in response to biotic and abiotic stresses. One of the defense mechanisms is systemic acquired resistance (Metraux *et al.*, 1990), which is a whole plant response to a localized attack by a pathogen and results in the induction of pathogenesis-related proteins (Malamy *et al.*, 1990). SA inhibits seed germination, and when applied with IAA, it stimulates adventitious rooting in mung beans (Raskin, 1995).

USE OF PGRS IN TISSUE CULTURE

Cultures of plant tissues may require a certain PGR, or a combination in one stage and then a completely different one or the absence of all PGRs in another for growth (Chapter 30). For example, in micropropagation stage 1, the initiation phase, a combination of auxins and cytokinins may be needed. In stage 2, the multiplication stage, most typically, cytokinins are used to stimulate shoot proliferation. In stage 3, the root formation stage, most often, auxins are used to induce roots. In stage 4, acclimatization, the herbicidal triazoles, paclobutrazol and uniconazole, have been shown to enhance survival in the case of *Prunus serotina*.

Much of the research literature on the *in vitro* culture of plant tissues reports on experimentation to determine what types and concentrations of PGRs will induce specific responses, such as adventitious shoot proliferation or root formation, or embryogenesis. Some species are quite amenable to clonal propagation via tissue culture and require only low concentrations of phytohormones to elicit the desired effects, and others, such as some woody species, require either higher concentrations or inherently more active phytohormones to obtain results. Sometimes, a culture that has previously required a specific growth regulator for growth no longer responds the same way in its presence, and this is called "habituation." This means that the culture either no longer needs it or cannot use it if it is in the medium. Many different facets of the *in vitro* environment and the explants affect success *in vitro*, but the type, concentration, and duration of exposure to PGRs typically have the most profound effect.

Auxins are the PGR most often used to initiate callus from explants, and the most often used auxin for this is, undoubtedly, 2,4-D. Once the callus is initiated by 2,4-D, maintaining it on the same medium may cause genetic variability, so sometimes, callus is transferred to a medium containing an alternative auxin, such as NAA or IAA (Machakova *et al.*, 2008). In cell suspensions, auxins tend to favor cell dispersion, whereas cytokinins tend to favor cell aggregation (Machakova *et al.*, 2008). Higher auxin levels may also promote embryogenesis and tend to inhibit the formation of chlorophyll promoted by cytokinins. Maintenance of suspension cultures is usually sustained with 2,4-D, often with cytokinins, but 2,4-D may be replaced by NAA or IBA when the suspension is to be used for subsequent morphogenesis (Machakova *et al.*, 2008).

Adventitious shoot proliferation can occur directly on the explant or on the callus proliferated from the explants. Another route to obtain shoot proliferation is from the release of the apical dominance and growth of axillary buds or shoots. Cytokinins (most often KIN or BA) are the predominant PGRs for the promotion of shoot proliferation or the development of adventitious shoots, but sometimes, the response can be enhanced by the addition of low concentrations of an auxin or even the addition of another cytokinin. Different species may also have a preference for certain cytokinins and respond to one or more, but not all, with respect to shoot culture. Finding the best cytokinin for shoot proliferation may require trial and error.

The formation of roots is stimulated by auxins, and the most often used auxins for this purpose are IBA and NAA. Cytokinins tend to inhibit root formation, and some cytokinins such as TDZ tend to have persistent negative effects on root formation on shoots. For some plants, roots are easily induced merely by transferring *in vitro* shoots on regeneration medium to one that contains no PGRs. These include *Nicotiana* species, cucumber, squash, *Acacia*, and some *Rosa* (Gaba, 2005). The type of auxin used to stimulate shoot proliferation prior to rooting *in vitro* also affects how easily roots are induced. The shoots of American yellowwood stimulated in response to BA rooted easily with IBA *in vitro*. Too high a concentration of auxin used to induce rooting may result in callus formation and poor root induction/quality.

USE OF PGRS FOR BREAKING SEED DORMANCY

Much of the work done with PGRs to enhance seed germination has focused on breaking seed dormancy or shortening the length of time required for cold stratification to be effective in enhancing the germination of woody tree species. GA has been used effectively to promote the germination of many woody trees, including angiosperm species *Acer*, *Fagus*, *Fraxinus*, and *Liquidambar*, and gymnosperm species *Picea*, *Pinus*, and *Taxodium* (Leadem, 1987). In some cases, cytokinins have been effective in improving germination. For *Acer pseudoplatanus*, cytokinins increased radicle emergence and stimulated the germination of unstratified seeds (Pinfield and Stobart, 1972), and for *Pinus taeda*, germination was increased (Biswas *et al.*, 1972). For *Cercis canadensis* and *Eleagnus augustifolia*, soaking unstratified seeds in ethrel, a compound that releases ethylene, increased germination (Hamilton, 1972). Ethylene pretreatment was even effective in promoting the germination of seeds of *Picea glehnii* stored for 20 years (Shibakusa, 1980).

The use of PGRs to enhance the germination of a field crop has been hampered by the need to soak seeds to gain a positive effect, and imbibed seeds are too soft to be mechanically planted. GA as a 6 to 24 h soak was effective in enhancing cotton seed germination over a range of temperatures, even those at which a chilling inhibition of germination occurred (Cole and Wheeler, 1974). Watermelon seeds are also sensitive to cold temperatures and germinate poorly at temperatures below 20°C and, basically, not at all below 15°C. When seeds were primed with either 1.0- or 3.0-μM meJA, they germinated significantly better at cooler temperatures than primed seeds without the growth regulator (Korkmaz *et al.*, 2004).

USE OF PGRS IN CUTTING PROPAGATION

Many herbaceous cuttings can be rooted without auxins, but cuttings often root faster and more extensively with the growth regulator applied. The most commonly used auxins for rooting cuttings are IBA and NAA. IAA degrades easily in light (UV) and in the presence of oxygen, and IBA and NAA are less prone to degradation.

Commercially, auxins are available either as a dry powder, in a gel formulation, or as a liquid. Talc is used for the dry powder formulations, with IBA or NAA as the active ingredient. Gel formulations use various materials to suspend IBA, NAA, or both, including the polymer carboxymethyl cellulose, 2-hydroxyethyl cellulose, propylene glycol, and acrylamide–acrylic acid hydrogels. Liquid formulations typically consist of the active ingredients (IBA or NAA) dissolved in an aqueous alcohol solution, such as ethanol or isopropyl alcohol. Several brand-name products are Rhizopon™, Dip'N Grow™, Vita Grow™ (formerly known as "Wood's™ Ready-to-Use Rooting Hormone"), Hormex™, Hormodin™, Rootone™, Clonex™, and Dynagro™ (Figure 4.8). Commercial hormone products are composed of a variety of different kinds and concentrations of auxins (Table 4.1), as well as other compounds, and sometimes include antibacterial and antifungal compounds.

As an alternative to using a preformulated commercial product, a propagator can prepare a formulation of auxin from the pure compound of choice. Both IBA and NAA can be purchased from chemical supply houses, such as Sigma Chemical Corp., Carolina Biological Supply, Phytotechnology Labs, and Caisson Labs. The powders are available in acid form as IBA or as the potassium salt of the acid.

There are advantages and disadvantages to both forms (free acid vs. potassium salt) of the auxins. The free acid form of auxin is only soluble in water in extremely low concentrations. Thus, ethanol or dilute potassium hydroxide must be used for dissolving the IBA or NAA.

FIGURE 4.8 Several commercial auxin products used for rooting cuttings. (a) Clonex, (b) Rhizodon, (c) Hormex, (d) EZ-Clone, (e) Dip'N Grow, and (f) Hormodin 1, 2, and 3. (Courtesy of Clonex—http://www.growthtechnology.com/; Rhizopon—http://www.rhizopon.com/; Hormex—http://hormex.com/; EZ-Clone—http://www.ezclone.com/; Dip'N Grow—http://www.dipngrow.com/; Hormodin 1, 2, and 3—http://www.ohp.com/.)

TABLE 4.1
Some Commercial Rooting Products Containing Auxins

Commercial Product	Manufacturer	Formulation	Active Ingredients
Clonex™	Growth Technology	Gel	0.3% IBA
Dip'N Grow™	Dip'N Grow, Inc.	Liquid	IBA and NAA
EZ-Clone™	EZ-Clone Enterprise, Inc.	Gel	0.3% IBA
Hormex™	Brooker Chemical Corporation	Powder	0.1% IBA 0.3% IBA 0.8% IBA
Hormodin™	OHP, Inc.	Powder	0.1% IBA 0.3% IBA 0.8% IBA
Rhizopon™	Rhizopon BV	Powder and water-soluble tablets	0.1% IBA 0.3% IBA 0.8% IBA
Vita Grow™ (formerly known as Wood's™ Ready-to-Use Rooting Compound)	Vita Grow, Inc.	Liquid	IBA and NAA

Ethanol, which generally makes up 50% of the final solution (see "Using Pure IBA" for instructions on how to prepare the compound), may be toxic to plants and cause the base of the cuttings to die. On the other hand, in some cases, alcohol is thought to enhance the uptake of auxin into the base of the cutting and is used preferentially in such situations. The potassium salt of the auxin is freely soluble in water and mixes easily. Since there is no alcohol present, the solution is less damaging to some plants.

USING COMMERCIAL FORMULATIONS OF AUXIN

Auxins can be applied in a variety of ways. The dilute soak method is an older method of using a dilute aqueous concentration (usually 50–150 ppm) of IBA to soak cuttings, most often hardwood, for periods of time of up to 24 h. This can be quite effective, but awkward due to the long soak times. The quick dip method (Figure 4.9) typically uses concentrations from 500- to 10,000-ppm IBA either in 50% alcohol or the potassium salt in water. Each species and type of cutting has its own optimum concentration for rooting (Figure 4.10). Just as the technique implies, the cutting bases are dipped into the solution for about 3 to 5 sec. Some cuttings may be sensitive to the alcohol solvent used, but some respond better to solutions made with alcohols or other solvents, perhaps due to better penetration. Talc and gel application methods both use concentrations typically ranging from 0.1% to 0.8% and vary only in the carrier of the growth regulator. Cuttings dipped in talc formulations (Figure 4.11) benefit from wetting to help the talc formulation adhere better. Gel applications often adhere better to the cut surface of the cutting and persist increasing effectiveness. The spray method involves using an applicator like an atomizer to spray concentrations of 500- to 5000-ppm potassium IBA in water over the leaves of the entire plant until drip off to stimulate rooting, but the effectiveness varies greatly by species.

Auxin formulations that are manufactured as powders are very easy to use. To use, transfer a small amount of the powder to a container, such as a weighing tin, weighing boat, or a small jar about 2 in. in diameter. The powder should always be transferred from the original container to avoid contamination of the remaining compound. Do not return any powder that has been used back into the original container. Take cuttings, 2.5 to 4 in. long, and dip the cut end in tap water. Then, dip it into the powder. Cuttings should be tapped to remove excess powder. The cuttings are then stuck into a flat or cell pack

FIGURE 4.9 Semi-hardwood cuttings of Jojoba being treated with IBA in 50% alcohol using the quick dip method prior to being stuck into the rooting medium. (Courtesy of Chiwon Lee, North Dakota State University.)

FIGURE 4.10 Effect of different concentrations of IBA in a 50% ethanol solution on the rooting of *Cordia* semi-hardwood cuttings. (Courtesy of Chiwon Lee, North Dakota State University.)

containing a mixture of soilless components (see below). It is very difficult to change the concentration of powders. If they are mixed with talc, one must be sure that the original and additional talc are well mixed and homogeneous so that the concentration of IBA throughout the powder is uniform. Powders are usually purchased and used in their original concentrations.

A different approach is taken by Hortus Inc., which sells potassium IBA under the name Hortus IBA Water Soluble Salts™. They include instructions on how to measure the proper amount of the salt. Potassium IBA (K-IBA) is freely soluble in water. Hortus also manufactures K-IBA in tablets (Rhizopon Rooting Hormone™).

FIGURE 4.11 Cuttings dipped into the talc formulation of IBA. Note the white powder adhering to the cut base of the cutting. (Courtesy of Chiwon Lee, North Dakota State University.)

This allows the propagator to dissolve a specified number of tablets in water to obtain the desired concentration of IBA. For liquid-prepared formulations (such as Hormex™, Dip'N Grow™, Vita Grow™), directions on the container indicate the proper dilution and liquid to use for plants that are easy-, moderate-, or difficult-to-root. Gel formulations, such as Clonex™, are typically purchased ready to use in the concentration that is desired depending on how difficult it is to stimulate adventitious roots.

USING PURE IBA

IBA is insoluble in water at the concentrations required for rooting cuttings. Thus, IBA is first dissolved in 100% or 95% alcohol (propanol is fine, but do NOT use methanol) and then diluted by half with water to obtain a 50% alcohol solution, which is then used for dipping the cuttings. Calculations for several different concentrations are provided in Figure 4.12.

The preparation of K-IBA is much more straightforward as it is readily water soluble. The proper amount of K-IBA is weighed and added directly to water. It will dissolve in extremely high concentrations, up to 40,000 ppm. A common recommendation is for all liquid auxin solutions to be kept in a dark container at 4°C to avoid degradation of the compound; however, IBA solutions in 50% isopropyl alcohol at concentrations as high as 5000 ppm were kept in a range of temperatures, including room temperature for 4 to 6 months, without any loss of biological activity (Robbins *et al.*, 1988).

Preparing different concentrations of IBA in 50% alcohol depending on the expected ease of rooting

For easy-to-root cuttings, use 1000 ppm (0.1%) IBA

1. • Measure out 500 mL alcohol.
2. • Add 1 g IBA and dissolve completely.
3. • Add water to a total volume of 1000 mL.

For moderate-to-root cuttings, use 3000 ppm (0.3%) IBA

1. • Measure out 500 mL alcohol.
2. • Add 3 g IBA and dissolve completely.
3. • Add water to a total volume of 1000 mL.

For hard-to-root cuttings, use 8000 ppm (0.8%) IBA

1. • Measure out 500 mL alcohol.
2. • Add 8 g IBA and dissolve completely.
3. • Add water to a total volume of 1000 mL.

FIGURE 4.12 Procedure for preparing different concentrations of indolebutyric acid (IBA) in 50% alcohol depending on the expected ease of the rooting of cuttings.

Auxin concentrations of 1000, 3000, or 8000 ppm (mg/L) will satisfy most rooting needs. Commercial powders and gels are sold in these three commonly used concentrations, and dilutions of liquid commercial formulations and IBA can easily be prepared in the same concentrations.

EXERCISES

With this background, several laboratory exercises can be developed to study the effects of the following on rooting:

1. Different formulations of auxin
2. Different concentrations of auxin
3. Same auxin concentration on different parts of a stem
4. Different auxin concentrations on different plant species

SELECTING A CUTTING TYPE AND A ROOTING MEDIUM

One of the most important aspects of rooting with auxins is matching the type of cutting with the appropriate concentration and type of auxin. Some kinds of cuttings do not need auxin to form adventitious roots. These generally are herbaceous and known as "easy to root." Cuttings in this group include *Salvia officinalis* (sage), *Pelargonium* cultivars (especially some scented geranium varieties), *Coleus blumei* (coleus), and *Penstemon digitalis* 'Husker Red' or other garden cultivars. Many houseplants, such as Pothos and succulents *Kalanchoe blossfeldiana* and *Echeveria*, fall in this category.

Most stem cuttings do need auxin; however, auxin may not be sufficient by itself to induce rooting, and other cultural practices may have to be performed either to the cutting or to the stock plant prior to taking the cutting. Selecting the proper part of the stem may also have a significant effect on rooting. Stems taken from a shrub or a perennial plant are composed of different kinds of wood (Figure 4.13).

Cuttings from the first 3 to 5 cm of the tip, known as "terminal cuttings," are considered to be softwood. They bend easily and are quite pliable. The section of the stem below the terminal cutting is known as the "subterminal." It is stiffer than the terminal section but still can be bent. The lower part of a stem is the basal section and is often woody and difficult or impossible to bend. A cutting taken from this section is known as a "basal or hardwood cutting." Stems of deciduous trees and shrubs taken during winter, when the leaves have abscised, are also known as "hardwood cuttings." Depending on the plant, the specific kind of stem cutting may be critical for success. In addition, the time of the year during which a cutting is taken may also be a crucial factor influencing the success in obtaining rooted cuttings. Unfortunately, there are no rules at present that will predict the kind of cutting that will best form roots. A suitable method for determining the most appropriate kind of stem to use for rooting as a cutting is to research the requirements of closely related species. The text *Plant Propagation: Principles and Practices* by Hartmann *et al.* (2010) contains a great deal more information on this subject as does the *Reference Manual of Woody Plant Propagation* by Dirr and Heuser (1987). In addition, a rooting database was developed by Burger and provides specific protocols for inducing adventitious rooting, including growth regulators and concentrations (http://rooting.ucdavis.edu/Pchome.htm).

A rooting medium is also necessary after the kind of cutting and the appropriate auxin concentration are chosen. An excellent rooting medium consists of 3 to 4 parts perlite to 1 part peat moss or vermiculite. This medium might appear to be very "light," but it retains the proper amount of moisture without becoming too wet, which would result in the base of the cuttings rotting.

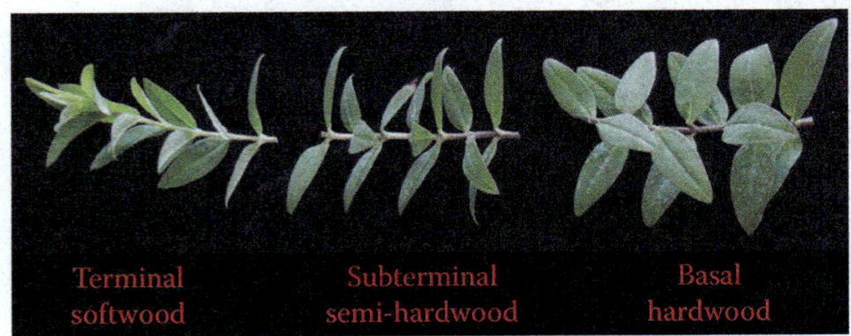

FIGURE 4.13 Portions of the stem illustrating terminal softwood, subterminal semi-hardwood, and basal hardwood cuttings.

Experiment 1. Effects of the Different Formulations of Auxin on the Rooting of Familiar Houseplants

MATERIALS

The following materials will be required for each student or team of students to complete this study:

- Plants that can be used include coleus, scented geranium, pothos, and begonia (Angel Wing type)
- Healthy stem cuttings from the terminal and the subterminal sections of stems
- Market pack, 3 × 5 × 8 in. or larger flats, depending on the size of the class
- Rooting medium—4 parts perlite, 1 part peat moss. Be sure that the peat moss is screened and moistened prior to use. Perlite dust is dangerous to inhale. One way to avoid dust is to pour water into the plastic bag of perlite first and mix it, prior to scooping it out. Another method is to have one person, wearing a mask, mix water into the dry perlite, which has been poured out onto a bench or mixing table
- Powder formulation of IBA
- Gel formulation of IBA in the same concentration
- Liquid formulation of IBA in the same concentration
- Clippers
- Water for dipping cuttings

Follow the instructions in Procedure 4.1 to complete this experiment.

ANTICIPATED RESULTS

There may or may not be a difference in the treatments. The terminal cuttings treated with 50% alcohol are more likely to rot if they are very succulent. The woody basal cuttings should root less well with the powder treatment. This experiment may also be performed using *Buddleia davidii* (butterfly bush). This plant is not available in the northern part of the country. Other possible plants to use include succulents such as those in the family Crassulaceae.

QUESTIONS (WILL DEPEND ON WHICH OF THE FACTORS WERE EXAMINED)

- Was there any difference between the powder, gel, and liquid form of the auxin used?
- Was there a difference in the rooting response with different concentrations of auxin? If so, what was the difference? In rooting percentage? In the morphology of roots formed?
- Did different types of cuttings (e.g., terminal, subterminal, basal) respond differently?
- Were there differences among different species tested?

Procedure 4.1

Effect of Different Concentrations of Auxin on the Rooting of *Dianthus caryophyllus* (Carnation) and *Dendrathema morifolium* (Chrysanthemum) Shoots

Step	Instructions or Comments
1	Fill the flats with the rooting mix to within 0.5 in. of the top. Water the mix and let it drain.
2	For coleus, use a 1000-ppm concentration of IBA in liquid, gel, and powder formulations. Take half of the terminal, subterminal, and basal cuttings and dip it in water (Figure 4.13). Then, dip in the powder and stick in the flat filled with the rooting mix. Take the other half of the cuttings and dip it in the IBA solution for 5 sec (quick dip). Stick the cuttings in the flat as well. Be sure that all treatments are well labeled and that the date that the experiment was done is also on the label.
3	Take coleus plants and clip off the terminal 2.5 to 3 in. Cut again 2.5 to 3 in. lower on the stem and then make a cutting from a woody part of the stem. There are now three types of cuttings: terminal, subterminal, and basal. Prepare 10 to 20 sets of cuttings per student or group depending on the number of cuttings available.
4	Take one-third of the terminal, subterminal, and basal cuttings and dip it in water. Then, dip in the powder and the stick in the flat filled with the rooting mix. Take another third of the cuttings and dip it in the IBA solution for 5 sec (quick dip). Stick the cuttings in the flat as well. Take the final third of the cuttings, dip it into the gel formulation of IBA, and insert it into the flat. Be sure that all treatments are well labeled and that the date that the experiment was done is also on the label. Data should be recorded in 2 to 4 weeks.

Experiment 2. Effect of Different Concentrations of Auxin on the Rooting of *Dianthus caryophyllus* (Carnation) and *Dendrathema morifolium* (Chrysanthemum) Shoots

Different plant species have different responses to auxin applied for rooting. A concentration that may be ideal for one species may be too high or too low for another. In this experiment, the effect of two concentrations of IBA on the rooting of the terminal cuttings of carnation and chrysanthemum will be determined.

MATERIALS

The following items are required for each team of students to complete this experiment:

- Chrysanthemum cuttings from a commercial grower
- Carnation cuttings from a commercial grower
- Cuttings (60) each of carnation and chrysanthemum are required for each group of students. It is best to have several extra in case some cuttings are damaged or destroyed
- Flat large enough to hold 120 cuttings
- Rooting medium of 4 parts perlite to 1 part peat moss or vermiculite. Be sure that the peat moss is screened and moistened prior to use. The dust of the perlite is dangerous to inhale. One way to avoid dust is to pour water into the plastic bag of perlite first and mix it, prior to scooping it out. Another method is to have one person, wearing a mask, mix water into the perlite, which has been poured out dry
- IBA in powder formulation at 1000 and 3000 ppm
- Water for dipping cuttings
- Clippers
- Talc

Follow the instructions presented in Procedure 4.2 to complete this exercise.

Procedure 4.2

Effect of Different Concentrations of Auxin on the Rooting of *Dianthus caryophyllus* (Carnation) and *Dendrathema morifolium* (Chrysanthemum) Shoots

Step	Instructions and Comments
1	The cuttings are used just as they come. Do not trim the bases. It is preferable to use the cuttings within 1 week of their delivery. Fill the flats with the rooting mix to within 0.5 in. of the top. Water the mix and let it drain. Place the wooden or plastic stakes in the flat with the proper treatment date and species.
2	The treatments are the same for carnations and chrysanthemums and are as follows: (1) control (no IBA), (2) 1000-ppm IBA, and (3) 3000-ppm IBA.
3	Take 20 cuttings of carnation, and dip them in water and then in the 1000-ppm hormone powder. Stick in the flat in the proper row. Repeat the procedure with another 20 cuttings for the 3000-ppm treatment. For the control, the cuttings are dipped in water then in talc, and stuck in the flat. Repeat with the chrysanthemum cuttings. Place the flat in the greenhouse.
4	After 2 weeks, 10 cuttings of each treatment are removed, and the amount of rooting is recorded on a data sheet (see example, Table 4.2). One method of quantifying root formation is to set up a scale ranging from 0 to 5, with 0 being no roots and 5 being a large number of well-developed roots.
5	After an additional 2 weeks, the remaining cuttings are removed, and the amount of rooting is recorded. At this time, it is possible that students have already rated some root masses as 5 and now see even larger root masses. They can use higher ratings without affecting the results of the experiment. Note, however, that the same student or students must do the ratings for them to be reliable over time. The root ratings are combined with the percentage of shoots rooted to produce a rooting index that will enable the students to compare the results across treatments and species (Table 4.3).

TABLE 4.2
An Example of Data Collection

Cutting Number	Week 2 Rating	Week 4 Rating
1	2	4
2	1	6
3	0	4
4	1	4
5	3	6
Mean	1.4	4.8
Rooting percentage	80%	100%

TABLE 4.3
Calculating the Weighted Root Rating That Takes into Account the Number of Cuttings at Each Root Rating

A. Rating	B. Week 2 No. of Cuttings	C.	D. Rating	E. Week 4 No. of Cuttings	F.
0	1	0	0	0	0
1	2	2	1	0	0
2	1	2	2	0	0
3	1	3	3	0	0
4			4	3	12
5			5	0	0
6			6	2	12
Rooting index		1.4			4.8

ANTICIPATED RESULTS

Roots will likely begin to appear within 7 to 10 days. Soon after that, rooting percentages and individual root lengths can be measured. With these species, it is likely that you will see increasing rooting percentages with increasing auxin concentrations. At high auxin concentrations, phytotoxic symptoms (browning stems and/or roots, stunted root elongation) may be apparent.

QUESTIONS

- Did the two species (carnation and chrysanthemum) respond differently to the auxin treatments?
- Were any phytoxicity symptoms observed? If so, describe them and speculate about what might have caused the symptoms.

LITERATURE CITED

Addicott, F. T., J. L. Lyon, K. Ohkuma, W. E. Thiessen, H. R. Carns, O. E. Smith, J. W. Cornforth, B. V. Milborrow, G. Ryback, and P. F. Wareing. 1968. Abscisic acid: A new name for abscisin II (dormin). *Science* 159:1493.

Aloni, R. 1995. The induction of vascular tissues by auxin and cytokinin. In Davies, P. J., ed., *Plant Hormones, Physiology, Biochemistry, and Molecular Biology*, Second Edition. Dordrecht, The Netherlands: Kluwer Academic Publishers, pp. 531–546.

Amasino, R. 2005. 1955: Kinetin arrives—The 50th anniversary of a new plant hormone. *Plant Physiol* 138:1177–1184.

Arteca, R. 1995. Brassinosteroids. In Davies, P. J., ed., *Plant Hormones, Physiology, Biochemistry, and Molecular Biology*, Second Edition. Dordrecht, The Netherlands: Kluwer Academic Publishers, pp. 206–213.

Bandurski, R., J. D. Cohen, J. P. Slovin, and D. M. Reinecke. 1995. Auxin biosynthesis and metabolism. In Davies, P. J., ed., *Plant Hormones, Physiology, Biochemistry, and Molecular Biology*, Second Edition. Dordrecht, The Netherlands: Kluwer Academic Publishers, pp. 39–65.

Biswas, P. K., P. A. Bonamy, and K. B. Paul. 1972. Germination promotion of loblolly pine and bald cypress seeds by stratification and chemical treatments. *Physiol Plant* 27:71–76.

Bleecker, A. and H. Kende. 2000. Ethylene: A gaseous signal molecule in plants. *Annu Rev Cell Dev Biol* 16:1–18.

Brown, C., F. J. Brooks, D. Pearson, and R. J. Mathias. 1989. Control of embryogenesis and organogenesis in immature wheat embryo callus using increased medium osmolarity and abscisic acid. *J Plant Physiol* 133:727–733.

Carlson, R. D. and A. J. Crovetti. 1988. Commercial use of GA and cytokinins and new areas of applied research. In Pharis, R. P. and S. B. Roods, eds., *Plant Growth Substances*. New York: Springer-Verlag, pp. 605–610.

Cheng, Z.-M., W. H. Dai, M. J. Bosela, and L. D. Osburn. 2005. Genetic engineering approach to enhance adventitious root formation of hardwood cuttings. *J Crop Improv* 17:211–225.

Clark, D. G., E. K. Gubrium, J. E. Barrett, T. A. Nell, and H. J. Klee. 1999. Root formation in ethylene-insensitive plants. *Plant Physiol* 121:53–60.

Cole, D. F. and J. E. Wheeler. 1974. Effect of pre-germination treatments on germination and growth of cotton seeds at suboptimal temperatures. *Crop Sci* 14:451–454.

Darwin, C. R. 1880. *The Power of Movement in Plants*. London, United Kingdom: Murray, 592 pp.

Davies, P. J. 2004a. The plant hormones: Their nature, occurrence, and functions. In Davies, P. J., ed., *Plant Hormones, Biosynthesis, Signal Transduction, Action!*, Third Edition. Dordrecht, The Netherlands: Kluwer Academic Publishers, pp. 1–15.

Davies, P. J. 2004b. Regulatory factors in hormone action: Level, location, and signal transduction. In Davies, P. J., ed., *Plant Hormones, Biosynthesis, Signal Transduction, Action!*, Third Edition. Dordrecht, The Netherlands: Kluwer Academic Publishers, pp. 16–35.

Dirr, M. A. 1998. *Manual of Woody Landscape Plants*, Fifth Edition. Champaign, Illinois: Stipes Publications, 1187 pp.

Dirr, M. A. and C. W. Heuser. 1987. *The Reference Manual of Woody Plant Propagation*. Athens, Georgia: Varsity Press.

Drew, M. C., M. B. Jackson, and S. Giffard. 1979. Ethylene-promoted adventitious rooting and development of cortical air spaces (aerenchyma) in roots may be adaptive responses to flooding in *Zea mays L. Planta* 147:83–88.

Gaba, V. P. 2005. PGRs in plant tissue culture and development. In Trigano, R. N. and D. J. Gray, eds., *Plant Tissue Culture and Development*. Boca Raton, Florida: CRC Press, pp. 87–99.

Galston, A. W. and R. Kaur-Sawhney. 1995. Polyamines as endogenous growth regulators. In Davies, P. J., ed., *Plant Hormones, Physiology, Biochemistry, and Molecular Biology*, Second Edition. Dordrecht, The Netherlands: Kluwer Academic Publishers, pp. 158–178.

Gaspar, T., C. Kevers, C. Penel, H. Greppin, D. M. Reid, and T. A. Thorpe. 1996. Plant hormones and PGRs in plant tissue culture. *In Vitro Cell Dev Biol (Plant)* 32:272–289.

Geneve, R. L. and C. W. Heuser. 1983. The relationship between ethephon and auxin on adventitious root initiation in cuttings of *Vigna radiata* (L.) R. Wilcz. *J Amer Soc Hort Sci* 108:330–333.

Gonzalez-Guzman, M., D. Abia, J. Salinas, R. Serrano, P. L. Rodriguez. 2004. Two new alleles of the abscisic aldehyde oxidase 3 gene reveals its role in abscisic acid biosynthesis in seeds. *Plant Physiol* 135:325–333.

Gray, D. 2004. Propagation from non-meristematic tissues: Non-zygotic embryogenesis. In Trigiano, R. N. and D. J. Gray, eds., *Plant Development and Biotechnology*. Boca Raton, Florida: CRC Press, pp. 187–200.

Hamilton, D. F. 1972. Ethrel for breaking dormancy in seeds of woody plants. *Intl Plant Prop Soc Proc* 22:368–373.

Hartmann, H. T., D. E. Kester, F. T. Davies Jr., and R. T. Geneve. 2010. *Plant Propagation: Principles and Practices*, 8th Edition. Upper Saddle River, New Jersey: Prentice Hall, 928 pp.

Hu, Y. X., F. Bao, and X. Y. Li. 2000. Promotive effect of brassinosteroids on cell division involves a distinct Cyc D3-induction pathway in Arabidopsis. *Plant J.* 24:693–701.

Kitto, S. L. and J. Janick. 1985. Hardening treatments increase survival of synthetically coated asexual embryos of carrot. *J Amer Soc Hort Sci* 110:283–286.

Klee, H. J., R. B. Hoesch, M. A. Hinchee, M. B. Hein, and N. L. Haffmann. 1987. The effects of overproduction of two *Agrobacterium tumefaciens* TDNA auxin biosynthetic gene products in transgenic petunia plants. *Genes Dev* 1:8696.

Koiwai, H., K. Nakaminami, M. Seo, W. Mitsuhashi, T. Toyomasu, T. Koshiba. 2004. Tissue-specific localization of an abscisic acid biosynthetic enzyme, AAO3, in Arabidopsis. *Plant Physiol* 134:1697–1707.

Korkmaz, A., I. Tiryaki, M. N. Nas, and N. Ozbay. 2004. Inclusion of plant growth regulators into priming solution improves low-temperature germination and emergence of watermelon seeds. *Can J Plant Sci* 84:1161–1165.

Kumar, S. V. and V. Rajam. 2005. Polyamines enhance *Agrobacterium tumefaciens vir* gene induction and T-DNA transfer. *Plant Sci* 168:475–480.

Kurosawa, E. 1926. Experimental studies on the nature of the substance secreted by the "bakanae" fungus. *Nat Hist Soc Formosa* 16:213–227.

Leadem, C. L. 1987. The role of plant growth regulations in the germination of forest tree seeds. *Plant Growth Regul* 6:61–93.

Letham, D. S. 1963. Zeatin, a factor inducing cell division isolated from *Zea mays*. *Life Sci* 2:569–573.

Liu, J.-H. and D. M. Reid. 1992. Auxin and ethylene-stimulated adventitious rooting in relation to tissue sensitivity to auxin and ethylene production in sunflower hypocotyls. *J Exp Bot* 43:1191–1198.

Ludwig-Muller, J. and E. Epstein. 1994. Indole-3-butyric acid in *Arabidopsis thaliana*: III. *In vitro* biosynthesis. *J Plant Growth Regul* 14:7–14.

Machakova, I., E. Zazimalova, and E. F. George. 2008. PGRs: I. Introduction: Auxins—Their analogs and inhibitors. In George, E. F., M. A. Hall, and G. de Klerk. *Plant Propagation by Tissue Culture*, Third Edition. Dordrecht, The Netherlands: Springer, pp. 175–204.

Malabadi, R. B. and K. Nataraja. 2007. Brassinosteroids influence *in vitro* regeneration using shoot tip sections of *Cymbidium elegans* Lindl. *Asian J of Plant Sci* 6:308–313.

Malamy, J., J. P. Carr, D. F. Klessig, and I. Raskin. 1990. Salicylic acid: A likely endogenous signal in the resistance response of tobacco to viral infection. *Science* 250:1001–1004.

Metraux, J. P., H. Signer, J. Ryals, E. Ward, M. Wyss-Benz, J. Gaudin, K. Raschdorf, E. Schmid, W. Blum, and B. Inverardi. 1990. Increase in salicylic acid at the onset of systemic acquired resistance in cucumber. *Science* 250:1004–1006.

Miller, C. O., F. Skoog, M. H. von Saltza, and F. M. Strong. 1955. Kinetin: A cell division factor from deoxyribonucleic acid. *J Amer Chem Soc* 77:1392.

Murthy, B. N. S., S. J. Murch, and P. K. Saxena. 1998. Thidiazuron: A potent regulator of *in vitro* plant morphogenesis. *In Vitro Cell Dev Biol (Plant)* 34:267–275.

Oh, M. H. and S. D. Clouse. 1998. Brassinolide affects the rate of cell division in isolated leaf protoplasts of Petunia hybrid. *Plant Cell Rep* 17:921–924.

Pinfield, N. J. and A. K. Stobart. 1972. Hormonal regulation of germination and early seedling development in *Acer pseudoplatanus* (L.). *Planta* 104:134–145.

Preece, J. E., C. A. Huetteman, W. C. Ashby, and P. L. Roth. 1991. Micro and cutting propagation of silver maple: I. Results with adult and juvenile propagules. *J Amer Soc Hort Sci* 116:142–148.

Raskin, I. 1992. Role of salicylic acid in plants. *Ann Rev Plant Physiol Mol Biol* 43:439–463.

Raskin, I. 1995. Salicylic acid. In Davies, P. J., ed., *Plant Hormones, Physiology, Biochemistry, and Molecular Biology*, Second Edition. Dordrecht, The Netherlands: Kluwer Academic Publishers, pp. 188–205.

Reid, J. B. and S. H. Howell. 1995. Hormone mutants and plant development. In Davies, P. J., ed., *Plant Hormones, Biosynthesis, Signal Transduction, Action!*, Third Edition. Dordrecht, The Netherlands: Kluwer Academic Publishers, pp. 448–485.

Rey, M., C. Diaz-Sala, and R. Rodriguez. 1994. Polyamines as markers for juvenility in filbert. *Acta Hort* 351:233–237.

Robbins, J. A., M. J. Campidonica, and D. Burger. 1988. Chemical and biological stability of indole-3-butyric acid (IBA) after long-term storage at selected temperatures and light regimes. *J Environ Hort* 6:33–38.

Ronsch, H., G. Adam, J. Matschke, and G. Schachler. 1993. Influence of (225,235)-homobrassinolide on rooting capacity an survival of adult Norway spruce cuttings. *Tree Physiol* 12:71–80.

Salisbury, F. B. and C. W. Ross. 1985. *Plant Physiology*, Third Edition. Belmont, California: Wadsworth Publishing, 540 pp.

Schwarz, O. J., A. R. Sharma, and R. M. Beaty. 2004. Propagation from non-meristematic tissues: Organogenesis. In Trigiano, R. N. and D. J. Gray, eds., *Plant Development and Biotechnology*. Boca Raton, Florida: CRC Press, pp. 159–172.

Shibakusa, R. 1980. Effect of gibberellin A_3, IAA, ethrel, and 6-benzyl-adenine on seed germination of *Picea glehnii*. *J Japan For Soc* 62:440–443.

Singha, S. and L. E. Powell. 1978. Response of apple buds culture *in vitro* to abscisic acid. *J Amer Soc Hort Sci* 103:620–622.

Sitbon, F., S. Hennion, B. Sundberg, C. H. A. Little, O. Olsson, and G. Sandberg. 1992. Transgenic tobacco plants co-expressing the *Agrobacterium tumefaciens iaa*M and *iaa*H genes display altered growth and indoleacetic acid metabolism. *Plant Physiol* 99:1062–1069.

Skoog, F. and C. O. Miller. 1957. Chemical regulation of growth and organ formation in plant tissue cultured *in vitro*. *Symp Soc Exp Biol* 11:118–131.

Sponsel, V. M. 1995. Gibberellin biosynthesis and metabolism. In Davies, P. J., ed., *Plant Hormones, Physiology, Biochemistry, and Molecular Biology*, Second Edition. Dordrecht, The Netherlands: Kluwer Academic Publishers, pp. 66–97.

Thomas, S. G. and T.-P. Sun. 2004. Update on gibberellin signaling: A tale of the tall and the short. *Plant Physiol* 135:668–676.

Visser, E., J. D. Cohen, G. Barendse, C. Blom, and L. Voesenek. 1996. An ethylene-mediated increase in sensitivity to auxin induces adventitious root formation in flooded *Rumex palustris*. *Sm Plant Physiol* 112:1687–1692.

Went, F. W. 1926. On growth-accelerating substances in the coleoptiles of *Avena sativa*. *Proc Kon Ned Akad Wet* 30:10–19.

Yabuta, T. and Y. Sumiki. 1938. On the crystal of gibberellin, a substance to promote plant growth. *J Agric Chem Soc Japan* 14:1526.

Zhang, Z. and Z.-M. Cheng. 1996. The effect of jasmonic acid on *in vitro* nodal culture of three potato cultivars. *Hort Sci* 31:631.

5 Sexual Reproduction and Breeding

Timothy A. Rinehart, Robert N. Trigiano, Phillip A. Wadl, and Haley S. Smith

Concept Box 5.1

- Most plants exhibit alternation of generations. In this case, the plant generations are the sporophytic or diploid (2n) phase that produces spores and the gametophytic or haploid (n) phase that produces gametes.
- The four parts of flowers are sepals, petals, anthers, and pistils. Flowers of some species are composed of all four parts, whereas flowers of other species lack one or more of the parts.
- Angiosperms exhibit double fertilization in which one sperm nucleus (n) unites with the egg (n) to form the zygote (2n), and the other sperm nucleus fuses with two polar nuclei to become the primary endosperm nucleus (3n or 5n). Water is not necessary for fertilization.
- Zygotic embryogenesis for monocotyledonous and dicotyledonous plants typically follows a predetermined developmental pattern, although there are many variations between species.
- Sexual reproduction is a major source of genetic diversity in populations and individuals.
- Plant breeding involves the selection of plants with combinations of improved traits that are inherited in a predictable manner.
- Collecting, understanding, and incorporating genetic variation into a horticultural breeding program are critical to success.
- Clearly defined goals help plant breeders choose an appropriate plant breeding strategy to avoid problems such as inbreeding depression.
- Manipulating chromosome numbers is an effective strategy to change plant morphology and/or overcome fertility barriers.
- Mutation breeding and biotechnology can be used to breed new cultivars from clonal material without sexual reproduction.

In this second edition of *Plant Propagation Concepts and Laboratory Exercises*, we have combined the first edition Chapters 36, "Sexual Reproduction in Angiosperms," and 37, "Breeding Horticultural Plants," into this single chapter, "Sexual Reproduction and Breeding." These topics are so closely related that we believe that concepts are best presented as one unit. In addition to combining the aforementioned chapters, we have added some high-resolution microscopic photographs of mitotic events that are comparable to the line drawings depicting the various phases of mitosis. We have maintained some black-and-white figures for clarity of presentation. Lastly, we have included in this edition a simple laboratory exercise involving the observation and frequency of mitotic phases in the cells of onion as seen in longitudinal sections of root tips on prepared slides.

Many horticultural crops are produced exclusively by vegetative or asexual means. Asexual reproduction using grafting, rooted cuttings, and various other methods ensures clonal fidelity of a cultivar. Phenotypic and genetic variation in these crops is undesirable, especially if the consumer is expecting to purchase a plant that they have seen growing somewhere else or pictured in a catalog. For example, all *Cornus florida* L. (flowering dogwood) 'Cherokee Sunset' plants have yellow variegated leaves and pink bracts subtending the floral disk, and are produced exclusively by a grafting technique (Chapter 25). There is little variation using this technique, and it produces thousands upon thousands of exact genetic copies of 'Cherokee Sunset.' In the end, the consumer receives the desired product with the expected characteristics.

Other horticultural plants, especially many bedding species and vegetable crops, are propagated mainly by seeds. This is a sexual process that entails meiosis in both the male and the female cells to form haploid gametes. The original diploid number of the chromosome is restored by the fusion of the haploid gametes from each parent. As will be discussed below in more detail, sexual reproduction in angiosperms involves pollination, in which the pollen or the male gametophyte lands on receptive female surfaces, usually the stigma of the ovary. Eventually, a process called "double fertilization" takes place, in which one male sperm nucleus fuses with the female egg to create the zygote and another sperm nucleus fuses with two other nuclei of the female gametophyte to form the primary endosperm nucleus. The result of sexual reproduction is a seed containing an embryo and everything necessary to grow a new seedling.

ALTERNATION OF GENERATIONS

The concept of alternation of generations is central to understanding sexual reproduction in angiosperms as well as most other organisms. In this case, generations do not refer to parents and children, but rather to the nuclear condition or the ploidy number of the plant's cells during the life cycle of the plant. Ploidy level is the number of copies of chromosomes in the nucleus. For all organisms that reproduce sexually, including angiosperms, one part of the life cycle is represented by haploid (n) cells (gametes that only have one-half the number of chromosomes), and the rest of the life cycle are plant tissues with diploid (2n) cells, which have the full complement of chromosomes. It is important to mention here that plants may contain single or multiple sets of chromosomes (polyploidy). For example, flowering dogwood haploid (n) cells contain one set of 11 chromosomes and, therefore, can be represented by the equation: $n = x = 11$, where x is a set of 11 chromosomes. Therefore, the diploid (2n) cells of flowering dogwood all contain 22 chromosomes or $2n = 2x = 22$. In contrast, other plants, such as the common chrysanthemum, have multiple sets of chromosomes. The haploid gametes of this plant each contains three sets of nine chromosomes, expressed in the above equation form: $n = 3x = 27$. Each diploid cell contains six sets, each consisting of nine chromosomes or $2n = 6x = 54$.

The life cycle phase of plants that produces gametes (n) is called the "gametophytic phase." For example, the green, "leafy" moss growing on a log or a rock is the gametophytic phase. Oftentimes, individual plants produce either

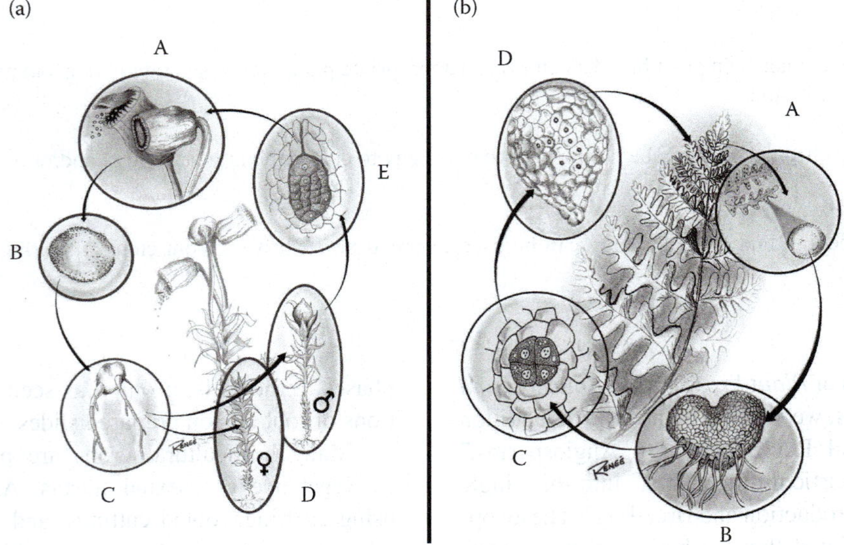

FIGURE 5.1 Alternation of generations. (a) Moss. The most notable phase of moss is the green, "leafy" gametophyte (n), the plant that produces gametes. The gametophytic phase begins with the formation of n meiospores (product of meiosis) in the sporophyte capsule (A). Meiospore are released from the capsule (B) and germinate (C). The gametophyte grows by typical cell division and differentiation and ultimately produces male and female gametangia containing gametes (D). Sperm from the male gametangia swims in a film of water to the female gametangia and fuse with the egg to produce a 2n zygote (E). The zygote develops and produces the mature sporophyte, which remains attached to the living gametophytic phase (A). (b) Fern. The most conspicuous phase of ferns is the sporophyte, which is green and can be very large. Meiosis occurs in sporangia and heralds the beginning of the gametophytic phase of the life cycle (A). The haploid meiospore germinates and grows into a very small and inconspicuous heart-shaped gametophyte containing both female and male gametangia (B). Fertilization requires water and sperm and egg fuse, creating the zygote and restoring the diploid (2n) or sporophytic phase (C). The zygote grows by mitotic divisions and differentiates into a mature sporophyte, which is typically thought of as a fern (D).

sperm v(male) or egg (female) gametes (Figure 5.1a). All of the cells of the gametophyte are haploid. Using this same example, the green (turning brown with age) hooked stalk with a terminal capsule growing from the leafy gametophyte of mosses is the "sporophyte" or "sporophytic phase," which will eventually produce spores. All of the cells of the sporophyte are diploid. The gametophytic phase is the most conspicuous phase of many "lower" plants, such as algae, mosses, and horsetails, whereas the gametophytic phase is reduced and is much less conspicuous in higher plants, such as ferns and gymnosperms (Figure 5.1b). In the case of ferns, we usually see the green and leafy, and seldom observe the much reduced in size gametophytic phase, which typically contains both male and female structures.

Angiosperms, or flowering plants, represent the epitome of gametophyte reduction. In this group, the female and male gametophytes are represented only by a few cells in the flower, and will be discussed later on in this chapter. The sporophyte of angiosperms is typically what you see in forests, fields, and landscapes, whereas the gametophyte is very small, and for the most part, we are unaware of its physical existence except perhaps as a coating of "dust" on a window or car. However, many of us are allergic to pollen (runny noses, itchy eyes, etc.), which is the unseen male gametophyte of angiosperms.

MITOSIS AND MEIOSIS

So, how do alternation and development of generations occur in angiosperms? The driving mechanisms for the alternation of generations are meiosis, which reduces the ploidy number from diploid to haploid, and the sexual fusion of gametes to form the zygote, which restores the diploid number. To understand the maintenance and reduction of ploidy levels, we need to consider the following two cellular processes that are related: mitosis (Figure 5.2) and meiosis (Figure 5.3).

Let us consider mitosis first because it is less complicated than meiosis. This process is integral in cell division and allows cells to maintain the ploidy level of the original cell, regardless of whether it was haploid or diploid. In other words, the number of chromosomes or sets of chromosomes will be the same in the daughter cells resulting from a mitotic division as it was in the mother

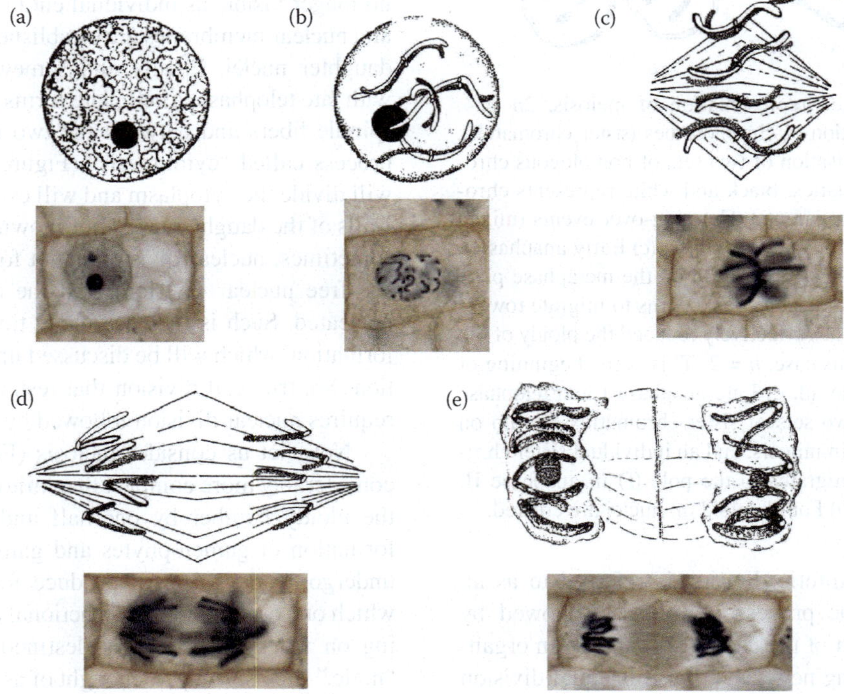

FIGURE 5.2 Idealized representation and photomicrographs of mitosis in onion root tips. In this figure, the idealized line drawing appears above, and the photomicrograph, below, for each phase of mitosis. All photomicrographs were taken from longitudinal sections of onion root tips. (a) Interphase nucleus. In this phase, chromosomes were replicated and remain uncondensed. Nucleoli are prominent. (b) Prophase nucleus. In this example, the set of four sister chromatids (two copies of each chromosome) has condensed (thickened and shortened). Note that the sister chromatids cannot be distinguished in the photomicrographs provided. In late prophase, the nuclear membrane and the nucleolus disappear. (c) Metaphase. The sister chromatids have aligned on the equatorial (metaphase) plate and the spindle fibers attached to centromeres. The individual sister chromatids (now chromosomes) begin to move toward the two poles. (d) Anaphase (late). The chromosomes have migrated to the poles. (e) Telophase. The chromosomes become long, threadlike (uncondensed) molecules, the nuclear membrane and nucleoli are reformed, and the wall between the two daughter cells is forming (cytokinesis). (Courtesy of Triarch Biology, Ripon, Wisconsin.)

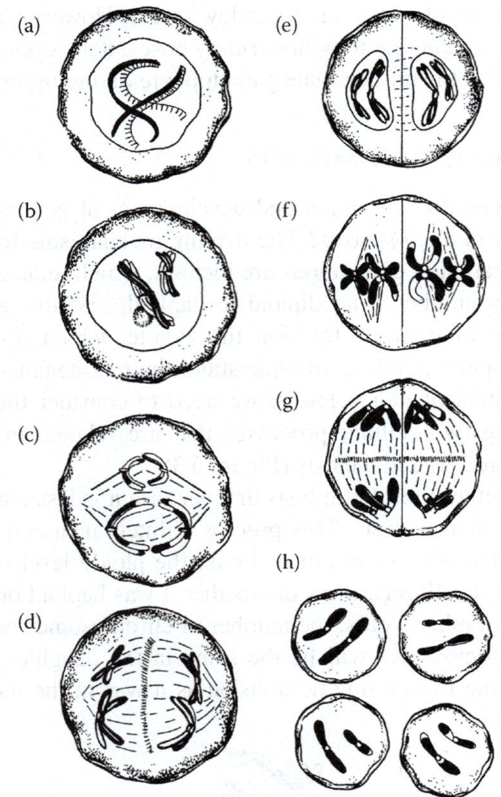

FIGURE 5.3 Idealized representation of meiosis. $2n = 4$. (a) Interphase. Replication of chromosomes (sister chromatids). (b) Prophase I. Condensation of two sets of homologous chromosomes (sister chromatids: black and white represents chromosomes from the two parents). Crossing-over events (mixed black and white) occur during this phase. (c) Early anaphase I. Homologous chromosomes are paired on the metaphase plate and one-half of the homologous pair begins to migrate toward each of the poles. This has effectively reduced the ploidy of the cell to haploid or, in this case, $n = 2$. This is the beginning of the gametophytic phase. (d, e) Late anaphase I and telophase I. (f) Metaphase II. Two sets of sister chromatids line up on the equatorial plate as in mitosis, and an individual sister chromatid (chromosome) migrates to the pole (f) in anaphase II. (g) Late telophase II. (h) Four haploid (n) nuclei are created.

cell. Sometimes, a mitotic division is referred to as an "equal division." The process of mitosis (followed by cytokinesis—division of the cytoplasm) allows an organism to grow by adding new cells. For this equal division to happen, duplication of the chromosomes must occur sometime during the cell cycle, which can be conveniently organized into the following five phases: interphase, prophase, metaphase, anaphase, and telophase. These phases are continuous with one another in the cell, but are demarked by more or less recognizable characteristics of the chromosomes. The nucleus of a cell remains in interphase for the majority of the cell cycle (Figure 5.2a). The nucleus appears granular after staining, and the nucleolus is apparent as a more densely staining sphere within the nucleus. At this time, the DNA molecules that the chromosomes are made of are very thin and long, and cannot be recognized as separate structures within the nucleus. The chromosomes are replicated or copied during this time. The template (the original DNA strand) and the replicated chromosome copy do not separate, but stay joined together and are now called "sister chromatids" instead of "chromosomes." Prophase (Figure 5.2b) is marked by the sister chromatids shortening and thickening. This process is also referred to as "condensation." The end of prophase is marked by the disappearance of the nucleolus and the nuclear membrane. The next phase is metaphase, in which the sister chromatids line up on the equatorial (metaphase) plate of the cell and spindle fibers growing from the poles attach to the centromeres (Figure 5.2c). These fibers are made of microtubules that contract, pulling one of each of the sister chromatids toward the opposite poles. Individual sister chromatids are now termed "chromosomes." The migration of the chromosomes signals the beginning of anaphase (Figure 5.2d). Telophase (Figure 5.2e) occurs when the chromosomes reach the poles and start to become thin and thread-like as just they were before prophase. They continue this process until they are no longer visible as individual entities. Nucleoli reappear and nuclear membranes are established around each of the daughter nuclei. Finally and somewhat simultaneously, with late telophase, a cell plate forms perpendicular to the spindle fibers and between the two nuclei, and begins a process called "cytokinesis" (Figure 5.2e). This process will divide the cytoplasm and will eventually result in cell walls of the daughter cells (not shown). One final thought: sometimes, nuclear division is not followed by cytokinesis (free nuclear division), and the cell becomes multinucleated. Such is the case many times with endosperm formation, which will be discussed under double fertilization. So, true cell division that results in two equal cells requires nuclear division followed by cytokinesis.

Now, let us consider meiosis (Figure 5.3), which is considerably more complex than mitosis. Meiosis reduces the ploidy number by one-half and is essential to the formation of gametophytes and gametes. A diploid cell undergoing meiosis will produce four haploid cells, of which one or more will be functional as a gamete depending on whether the cell is destined to be "female" or "male." Meiosis may be thought of as a reduction division (diploid to haploid) followed by an equal division, similar to that of mitosis. As in mitosis, the chromosomes are replicated during interphase (Figure 5.3a) and are now in the form of sister chromatids. The chromosomes thicken and become visible in prophase, as discussed in mitosis. However, at the end of this phase and unlike in mitosis, chromosomes are paired (Figure 5.3b), and each pair now consists of four sister chromatids. Another phenomenon happens in this stage as well. Segments from sister chromatids of pairs of chromosomes are exchanged. In other words, a portion of one chromosome is swapped with the corresponding site on the other chromosome and vice versa. This is called "crossing over" and represents a rearrangement of genes that were contributed by each parent in the previous generation. Before crossing over, each pair of chromosomes in the diploid cell, called "homologous chromosomes," contained one chromosome from the mother and one from the father. After crossing over during meiosis, the maternal and paternal genetic material has been recombined into a totally new arrangement that will be passed on to the next generation through the formation of gametes and subsequent sexual reproduction. Recombining the genomes before sex contributes greatly to the genetic diversity of the species.

The pairs of homologous chromosomes line up on the equator in metaphase I (the "I" refers to the first division of meiosis), and spindle fibers are formed. However, instead of the sister chromosomes migrating to opposite poles as in mitosis, one-half of the pair of homologous chromosomes (each consisting of two sister chromatids) moves toward each pole during anaphase I (Figure 5.3c). Cytokinesis occurs at the end of telophase I, and each of the daughter nuclei contains one-half of the number of chromosomes (Figure 5.3d, e). The second (II) stage of meiosis begins as metaphase II in the daughter cells, where the sister chromatids align on the equatorial plate (Figure 5.3f). Just like in mitosis, spindle fibers form and attach to the centromeres. During anaphase II (Figure 5.3g), the sister chromatids (now called "chromosomes") separate, and one of each is drawn to the opposite pole (telophase II). Meiosis results in four haploid cells, each with one-half of the diploid complement of chromosomes (Figure 5.3h). Fusion of these haploid cell nuclei during sex, one from the male parent and one from the female parent, results in the formation of the zygote—a new cell with a diploid number of chromosomes that eventually becomes a new seedling plant.

FLOWERS

In angiosperms, sexual reproduction takes place in highly evolved, specialized structures called "flowers." Flowers come in many shapes, colors, and arrangements, and depending on the species of plants, they may appear at any time during the growing season. Flowers may be borne singly as shown in Figure 5.4 or many flowers may be clustered on a common axis (inflorescences). There are many types of inflorescences, and these arrangements are sometimes distinguishing characteristics of certain groups or plant families. Some species of plants produce flowers that may lack one or more basic parts (incomplete), whereas others have all of the structures (complete). Flowers may contain both male and female structures (bisexual), whereas other flowers contain only female, or pistillate, or male, or staminate, parts. Individual plants may have both male and female flowers on the same plant

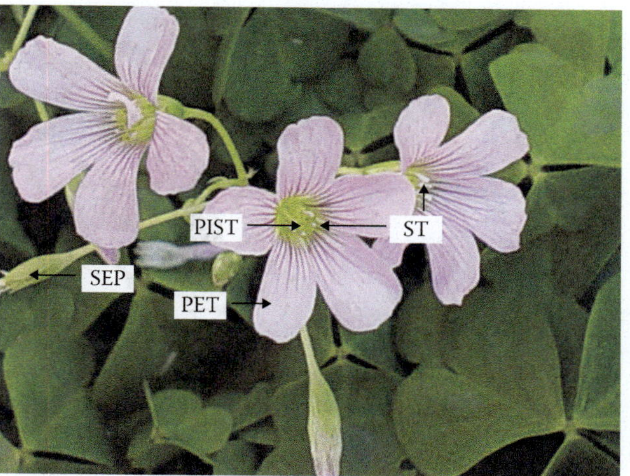

FIGURE 5.4 Morphology of a dicot flower. The perfect flower of this *Oxalis* species has five sepals (SEP) that are hidden from view in the open flowers by the five petals (PET), five stamens (ST), and one pistil (PIST) (cf. Figure 5.5).

(monoecious—literally, "one house"), whereas other plants may only possess flowers that are either male or female (dioecious—literally, "two houses"). We could go on to describe many other differences between flowers of different species, some obvious and others subtle, but one unifying function of flowers among the diverse species of angiosperms is that sex, including meiosis, the production of gametes, and the fusion of gametes to form the zygote, always occurs within the flowers.

It is not our intention to explore all of the different variations of flower morphology, so we will restrict our discussion to the very simple case of a perfect flower—a flower that has all four basic parts with a superior ovary (Figure 5.5).

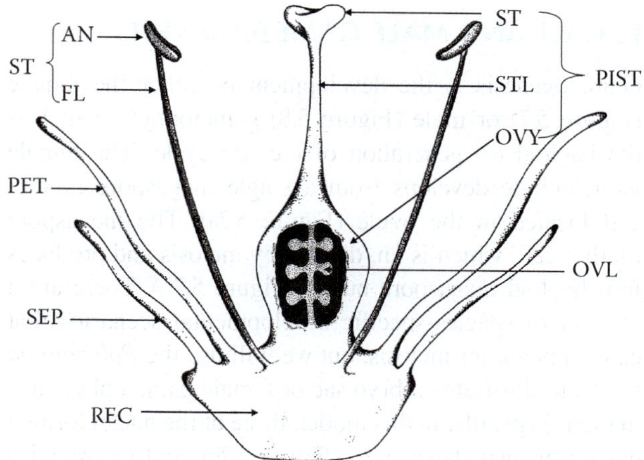

FIGURE 5.5 Idealized diagram of a perfect flower. The four parts of a flower are sepals (SEP), petals (PET), stamens (ST), and pistil(s) (PIST). In this case, all parts are mounted on a receptacle (REC). Stamens are composed of filaments (FL) that support the anthers (AN), which contain pollen. A pistil has three parts: a stigma (ST), a knob-like structure where pollination occurs; a style (STL); and an ovary (OVY), which contains ovules (OVL).

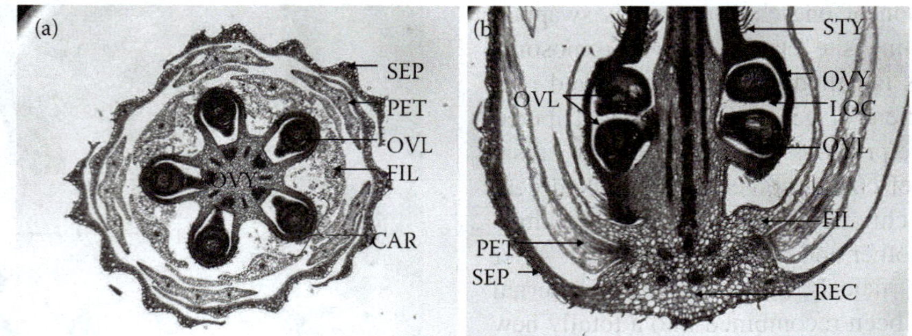

FIGURE 5.6 Flower anatomy. (a) Cross section through a *Geranium* species (Stork's bill) flower. Progressing from the outside inward, sepals (SEP), petals (PET), and filaments (FIL). Occupying the center of the flower is the ovary (OVY), which is composed of five carpels (CAR) in which two ovules (OVL) are borne. Note that only a single ovule can be seen in each locule (space) with this section. (b) Longitudinal section through the Stork's bill flower seen in Figure 5.6a. In this view, two carpels (CAR) and locules (LOC) of the ovary (OVY) are visible, but each contains two ovules (OVL) of which only one will mature. Note the trichomes (hairs) on the style (STY) portion of the pistil. The sepals (SEP), petals (PET), and filaments (FIL) are inserted below the ovary on the receptacle (REC). (Adapted with permission from Trigiano, R. N. and D. J. Gray, *Plant Biotechnology and Development*. Boca Raton, Florida: CRC Press, 2005.)

The four basic building units of flowers working from the outside to the inside are the sepals, green, leaf-like appendages at the base; the petals, which are often pigmented; the stamens, consisting of filaments (stalk) and anthers; and the pistils, which are made up of the stigma, the style, and the ovary. All of the floral components in this case are mounted on a receptacle (Figures 5.5 and 5.6b). The male portions of the flower are the stamens (Figure 5.5), which contain many cells (microspore mother cells) that will undergo meiosis and will eventually form the male gametophytes or pollen. The female part of the flower is the pistil. The ovary contains one to many ovules (Figures 5.5 and 5.6a, b). Each ovule contains one megaspore mother cell that will give rise to the female gametophyte through the process of meiosis.

FEMALE AND MALE GAMETOGENESIS

Gametogenesis is the development of either the female (Figure 5.7) or male (Figure 5.8) gametophyte, which is the haploid (n) generation of the life cycle. The female gametophyte develops from a single megaspore mother cell located in the ovule (Figure 5.7a). The megaspore mother cell, which is 2n, divides by meiosis and produces four haploid megaspore nuclei (Figure 5.7b). There are a number of species-specific developmental scenarios that can happen after meiosis, but we will use the *Polygonium* model to illustrate embryo sac or female gametophyte formation. Typically, in this model, three of the nuclei formed during meiosis degenerate (Figure 5.7c), and the remaining megaspore divides via mitosis three times to produce an eight-celled (nucleated) female gametophyte bounded by the nucellus. The nucellus is a sporophytic tissue and is not formed from the megaspore mother cell. The solitary egg (n) or female gamete, flanked on either side by synergids (n), is located at the micropylar end of the ovule (Figure 5.7a, d). The micropyle is the opening formed by

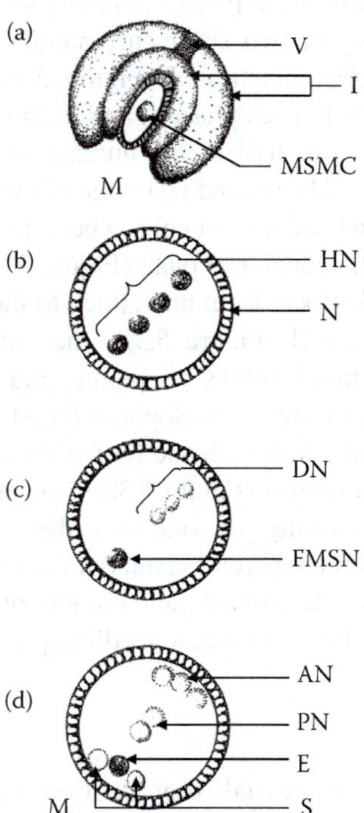

FIGURE 5.7 Idealized diagrammatic representation of female gametophyte development. (a) A single diploid (2n) megaspore mother cell (MSMC) resides within each ovule contained in the ovary. I = integuments (may be one to several); M = micropyle; V = vascular tissue at the opposite end of the ovule. (b) The MSMC undergoes meiosis (see Figure 5.3) to produce four haploid (n) nuclei (HN) within the nucellus (N). (c) In this example, three of the HN degenerate (DN), leaving one functional megaspore nucleus (FMSN). (d) The FMSN will divide via mitosis three times, and the eight haploid nuclei migrate to form the complete female gametophyte or embryo sac consisting of an egg (E); two synergids (S), which lie on either side of the egg at the micropylar end; two polar nuclei (PN) in the center; and three antipodals (AN) at the distal portion of the embryo sac. M = micropylar end of the ovule.

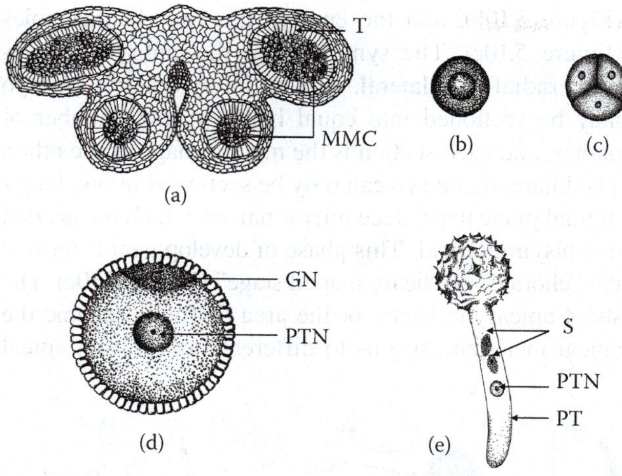

FIGURE 5.8 Idealized diagrammatic representation of male gametophyte development. (a) Many diploid (2n) microspore mother cells (MMC) located in the anther portion of the stamens are surrounded by the tapetum (T), the cell layer derived from the sporophyte that surrounds the MMC and, later, the pollen. (b) A single MMC undergoes meiotic division to produce four haploid (n) microspores, which typically are found in a group called a "tetrad," shown in (c). (d) After the microspores separate (while still in the anther), the nucleus divides via mitosis to produce a generative nucleus (GN) and a pollen tube nucleus (PTN). Hereafter, the binucleated cell is termed "pollen." (e) After pollination, the pollen grain germinates via a pollentube (PT), and the generative nucleus divides to produce two sperm nuclei(s) that are haploid.

the immature integuments (later in development, the seed coat or testa) of the ovule. The polar nuclei, both haploid, are located in more or less the center of the embryo sac. Lastly, there are three antipodal nuclei (n) located at the end opposite of the micropyle (Figure 5.7d).

Male gametogenesis occurs from hundreds of microspore mother cells, which are all 2n, located in the anthers (Figure 5.8a, b). Each microspore mother cell undergoes meiosis to form four (a tetrad) microspores, which are haploid (Figure 5.8c). These microspores, while in the anther, divide via mitosis once again to produce binucleated (two nuclei) pollen grains (Figure 5.8d). Each grain has a tube nucleus and a generative nucleus, which will give rise to two sperm nuclei—the male gametes (Figure 5.8e). The male gametophyte usually matures (produces sperm nuclei) only after dehiscence of the anther, which liberates the pollen and allows subsequent pollination.

POLLINATION AND DOUBLE FERTILIZATION

Now that the male (typically three cells) and female gametophytes (typically eight cells or nuclei) have been produced, how do the male and female gametes come into close proximity? The act of pollen grains being transferred from the anthers to the stigma of a flower is called "pollination." Pollen can be delivered to the stigma of the pistil either via wind currents or by a vector. Those plant species that rely on wind dissemination usually produce copious amounts of light, dry pollen. Some examples of plants depending primarily on wind are some trees, such as oaks and maples; corn and wheat; and perhaps the most notorious of them all, ragweed, a cause of "hay fever" in people during the fall months. Pollen may be vectored to the stigmatic surface of the pistil by numerous types of insects, including honeybees, bumblebees, moths, and beetles; other insects; and mammals, such as bats and birds. In general, those species relying on vectors for pollination produce more conservative amounts of heavy, sticky pollen than those species that depend on wind pollination.

Pollination may occur between the pollen and the stigma on the same plant (self-pollination) or may happen between two different plants (cross-pollination). The surface of the stigma is typically covered with exudates (usually some type of sugar) that help the pollen grain adhere to the tissue and provide physical and physiological cues for germination. Remember, pollination does not mean fertilization—they are two different processes. After pollination occurs, the pollen grain will germinate, producing a pollen tube, which grows through the style toward the ovules in the ovary. Shortly after germination, the generative cell divides, and two sperm nuclei are produced. There are a number of factors that will affect the germination of the pollen and the growth rate of pollen tubes. For example, genetic incompatibility factors exist in some plants, which prevent the germination of the pollen and/or retard pollen tube growth. These factors control mating and thus direct recombination and genetic variability. Assuming that all physical, environmental, and genetic factors are favorable for pollen germination, pollen tubes grow through the length of the style to the micropylar end of the ovule in the ovary and deliver two nonflagellated sperm (Figure 5.9a, b). Since the sperm (nuclei) are not required to swim, free water is not necessary for fertilization in angiosperms and represents a significant advancement over most other plants, which require free water for fertilization. Another process that is unique to the life cycle of angiosperms is double fertilization. Double fertilization entails one of the sperm nuclei uniting with the egg to form the zygote, which is the first cell of the 2n sporophytic phase, and the other sperm nucleus fusing with both of the polar nuclei to form the primary endosperm nucleus, which is triploid (3n) in our *Polygonium* example (Figure 5.9c, d). Note that the endosperm can be pentaploid (5n) in some species of plants. The primary endosperm nucleus divides repeatedly to form the endosperm, which functions as food and provides other physiological and physical factors for the developing sporophyte and/or will be typically absorbed into the cotyledons of the embryo as the seed matures.

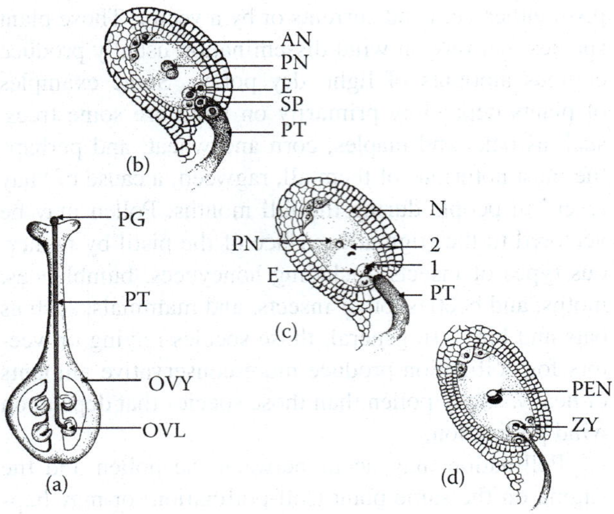

FIGURE 5.9 Idealized diagrammatic representation of pollination, pollen tube growth, and double fertilization. (a) Pollen grain (PG) germinates on the stigma of the pistil and grows via a pollen tube (PT) toward an ovule (OVL) containing a mature female gametophyte. OVY = ovary. (b) The pollen tube (PT) grows through the micropylar end of the ovule toward the egg (E). AN = antipodal nuclei; PN = polar nuclei; SP = sperm nuclei. (c) Double fertilization occurs when one sperm nucleus fuses with the egg (1) and the other sperm nucleus unites with the polar nuclei (2). N = nucellus. (d) Double fertilization results in the formation of zygote (ZY), which is diploid (2n), and the primary endosperm nucleus (PEN), which, in our example, is triploid or 3n.

ZYGOTIC EMBRYOGENESIS

In this section, we will describe only zygotic embryogenesis in dicotyledonous plants, and our example will be redbud (*Cercis canadensis* L.). Zygotic embryogenesis begins with the first mitotic division of the zygote to form the embryo proper (the cell that will form the embryo) and the suspensor. The cells of the suspensor continue to divide and push the developing embryo into the endosperm. The suspensor may help transfer nutrients and water to the developing embryo but does not have any vascular tissue (xylem and phloem—see Chapter 2). The mature suspensor may be a long single file of cells (uniseriate) as in flowering dogwood or may have multiple files of cells (multiseriate) as in redbud (Figure 5.10a, b). The cell that becomes the embryo proper divides via mitosis many times and produces a more or less round (isodiametric) ball of cells with radial symmetry (Figure 5.10a, b). This is the globular stage of embryogenesis, and at this point, there is very little morphological or physiological specialization evident. The protoderm, the cells on the perimeter of the embryo that will form the epidermis, may be distinguishable from other cells and tissues during this stage. With continued divisions and development, the distal portion of the globular embryo begins to flatten (Figure 5.10b), and the embryo has two distinct poles (Figure 5.10c). The symmetry of the embryo changes from radial to bilateral. The radial symmetric embryo may be sectioned into equal halves in any number of planes, and each section is the mirror image of the other. The bilateral embryo can only be sectioned in one longitudinal plane to produce mirror halves, which means that it is bisymmetrical. This phase of development is termed the "chorate" or "heart-shaped stage" (Figure 5.10c). The shoot apical meristem, or the area that will become the apical meristem, begins to differentiate; the root apical

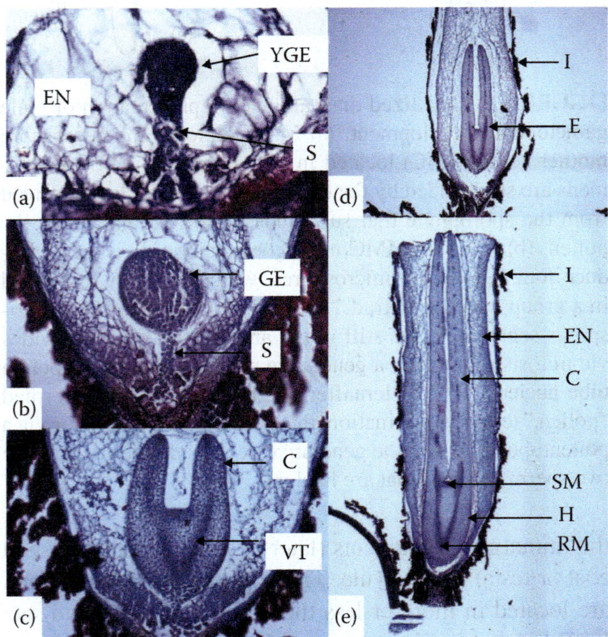

FIGURE 5.10 Embryo development in *Cercis canadensis* (redbud), a dicotyledonous plant. (a) A longitudinal section through a very young globular-stage embryo (YGE). Note the suspensor (S) and the cellular endosperm (EN). (b) Longitudinal section through a late globular-stage embryo (GE) with characteristic radial symmetry. S = suspensor. (c) A median longitudinal section through a bipolar, late heart-shaped embryo exhibiting bilateral symmetry. The shoot and the root meristems (arrows) are not well-differentiated at this stage. C = cotyledon; VT = vascular tissue development. (d) A median longitudinal section through an early torpedo-stage embryo (E). Meristems, cotyledons, and the vascular tissue are well differentiated. I = integuments. (e) A median longitudinal section through a cotyledonary-stage embryo with distinct cotyledons (C), a shoot meristem (SM), and a less conspicuous root meristem (RM). H = hypocotyl; I = integuments; EN = endosperm. Compare to a mature monocotyledonous embryo shown in Figure 5.11a. ((b) and (c) modified and adapted with permission from Jain, S. M., P. K. Gupta, and R. J. Newton (eds.), *Somatic Embryogenesis in Woody Plants*. Dordrecht, The Netherlands: Kluwer Academic Publishers, 1999; (e) modified and adapted with permission from Trigiano, R. N. and D. J. Gray (eds.), *Plant Biotechnology and Development*. Boca Raton, Florida: CRC Press, 2005.)

meristem remains obscure and relatively undifferentiated. Cotyledons are initiated, and the vascular tissue in the hypocotyl (below where the cotyledons join the long axis of the embryo) starts to form (Figure 5.10c). The embryo continues to grow through the "torpedo stage" (Figure 5.10d) and the "cotyledonary stage" (Figure 5.10e). At this time, the ovule containing the mature embryo becomes a seed, and the integuments harden to form the seed coat. Further maturation usually includes accumulation of storage food materials and loss of moisture in preparation for dormancy.

Seeds may or may not be able to germinate immediately. They often serve as overwintering or survival structures for many plants and, therefore, may remain dormant for extended periods of time. Dormancy assures that seeds will not germinate until favorable conditions for growth are present. Many species require that specific conditions be met to break dormancy and encourage germination. For example, a prolonged (4-month) cold-moist treatment of redbud and flowering dogwood seeds in soil is essential for germination in the laboratory. One might easily equate this treatment to what the winter environment is like in soil. In fact, nursery producers growing these two woody species plant the seeds in the fall to ensure germination in the spring.

Remember that the young ovules were located in the ovary of the flower. In many species, the ovary and/or the receptacle will continue to develop and form a conspicuous fruit. Seeds produced inside the fruit are characteristic of only the angiosperms—no other group of plants forms fruit. Fruit morphology and composition is extremely variable, ranging from the simple pod of legumes to the aggregate fruit of a blackberry to the kernel of a corn. For additional information on fruit types, consult any number of introductory botany textbooks. Fruit, besides being good to eat, serves to protect and aid in the dispersal of seeds.

SEX AND GENETIC DIVERSITY

Sexual reproduction typically increases genetic variability because each gamete contains half of each parent's genetic material. Fusing the gametes during sex creates a seedling where each cell contains copies of both the paternal and the maternal genomes. So, how do seed producers maintain the desirable characteristics of a cultivar when propagating a cultivar through seed production? After all, when we plant a certain variety of petunia or tomato, we have expectations of growing a uniform crop. The answer is to use parents with highly similar, or even identical, genetic material so that any progeny created is essentially identical to the parents. Based on what you now know about plant sexual organs and reproduction, it can be relatively easy to use the same plant as both the male and the female parent. Self-fertilization, or selfing, is the process where plants can mate with themselves or other plants in the population, which are essentially identical, and produce seed with nearly identical genetic material and uniform traits in the seedlings. In this manner, seed producers can propagate a cultivar from seed for many generations, each time essentially recreating almost exact copies of genetic material from the parent in all the seeds.

Not all cultivars can be propagated by seed. There are many instances where plants cannot mate with themselves or with similar genotypes in the population. This reproductive limitation is called "sexual incompatibility" and is found commonly in many woody ornamental plants, such as flowering dogwood. Sexual incompatibility can manifest as the failure of the pollen to germinate and form germ tubes, or even the poor (slow) growth of pollen tubes. Selfing may be prevented by incompatible genetics, either one allele (gametophytic) or two alleles (sporophytic). Generally, self-incompatibility creates a reproductive barrier that forces a plant to mate with a different, unrelated genotype (obligate outcrossing) and encourages increasing the genetic diversity within the population.

There are other mechanisms that plants have to effectively reduce selfing or even prevent hybridization between closely related groups of plants. Timing of the maturity of the sexual floral organs may play an important role on who is available to have sex, including oneself. The pollen (male gametophyte) in the perfect flower (having both functional male and female floral organs) may not be mature when the stigma (the receptive surface of the female floral organ) is receptive. In other words, pollination may occur in this instance, but the pollen will not germinate and, therefore, cannot complete self-fertilization. The opposite may be true as well. The stigma may be receptive either before or after the anther (a part of the male floral organ) dehisces and self-pollination is thwarted. Another simple solution that plants utilize to eliminate selfing is to have male and female floral organs on separate plants (dioecious = two houses). In this case, there is no alternative except to cross with another plant. This forces individual plants to mate with other plants that are genetically different, presumably to increase the genetic diversity and survival of the species.

So far, in this chapter, we have described the biological principles behind sexual reproduction, which can be used to propagate cultivars with desirable traits. Sex can also be used to transfer desirable traits from one plant to another plant. It was Gregor Mendel (1822–1884) that discovered that traits are determined by elementary units transmitted between generations in a uniform and predictable fashion. Each genetic unit is called a "gene" and must adhere to two principles. First, it must be inherited between generations such that each descendent has a physical copy of the material. We have already discussed the development of gametes and subsequent fusion (sex) in angiosperms. Second, it must provide information regarding structure, function, or other biological attribute. Thus, there are two ways to think about genes: one is to identify the specific genetic material, or the DNA, that is transferred between generations, and the other is to focus on the way that a specific piece of DNA manifests itself or the traits that are inherited. Both concepts are critical to the successful breeding of horticultural plants.

Mendel's experiments with plant hybrids effectively disproved the notion that heredity was a blending process and the dilution of parental characteristics. He also cast serious doubt on the idea of pangenesis, that physical traits acquired during a parent's lifetime could be inherited in their progeny. Several biologists performed experiments with plant hybrids, comparing the similarities and dissimilarities between offspring of different generations and parental combinations. Kolreuter (1733–1806), Gartner (1772–1850), Naudin (1815–1899), and Darwin (1809–1882) each observed that hybridization often produces progeny with considerable diversity compared to parental stock. However, it was Mendel who paid attention to the numerical ratio in which the different parental characteristics appeared in the offspring. It was Mendel who demonstrated that the appearance of traits in the offspring followed specific mathematical rules, which he deduced by counting the diverse kinds of offspring produced from a particular set of crosses. Although Mendel published the basic groundwork for modern genetics in 1866, his experiments were not recognized until the 1900s after it was shown that chromosomes are the hereditary material passed from one generation to the next (Olby, 1966).

Modern genetic technology can clearly visualize chromosomes and has confirmed that DNA is inherited as discrete, physical units. Statistical tools predict the behavior of DNA from one generation to the next, and gene expression assays can confirm the relationship between genotype and phenotype. However, the basic principles of inheritance remain unchanged from Mendel's day. Plant breeding relies heavily on identifying plants that exhibit desirable traits from among the morphological and physiological variation that exists in natural populations. These plants are used as parents to produce seedlings, some of which inherit the desirable traits. Individuals with improved combinations of traits are selected by plant breeders and become the parents for the next generation until a superior plant is chosen for mass production. Selecting plants with improved traits is a central theme of modern plant breeding and is critical to selecting a breeding strategy to create a new cultivar.

We already know that sexual reproduction is intrinsic to the transfer of genes from one generation to the next. We also covered mitosis, meiosis, and possible differences in chromosome numbers (ploidy). To discuss Mendel's laws of genetics, we will assume that the plants in each example are diploid, which means that they contain two copies of each gene. The paternal copy of the gene came from the pollen and the maternal copy from the egg. When fused, they reconstitute the diploid genome in the progeny. This splitting and uniting of genetic material during sexual reproduction is the basis for segregation, which is Mendel's first law of genetics. Simply put, the mother and father plants contain two copies of each gene, called "alleles," but they only contribute one copy to each progeny. This is because each member of an allelic pair separates, or segregates, from the other during meiosis and gamete formation. The unique combination of one allele from your mother (maternal contribution) and one allele from your father (paternal contribution) is a significant source of genetic variation.

We can visualize the segregation of alleles using some examples. In their simplest form, alleles are either dominant or recessive. Dominant alleles (denoted by uppercase letters) can be observed in the plant even when there is only one copy. Recessive alleles are only observed in the phenotype when the dominant allele is not present. In a diploid plant, there must be two copies of the recessive allele to see the recessive phenotype. Figure 5.11 is a Punnett square depicting all the possible allele combinations of a gene for flower color when you cross two heterozygous parents. Segregation of genotypes is different from segregation of phenotypes because of the dominant and recessive allele combinations. You can follow this gene transfer forward an infinite number of generations, with each generation segregating for the maternal or the paternal contribution of the previous parents.

The second predictable source of genetic variation is Mendel's law of independent assortment. This law states that the segregation of one gene is independent of the segregation of a second gene. It can also be visualized by a Punnett square (Figure 5.12). As more desirable traits are selected, the number of progeny to be evaluated increases

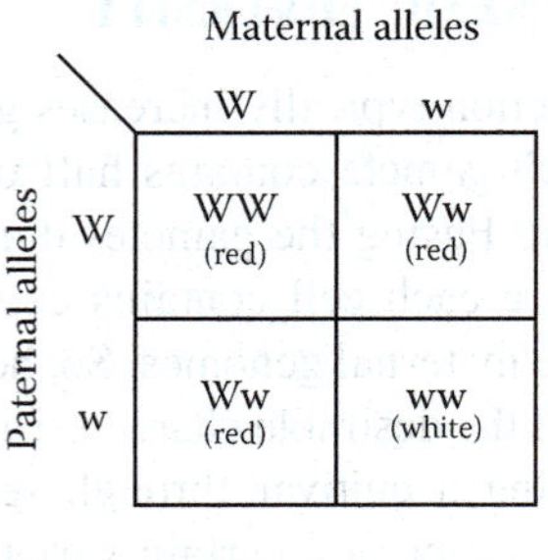

Paternal alleles \ Maternal alleles	W	w
W	WW (red)	Ww (red)
w	Ww (red)	ww (white)

FIGURE 5.11 Punnett square depicts the possible allele combinations found in the progeny of two parents that are heterozygous for a gene controlling flower color. As shown in the diagram, 75% of the offspring are expected to produce red flowers because the homozygous (WW) and heterozygous (Ww) allele combinations produce the red phenotype. The remaining 25% are expected to produce white flowers because they contain two copies of the recessive (w) allele.

Possible paternal allele combinations \ Possible maternal allele combinations	DW	Dw	dW	dw
DW	DDWW (tall) (red)	DDWw (tall) (pink)	DdWW (tall) (white)	DdWw (tall) (pink)
Dw	DDWw (tall) (pink)	DDww (tall) (white)	DdWw (tall) (pink)	Ddww (tall) (white)
dW	DdWW (tall) (red)	DdWw (tall) (pink)	ddWW (dwarf) (red)	ddWw (dwarf) (pink)
dw	DdWw (tall) (pink)	Ddww (tall) (white)	ddWw (dwarf) (pink)	ddww (dwarf) (white)

FIGURE 5.12 Punnett square depicts the possible allele combinations in progeny from two parents that are heterozygous for two different genes. In this case, flower color is controlled by a gene exhibiting incomplete dominance with heterozygous (Ww) progeny producing intermediate or pink flowers. The gene for plant height is completely dominant, and heterozygous progeny (Dw) is indistinguishable from homozygous progeny for the dominant trait (DD). As diagrammed, a plant breeder would expect to observe white-flowered dwarf plants in one-fifteenth of the progeny or 5.25% of the population. The percentage of plants expected to be homozygous recessive for all alleles is $(1 \times 4^n) \times 100$, where n is the number of gene pairs.

due to the smaller ratio of progeny expected to contain rare combinations of alleles.

There are exceptions to independent assortment. Genes that are physically located next to each other on a chromosome may experience linkage disequilibrium. The close physical connection between genes reduces the likelihood of crossing over during meiosis and increases the probability that both paternal or both maternal alleles will be found in individual progeny. On the other hand, genes on different chromosomes are always unlinked. Linkage distorts the ratio between the observed and the expected number of progeny with a particular allele combination. The more tightly that genes are linked, the more that the ratio is skewed from the expected 50% probability that two genes in any given individual hybrid came from different parents. Genetic linkage maps are based on the recombination frequencies observed between genes on the same chromosome and reveal the order of genes along a chromosome. Linkage between genes often results in unwanted or unrelated traits being linked to desirable traits. Breeding strategies can break the linkage between genes and/or make use of the linkage between genes to indirectly select for traits.

All of the genetic traits that we have talked about so far have been qualitative, that is, the phenotype is controlled by a single or few genes, each usually having a dominant and a recessive phenotype. We have also explored the mathematical repercussions of selecting more than one qualitative trait at a time. However, some traits are controlled by multiple genes, each contributing unequally to the observed phenotype. Thus, the addition or subtraction of a particular allele does not have the dramatic, qualitative effect seen in previous examples. Instead, allele combinations display incremental or quantitative differences. Quantitative phenotypes are often measured on a continuous scale, where progeny displays a distribution of phenotypes based on the allele combinations present in the population. The frequency of a particular phenotype depends on the number of genes involved and the allele combinations present in the parental genomes (Figure 5.13). Breeding strategies for quantitative traits generally focus on increasing the frequency of favorable alleles and moving the distribution of phenotypes toward one end of the curve.

The act of evaluating plants for desirable traits and selecting superior plants results in plants with exaggerated traits in combinations that would not typically be seen in natural environments. Natural selection occurs over evolutionary time and reflects the effects of pressure from environmental conditions and competition for resources. Natural selection can be disruptive, stabilizing, or directed depending on the trait. Selection during plant breeding, however, is directional, acts on only a few traits at a time, and is accelerated by applying selection over multiple generations, typically with much smaller populations that are found in the wild, and by controlling hybridization between selected plants.

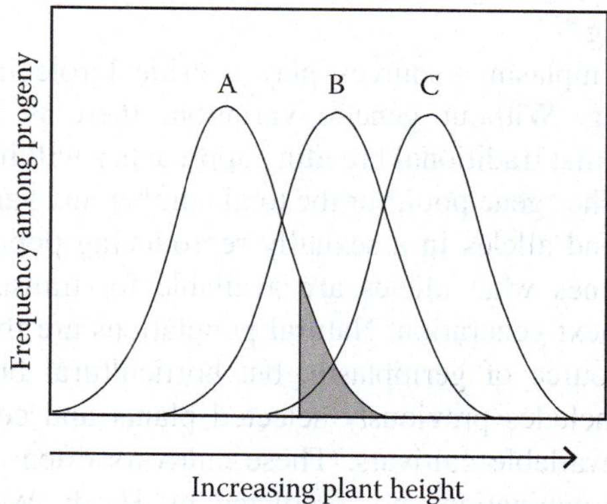

FIGURE 5.13 Graph of plant height from three seedling populations created from three different sets of parents. The distribution of plant heights from populations A, B, and C overlap, but the mean plant height for population C is larger than that of A or B, presumably due to the increased number of favorable alleles affecting plant height. Breeding for increased plant height with population A would be inefficient even if the tallest plants on the most extreme right side of the curve (shaded area) were selected as parents for subsequent generations.

Some traits, such as increased yield or number of flowers, are easily quantified, and improvement, no matter how incremental, can be objectively measured. Other traits are not easily quantified. For example, insect resistance, disease resistance, and drought tolerance may be hard to objectively measure since they involve environmental interactions or the specific behavior of diseases or insects. Some traits are selected based entirely on aesthetics. Flower color and shape, plant growth habit, bark characteristics, and many other horticultural traits have aesthetic components based on consumer appeal. An "eye for plants" is often considered as one of the most important qualities of a horticultural plant breeder since breeding necessarily produces hundreds to thousands of seedlings, out of which only a few are expected to become new cultivars. It is not surprising that commercial plant breeding efforts often include test marketing. Modern plant breeders rely less on intuition and more on science, but a keen sense of observation will always be critical for successful breeding.

Clear objectives, whether based on production or on consumer interests, should be defined before initiating a breeding program since unfocused breeding can result in an exponential increase in the numbers of plants to be evaluated. Crossing every conceivable plant and evaluating tens or hundreds of thousands of progeny produced year after year is simply not physically possible or financially feasible. Luther Burbank (1849–1926) wrote that "Plant breeding, to be successful, must be conducted like architecture. Definite plans must be carefully laid for the proposed creation; suitable materials, selected with judgment; and these must be securely placed in their proper order and position. No occupation requires more accuracy, foresight, and skill than does scientific plant breeding."

Germplasm resources play a critical role in plant breeding. Without genetic variation, there is a little chance that traditional breeding approaches will improve traits. The "gene pool," or the total number and variety of genes and alleles in a sexually reproducing population, determines what alleles are available for transmission to the next generation. Natural populations are the most basic source of germplasm, but horticultural breeding often includes previously selected plants and commercially available cultivars. These cultivars often contain allele combinations not predicted by Hardy–Weinberg equilibrium equations, which are used to estimate the frequency of alleles in natural populations (Weinberg, 1908). Hardy–Weinberg is a mathematical model for the gene pool and is based on the following five factors: population size, migration between populations, mutation within a population, differences in fertility, and mating systems for a specific crop. We can calculate theoretical allele frequencies using Hardy–Weinberg equations but, for our purposes, it is sufficient to understand that allele frequencies are generally balanced for large, random mating populations in nature, where the effects of mutation, migration, and selection are minimal. In the altered framework of plant breeding, however, the directed evolution of small breeding populations is mainly due to the parents, the type of cross, and the artificial selection of desired traits. Therefore, careful consideration must be given to the selection of parents since it has a profound effect on the results. Crossing and selection schemes cannot be successful if the desired alleles are not present in the parents.

Two types of variation are observed when evaluating germplasm. Environmental variation is the result of differences in nutrients, moisture, light, temperature, and other factors that affect the growth of plants, even when comparing plants that have identical genotypes. Environmental conditions may even create situations where inferior genotypes outperform superior genotypes. This can be demonstrated by growing two different cultivars under similar conditions but only applying fertilizer to one cultivar. The cultivar with fertilization may grow taller than the unfertilized cultivar even if it is known to be shorter when grown under identical conditions. In this way, environmental variation can significantly impact the selection of heritable traits. To offset the confounding effects of environment, plants are evaluated under uniform conditions that approximate production or landscape use and in multiple locations and/or over several years.

Genetic variation is heritable and can be detected at the molecular level using biotechnology. Plant breeding is designed to take advantage of genetic variation, but to do so, it must first be separated from environmental variation. Statistical tools were developed to estimate the relative importance of genotypic (G) and environmental (E) effects on phenotype (P), or $P = G + E$. Plant breeders measure the environmental effects that cause predicted phenotypes to deviate from expected genotypes using a statistical tool called "variance." Phenotypic variance (pe), genotypic variance (ge), and environmental variance (ee) all play a role, especially in analyzing quantitative traits where genotypic variance is expected to be significantly larger than with qualitative traits, which have essentially no genotypic variance. The ratio of genotypic to phenotypic variance, called "heritability," represents the proportion of trait variation that is actually due to DNA. A plant breeder's ability to make progress increases as this ratio increases.

There is one more critical factor that needs to be revisited before discussing breeding strategies. In natural populations, plants are either self-pollinating or cross-pollinating. While these naturally occurring reproductive systems can be artificially changed (self-pollinating species can be forced to outbreed, and vice versa), the underlying genetic structure that these reproductive systems impose cannot be changed. This "genetic architecture," along with dominance of traits, gene interactions, and heritability, plays an important role in the decision to use a breeding strategy. For example, some breeding strategies include the creation of pure lines, which are the products of repeated self-fertilization, the most extreme form of inbreeding. With each generation of self-fertilization, heterozygosity decreases by 50%. The coefficient of inbreeding (F), which can be plugged into the Hardy–Weinberg equilibrium equations, suggests that plants approach 100% homozygosity in just eight generations of self-fertilization. Full-sib and half-sib mating schemes also significantly increase homozygosity over time (Figure 5.14). Thus, crops that are self-incompatible are not good candidates for breeding strategies that require pure lines. Crops that outbreed in nature may suffer from "inbreeding depression" when forced to self-fertilize. Inbreeding depression varies among outcrossing species, but in its extreme form, it can be lethal after just a few generations. In other outcrossing species, the consequence of inbreeding is observed as reduced vigor, typically most pronounced in the first four generations. This is due to the increasing homozygosity of deleterious recessive alleles whose effects do not manifest in the heterozygous state (see Figures 5.11 and 5.12). As discussed previously, self-fertilization is also an efficient method to propagate some crops through seed production in species where self incompatibility is not an issue.

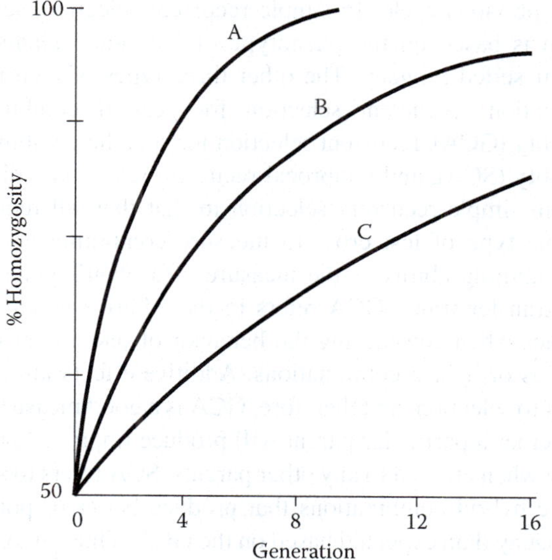

FIGURE 5.14 Effects of inbreeding on the reduction of heterozygosity. A is self-fertilization, B is full-sib mating, and C is half-sib mating. Full-sib mating is the hybridization between progeny from a single cross (siblings), whereas half-sib mating is the crossing of plants that only share one parent. Over time, all three crossing schemes result in high levels of homozygosity.

BREEDING SELF-POLLINATED PLANTS

Several breeding strategies have been developed for self-pollinated plants, and all involve the development of pure, or highly homozygous, lines. Most begin with controlled pollinations between two highly homozygous plants, followed by self-pollination to produce the F_2 population. Parents are often chosen because they have complementary traits. For example, if the goal of a breeding program is to develop a high-yielding cultivar that matures early, an early-season, low-yielding cultivar may be hybridized to a cultivar that matures later but has high yields. Alternatively, parents that possess the same desirable trait may be hybridized if it is thought that they possess different genes for the trait of interest. As an example, two cultivars that mature fairly early in the season because of two different genes may be hybridized in an effort to breed for an even earlier maturity.

In the classical pedigree method, superior F_2 plants are selected and self-pollinated. From the F_3 to F_5 or F_6 generation, both individual plant performance and performance of families (e.g., all F_3 plants derived from a single F_2 plant) are evaluated. Superior plants from the best performing families are advanced to the next generation. In early generations, individual plant performance plays a greater role than family performance in determining which plants are self-pollinated and advanced to the next generation. As heterozygosity declines, selection is based more on family than on single plant performance. Following the F_5 or F_6 generation, selection is based solely on family performance. Because the pedigree method is extremely labor intensive—requiring a detailed record keeping and extensive evaluations of families in early generations when heterozygosity is high and heritability is low—it has been replaced by other breeding methods.

Most of today's breeding schemes for self-pollinated plants are based on bulk population methods. In these breeding schemes, selection in early generations (F_2–F_5) is based solely on single plant performance. Since superior plants are self-pollinated and their seed bulked to produce the next generation, no progeny rows or pedigree records are required in early generations. Selection among family lines is delayed until homozygosity and, therefore, heritability is high. Various bulk population methods have been proposed, most of which differ in terms of the number of plants grown and the stringency of selection in the early generations. For example, single-seed descent breeding schemes require no selection during early generations where only a single self-seed of each individual plant is carried forward to the next generation. The advanced generations produced by the single-seed descent method are expected to be a general representation of a random sample of the pure lines that any given F_2 hybrid is capable of producing. Only the advanced lines are evaluated to make sure that no lines are eliminated early in the breeding process. This bulk population method increases the chances of a superior, true-breeding individual being present, although it significantly limits the number of plants that can be evaluated for each line.

Backcross breeding is a modified form of inbreeding and is generally applied to situations where existing cultivars are only deficient in one or two traits. Backcross breeding involves repeated crossing back to one original parent, called the "recurrent parent," to produce progeny containing the desirable traits. When the trait of interest is controlled by a recessive gene, each backcross will be followed by self-pollination. The recurrent parent should not contain other deleterious traits since the resulting progeny will likely also contain those inferior traits. The parent containing the trait missing from the recurrent parent is called the "donor parent." Traits that are qualitative and dominant, and produce a readily observed phenotype are the easiest to transfer from the donor parent. Transferring quantitative traits through backcross breeding is generally not productive. Depending on the crop and the trait, backcross breeding usually takes at least five generations since the proportion of donor genes is reduced by 50% with each generation.

As mentioned previously, pure lines of self-pollinated crops are true breeding, and seed from a pure line cultivar can be saved and used to generate the next season's crop. However, seed companies have often invested considerable resources into developing pure lines and are not likely to market and sell pure lines as new cultivars since the resulting plants will continue to produce seeds with the same desirable traits for many generations. Thus, resourceful farmers could save a portion of the seed from their crops each year and would not have to purchase more from the seed company. Instead, seed companies use two pure or inbred lines to produce an F_1 hybrid seed that is sold as a new cultivar. Because it is heterozygous, the hybrid seed will not breed true in the next generation (see Figure 5.12 for the segregation of traits when crossing heterozygous parents). In this way, farmers have to return to seed companies each season to purchase a new hybrid seed for each crop that they want to plant.

BREEDING CROSS-POLLINATED CROPS

As mentioned earlier, using self-fertilization to make pure lines of cross-pollinated crops is usually a bad idea because of inbreeding depression. However, modified versions of the breeding schemes listed above can be used to breed cross-pollinated crops as long as the negative effects of inbreeding are mitigated by regular outcrossing to other genetic backgrounds. In this way, inbreeding depression is reversed by heterosis, otherwise known as hybrid vigor. Elite breeding lines, each created by self-fertilization, may display the ill effects of inbreeding depression, but they can be used to make a final hybridization prior to cultivar selection as long as each pure line was derived from a distinct genetic background. In addition to uniting the desirable traits found in each breeding line, mixing the genetic backgrounds can restore vigor and even produce offspring that exhibit improvement in traits beyond what either parent displays. Hybrid vigor is presumably due to the new combination of hundreds of unselected genes that have small effects. The few genes responsible for the negative inbreeding phenotype are reshuffled when the genetic backgrounds are mixed.

The oldest method of breeding cross-pollinated crops is mass selection, where diverse populations of heterogenous plants are evaluated. Individual plants with the best performance each season are used as the parents for the next generation. Generation after generation, seed is collected only from those plants having superior traits. Assuming that the selection criteria are effective, the end result of mass selection is an increase in the frequency of desirable genes in a population where the overall genetic diversity remains high. In this way, there is little chance of inbreeding depression. This method is usually only effective for traits that have high heritability.

Several recurrent selection schemes have been developed for cross-pollinated plants and are used to concentrate genes for quantitative characters without incurring a loss in genetic variability. All recurrent selection schemes involve repeated (or recurrent) cycles of (1) self-pollinating plants from a heterozygous source, (2) saving self-seed from only those plants that are judged to be superior, and (3) intercrossing the self-progeny from the previous cycle. In simple recurrent selection, selection is based on the phenotype of individual plants or their selfed progeny. The other three types of recurrent selection—recurrent selection for general combining ability (GCA), recurrent selection for specific combining ability (SCA), and reciprocal recurrent selection—differ from simple recurrent selection in that they all require some type of test cross to measure combining ability. Combining ability is the measure of a plant's potential to transfer traits. GCA refers to the additive actions of genes when considering the behavior of one parent in a series of hybrid combinations. Additive effects are subject to selection, and therefore, GCA is a good measure of whether a particular parent will produce improved progeny when crossed to any other parents. SCA refers to specific hybrid combinations that produce better or poorer progeny than expected based on the GCA. Thus, SCA is a good measure of nonadditive genetic effects (Hull, 1945). Reciprocal recurrent selection provides for simultaneous selection for both GCA and SCA. Backcross breeding, as described for self-pollinated crops, is sometimes used for improving a specific trait in an elite inbred line. For more details related to the different recurrent selection schemes, see Allard (1999) or other plant breeding textbooks.

OTHER BREEDING CONSIDERATIONS

Linkage

Genes for desirable traits are sometimes physically located next to genes responsible for undesirable phenotypes. Getting rid of this unwanted genetic baggage requires large populations because selecting for the desired trait has a high probability of dragging along an undesirable gene. Recombination events that separate tightly linked genes are rare, so large numbers of progeny need to be evaluated to find the one or few plants where the two genes have been separated by crossing over. The frequency of recombination is proportional to the distance between the genes. The closer the physical association, the less likely that they are to be split apart during meiosis. Breeders create linkage maps that are based on the empirically determined rate at which meiotic recombination separates genes and/or traits. For example, traits that always show up together when evaluating progeny are 100% linked. Data from enough progeny and traits can be used to describe the order and the distance between genes on each chromosome.

The good news about genetic linkage is that indirect selection can be used for traits that are difficult to evaluate. For example, a particular type of insect resistance may be tightly linked to a unique leaf shape due to the proximity of the genes on the chromosome, despite the fact that there is no physiological or mechanistic association between the two traits. Direct screening for insect resistance may be very difficult, in either getting the insects or analyzing the results of feeding. On the other hand, leaf shape is easily evaluated, and hybrids showing the correct leaf shape will have a high probability of having the desired insect resistance. These linkages are rare but easily identified when recording the segregation of multiple traits in a single population. In this way, genetic linkage maps are useful tools for plant breeders.

Non-Mendelian Inheritance

Extrachromosomal inheritance may affect the selection of some traits. Nuclear DNA is evenly contributed by maternal and paternal sources at fertilization because meiosis evenly divides genomes during gametogenesis. However, only the egg contains cytoplasm, so plastid DNA from chloroplasts and mitochondria are contributed exclusively from maternal sources. This maternal bias influences some traits, particularly those that involve chlorophyll, such as variegation in foliage. Most horticultural breeding strategies mitigate possible maternal effects by including reciprocal crossing schemes where the same parent is used as the female for one population and as the pollen donor for the other. Thus, differences in traits that are only observed between reciprocal populations are evidence of maternal or paternal inheritance for those traits.

Other factors affecting inheritance include transposable elements, which are found in all eukaryotic genomes. Transposable elements, also known as jumping genes, are small, mobile regions of DNA that can increase the rate of mutation during plant breeding by inserting into genes (McClintock, 1950). Some transposable elements can also jump back out of genes, causing them to revert to their wild type state or nonmutated phenotype. Phenotypes affected by transposable element movement are not inherited in a Mendelian fashion, and traits are often unstable during vegetative propagation. Traits displaying non-Mendelian inheritance and instability may also be under the influence of other epigenetic phenomenon such as chromatin remodeling, paramutation, imprinting, position effects, or sex-chromosome inactivation (for review, see Chandler and Stam, 2004).

Wide Hybridization, Incompatibility, and Polyploidization

Intergeneric or interspecific hybrids are important tools in breeding strategy since they broaden the genetic base, introduce new genes, and can bridge gaps between species not sexually compatible. Backcross breeding is often used to transfer desirable traits from one species to another after a wide hybrid is successfully produced. Unfortunately, incompatibility may prevent successful wide hybridization. Prezygotic incompatibility is associated with mechanisms that keep the pollen and the egg from ever fusing. Chemicals secreted on the stigma may prevent the germination of pollen grains once deposited. Pollen tube growth may be inhibited before reaching the ovule. Even if the pollen cells reach the egg, the embryo sac may deteriorate before fertilization can occur. Late acting, or post-zygotic incompatibility, includes embryo lethality, halted growth, and aberrant development.

Fertility issues in wide hybrids are also associated with ploidy differences because chromosome number often varies between different genera and species. Differences in chromosome number do not necessarily prevent fertilization, but zygotes are often aborted or show abnormal growth as the two different sets of haploid chromosome attempt to establish new gene expression patterns. Wide hybrids often display unstable phenotypes. Even if embryos grow into flowering plants, they may not produce viable gametes if chromosomes are unable to recognize and pair with their diploid partners during meiosis. Sterile hybrids are a breeding dead end.

One method used to recover fertility in wide hybrids is the deliberate doubling of chromosome number in the sterile plant, thus providing a diploid partner for every chromosome before gamete formation. Mitotic inhibitors, such as the chemicals colchicine and oryzalin, are known to arrest the mitotic cell cycle after DNA is replicated but before the cell divides (Nakasone and

Kamemoto, 1961). Plant tissues treated with these chemicals contain higher proportions of cells with doubled genomes. If these doubled cells go on to form the sexual reproduction structures and gametes, then the process of meiosis, gamete formation, and fusion can occur as outlined previously in this chapter. Ploidy manipulation can also be an effective strategy to change the plant phenotype without sexual reproduction. Cells with doubled genomes can be clonally propagated to produce complete plants that are tetraploid. These plants generally display shorter internode lengths, larger flowers, increased leaf and stem thickness, and other exaggerated morphology when compared with diploid starting material. Because they are clonally produced, ploidy manipulation does not usually affect flower color, disease or insect resistance, and other environmental adaptations present in the starting material.

It is worth noting that the spontaneous generation of unreduced gametes, usually pollen containing the diploid complement of genes rather than the haploid number expected, occurs naturally in some horticultural crops. Plants produced by fertilization with an unreduced gamete are usually triploid, sometimes tetraploid if both gametes are unreduced. Triploids are generally sterile, but not always. While sterility in breeding lines is ruinous, sterility in released cultivars can be beneficial. Terminal crosses between diploid and tetraploid breeding lines to produce sterile triploid forms are one strategy to overcome plant invasiveness. Triploidy has also been used as a way of developing seedless fruit crops.

Mutagenesis Breeding

Mutagenesis breeding is another strategy that does not necessarily involve sexual reproduction. Spontaneous mutant phenotypes seen in vegetative tissues, called "sport mutations," generally retain all the characteristics of the original plant except for the single, mutated trait. For example, four variegated *Hydrangea macrophylla* cultivars are all naturally occurring sport mutations of a very old, non-variegated cultivar (Reed and Rinehart, 2007). Stems with differences in leaf color were observed in the original plant and were removed for clonal propagation. These *H. macrophylla* sport mutations are easily distinguished by their different patterns of leaf color, but they are indistinguishable at the genetic level because they were not produced by sexual reproduction. Mutations can be induced to increase the odds of observing sport mutations. Ethyl methane sulfonate is a chemical mutagen that produces single nucleotide changes, typically affecting only a few genes in the exposed plant. On the other hand, mutations caused by x-ray and gamma-ray sources can cause chromosomal rearrangements, breaks, and aneuploidy (loss of a chromosome). Because these events are random, mutation breeding typically involves screening large numbers of plants or seeds that have been mutagenized to find novel phenotypes. Once identified, a stable sport mutant is a potential new cultivar, especially if the crop is clonally propagated. Mutation breeding often starts with a proven cultivar that shows superior ornamental features in an effort to expand the range of flower or leaf color and/or shape; variegation; and, sometimes, environmental tolerance. Improvements in disease and insect resistance or other complex horticultural traits are generally not found in mutation populations.

Other natural phenomenon can produce new, improved plants without sexual reproduction. There is a special case of embryogenesis called "non-zygotic (somatic or adventitive) embryogenesis" in plants that does not rely on the fusion of gametes and the formation of endosperm. In nature, it is called "apospory" and occurs most notably in citrus seeds from the nucellus, which is sporophytic tissue. However, the process is relatively common in tissue culture (*in vitro*), which is a way to clonally propagate plants using small amounts of vegetative tissue or even regenerate whole plants from single cells. Non-zygotic embryogenesis and tissue culture take advantage of the totipotency of some plant cells. "Totipotency" is the word that describes the ability of a single cell to reproduce the entire organism and, in our case, the entire plant. Not all cells in the plant are naturally totipotent, but with certain treatments, such as exposure to specific growth regulators (synthetic auxins [2,4-D, dicamba, etc.] and cytokinins [benzyladenine]), some cells can be induced to a less-differentiated state that are capable of forming non-zygotic embryos. Non-zygotic embryos are derived from single cells and generally follow the same ontological or developmental sequences that were discussed for their zygotic counterparts by first producing a root and then a shoot. When compared side by side, the morphology of zygotic embryos (Figure 5.15a, c) are remarkably similar to the corresponding non-zygotic embryos (Figure 5.15b, d). Of course, non-zygotic embryos are without a seed coat and have sometimes been called "naked embryos." Much like sport mutants, non-zygotic embryos can exhibit aberrant development or novel traits, presumably due to differences in gene expression or genetic mutation during the regeneration process. Somatic mutants that show desirable or novel traits can be propagated as new cultivars. Screening a large number of plants for somatic mutants using *in vitro* methods takes much less space than traditional mutagenesis.

Biotechnology

Molecular markers, which are generally random regions of DNA that show variation, can also be indirectly associated with phenotypic traits. Selection for plants containing molecular markers linked to desirable traits has a probability of containing the gene for the desired

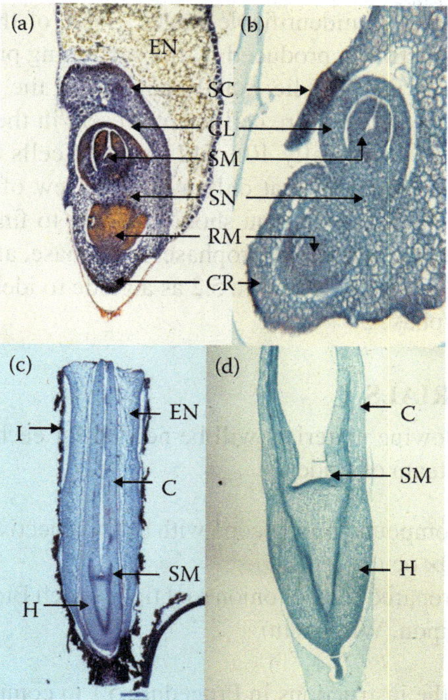

FIGURE 5.15 Comparison of zygotic and non-zygotic embryogenesis in monocotyledonous and dicotyledonous plants. (a) Median longitudinal section through a mature zygotic embryo of *Dactylis glomerata* (orchardgrass) showing the endosperm (EN), the scutellum (SC) or cotyledon coleoptile (CL), the shoot meristem (SM), the scutellar notch (node) (SN), the root meristem (RM), and the coleorhizza (CR) (cf. Figure 5.15b). (b) Median longitudinal section through a morphologically mature non-zygotic embryo of *Dactylis glomerata* (orchardgrass). The non-zygotic embryo is morphologically very similar to its zygotic counterpart shown in Figure 5.15a, except for the lack of endosperm. (c, d) Median longitudinal sections through a morphologically mature zygotic embryo (c) and a morphologically mature non-zygotic embryo (d) of *Cercis canadensis*. The two embryos are very similar except that the hypocotyl (H) of the non-zygotic embryo is swollen and the meristem of the non-zygotic embryo is malformed. The cotyledons (C) of zygotic and non-zygotic embryos have similar morphology and anatomy. The non-zygotic embryo also lacks endosperm (EN) and integuments (I). ([a] With kind permission from Springer Science+Business Media: *Protoplasma*, A histological study of tissue proliferation, embryogenesis, and organogenesis from tissue cultures of *Dactylis glomerata* L., 110, 1982, 121–128, McDaniel, J. K., B. V. Conger, and E. T. Graham. [d] Modified and adapted from Jain, S. M., P. K. Gupta, and R. J. Newton [eds.], *Somatic Embryogenesis in Woody Plants*. Dordrecht, The Netherlands: Kluwer Academic Publishers, 1999.)

trait that is proportional to the genetic distance between the two. When using multiple, tightly linked molecular markers, plant breeders can indirectly select for genes with relative confidence. While this type of biotechnology, called "marker-assisted selection" (MAS), is commonly associated with agronomic crops, it is becoming more feasible to use DNA-based tools in the plant breeding of horticultural crops. MAS is typically associated with breeding for disease and insect resistance, especially where multiple forms of resistance to a single pest need to be incorporated into a single genotype. Without indirect selection, one gene for pest resistance would hide the effects of additional pest resistance genes, making it impossible to select for both genes in a single population using traditional phenotype observations. Using molecular markers, multiple genes can be indirectly selected regardless of phenotype. In this manner, agronomic crops such as barley contain multiple genes for disease resistance and are less likely to succumb to new pathogen variants when they arise (Castro *et al.*, 2002).

Molecular markers have been extensively analyzed for association with quantitative traits, or traits controlled by multiple genes, and those that show linkage are called "quantitative trait loci" (QTL). QTL markers can be genetically mapped based on the degree of co-segregation with the desired phenotype and are often associated with the percentage that they contribute to the desired phenotype (Lynch and Walsh, 1998). This detailed information allows plant breeders to select for genes that represent minor contributions to a quantitative trait. The impact of these genes would not be observed in traditional plant evaluations, but when assembled together in the final cultivar, they can contribute significantly to the final phenotype. For example, a gene that adds 10% more height might not be selected in a population that is 50% shorter than the rest, but it might still boost plant height at 10% when combined with the tallest phenotypes. QTL markers are often associated with agronomic traits such as increased yield.

Biotechnology has opened new doors in plant breeding, such as gene transfer and the creation of genetically modified organisms. Gene transfer requires tissue culture or plant regeneration systems, which have not been developed for all horticultural crops. Foreign DNA with proven usefulness must also be available. A few successful examples of genetically modified horticultural crops exist, including blue roses developed by Florigene, Inc. Given the legal complications associated with genetically modified organisms and the biotechnology used to create them, it remains to be seen whether these methods will show widespread benefit. It is sufficient to say that the traditional horticultural breeding methods described in this chapter will remain relevant and useful in the near future, as they have for the past 100 years.

In this chapter, we have attempted to describe the general processes and structures involved in sexual reproduction, embryo development in angiosperms, and the usefulness of sexual reproduction for seed propagation and breeding. As the reader, you should be aware that there are many species-specific variations on the general themes explored here and that, although we have tried to be accurate, many details have been omitted. We sincerely hope that the preceding pages have whetted your appetite for a greater understanding of not only angiosperm sexual reproduction, but also sexual reproduction in other broad groupings of plants. There are many good introductory botany texts that provide fantastic information on this subject. If you have the opportunity, we encourage you to further your formal education by enrolling in additional courses such as plant developmental anatomy, biotechnology, and/or genetics.

EXERCISE

Experiment 1. Observing the Stages of Mitosis in Prepared Slides of Onion Root Tip

The various stages of mitosis can be observed easily by using prepared (purchased) slides of longitudinal sections of onion root tips. The manufacturer (Triarch Biology, Ripon, Wisconsin) has timed the sampling of root tips to correspond with the high frequency of mitotic cell divisions. Some cells will have "perfect or textbook" depictions of the stages, whereas others will depict almost unidentifiable stages. Some of the variations are artifacts produced by the sectioning process—some cells may not lie exactly parallel to the plane of sectioning or the entire cell is not present in the section (sections are typically 10 to 12 μm, and cells are considerably larger) so that only a partial view of the cell is possible. However, you should be able to find ample examples of interphase, prophase, metaphase, anaphase, and teleophase. Use Figure 5.2 as a guide to identify the various phases.

MATERIALS

The following materials will be needed for each student or each team of students:

- Compound microscope with a 40× objective and 10× ocular lens
- Prepared slide of onion root tip (Triarch Biology, Ripon, Wisconsin)

Follow the instructions in Procedure 5.1 to complete this exercise.

Procedure 5.1
Observing and Counting Mitotic Figures in Onion Root Tips

Step	Instructions or Comments
1	Obtain a compound microscope from the instructor and carefully transport it to your work area. Be sure to keep one hand under the base of the microscope. Also, obtain a prepared slide of onion root tip.
2	Before beginning the observation, clean the surface of the slide with either 70% or 95% ethanol and a clean cloth. This process should remove any dirt and debris, including oils from hands, from the slide surface and make viewing better.
3	Be sure that the low-power (4× or 10×) objective is in place before any viewing. Mount the slide on the microscope stage with the coverslip closest to the objective lens. Lower the 4× or 10× objective lens while looking through the ocular lens until the section on the slide is in focus.
4	Move the slide so that you can identify the apical meristem (see Figure 2.1a). There will be no cells undergoing mitosis here, but this will serve as a point of reference. Move the slide away from the meristem (up the root) until you can identify a cell that is undergoing mitosis. If needed, swing the 40× objective into place, and using the fine adjustment knob, bring the image into focus. Never use the course adjustment knob for the 40× objective. Identify the stage of mitosis seen. This is the starting point for counting 100 cells in the section.
5	Return to the lower-power objective and, moving up the root, count the cells in the various phases of mitosis (Figure 5.2). You might need to use high power to identify some stages. Some cells may be in stages of mitosis that do not exactly conform to the five stages listed, for example, partway between prophase and metaphase. Do your best to classify, but do not be overly concern with getting "the right phase."
6	Record the number of cells and frequency (%) in each phase of mitosis on Table 5.1. Collect the class data, and recalculate the overall frequency of the different stages of mitosis.

TABLE 5.1
Observations of Mitotic Figures in Onion Root

Student or Group Number	Interphase	Prophase	Metaphase	Anaphase	Telophase
1[a]	94 (94%)	2 (2%)	1 (1%)	1 (1%)	2 (2%)
Means of class					

[a] Sample data—do not include in your analysis.

ANTICIPATED RESULTS

Mitotic figures should be easily seen by all students. The combined frequencies of cells in the different stages of mitosis should be easily calculated and provide a very rough estimate of the time spent in the individual phases of mitosis and the cell cycle.

QUESTIONS

- Which phase of the cell cycle is the most commonly observed? Why?
- Which phase is least common? Why?

LITERATURE CITED AND SUGGESTED READING

Allard, R. W. 1999. *Principles of Plant Breeding*, Second Edition. New York: John Wiley, 584 pp.

Castro, A., P. Hayes, T. Filichkin, and C. Rossi. 2002. Update of barley stripe resistance QTL in Calicuchima-sib X Bowman mapping population. *Barley Genetics Newsletter* 32:1–12.

Chandler, V. L. and M. Stam. 2004. Chromatin conversations: Mechanisms and implications of paramutation. *Nat Rev Genet* 5:532–544.

Hayes, P., A. Castro, A. Corey, T. Filichkin, C. Rossi, J. Sandoval, M. Vales, H. Vivar, and J. von Zitzewitz. 2001. Collaborative stripe rust resistance gene mapping and deployment efforts. In Vivar, H. and A. McNab, eds., *Breeding Barley in the New Millennium: Proc Intl Sym,* 82 pp.

Hull, H. 1945. Recurrent selection for specific combining ability. *J Amer Soc Agron* 37:134–145.

Jenkins, M. 1940. The segregation of genes affecting yield of grain in maize. *J Amer Soc Agron* 32:56–63.

Lynch, M. and B. Walsh. 1998. *Genetics and Analysis of Quantitative Traits*. Sunderland, Massachusetts: Sinaue Association, Inc, 980 pp.

McClintock, B. 1950. The origin and behavior of mutable loci in maize. *Proc Natl Acad Sci (USA)* 36:344–355.

McDaniel, J. K., B. V. Conger, and E. T. Graham. 1982. A histological study of tissue proliferation, embryogenesis, and organogenesis from tissue cultures of *Dactylis glomerata* L. *Protoplasma* 110:121–128.

Nakasone, H. and H. Kamemoto. 1961. Artificial induction of polyploidy in orchids by the use of colchicine. *Hawaii Agric Exp Stn Tech Bull* 42:32.

Olby, R. 1966. *Origins of Mendelism*, Second Edition. Chicago, Illinois: University of Chicago Press, 204 pp.

Reed, S. M. and T. A. Rinehart. 2007. Simple sequence repeat marker analysis of genetic relationships within *Hydrangea macrophylla. J Amer Soc Hort Sci* 132:341–351.

Sprague, G. and L. Tatum. 1942. General vs. specific combining ability in a single cross of corn. *J Amer Soc Agron* 34:923–932.

Trigiano, R. N., L. G. Buckley, and S. A. Merkle. 1999. Somatic embryogenesis in woody legumes. In Jain, S. M., P. K. Gupta, and R. J. Newton, eds., *Somatic Embryogenesis in Woody Plants*, Volume 4. pp. 189–208.

Trigiano, R. N. and D. J. Gray. 2005. A brief introduction to plant anatomy. In Trigiano, R. N. and D. J. Gray, eds., *Plant Biotechnology and Development*. pp. 73–85.

Weinberg, W. 1908. Über den nachweis der vererbung beim menschen. *Jahreshefte des Vereins für Vaterländische Naturkunde in Württemberg* 64:368–382.

Wilson, C. L., W. E. Loomis, and T. A. Steeves. 1971. *Botany*, Fifth Edition. New York: Hort, Rinehart and Winston, 752 pp.

6 Juvenility and Its Effect on Macropropagation and Micropropagation

Caula A. Beyl

Concept Box 6.1

- As plants transition from juvenile to mature, they may exhibit changes in leaf shape, pigmentation, stem characteristics, and rate of growth. The most important change for propagators is that, as a plant matures, it becomes increasingly more difficult to induce adventitious roots.
- The area of the tree containing tissue that is the most juvenile is called the "cone of juvenility" and encompasses a vaguely cone-shaped region broadest at the crown and diminishing as the distance from the crown increases for both the aboveground portion and the root system.
- Various techniques have been used to rejuvenate mature plants so that they are more amenable to clonal propagation, including hedging, serial propagation, serial grafting, shoot forcing, etiolation and blanching, and tissue culture.
- Shoots at the top of the tree are the youngest chronologically, but from an ontogenetic standpoint, are the most mature. Those at the base of the tree or closest to the crown are the oldest chronologically, but the most juvenile ontogenetically.
- The transition from juvenile to adult is believed to be epigenetic in nature with internal or external cues changing gene expression, but not the genes themselves, in a stable and persistent manner.
- Juvenile tissue is not only the most valuable from a macropropagation point of view but is also more amenable to the establishment in tissue culture (micropropagation) and typically exhibits better shoot proliferation rates, and the *in vitro* shoots are more amenable to rooting than mature tissue.

Woody and herbaceous plants grow from a seedling to a fully mature flowering plant, a process occurs that is much akin to the changes that we go through as we transition through childhood, adolescence, and adulthood. Maturity is attained when the apical meristem changes from vegetative to flowering in response to some environmental stimulus to flower, usually day length and/or chilling. In its juvenile phase, a plant is incapable of flowering or sexual reproduction even if the proper environmental cues to flower are given. This process is of profound impact to plant propagators because one of the characteristics of the mature phase is the difficulty encountered with vegetative propagation either by traditional means or by tissue culture. Oftentimes, the true value of an ornamental, forest, or fruit selection cannot be determined until it is fully mature and has exhibited flowers, form, or fruit characteristics that make it unique and prized. This is also the time when it is most difficult to propagate either by conventional cuttings or *in vitro*.

Plant scientists often use terms such as "chronological age," "physiological age," and "ontogenetic age" to describe plants as they go through this "change of phase." The chronological age of a tree refers to how many years it has been growing. The chronological age of a tissue refers to how long it has been in existence. As time passes, the chronological age increases. A new shoot at the top of the tree, which has developed in the current growing season, is chronologically young. Plant scientists also talk about biological or physiological age, which refers to the degree of maturation caused by hormonal or environmental and nutritional changes. The tissues formed when the tree was a seedling or very young are the most juvenile, and these are located in the part of the tree that is closest to the crown (the point where the trunk meets the roots). This also means that the new shoot at the top of the tree which is "young" chronologically is the most "mature" physiologically. Another term important in describing the processes involved in maturation phenomena is "ontogenetic aging," which refers to the relatively irreversible changes in gene expression resulting in maturation.

Typically, the transition to the mature form is not an abrupt one. Individual traits that are characteristic of the juvenile phase may disappear at different rates as the plant ages, and new traits characteristic of the mature phase, such as a different leaf form, flowering, and loss of rooting ability, may appear at different times during ontogenetic development. The change of phase from juvenile to mature is not genetic but is believed to be epigenetic. Epigenetic changes are caused by the differential expression of genes or genes being turned on and off resulting in cells and tissues with distinct attributes (see Chapter 40). Some features of the juvenile phase are characteristics such as a tendency toward rapid growth; different forms for leaves and stems; thorniness; leaf retention; and, most important to propagators, easier adventitious rooting than in the mature phase. In some plants, the ability to root easily is compromised very quickly, and in others, the difference in the ability to root between juvenile and mature forms may not be as extreme. Maturity and the loss of the rooting competence of woody shrubs and trees is a daunting challenge to propagators, and for some species such as the fringe tree (*Chionanthus virginicus* L.) and the pawpaw (*Asimina triloba* [L.] Dunal), inducing adventitious roots may continue to be a challenge for the next generation of propagators.

EPIGENETIC NATURE OF THE JUVENILE-TO-MATURE TRANSITION

The nature of new cells, tissues, and organs is determined by the master architects, the apical meristems. The process of ontogenetic aging is dependent on the meristems or, in actuality, the number of cell divisions behind each meristem. If you look closely at the juvenile apical meristems, you will find that they are smaller than their mature counterparts and contain cells with smaller nuclei. Architects direct the construction of new buildings based on instructions, and in the apical meristem, these instructions reside in the genes and the way that they are expressed. The juvenile-to-mature transition is said to be epigenetic in nature because factors such as the environment influence how a cell behaves as a result of its genetic programming (genotype), not by changing the DNA directly, but by changing how it is expressed. This gene expression is stable and tends to persist with each successive cell division occurring in the meristem, thus resulting in juvenile or mature tissue in propagules. An architect must sometimes alter the plans when the environment changes. So too does the apical meristem change when gene expression is altered in response to internal factors (ontogenetic or hormonal) or external factors (photoperiod or chilling), ultimately resulting in either the mature phenotype or a transitional form in between juvenile and mature. These changes are described as epigenetic and once induced, they are perpetuated in the clones.

JUVENILE TRAITS

Many woody plants have different leaf forms while they are juvenile than they exhibit when they are mature. Sometimes, the change is expressed as a difference in the degree of lobing, and in others, juvenile leaves may be simple, whereas the mature leaves may be compound. One of the plants most often cited when the differences between juvenile and mature plants are related is the English ivy (*Hedera helix* L.). Juvenile leaves are lobed (Figure 6.1a), whereas mature leaves are entire (lack lobing; Figure 6.1b). Differences between the two forms can also be seen with the stem which, when juvenile, supports many adventitious roots. Shoot growth of the

FIGURE 6.1 *Hedera helix* in its juvenile form (a) is a vine bearing lobed leaves, whereas its mature form (b) has a more compact and shrub-like growth habit than the juvenile form as well as a more rounded leaf shape with no pronounced lobing.

juvenile form is vine-like, whereas the more mature form is more like an upright shrub. The juvenile passion flower (*Passiflora*) has cordate (heart-shaped) leaves, but the mature passion flower has very divided leaves, as does *Pothos*, a tropical plant often used for interior landscaping. For conifers such as Japanese cedar (*Cryptomeria japonica* D. Don), juvenile foliage is awl-shaped, and the more mature foliage is scale-like. Juvenile leaves of many species are pubescent or hairy, such as the leaves of juvenile pecan trees (*Carya illinoinensis* [Wang] K. Koch.), which have a great deal of pubescence particularly on the lower surfaces. Mature pecan leaves are shiny and lustrous. The arrangement of leaves on the stem or "phyllotaxy" may also be different between juvenile and mature stems. For example, juvenile grape (*Vitis vinifera* L.) stems have a 2/5 phyllotaxy (five leaves arranged in two spirals around the stem before a leaf is directly above the leaf used as the starting point) and no tendrils. Mature grape stems have a 1/2 phyllotaxy and tendrils.

Just as traits like leaf shape and pubescence may vary with the ontogenetic phase, so may pigmentation. For some species, such as pecan, young leaves and even shoots of juvenile plants have a reddish pigmentation in the spring. Leaf retention is another characteristic of juvenile trees. This characteristic can be seen most vividly in beech (*Fagus* spp.), oaks (*Quercus* spp.), and some maples (*Acer* spp.). In older trees, the pattern of leaf retention is a reflection of the portions of the tree that are the oldest chronologically, but are the most juvenile—those areas nearest the crown and the center of the tree (Figure 6.2)—and also gives an idea of the shape of the tree when it was younger. Leaves of juvenile trees are also able to tolerate shade, whereas leaves of mature forms may not. Not only leaf characteristics can be different between juvenile and mature forms, but also stem characteristics. The juvenile seedling trees of honey locust (*Gleditsia triacanthos* L.) and *Citrus* spp. have an abundance of thorns relative to their mature counterparts.

FIGURE 6.2 Leaf retention is a juvenile characteristic in deciduous trees. The pattern of leaf retention in this oak tree vividly illustrates the branches within the cone of juvenility.

As the age of the plant increases, it begins to lose juvenility, which has an impact on clonal propagation success in terms of root induction on stem cuttings, shoot formation on root cuttings, and either organogenesis or embryogenesis in tissue culture. In traditional propagation systems, this loss of juvenility is expressed in terms of a lower percentage of cuttings that form roots or the length and number of roots formed. Sometimes, it may even be expressed in terms of how well the cuttings survive and grow after forming roots.

JUVENILITY AND ITS EFFECT ON MACROPROPAGATION SUCCESS

Loss of rooting competence presents a difficult problem for horticulturists and foresters as they normally select superior clones based on mature characteristics such a vivid flower color; fruit color, shape, or taste; shoot architecture or form; and wood characteristics. However, once the candidate for cloning has attained maturity and displayed its characteristics, it has now also attained the dubious distinction of losing the ability to form adventitious roots and shoots easily.

This loss of ability for adventitious root formation can be quite dramatic, for example, the cotyledonary node of Eucalyptus (*Eucalyptus grandis* W. Hill ex Maiden) has good rooting potential, but by the time the 15th node has developed, the rooting potential is completely lost. Loss of the ability to root as the stock plant ages is expressed as a low percentage of rooting and/or longer rooting times. In some cases, higher concentrations of indolebutyric acid (IBA) are needed to induce roots from mature cuttings than what are needed for juvenile ones.

Taking cuttings from the more juvenile portions of a mature tree, often referred to as the "cone of juvenility" (Figure 6.3), increases the chances of obtaining adventitious roots. This has been demonstrated with many species, such as olive (*Olea europaea* L.), Norway spruce (*Picea abies* [L.] Karst), and apple (*Malus domestica* Borkh.). These juvenile tissues are those that were formed when the tree was young and include the crown of the tree where the shoot joins the root, the lower trunk, and the

FIGURE 6.3 A stylized tree showing regions both aboveground and belowground, which contain the most juvenile tissues—those tissues nearest to the crown. This is referred to as the "cone of juvenility." These juvenile areas of the tree are shaded darkly, and the transition to more mature tissue is indicated by the gradual loss of shading. (Drawing by Collier Collins.)

interior portion of the lower limbs. In a similar pattern, the most juvenile portions of the root are those nearest the crown. The closer the apical meristem is to the roots of the plant, the more juvenile it is likely to be. In a study done with *Acer pseudoplatanus* (Girma, 2004), cuttings were taken from the top, the middle, and the bottom portions of the five-year-old stock and were compared to those taken from the same three portions of the tree on 15-year-old stock plants. On the younger stock plants, the percentage of rooting was decreasing the farther away from the crown that the cuttings were taken, with bottom cuttings rooting at 71.5%; middle cuttings, at 56%; and top cuttings, at 45.5%. On the older stock plants, bottom cuttings rooted at 28%; middle cuttings, at 22%; and top cuttings, at only 9.5%. The effect of both stock plant age and the portion of the tree used as the source of cuttings can be clearly seen.

This greater potential of tissues nearest the crown for adventitious rooting is exploited by techniques such as hedging or stool bedding, which employ severe pruning to decrease the distance between the new growth and the root system, thus acting to rejuvenate the plant and benefit from the ease of rooting that is characteristic of the juvenile phase. The outside of the mature tree or the area farthest away from the crown and root system is also the most mature or adult and, hence, the hardest to propagate via cuttings. Root suckers, because of their proximity to the roots, are often very easy to root adventitiously, a trait put to use by propagators of hard-to-root species.

JUVENILITY AND ITS EFFECT ON MICROPROPAGATION SUCCESS

The mature or adult form also represents a challenge to successful micropropagation, as well as macropropagation. The first difficulty is often disinfesting the explant. Explants taken from mature material and disinfested with bleach or hydrogen peroxide solutions are prone to necrosis (death of cells and tissues), whereas juvenile material is less sensitive. With *in vitro* propagation, just as in conventional macropropagation, the juvenility of the plant that is the source of the explant tissue profoundly influences how successful the tissue culture propagation will be, whether an organogenesis route or an embryogenesis route is intended. Mature explants also have a greater tendency to produce phenolic compounds, which leach into the medium, giving it a dark appearance and often becoming toxic to the explant. This is a serious limitation with the establishment of some woody species such as black walnut (*Juglans nigra* L.). Even the callus derived from explants taken from juvenile plants may be different with respect to cell size, rate of growth, and ability to form roots.

TECHNIQUES TO CIRCUMVENT MATURITY-RELATED LOSS OF ROOTING COMPETENCE

Propagators faced with recalcitrant species that are difficult to root can experiment with various techniques that reinvigorate or partially rejuvenate the plant, thereby making it easier to root adventitiously. Some of these techniques have become mainstays of the commercial nursery industry for the propagation of certain species. These include hedging and stool bedding, etiolation and blanching, serial propagation or serial grafting, application of growth regulators, establishment and culture *in vitro*, and the use of specialized structures such as lignotubers.

Hedging and Stool Bedding

Severe pruning can be used to rejuvenate hard-to-root species and make them more amenable to cutting propagation. Repeated shearing or hedging has been a useful technique to maintain the juvenility of Monterey pine (*Pinus radiata* D. Don) for up to 6 years and yellow cedar (*Chamaecyparis nootkatensis* [D. Don] Spach) for up to 10 years. When a stock plant is hedged, terminal buds on the periphery of the tree, which are also the most mature, are removed. The new shoots that form from dormant buds nearer to the crown are more juvenile. With stool bedding, the plants are pruned back to the ground in the dormant season, and the shoots that form in the spring have juvenile characteristics and are called "juvenile reversion shoots." Stool bedding or stool bed layering is a common practice for the production of the rootstocks of apple. The same severe pruning techniques can be applied to other difficult-to-root species. If potentially valuable selections are divided into two groups, one group can be maintained in the juvenile state as a source of cutting material, and the other group can be allowed to mature normally for the evaluation of adult characteristics.

Etiolation and Blanching

Exclusion of light from green growing shoots using a light barrier results in etiolation and, for many plants, promotes greater adventitious root formation. The practice of applying a band of black tape or Velcro tape around a green shoot to exclude light is called "blanching." Both practices of light exclusion have been useful in promoting greater rooting frequency. When McIntosh apple shoots were blanched until they had reached 7.5 cm and the basal portion of the blanched shoot was kept in darkness with a tape, cuttings taken from these shoots rooted at very high percentages. The effect of etiolation in promoting the rooting of apple shoots tends to persist even if the etiolated shoots are allowed to become green for several weeks and, even up to 9 months later, an enhancement of rooting can be seen. Avocado (*Mangifera indica* L.) shoots lose the etiolation-induced enhancement of rooting success with light exposure gradually over a week, but if the base is kept in the dark longer, for up to 5 weeks, although they have been exposed to light for 7 days prior to taking the cuttings, they root with a high success rate. This same promotive effect of light exclusion on adventitious root formation has been documented in many species and across family lines. Insertion of cuttings into a rooting medium also has the effect of excluding light to the basal portion of the cutting. Although exclusion of light affects auxin and other rooting cofactors, that effect is transitory and does not explain the persistence of the light exclusion effect. The changes that are induced, which enhance rooting, may be epigenetic, acting directly on the meristem and the cells derived from it and resulting in the altered expression of the genes that regulate the adventitious root formation process.

Exclusion of light is a technique with value for micropropagation as well. Shoot explants taken from sweet chestnut shoots that had been stripped of leaves and then wrapped in foil to exclude light for 4 months not only had better establishment in culture (79% relative to 22%), but also had enhanced shoot proliferation.

Serial Propagation and Grafting

Serial propagation by taking cuttings at intervals, rooting them, and using them to replace the original stock plant helps maintain juvenility or, if the plant is already mature, helps rejuvenate it. The ontogenetic maturity, in this case, is influenced by the proximity of the meristem to the new adventitious root system or perhaps by some physiological influence related to the newly generated roots. Juvenility can be maintained for many years using serial propagation. Grafting a scion onto juvenile rootstocks also acts to partially reverse ontogenetic aging, an effect that is compounded when the grafting is repeated several times. It has been proposed that the endogenous growth regulators and concentrations being produced by the roots are responsible. The rejuvenation effect is also enhanced by minimizing the size of the mature scion and allowing leaves to remain on the juvenile rootstock. The presence of leaves on the adult scion can inhibit the process.

Many of the same treatments that have been developed to circumvent the maturity-related loss of rooting competence for cuttings also benefit tissues to be cultured *in vitro*. For example, when plants of sweet chestnut were serially grafted four times and then compared to adult trees, not only did explants of those that had been serially grafted had better establishment and excellent shoot proliferation, but also the microshoots derived from them rooted better than their counterparts from adult explants.

High-Temperature Exposure

When mature plant material is exposed to high temperature, growth of juvenile shoots results. If potted plants are exposed to bottom heat, juvenile shoots tend to form at the bases. This technique was applied to plants of Japanese witch hazel (*Hamamelis japonica* Sieb. & Zucc.), saucer magnolia (*Magnolia soulangiana* Soul.-Bod.), and star magnolia (*M. stellata* Maxim.) to obtain explants that were amenable to *in vitro* culture. In another example, temperatures of 27°C enhanced the juvenile reversion effect of grafting mature Algerian ivy (*Hedera canariensis* Willd.) onto juvenile rootstock.

Growth Regulator Treatments

The role of various growth regulators in the promotion of rooting is well recognized. Application of exogenous auxins, particularly IBA or 1-naphthaleneacetic acid (NAA), to the bases of cuttings is a well-accepted horticultural practice that promotes rooting. This clear-cut benefit of the localized application of a growth regulator to promote rooting is in contrast to the much less clear effects of applying growth regulators to plants to rejuvenate them and thus promote rooting. Treating mature shoots of English ivy with gibberellins can cause them to develop juvenile characteristics again. For other plants, either spraying the entire plant or injecting it with cytokinins provides the stimulus for a reversion to a juvenile-like state. For some conifers, flowering, the signature characteristic of maturity, can be induced with auxin/gibberellin combinations or gibberellin alone.

For some woody plants, spraying with combinations of a cytokinin and a gibberellin can result in vigorous new juvenile growth that could serve as a source of cuttings. When propagators cut dormant shoots and bring them indoors to force budbreak and growth, if they include a cytokinin in the forcing solution, the forcing can be more successful. For narrow-leaf ash (*Fraxinus angustifolia* Vahl), the presence of the cytokinin, benzyladenine, in the forcing solution promoted more bud sprouting than the inclusion of gibberellic acid (GA_3) or the auxin, NAA.

Application of growth regulators can sometimes aid in producing shoots that have more juvenile characteristics and serve as a better source of explant material, but if gibberellins are used, they can result in poor establishment, interference with root formation, and inhibition of shoot formation.

Tissue Culture

The tissue culture environment can serve as a means of reinstating the juvenile state or a pseudo-juvenile state. Because the mature influence is not erased completely as evidenced by the fact that plantlets can sometimes be induced to flower *in vitro* or that plants that had been developed from *in vitro* material seem to be able to flower sooner than if they had been a seedling material, some scientists prefer the terms "apparent rejuvenation" or "reinvigoration" rather than rejuvenation. If mature tissues are established *in vitro*, over time, they begin to exhibit more juvenile characteristics, such as faster growth rate of shoots, smaller leaves on shoots, and easier induction of adventitious roots. The advantages of going through a tissue culture cycle also extend to when the plantlet has been acclimatized and established. When tissue-cultured plants are used as stock plants, the shoots taken from them root more easily and at higher rates than those taken from conventional stock plants. This technique of using tissue culture–derived stock plants has worked well with common lilac (*Syringa vulgaris* L.) and 'Kwanzan' flowering cherry (*Prunus serrulata* Lindl.).

Rooting may still be challenging in the *in vitro* environment if mature explants are used to establish the cultures. With mature cultures of *Eucalyptus grandis*, microcuttings taken from juvenile shoots rooted with 60% success as opposed to those from mature shoots that only rooted with 35% success. In other studies with several *Eucalyptus* species, roots could not be induced on nodes from mature trees, although they could be induced on those from one-year-old seedling trees. Not all species exhibit the same difference as evidenced by the rooting of microshoots from both mature and juvenile silver maple (*Acer saccharinum* L.) and shoots from mature and juvenile apple rootstock.

Lignotubers and Burls

Some trees and woody shrubs form specialized woody swollen structures near the root crown call "lignotubers." Sometimes, these are referred to as "burls," but the term "burl" can refer to any swelling that has swirled woody grain or that is the result of injury or infections, and technically, these are not the same as lignotubers. Burls form higher on the plant than lignotubers. What makes lignotubers different and important from a propagation standpoint is that, ontogenetically, they are juvenile. Lignotubers contain buds and meristematic tissue that developed early in the life of the seedling in the axils of the cotyledons or the first few seedling leaves and retain their juvenile nature. When a lignotuber is induced to form shoots, the shoots exhibit juvenile characteristics, including the ability to root easily. Lignotubers evolved as a mechanism for survival. If the original shrub or tree is damaged, the dormant buds become active generating suckers, one of which usually ends up becoming dominant. The Mediterranean shrub tree heath (*Erica arborea* L.) regenerates from the lignotuber after fires; in fact, the fire-resistant lignotubers are used to make pipes. Examples of other species that form lignotubers are *Eucalyptus* spp. (Figure 6.4), many of the Ericaceae family such as mountain laurel (*Kalmia latifolia* L.) and *Rhododenron* spp., creosote bush (*Larrea tridentata* [D. C.] Colville), and redwood (*Sequoia sempervirens* [D. Don] Endl.). Aerial lignotubers of maidenhair tree (*Ginkgo biloba* L.) are abundant on older trees. In

FIGURE 6.4 The large swellings just beneath the ground are lignotubers of Eucalyptus. These are thought to be a means of ensuring survival since new shoots sprout from the fire-resistant lignotuber that survives fire events. (Courtesy of Dr. Ed Barnard.)

FIGURE 6.5 Venerable ancients like the "Ginkgo of Kitakanegasawa" in Aomori Prefecture, the oldest and biggest Ginkgo in Japan, bear numerous "chichi" on their trunks. These pendulous formations can grow into the ground, root, and form shoots. Bonsai trees can be developed from the severed, inverted chichi. (Courtesy of Tatsuo Komori.)

Japan, these swellings on the trunk are called "chichi," which means "mothers' breasts," because of their pendulous appearance (Figure 6.5). The chichi can grow into the ground, root, and form shoots. Instant bonsai can be created by severing the lignotuber of Ginkgo in the dormant season and placing it upside down in the growing medium. New shoots quickly develop from the lignotuber, resulting in the "stalactite trunk" bonsai style.

SUMMARY

As you have discovered, the age and state of maturation of a plant has a profound effect on how easily it can be propagated vegetatively. Through observation and experimentation, propagators have developed various techniques and practices that allow them to optimize the chances of rooting cuttings successfully. These include etiolation and blanching, hedging and stool bedding, serial grafting, high temperatures, growth regulator treatments, *in vitro* culture, and the use of specialized structures. Successful plant propagators have found some techniques more successful with some species than others. This is why patience, experience, and education are all the hallmarks of a good propagator.

EXERCISES

Experiment 1. Differences in Rooting Response of Leaf Petiole Cuttings of Juvenile and Mature Forms of Hedera helix

A classic example of the difference between juvenile and mature forms of the same species exists for *Hedera helix* (English ivy). Not only is leaf shape and growth habit different for the two phases, but also phyllotaxy (the way leaves are arranged on the stem) changes from alternate in juvenile stems to spiral in adult stems. Stem pigmentation and whether the stem has adventitious roots also varies with maturity. This species represents an excellent system for examining the effects of maturation on the ease of rooting because English ivy is readily available, and the differential response of the juvenile versus the mature phase is so well documented. Juvenile English ivy cuttings root easily with extremely high success rates, but mature English ivy roots, at much lower percentages. The manner in which adventitious roots are initiated also differs in the juvenile versus the mature form. For juvenile cuttings, adventitious roots arising from the phloem and the ray parenchyma of the stem appear at about 2 weeks. For mature cuttings, wound callus forms from cells in the pith, secondary xylem, phloem, cambium, and cortex first. It takes about 4 weeks from the time the cutting is taken before roots begin to emerge from the callus at the base of the cutting. If leaf petiole cuttings and some form of auxin are used to initiate roots, a similar pattern is seen. Juvenile leaf petiole cuttings root easily, but mature cuttings root more infrequently and slowly because the roots form from tissue that must dedifferentiate into callus and then redifferentiate again into root primordia. This adds to the time required for a rooting response to be seen.

MATERIALS

The following items are needed for each student or group of students. Each student or group may act as one replication of the treatment.

- Ten leaf petiole cuttings from juvenile and ten leaf petiole cuttings from mature English ivy. A leaf petiole cutting consists of the leaf with most of the petiole still attached.

- 0.1% (1000 ppm) IBA. Commercial formulations in talc can be used for convenience.
- Weighing boats or Petri dishes to hold the rooting powder
- Sharp knife or pruning shears for severing the petioles and blade from the stem
- Labels and markers
- Flats containing 1 peat: 3 perlite (by volume), one per student or group. Dry peat is sometimes resistant to wetting. Adding a little wetting agent to the water helps.
- Intermittent mist system

Follow the experimental protocols outlined in Procedure 6.1 to complete this experiment.

ANTICIPATED RESULTS

Petioles of the juvenile leaf cuttings will form roots more rapidly in response to auxin treatment, because they contain preexisting competent cells that are able to respond to the auxin and then form roots. Petioles of the adult leaf will form callus in response to auxin, but then some of the callus cells divide to form root primordia. This process takes a longer time.

QUESTIONS

- What morphological differences did you observe between juvenile and mature English ivy specimens from which you took the leaf petiole cuttings?
- Would results have differed if leaf-bud cuttings had been used? Stem cuttings?
- In this experiment, you measured the percentage of cuttings forming roots, the time required until the emergence of the first roots, and the number of roots formed per cutting. What other parameters could you have observed to determine the response to the IBA treatment as a function of maturity?
- What variations of this experiment could have been conducted to explore more fully the concept of change of rooting competence?

Procedure 6.1

Comparison of the Rooting Response of Leaf Petiole Cuttings from Juvenile and Mature Forms of English Ivy (*Hedera helix*)

Step	Instructions and Comments
1	Enough shoots of both mature and juvenile forms of English ivy should be obtained to yield at least 10 leaf petiole cuttings of each type for each student or group of students.
2	Before the shoots are used to obtain cuttings, students should note the characteristics that indicate the state of maturity in each of the forms.
3	Leaves that are healthy and have undamaged blades should be used. Using a knife or pruning shears, students should sever the petioles of chosen leaves at a position on the petiole just above where it connects to the stem.
4	About 1 cm of the base of the leaf petiole should be dipped into the 0.1% IBA in talc. When using IBA in talc, always use a clean scoop to remove some from the original container and place the powder in a container such as a plastic weighing boat or a Petri dish. That way, when cuttings are dipped into the powder, the material in the original container is not contaminated. The petiole should be tapped to remove the excess. When done, do not put the unused IBA in talc back into the container.
5	Place the cuttings in rows in the plastic flat containing 1 peat: 3 perlite (by volume) and then label the rows with the name of the treatment. Label each flat with the name of the student or group that has performed the treatment.
6	After 2 weeks, students should begin checking periodically the cuttings for root formation. Carefully record the date that the first roots emerge so that elapsed time to first root emergence can be compared between the mature and juvenile cuttings.
7	At the end of the 6 week period, terminate the experiment and collect data on the percentage of cuttings rooted and the number of roots induced.
8	The two treatments can be statistically compared if the cuttings prepared by each student or group are treated as replications. Alternately, comparisons can be made within each group if each leaf petiole cutting is considered as one experimental unit or replication.

Experiment 2. Forcing and Rooting (1) Epicormic Shoots from Detached Stem Segments or (2) Suckers from Root Segments

On mature trees, the easiest material to take for cuttings is on the periphery of the tree, but that propagation material is more mature than the tissues of the main trunk and the interior portions of the lower branches. The area within the cone of juvenility has the best potential for rooting, but may not offer a large supply of suitable shoots. Fortunately, epicormic shoots can be forced from branch segments that are cut, laid horizontally in a well-drained substrate like perlite, and placed under mist. These shoots can then be used for shoot cuttings and rooted relatively easily. With species that sucker readily, belowground, within the cone of juvenility, large roots can be severed near the crown and cut into lengths, and suckers or shoots can be forced as well. These also serve as a very good source of cuttings and can be rooted with IBA treatment and intermittent mist.

MATERIALS

The following items are needed for each student or group of students:

- Pieces of the main stem or branches, approximately 40 to 50 cm long, harvested from within the cone of juvenility of a mature silver maple (*Acer saccharinum* L.). Other species may prove suitable for this exercise such as *Eucalyptus* spp., *Quercus* spp., *Betula* spp., and *Pyrus* spp. Ideally, the branches should be about 5 to 10 cm (~2–4 in.) in diameter, but larger diameters will work as well. One piece may be used for each student or group representing one replication of the treatment.
- Pieces approximately 20 cm (~8 in.) long, harvested from thick horizontal roots near the crown of trees or shrubs that are known to sucker can

Procedure 6.2

Forcing and Rooting: (1) Epicormic Shoots from Detached Stem Segments or (2) Suckers from Root Segments

Step	Instructions and Comments
1	Each student or group of students should prepare one stem segment that is 40 to 50 cm (~20 in.) long. The stem segment should be embedded horizontally into the flat containing perlite so that only about half of its diameter is exposed. The student or group of students should then prepare a root segment that is 20 cm long and embed the piece into the flat in the same way as was done with the stem segment.
2	The flat should be placed into a greenhouse at 25°C and the medium watered well. The flats should be misted frequently. A cycle of 6 sec every 6 min can be used. If necessary, a broad spectrum fungicide/algaecide like Zerotol can be applied as a drench to discourage fungal and algal growth in the flats.
3	Begin examining the stem and root segments for shoot induction weekly and record the dates when the shoots first emerged.
4	Four to six weeks after forcing the stem and root segments, shoots that are large enough to be used as cuttings (≥5 cm long) may be measured for shoot length, harvested, and then used in the next phase of the experiment.
5	Immediately after collection, shoots should be trimmed to uniform lengths, their basal ends dipped into 0.1% (1000 ppm) IBA in talc and tapped to remove the excess, and then inserted into the flats containing rooting medium, 1 peat: 3 perlite (by volume).
6	The flats should be watered thoroughly and then placed under intermittent mist on a schedule of about 4 sec every 6 min. This is also a good opportunity to acquaint students with how intermittent mist timers are set to obtain various schedules of misting. Mist is only necessary during daylight hours.
7	After 6 weeks, the cuttings can be examined for rooting. Collect data on the percentage of cuttings rooted, the number of roots, and the length of the longest root. What other kinds of data can be collected?
8	The two treatments can be statistically compared if cuttings prepared by each student or group are treated as replications.
9	Variations of this experiment can include comparisons of different cultivars within a species, comparisons among several species, using different lengths of stem or root segment, and using segments of stems farther up the tree or roots farther away from the crown.

be used. Suggested species that are suitable for forcing shoots are aspens (*Populus* spp.), oak leaf hydrangea (*Hydrangea quercifolia* Bartr.), yaupon holly (*Ilex vomitoria* Ait.), and lacebark elm (*Ulmus parvifolia* Jacq.), although any species which naturally produces abundant suckers will work well.

- Standard flats filled with coarse perlite for forcing stem and root segments; one for each student or group of students
- Standard flats filled with 1 peat/3 perlite (by volume) for rooting of the forced shoots
- Intermittent mist
- Metric ruler

Follow the experimental protocols outlined in Procedure 6.2 to complete this experiment.

ANTICIPATED RESULTS

Within 28 days, shoots should have developed from the stem segments. Depending on the species chosen for the exercise, root segments may begin yielding root suckers at about the same time or a little bit later. Both sources of shoots should provide cuttings that root more easily than cuttings taken from the mature portions of the tree, although depending again on the species chosen, shoots derived from root cuttings may exhibit a better rooting response.

QUESTIONS

- What problems could occur with the protocol used above if the shoots being forced were intended to be used as a source of explant material for *in vitro* culture?
- How can the diameter and the length of either the stem or the root segment being used affect the initiation and growth of shoots?
- Would the response in terms of the number of shoots forced be different if the stem segments were taken from farther up into the tree or if the root segments were taken from farther away from the crown?
- Will all cultivars of the same species respond in the same way to the forcing and root-inducing conditions?

LITERATURE CITED AND SUGGESTED READING

Abo El-Nil, M. 1982. Method for asexual reproduction of coniferous trees. U.S. Patent No. 4353184.

Ballester, A., M. C. Sanchez, and A. M. Vieitez. 1989. Etiolation as a pretreatment for *in vitro* establishment and multiplication of sweet chestnut. *Physiol Plant* 77:395–400.

de Fossard, R. A., P. K. Barker, and R. A. Bourne. 1977. The organ cultures of nodes of four species of Eucalyptus. *Acta Hort* 78:157–165.

Debergh, P. and L. Maene. 1985. Some aspects of stock plant preparation for tissue culture propagation. *Acta Hort* 166:21–23.

Giovannelli, A. and R. Giannini. 1999. Effect of serial grafting on micropropagation of chestnut (*Castanea sativa* Mill.). *Acta Hort* 494:243–245.

Girma, A. 2004. Effects of cyclophysis, topophysis, and length of cuttings on rooting and early field performance of *Acer pseudoplatanus* L. Thesis. University of Hannover, Hannover, Germany, 94 pp.

Hackett, W. P. 1987. Juvenility and maturity. In Bonga, J. M. and D. J. Durzan, eds., *Cell and Tissue Culture in Forestry*, Volume 1. Dordrecht, The Netherlands: Martinus Nijhoff, 422 pp.

Hackett, W. P. and J. R. Murray. 1997. Approaches to understanding maturation or phase change. In Geneve, R. L., J. E. Preece, and S. A. Merkle, eds., *Biotechnology of Ornamental Plants*. Wallingord, United Kingdom: CAB International, 416 pp.

Henry, P. H. and J. E. Preece. 1997. Production and rooting of shoots generated from dormant stem sections of maple species. *HortScience* 32:1274–1275.

Libby, W. J. and J. V. Hood. 1976. Juvenility in hedged radiata pine. *Acta Hort* 56:91–98.

Perez-Parron, M. A., M. E. Gonzalez-Benito, and C. Perez. 1994. Micropropagation of *Fraxinus angustifolia* from mature and juvenile plant material. *Plant Cell Tiss Org Cult* 37:297–302.

Pharis, R. P. and R. W. King. 1985. Gibberellins and reproductive development in seed plants. *Annu Rev Plant Physiol* 36:517–568.

Plietzsch, A. and H. H. Jesch. 1998. Using *in vitro* propagation to rejuvenate difficult-to-root woody plants. *Comb Proc Intl Plant Prop Soc* 48:171–176.

Preece, J. E., C. A. Huetteman, W. C. Ashby, and P. L. Roth. 1991. Micro- and cutting propagation of silver maple: I. Results with adult and juvenile propagules. *J Amer Soc Hort Sci* 116:142–148.

Sankara Rao, K. and R. Venkateswara. 1985. Tissue culture of forest trees: Clonal multiplication of *Eucalyptus grandis*. *Plant Sci* 40:51–55.

Stoutmeyer, V. T. and O. K. Britt. 1961. The behavior of tissue cultures from English and Algerian ivy in different growth phases. *Amer J Bot* 52:805–810.

7 Chimeras

Robert M. Skirvin and Margaret A. Norton

Concept Box 7.1

- A chimera is a plant with two or more genetically dissimilar tissues growing side by side.
- Higher plants have layered meristems that originate from a few cells in the central zone of the shoot apical meristem. The outer layers maintain their integrity because they divide anticlinally.
- Chimeras arise from genetic changes in one or more of the layers in the apical meristem.
- There are three types of chimera: mericlinal, in which the genetically different tissue is found in part of a single meristem layer; sectorial, in which the genetically different tissue is found in part of all meristem layers; and periclinal, in which the genetically different tissue makes up one entire meristem layer.
- Chimeras can be reliably propagated only from axillary buds.
- Mericlinal and sectorial chimeras are unstable in propagation. Only periclinal chimeras can be reliably propagated.
- Mericlinal and sectorial chimeras can be stabilized as periclinal chimeras by the selection of axillary shoots.

Plant chimeras have intrigued scientists and plantsmen for centuries. Their sometimes odd appearance and unpredictable behavior during propagation have made them the object of speculation and experimentation. Chimeras have been the source of numerous attractive and desirable ornamental plants, and they have given rise to valuable new fruit and vegetable cultivars. Chimeras are surprisingly common. A visit to the produce department at your local grocery store will likely turn up one or two chimeras in the citrus bins. Others may lurk in among the apples and pears. A walk through a residential neighborhood will take you to many more (Figure 7.1).

What is a chimera? A chimera is a plant with two (or more) genetically dissimilar tissues growing side by side. The term "chimera" was borrowed from mythology: the chimera was a mythical beast with a head like a lion, a body like a goat, and a tail like a serpent. In this chapter, we will discuss the origin of plant chimeras, their importance in horticulture, and the methods used both to maintain and to stabilize chimeras.

HISTORY

Some of the first known chimeras arose as a consequence of grafting. Probably the oldest chimera on record is the bizzarria orange found in a garden in Florence, Italy, in 1644. It is believed to have originated from an unsuccessful graft of a sour orange scion (*Citrus aurantium* Linn.) onto a citron rootstock (*C. medica* Linn.). The scion died, but a shoot grew from an adventitious (not preformed) bud at the graft union. The resulting branch produced different kinds of fruit: some resembled a smooth-skinned orange; others, a rough-skinned lemon; and occasionally a fruit that seemed to be part orange and part lemon. Similar plants caught the attention of botanists: a shoot that seemed to be part *Cytisus* and part *Laburnum*, arising from a graft of *Cytisus purpureus* Scop. onto *Laburnum vulgare* Berch. & J. Presl; a branch from the graft union of medlar (*Mespilus germanica* L.) on hawthorn (*Crataegus monogyna* Jacq.) that had characteristics intermediate between the two genera; and a shoot from a pear (*Pyrus communis* L.) on quince rootstock (*Cydonia oblonga* Mill.) that seemed to be a blend of both plants.

Originally, botanists thought that these plants were graft hybrids that had formed when cells from the rootstock fused with cells from the scion to form new cells, combining the genetic traits of both plants. Hans Winkler, a scientist, attempted to prove that graft hybridization was possible. He grafted tomato (*Lycopersicum esculentum* Mill.) scions onto black nightshade (*Solanum nigrum* L.) stocks and vice versa, allowed the union to heal, and then cut the stem across the graft union. Callus

FIGURE 7.1 Chimeras. (a) Chimeral apple with light sector. (b) Chimeral orange with light sector. (c) Chimeral orange with thick rind sector. (d) Chimeral rubber plant with albino variegation. (e) Chimeral *Hoya* with albino variegation. (f) Chimeral *Ficus* with albino variegation.

cells from both stock and scion formed over the cut surface, and eventually, adventitious shoots formed from this callus. Winkler chose to use tomato and nightshade because they are related but have distinctive leaf and fruit characteristics. Tomato has a lobed, compound leaf, whereas nightshade has an entire leaf. The tomato that Winkler used had a plum-shaped red fruit; the nightshade had a small, round, dark purple fruit.

One of the adventitious shoots that formed resembled tomato on one side and nightshade on the other. Winkler called this shoot a "chimera" after the mythical beast. Eventually, Winkler produced a series of shoots that he thought were graft hybrids. The leaves of Winkler's graft hybrids were intermediate in morphology between tomato and nightshade. Some resembled a tomato leaf with less dissection, whereas others resembled a nightshade leaf with lobes (Figure 7.2a).

Another botanist, E. Bauer, developed a different hypothesis. Working with a variegated zonal geranium (*Pelargonium zonale* L. "albomarginata"), he concluded that the leaf pattern, a green center surrounded by an albino margin, was the result of genetically different tissues growing adjacent to each other in layers. Instead of fusing, the cells maintained their genetic integrity. The

FIGURE 7.2 Leaf, flower, and fruit morphologies of Winkler's "graft hybrids": (from left to right) *Solanum nigrum*, *S. tubingense*, *S. proteus*, *S. Gaertnerianum*, *S. Köelreuterianum*, and *S. Lycopersicum*. At the bottom are graft chimera explanations for the "graft hybrids," with layers derived from the *S. nigrum* colored in green, and those from the tomato, in yellow. (Reprinted from Fitting, H. *et al.*, *Strasburger's Textbook of Botany*, 5th English Edition. New York: Macmillan and Co., p. 302, 1921. With permission.)

outer layers of albino tissue were wrapped completely around a green central core in what Bauer termed a "periclinal arrangement." He linked this layering in mature tissues to the layering found in plant meristems. In his chimera hypothesis, he concluded that the epidermis was formed from the outer cell layer of the apical meristem, which he termed the "LI"; another layer of cells formed from the second layer of the meristem, which he called the "LII"; and the central part of the meristem, the "LIII," formed the core of the plant. He suggested that Winkler's graft hybrids were actually graft chimeras, with an epidermis formed from the tissue of one plant, and a core, from the tissues of the other (Figure 7.2b). Bauer's concept of chimeras laid the groundwork for our understanding of chimeras.

PLANT SHOOT TIP MERISTEMS

To understand chimeras thoroughly, we need to take a closer look at how a plant's shoot tip meristem is structured and how it behaves when it divides. This will help us choose the best methods to propagate chimeras.

Meristem Structure

A meristem is an area of actively dividing cells. The shoot apical meristem is a region located at the topmost growing point of a plant's shoot. The apical dome is a small group of cells located above the first leaf primordium, a swelling on the flanks of the meristem that will eventually become a leaf (Figure 7.3).

In longitudinal section, the most striking characteristic of a shoot meristem, at least in higher plants, is layering. Ferns and lower plants do not have layered meristems; their tissues arise from one or more apical initials, which can contribute to any part of the plant (Figure 7.4a). Gymnosperm and angiosperm meristems, however, exhibit distinct layers. The pattern of layering differs among gymnosperms and angiosperms. Gymnosperms

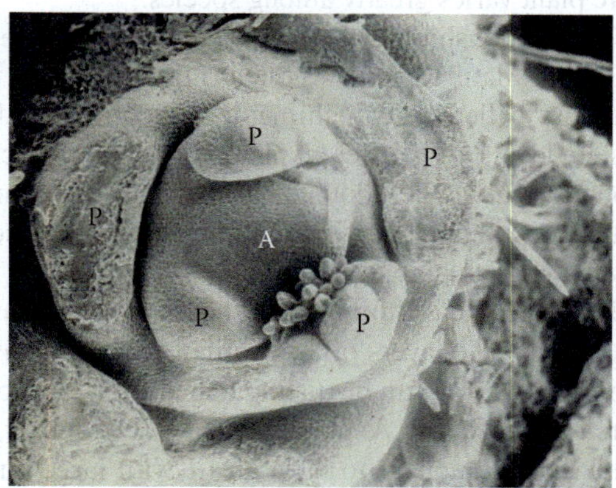

FIGURE 7.3 Scanning electron micrograph of a shoot tip meristem showing the apical dome (A) and leaf primordia (P).

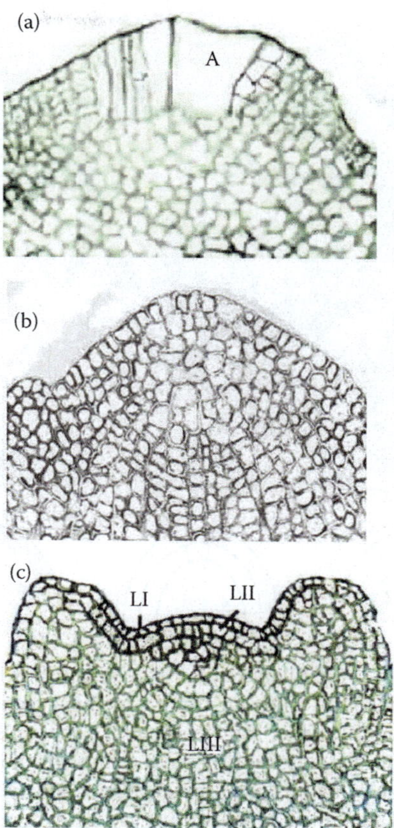

FIGURE 7.4 (a) Diagram of a longitudinal section through a fern shoot apical meristem, showing the apical initial (A). (b) Diagram of a longitudinal section through a gymnosperm shoot apical meristem, showing the apical initials and layering. (c) Diagram of a longitudinal section through an angiosperm shoot apical meristem, showing layering.

generally have a group of apical initials that form both the surface layer of the meristem (tunica) as well as the central core (corpus) (Figure 7.4b). Angiosperms have the most complex meristem structures. They have at least one tunica layer and often have many more (Figure 7.4c). Unlike gymnosperms, each angiosperm tunica layer has its own set of apical initials. As a result, the tunica layers are maintained distinct from one another.

The outermost layer in the meristem is designated LI, and the next layer, LII (Figure 7.4c). Numbering continues, with the corpus being given the highest number. Most angiosperms have an LI, LII, and LIII, with the LIII being the corpus.

Cell Division in the Shoot Tip Meristem

All the cells that make up a plant's shoot come from a few cells in the meristem. These cells are found in the central zone of the apical dome (Figure 7.5a, b). Cells in the central zone divide slowly. One cell will remain in the central zone as an initial, whereas the other is pushed away from the central zone (Figure 7.5c).

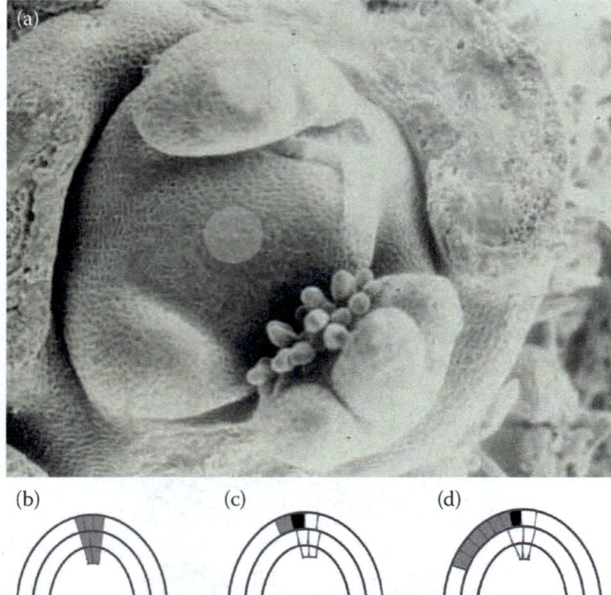

FIGURE 7.5 (a) Shoot apical meristem, showing central zone (in circle). (b) Diagram of a longitudinal section through the shoot apical dome, showing central zone cells (in gray). (c) Diagram of a longitudinal section through the shoot apical dome, showing an apical initial (black) and daughter cells (gray). (d) Diagram of a longitudinal section through the shoot apical dome, showing an apical initial (black) with daughter cells and their derivatives (gray).

This second cell (daughter cell) will continue to divide for some time, producing a patch of derivative cells (Figure 7.5d). Eventually, these derivative cells cease to divide. All cells of a plant's body can be traced back to one particular central zone cell.

Because the cells in the central zone divide slowly, they undergo fewer divisions than the other cells in the apical dome. As a result, there is less opportunity for an error to occur during cell division, making the central zone cells a kind of reservoir to preserve the genetic integrity of the plant. Occasionally, a daughter cell will displace a central zone cell instead of being pushed outside the central zone. In this case, the daughter cell will take over the function of the displaced central zone cell, and the displaced cell will act as a daughter cell.

Cells can divide in many directions, but in the meristem, they generally divide in two planes: anticlinal (perpendicular to the central axis of the plant) and periclinal (parallel to the central axis of the plant) (Figure 7.6). Anticlinal divisions result in a sheet of cells one layer thick, whereas periclinal divisions result in an increase in plant girth.

Tunica cells in the angiosperm meristem divide almost exclusively by anticlinal divisions. It is this pattern of cell division that causes the layering in meristems. As a result of these anticlinal divisions, each layer is maintained one cell thick and remains distinct from the adjacent layer. The layers in gymnosperm meristems are less distinct because the tunica occasionally divides periclinally and its cells mingle with cells in the corpus. Cells in the corpus divide both anticlinally and periclinally, resulting in an increase in stem length and girth.

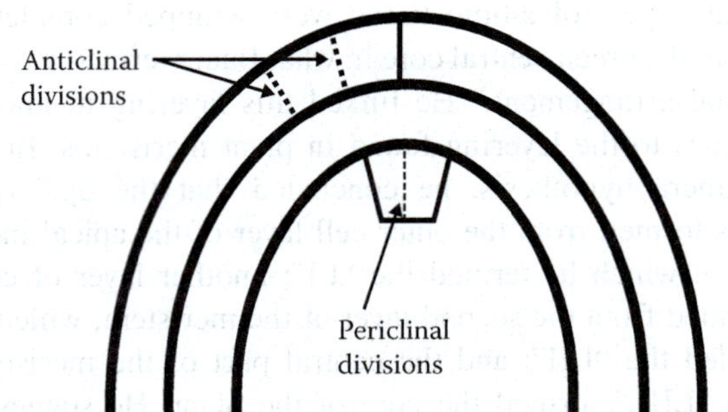

FIGURE 7.6 Diagram of a longitudinal section of shoot tip meristem showing anticlinal cell division and periclinal cell division.

Ultimate Fate of Cells in the Shoot Tip Meristem

The kind of mature, differentiated cell that a daughter cell becomes depends on the meristem layer from which it originates. Cells from each layer in the meristem tend to contribute to certain tissues because of their relative position in the maturing plant. For example, in plants with a three-layered meristem, LI cells will form the epidermis of the shoot; LII cells will form tissue just beneath the epidermis, including much of the leaf lamina and the gametes in the flower; and LIII cells will form the bulk of the structure of the plant, including adventitious roots, which nearly always form from cells in or around the vascular tissue. In a plant with a two-layered meristem (LI and LII), the LI will become the epidermis (possibly many layers thick) and part of the leaf lamina, and the LII will become the remainder of the plant, including the gametes. The precise contribution that a particular meristem layer makes to the mature plant varies greatly among species.

The relationship of the meristem layers to the tissues in the mature plant was determined using a type of chimera called a "cytochimera." Cytochimeras have meristem layers with different ploidy or numbers of chromosomes. When a cell doubles its chromosome number (spontaneously or following treatment with a chemical called "colchicine"), the volume of its nucleus also doubles. This is easy to visualize because nuclei can be stained dark. The layers of a periclinal cytochimera that vary in ploidy level can be distinguished from each other under a microscope. A plant with a polyploid LI, for example, would appear to have an LI with large nuclei, and LII and LIII cells with normal-sized nuclei; such a chimera could be labeled a 4-2-2 cytochimera (tetraploid LI, diploid LII, and diploid LIII). By examining the tissues in a mature plant derived from a 4-2-2 cytochimera, we learn that all epidermal tissues of the shoot are tetraploid and that all internal tissues are diploid, and we can deduce that the epidermis is derived from the LI. Similarly, analysis of a 2-4-2 cytochimera reveals that much of the internal body of leaves and fruits as well as all gametes (male and female) are derived from the LII. Analysis of a 2-2-4 cytochimera would show that most of the internal tissue of the plant and all roots are tetraploid and, thus, derived from the LIII.

HOW CHIMERAS FORM

Most people assume that all of a plant's cells are genetically identical. After all, most plants and animals begin their lives as a single cell derived from the union of male and female gametes, so theoretically, all cells of an organism should be identical. However, genetic changes occur all the time in plants. Cells mutate spontaneously following exposure to cosmic rays, ultraviolet light, or intentional mutagenic treatments. These mutant cells can be maintained by ordinary cell division. Most of these cells are never observed because they occur away from the central zone of the meristem, in a region of the plant, which will eventually become non-meristematic. A mutant cell outside the central zone may divide to form an isolated patch of clonal cells, but as the plant ages, this patch of mutant cells may be shed as a leaf dies or a layer of bark falls off, or it may be buried within the plant as its girth increases. Some scientists call a plant such as this a "mosaic" rather than a "chimera," preferring to reserve the term "chimera" for plants with mutations in the central zone cells, which give rise to more or less permanent streaks or layers of genetically different tissue.

Chimeras arise from spontaneous mutations as well as from grafting. Chimeras are most likely to be observed if they affect a color or morphologic characteristic. For this reason, the best-known chimeras are those that express variegation in leaf color. This is most often seen as green (chlorophyll) and albino tissues growing together (Figure 7.1d–f). However, just because a plant is variegated does not mean it is a chimera. Some plants exhibit genetic patterns that are passed on through the sexual cycle (Figure 7.7). Other chimeras can include observable changes in growth habit, thorniness, leaf shape, and fruit color (Figure 7.1a–c) or less apparent changes in metabolic products, enzyme levels, or other biochemical processes.

KINDS OF CHIMERAS

Chimeras can be classified into three forms. These include mericlinal chimeras and sectorial chimeras, which are marked by instability, and the more stable periclinal chimeras, in which one tissue completely encloses another like a hand in a glove.

FIGURE 7.7 Rex begonia (*Begonia* Rex-cultorum) is a variegated plant that has a heritable pattern in its leaves. It is not a chimera because its pattern arises from differential gene expression, not mutation.

MERICLINAL CHIMERA

Mericlinal chimeras are chimeras in which a genetically different tissue occurs in a part of one meristem layer. Mericlinal chimeras are common, arising when a central zone cell in one layer of the meristem mutates (Figure 7.8a). It divides to form a daughter cell, which, in turn, divides to form derivatives (Figure 7.8b). As the

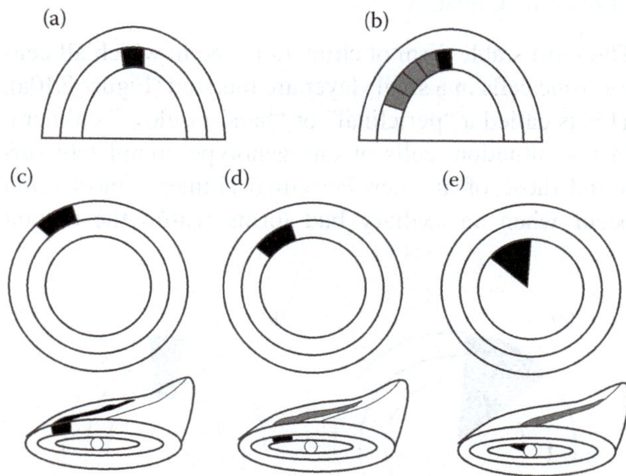

FIGURE 7.8 (a) Diagram of a longitudinal section of shoot tip meristem showing mutant central zone cell in LII. (b) Diagram of a longitudinal section of shoot tip meristem showing mutant central zone cell in LII, with its derivatives. (c) Diagram of a longitudinal section of shoot tip meristem showing mutant central zone cell and its derivatives in the LI, extending into the leaf primordium and the mature leaf. (d) Diagram of a longitudinal section of shoot tip meristem showing mutant central zone cell and its derivatives in the LII, extending into the leaf primordium and the mature leaf. (e) Diagram of a longitudinal section of shoot tip meristem showing mutant central zone cell and its derivatives in the LIII, extending into the leaf primordium and the mature leaf.

mutant cell in the central zone continues to form more daughter cells, the continual production of cells forms a streak of mutant tissue, extending the length of the shoot. If the mutant central zone cell is in the LI layer (Figure 7.8c), its progeny will appear as a streak of mutant epidermis in the mature shoot. If the mutant central zone cell is in the LII, its progeny will lie just beneath the epidermis in the shoot and in the lamina of the leaf (Figure 7.8d). If the mutant central zone cell is in the LIII, its progeny will form a wedge of tissue in the central part of the stem and may also be seen around the midrib of the leaf (Figure 7.8e).

Sectorial Chimera

A sectorial chimera is similar to a mericlinal chimera, except that the mutation occurs in part of each layer of the meristem (Figure 7.9a). The result is a wedge of mutated tissue in the mature stem, which extends from the axis of the plant to the epidermis. Sectorial chimeras often occur in seedlings early during embryo development, before the meristem layering has developed, but they may also arise in mature plants when daughter cells from a single mutant central zone cell divide periclinally and displace adjacent central zone cells in the other layers of the meristem (Figure 7.9b).

Periclinal Chimera

The most stable form of chimera is one in which all central zone cells in a single layer are mutated (Figure 7.10a). This is called a "periclinal" or "hand-in-glove" chimera. In this situation, cells of one genotype completely surround those of another. Periclinal chimeras most often occur when an axillary bud forms within the mutant

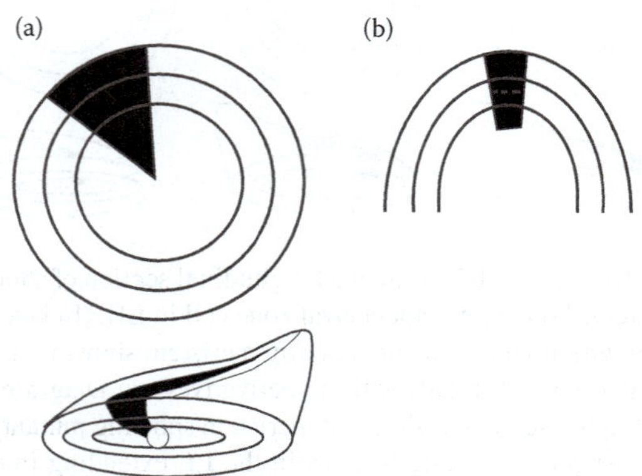

FIGURE 7.9 (a) Diagram of a longitudinal section of shoot tip meristem showing mutant central zone cells in LI, LII, and LIII, with their derivatives, extending into the mature leaf. (b) Diagram of a longitudinal section of shoot tip meristem showing mutant central zone cell in LII dividing periclinally to replace central zone cells in LI and LIII.

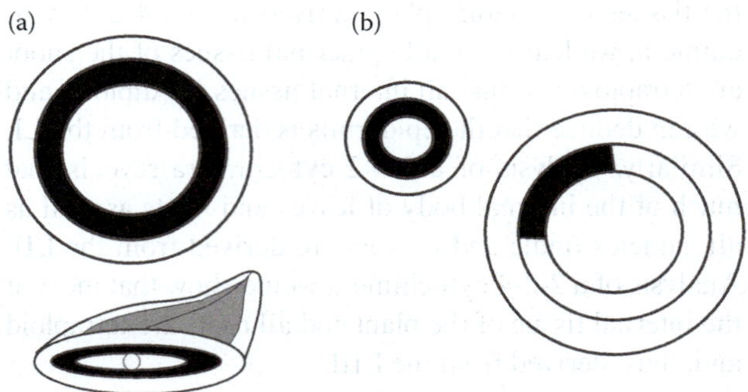

FIGURE 7.10 (a) Diagram of a longitudinal section of shoot tip meristem showing mutant central zone cells in LII and their derivatives, extending into the mature leaf. (b) Diagram of a transverse section across a shoot tip meristem (large circle) showing mutant central zone cells in LII (mericlinal chimera) with an axillary shoot (small circle) originating in the center of the mutant streak, forming a periclinal chimera.

streak of tissue in a mericlinal chimera (Figure 7.10b). Daughter cells from a single mutant central zone cell might possibly displace adjacent central zone cells in the same layer of the meristem, but this is a less likely scenario because of the multiple steps involved.

When all central zone cells are replaced with mutant tissue, the result is an entire layer (LI, LII, or LIII) of mutant tissue growing adjacent to other layers of nonmutant tissue. The periclinal chimera is called a "hand-in-glove" chimera because a complete layer of one tissue (the "glove") can surround a core of genetically different tissue (the "hand"). In some cases, the phenotype of the glove can mask the phenotype of the hand.

BRANCHING AND STABILITY IN CHIMERAS

Sectorial and mericlinal chimeras are less stable than periclinal chimeras during growth. Depending on where an axillary bud forms relative to a sector of mutant tissue, the meristem of the resulting shoot will maintain, lose, or change its chimeral nature. A mericlinal chimera, for example, might remain a mericlinal chimera if an axillary bud forms at the junction of mutant and nonmutant tissues (Figure 7.11a); it might become a periclinal chimera if an axillary bud forms in the center of the mutant streak (Figure 7.11b), or it might lose its chimeral status if it forms in a region that contains no mutated tissue (Figure 7.11c). Sectorial chimeras will either remain sectorial or become non-chimeral (Figure 7.12a–c). Periclinal chimeras, however, will reproduce themselves reliably because the layer of mutant tissue is continuous (Figure 7.13).

EXAMPLES OF CHIMERAS

There are numerous examples of chimeras in horticulture. Many variegated plants are chimeral. In this case,

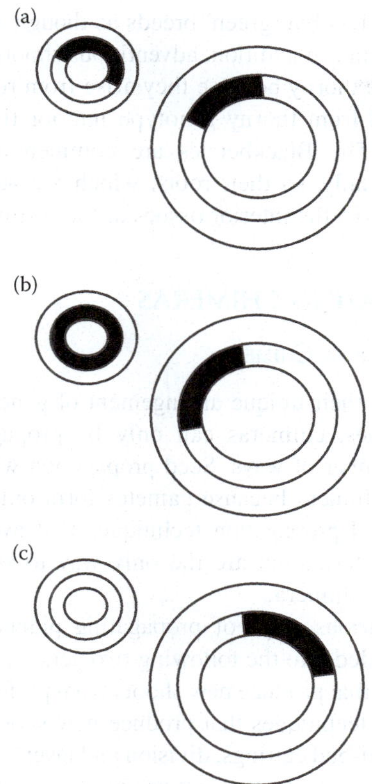

FIGURE 7.11 (a) Diagram of a transverse section across a shoot tip meristem (large circle) showing mutant central zone cells in LII (mericlinal chimera) with an axillary shoot (small circle) originating at the junction of mutant and nonmutant tissue. The resulting shoot will maintain its mericlinal nature. (b) Diagram of a transverse section across a shoot tip meristem (large circle) showing mutant central zone cells in LII (mericlinal chimera) with an axillary shoot (small circle) originating in the center of the mutant streak, forming a periclinal chimera. (c) Diagram of a transverse section across a shoot tip meristem (large circle) showing mutant central zone cells in LII (mericlinal chimera) with an axillary shoot (small circle) originating in a region with no mutant tissue. The resulting shoot will lose its chimeral nature.

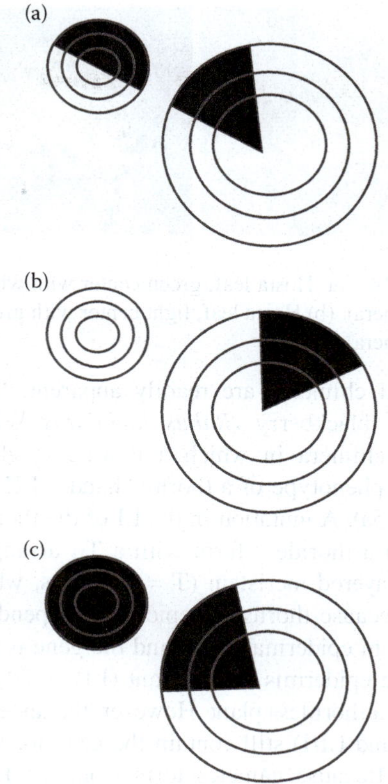

FIGURE 7.12 (a) Diagram of a transverse section across a shoot tip meristem (large circle) showing mutant central zone cells in LII (mericlinal chimera) with an axillary shoot (small circle) originating at the junction of mutant and nonmutant tissue. The resulting shoot will maintain its sectorial nature. (b) Diagram of a transverse section across a shoot tip meristem (large circle) showing mutant central zone cells in LII (mericlinal chimera) with an axillary shoot (small circle) originating in nonmutant tissue. The resulting shoot will become nonchimeral. (c) Diagram of a transverse section across a shoot tip meristem (large circle) showing mutant central zone cell in LII (mericlinal chimera) with an axillary shoot (small circle) originating in the center of the mutant tissue. The resulting shoot will be composed of entirely mutant cells.

the chimera is easily detected because the mutated tissue is a different color than the original tissue. Hostas are a good example of variegated plants that are chimeras (Figure 7.14a, b). Hostas are monocots and have two-layered meristems. The LI forms the epidermis of the leaf as well as a border around the edge of the leaf, whereas the LII forms the central part of the leaf. A leaf that has a green center (Figure 7.14a) surrounded by a white margin can be termed a "WG chimera" (W or white comes from the LI, whereas G or green comes from the LII). A leaf having a green margin around a white center (Figure 7.14b) is termed a "GW chimera."

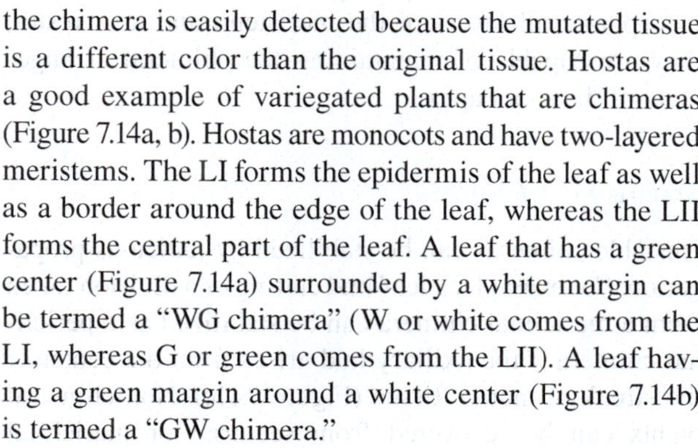

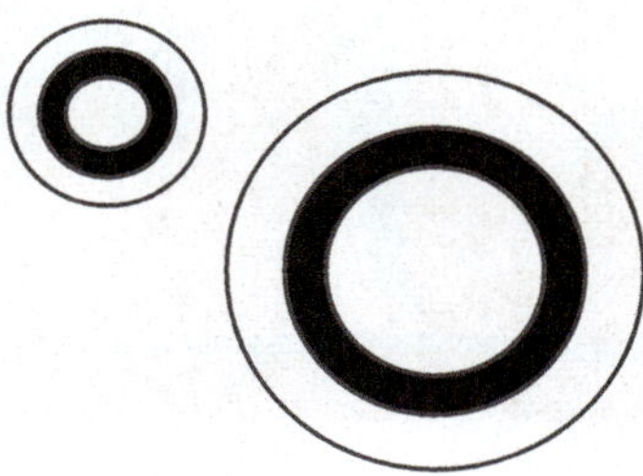

FIGURE 7.13 Diagram of a transverse section across a shoot tip meristem (large circle) showing mutant central zone cells in LII (periclinal chimera) with an axillary shoot (small circle). No matter where the axillary bud forms, the resulting shoot will always be a periclinal chimera.

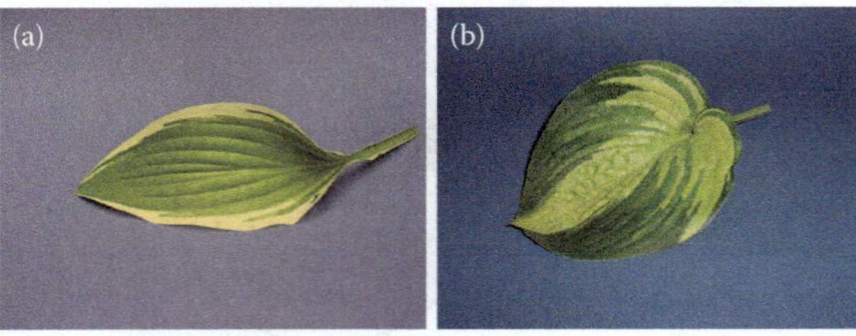

FIGURE 7.14 (a) Hosta leaf, green center with white margin (a WG chimera). (b) Hosta leaf, light center with green margin (a GW chimera).

Not all chimeras are readily apparent. 'Thornless Evergreen' blackberry (*Rubus laciniatus* Willd.) is a periclinal chimera in which a thornless "glove" (LI) masks the phenotype of a thorny "hand" (LII and LIII) (Figure 7.15a). A mutation in the LI of the thorny parent resulted in a thornless form with a Ttt arrangement in the three-layered meristem (T = thornless, whereas t = thorny). Because thorns are epidermal appendages (i.e., derived from epidermal cells) and the gene is expressed only in the epidermis of the plant (LI), a Ttt periclinal chimera is a thornless plant. However, the underlying tissues (LII and LIII) still contain the gene for the thorny character. Because gametes form from the LII, which is t, 'Thornless Evergreen' breeds as though it were still a thorny plant. In addition, adventitious shoots from root cuttings are thorny because they arise from roots, which are formed from thorny genotype interior tissue (LIII) (Figure 7.15b). Blackberries are commercially propagated asexually, so their roots, which are adventitious, originate from the interior tissues of the cuttings.

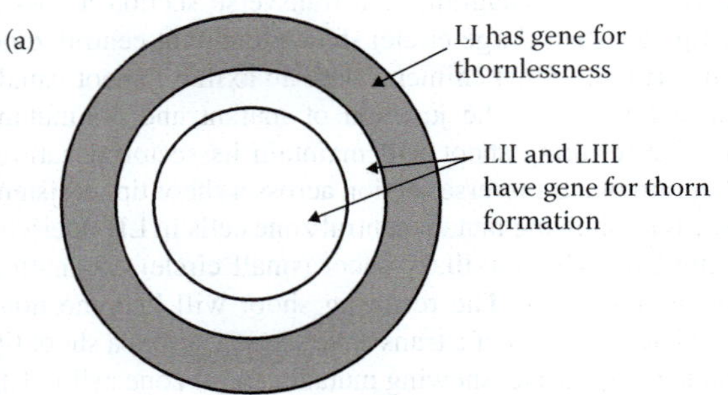

FIGURE 7.15 (a) 'Thornless Evergreen' chimeral structure. The LI contains the mutant gene for thornlessness. The LII and LIII retain the wild-type gene for thorn formation. (b) 'Thornless Evergreen' chimeral cane (bottom) and 'Thornless Evergreen' root sucker, which has reverted to a non-chimeral state.

PROPAGATING CHIMERAS

Maintaining a Chimera

Because of their unique arrangement of genetically different tissues, chimeras can only be propagated in a limited number of ways. Seed propagation will not perpetuate a chimera because gametes form only from the LII. Asexual propagation techniques that avoid adventitious bud formation are the only way to successfully propagate a chimera.

The various ways of propagating plants asexually can be divided into the following two general categories: techniques that produce new shoots from preformed meristems and techniques that produce new shoots adventitiously. Leaf-bud cuttings, division and layering, grafting, shoot cuttings, and tissue culture methods, which multiply plants from axillary buds, are techniques by which shoots are produced from preformed (axillary) buds. Leaf cuttings, root cuttings, and tissue culture processes involving callus, single-cell organogenesis, or somatic embryogenesis result in adventitious shoot formation. Adventitious buds tend to arise from one to a few adjacent cells, and as a result, they generally do not maintain the chimeral arrangement of tissues. Techniques that use preformed buds are the only methods that will reliably reproduce a chimera.

Axillary buds are fully formed meristems that include all the layers found in the apical meristem and develop very early in the differentiation of a shoot tip. Recall that the first signs of leaf formation occur at the base of the apical dome (Figure 7.3). The first leaf primordium appears as a slight bulge near the apical dome. Its accompanying axillary bud begins to form shortly thereafter. Axillary buds originate from many cells and incorporate all layers of the meristem, and as a result, they are capable of maintaining the chimeral arrangement of the shoot apex.

Stabilizing a Chimera

Unstable chimeras can be stabilized by judicious propagation. For example, a leaf-bud cutting or a division taken from a mericlinal chimera can be stabilized as a periclinal chimera if the axillary bud arises from the center of the streak of mutant tissue (Figure 7.11c). Non-chimeral plants can be recovered from sectorial or mericlinal chimeras (Figures 7.11b and 7.12b, c). This is a technique frequently utilized in hosta breeding and propagation. Many variegated hostas begin life as "streaky" seedlings. Some of their cells contain plastids incapable of forming chlorophyll, and these patches of albino tissue form sectors or "streaks" that can be seen on the leaves. As the seedling matures, offsets form from axillary buds. Some of these offsets will arise from the center of a streak, forming a stable periclinal chimera.

Destabilizing a Chimera

Destabilization of a chimera can be desirable. 'Thornless Evergreen' blackberry, which was discussed previously, is a stable periclinal chimera. However, the plant produces suckers (LIII tissue) during the growing season. Since these suckers arise adventitiously from roots, the new growth is thorny (Figure 7.15b). Growers find it labor intensive to remove these thorny suckers. In the case of a harsh winter during which canes freeze back to the ground, all new growth from the previously thornless plant will be thorny. A grower may find that his/her entire planting of thornless blackberries has become thorny over a single winter.

Using tissue culture techniques, the 'Thornless Evergreen' periclinal chimera was induced to form adventitious shoots. Thornless shoots were selected and tested for their chimeral status. Eventually, a non-chimeral thornless plant was selected and named 'Everthornless.' Because it is no longer a chimera, all its root suckers are also thornless.

Tissue Culture

Tissue culture has become a standard tool for the nursery trade. Plants produced in tissue culture arise in the following two different ways: by the growth and enlargement of preformed shoots (axillary buds and apical meristems) and by adventitious shoot formation. Chimeral shoots propagated at moderate to slow rates in tissue culture on a medium that minimizes callus formation tend to be relatively stable because they arise from intact preformed buds that include all the layers of the apical meristem (LI, LII, and LIII). When shoots are propagated rapidly or on a medium with high levels of growth regulators, a portion of the shoots that develop *in vitro* arise adventitiously. Adventitious shoots develop from either single cells or small groups of cells. Chimeras propagated on a high growth regulator medium tend to segregate into pure types (e.g., green and albino plants) and off types (rearranged chimeras).

Rearranged chimeras may have a value as a new cultivar, but for a nursery grower interested in clonal integrity, such off types can lead to loss of revenue, distrust among customers, and even lawsuits. Because the danger of chimeral segregation is high, micropropagation nurseries maintain clonal stability by propagating their plants at moderate to slow rates on a medium that minimizes callus formation and by selecting only shoots originating from axillary buds.

Breakdown of formerly stable chimeras may be one source of somaclonal variation or tissue culture–induced variability. For example, some hostas have a variegation pattern that is difficult to see in tissue culture, and the variegation can be lost during prolonged culture. This is an example of undesirable destabilization of a chimera.

EXERCISES

Experiment 1. Propagation of Color-Variegated Chimeral Plants

Color-variegated chimeras are the most readily identifiable chimeral plants because the source of the different tissues is identified by the tissue color. In this exercise, we will propagate the green-and-white–variegated oval-leaf peperomia (*Peperomia obtusifolia* L. 'variegata'; Figure 7.16) by the following two different methods: leaf cuttings, which produce new shoots and new roots adventitiously, and two-node stem cuttings, which produces shoots from preformed meristems (in this case, an axillary bud). *Peperomia obtusifolia* 'variegata' is a GWG periclinal chimera, that is, the LI and LIII layers of the meristem have normal, green tissue, but the LII tissue has mutated to albino tissue. The leaves of this plant have a white margin with a green center. The LI tissue is very thin and contributes little tissue to the mature leaf. The contributions of the LII and LIII layers of the meristem are apparent by the pattern of variegation.

The objective of this exercise is to demonstrate the effect of two different propagation techniques, leaf cuttings and stem cuttings, on the stability of the periclinal chimera *P. obtusifolia* 'variegata.'

Variegated *Sanseveria trifasciata* Prain. may be used instead of variegated peperomia, if desired. *Sanseveria*

FIGURE 7.16 Segregating adventitious shoots from a leaf cutting of the periclinal chimera *Peperomia obtusifolia* L. 'variegata.'

trifasciata can be propagated from leaf cuttings (adventitious shoots) or by the division of offsets (preformed shoots). Nurseries selling *P. obtusifolia* 'variegata' and several variegated forms of *S. trifasciata* can be easily located on the Internet.

MATERIALS

The following materials will be required to complete this experiment:

- Peperomia plants (the number of plants depends on the class size and the number of cuttings used)
- Knife
- Commercial rooting powder (0.1–0.3% auxin)
- Propagation mix and trays
- Humidity dome and greenhouse

Follow the experiment protocol outline in Procedure 7.1 to complete this experiment.

ANTICIPATED RESULTS

Two-node cuttings should root readily. Because the shoot tips have been removed, apical dominance has been broken and axillary buds will break. Shoots from the axillary buds would be expected to maintain their chimeral arrangement. Leaf cuttings will take it longer to root and form shoots because the shoots must form adventitiously. Shoots would be expected to be either completely albino or completely green because of their adventitious origin.

QUESTIONS

- Where did new shoots form on leaf cuttings? On stem cuttings? Were these axillary or adventitious shoots?
- Were there any non-variegated shoots formed on the stem cuttings? How might this happen?
- Were there any variegated shoots formed on the leaf cuttings? How might this happen? Is the pattern of variegation the same as that of the mother plant?
- If a unique chimeral plant appeared in a propagation bench, how would you stabilize it?

Experiment 2. Separation of Phenotypes in *Pelargonium graveolens* L'Hertier ex. Aiton 'Rober's Lemon Rose'

The scented geranium 'Rober's Lemon Rose' (Figure 7.17) is a periclinal chimera in which the morphological

Procedure 7.1
Propagation of the Color-Variegated Periclinal Chimera *Peperomia obtusifolia* L. 'Variegata' by Leaf and Stem Cuttings

Step	Instructions and Comments
1	Obtain a single, multi-stemmed plant of variegated oval-leaf peperomia. Make 10 two-node cuttings, trimming the lower part of the stem to just below the first node and the upper part of the stem to just above the second node.
2	Remove the lower leaf on each cutting by snapping off the petiole close to the stem. These leaves will be used for leaf cuttings. The upper leaf should remain on the stem cutting.
3	Plant the stem cuttings by dipping the lower part of the stem in a commercial rooting powder containing a low concentration (0.1–0.3%) of auxin. Make sure that the lower node is coated with powder. Tap off the excess and stick the cutting into the pre-moistened propagation medium. Cover with a humidity dome and move to a greenhouse. The cuttings should be kept moist.
4	Plant the leaf cuttings by sticking the petioles into the pre-moistened propagation medium. The entire petiole up to the base of the leaf should be in the soil. Do not use rooting powder for the leaf cuttings. Cover with a humidity dome and move to a greenhouse. The cuttings should be kept moist.
5	Observe the cuttings for signs of shoot formation. This will take several weeks, especially for the leaf cuttings. Note the location where new shoots form and whether the shoots are variegated, albino, or entirely green.
6	Graph your results using a bar graph, comparing the number of variegated, albino, and green shoots for each cutting type.

FIGURE 7.17 Leaf forms of segregating adventitious shoots from the periclinal chimera *Pelargonium* X *domesticum* Bailey 'Rober's Lemon Rose.' Top to bottom, left to right: parental lobed, segregant dentate, segregant hairy, segregant dwarf, segregant flat leaf, segregant hairy dwarf, and segregant lobed.

phenotype of the internal tissue is masked by that of the outer tissue. Normally, the cultivar has lobed leaves and small, asymmetrical violet flowers with a two-parted stigma and crinkled petals. The flowers are sterile. However, plants forming from internal tissues have highly dentate leaves and flowers with smooth petals and a five-parted stigma.

Separation of the inner phenotype is possible by propagation from regions of the plant, which are derived from internal tissue. Adventitious shoots from root or leaf cuttings will express the internal phenotype.

The objective of this exercise is to separate the inner phenotype in 'Rober's Lemon Rose' by forcing adventitious shoots from the internal tissue by means of root and leaf cuttings.

'Rober's Lemon Rose' is readily available by mail order. An Internet search will reveal numerous nurseries that carry this cultivar.

MATERIALS

The following materials will be required for this experiment:

- Scented geranium 'Rober's Lemon Rose' plants (the number of plants depends on the class size and the number of cuttings used)
- Knife
- Propagation mix and trays
- Humidity dome and greenhouse

Follow the experimental outline in Procedure 7.2 to complete the experiment.

ANTICIPATED RESULTS

All shoots arising from root and leaf cuttings are adventitious. Adventitious shoots from the leaves can arise from either epidermal (lobed phenotype) or internal (dentate phenotype) tissues and may exhibit either morphology. 'Rober's Lemon Rose' stock plants are vegetatively propagated from cuttings, and their roots (which are adventitious in origin) come from LIII tissue. Therefore, all shoots arising from root cuttings would be expected to show the dentate phenotype only.

Procedure 7.2

Separation of Phenotypes in the Scented Geranium 'Rober's Lemon Rose'

Step	Instructions and Comments
1	Remove 25 leaves with attached petioles from a stock plant of 'Rober's Lemon Rose.' Make sure that all traces of the axillary bud are removed. The stock plant should be retained to serve as a control.
2	Stick the leaf cuttings into a flat of pre-moistened propagation mix and cover with a humidity dome. Place in a greenhouse. The cuttings should be kept moist at all times.
3	Make 10 root cuttings from the stock plant. Cuttings should be at least 3 mm in diameter and about 5 cm long. Maintain root polarity by cutting the proximal end of the root straight across and the distal end at a slant.
4	Plant root cuttings horizontally in a flat of pre-moistened propagation medium. Cover the flat with a humidity dome. Place in a greenhouse. The cuttings should be kept moist at all times.
5	Observe cuttings for the appearance of adventitious shoots. Note the leaf morphology of each new shoot and compare it to the stock plant.

QUESTIONS

- All shoots arising from leaf and root cuttings are adventitious. Were there any shoots produced that resemble the mother plant? How do you account for this?
- If a plant is non-variegated and shows no obvious differences in morphology, how could you determine if it is a periclinal chimera?

LITERATURE CITED AND SUGGESTED READING

Dermen, H. 1960. Nature of plant sports. *Amer Hortic Mag* 39:123–173.

Fitting, H. et al. 1921. *Strasburger's Textbook of Botany*, 5th English Edition. New York: MacMillan and Co., 302 pp.

Mauseth, J. D. 1988. *Plant Anatomy*. Menlo Park, California: The Benjamin/Cummings Publishing Co., Inc., 560 pp.

McPheeters, K. and R. M. Skirvin. 1983. Histogenic layer manipulation in chimeral 'Thornless Evergreen' trailing blackberry. *Euyphytica* 32:351–360.

Neilson-Jones, W. 1934. *Plant Chimeras*. London, United Kingdom: Methuen & Co. Ltd., 123 pp.

Skirvin, R. M. and J. Janick. 1977. Separation of phenotypes in a periclinal chimera. *J Coll Sci Teach* 7:33–35.

Steeves, T. A. and I. M. Sussex. 1989. *Patterns in Plant Development*. Cambridge, United Kingdom: Cambridge University Press, 388 pp.

Tilney-Bassett, R. A. 1986. *Plant Chimeras*. London, United Kingdom: Edward Arnold, Ltd., 199 pp.

Part III

Plant Propagation Structures, Media, and Containers

8 Propagation Structures
Types and Management

Gerald L. Klingaman

Concept Box 8.1

- **A variety of structures, such as greenhouses, cold frames, polytunnels, and hot beds are used to control the propagation environment.**
- **Facilities should be designed to allow for maximum flexibility in terms of usage and potential expansion.**
- **Propagation sites should have access to large volumes of high-quality water, which should be tested for salinity, alkalinity, pH, hardness, electrical conductivity, and mineral content.**
- **Greenhouse coverings (roofs) are made from polyethylene or other films of various compositions, fiberglass-reinforced plastic, rigid acrylic or polycarbonate resins, and glass. Each of these types of coverings offers advantages and disadvantages.**
- **Bench systems can be arranged to maximize space utilization—rolling benches use as much as 90% of the usable space, whereas peninsular arrangements are about 75% efficient. Many growers also use floor space for growing plants.**
- **Heating of greenhouse structures can be accomplished using gas and oil as fuels and passive solar irradiation; electrical heating systems are typically only used to power heat mats and/or heating cables.**
- **Greenhouses are cooled by ventilation (open roofs or side vents), wet pads and cooling fans (evaporative cooling), and fog systems and, most commonly, by reducing solar heating with shade cloth.**

An environment suitable for plant growth must be provided during propagation, taking into account the reduced capacity of the plant to take up water without a root system. Of primary concern is that desiccation must be prevented until the plant becomes sufficiently established to endure the rigors of the real-world environment. By thoroughly understanding seasonal growth patterns and weather cycles, and with a good measure of luck, people have propagated plants for centuries with little or no special equipment. In today's economy, a variety of structures, such as greenhouses, cold frames, and hot beds, are employed to increase control over the environment to maximize the likelihood of success.

SITE SELECTION

Layout

The type of propagation dictates the facilities that are necessary. Facilities should be designed to allow for maximum flexibility in terms of use and potential for expansion because businesses may grow and they may often have to respond to changes in market demand. Because propagation is the most basic part of a greenhouse or nursery operation, these areas should be easily accessible to the main growing facilities. If additional space is needed for propagation, it can be taken from the growing facilities. For example, a container nursery operation could give up some outdoor growing area for additional propagation facilities, whereas a greenhouse operation could turn over some of its production ranges for propagation. For efficient management, propagation areas should be adjacent to each other and not scattered throughout a facility.

Many nursery and greenhouse businesses now have regional production facilities situated in several locations, enabling the operation of the opportunity to take advantage of unique climatic conditions or improved access to major or new emerging markets. In most cases, it is advantageous for each facility to have its own propagation area. Because climates and local weather conditions vary, even within a given hardiness zone, it is easier to manage propagation schedules and have liners or transplants ready to move into the production mode if they are grown locally.

Topography

The physical topography of the site must be considered for both propagation and production purposes. Ideally, gently sloping but essentially level land is preferred for nursery and greenhouse purposes; however, there are many examples of producers that have made do with less than ideal sites, often using extremely sloping and rolling land.

Propagation facilities use a lot of water and some provision must be made to assure adequate water drainage away from the site. Drainage issues are easy to address before structures are built, but are difficult and expensive to deal with following construction. Excess surface water not only creates an unpleasant work environment, but also serves as a breeding ground for insects such as shore flies and fungus gnats, pests such as slugs and snails, and root rot pathogens such as *Phytophthora, Rhizoctonia,* and *Pythium* species (Chapter 11). Ideally, greenhouses should have approximately 1% slope along the length of the span, with drainage swales outside to move rainwater away from the site. The floor of single greenhouses should be crowned (higher in the center) to move surface water to the outer edge of the house. Areas with less than 1% slope should have in-ground tile installed to assure adequate water removal. For gutter-connected structures with concrete floors, the floor should be sloped into drainage catch basins to move excess water. Some governments now regulate water discharge from nursery and greenhouse ranges. Water catchments are designed to control storm water runoff and reduce the amount of non-point source pollution leaving the site. Most nurseries and greenhouses use the water in these catchment ponds to irrigate their growing crops, but not for propagation.

WATER

Perhaps the most important consideration before locating any horticultural operation is to assure that the site has access to large volumes of high-quality water. Propagation is no exception. In fact, water used for propagation must be of higher quality than that used to grow the finished crop because young tender plants are more susceptible to stresses such as high salts and attack by pathogens. Before a site is selected for a propagation facility, the quality of water should be tested. The most important water quality characteristics are salinity, alkalinity, pH, hardness, electrical conductivity, and mineral content (Table 8.1).

Well water is preferable to surface water, provided that the quality parameters are the same. Surface water supplies are more variable than well water during the course of a growing season because they are affected by heavy rains, droughts, point-source and non-point source pollutions, and complications caused by algae and disease organisms. A multistage treatment facility may be needed to assure high-quality water for propagation if surface water is the only available source. This is accomplished by passing the water through a sand filter to remove as many of the particulates as possible. In-line canister filters in the propagation house may also be

TABLE 8.1
Water Quality Guidelines for Use in Propagation

Water Characteristics	Optimum	Maximum Usable
Electrical conductivity (EC)	0–0.3 dS/m	1.5 dS/m
Alkalinity	0–60 ppm as $CaCO_3$	150 ppm as $CaCO_3$
Hardness	20–60 ppm	200 ppm
pH	5–7	8.5 upper limit; 3.5 lower limit
Calcium (Ca)	20–40 ppm[a]	120 ppm[a]
Magnesium (Mg)	5–10 ppm[a]	25 ppm[a]
Sodium (Na)	0–10 ppm	50 ppm
Chloride (Cl)	0–40 ppm	140 ppm
Iron (Fe)	1–3 ppm	5 ppm
Manganese (Mn)	0.2–0.5 ppm	1 ppm
Zinc (Zn)	0–0.1 ppm	0.2 ppm
Copper (Cu)	0–0.05	0.2 ppm

[a] The ratio of calcium to magnesium should remain between 4:1 and 6:1.

required to prevent the plugging of mist and fog nozzles. Surface water destined for use in propagation can be disinfested using either a chlorination system or ultraviolet (UV) lights.

Of these, chlorination is currently most common and effective. Many municipalities use chlorine gas as a disinfestant, but this corrosive, deadly gas is difficult to handle, so most propagators use either calcium hypochlorite ($Ca(OCl)_2$)—the material used to treat swimming pools—or sodium hypochlorite (NaOCl)—the component found in household bleach products. The amount of chlorine required to treat the water depends on the amount of organic residue, the temperature, the pH, and the amount of time allowed for the treatment. Most water supplies have a free residual level of about 1 ppm chlorine at the faucet, which is achieved by adding about 10 ppm at the treatment plant.

Well water that does not meet all of the quality parameters can still be used, provided that a means of correcting the specific problem can be found. Some problems, such as a high level of alkalinity, can easily be corrected by injecting acid into the water line. High electrical conductivity and high total dissolved solids, common problems in coastal areas or arid regions, can be alleviated by processing through reverse osmosis treatment plants. Because of the importance of a sustainable, high-quality supply of water, it is critical to explore the economics of water treatment before selecting a site for propagation.

Municipal water supplies can be used for propagation and may be cost competitive if extensive water treatment is necessary. Chlorine is added by most municipalities to disinfest the water supply and is perfectly safe for use in propagation. Most communities now fluoridate their water supply to help reduce tooth decay. The fluoride added does no damage to most plants, but a few species in the Agavaceae (*Dracaena* and *Cordyline*) and Liliaceae (*Chlorophytum* and *Lilium*) accumulate fluorides in their leaf tissues, and the leaf tips may become necrotic. This problem usually only occurs when the fluoride concentration in the water is above 1 ppm.

As populations have grown, water demand has increased, resulting in ordinances during periods of drought. Some communities treat agricultural usage favorably, whereas others equate it with lawn watering, which frequently goes on even–odd watering cycles at the first hint of drought. Entire days without water could result in stress or death of young propagules.

FACILITY NEEDS

The size and nature of the propagation facility depend on the kind of propagation to be conducted. In almost all instances, greenhouses will satisfy most of the needs of the propagator, but support facilities are also needed. Headhouse facilities for propagation ranges usually amount to about 5 to 10% of the allocated greenhouse space. The headhouse facilities might be used for offices, cutting preparation, potting, cold storage, general storage, and restrooms. Some of the functions of the headhouse can be done in the field, such as trimming the cuttings as they are removed from the stock plants. Transplanting can sometimes be accommodated by using potting wagons, thus facilitating the flow of liners from propagation to the next growing stage. As the size of the propagation area increases, the percentage area devoted to headhouse facilities will decrease. If tissue culture propagation is anticipated, an isolated area that can be kept clean should be used.

The amount of space required for propagation depends on the number of plants to be produced, the number of cuttings stuck (planted) per given area (plant density), and the length of time that the cuttings will occupy the propagation space. Additional propagules may be planted to compensate for the losses caused by poor rooting or low germination percentages, and a myriad of other biotic and environmental factors. Planning for the required amount of space requires a thorough understanding of plant propagation characteristics and careful attention to the details that can make or break a crop. Good propagators are detail oriented, observant, and consistently looking for ways to minimize loss during propagation.

In a perfect world, one would stick the number of cuttings desired for a particular crop, and it would be easy to estimate space requirements. For example, only 50% of the cuttings of *Juniperus virginiana* L. 'Skyrocket' (Skyrocket Juniper) usually root, and of those, only half of the cuttings that root will be vigorous enough to make a good liner. So, if you wanted 1000 usable liners, you would have to stick 4000 cuttings [number of cuttings to stick = number of liners desired/(percent rooting × percent vigorous liners) or $4000 = 1000/(0.5 \times 0.5)$]. Not only does the propagator have to acquire and stick the extra cuttings, but also the room must be provided for them in the propagation area. Obviously, anything that can be done to improve the rooting and the vigor of cuttings will increase the efficiency of the operation.

To determine the amount of space required for a given crop, you must decide on the density of the propagules in the propagation area and whether they will be in open beds or in flats. Open bed systems are often used for the most difficult species and cultivars because they assure optimum drainage and aeration for the propagules.

Higher plant densities are desirable; however, if planting densities are too high, lower-quality plants could result. The two most common problems with high plant densities are legginess (stretching) and *Botrytis* infection (Chapter 12). Density is also determined by the length of time that the cuttings will be in the propagation area, the leaf size and the shape of the species, and the inherent characteristics of the plant that allow it to grow close together in a mist bed and still produce a high-quality liner. Plants such as some cultivars of chrysanthemum root quickly and can be placed at high densities with as many as 800 cuttings per square meter (72/ft^2), the same density that can be used for slow-to-root *Juniperus horizontalis* Moench. 'Wiltonii' (Blue Rug Juniper). The cutting spacing required to achieve these densities is 2.5 × 5 cm (1 × 2 in). Chrysanthemum cuttings will remain in the propagation area for about 2 weeks, whereas the juniper cuttings may occupy their space in the propagation bench for 8 to 12 months, depending on the production schedule. Plants with larger leaves, such as *Euphorbia pulcherrima* Willd. ex Klotzsch (Poinsettia) and *Ilex cornuta* Lindl. & Paxt. 'Burfordii' (Burford Holly), are usually spaced 5 × 5 cm (2 × 2 in) and have a density of 400 cuttings per square meter (36/ft^2). Poinsettias will be under mist for about 3 weeks, whereas the holly may occupy the propagation area for 3 or more months.

For easy-to-root species and for bedding plant growers, various containers have become popular. Plug production of bedding plants permits growers to produce plants at very high densities during the initial stages of seedling growth when the rate of growth is slow, thus increasing the efficient use of the greenhouse space. Propagation containers vary dramatically from crop to crop, but they all share one feature, namely, a specific spacing arrangement per given area. In the open bed, the propagator can determine the spacing to be used, but when rooting or plug trays are used, the choice of container establishes the density of propagules per unit area.

Rooting trays are plastic inserts designed to fit into a nursery carrying tray (Figure 8.1). These have a given number of planting cells per sheet—often 18, 36, or 72—giving 102, 205, or 411 planting cells per square meter (13, 26, or 52/ft^2). Plug trays are even smaller, with individual trays having from 188 to 480 cells per tray and plant densities of 800 to 4000 seedlings per square meter (100–500/ft^2). Individual plastic pots may also be used for propagation with densities per square meter similar to those given for the rooting trays. For easy-to-root species, such as poinsettia, growers often use a technique called "direct sticking," where they root directly in the pot in which the plant will be sold.

FIGURE 8.1 A polytunnel propagation house filled with 25,000 rooted barberry cuttings in rooting trays.

As you note from the above discussion, propagation facilities may be in use for a long time or for only a matter of weeks. For crops such as chrysanthemum that have a year-round demand, propagators use the same space over and over during the course of a year. A chrysanthemum propagator may have as many as 25 "turns" or cycles of the bench space per year. On the other hand, woody plants may have only one turn per year because of the longer times required for rooting and subsequent growth. Easy-to-root woody nursery crops such as *Forsythia* × *intermedia* Zab. or *Weigela florida* (Bunge) A. DC. often occupy propagation space for a few months, but seasonally related planting cycles may dictate that the plants be held for several more months before they are planted. One of the principal roles of the propagator is to take the mosaic of several hundred to a thousand or more species or cultivars that are grown by the nursery and schedule the time and space allotment for the crop in propagation. A spreadsheet can be used to calculate the space requirement for each crop. Data in the spreadsheet can be based on the propagation requirements for the species as well as information from past years' performances such as plant spacing, rooting percentages, and other modifying factors. Successful propagators maintain accurate records of propagation performances from past years, and they use the experience gained to increase their efficiency in future years.

ENVIRONMENTAL REQUIREMENTS FOR PROPAGATION

This book discusses a number of different propagation techniques, but central to all of them is the basic requirement that the plant tissue be maintained in a turgid condition until the new propagule is able to support itself in an open environment. Preventing desiccation is the most crucial factor involved in propagation, and a number of techniques such as intermittent mist and high-humidity chambers have been developed to forestall this possibility. Water loss from tissue increases as sunlight, air movement, and temperature increase. At 100% humidity, water loss through transpiration stops.

Growth and rooting, though, are temperature-related phenomena occurring between approximately 10 and 32°C (50 and 90°F), depending on the sensitivity of the crop. Most greenhouse and nursery plants root best when the rooting medium is between 21 and 27°C (70 and

81°F). These kinds of environmental conditions can be satisfied without much trouble in a steamy tropical jungle, but in temperate climates, we need help—a greenhouse.

GREENHOUSE STRUCTURES

Greenhouses are a mainstay in plant propagation, permitting the easy manipulation of all of the environmental parameters that affect tissue desiccation. Small growers may use just a portion of a greenhouse for propagation, with the rest of the house devoted to production. Large growers have ranges of greenhouses devoted exclusively to propagation. In the United States, approximately 63% of all greenhouse space is covered with polyethylene sheeting, glass is approximately 14%, and the remainder is about equally divided between fiberglass-reinforced plastic (FRP) and rigid plastic panels. Polyethylene-covered greenhouses are the most common propagation structures because they are inexpensive, double polyethylene offers higher insulation values, and although light levels are lower than under glass, they are more than adequate for propagation.

Greenhouses frames were initially built from decay-resistant woods, such as redwood or cypress. These eventually gave way to steel pipe frames supporting wooden sashes to hold the glass. This technology lasted through World War II when new construction techniques and materials began to enter the market. Extruded aluminum structural components began to replace wooden frames, providing a low-maintenance alternative to the laborious task of stripping wooden sills, repainting, and reglazing every few years. When polyethylene sheeting first became available in the mid-1950s, wooden A-frame houses became popular. These all-wood houses had a short useful life and were quickly phased out as the growers' financial status improved. The wood houses were replaced primarily with tubular galvanized metal Quonset houses. Today, greenhouse frames are most commonly made from galvanized steel that is fabricated in a variety of shapes and sizes. Extruded aluminum is still available and commonly used for sash bars to support glass or rigid plastic panels.

Cold Frames and Hot Beds

At the turn of the 20th century, cold frames and hot beds were the primary structures used to propagate plants. These frames were built low to the ground, with the northern side slightly higher so that the glass sash covering the structure had about a 15° slope to the south to intercept more sunlight in the winter months. While they served their purpose, they were difficult to automate. Because the air space inside the structure was small, cold frames and hot beds heated or cooled quickly as weather conditions changed, necessitating constant attention. Because of these limitations and the amount of hand labor required to check constantly and maintain the frames, they are used infrequently today for commercial purposes. Cold frames and hot beds are easy to build, inexpensive, and unobtrusive in the landscape, so hobby gardeners still find uses for them.

Polytunnels

Today, the term "cold frame" is being used more often to describe an unheated polytunnel or Quonset greenhouse (Figure 8.2). These structures can be usually about 2.5 m (7 ft) tall, 4 m (13 ft) wide, and 30 m (97 ft) in length, but they can be much longer if the need dictates. The structures are made from pieces of pipe tubing bent in a bow placed 1.3 m (4 ft) to 2 m (6.5 ft) apart, depending on the anticipated snow load for the area. There is usually little or no internal bracing, except for a purlin that runs the length of the house at the ridge line. They are anchored by inserting the bows into a larger piece of pipe that is driven into the ground. Most growers use only a single layer of polyethylene when covering unheated tunnels.

The preferred orientation of unheated polytunnels is north and south. With a north–south orientation, the sun passes over the house and uniformly heats the structure as it passes in its arc through the sky. If the house is oriented east and west, especially during the late winter months, the southern side of the house will be much warmer than the northern side. This can cause plants to come out of dormancy at different times. Environmental control in polytunnel structures is minimal, usually limited to the use of shade cloth or ventilation by opening the doors and/or rolling up the sides. Many growers in warm-temperate climates use the protection afforded by the poly during the cold winter months but then cut progressively larger vent holes in the plastic as the temperature

FIGURE 8.2 Polytunnel greenhouses that will be used for the in-ground propagativon of woody nursery stock. Each greenhouse can hold in excess of 50,000 cuttings.

FIGURE 8.3 The basic Quonset design is versatile and can be adapted to a wide number of uses. In this small propagation frame, PVC tubing is used to support the plastic that will be added as winter approaches.

FIGURE 8.4 The polyethylene-covered, stand-alone Quonset greenhouse is the most common greenhouse structure in the United States because it is inexpensive and versatile.

increases during the summer. Propagators sometimes install heaters in these houses but keep them set just above freezing to prevent the plumbing from freezing and reduce the possibility of winter burn on broadleaf plants. However, these houses are not warm enough to encourage mid-winter growth.

A scaled-down version of these hoop houses, only wide enough to cover an outdoor bed, can be constructed for small producers or hobbyists (Figure 8.3). In Europe, the term "cloche," originally used to describe a glass bell jar to cover a single plant, is used to describe these small poly structures, especially when they are used in the vegetable garden. Unheated hoop houses (high tunnels) are also used for the production of vegetables, flowers, and fruits directly in the ground during cooler seasons.

Stand-Alone Greenhouses

A number of different greenhouse designs have been used over the years. The first to be employed on a large scale by commercial growers were glass houses using typical gable construction and looking much like a contemporary ranch style house, but built of glass. These first houses were built with either cypress or redwood rafters supported by a steel pipe frame. This system remained largely unchanged in design from the end of the Victorian era until the 1950s when new materials, especially plastic coverings, extruded aluminum sashes, and galvanized tubing became more readily available. Few of these stand-alone glass houses are used today, except for hobby greenhouses and some institutional construction. Curved surfaces, including the gothic-arch style, are popular for these high-end greenhouses.

The workhorse of the stand-alone greenhouse is a larger version of the polytunnel, the Quonset greenhouse, which became popular after 1970 because of its low cost and relatively simple construction (Figure 8.4). The Quonset design was developed by American forces during World War II as a quick way of building portable buildings and hangers. As greenhouses, these free-standing structures are usually about 10 m (32 ft) wide and 4 m (10 ft) tall, with the length usually 30 to 120 m. They are more substantially built than polytunnels, with raised sidewalls, internal bracing, heating and cooling systems, and various environmental control systems. These structures can be equipped with an open-roof design where a portion of the roof lifts up to facilitate natural ventilation. Stand-alone greenhouses are usually oriented north to south in latitudes south of 40E and east and west north of that line. The east–west orientation permits more wintertime light to enter the house. Because of the environmental controls in these houses, heat buildup on the southern side is not an issue.

Gutter-Connected Greenhouses

The cost of heating a greenhouse in a given location is a function of the thermal characteristics of the covering and the amount of exposed surface area. By connecting greenhouses together, much of the exposed surface area disappears, and the structure becomes more efficient to heat. (These gutter-connected houses may be Quonset style [Figure 8.5] if poly covering is used or ridge-and-furrow design if glass or rigid plastic sheeting is chosen.) Gutter-connected houses are composed of a number of bays, with the width of the span between the gutters on the roof determined by the type of structure selected. The old-style ridge-and-furrow houses were high-profile units with wide individual bays. These were followed by low-profile Dutch-style, ridge-and-furrow houses with narrow bays but low roof lines that decreased the surface area

FIGURE 8.5 A gutter-connected, polyethylene-covered propagation house where cuttings are rooted and grown through their liner stage in rooting trays.

of the greenhouse and kept the warm air closer to the plants. Today, the width of individual bays varies from 4 to 12.2 m (13–40 ft), with most designs maintaining the low-profile roof and as wide a bay as possible. Some of the wide-span, low-profile designs employ a truss system that requires support posts for only every other gutter, thus maximizing the amount of post-free growing space.

As mechanization has become more important in greenhouses, the height of the sidewall has increased to accommodate the curtains, trolley, plumbing, lighting, and other equipment positioned above the growing space. Four-meter sidewalls are common, with many southern growers using even taller houses.

The compass arrangement of gutter-connected houses is less critical than for unheated polytunnels. Some authorities recommend the east–west arrangement in northern areas, but there seems to be less consensus with this style house. More frequently, since these units have such a large footprint, grading considerations often take precedence. Because bays within gutter-connected structures share some structural elements, the cost of construction and operation per unit area is less for these houses than for stand-alone houses equipped in a similar fashion. Provided that the environmental controls are installed in a series of zones, gutter-connected greenhouses can be maintained with different growing conditions by dropping curtains between adjacent bays.

Retractable-Roof Greenhouses

Starting about 1990, several competing firms began producing greenhouses with retractable roofs. These were primarily developed for bedding plant growers, permitting operators to open the houses during good weather and quickly cover them if bad weather threatened. Radically different flat-roofed designs were employed by some firms, whereas others kept the traditional shape of existing structures. Different manufacturers have employed a variety of methods to achieve the open-roof goal. Several companies roll up the greenhouse covering, whereas others bunch up the pleated roof covering just as a drapery is compressed as the curtains are opened. Some use a Venetian blind approach, constructing the roof as a series of heavy plastic slats that open and close. One style has a ridge-and-furrow construction, with the panels covered with lightweight rigid plastic sheets. The entire roof of the greenhouse opens like a crank-out window (Figure 8.4). Depending on the type of design chosen, these houses cost about the same as traditional gutter-connected structures.

While retractable roof greenhouses have not been used much in cutting propagation yet, there is little doubt that they will be in the future. They provide the same benefit to the cutting propagator as the plug grower, providing a more natural environment for plant growth. Because the plants are exposed to elements, especially wind, fluctuating temperatures, and bright sunlight, in a controlled manner, they are more hardened than the same plant grown in a completely enclosed environment. Operational costs, especially if the greenhouse is used with minimum heat, are comparable to other structures. Summertime cooling costs are eliminated by this design. Because of the increased use of electric motors, pulleys, cables, and other assorted machinery necessary to open and close the roof, maintenance costs have been higher than those of conventional greenhouses.

GREENHOUSE COVERINGS

The choices available for greenhouse covering have changed considerably since World War II when glass was the only possibility. Today, growers choose coverings based on an evaluation of what they need from the covering, effective life, initial cost, long-term cost, tax benefits, operational costs, the type of structure, and convenience. While glass is still the standard bearer for durability—provided that a hailstorm does not turn it into a truckload of shards—it is also expensive. Business decisions, especially for nursery production that may be made 5 to 8 years before a crop is salable, are fraught with uncertainty. Many nursery managers are reluctant to face a 20-year payout for a more expensive glass structure when a less expensive, more disposable propagation structure reduces their initial capital outlay and makes it easier to amortize over a shorter period.

Polyethylene Film

Polyethylene, or "poly" in the jargon of the trade, is the dominant covering for greenhouses in most parts of the world. While it is petroleum based, its cost is always proportionally less than the other coverings because even glass has a high energy component in its manufacture and cost. The polyethylene sheeting employed for greenhouse use has a UV inhibitor mixed with the plastic resin, permitting 3 to 4 years of use before it must be replaced. Without the UV inhibitor, poly sheeting may not last through a single winter. Polyethylene is sold according to thickness, and most growers use 0.15-mm (6-mil)–thick material for the outer layer and 0.1-mm (4-mil)–thick material for the inner layer. Most growers that maintain temperatures warm enough to permit wintertime growth use two layers of poly to achieve an air-inflated insulation barrier that reduces heating costs by as much as 40% (the exception is those in warm areas, such as zones 9 and 10 in the southern United States). The insulation air layer is maintained by means of a small fan that inflates the poly sheets, maintaining a distance between layers of 50 to 70 mm. The air-filled space also cushions the outer layer of plastic, reducing abrasion and helping extend its life. Polyethylene sheets are available in widths to accommodate all standard greenhouse designs. Poly tubes are also available, making the job of installing a double layer of poly much easier.

Designer polyethylene sheets are now available, which have a variety of benefits, thanks to a new technique that permits manufacturers to extrude up to three layers at one time. All greenhouse polyethylene is manufactured for UV resistance, but in addition, it may be imbued with a surfactant to reduce the drip problem during the winter, compounds to reduce electrical charge to prevent dust from clinging, or infrared (IR)-absorbing materials that reduce the amount of radiational cooling that occurs at night. Light transmission through a double layer of polyethylene film is 76%, compared to 88% for glass. The addition of IR-absorbing compounds reduces light transmission to 67%.

Other Sheet Plastics

While polyethylene dominates the plastic film market in most parts of the world, other products are available and offer some advantages. UV-resistant polyvinyl chloride available in thicknesses of 0.2 and 0.3 mm (8 and 12 mil) are the most popular in film covering for greenhouses in Japan. However, they cost about three times as much as polyethylene film while only lasting for 5 or 6 years. Because they carry a static electric charge, they also tend to collect dust, which reduces light transmission.

Another product, UV-resistant ethylene tetrafluoroethylene (ETFE) film sold under brand names, such as F-Clean and Tefzel T[2], has potential use as a greenhouse covering because it is purported to have a life expectancy of over 15 years. The product is extremely clear with 95% light transmission, making it a popular choice for covering the surface of photoelectric cells. It also has a good IR transmission; thus, more heat is radiated back outside on sunny days. Currently, ETFE is over 10 times more expensive than polyethylene and not available in widths needed for commercial greenhouse applications, but if these difficulties can be resolved, it could become a useful addition for covering greenhouses.

Fiberglass-Reinforced Plastic (FRP)

This corrugated rigid plastic has been used as a greenhouse covering since the 1950s, but its popularity has declined in recent years. It has the advantage of being relatively long lived and, while new, to have a light transmission coefficient of at least 88%. Because of the fiberglass particles embedded in the plastic, it tends to scatter light and gives a uniform light distribution pattern throughout the greenhouse. These corrugated sheets are strong and more resistant to hail and vandalism than other greenhouse coverings. FRP is available in a variety of widths and lengths and is flexible enough to cover a Quonset greenhouse. They tend to be tight greenhouses and slightly more efficient to heat than glass houses. The cost of FRP is intermediate between those of polyethylene and glass.

FRP does have significant drawbacks. It is combustible and tends to age poorly, turning yellow or brown over time and becoming brittle. Greenhouse grades of FRP are usually guaranteed to last for 10 to 15 years, but as the acrylic or polycarbonate covering erodes from the surface, the fiberglass particles are exposed. These capture dust and debris and further darken the surface. Resurfacing products are available but, because of the amount of labor involved in the process, are seldom used. The shading problem is not as serious for cutting propagators as for conventional greenhouse growers because most propagators use shade cloth during at least part of the propagation cycle. Many growers used FRP to cover the ends of their greenhouses where shading is not a serious concern.

Rigid Plastic Panels

Since about 1990, a new category of greenhouse covering has gained popularity—the twin-walled rigid plastic panels. The unique twin-walled construction and large panel size makes these very tight houses, reducing heat loss by about one-half when compared to a single layer of glass. These panels are made from either acrylic or polycarbonate resins—chemicals that give the panel unique characteristics.

Acrylic panels are available in 8 or 16 mm in thicknesses, with widths of 120 cm (47.25 in) and a variety of lengths. Light transmission is 83%, and that level can be maintained for at least 15 years. Some of the original greenhouses covered with rigid acrylic sheets have shown almost no loss in light transmission as they have aged. Acrylic panels are inflexible and must be used on a flat surface. They are flammable and more subject to damage by hail than polycarbonate panes. This covering is especially popular where high light issues are a concern and cost is not, for it costs more than twice as much glass (Figure 8.6).

Polycarbonate panels have a similar appearance to acrylic panels but, unlike acrylic panels, they darken with age (Figure 8.7). New 8-mm–thick panels have a 79% light transmission coefficient, with about a 1% reduction expected each year. They are available in thicknesses of 4, 6, 8, 10, and 16 mm and have similar insulation advantages when compared to glass. The thinner sheets are flexible enough to be used on a Quonset style greenhouse. Polycarbonate panels are noncombustible. They cost about twice the current cost of glass. Polycarbonate panels are more popular with growers than acrylics and are often used for side and end walls.

FIGURE 8.6 A low-profile, open-roofed greenhouse range covered with acrylic panels. One hundred percent of the roof opens to achieve near-ambient temperature conditions, but still allows quick protection of the crop if weather conditions change.

FIGURE 8.7 Cooling fans protrude from these gutter-connected Quonset greenhouses covered with 8 mm polycarbonate sheets. The fans are designed to replace the entire volume of air inside the greenhouse once every minute. The pipe protruding from near the ridge of the houses is the exhaust port for the ceiling-mounted radiant heater used in the houses.

Glass

Greenhouse growers with sufficient capital or who can spread the cost of their facility over a number of years still prefer glass greenhouses. Unless some calamity befalls it, glass is a permanent greenhouse structure. Glass has 88% light transmission and is easy to maintain, and its qualities do not change over time. As greenhouse structures have become more sophisticated, the size of glass panes has gotten wider and longer. Some panes are up to 1 m wide and 2 m long. Most commonly, double-strength, 3 mm (0.125) float glass is used in the United States, whereas tempered, 4 mm (0.16) panes are used in Dutch houses. Because glass is heavy, the cost of structural members is higher for glass houses than for lighter-weight rigid plastic panels. Glass greenhouses are popular for plug growers and propagators of high-value, vegetatively propagated, and herbaceous liners.

BENCHING SYSTEMS

Good plant propagators have developed their own favorite way of increasing their plants. If given the choice, most would prefer to use some type of benching system in the propagation house because the plants are at a convenient height for work and inspection. Crops produced in benches usually have less incidence of disease for several reasons. Disinfestation between crops is easier, contamination by soil-borne organisms is less likely, and air circulation is better, thus reducing the incidence of foliar disease. Also, in a conventionally heated greenhouse, the floor is the coldest location in the house. Cold soil slows root development and increases the possibility of disease.

Bench Arrangement

Benching systems can be arranged to maximize the use of floor space, with rolling benches utilizing as much as 90% of the total floor space. Peninsular benches are arranged so that each main bench extends from the greenhouse wall to the center aisle. The ends of these benches are connected by crossbenches at the walls. Peninsular bench arrangements utilize about 75% of the floor space and, while convenient, waste a lot of space. Some growers who produce large numbers of plants have modified the peninsular system to gain efficiency by using a series of inverted T rails at right angles to the main isle. These rails are as far apart as the length of carry flat. As the greenhouse is filled with cuttings, flats are slid down each row of rails, completely filling the sides of the greenhouse and leaving the central aisle for access.

Benching systems are used mostly to accommodate plants grown in individual containers nested in carry flats, or plug or rooting trays. The size and type of benching material should be strong enough to hold the flats level and constructed of a material that can be effectively disinfested between crops. Expanded metal benching is perfect for this use but expensive. Various kinds of wire benching systems are also effective and more economical. Bench widths of up to 2 m are acceptable so long as the bench can be accessed from both sides. Wider benches reduce the number of isles and increase the efficiency of the floor plan. However, the tradeoff between space utilization and efficiency must be considered. If the grower cannot conveniently access the plants, any gains made in floor space utilization may be lost in the overall efficiency of the operation. Bench height should be about 750 mm (30 in).

Growing on the Floor

Although most propagators would prefer producing plants on raised benches, most choose instead to produce on the floor because the cost of the benches is eliminated and it is possible to use a higher percentage of the floor space to grow plants. The major disadvantage of growing on the floor is that it is the coldest location in the greenhouse; thus, rooting is slowed. This disadvantage can be rectified by the use of ground bed heating systems, an expensive, but effective way of dealing with cold soil. The ideal surface for growing on the ground is concrete because it is easy to disinfest between uses and is clean and uniform, and makes a generally agreeable working surface. It is also costly, so many growers opt for less expensive ground treatment. The most common alternative to concrete is a gravel base, sometimes with concrete isles. Most growers cover the gravel with a woven ground cloth that serves as a weed barrier. Bare soil should never be used because of the weed and disease pressure that can be encountered. The carry or rooting trays are arranged to achieve the best utilization of space and to accommodate any mechanization equipment that is available. Some growers have adapted systems of mechanization that place plants onto large, mobile benches. These benches are then transported to the growing area where they remain until their production cycle is finished, and then the bench is retrieved for shipping or growing on.

Ground Beds

Floor production may be done in open ground beds where pine bark, sphagnum peat, sand, vermiculite, perlite, pumice, and a host of other ingredients are mixed to produce a suitable rooting medium (Figure 8.2). Most propagators use the medium once and then disinfest it before using it again. Because of the possibility of developing wet spots in these ground beds, careful attention for surface and subsurface drainage should be considered.

HEATING SYSTEMS

The greenhouse grower and the propagator have different needs in terms of a heating system. Greenhouse growers use heat to keep their crop growing during the winter months. Propagators may or may not need to heat their crop during the winter. In some cases, it may be more advantageous to keep the crop dormant until it can be planted in spring. Before deciding on a heating system, the decision must be made on exactly how the facilities will be used.

Fuel

The array of fuels available for heating greenhouses is the same for heating a home, and the same considerations come to bear on the decision of which to use. In a greenhouse or a home, the affordability, reliability, and convenience of a fuel are desirable. Greenhouses, unlike homes, are poorly insulated and lose a tremendous amount of heat. Also, greenhouses may be set up with a central heating system (a heating plant) that delivers hot water to individual houses. Some fuels, such as coal and oil, are appropriate to fire boilers but not suited to unit heaters in individual greenhouses. Boiler systems are more popular and cost effective in colder climates than in warmer regions.

Electrical resistance heating is an expensive way of heating and is never used for space heating in the greenhouse. If used in the greenhouse, electrically generated heat is used only to power heat mats or heating cables.

Solar heating seems like an obvious way of reducing heating costs—use the heat gain by the greenhouse during the day to heat the crop at night. Following the high energy costs of the 1970s, solar heating systems were thoroughly studied for their potential to heat greenhouses. Greenhouses trap a tremendous amount of heat during the day, so relatively little supplemental heat is needed on sunny days. The largest heat demand comes during the night or on cloudy days. Most of these heating systems that have been developed use solar energy to heat water. During periods of demand, the hot water is circulated through some type of heat exchanger to extract the heat. In colder areas, the solar collector surface area must be equal to the surface area of the greenhouse. If less collector space is used, the system cannot store sufficient heat in the water to keep up with the demand. Based solely on the daily British thermal unit (BTU) output, about 42 m^2 of collector surface area must be provided to equal the heat output of 3.8 L (1 gal) of heating oil. Because of the extensive areas of solar collector arrays necessary, most attempts at using solar heating for greenhouses were abandoned by the 1990s when fuel prices stabilized.

Natural gas is the least expensive and easiest to use heating fuel and, therefore, is the most popular choice for greenhouse heating. Heaters outfitted to burn natural gas are less expensive than equipment adapted to burn other fuels. Natural gas is delivered via buried supply lines, but these lines do not service every area. Access to natural gas should be one of the considerations weighed before deciding on a site for a new greenhouse location. If natural gas is not available, many growers use liquid propane (LP) gas. It is an easy-to-use and readily available compressed gas; however, it is expensive, costing about twice as much as natural gas. It is stored in aboveground tanks under pressure, where it exists as a liquid. LP prices vary considerably from summer to winter, with winter prices oftentimes double that of summer prices. Some growers find that the investment in extra storage tanks is cost effective, allowing much of the fuel to be purchased during the summer when prices are low.

Heating oil is used, for example, in areas of the northeastern United States, but is less commonly available in other areas. Heating oil comes in six grades, with the higher grades being more viscous and difficult to use. Grades number 1 and 2 can be pumped to individual unit heaters, but because they are liquid, they must be moved under pressure. The heavier grades are slightly less expensive and are used to fire central heating plants. Coal systems, while once used to fire central heating systems, are seldom seen today.

Wood heat has been used to fuel central boilers and may be cost effective if a dependable, cheap source of wood can be found. Some small growers are attracted to the low startup cost of individual fireboxes in a greenhouse but quickly discovered that feeding a fire is like having a baby, always wanting to be fed when sleeping is best. If wood heat is used, care must be taken to assure that no combustion gases enter the greenhouse. Wood smoke is a ready source of ethylene gas, which can cause serious crop injury. The most efficient wood-fired heating plants use either wood chips or 1- to 2-m–long logs that can be automatically fed into the furnace as the system requires them. The fire heats water that is then piped to the heat exchange device. Because these systems are not pressurized, they are less expensive to operate, install, and own.

Heat Delivery

In regions with cold winters and for large greenhouse ranges, central heating plants are more economical than unit heaters. These central heating plants, most commonly, are hot water systems that deliver the heat via finned pipes, small fan–powered heat exchangers, or, more recently, underground pipes buried in the floor (Figure 8.8). The first two require a steel pipe and have high startup costs, while the piping used to heat the floor is plastic and correspondingly less expensive.

Unit heaters are most commonly fueled with natural gas. These unit heaters heat the air and give about 80% overall fuel efficiency. The exhaust gases must be vented outside. Some companies promote high-efficiency heaters, claiming 95% efficiency. These heaters are mounted outside the greenhouse and blow the heat and exhaust gases directly into the greenhouses. Ethylene injury can be a major problem with these types of heaters unless fresh air is brought into the houses, negating the claimed gain in efficiency. Unit heaters can be equipped with polyethylene tubes to distribute the warmed air uniformly through the greenhouse. If propagation is being done on benches, it is possible to duct the heated air beneath the tables and warm the soil. Horizontal airflow fans are now more commonly used than poly tubes to move the heated air around the greenhouse.

Radiant heaters have been used in some greenhouses because they claim a higher fuel efficiency than unit heaters. Radiant heaters burn natural gas inside a long pipe, which turns cherry red. Above the pipe, a stainless steel reflector deflects the IR heat downward to the crop. Radiant heaters warm plants, soil, pots, and walks, not the air. Because the air is not heated, radiant heaters have a higher heating efficiency. The air temperature in a greenhouse heated by this method is usually 3 to 6°C (5–10°F) cooler than the air temperature inside a greenhouse with a forced air heating system. However, because they have a higher initial cost, they are not yet widely used in greenhouses.

Heated floor systems are especially well-suited for propagation purposes. Because it is more important to

FIGURE 8.8 Plastic tubing being installed in the porous concrete floor of a greenhouse range to provide bottom heat.

FIGURE 8.9 Bottom heat is provided for these propagation ground beds using a homemade circulating hot-water heating system that uses a conventional water heater as a source of heat.

keep the soil warm than the air around the crop, these systems deliver the heat where it is needed most. Cross-linked polyethylene (PEX), ethylene propylene diene terpolymer (EPDM) (a synthetic rubber), and polybutylene pipes have been used most commonly in the heated floor systems. The piping is usually 19 mm (0.75 in) in diameter and placed 15 to 30 cm apart in the floor (Figure 8.8). Hot water is delivered to the floor pipe at about 50°C (122°F). A portable, table-top heating system employing EPDM tubing is readily available and has become popular with plug growers and vegetative cutting propagators. It uses hot water heaters to provide the heat for a network of tubing that is placed beneath the plants. Some growers devise their own hot water systems to heat rooting beds (Figure 8.9).

COOLING SYSTEMS

Cooling greenhouse space becomes increasingly important as one heads toward the equator; conversely, heating is more important in northern latitudes. The greenhouse is an excellent solar collector, with temperatures inside a greenhouse often 15 to 20°C (28–36°F) warmer than the outside air temperature. In the winter months, this is not a problem, but during late spring and summer months, when warm ambient conditions prevail outside, inside temperatures become too hot for most crops. Some growers in warmer climates routinely stop growing in the greenhouses during the summer and, where possible, move the production to outdoor gravel beds during hot weather.

The traditional approach for greenhouse cooling in colder areas relied on natural ventilation. Roof and side vents were opened, and the natural chimney effect (convection) did the rest. However, changes in greenhouse design, especially the introduction of poly houses, saw natural ventilation systems disappear in favor of fan systems, which pull fresh air through the greenhouse. As electric energy costs have increased, passive cooling systems that rely less heavily on fans have become more practical. Higher greenhouse sidewalls are a feature of many southern greenhouses to take advantage of the fact that hot air rises in the greenhouse, thus keeping it away from the crop (Figure 8.10).

FIGURE 8.10 Sidewalls 4 m or more in height are common in modern greenhouse construction. Higher sidewalls keep the heat away from the crop and allow room for the various kinds of mechanization suspended above the crop.

Open-Roof Greenhouses

Several manufacturers are designing greenhouses with retractable, roll-up, or stand-up roofs that allow for efficient crop cooling. The original passively cooled greenhouses relied on roof vents of only about 10% of the roof area. These new open-roof designs have almost 100% of the roof opened. These maintain the crop temperature at or near-ambient conditions and permit year-round use of the greenhouse space. This further increases efficiency because mechanization can be more easily justified if the space is used on a continual basis. The opening and closing of the houses occurs within a matter of minutes and is controlled by the greenhouse environmental control system. Polyethylene-covered Quonset greenhouses can now be equipped with lift-top attachments that permit lifting about half of the roof off of the structure. These hinged roofs are positioned so that prevailing winds pass over the hinged side of the house, thus pulling the heat out as the air passes the opening. Open-roof greenhouse design seems to be the type of cooling system that will dominate in the immediate future.

Pad and Fan Cooling

Pad and fan cooling systems rely on rapid air exchange, essentially exchanging the volume in the greenhouse once every minute. The number and placement of fans is critical for the proper design of these systems. Because the air continues to be heated by the sun as it is pulled through the greenhouse, the effective distance from the air intake to the exhaust fans is about 70 m (200 ft). Simple air exchange works effectively when the outside air is cool, but when the ambient temperature is already warm, further cooling is needed (Figure 8.7).

An additional step in cooling relies on evaporative pads to remove heat energy from the incoming air, thus introducing cooler air into the greenhouse. As air passes through the pads, water changes from the liquid to the gaseous phase, extracting heat energy from the air for the phase change. In an environment with no water vapor present, air temperature can be reduced as much as 22°C (40°F) by this method. However, as the relative humidity increases, less water can be evaporated from the cooling pads, thus reducing their effectiveness. At 100% relative humidity, there is no cooling effect. Pad and fan cooling systems work effectively in areas with low summertime humidity but less well in areas where the humidity often exceeds 70% during the day. At 70% humidity, the incoming air temperature can only be reduced 3°C (7°F). Aspen pads were used for greenhouse cooling but are seldom seen today. Cross-fluted cardboard pads that are 10 cm (4 in) thick are the norm today for greenhouse pads.

Fog Cooling Systems

Greenhouse fog systems have been used in greenhouse cooling and rely on the same principle of the evaporative cooling used in the pad and fan cooling system. However, instead of using saturated pads as the source of water, fog is pumped directly into the greenhouse space where it emerges from the fog generator as a 10- to 40-μm drops. These drops are small enough that they will remain suspended in the air. Fog cooling systems cannot be used with fan systems because the fog would simply be pulled from the house. When used, they are combined with a passive ventilation system. Fog systems can be used in propagation, so a system using the fog to prevent desiccation of the cuttings and provide cooling might be feasible.

Shading

Shading has always been a part of greenhouse cooling. Even in northern areas, sunlight is not a limiting factor during the summer, so shading is used to reduce the amount of radiant energy entering the structure. Lime- or latex-based whitewash is applied to the greenhouse roof in the summer and removed during the winter. With the advent of plastic greenhouses, painting on shading compounds was not feasible because it could not be removed in the winter, so various types of shade fabrics became available. These reduced the amount of incoming radiation from 20 to 80%, with 40 to 50% shade reduction being most commonly used (see Figure 8.4). As greenhouses became mechanized, shading was automatically added or removed as needed to maintain crop temperature. Originally, shade cloths were black or green woven plastic material. Today, aluminized screens and white fabric cloths are used in shading. The newer products have several functions: they keep the crop cooler during the day, they may reduce crop water use, and they can be used as heat blankets at night to reduce nighttime heating costs.

Environmental Control Systems

Greenhouses are much more sophisticated than in the past, so most new structures combine computer-based technology with an integrated environmental control system that routinely monitors the crop environment and then responds as needed. These systems control heating, cooling, photoperiod, shading, and other environmental parameters as preset by the grower.

LITERATURE CITED AND SUGGESTED READING

Agricultural Statistics Board. 1996. *Floriculture Crops: 1995 Summary*. Washington, D.C.: USDA National Agricultural Statistics Service, 108 pp.

Aldrich, R. A. and J. W. Bartok, Jr. 1994. *Greenhouse Engineering*, 3rd Revision. Ithaca, NY: Cornell University, 203 pp.

Boodley, J. W. 1997. Greenhouses. In: *The Commercial Greenhouse*, 2nd Edition. Albany, NY: Delmar Publishers, pp. 28–56.

Landicho, S. 2003. The Dark Ages: From black woven cloth to high-tech shade systems. In: *Amer Nurseryman*. Chicago, IL, pp. 26–28.

Nelson, P. V. 1998. Greenhouse construction, heating, and cooling. In: *Greenhouse Operation and Management*, 5th Edition. Upper Saddle River, NJ: Prentice-Hall, Inc., pp. 35–170.

Robbins, J. and G. Klingaman. 2000. Irrigation water for greenhouses and nurseries. *Arkansas Coop Ext Fact Sheet FSA6061*, 6 pp.

Simeonova, N. 2003. This year in greenhouse structures. *Greenhouse Product News* 13, pp. 40–43.

Yamada, Y. 2003. Choosing a covering. *Greenhouse Grower* 21, pp. 38–40.

9 Intermittent Mist Control for Plant Propagation

David W. Burger

Intermittent mist systems provide the necessary environment for the vegetative propagation of herbaceous, softwood, and semi-hardwood cuttings (Figures 9.1 and 9.2). Mist controls water loss from the cuttings by reducing leaf temperature via evaporative cooling and by raising the relative humidity around the cuttings.

An intermittent mist system can be constructed inside a greenhouse, where many aspects of the environment (light intensity, day length, and temperature) can be controlled. However, intermittent mist systems can also be constructed outside. This is often advantageous if the plant material being propagated is adapted to higher light intensities and temperatures. Whether constructed inside or outside, intermitted mist systems require a source of high-quality water under pressure (at least 40 pounds per square inch) and electricity. The basic intermittent mist system includes a daytime/nighttime clock that turns the system on (during the day) and off (during the night). When the system is on during the day, a second clock controls the frequency and the duration of misting events. This second clock is typically set so that misting occurs for 5 to 10 sec every 5 to 10 min. However, many plants form adventitious roots more quickly when the misting is kept to a minimum. This is difficult to do with time clocks since they do not respond to changes in the surrounding environment. For example, on overcast or rainy days, when the light intensity and temperature are lower, mist systems controlled by the time clock still run as if it were a bright, hot day. Therefore, alternatives have been sought to replace time clocks with other mist-controlling mechanisms that respond to changes in the environment. These devices, often called "electronic leaves," work by either depending on water to complete a circuit to activate/deactivate the mist system or on the weight of water on a simulated leaf surface (e.g., stainless steel screen). In the electric circuit completion case, while the surface of the electronic leaf is covered with a film of water, a circuit is completed, keeping the normally open solenoid valve closed. When the water evaporates from the surface of the sensor, the circuit is broken and the valve opens, allowing water to be distributed through the mist nozzles. When mist again completely covers the sensor's surface, the circuit is completed and the valve closes. Of course, this type of electronic leaf is not useful when using deionized water that does not conduct electricity well.

In the screen (balance type) (Figure 9.3), as water evaporates from the surface of the screen, it slowly rises until a threshold is reached, whereby the normally closed valve is opened. As mist is deposited on the screen's surface, it once again gains weight and lowers to a point where the valve will be turned off. Both of these electronic leaves are more responsive to changing environmental conditions and help improve the cutting's ability to form adventitious roots and reduce the use of water. Another type of intermittent mist control involves photovoltaic (solar) cells (Figure 9.4). They work by converting solar energy into electrical energy and storing the energy in a capacitor until needed. During the day, as the light intensity and temperature increase, the rate of energy conversion/storage increases, and the frequency of misting increases concomitantly. Later in the day, as the light intensity and temperature decrease, the reverse happens. Of course, during the night, the mist system is off. An advantage of using solar cells is that no additional electricity is required since the photovoltaic cell generates electricity that is stored in a capacitor until a threshold is reached. Once that threshold is reached, the stored electricity can be used to activate the solenoid valve.

EXERCISES

Two important controls that the plant propagator has over the use of intermittent mist systems are: (1) water distribution uniformity and (2) misting frequency. The following laboratory exercises will demonstrate the effects of each on the success/failure of plants to form adventitious roots.

Experiment 1. Water Distribution Uniformity

Uniform water distribution is a necessity during adventitious root initiation on herbaceous, softwood, and semi-hardwood cuttings. If water distribution from the mist system is not uniform, some plants will receive too much water, and others will receive too little. This may lead to delays and/or variation in root initiation and rooted

FIGURE 9.1 An example of an intermittent mist system while off.

FIGURE 9.2 An example of an intermittent mist system while on.

FIGURE 9.3 An example of a balance-type electronic leaf.

FIGURE 9.4 An example of a photovoltaic mist controller.

cuttings that are not uniform. A simple "catch can" test can determine how uniform the intermittent mist system is. It also serves to identify areas that receive too much/too little water so that they can be corrected or avoided and to help determine the application rate of the mist system.

MATERIALS

The following materials will be required for the experiment:

- Enough uniform containers (plastic cups, cat food cans, etc.) to place on each square foot (in a grid pattern) on the mist bench
- 25-mL graduated cylinder
- Metric ruler
- Tape measure for laying out the grid

Follow Procedure 9.1 to complete this experiment.

ANTICIPATED RESULTS

The coefficient of uniformity (CU) will be between 0 and 100. If it is less than 80, then the distribution of mist over the evaluated area is less than satisfactory. The application rate will be in units of inches per hour or centimeters per hour and can give the propagator an idea of whether they are applying too little or too much water to the cuttings. This decision will be helped by including results from the second experiment focused on misting frequency.

QUESTIONS

- What was the calculated CU? If it were less than 80, what do you think was the cause? Are any of the mist emitters plugged or misaligned?

Procedure 9.1
Evaluating Water Distribution Uniformity in a Mist System

Step	Instructions and Comments
1	Place the uniform containers in a grid pattern over the entire surface of the mist bench, spacing them approximately 1 ft apart (30.5 cm). A regular pattern helps determine high/low water delivery areas after the test.
2	Activate the mist system for a known length of time. This will probably be between 5 and 15 min. The goal is to catch more than a few milliliters of water, but not more than 25 mL in the containers spaced over the bench.
3	Stop the system after a known period of time and measure the volume of water in each container. Calculate the mean volume (*M*) in the containers by adding all the individual volumes and dividing by the number of containers (*n*).
4	Calculate the surface area of the container (*A*) by first measuring its diameter using a metric ruler. Since the radius is one-half of the diameter, you can then calculate the surface area by using the following formula: surface area = $\pi \times$ radius^2, where the value of π is 3.14. By knowing the surface area of the container in square centimeters, and by measuring the volume of water in it in cubic centimeters (remember 1 cm^3 = 1 mL), you can calculate the depth of the water in centimeters.
5	Next, calculate the average absolute deviation from the mean (*D*) by taking the volume of each container and subtracting from it the mean volume (*M*) determined in step 3. Take the absolute values of these deviations from the mean, add them together, and divide the sum by the number of containers (*n*) to get the mean deviation (*D*).
6	Now, you can calculate the CU for the mist system, which gives you a measure of how uniformly your mist system is distributing water, using the following formula and values you calculated in steps 3 and 5: $$\text{CU} = 100\,(1 - D / M)$$ where *D* is the average absolute deviation of the mean (calculated in step 5), and *M* is the mean volume of the containers (calculated in step 3).
7	The application rate for the mist system can now be calculated by dividing the average volume of water (*M* from step 3) from the test (in mL or cm^3) by the surface area (*A* from step 4) of the container (in cm^2) and then dividing that result by the number of minutes the mist system was on. The result is the rate of mist application (in cm/min). Multiply this value by 60 to obtain the mist application rate per hour.

- How do you think air currents in/around the mist bench could affect the CU?
- What effect do you think water pressure would have on droplet size from the emitter and the resulting mist distribution?

Experiment 2: Misting Frequency and Its Effect on Rooting

In general, most plants root better when misted just enough to prevent desiccation and maintain an optimum air temperature around them. All too often, cuttings under intermittent mist systems are given too much water either by having the mist on too long or too often (or both), or left on during nighttime hours. Interesting results can be obtained by experimenting with different mist frequencies and times.

MATERIALS

The following materials will be required for the completion of the experiment:

- At least two mist benches under separate timer control
- Cuttings from plants that root easily (e.g., coleus, chrysanthemum, etc.)
- Ruler

Follow Procedure 9.2 to complete this experiment.

Procedure 9.2
Misting Frequency and Its Effect on Rooting

Step	Instructions or Comments
1	Place cuttings in flats containing a porous rooting medium, such as 2 perlite:1 peat, on mist benches of varying misting frequencies and times.
2	One such combination of treatments could be the following: one mist bench set at 5 sec every 5 min and another set at 5 sec every 10 min.
3	After 1 to 2 weeks, assess the rooting response (percentage of cuttings rooted, number of roots per cutting, and mean root length).

ANTICIPATED RESULTS

Calculate the rooting percentage by dividing the number of cuttings rooted by the number of cuttings started and multiplying the result by 100. Calculate the average root length for the cuttings of each species under each misting frequency tested. Cuttings rooted under a higher misting frequency (mist applied for longer periods of time and/or more often) will likely have a lower rooting percentage and average root length. This is most likely due to too much water in the rooting medium, which interferes with oxygen diffusion to the stem/root surface. This effect is more pronounced when a rooting medium with a high water holding capacity is being used.

QUESTION

- Were there differences in rooting percentage and root length from the species and frequencies tested? If so, what are some of the possible causes? If no, why?

10 Substrates and Containers for Seed and Cutting Propagation and Transplanting

Neil S. Mattson and Amy Fulcher

Concept Box 10.1

- Substrates facilitate the growth and development of plant roots. As such, substrates must store and deliver water, nutrients, and air to the roots.
- Several physical, chemical, and biological properties of substrates can be determined and used to assess the potential suitability for plant propagation and transplanting.
- Containers, along with substrates, substrate handling, and irrigation must be managed properly to produce healthy, high-quality greenhouse and nursery crops.
- Container size and shape vary with the market. The use of larger individual containers has increased for bedding plants, herbs, and vegetable transplants.
- Plastic flats, cell packs, and individual pots have dominated the market in recent times, although wood fiber, peat (mostly Jiffy Pot® products), and other non-plastic containers have been used.
- The interest in biodegradable containers and other alternatives to plastic containers is increasing.

HISTORICAL CONTEXT/INTRODUCTION

The earliest agricultural efforts by humans were undertaken in field soils. In the field, soil serves as a large reservoir for moisture, air, nutrients, and anchorage for the plant. More recently, plants have been cultivated in containers. Containers were used for growing certain plants in ancient Egypt. In Athens, Greece, 2500 years ago, it was common to grow plants in containers, and several fast-growing herbaceous plants were grown in containers for festivals. Containers restrict roots to a potentially very small volume as compared with field soils, thereby leading to a smaller reservoir for moisture, air, and nutrients. As we will see later in this chapter, this has a large effect on moisture dynamics and also means that conditions can fluctuate much more rapidly in a container than in the field.

The term "substrate" refers to the components that are used to fill the container and surround plant roots. Alternative terms are "container medium" (plural is media), "potting medium," and "potting mix." Many soils used by themselves make poor container substrates. Over time, growers made improvements to container substrates initially by adding amendments to soil such as aged animal manure, decomposing leaves, sand, and wood ashes. In the mid-19th century, scientists began to determine the characteristics that were important for container substrates. For example, in 1855 John Lindley stated that potting soil should have satisfactory water holding capacity, aeration, water permeability, and fertility, and should be lightweight. Because the original container substrates were soil based, there were problems with uniformity. What worked well for one person may not have work well for others who used a different type of soil. This is because soils vary in their chemical and physical properties. Also, diseases and weeds can be spread easily in the soil.

In the late 1940s, containers, rather than in-ground production, became commonplace for nursery crops. Pine bark became the primary component used in nursery mixes by the mid-1960s. In the 1950s, researchers at the University of California at Los Angeles developed a widely used soilless container substrate, called

"UC Mix," which was initially comprised of 50% fine sand and 50% peat amended with fertilizer nutrients. In the 1960s, scientists Boodley and Sheldrake at Cornell University developed a series of recipes with a base of 50% peat and 50% perlite or vermiculite. Many variations of the UC and Cornell mixes were developed. These components have formed the basis for substrate components used in greenhouse and nursery production. As we will see later in this chapter, the type of materials currently used in substrates is much broader than the few components noted above.

SUBSTRATES

What Functions Do Substrates Serve?

The substrate in the root zone of plants must serve both as a reservoir and as a conduit of water, oxygen, and 14 essential mineral nutrients. The substrate also anchors the plant into the container. In addition, the biological community associated with the substrate can also affect the plant's ability to extract water or nutrients from the substrate or help/hinder the plant with regard to root diseases. Field soils supply a relatively large volume for the roots to access these important resources, whereas container-grown plants typically have a much smaller root volume. Plants are often (although not exclusively) propagated in containers with soilless substrates. Following propagation, plants are often grown in the greenhouse and nursery in containers with soilless substrates. Many greenhouse and nursery operations choose to purchase commercially prepared "off-the-shelf" potting mixes. Larger operations may create their own potting mix by blending several components together on a "soil mixing line." This section will cover the common components of soilless substrates, the characteristics to consider when choosing and mixing substrates for a particular application, and how substrate choice impacts irrigation and fertilization.

Important Characteristics of Substrates

Substrates can be examined based on several measurable properties to determine their suitability for plant propagation and growth. These properties can be divided into the following three broad categories: physical, chemical, and biological properties.

Substrate Physical Properties for Propagation and Transplanting

The bulk density (ρ_b, g/cm^3) of a substrate is defined as its dry mass (g) per unit of volume. The volume of the substrate is measured in its moist state; this is because wet substrate may settle or compress within a container. Therefore, to measure bulk density, moist substrate is used to fill a container with known volume; the substrate is later dried down completely and weighed. The typical ρ_b of sands varies from 1.4 to 1.8 g/cm^3; organic substrate amendments, from 0.1 to 0.5 g/cm^3; and common substrate container mixes, from 0.5 to 1.1 g/cm^3. The bulk density has practical implications in terms of transporting substrates or containers with substrates. A substrate with a high ρ_b will be heavier and will require more fuel to transport/lift.

The ability of a substrate to hold water and air is dependent on the size of the particles that make up the substrate. A finely textured substrate (that is, one with small particle size) tends to retain more water, whereas a coarsely textured substrate (one with large particle size) will have a large amount of air-filled pores. When a substrate is used to fill a container, only a portion of the container volume is filled by the substrate particles themselves; the remaining space is filled with water or air. The total porosity (E, %) is the percent by volume of a substrate that is occupied by water or air (i.e., not by substrate particles). When a container is watered, much of the remaining substrate void space becomes filled with water, and the remaining space is occupied by air. Two opposing forces act on the water in a container. Gravity acts to pull water down and out of the holes at the bottom of the container, whereas the capillary action of the substrate acts to pull water up. "Capillary action" refers to the ability of water to move into narrow spaces due to the forces of cohesion and adhesion. The force of gravity is dependent on the height of a container and, thus, the height of the substrate column. The capillary rise is a function of the pore size, with a greater force as pore size decreases. When a container is watered to saturation, water stops flowing out when the gravity pushing the water down is balanced by the capillary action pulling the water up. At this point, the substrate is said to have reached container capacity, which is defined as the maximum amount of water that a substrate can hold in a container with a given height following drainage.

At container capacity, the total porosity (E) of a substrate can be separated into the fractions comprised of water and air. Volumetric moisture content (P_v, %) is the percent by volume of water in the substrate at container capacity. The air-filled porosity (E_a, %) is the percent by volume of air in the substrate at container capacity. To summarize, $E = P_v + E_a$. For example, let us say that a given container holds 1 L (1000 mL or 1000 cm^3) of substrate. When the container is filled with a particular substrate and watered to container capacity, it holds 400 mL of water and 150 mL of air in the pore space. In this example, P_v = 40% (400 mL water/1000 mL substrate volume), E_a = 15% (150 mL air/1000 mL substrate), and E = 55% (40% [P_v] + 15% [E_a]). Table 10.1 provides the definitions and units for these terms.

TABLE 10.1
Abbreviations and Definitions for Substrate Physical Properties

Abbreviation	Definition
E	Total porosity (%); the percent by volume of pore space in a given volume of substrate at container capacity; $E = P_v + E_a$
P_v	Volumetric moisture content (%); the percent by volume of pore space that is filled with water at container capacity
E_a	Air-filled porosity (%); the percent by volume of pore space in a substrate filled with air at container capacity
ρ_b	Bulk density (g/cm^3); the dry weight of the substrate per unit of volume

At container capacity, there is a zone at the bottom of the container in which all pores are filled with water. The saturation zone can be thought of as a perched water table in the container. Within the saturation zone, there is very little air for roots; above this point, pores hold a mixture of air and water. Gradually, air can enter the saturation zone as plant roots absorb water and air fills the once water-filled pore space. The height of this saturation zone depends only on the pore size, which is a function of the particle size (Table 10.2; Figure 10.1). For example, when very fine sand with a diameter of 0.002 cm is used to fill a container, there is enough capillary rise to create a saturation zone of up to 147 cm (Table 10.2), whereas when very coarse sand is used (0.2 cm diameter), the saturation zone will be at 1.47 cm. The height of the saturation zone (or perched water table) does not depend on the container size or height (Figure 10.2). Above the saturation zone, the proportion of air and water in the pore space depends on the container height. For example, think about the sponge exercise (Experiment 1); the saturation zone does not change regardless of whether the sponge is lying flat versus upright at its tallest length. However, the proportion of air and water above the saturation zone does change. At the tallest sponge configuration, there is a greater force of gravity pulling the water down, so there is a greater air space above the saturation zone.

It is important to select a substrate with an appropriate balance of E_a and P_v for the height of the container to be used (Experiment 2 describes how to measure physical properties). The air space is important as roots require oxygen for respiration and growth. If normal root growth cannot occur, the plant will have problems taking up enough water and nutrients to support the shoot. Symptoms of poor aeration may initially include less vigorous plant growth and smaller leaves. More advanced symptoms of poor aeration include yellow chlorotic leaves (symptomatic of nutrient deficiencies), poor root growth/shallow growth (root growth only toward the top of the container where there is air), leaf abscission, or wilting of leaves in the early afternoon when leaf transpiration is the greatest. On the other hand, if a substrate holds too much air, then its water holding capacity (P_v) will be low, which means that a container must be rewatered frequently; otherwise, the plant will dry out. A substrate with too low of a water holding capacity may also more readily leach nutrient and pesticide-laden water. A general guideline is that a substrate in a particular container should have a total porosity greater than or equal to 50%, a volumetric moisture content greater than or equal to 40%, and an air-filled porosity greater than or equal to 10%. Substrates that have high air-filled porosity and low volumetric water content will need to be irrigated frequently to supply the plant with enough water. Substrates with low air-filled porosity may need to be irrigated sparingly so as to avoid waterlogging of the substrate. A common misconception is that a poorly drained substrate can be improved if a layer of coarse gravel or rocks is placed at the bottom of the container. Ironically, this just exacerbates the situation. The layer of gravel/rock raises the bottom of the container, in effect reducing the height of the container and raising the saturation zone higher in the container.

Many soils drain adequately in a field setting yet drain poorly in a container. This is because, in a field setting, there may be a much greater soil depth (and, thereby, force of gravity) than in a container. Similarly, a deep soil

TABLE 10.2
The Height of the Saturation Zone in a Container Depends Only on the Pore Size (Which Is Determined by Particle Size)[a]

Pore Radius (cm)	Equivalent Particle Diameter (cm)	Height of Saturation Zone (cm)
0.1	0.2 (very coarse sand)	1.47
0.01	0.02 (fine sand)	14.7
0.001	0.002 (very fine sand)	147

[a] This equation is height of saturation zone (in cm) = 0.147 / pore radius (in cm).

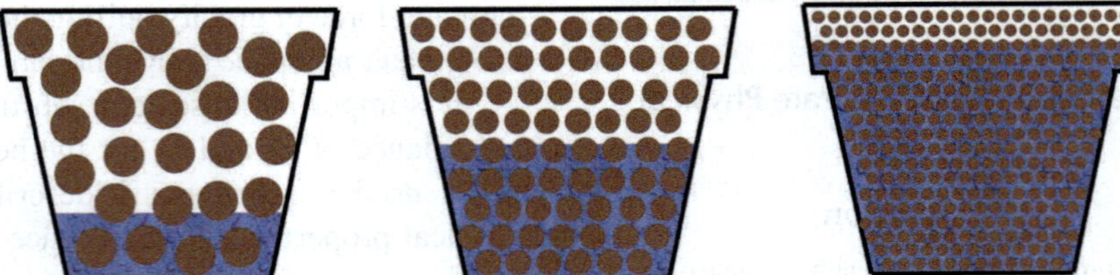

FIGURE 10.1 The particle size (pore size) of the substrate, not the container height, affects the height of the saturated zone in a container.

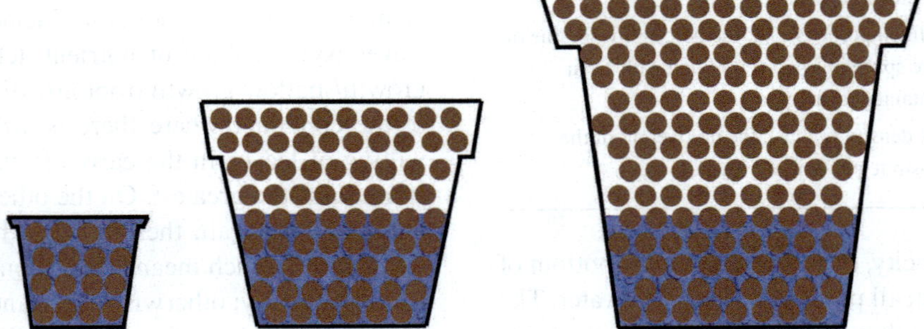

FIGURE 10.2 The saturated zone height is not influenced by the container height (rather, it depends on pore size).

profile means that there will be a substantial capillary action pulling the water down and away from the surface. Although in a container, once the water reaches the bottom, there is no additional capillary force to pull the water down, and so pores in the substrate may only pull the water up rather than aid drainage.

Substrate Chemical Properties for Propagating and Transplanting

The term "substrate solution" (similar to the term "soil solution" in field soils) refers to the water and its solutes (consisting primarily of mineral ions as well as dissolved gases such as O_2 and CO_2, and organic compounds) within the substrate. The surface of substrate particles carries electrical charges that have the ability to interact with ions in the substrate solution. These surface sites tend to be negatively charged so that they primarily bind with cations such as H^+, K^+, Ca^{2+}, Mg^{2+}, and NH_4^+. The binding or exchange sites typically do not permanently bind with a particular molecule/atom; rather, this can change over time in response to the ion composition of the substrate solution. The cation exchange capacity (CEC, meq/100 cm^3) of a soilless substrate is a measure of the maximum quantity of cations (expressed in milliequivalents) that can be held per volume of substrate (e.g., meq per 100 cm^3). A high CEC is useful in a substrate as the substrate will be more buffered from rapid changes in pH or loss of cations through leaching.

The pH of the substrate is important as it impacts nutrient availability/solubility. The availability of phosphorus, iron, manganese, zinc, copper, and boron decreases markedly as pH increases above 6.5. On the other hand, calcium and magnesium availability declines at pH below 5.5. Based on the above problems, excessively high and excessively low substrate pH should be avoided. For most plants, a substrate pH of 5.5 to 6.5 is ideal for rooting, seedling growth, and continued production in larger containers. Certain plants such as *Ericaceous* species (cranberry, blueberry, azalea, rhododendron, etc.) are adapted to low-pH soils. For these species, a lower substrate pH such as 4.5 to 5.5 is necessary for optimum growth. The pH and CEC of common substrate components are listed in Table 10.3. Some substrate particles such as sand or perlite are nearly chemically inert, which means that they have very few exchange sites and contribute very little to the pH or CEC of the substrate.

Soluble salts include all of the mineral ions in the substrate, which are dissolved in water. This includes fertilizer nutrients, as well as salts that are impurities in water or fertilizers (such as sodium or chloride). Soluble salts are typically measured indirectly using an electrical conductivity (EC) meter. EC meters are handheld probes that measure the passage of electrical current through a solution. The more soluble salts present in the water, the greater the conductance of an electrical current through the solution. Low EC indicates a potential problem with low fertility. High EC indicates excess fertility or a high amount of impurities in the substrate or water source. Several waste-derived substrate components may be high in salts such as poultry litter composts or spent mushroom compost. High salts can inhibit or reduce germination. Roots of young seedlings and cuttings, especially herbaceous cuttings or

TABLE 10.3
Chemical and Physical Properties of Some Common Substrate Components

Property	pH[a]	EC[a] (dS m^{-1})	CEC (meq· 100 cm^{-3})	Bulk Density (g cm^{-3})	Air-Filled Porosity (%)	Volumetric Moisture Content (%)
Sand	7.8[b]	0.10	Negligible	1.70	27	9
Perlite	7.5[b]	<0.10	0.15	0.12	32	24
Vermiculite	Variable	0.10	2–3	0.10	28	46
Sphagnum peat	3.5–4.2	0.10	15	0.11	30	58
Pine bark	4.0	0.1	8	0.20	46	32
Coir	5.0–7.0	1.0	>20	0.1	29	56
Turkey litter compost	8.7	4.1	30	0.31	45	28

[a] pH and EC measured in 1 substrate: 2 water (v/v) extracts (see Experiment 4).

[b] Perlite and washed sand free of carbonates are relatively inert materials that will not contribute greatly to substrate pH when blended with other components.

slow-rooting hardwood cuttings, are also typically sensitive to high salts. High salt concentrations can cause cell desiccation from osmotic stress or cell damage due to salt accumulating to toxic levels; informally, this is called plant "salt burn." High EC can be corrected by drenching the substrate thoroughly with clear water to leach out excess salts. Monitor EC and repeat leaching as necessary.

EC and pH are not only a function of the substrate components, but can change during production as a function of the water source or fertilization practices. Besides testing substrate components or potting mixes prior to planting, it is also a good practice to test the pH and EC periodically during crop development. There are several methods for testing substrate pH and EC. The methods utilize different volumes of distilled water, and therefore, EC guidelines specific to the testing method must be used (Table 10.4). Detailed information on the pour thru method for pH and EC testing and interpretation is available from Cavins *et al.* (2000). More information on the 1 : 2 dilution testing method is given in Experiment 4.

Substrate Biological Properties for Propagating and Transplanting

Several types of organisms may also be associated with substrates; their presence is referred to as the "biological activity" of the substrate. Bacteria, fungi, and water molds in the substrate may be beneficial, harmful, or neutral to plants being propagated. Water molds (oomycetes) such as *Pythium* or *Phytophthora* can infect roots, leading to poor health and, oftentimes, mortality of infected plants. Other microorganisms associated with the substrate may be beneficial, producing, for example, antibacterial or antifungal agents or hormones that induce rooting (auxins), as well as other vitamins and growth-promoting substances (Hoitink *et al.*, 1997). Mycorrhizal fungi can

TABLE 10.4
EC Interpretation Values for Three Different Substrate Extraction Methods[a,b]

1:2 Dilution	Pour Thru	Saturated Media Extract	Interpretation
0–0.25	0–1.0	0–0.75	*Very low*. Nutrient levels may not be sufficient to sustain rapid growth.
0.26–0.75	1.0–2.6	0.76–2.0	*Low*. Good for germination. Suitable for light feeders and salt-sensitive plants.
0.76–1.25	2.6–4.6	2.0–3.5	*Normal*. Standard range for most established crops actively growing. Upper range for salt-sensitive species.
1.26–1.75	4.6–6.5	3.50–5.0	*High*. Reduced vigor and growth may result, particularly during hot weather.
1.76–2.25	6.6–7.8	5.0–6.0	*Very high*. Salt injury may occur; reduced growth rates are likely.
>2.25	>7.8	>6.0	*Extreme*. Most crops will suffer salt injury at these levels. Immediate leaching is required.

[a] Units: mS/cm = dS/m = mhos/cm (these three units are equivalent to each other).

[b] Adapted from Cavins *et al.*, 2000. Monitoring and managing pH using the pour thru extraction method. Horticulture Information Leaflet 590:17. Available online: http://www.ces.ncsu.edu/depts/hort/floriculture/hils/HIL590.pdf.

associate with plant roots and can more efficiently extract water and nutrients that are then supplied to the plant. The plant, in turn, provides carbohydrates and sugars to the fungi. Microbes may be already established in substrate components, or they may become introduced over time. The presence of a strong microbial community can make it difficult for pathogenic organisms to become established in the substrate. For example, the process of composting can create a diverse microbial community that may be disease suppressive (Jack *et al.*, 2011). An increasing number of commercial products are also available to introduce microbes to substrates. Examples include bacteria (such as *Streptomyces*) or fungi such as (*Trichoderma*), which can suppress plant pathogens.

Beyond microorganisms, other biological components of substrates include insects, weed seeds, and nematodes. When soil or other substrate components that may contain harmful organisms are used, they should be pasteurized by heating to 60°C for 30 min. This temperature kills most pathogens, weed seeds, and insects but leaves behind beneficial fungi. Higher-temperature sterilization kills all organisms in the substrate, but this can lead to the rapid subsequent proliferation of disease organisms. Steam pasteurization is most commonly used, but is expensive due to the heating energy required. Problems that can occur with pasteurization include the toxicity of manganese and ammonium.

Common Substrate Components and Their Qualities

To achieve a balance of desired physical, chemical, and biological properties, most container substrates are a combination of several components (amendments). The components listed below are combined to form the primary bulk (or volume) of the substrate. The ideal bulk components should be locally available, inexpensive, and uniform from batch to batch. Listed below are some of the more common organic and inorganic amendments to substrates. Here, "organic" refers to those components that are carbon based; most are essentially some form of plant or animal by-product. One problem with organic amendments is that, over time, they may break down and take up less volume, which is termed as "substrate shrinkage." Besides being a nuisance, this effectively reduces the substrate air-filled porosity, which can lead to inadequate substrate aeration. Inorganic amendments are primarily materials that are mined from the earth and then further processed. Many inorganic materials are added to substrates because they improve the air-filled porosity of the substrate and do not break down over time.

ORGANIC AMENDMENTS

Peat is decomposed sphagnum moss that forms in bogs where plant residue accumulates faster than it decomposes (Figure 10.3). As plant material accumulates, it sinks into the water, where low oxygen levels slow the rate of decomposition. Most horticultural peat comes from sphagnum moss, but other types come from hypnum moss, reeds, sedges, and other slowly decomposing organic matter. Sphagnum peat is acidic (pH, 3.5–4.0). Peat is among the most commonly used ingredients in container substrates because it has high water holding capacity, total porosity, and CEC. It is lightweight and can be compressed for shipping. Mechanical damage from overgrinding/overhandling peat can break apart fibers; this decreases its porosity in a container.

FIGURE 10.3 Sphagnum peat moss.

Coconut coir is derived from the husks of the coconut fruit and is a by-product of coconut production (Figure 10.4). The husks are shredded, and the long fibers are used for the production of brushes, mats, and twine, whereas the shorter fibers, previously a waste material, are now used as container substrates. The physical properties of coir are similar to peat, but the pH is higher (5–7). Coir often has a high concentration of sodium and potassium salts, which may need to be leached out prior to use. Coir is often shipped in compressed bricks that expand 6 to 8 times (by volume) when water is added.

Tree bark is a by-product of the lumber industry and has long been used as a substrate component in nursery and greenhouse production. The physical and chemical properties of bark and wood amendments vary based on the particular source as well as on how finely ground

FIGURE 10.4 Coir is derived from shredded coconut husk.

it is and how long it has been aged/composted. The pH of coniferous bark is typically quite low (Table 10.3). While bark is relatively slow to decompose due to lignin in the bark cells, wood products can decompose rapidly. During decomposition, microorganisms can consume much of the substrate nitrogen, thereby limiting nitrogen availability to plants. To get around this problem, wood products should be composted or aged before use. Due to space and time requirements, some operations will forego this process and instead add additional nitrogen to the substrate prior to planting. Irrigation practices may also have to be adjusted for fresh pine bark–based substrates. Fresh and aged materials can have substantially different chemical properties; therefore, testing materials when a new shipment arrives as well as after prolonged storage is recommended. A relatively recent development in the use of wood materials is whole tree (WT) substrates. WT is composed of the entire aboveground parts of a pine tree, including wood, bark, needles, and branches, which are chipped and then ground with a hammer mill and screened to the desired size. WT has been used successfully as the sole bulk component of substrates, along with some fertilizer amendments. WT that has been aged for at least 90 days has been reported to lead to improved plant growth as opposed to freshly used material.

Rice hulls are a by-product of threshing rice. Fresh rice hulls can be contaminated with weed seeds. Therefore, rice hulls have been traditionally composted or burnt prior to their incorporation in a substrate. More recently, parboiled (or high water temperature–treated) rice hulls have been used in substrate mixes. The hulls are 6 to 10 mm long by 1 to 2 mm wide (Figure 10.5). The relatively large particle size of the hulls makes them a useful amendment to improve substrate air porosity. When used at 20 to 25 percent by volume in the substrate, they impart a container substrate with physical properties similar to perlite (Evans, 2008).

Compost is a humus-like product that develops when organic matter is subjected to a controlled decomposition process. During composting, organic matter increases in temperature, killing most pathogens, weed seeds, and arthropods. Common feedstocks for composts include animal manure and bedding, tree leaves, grass clippings, food wastes, and many other plant- and animal-based by-products. When used in container substrates, composted materials should be composted to maturity, in the sense that they are well-processed and decomposed by microbes so that they are relatively stable in structure. Immature composts can contain high salt levels as well as give off ammonia and volatile organic compounds that can injure plants. A broader problem with the use of some compost is the inconsistency of the material from batch to batch based on the feedstocks used and the procedures followed for compost. Nevertheless, many professional composting operations exist, and the use of composts to replace a portion of the peat or other organic components is becoming increasingly common. Testing each batch of compost is prudent to ensure its suitability for propagation and transplanting. An ideal compost for container use would have a C:N ratio of 15 to 25 : 1 (indicating that it is relatively mature), an EC of 1 to 3 dS/m as measured by the saturated media extract method, and a pH of 6 to 6.5 (Sideman, 2007). A relatively new tool for growers is the Solvita® compost maturity test kit, which determines a maturity index based on carbon dioxide and ammonia release from the compost sample. University and commercial laboratories can also test compost to determine its suitability for use in container substrates.

FIGURE 10.5 Rice hulls (elongated tan particles) are used to improve air porosity in potting mixes.

Most composts are finely textured and have low air-filled porosity. Therefore, they should not be used as the sole component of a substrate as they will be too waterlogged. Aggregates such as perlite or rice hulls are necessary to improve the air-filled porosity of compost-based substrates. One compost-based substrate recipe that has been successfully used by the University of Maine (Burnett, 2009) blends 1 : 1 : 1 (by volume) sphagnum peat:coarse perlite:compost along with a fertilizer starter charge incorporated into the mix. The composting processes usually raises the pH, which can reduce the need to add lime to the substrate potting mix when compost is added. The use of vermicomposts, or worm-processed composts, is becoming increasingly popular in organic substrates. The worms digest and fragment the feedstock, leading to a nutrient-enriched material that can also foster a beneficial microbial community (Jack *et al.*, 2011).

INORGANIC AMENDMENTS

Perlite is a mined volcanic material that is prepared by crushing into small particles and heating to very high temperatures. As the water inside the particle vaporizes, the particle puffs up to create a lightweight expanded particle. Perlite is used to improve the air-filled porosity of substrates. Perlite is impermeable; there is no exchange of air or water with the inside cavities of perlite particles. Therefore, the particles do not add to the total porosity or volumetric water content of substrates. Perlite is chemically inert, that is, it does not contribute greatly to substrate CEC or pH. Perlite is physically stable, which means that its volume does not decrease over time unless it is subjected to severe physical compression.

Vermiculite is a mined mineral material composed of sheets of alumina held together by calcium or magnesium. The material is ground, and particles are heated to very high temperatures. Water trapped between the layers is turned into vapor, which expands the layers. As compared to perlite, vermiculite has a relatively high CEC. In addition, water is adsorbed by alumina layers; therefore, vermiculite increases substrate total porosity and volumetric water content. Coarse vermiculite is used to increase substrate porosity, whereas finely ground vermiculite is often applied as a thin layer to cover seeds in germination trays, thereby keeping the seeds uniformly moist.

Expanded clay pellets (grow rocks) are produced by heating naturally mined clay in a rotating kiln to temperatures of about 1200°C. The process produces round pellets of about 1 cm in diameter. Similar to perlite, the pellets are hollow and do not interact with outside air or water. The pellets are chemically and physically inert. Due to their large size, the pellets provide great aeration but low moisture holding capacity. Frequent watering is required to keep roots from drying. Expanded clay pellets are primarily used in hydroponic production.

Rockwool is a mixture of coke, basalt, and limestone heated to 1600°C and then spun into fibers similar to fiberglass. The fibers can be spun into different shapes, and the fineness of spinning can be adjusted to influence its physical properties. Rockwool is essentially chemically inert but can be engineered to have excellent physical properties, including both high water holding capacity and aeration due to the portion of micro- and macropore spaces. Rockwool cubes are often used to germinate the seeds or root cuttings of plants that are intended for hydroponic production (such as greenhouse vegetables). The cubes are then placed onto larger rockwool slabs that are drip irrigated with a nutrient solution.

Several other components may be added to substrates, such as powdered limestone to raise the pH of the substrate, if desired; fertilizers to provide a "starter nutrient charge"; and sometimes, microbial inoculants to facilitate plant nutrient availability and absorption or to reduce the incidence of root-borne pathogens. In addition, wetting agents are often added to peat-based substrates. When peat becomes dry, the fibers are strongly hydrophobic or water repellant. Wetting agents are surfactants that decrease the surface tension of water, making it easier to rewet dry peat.

SUSTAINABLE AND ORGANIC SUBSTRATES

In the last decade, there has been an increased interest in containerized plants that are grown sustainably or organically. Many of the commonly used substrate amendments can be used in the U.S. Department of Agriculture (USDA)–certified organic production, including, but not limited to, peat, bark, perlite, vermiculite, limestone, and compost prepared following USDA National Organic Standards. However, conventional fertilizers and wetting agents are not allowed.

Regarding sustainability, some operations are actively seeking to reduce the use of mined or nonrenewable substrate components. The sustainability of peat in substrates is in question. The development of peat bogs requires hundreds or thousands of years of plant growth and may be considered nonrenewable. Bogs must be partially drained to harvest peat, which leads to a more rapid decomposition and release of carbon dioxide to the atmosphere. On the other hand, only a very small proportion of peat land is harvested for horticultural use. For example, most U.S. peat comes from Canada. Canada is estimated to contain 280 million acres of peat land, whereas the peat industry harvests on only 42,000 acres (Daigle and Gautreau-Daigle, 2001).

The use of organic substrate components with suitable physical or chemical properties to replace peat remains an area of active research. Another aspect of substrate sustainability relates to the leaching (or release) of nutrients such as nitrates and phosphates into the environment. Nutrient leaching can be reduced by using substrates with a high CEC, fertilizing in accordance with plant nutrient demands, and avoiding excessive irrigation.

CONTAINERS

Containers, along with substrates, substrate handling, and irrigation, are the primary factors that affect air and water status within a container and are critical factors that must be managed to profitably produce plants. Container selection is also important because containers hold substrate, dictate minimum plant spacing, and facilitate the transportation of plants. Container dimensions and the environment within the container affect root growth. Containers must maintain integrity throughout propagation and/or production and marketing. Containers must be lightweight, made from a material that is readily available, and inexpensive. For containers used for both propagation and sales, containers must hold up to loading, shipping, unloading, and the retail sales period. Consumers are increasingly interested in biodegradable containers and other sustainable container choices. Container color and design are an integral component of marketing and branding programs.

History of Containers

Until as recently as the 1970s, commercial growers used a wide array of containers, including clay pots; institution-sized tin cans, often from canned fruit, with holes punched at the bottom for drainage; and tar paper containers (Avent, 2003). The disadvantages of each varied with container type; problems included the following: being heavy and fragile, rapid moisture, loss, availability, difficulty in extracting plants, and expense. Pots were filled with soil, and general production practices were adaptations from field production, not the highly specialized and researched approach used in many greenhouses and container nurseries today. For propagation, trays were 22 to 24 in; constructed from wood, metal, or Styrofoam, with openings to allow drainage; and used in protected and unprotected outdoor growing systems (Hartmann *et al.*, 1997).

Types of Containers

Flats

Currently, container trays are almost exclusively plastic (polyethylene or polystyrene), 11 by 22 in, and are referred to as "flats." Flats are easily cleaned, are reusable, and can be easily nested and stacked when empty for space-efficient storage. Characteristics of trays include raised ridges on the bottom surface for enhanced water drainage and gas exchange, a curved top edge to assist in carrying flats and separating nested flats, and a reinforced rib to extend the lifespan and facilitate reuse. Cuttings are direct stuck into flats and transplanted into individual containers once sufficiently rooted to survive without mist. Seeds are direct sown into flats, plug trays (see "Plug trays" subsection), or multi-cell inserts (see "Multi-cell (cell packs) inserts" subsection) that are made to fit within trays. Directly sown seeds are often planted in rows and transplanted from the flats into larger containers. As compared with using plug trays or multi-cell inserts, transplanting from direct seeded flats can require added labor and may lead to transplant shock because the seedlings' delicate root systems are intertwined and are easily damaged when they are teased apart. The advantage of direct seeding into a flat or using plug trays is that many seeds are concentrated in a small area; thus, very specific germination conditions can be met for many seedlings in a precisely controlled growth chamber or greenhouse.

Plug Trays

Since the early 1990s, it has become common to sow into plug trays rather than open flats (see Chapter 37 for more information on seed propagation). Starting with plugs versus seeds shortens production time, which minimizes the amount of time that growers must heat the greenhouse, control pests, hire labor, etc. As in all cases of placing a single seed or plant per container, transplant shock is

FIGURE 10.6 (a) Various shallow injection-molded propagation plug trays include 40-cell square (top) and 72-cell round (bottom) plug flats, and six-strip 60-cell sheets with matching ribbed trays to facilitate in spacing the strips (center). (Photo courtesy of Dillen Products.) (b) A line of plug flats (square, 50–512 cells; round, 72–288 cells) designed to be rigid enough to use without a carry tray (a), and compatible with most, if not all, automatic seeders and transplanting equipment (b). (Photo courtesy of ITML Horticultural Products Inc.)

reduced. Also, some growers have indicated that they can hold plugs for a short period and transplant to larger single or multi-cell containers on demand, which is not possible with flat-sown seedlings. Holding plugs too long will halt development, and the roots will become pot bound. Using plug cells can also allow for automation, that is, transplanting machines, whereas this would be impossible for seeds scattered across a single tray. Whether a transplanting machine or manual labor is used, transplanting is faster from a plug tray. See "Single cell" subsection for more advantages of producing plants in individual cells or containers. Disadvantages of plug trays are the added expense and that it may require more space to accommodate the same number of seedlings. However, because the crop finishes faster and is easier to transplant, the advantages of plugs generally outweigh the negatives (Figure 10.6).

Individual Containers

Single Cell

Single-cell containers, generally 2.25 in and larger in size, offer the advantage of eliminating plant competition and flexibility to space containers as the crop grows. Like with plug trays compared to sowing into flats, single-cell containers often lead to greater plant growth and reduced transplant shock. Also, successful transplanting is often possible over a range of weather conditions because roots are relatively undisturbed. When plants are grown in individual containers, germination percentage can be higher, and growth is more uniform. However, substantially greater space is taken up by the containers long before the plant dimensions require the space, and small transplants can stay too wet in large containers, a condition that is conducive to root rot infection. Single-cell containers are not often used if the cull percentage is high as it is best to eliminate dead and inferior plants in the seedling transplant stage before large resources (space, container, and substrate) are expended.

Multi-Cell (Cell Packs) Inserts

Advantages are similar to single-cell containers. A disadvantage of cell packs when direct seeding is that "misses" or runts cannot be easily removed and replaced prior to sales, diminishing the value of the cell pack. While it involves an extra transplanting step compared to direct seeding, transplanting seedlings from plug trays into multi-cell inserts can help ensure uniformity.

Hanging Baskets

Hanging baskets are generally grown in 10 to 12 in containers; however, premium markets may dictate a 16 in container, and 8 in baskets have been marketed in regions where a lower price point is desirable. Hanging baskets were traditionally planted with ferns or a single type of flowering plant with a trailing habit and sold in the spring and summer. In the past several years, mixed color baskets have been offered as well as baskets filled with a single type of flowering plant such as petunia, impatiens, or calibrachoa. Hanging baskets are often plastic pots. Fiber-lined wire baskets are also offered pre-planted, whereas 15 years ago, fiber-lined baskets were only available separately, where liner, basket, and transplants were purchased individually and assembled by the consumer.

Container Composition

Petroleum Based

Plastic containers have been traditionally manufactured from virgin petroleum-based resins such as polyethylene, polypropylene, or polystyrene resins. Currently, recycled plastics are often incorporated, and some containers are made from 100% recycled plastics. Plastics may be pressure formed, thermo formed, injection molded, or blow molded. Injection-molded containers have thicker sidewalls and can be reused more than blow-molded containers, but require more plastic to manufacture and, therefore, are more expensive. Blow-molded containers are the most common in nursery production.

Alternatives to Petroleum-Based Containers

Consumers are increasingly interested in plastic alternatives and other sustainable containers. Biocontainers are not manufactured from petroleum; instead, they are most commonly fiber or paper-based, or based on biopolymers, which are generally derived from starch, cellulose, soy protein, or lactic acid (Figure 10.7). Types of bio-plastic containers include starch and starch blends, polylactide, and poly-3-hydroxybutyrate, all of which are derived from plant products. Commercially available biodegradable containers are largely made from pressed fiber and wood pulp, but rice hull, cow manure, and other biodegradable byproducts are being manufactured into containers. Consumers have a preference toward and are willing to pay more for containers made from sustainable materials (Hall *et al.*, 2010; Yue *et al.*, 2010). Another advantage of biocontainers is that, depending on the raw materials and manufacturing process, they may not be subject to the volatility in petroleum pricing.

FIGURE 10.7 Ellepots™ are one example of an alternative to petroleum-based containers; they are made from a biodegradable paper sleeve.

Biopolymer-based containers are compostable, either through traditional home composting or via industrial composting, which involves applying heat in order for composting to occur within an acceptable time frame. Fiber or paper biocontainers are home compostable or plantable. Compostable containers are removed prior to planting. Plantable containers remain on the plant during transplanting and degrade in the planting hole. Peat pots and Jiffy Pot® are commonly recognized examples of plantable containers.

Substrate-Based Containers

Substrate-based containers are another choice that growers have; the substrate binds together so that an additional outer container is not needed. Examples are Oasis® and soil blocks. Substrate-based containers are planted even if they do not readily degrade. An advantage of substrate-based containers is that they can save labor and reduce waste; however, they cannot be reused.

Implications of Using Biocontainers

Fiber- and paper-based containers are porous, unlike conventional plastic containers, and this has a widespread effect on propagation and production. The use of biocontainers or other containers with unusual features may affect irrigation. For example, more water is lost through evaporation from the container sidewalls of biocontainers than from solid sidewall plastic containers. Therefore,

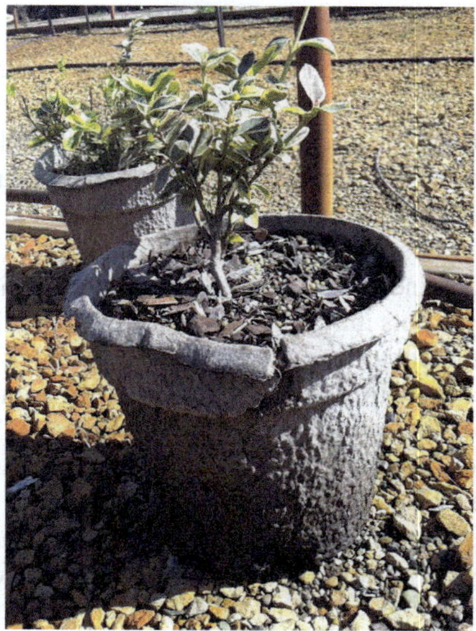

FIGURE 10.8 Before adopting a new biocontainer, test on a small scale for the ability to withstand the stress of production, transportation, and retail display. This biocontainer is beginning to crack after a season of production. (Photo courtesy of Xueni "Vickie" Wang.)

greater irrigation frequency and/or irrigation volume may be needed to produce plants in biocontainers due to greater evaporation (Nambuthiri *et al.*, 2012). Consequently, conventional plastic containers and biocontainers may not be able to be mixed within the same irrigation zone. The root zone within a biocontainer is cooler by as much as 10.8°F (Nambuthiri *et al.*, 2012), and this has been linked to increases in plant growth and survival (Fulcher *et al.*, 2012). Fiber containers are also thought to have greater air exchange than solid, plastic containers and a reduced zone of saturation at the bottom of the container (Ruter, 2000; Mathers *et al.*, 2007). Biocontainers may not withstand the force of potting machines and other automation or may begin to break down when used for crops with a long production period. A range of production tests should be done to test both wet and dry container strength before integrating a new biocontainer into production (Figure 10.8). Also, algae may grow on the pot surface, affecting aesthetics in the retail setting and container integrity.

Container Dimension

Container dimension can be designated in a number of ways. Plug trays are referred to by the number of cells per tray. Larger containers used in woody plant production often use volume to designate the container size. Historically, a 1-gal pot held 1-gal of substrate, but with standardization and to conserve substrate, they now hold somewhat less volume. Therefore, a "3-gal" pot does not technically hold 3 gal; reference to historical container size must be avoided to prevent false advertising and misrepresenting products to consumers. A more appropriate designation of container size is the standard #3 pot. Other pot sizes are designated based on the top diameter of a round container or the width of the top of a square container.

Transplanting Schedules and Matching Final Container Size to Markets

Different markets demand different container sizes. Upscale markets often prefer individual pots such as 3 to 6 in and larger containers for seasonal plants such as beddings plants, herbs, and vegetables, whereas, historically, discount stores and mass merchandizers often carry cell packs in which there is one smaller plant per cell and some number of cells per pack (typically, 4–9). This is rapidly changing as some national chains recognize consumer demand for larger plants. Container size affects not only the overall production cycle, but also production cycle components, for example, how quickly a plug is ready to transplant into the next size container. Once the plug cell is filled, the plugs are transplanted to larger, individual containers. Therefore, 128s require about 4 to 5 weeks to finish, whereas 512s, a smaller plug cell, require less time. Common seed production plug cell sizes for vegetable transplants are 128s and 288s; however, 72s are common for tomatoes and peppers, and larger seeded plants such as cucurbits are often seeded into 48s.

Timely transplanting to the appropriately sized container is essential to crop success (see "Circling Roots" section). Remaining in a small container too long, that is, becoming "pot bound," can lead to small, constricted root systems and delayed or reduced aboveground growth, as well as reduced growth and girdling tree roots, once transplanted. Roots of pot-bound plants fill the pore space in the substrate, reducing the ability of the substrate to hold enough water (Figure 10.9). Transplanting directly into a larger container or transplanting earlier can lead to larger plants that may establish more readily in the landscape.

FIGURE 10.9 The root system of a pot-bound plant fills much of the substrate pore space that previously held water, reducing the ability of the substrate to serve as a reservoir. Pot-bound root systems often contain circling roots.

Container height will affect the substrate, air, and water status of the root zone. A taller container will have a greater volume that is not within the saturated zone and, thus, available for root growth. A squat or short pot, often designated with an "S," has a volume similar to the corresponding conventional size container, but a larger portion of the container volume is within the saturated zone. These shorter-profile containers are not recommended for crops intolerant of high moisture.

Both round and square containers (as viewed from above) are used. Both may have specialized flats to help support the containers during propagation and production, and in the retail marketplace. Round containers are popular, but square containers allow more root volume per flat or bench space, are better adapted to mechanization, and permit less waste from automated or manual potting.

Container Features and Root Growth

Circling Roots

Circling roots are a problem in container production and must be prevented to maximize the advantage of container-grown plants. Roots grow where there is sufficient oxygen and water. In containers, these conditions are often met at the interface between the substrate and the sidewalls. Roots often reach the sidewalls and then follow the sidewall horizontally. Roots circle the interior of the container sidewall in both round and angular-shaped containers. During propagation, extra effort must be made to prevent circling roots as they persist in later phases of propagation and production (Figure 10.10). This circular pattern of root growth is especially significant in woody plants because the trunk can eventually become girdled, affecting growth rate, plant health, and even survival. Also, roots at the interface are particularly susceptible to heat damage when containers are moved and previously shaded sides of the containers are suddenly exposed to high radiant heat loads.

Root Pruning Containers

Without root pruning, roots often circle the container and have little branching or secondary roots. Some containers use design features, and others are chemically treated to prune roots and prevent circling. Traditionally, container sidewalls have been solid; however, containers may have sidewalls with pores or slots that prune the roots. These containers are designed to prune roots in one of two ways. Roots may grow through small, circular openings in the container and are girdled, pruning the root as it increases in width, or roots are air pruned when they grow through the openings in sidewalls. The root tips desiccate and die because of exposure to air. When roots are pruned, secondary roots form behind the dead root tip, which increase water and nutrient uptake and facilitate the establishment in the landscape or larger container. Accelerator® and RootMaker® containers are readily available root pruning containers (Figure 10.11).

Containers can also be treated to chemically prune roots (Figure 10.12). For example, the inside of the container can be coated with latex containing copper hydroxide, widely sold as SpinOut®, to prevent circling roots and to encourage root branching. The coating thickness is adjusted based on how aggressively rooting a crop is. Flats are often coated with a slipping agent to facilitate pulling nested trays apart. This coating may interfere with the adhesion of copper hydroxide products.

Container Temperature and Color

In black containers facing the sun, substrate temperatures can be greater than 120°F, yet temperatures as low as 100°F can damage roots. The color and composition of the container can affect root zone temperature (Markham *et al.*, 2011; Nambuthiri *et al.*, 2012). While color plays the greater role, the effect of radiant heat is mitigated by porous containers due to greater cooling by evaporation or the latent heat of vaporization. The ability of containers to absorb or reflect radiation is an important factor in heat buildup in a substrate within a container. Light-colored

FIGURE 10.10 Circling roots that form during propagation persist into the later phases of production and into the landscape where they can become stem-girdling roots.

FIGURE 10.11 (a) Examples of root pruning propagation trays, top to bottom: RootMaker® 60 cell tray, RootMaker® 32 cell tray, RootMaker® 18 cell tray, and the RootMaker® Express 18. (b) Oak seedlings were transplanted from a traditional plug cell (left) and a RootMaker® 18 cell root pruning propagation tray (right) to a 11.4 liter container (#3). After 20 days seedlings were harvested to reveal the pattern of root growth. Notice how the dimension of the original propagation container continue to influence root development following transplanting. (Photo courtesy of Dr. Carl Whitcomb, Lacebork, Inc.)

containers have a greater ability to reflect radiation than darker containers, so less heat from solar radiation is absorbed. Some black containers are coated with white paint or placed within white fabric sleeves to reduce the amount of radiant heat that is absorbed by the containers.

FIGURE 10.12 SpinOut was used to chemically prune the roots of this #15 red maple, creating fibrous root mass.

Containers covered in white are approximately 1 to 7°C (2–13°F) cooler than black containers (Whitcomb and Whitcomb, 2006). Markham *et al.* (2011) found that red maples grown in white containers had 2.5 times greater root density than those grown in black containers.

Sanitizing Containers

Large greenhouse operations often do not reuse plastic trays or containers because they are sold with the plants. However, for specialty containers, some propagators may offer a credit for returned containers, and vertically integrated propagation/retail/landscape operations may reuse trays and containers. Sanitizing containers before reusing is a key component of Integrated Pest Management (IPM) and is critical to preventing root rot pathogens from persisting from one crop to another (see Chapter 13 for substrate sterilization information and Chapter 15 for more on IPM). Several compounds effectively sanitize surfaces, flats, and reused containers. These include alcohol, hydrogen peroxide and peroxyacetic acid, quaternary ammonium, sodium hyperchlorite, steam, and solarization. Read product labels for safety precautions, exposure time, and if organic matter deactivates the product.

Container Handling and Storage

Irrigation, high temperatures and humidity, high irradiance, and frequent handling can affect the lifespan of any container. Container composition may impact container stability during propagation, its ability to withstand transportation stresses, and shelf life during retail sales. Once filled with substrate, flats and plug trays must be handled carefully as stacking can lead to compressed substrate. Excessively or unevenly compacted substrate is a major source of variation in plant growth. Buying containers in large quantities can save money, but appropriate storage conditions are necessary. High light and heat environments cause plastic containers, trays, and flats to become brittle. These items should be stored out of sunlight to extend their lifespan. Containers made of biodegradable materials require a dry environment with a relatively low humidity to extend the life of the products and protection from rodents. Containers should be protected from weed seeds, insects, and disease pests. Containers should be stored on pallets or, otherwise, protected from standing water, which can contain pathogens.

Recycling Plastic Containers

Levitan and Barros (2003) estimated that 1.7 billion pounds of plastic were used in agriculture in 2002. Recycling plastic containers is becoming more common both in municipalities and via coordinated programs to recycle agricultural plastics. However, many barriers to recycling greenhouse and nursery containers still exist. Degradation from ultraviolet (UV) light and potential pesticide residue are two barriers to recycling containers. UV light degradation reduces the value of plastic because it reduces the flexibility and ability to reuse the plastic.

Trends in Containers

Container composition, design, and color are all subject to popular trends as well as advancements through science. Currently, petroleum-based plastic containers dominate the industry, but there is interest in more sustainable container materials. Recent interest in sustainability and increases in petroleum prices fuel the demand for biocontainers and other alternatives to plastic containers. The convention of solid sidewalls has been challenged in the last two decades by air pruning technology. Also, containers are now used as marketing tools with color and print to distinguish brands and help products stand out in the retail marketplace. Even standard container sizes change with marketplace demands. For bedding plants, larger cell packs and larger individual containers rather than multi-cell packs are increasingly prevalent. Traditionally, bedding plants were commonly sold in flats consisting of 12-cell packs each with six plants (called 1206s). Over time, the trend shifted to the production of larger, premium transplants such as eight-cell packs of four plants each per flat (called 804s). Many other configurations are used, but the general trend is to market larger plants in a larger cell size. More retail markets are selling mixed-color containers with multiple plant species per pot. These are finished products (often round, 10 in or larger plastic containers or hanging baskets) and can be placed for instant color in decorative pots, planters, or hanging baskets by the final consumer. The recent economic downturn led to the production of smaller-sized cell packs in some markets. Mechanization also affects container trends. As container-filling machines and other mechanization such as transplanting machines are utilized, the range of container sizes is limited.

Chapter 10 was written in remembrance of Dr. Calvin Chong, University of Guelph, who contributed to Chapter 5, “Media and Containers for Seed and Cutting Propagation and Transplanting,” and Chapter 6, “Physical Properties and Other Factors When Selecting Propagation Media,” for the first edition of *Plant Propagation Concepts and Laboratory Exercises.*

EXERCISES

Experiment 1. Using a Rectangular Sponge to Illustrate the Saturated Zone and the Effect of Substrate Depth on Saturated Zone

The taller the substrate column, the greater the influence of gravity on the water in the substrate, and thus, the taller the container, the greater the drainage, all other variables being equal. However, the saturated zone is not affected by container height. The saturated zone height is easy to demonstrate using a rectangular sponge at three different orientations, which mimic three different substrate depths or container heights.

MATERIALS

The following materials are needed for each team of four students:

- 3 to 2 in. × 4.25 in. × 8.5 in. sponges (or similar dimension, light color)
- Water dyed with food coloring
- Flat, shallow tray at least 22 in. wide
- Measuring cup
- Spare sponge, any dimension

Follow the protocols in Procedure 10.1 to complete the experiment.

Procedure 10.1
Using a Rectangular Sponge to Show the Height of the Saturated Zone

Step	Instructions and Comments
1	Color approximately 3 cups (708 mL) of water with food coloring, very darkly.
2	Wet three sponges and wring all water from them.
3	Carefully place the three sponges next to one another in a shallow tray. Place one sponge at the shortest orientation (2 in. tall), place the second sponge at the intermediate orientation (4.25 in. tall), and place the third sponge at its tallest orientation (8.5 in. tall).
4	Pour the dyed water over each sponge very slowly, allowing it to be absorbed by the sponge.
5	Allow the water to drain to the bottom of the sponge. Use the fourth sponge to soak up any gravitational water that drains from the base of the sponges.
6	Observe the zone of saturation forming at the base of each sponge. Calculate the percentage of total sponge height that the zone of saturation represents for each of the sponges. Calculate what volume, in milliliters, is unsaturated for each of the sponges. *Note:* Placing a few drops of undiluted food coloring at the base of each sponge can enhance the visibility of the saturated zone. Optional: Consider modifying the experiment to illustrate the effect of substrate column/container height on total air space and water holding capacity. After steps 1 to 8, squeeze a sponge underwater and allow it to fully saturate. Hold it so that it has a 2 in. height and collect the water that drained from it. Now, turn the sponge in its side so that it has a 4.25 in. height. Now, hold the sponge so that the height is 8.5 in. Notice as the height changes the volume of the water draining and the total air space changes.

ANTICIPATED RESULTS

The height of the zone of saturation does not change with differing container heights.

QUESTIONS

- Does the zone of saturation have a greater potential influence for plug production, where containers are approximately 1 in. tall, or nursery production, where containers are approximately 8 in. or taller?
- How would the height of the zone of saturation change for a more coarsely textured substrate? A finely textured substrate?

Experiment 2. Measuring Substrate Physical Properties

Physical properties depend on the container height and substrate pore size. These properties can be readily measured with simple tools and can help indicate whether the substrate is appropriate for a particular container and/or crop.

MATERIALS

The following materials are needed to complete the following experiment. We suggest that either each student or a small team of students (2 to 3) is given one substrate to measure with two different container heights:

- Containers with two different heights, for example, a cell pack, and a 6 in. or one gallon pot. Larger containers will require more time for drainage, lengthening the time required for the exercise
- Several substrates that vary in texture (e.g., vermiculite, coarse perlite, peat, fine sand or soil, a peat/perlite-based potting mix, and a bark-based potting mix)
- Duct tape
- Scale, suitable for mass greater than 1000 g, with precision from 1 to 5 g. The balance will be used for several measurements and takes advantage of the property of water that 1 g of water = 1 cm^3 = 1 mL
- Oven for drying substrates
- Oven-safe containers

Follow the steps in Procedure 10.2 to complete this experiment.

ANTICIPATED RESULTS

Finely textured substrates should have a lower E_a and a larger P_v than coarsely textured substrates. In a taller container, a given substrate will have a larger E_a and a lower P_v than the same substrate in a shorter container.

Procedure 10.2
Measuring Substrate Physical Properties

Step	Instructions and Comments
1	Determine the container volume (*V*). Securely duct tape the drainage holes of the container. Weigh the empty container after the duct tape is in place. Fill with water to the collar. The difference in weight between the full container and the empty container is the container volume (express units in cubic centimeters [cm^3]).
2	Pour out the water from the container. Fill the container with moist substrate (keep the duct tape in place). Very slowly add water to the substrate until it is saturated; this will allow for the calculation of *E*. Allow time for it to reach the bottom of the container and fully saturate the pores. Continue adding water until there is a thin film of water at the top of the substrate. Weigh the filled container and subtract the empty container weight; this is the saturated substrate weight (express units in cubic centimeter [cm^3]).
3	Remove the duct tape from the drainage holes; place the tape on the side of the container (thereby keeping the empty container weight the same). Allow the substrate to drain completely—this may take up to half an hour. Wait until the water stops dripping from the container. Weigh again and subtract the empty container weight; this is the container capacity weight (express units in cubic centimeter [cm^3]).
4	Place the substrate in a pan (keep substrates from each container separate). Place in an oven to dry at 160 to 200°F until all moisture is removed.
5	Cool the substrate to a workable temperature and weigh; this is the dry substrate weight (express in gram [g]).
6	Conduct the following calculations (abbreviations noted in 10.1): a. ρ_b (g/cm^2) = dry substrate weight / container volume b. E (%) = (saturated substrate weight – dry substrate weight) / container volume × 100 c. P_v (%) = (container capacity weight – dry substrate weight) / container volume × 100 d. E_a (%) = (saturated substrate weight – container capacity weight) / container volume × 100

QUESTIONS

- Did you encounter any difficulties during step 2—saturating the substrate? How could these same difficulties be encountered and managed by commercial propagators attempting to irrigate crops?
- Which of the measured substrate(s) would be best suited for rooting cuttings in a tray under mist? Which substrate(s) would be best for growing a nursery crop with large water demand in a 6-in. or one gallon pot?
- Why are peat and perlite combined together as the basis for many commercial potting substrates for floriculture crops?

Experiment 3. Growing Seedlings/Cuttings in a Finely Textured and a Coarsely Textured Substrate

Grow bedding plants/vegetable transplants in a container filled with a fine-textured (heavy) or coarse-textured substrates. Fine-textured substrates will retain more water and have lower air porosity. Note: Soil should always be steam pasteurized before incorporating into a substrate to eliminate pathogens.

MATERIALS

The following materials are needed for each team of four students:

- 20 seedlings sown in plug trays 4 to 6 weeks prior to the beginning of the experiment
- 20 6-in. pots
- Soil (steamed) or another of the fine-textured substrates used in Experiment 2
- Promix 280 or another coarse substrate
- Fertilizer
- Plastic trays (solid, without holes)
- PVC pipe, 2 in., cut into 1 in. sections to serve as a spacer between the container and the plastic tray
- Ruler

Complete the protocol in Procedure 10.3 to complete this experiment.

ANTICIPATED RESULTS

Plant growth will be poorest in the soil-based substrate. The substrates will hold different volumes of water, and the soil-based substrate will dry out more slowly. Plants will attain greater biomass in the peat/pine bark–based substrate.

Procedure 10.3
Growing Seedlings/Cuttings in a Heavy and a Coarse Substrate

Step	Instructions and Comments
1	Transplant seedlings into 6 in. pots. Fill half of the containers with Promix 280, and the other half, with soil. Place each container on three to four PVC spacers on a plastic tray.
2	Water/fertigate when needed, but irrigate every container with the same volume of water.
3	Record the frequency of irrigation for each container and leachate.
4	Grow plants for 4 to 6 weeks.
5	Record mortality.
6	Measure and record water use and leachate (collected in the plastic trays) per container for each irrigation event.
7	Measure and record plant height and growth index.
8	Rinse roots, observe differences, oven dry at 55°C, weigh, and record.

QUESTIONS

- What do you think would happen if the soil substrate was used in a taller container of the same volume? Where do you think roots would grow the most?
- If you partitioned the container volume into quarters from top to bottom, where would you find the greatest root biomass in each substrate? Why would the root distribution differ based on the substrate?

Experiment 4. Measuring Substrate pH and EC Following the 1:2 Dilution Method

Substrate pH and EC can be readily measured with handheld meters following a simple extraction with one part substrate and two parts deionized water (termed "1:2 dilution method"). This is a useful tool for determining the suitability of substrate components or a pre-made potting mix for plant propagation and growth. Note that using the 1:2 dilution method is ideal for measuring substrate batches prior to planting. For monitoring a crop sequentially over time, the method is somewhat destructive; you may consider taking small pinches of substrate from the middle of the root zone for several plants and pooling these to make a sample. Alternatively, it is common to follow the pour thru method, which is not destructive, when many sequential measurements will be taken for long-term crops. The pour thru method for substrate sampling is recommended when controlled-release fertilizers are in the substrate because, in the 1:2 dilution method, the disruption of prills may lead to artificially high EC readings. Pour thru methods are described in depth by Cavins *et al.* (2000).

MATERIALS

The following materials are needed; the same substrates from Experiment 2 and/or additional materials may be used.

- Several substrate components or pre-made potting mixes
- Deionized water
- Handheld pH and EC meters and calibration solutions
- Plastic cups or glassware
- Cone coffee filters

Use the protocol in Procedure 10.4 to complete this experiment.

ANTICIPATED RESULTS

Commercially available substrates are often formulated to a pH of around 6.0 and an EC of 0.25 to 1 dS/m. Substrate components vary greatly in their pH and EC—manure-based composts often have both high pH and high EC; peat or aged bark often have low pH and EC values.

QUESTIONS

- Would any of the substrate components you measured be suitable for use as a substrate by themselves with regard to pH and EC properties?
- Assuming that you measured a high pH and EC for a manure-based compost, how might you blend this with other components to make a suitable potting mix?

Procedure 10.4
Measuring Substrate pH and EC

Step	Instructions and Comments
1	Take a representative substrate sample. The sample should be relatively dry. Otherwise, the sample may be air dried for 24 h in a shallow pan.
2	In a cup, combine one part substrate and two parts deionized water (by volume) (e.g., 1/2 cup of substrate and 1 cup water).
3	Stir thoroughly and allow the mixture to stand for 30 min to allow time for the substrate to interact with the water.
4	Filter the mixture by pouring it to two to three layers of coffee filter. Capture the water in a clean cup.
5	Calibrate the pH and EC meters. Measure the pH and the EC of the water sample.
6	Optional: Consider setting up experiments on the effects of high pH or EC on the growth of transplanted seedlings. For a high-pH treatment, prior to transplanting, amend the substrate with 0 or 12 g of limestone (calcium carbonate) per 1 L of substrate (this is equivalent to limestone incorporation of 0 or 20 lbs per cubic yard). For a high salt treatment, amend the irrigation water with 0 or 50 mM NaCl. At each watering, plants receive the salt treatment. Periodically monitor the substrate pH and EC. After several weeks, make final observations on plant growth in response to high pH or high EC.

LITERATURE CITED

Avent, T. 2003. *So You Want to Start a Nursery?* Portland, Oregon: Timber Press, 340 pp.

Burnett, S. E. 2009. Making the switch to organic substrates. *Greenhouse Mgt and Prod* 29:22–26.

Cavins, T. J., B. E. Whipker, W. C. Fonteno, B. Harden, I. McCall, and J. L. Gibson. 2000. Monitoring and managing pH and EC using the pour thru extraction method. *Horticulture Information Leaflet* 590:17. Available online: http://www.ces.ncsu.edu/depts/hort/floriculture/hils/HIL590.pdf.

Daigle, J.-Y. and H. Gautreau-Daigle. 2001. *Peat Harvesting and the Environment*, 2nd Edition. Issues Paper 2001-1. Published in Partnership with the North American Wetlands Conservation Council Committee, 41 pp.

Evans, M. 2008. Rice hulls 101. *Grower Talks* 71:61–64.

Fulcher, A., G. Niu, G. Bi, M. Evans, T. Fernandez, R. Geneve, A. Koeser *et al.* 2012. Assessing biocontainers and a sustainable irrigation regime for the U.S. nursery industry. *Proc So Nursery Assoc Res Conf* 57:73–77.

Hall, C., B. L. Campbell, B. Behe, C. Yue, R. G. Lopez, and J. H. Dennis. 2010. The appeal of biodegradable packaging to floral consumers. *HortScience* 45:583–591.

Hartmann, H. T., D. E. Kester, F. T. Davies, and R. L. Geneve. 1997. *Plant Propagation: Principles and Practices*. Upper Saddle River, New Jersey: Prentice Hall, 877 pp.

Hoitink, H. A. J., D. Y. Yan, A. G. Stone, M. S. Krause, W. Zhang, and W. A. Dick. 1997. Natural suppression. *Amer Nurs* 186:90–97.

Jack, A. L. H., A. Rangarajan, S. W. Culman, T. Sooksa-Nguan, and J. E. Thies. 2011. Choice of organic amendments in tomato transplants has lasting effects on bacterial rhizosphere communities and crop performance in the field. *Appl Soil Ecol* 48:94–101.

Levitan, L. and A. Barros. 2003. *Recycling Agricultural Plastics in New York State: Environmental Risk Analysis Program*. Ithaca, New York: Cornell Center for the Environment. Available online: http://cwmi.css.cornell.edu/recyclingagplastics.pdf.

Markham III, J. W., D. J. Bremer, C. R. Boyer, and K. R. Schroeder. 2011. Effect of container color on substrate temperatures and growth of red maple and redbud. *HortScience* 46(5):721–726.

Mathers, H., S. B. Lowe, C. Scagel, D. K. Struve, and L. T. Case. 2007. Abiotic factors influencing root growth of woody nursery plants in containers. *HortTechnology* 17:151–162.

Nambuthiri, S., R. Geneve, T. Fernandez, A. Fulcher, A. Koeser, G. Bi, M. Evans *et al.* 2012. Substrate heat buildup and evaporation rate differs between plastic and alternative one-gallon nursery containers. *Proc So Nursery Assoc Res Conf* 57:60–62.

Ruter, J. M. 2000. Biodegradable fiber containers improve the growth of two daylily cultivars. *Acta Hort* 517:271–273.

Sideman, E. 2007. Soilless mixes for vegetable seedling production. *Maine Organic Farmers and Gardeners Association Fact Sheet* 9. Available online: http://www.mofga.org/Portals/2/Fact%20Sheets/FS%2009%20Soil-less%20Mixes%20web.pdf.

Whitcomb, C. E. and A. C. Whitcomb. 2006. Temperature control and water conservation in aboveground containers. *Proc Intl Plant Propagators' Soc* 56:588–594.

Yue, C., C. Hall, B. Behe, B. Campbell, J. Dennis, and R. Lopez. 2010. Are consumers willing to pay more for biodegradable containers than for plastic ones? Evidence from hypothetical conjoint analysis and nonhypothetical experimental auctions. *J Agr Appl Econ* 42:757–772.

Part IV

Plant Propagation Diseases and the Importance of Sanitation

11 Disease Management

Alan S. Windham

Concept Box 11.1

- Common plant pathogens include fungi, bacteria, viruses, mollicutes, and nematodes.
- The most common plant diseases observed during propagation are those caused by fungi and fungi-like organisms.
- *Rhizoctonia solani*, which causes damping-off, stem rots, and web blight, and *Botrytis cinerea*, which causes botrytis blight, (gray mold) are two of the most common fungal pathogens observed during the propagation of plants.
- The environment in which most plants are propagated is highly favorable for plant disease development.
- Plant disease management during propagation involves sanitation, disease-free stock plants, environmental and cultural controls, as well as chemical controls.
- The international trade in plant material results in diseases new to the United States, such as boxwood blight and downy mildew of impatiens.
- Viruses vectored by insects, such as impatiens necrotic spot virus, may be very difficult to manage.
- Fungicides should be used to keep healthy plants healthy, not cure diseased plants.

Disease management during plant propagation is critically important to ensure a disease-free plant for the consumer. In most production systems, there is no other time that so many immature plants are exposed to conditions so favorable for disease development. Seedlings or cuttings are grown essentially in monoculture in flats or beds that can account for the rapid spread of disease. Vegetatively produced cuttings are often grown under conditions of high relative humidity that favor several foliar diseases. Also, in many parts of the world, cool-season annual bedding plants are predisposed to some root rot diseases by exposure to high temperature. Anyone involved in plant production needs to realize that producing a healthy, saleable plant starts prior to and continues during plant propagation. At this point of plant production, you are most likely to lay the foundation for the health of the crop.

Seedlings or rooted cuttings may not die from infectious diseases, but they may be disfigured, discolored, stunted, or become generally unthrifty. Many plant diseases that occur during plant propagation have the ability to reduce the quality or grade of infected plants. There are some pathogens that infect vegetatively propagated plants that are not detected in the greenhouse, nursery, or garden center. This often occurs as the signs or symptoms of disease are not recognized or are confused with the symptoms of other maladies, such as nutrient deficiencies, pesticide phytotoxicity, or injury from abiotic stresses.

Diseases of plants in propagation systems may be caused by plant pathogens such as fungi, fungi-like organisms, bacteria, mollicutes, viruses, viroids, and nematodes. Fungi lack chlorophyll and are eukaryotic. They are generally filamentous, branched organisms that reproduce by spores and have cell walls made of chitin and other polymers. The fungi-like organisms such as *Phytophthora* and *Pythium* belong to a group of organisms that are responsible for causing very damaging diseases such as damping-off, root rots, wilting rots, foliar blights, and downy mildew. This group of pathogens (oomycetes) has filamentous growth and reproduces by spores, which is very similar to the true fungi. However, once considered fungi, these organisms are now thought to be more closely allied to the brown algae. Prokaryotic organisms, such as bacteria, are commonly found in propagation systems causing soft rots and leaf spot diseases. Bacteria are single-celled organisms that do not have nuclei, double membrane–bound organelles, but have a rigid cell walls. Plant pathogenic bacteria are normally rod-shaped or filamentous. Mollicutes are distinguished from bacteria by their smaller size and lack of a cell wall. They are surrounded by a plasma membrane. Since there is no rigid cell wall, these organisms may have a variety of shapes. Mollicutes are causal agents of yellow diseases in plants. Viruses and viroids are much smaller than bacteria and cannot be seen with light microscopy. Neither viruses nor viroids can reproduce alone as each needs the host plant's replication machinery for multiplication. Viruses are nucleoproteins with nucleic acid (DNA or RNA) surrounded by a protein coat. Viroids possess many of the attributes of viruses, but are essentially naked strands of RNA without a protein coat.

FIGURE 11.1 Zonate lesion typical of impatiens necrotic spot virus on impatiens leaf.

VIRUS DISEASES

Plant viruses represent one of the major threats to healthy ornamental crops. There are a variety of symptoms that are associated with viral diseases, including ring spots, oakleaf pattern, mosaic, stunting, stem cankers, leaf spots, flowering break, etc. Viruses may be spread by several different means. For example, tomato ring spot virus may be spread via an infected seed; tobacco mosaic virus may be spread mechanically by handling plants with tools or hands infested with the virus; impatiens necrotic spot virus (INSV) and the tomato spotted wilt virus (TSWV) are often transmitted when vegetative cuttings are taken from infected stock plants. Viruses may also be vectored by insects. Aphids often transmit cucumber mosaic virus, and western flower thrips (WFT; *Frankliniella occidentalis* Pergande) are common vectors of INSV and TSWV.

Impatiens Necrotic Spot Virus

It is critically important to purchase vegetative cuttings from a horticultural supplier that uses a virus indexing program to ensure the health of their stock plants. New Guinea impatiens cuttings should be purchased from suppliers indexing for INSV. This virus has more than 450 hosts representing more than 60 genera of plants. Although infected plants rarely die, they are often severely stunted if infected as seedlings or young cuttings. Other symptoms include leaf distortions, ring spots (Figure 11.1), oakleaf patterns, and stem lesions.

To prevent damage from INSV, monitor greenhouses for WFT by placing yellow or blue sticky cards among the crop. In a very short time, WFT can acquire INSV and transmit it to healthy plants. Weed management is important inside and around the perimeter of a greenhouse. Weeds may serve as asymptomatic hosts for viruses and the insects that act as vectors. If possible, isolate seedlings from vegetatively produced crops in different greenhouses. There is always danger of a virus entering the greenhouse in vegetatively produced cuttings and then spreading to seedlings via insects (WFT).

Rose Rosette

Rose rosette is quickly becoming the most damaging disease of landscape roses, particularly shrub roses. Common symptoms include witch's brooms (rosettes), furry canes (excessive thorns), new growth may be red to maroon, and death (Figure 11.2). The causal agent for rose rosette is a negative-sense RNA virus, the rose rosette virus, an *Emaravirus*. Rose rosette can be spread in the following two ways: (1) in propagation, through the vegetative propagation of roses infected with the virus; and (2) in landscapes, where the virus is vectored by an eriophyid mite *Phyllocoptes fructiphilus*. As soon as practical assays for the virus become available, propagators should index stock plants to ensure that they are free of the virus. In landscape beds, roses infected with rose rosette should be promptly removed.

FIGURE 11.2 Rose rosette on a landscape rose.

Rose Mosaic

Rose mosaic is a common disease on roses worldwide. The causal agents for rose mosaic are apple mosaic virus and *Prunus* necrotic ring spot virus. These viruses may occur together or individually and are often spread during the vegetative propagation of roses. Common symptoms in rose include vein clearing, mosaic, and line patterns such as "oak leaf symptoms." During hot weather, infected roses may be asymptomatic. Roses infected with rose mosaic may produce fewer and smaller flowers. Rose producers may procure virus-free root stocks and scions from propagators that have a virus indexing program.

BACTERIAL DISEASES

Bacterial diseases, such as soft rot, bacterial leaf spot, bacterial wilt, and crown gall, are commonly found in propagation systems. Soft rot caused by *Erwinia* spp. can turn cuttings under mist into a putrid, slimy mess during warm weather. Poinsettia (*Euphorbia pulcherrima* Willd. ex Klotzsch) is very susceptible to bacterial soft rot during propagation. Formerly in the genus *Erwinia*, *Pectobecterium carotovora* and *P. chrysanthemi* are both capable of causing soft rot in plants reproduced by vegetative cuttings. Wilting and the collapse of cuttings are common symptoms. *Erwinia chrysanthemi* may also be associated with vascular rot and discoloration of tropical foliage plants.

There are several bacteria that are causal agents of leaf spot diseases in ornamental crops. One of the most common is bacterial leaf spot of zinnia. *Xanthomonas campestris* pv. *zinnae* (Figure 11.3) may be found on seeds, and the bacterium spreads to the leaves of the seedlings just after germination. Leaf spots are angular and dark brown to black. Overhead watering spreads the bacterium to adjacent plants. This particular disease can be prevented by disinfecting seeds prior to planting.

One of the most damaging bacterial diseases is bacterial blight of geranium caused by *X. campestris* pv. *pelargonii*. It can cause not only leaf spots, but also V-shaped lesions, wilting, and death. Bacterial blight is very contagious and can spread rapidly among zonal and ivy geraniums. Other geranium species may serve as hosts. Zonal cuttings should be purchased from suppliers that index their stock plants for this disease. Isolate seed geraniums from those that are vegetatively propagated. Also, geraniums should not be carried over from one season to the next.

FIGURE 11.3 Xanthomonas leaf spot on zinnia.

Another bacterial wilt disease of geranium is caused by *Ralstonia solanacearum*. The symptoms of this disease look very much like those of bacterial blight, with the exception of leaf spots. Infected plants become chlorotic, wilt, and die. In recent years, *R. solanacearum* race 3 biovar 2 has been detected in geranium cuttings from major propagators. This pathogen has been imported into the United States in geranium cuttings from other countries. This pathogen poses a serious threat not only to ornamental plants, but also to important horticultural crops in the Solanaceae that are grown for food, such as potatoes and tomatoes.

FUNGAL DISEASES

Fungal diseases are the most common diseases of ornamental plants. Included in this group are powdery mildew, downy mildew, rust, fungal leaf spots, rhizoctonia stem rot, pythium root rot, phytophthora root and crown rot, black root rot, fungal leaf spot, and damping-off of seedling crops.

Damping-Off and Web Blights

Damping-off and stem rot are often caused by the fungus *Rhizoctonia solani*. This fungus is ubiquitous and may be found anywhere plants are grown. Young seedlings often fall over after their stems are girdled (damping-off), and older plants may be stunted and suffer stem breakage. This fungus is a soil inhabitant and may be spread in contaminated pots, flats, media, and tools. As it rarely sporulates in nature, *R. solani* spreads by mycelial growth, killing seedlings as it grows. In mist propagation or in polyhouses where temperatures and humidity are often high, "web blight" may develop at the base of cuttings or on the foliage. Growth resembling spider webs is often easily visible on infected plants.

Botrytis Blight

Botrytis blight, also called "gray mold," is probably the most prevalent of all ornamental diseases (Chapter 12). The causal agent, *Botrytis cinerea*, is favored by cool, wet conditions, and leaf wetness is important for infection. Older leaves or flowers are most frequently damaged, but under ideal conditions, such as stagnant air and high humidity, almost any plant part is at risk of damage. Botrytis blight can be particularly damaging on geranium, poinsettia, exacum, and begonia. *Botrytis cinerea* may exist in most greenhouse as a saprophyte on dead plant tissue. It sporulates readily on plant tissue, and almost any activity in the greenhouse may initiate the release of spores. Spores must come into contact with free water to germinate. Humidity reduction, increased air movement, good sanitation practices, and fungicide sprays are important management strategies.

Root Rot Diseases

Pythium root rot is one of the more common root rot diseases in floral crops. It can be found on chrysanthemums, poinsettia, and many bedding plants. Wilting is one of the first signs of a problem (Figure 11.4). Upon closer examination, diseased roots are dark brown and decayed. Infected plants may wilt rapidly and die. *Pythium* is a "water mold" and may be spread like *Rhizoctonia* or in irrigation water. High soluble salt levels in media favor damage from pythium root rot. Phytophthora crown or root rot is not as common as pythium root rot. Like *Pythium*, *Phytophthora* is a water mold and can be spread in irrigation water as well as from plant to plant in splashed rain or irrigation water. These fungi-like organisms can also be spread by contaminated media that have been improperly stored or become contaminated with soil. *Phytophthora*-affected stems and leaves may be dark brown to black. Symptoms include a root rot that extends into the stems of infected plants. Saturated, poorly drained media and overwatering favor both pythium and phytophthora root rots. *Phytophthora ramorum* is thought to be an exotic pathogen in the United States, which has been associated with the dieback and death of native oaks in coastal California, called "sudden oak death." Unlike many *Phytophthora* species that cause root rots, this fungus causes canker diseases, foliar blights, and leaf spots,

FIGURE 11.4 Pythium root rot on a rooted cutting of garden mum. (Reprinted from Trigiano, R. N. *et al.*, *Plant Pathology Concepts and Laboratory Exercises*, Second Edition, with permission of CRC Press, LLC, Boca Raton, Florida.)

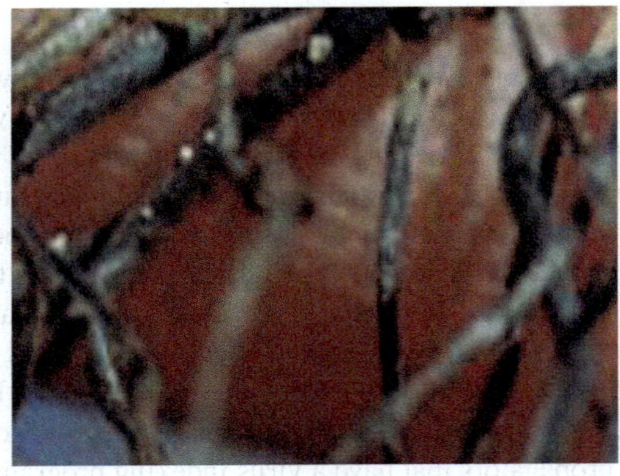

FIGURE 11.5 Blackened roots of Japanese holly infected with *Thielaviopsis basicola*.

depending on the plant species infected. The host range is growing larger as more is learned about this pathogen. Common hosts include rhododendron, viburnum, mountain laurel, pieris, and camellia. Diseases caused by *P. ramorum* are so difficult to control that a "systems approach" composed of an intense, integrated management plan has been devised to minimize the threat of this pathogen on woody ornamentals in production nurseries.

Black root rot (*Thielaviopsis basicola*) is a problem on pansy and may be found on plants as diverse as blue holly, Japanese holly, inkberry, fuchsia, verbena, vinca, petunia, and million bells. Infected plants generally look unthrifty with poor color and slow growth. Infected root systems may range in color from dirty off-white to brown and black (Figure 11.5). This disease may spread very rapidly and infect whole crops. Dark aleuriospores of *T. basicola* are often easily spotted inside infected roots. Media pH values that are near 7 or alkaline favor infection. Heat stress may also play a role in the infection process for cool season crops.

Fungal Leaf Spots/Blights

Fungal leaf spots are not common in greenhouses, but may appear on rooted cuttings and pot plants. Poinsettia scab is an example that was first noticed in 2000. Fungal leaf spots may be zonate, irregularly shaped, or angular. Cercospora leaf spot can be particularly damaging on pansy plugs. Avoid taking cuttings for propagation from plants infected with fungal leaf spots (Figure 11.6). Fungicide sprays are often used to protect stock plants from fungal leaf spot diseases.

Boxwood blight is a fungal disease of all cultivated boxwood species, new to the United States. It was first found in the United States in 2011. The causal fungus, *Cylindrocladium pseudonaviculatum* (syn. *C. buxicola*), attacks the foliage and twigs of boxwood. Leaf spots are circular and brown with white fungal growth on the underside of the leaves. Black lesions may be found on twigs and small branches. As the disease progresses, all of the

FIGURE 11.6 Avoid taking cuttings from plants damaged by fungal leaf spot diseases.

FIGURE 11.7 Boxwood blight caused by *Cylindrocladium pseudonaviculatum* on American boxwood.

leaves may become necrotic and drop from the infected boxwood (Figure 11.7). It is critical that cuttings are from stock plants that are free of disease as this disease can spread quickly in propagation and produce resting structures (microsclerotia) that make it difficult to eradicate. Check with your local extension office for new information regarding fungicides labeled for boxwood blight.

Powdery Mildew

Powdery mildew is easily recognized on most crops by the white, powdery fungal growth on leaves. These fungi may disfigure and stunt the leaves and flowers. The fungi that cause powdery mildew, such as *Erysiphe*, *Microsphaera*, and *Podosphaera*, are rarely found attacking plants in mist propagation systems as free water inhibits spore germination. Rather, they may be a problem on stock plants. While powdery mildews do not need free moisture for infection, high relative humidity is necessary.

Downy Mildew

Downy mildew is similar to powdery mildew in only one way, its name. The fungi-like organisms that cause downy mildew are actually more closely related to *Pythium* and *Phytophthora*, organisms that cause stem rots, root rots, and foliar blights, than to the fungi that cause powdery mildew. Downy mildew is becoming increasingly common on a variety of plants, including rose, garden impatiens, rudbeckia, salvia, snapdragon, and pansy. Infection is favored by cool, wet weather. On herbaceous plants, symptoms include stunting, foliar chlorosis, and leaf necrosis. On herbaceous plants such as garden impatiens, severe defoliation may occur (Figure 11.8). In the early stages, angular brown, black, or maroon leaf spots may appear on foliage. Also, infected shoots may become systemically infected and wilt rapidly. Downy mildews are often visible on the undersides of leaves. Off-white to gray fungal growth is often visible directly under leaf spots that are visible on the upper surface of leaves.

FIGURE 11.8 Defoliation of garden impatiens by downy mildew *Plasmopara obducens*.

Downy mildew of garden impatiens (*Impatiens walleriana*) has become a major disease in the floral crops industry in the United States and Europe. *Plasmopara obducens*, the causal organism, sporulates profusely on the underside of leaves. These spores (sporangia) may be windblown locally or to long distances to infect healthy plants. *Plasmopara obducens* also produces oospores in the stems of infected impatiens in late fall. These spores may overwinter in landscape beds and infect impatiens the next season. Downy mildew is not seedborne, but may spread on infected plants. It is important that growers of impatiens follow intense fungicide spray protocols to prevent infection in greenhouses. Growers should also scout for signs of disease (white mycelium) and discard infected plants. Resistant plants that can be used as replacements for impatiens include begonia, New Guinea impatiens, and coleus.

Rust

Rust diseases may appear on snapdragon and geranium and perennials, such as potentilla and daylily. By the time that yellow leaf spots are noticed, the fungus is usually sporulating and producing millions of yellow-to-orange or brown spores in pustules, which make the identification of the fungus easy. Symptoms, including angular leaf spots, yellowed foliage, and fungal growth, may be spotted on the undersides of leaves. In 2000, daylily rust caused by *Puccinia hemerocallidis* was first detected in the United States. This pathogen of daylily is now established in many areas of the southern United States and is endemic to Asia. The propagation and sale of daylily by hobbyists make the regulation of this disease nearly impossible.

NEMATODE DISEASES

Nematodes are microscopic, worm-like animals, and most are harmless to plants and animals. Plant parasitic nematodes have a mouthpart, called a "stylet," which is similar to a hypodermic syringe. It is used to rupture cell walls, inject enzymes into the cell that aid in digestion, and reabsorb the contents of the cell. Root knot nematode (*Meloidogyne* spp.) may be found in floral crops where the soil has contaminated media. Infected roots are swollen and knotted. A larger nematode problem are the foliar nematodes (*Aphelenchoides* spp.). They feed on the foliage of several floral crops such as African violet and begonia. Many shade perennials such as hosta (Figure 11.9), anemone, brunnera, and Lenten rose may be infected with foliar nematode. It is usually spread during the propagation of infected plants. Infected plants should be discarded and not propagated. Hot water treatment can be used to clean up plants, but it is time consuming and difficult, and there is a risk that the plant material will be killed.

DISEASE MANAGEMENT TACTICS

- *Sanitation*. Disinfestants that contain ammonium chloride, such as Physan, Consan, and Green Shield, are useful for killing bacteria and fungi on benches, pots, flats, tools, and walkways. Disinfect any pots or flats that are to be used more than once. Flats of plants susceptible to infection by soil-borne pathogens should be placed on clean ground cloth, gravel, or benches. Each greenhouse should have a trash can with a lid to hold plant debris until it can be destroyed. Rogueing, the removal of infected plants, can help slow the spread of diseases such as INSV and bacterial blight.

FIGURE 11.9 Angular lesions caused by an *Aphelenchoides* sp. on a *Brunnera* leaf.

- *Media smarts*. Soilless media are not sterile; therefore, you may occasionally recover some plant pathogens such as *Rhizoctonia*, *Thielaviopsis*, and *Pythium*. Media should be protected from soil contamination by storing and mixing on a clean bench or concrete pad. Keep hose nozzles off the ground to prevent the introduction of plant pathogens into clean media.
- *Disease-free plant material*. Take cuttings from healthy plants free of obvious disease. If available, take cuttings or budwood for fruit trees from a block of trees certified free of common plant viruses. Buy cuttings from horticultural suppliers that index their stock plants for INSV (impatiens, gloxinia, etc.) and bacterial blight (geranium). The added expense is worth the cost of not having to deal with these diseases. Check all incoming plants for symptoms or signs for insect and disease. Remember to check the root systems of incoming plants as well as the foliage. Black root rot can be easily detected if you will take the time to look at the roots of a few plants. If you have a problem with incoming plant material, contact the plant inspector with the state department of agriculture to document the damage. This can make things go more smoothly when dealing with your supplier.
- *Environmental controls*. Humidity control is important for managing diseases, such as powder mildew, downy mildew, and botrytis blight. Vent moisture-laden air in the late afternoon and heat incoming air in the morning to reduce humidity levels. This can be crucial in preventing condensation from forming on leaf surfaces at night when infection from *Botrytis* or mildew is likely to occur. Horizontal airflow fans can be used to reduce leaf wetness and condensation on plastic greenhouse coverings. Mount fans just above the crop and circulate air in a racetrack orientation. If possible, avoid overhead watering, especially on crops that are susceptible to botrytis blight. Sub- or drip irrigation can reduce bacterial and botrytis blight. Overwatering and saturated media favor the water molds *Pythium* and *Phytophthora* as well as insects such as

TABLE 11.1
Chemical Control of Diseases of Seedlings, Cuttings, and Liners of Ornamental Plants

Disease	Susceptible Plants	Chemical Control
Stem/crown rot *Southern blight*. Usually occurs in gardens, perennial borders, and nurseries during hot weather, near midsummer. Symptoms include wilting, leaf scorch, followed by plant death. Signs of disease include white mycelium on the stem of infected plants and tan to reddish-brown round, spherical resting structures of the fungus (sclerotia) on the stem and the soil surface. *Sclerotinia crown rot*. Unlike southern blight, this disease usually appears during mid-spring to early summer when conditions are cool and moist. Affected plants usually wilt and die. White mycelium may be visible on stems near the soil surface. Black, oblong sclerotia may be present on the outer surface of woody plants or in the stem pith of herbaceous plants. Diseased stems should be split lengthwise and examined for signs of sclerotia. *Rhizoctonia stem rot/damping off*. This disease is often the cause of damping-off (stem rot) of seedling plants. Seedling annual or perennial flowers or woody ornamentals may be killed by this fungus after it attacks the stem near the soil surface. Diseased seedlings often fall over and die. In the field, the fungus may move short distances down the row, killing several adjacent plants. In propagation beds or flats, diseased plants may be killed in circular areas as the fungus moves outward.	*Southern blight*: ajuga, apple, clematis, crabapple, forsythia, hosta, many annual and perennial flowers *Sclerotinia stem rot*: campanula, euonymus, several herbaceous flowers *Rhizoctonia stem rot*: many herbaceous plants and seedlings of woody plants and conifers	*Southern blight*: azoxystrobin, fludioxonil, flutolanil, fluoxinil, pentachloronitrobenzene (PCNB) *Sclerotinia crown rot*: thiophanate methyl *Rhizoctonia diseases*: azoxystrobin, fludioxonil, iprodione, thiophanate methyl, triflumizole
Gray mold (botrytis blight). Gray mold may be found on herbaceous and woody ornamentals usually during cloudy, cool, moist weather. Stems, leaves, and flowers may be attacked. Woody ornamentals in overwintering structures may become infected. Symptoms of infection are blighting of flowers, tan-to-brown leaf spots, shoot blights and stem rot. A sign of disease is gray-brown mold on diseased plant parts.	Almost any herbaceous or woody plant. Geraniums are particularly susceptible to gray mold.	Chlorothalonil, fenhexamid, fludioxonil, iprodione, mancozeb, thiophanate methyl, triflumizole, vinclozolin
Powdery mildew. Powdery mildew is easily identified by the presence of white-to-gray mycelium on affected leaves and/or flowers. The first sign of disease is usually isolated colonies of white fungal growth. With time, whole leaves may be totally covered with fungal growth. On some plants, such as pin oak, mildew may be present only on the undersides of leaves. On dogwood, crape myrtle and nandina infected leaves may be curled, twisted, or, otherwise, distorted. Leaves may be abnormally red with little mycelium visible; on sedum lesions are scabby and brown.	Amelanchier, apple, azalea, begonia, columbine, crabapple, crape myrtle, dogwood, euonymus, hydrangea, lilac, nandina, phlox, rhododendron, rose, sedum, tulip tree, zinnia	Azoxystrobin, chlorothalonil, kresoxim methyl, metconazole, myclobutanil, piperalin, propiconazole, tebuconazole, thiophanate methyl, triadimefon, trifloxystrobin, triflumizole

(*continued*)

TABLE 11.1 (Continued)

Chemical Control of Diseases of Seedlings, Cuttings, and Liners of Ornamental Plants

Disease	Susceptible Plants	Chemical Control
Downy mildew. Although this sounds similar to powdery mildew, the diseases are very different and caused by fungi from entirely different taxonomic classes. The fungi-like organisms that cause downy mildew are more closely related to the fungi-like organisms that cause phytophthora and pythium root rots than to the fungi that cause powdery mildew. Symptoms of downy mildew can range from leaf spots and defoliation to rapid blighting of diseased shoots. Angular leaf spots on rose may range from red to brown to black. Signs to look for include gray tufts of mycelium on the undersides of leaves, directly below chlorotic lesions. Look for mycelium early in the morning while the leaves are still wet.	Alyssum, brambles, grape, pansy, rose, salvia, snapdragon, tobacco, viburnum	Dimethomorph, fenamidone, fluoxastrobin, fosetyl-Al, mandipropamid, mefenoxam, potassium salts of phosphorous acid, boscalid + pyraclostrobin
Root rot diseases. Plants affected with fungal root rots may be stunted, wilted, look generally unthrifty (mimic nutrient deficiency), and eventually die. Discolored decayed roots are sure symptoms of root rot diseases. Poor drainage, standing water, improperly constructed landscape beds, planting infected plants, and excessive irrigation favor phytophthora and/or pythium root rots.	*Black root rot*: Japanese holly, blue holly, vinca, pansy, petunia *Phytophthora root rot*: azalea, dogwood, forsythia, fir, holly, juniper pieris, rhododendron, yew *Pythium root rot*: herbaceous ornamentals	*Black root rot*: fludioxonil, thiophanate methyl, triflumizole *Phytophthora and pythium root rot*: etridiazole, fosetyl-Al, fluopicolide, mefenoxam, potassium salt of phosphorous acid
Rust (leaf, stem, needle). Signs include bright yellow, orange, reddish-brown, or chocolate-brown raised pustules are visible usually on the undersides of leaves. Gelatinous tendrils of rust spores are produced from galls each spring on eastern red cedar infected with cedar apple rust. Pine needle rust produces pustules on pine during spring. Early symptoms on leaves are yellow leaf spots. Rust galls may appear on stems of pine, cedar, and hawthorn. Twig rust may cause branch dieback on plants as diverse as hawthorn and hemlock.	Amelanchier, apple, aster, azalea, buckeye, cedar, crabapple, daylily, fuchsia, geranium, grasses, hawthorn, hemlock, hollyhock, iris, jack-in-the-pulpit, juniper, mayapple, oak, pear, pine, potentilla, quince, snapdragon, sunflower	Azyoxystrobin, chlorothalonil, flutolanil, kresoxim methyl, mancozeb, myclobutanil, propiconazole, triadimefon, trifloxystrobin, triflumizole
Leaf spot diseases. Leaf spot diseases are usually caused by fungi, but a few may be caused by bacteria. These are among the most common plant diseases. Symptoms vary depending on the pathogen and the host. Some common symptoms include the following: frogeye or bull's eye spot marked with concentric rings; irregular, round tan spots with small black fruiting bodies; angular tan or black spots; black or tan spots surrounded by a yellow "halo"; oval-shaped leaf spots; and tan-to-gray spots with red or purple margins. Fungal leaf spot diseases are usually favored by wet seasons, high humidity, and/or frequent overhead irrigation. Many leaf spots cause premature defoliation.	*Alternaria LS*: aucuba, impatiens, marigold, zinnia *Bull's eye LS*: magnolia, maple *Cercospora LS*: buckeye, crape myrtle, leucothoe, laurel, red bud, rose *Entomosporium LS*: hawthorne, pear, photinia *Phyllosticta LS*: holly, magnolia, maple, witch hazel	Azoxystrobin, chlorothalonil, copper hydroxide, flutolanil, fludioxonil, iprodione, kresoxim methyl, mancozeb, metconazole, myclobutanil, propiconazole, tebuconazole, thiophanate methyl, triadimefon, trifloxystrobin, triflumizole

(*continued*)

TABLE 11.1 (Continued)
Chemical Control of Diseases of Seedlings, Cuttings, and Liners of Ornamental Plants

Disease	Susceptible Plants	Chemical Control
Shot-hole diseases. Some plants shed diseased leaf tissue in response to fungal or bacterial infections. Infected leaves are covered with circular, "shot" holes where diseased tissue has fallen out. Infected leaves may become chlorotic and drop prematurely. Shot-hole diseases may be caused by fungi or bacteria. Damage from shot-hole disease may be confused with insect feeding.	Almond, apricot, cherry, cherry laurel, peach, plum (plants in the genus *Prunus*)	*Bacterial shot hole*: copper hydroxide *Fungal shot hole*: myclobutanil, propiconazole
Anthracnose diseases. Anthracnose refers to diseases that cause leaf, stem, and/or fruit lesions. These diseases may appear as irregular leaf spots/lesions along leaf margins and across or between veins. Anthracnose may kill entire leaves, young shoots, and twigs, and cause premature defoliation. Diseased leaf tissue may fall out of leaf lesions. Stem cankers may form at the base of succulent shoots.	Ash, dogwood, euonymus, hosta, maple, oak, sycamore	Chlorothalonil, mancozeb, myclobutanil, propiconazole, thiophanate methyl
Crown gall. Rough-surfaced, hard or soft, spongy swollen tumors or galls up to several inches in diameter may form on stems or roots. Galls are induced by infection with *Agrobacterium tumefaciens*, a bacterium. Galls may be flesh colored, greenish, or dark. Galls are usually found near or below the soil line. Galls may form at wounds made during propagation. As galls continue to develop and enlarge, surface layers may become brown, woody, and roughened. Plants with crown gall usually become unthrifty and possibly stunted. Plant death may eventually occur.	Euonymus, holly, maple, peach, plum, rhododendron, rose, willow	No chemical controls. Use rogueing, crop rotation, and resistance. Clean pruning tools often or between plants.
Nematode diseases. Millions of nematodes may live in a square meter of soil; however, only few are parasites of plants. Most plant parasitic nematodes attack plant roots; some attack foliage. Nematode damage can be difficult to diagnose as most of the damage occurs belowground. Plants damaged by nematodes may appear stunted, unthrifty, and discolored, and have discolored roots with lesions or galls. One sure way to identify nematode problems is to submit a soil and/or root sample for analysis at a plant diagnostic laboratory; submit symptomatic foliage where foliar nematode is suspected.	*Root knot nematode*: abelia, aucuba, boxwood, dogwood, gardenia, holly, hydrangea, impatiens, ligustrum, nandina, photinia, rose *Foliar nematode*: African violet, begonia, hosta, many perennials *Lesion nematode*: boxwood, juniper	*Foliar nematode*: chlorfenapyr (greenhouse use only)

Note: For each fungicide, consult the pesticide label for the following information: registered for legal use in your locale, target disease, target host, cropping system (nursery, greenhouse, commercial, or residential landscape), rates, use intervals, compatibility with other pesticides, information on resistance management, mammalian toxicity and phytotoxicity, and other valuable information.

fungus gnats and shore flies. Space plants to allow for air movement and light penetration of the plants' canopies. Closely spaced plants lose lower leaves that are soon colonized by *Botrytis*.

- *Nutrient monitoring.* Monitoring the nutritional status of your crop along with media pH and soluble salts allow you to make adjustments to your fertility program before a major problem arises. Pythium root rot is favored by high soluble salt levels. At least 50% of the floral crop problems diagnosed by plant disease clinics are related to the lack of nutrient monitoring. Make sure that stock plants are well maintained and receive optimum amounts of nutrients. Fertilizer is not necessary for rooting cuttings or seed germination but will be needed shortly after.

- *Scouting.* At least one person should be designated as a scout for disease, insects, and early signs of poor plant growth. Spotting diseases early and rogueing infected plants is a sound management strategy. Scouts should note the plant species affected, make a tentative diagnosis, and note the location of the diseased plants.
- *Insect management.* Insects can vector plant pathogens that attack floral crops. WFT can transmit INSV; high thrips populations are sometimes responsible for the total losses of crops susceptible to INSV. Fungus gnats and shore flies can spread fungi such as *Pythium*, *Thielaviopsis*, and *Botrytis*. White flies can spread bacterial blight of geranium. Sticky cards that are monitored and changed frequently can give you important information about the insect populations in your greenhouse.
- *Weed control.* Weed-free greenhouse floors and a clean perimeter around the greenhouse can cut down on WFT and virus diseases such as INSV and TSWV. Weeds are often controlled by herbicides, hand weeding, or with a flame.
- *Fungicide drenches and sprays.* Most fungicides are not specifically labeled for use during plant propagation. If you are going to use a fungicide on young seedlings or unrooted cuttings, test the fungicide on a small group of plants and observe for obvious signs of phytotoxicity. Also, check to see if the fungicide has negatively affected root initiation, root number, or root mass on vegetatively produced cuttings.

Fungicides can be an important tool in the prevention (not curing) of fungal diseases (Table 11.1). You should be aware of which fungicides have efficacy against the target pest in question. Also, consider the use rate and interval prior to choosing a fungicide. The least expensive product is not always the product that is cheapest per pound. Many crops should be drenched with fungicides to prevent stem and root rots when rooted cuttings are potted. Fungicide sprays can be used to protect the foliage of plants susceptible to powdery mildew and botrytis blight. Rotate fungicides used for botrytis blight. Recent surveys have shown that, in many greenhouses, *Botrytis* isolates resistant to benzimidazole and dicarboximide fungicides are quite common.

Do not delay in asking for assistance. Acting quickly to get a pest or disease identified on the front end of an outbreak can save plants and your profits. Too often, growers wait too late in an epidemic to ask for assistance in identifying plant problems. Plant disease clinics at land grant universities and private laboratories can be important sources of information for solving disease problems.

EXERCISE

Experiment 1. Carrot Disc Assay for *Thielaviopsis basicola*, the Pathogen Causing Black Root Rot on *Ilex crenata*

Black root rot is commonly found on the roots of several herbaceous and woody ornamentals, especially cultivars of *Ilex crenata*. If Japanese holly is not available, other hosts are blue holly (*Ilex* × *Meserve*), inkberry (*Ilex glabra*), pansy, verbena, vinca, and *Calibrachoa*. Black root rot does not quickly kill the plant, but may cause stunting, chlorosis, and root death. It is important to take cuttings for propagation from plants that are free of the disease. One way to determine root health is to use bioassays for the presence of the pathogen. The aim of this experiment is to determine if the root system of Japanese holly is infected with *Thielaviopsis basicola*. This experiment will require about 1 to 2 h in the course of two laboratory periods.

MATERIALS

The following items are needed for this laboratory exercise:

- Whole, fresh carrots from a local grocery
- Two to three 1-gal containers of *Ilex crenata* 'Helleri' or 'Soft Touch' exhibiting chlorosis and discolored roots
- 2 to 3 pairs of scissors
- 2 to 3 paring knives

FIGURE 11.10 *Thielaviopsis* growing from infected root segments onto carrot discs.

Procedure 11.1
Carrot Disc Assay for *Thielaviopsis basicola,* the Pathogen Causing Black Root Rot on *Ilex crenata*

Step	Instructions and Comments
1	Moisten a paper towel with tap water and place it in the bottom half of a Petri dish.
2	Wash and blot carrots dry. Cut carrots into discs, 0.5 cm thick, and place on the moist paper towel in the Petri dish.
3	Remove plants from containers and gently remove the soilless mix from the root zone.
4	Use scissors to cut and remove dark, discolored root segments. Place root segments in a sieve and wash under tap water. Blot dry with paper towels.
5	With forceps, place two to three root segments on the upper surface of each carrot disc. Cover with the top of the Petri dish and incubate at room temperature for 5 to 7 days. For an untreated control, do not place roots on the carrot discs and incubate with other plates.
6	Examine plates at days 5 to 7 for signs of black mycelium (Figure 11.10). Place the Petri dish onto the stage of a dissecting microscope and look for aleuriospores of *Thielaviopsis* at 10 to 40×.
7	With forceps, transfer a small amount of the carrot tissue with aleuriospores to a microscope slide, add a drop of water, cover with a coverslip, and press gently to flatten the mount. View with a compound microscope at 40 to 100×. View aleuriospores and phialospores at this magnification.

- Forceps
- 4 plastic Petri dishes for each student
- Paper towels
- Sieve or colander to wash roots

Follow the instruction in Procedure 11.1 to complete this experiment.

ANTICIPATED RESULTS

After 5 to 7 days, mycelium of *T. basicola* will form a black mat on the carrot discs if root segments were positive for black root rot. Darkly pigmented, segmented aleuriospores will be visible with a dissecting microscope. Aleuriospores will form along the root segment, throughout the mycelia mat, and on the surface of the carrot disc. *Thielaviopsis* may also form phialides and small, hyaline phialospores. *Thielaviopsis* should not be present on the carrot discs that served as the untreated controls.

QUESTIONS

- What strategies could growers use to identify and minimize the incidence of black root rot?
- How do infected plants move through a production nursery and end up at a retail garden center? Why are diseased plants not detected and discarded?

SUGGESTED READING

Albajes, R., M. L. Gullino, J. C. van Lenteren, and Y. Elad. 1999. *Integrated Pest and Disease Management in Greenhouse Crops*. Kluwer Academic Publishers, 545 pp.

Daughtrey, M. L., R. L. Wick, and J. L. Peterson. 1995. *Compendium of Flowering Potted Plant Diseases*. St. Paul, Minnesota: APS Press, 90 pp.

Griesbach, J. A., J. L. Parke, G. A. Chastagner, N. J. Grünwalde, and J. Aguirre. 2012. *Safe Procurement and Production Manual: A Systems Approach for the Production of Healthy Nursery Stock*. Oregon Association of Nurseries, 108 pp. http://www.oan.org/associations/4440/files/pdf/SafeProduction.pdf. Accessed on March 31, 2014.

Horst, R. K. and P. E. Nelson. 1997. *Compendium of Chrysanthemum Diseases*. St. Paul, Minnesota: APS Press, 62 pp.

Jarvis, W. R. 1992. *Managing Diseases in Greenhouse Crops*. St. Paul, Minnesota: APS Press, 288 pp.

Jones, R. K. and D. M. Benson. 2001. *Diseases of Woody Ornamentals and Trees in Nurseries*. St. Paul, Minnesota: APS Press, 482 pp.

Laney, A. G., K. E. Keller, R. R. Martin, and I. E. Tzanetakis. 2011. A discovery 70 years in the making: Characterization of the rose rosette virus. *J Gen Virol* 92:1727–1732.

Parke, J. L. and N. J. Grunwald. 2012. A systems approach for management of pests and pathogens of nursery crops. *Plant Dis* 96:1236–1244.

Trigiano, R. N., M. T. Windham, and A. S. Windham. 2007. *Plant Pathology Concepts and Laboratory Exercises*, Second Edition. Boca Raton, Florida: CRC Press, 523 pp.

12 Botrytis and Other Propagation Pathogens

Mark P. McQuilken

Concept Box 12.1

- Botrytis gray mold (*Botrytis cinerea*) and damping-off (*Pythium*, *Phytophthora*, and *Rhizoctonia*) are common diseases during plant propagation, causing serious economic losses.
- Gray mold often affects the stem cuttings of ornamental shrubs and herbaceous plants during rooting.
- Damping-off prevents seed germination and kills young seedlings.
- A combination of fungicide sprays and cultural measures, such as good crop hygiene and increased ventilation, provides an effective control of gray mold.
- Commercial biocontrol products for incorporation into a seed sowing medium or applied as a drench (e.g., Prestop) are available for the control of damping-off.
- Three experiments described in this chapter show the typical symptoms of gray mold and damping-off. Fungicide sprays and increased ventilation are evaluated for the control of gray mold on heather cuttings, and Prestop®, for the biocontrol of damping-off in cress.

There is a great risk of attack by a range of diseases during plant propagation (Chapter 11). The environmental conditions are often conducive to serious outbreaks of *Botrytis* gray mold on rooting stem cuttings and damping-off during seed germination and seedling growth. Both of these major diseases are the subject of this experimental chapter.

Botrytis gray mold, caused by the fungus *B. cinerea*, remains one of the most damaging diseases affecting the stem cuttings of ornamental shrubs such as heather (*Calluna vulgaris* [L.] Hull), sunrose (*Helianthemum* nummularium [L.] Mill.), lavender (*Lavandula* spp.), *Rhododendron* spp., and *Hebe* spp. It also affects herbaceous plants such as common plants (e.g., *Pelargonium* × *hortorum* L. H. Bailey), *Cineraria*, *Cyclamen* spp., primroses (*Primula* spp.), and *Fuchsia* spp., as well as vines such as English ivy (*Hedera helix* L.) during rooting (Figures 12.1 and 12.2). It typically starts from the cut end or the foliage tips, and symptoms include dieback, severe browning followed by poor rooting, and cutting death. Characteristic furry, gray-brown masses of spores often develop on affected cuttings. The main sources of gray mold are diseased plants and crop debris. Infections can often remain in a latent state for several weeks, and cuttings taken from apparently healthy stock plants may be infected. Spores produced on diseased plants and on dead and decaying crop debris spread to other plants on air currents and by water splash. Spores landing on susceptible plant tissues can germinate and invade tissue immediately, or remain dormant for up to several weeks and infect plants when conditions become more favorable for infection.

Damping-off (Figures 12.3–12.5) is, by far, the most common disease during seed germination and seedling growth. Fungal species of *Rhizoctonia*, *Fusarium*, and even *B. cinerea*, as well as fungal-like pathogens, such as species of *Pythium* and *Phytophthora*, may be the cause of economically important diseases, resulting in a serious loss of seeds, seedlings, and young plants during propagation. These diseases occur at various stages of production. Seeds or seedlings can be attacked before emergence, resulting in decay and rotting (preemergence damping-off). Seedlings that have already emerged are attacked at the roots and stems at or below the surface of the propagation medium. The invaded tissue becomes water soaked and discolored, and the seedling collapses and dies (postemergence damping-off). Sometimes, the

FIGURE 12.1 Foliar lesions caused by *Botrytis cinerea* on cineraria. Note the necrotic lesion with gray "fuzz," which is sporulation of the pathogen.

FIGURE 12.2 Severe foliar browning, scorching, and sporulation of *Botrytis cinerea* on heather.

FIGURE 12.3 Diseases caused by *Rhizoctonia*. (a) Damping-off of impatiens. (b) Stem rot of oakleaf hydrangea. Note the discolored (red-to-brown) lesions. (Photographs courtesy of Alan S. Windham, University of Tennessee.)

seedling may remain alive and standing, but the stem becomes girdled and stunted and eventually dies (wire stem). In larger plants, the damping-off pathogens may attack and kill the rootlets. Such attacks cause plants to become stunted and eventually die (root rot). Species

FIGURE 12.4 Postemergence seedling damping-off. Note that the tissue is water soaked and discolored, and the seedlings have collapsed and died.

FIGURE 12.5 Damping-off. (a) Damping-off of Boston ivy. (b) Enlargement of roots and stems shown in (a). Note that the lateral roots are also affected by the pathogen. (Photographs courtesy of Alan S. Windham, University of Tennessee.)

susceptible to damping-off include the following: pansy (*Viola* spp.), *Lobellia* spp.; alyssum (*Lobularia maritima* [L.] Desv.); *Salvia* spp.; and many other common plants, such as pea (*Pisum sativum* L.), beet (*Beta vulgaris* L.), tomato (*Lycopersicon esculentum* Mill.), and cucumber (*Cucumis sativus* L.).

The following three experiments will show the typical symptoms of gray mold and damping-off, and demonstrate how these two important diseases can be controlled and managed during propagation. Teams of three to five students are recommended for Experiments 1 and 2, and teams of two to three students, for Experiment 3.

GENERAL CONSIDERATIONS

Growth of Stock Plants and Cutting Propagation

Stock plants of *Calluna vulgaris* are maintained to provide a source of cuttings for Experiments 1 and 2. Eight to ten stock plants will provide enough material for a class of 10 to 15 students to conduct both experiments. The cultivars Arran Gold, Dark Beauty, Flamingo, Robert Chapman, Silver Queen, Sun Rise, or Velvet Fascination are recommended as they are very susceptible to gray mold. Grow stock plants in 2-L pots containing sphagnum peat medium amended with dolomitic limestone (1.8 kg/m^3), superphosphate (0.75 kg/m^3), fritted trace elements (0.3 kg/m^3), and controlled-release fertilizer (1.0–1.5 kg/m^3, either 8–9- or 12- or 14-month formulation). Addition of a wetting agent will aid water absorption by the peat. It is a good idea to spread pine bark on the top of the medium to reduce moss growth. Ideally, grow the stock plants under a shaded polyethylene tunnel with well-spaced capillary sand beds to keep the foliage dry and reduce the risk of *Botrytis* and *Rhizoctonia*. Re-pot stock plants every year, and replace every third year. Once cuttings are taken, prune plants to produce even growth.

Small branches of semi-ripe material should be cut with a pair of scissors from March/April until September. Cuttings are prepared by breaking 15 to 20 mm of stem from the branch and drawing fingers down the stem to remove lower leaves. For both experiments, fill 15 plastic trays (~23 × 18 × 6 cm) with a peat, fine bark mixture (1:1 v/v) propagation medium without any lime or fertilizers. Insert 30 cuttings into each tray with pre-made holes. For rooting, place trays of cuttings on capillary sand beds, under low-polyethylene propagation tunnels within a glass-covered greenhouse. Alternatively, cuttings can be rooted in the moist atmosphere created under a polyethylene sheet supported just above the cuttings. A polyethylene sheet laid directly over the cuttings should be avoided because its removal lifts out the cuttings from the medium. The objective is to provide the high-humidity, stagnant-air environment that favors *Botrytis* growth; therefore, a clear plastic bag that is sealed at the top should also be sufficient. Suitable alternative cuttings to use include lavender, geranium, *Cyclamen* spp., primroses, *Hebe* spp., and *Fuchsia* spp., as well as vines, such as English ivy.

Botrytis cinerea Inoculum

Propagation beds and glasshouses that are used regularly for producing plants should contain sufficient amounts of *B. cinerea* inoculum (spores) to initiate the infection of cuttings. Natural inoculum can be supplemented by placing sporulating cultures (lids of Petri dishes removed) of *B. cinerea* within the experimental area. Using aseptic technique, sporulating cultures are produced by centrally inoculating plates of Difco 3% (w/v) malt extract agar, which should be prepared according to the manufacturer's instructions printed on the product label), in Petri dishes (9-cm diameter) with a 0.5-cm^2 agar plug of *B. cinerea*, and growing the pathogen for 7 to 10 days under fluorescent light (16-h photoperiod) at 20 to 22°C. Alternatively, commercial preparations of nutrient agar or potato dextrose agar (Fisher Scientific, Atlanta, Georgia) are suitable for growing inoculum.

Disease Assessment

Assess each cutting for incidence of *Botrytis* (dieback; browning; and/or the presence of furry, gray-brown mass of spores) and disease severity, 4 and 8 weeks after the start of each experiment. Disease severity is rated on a 0 to 5 scale (% proportion of cutting with browning), where 0 = no visible symptoms/healthy cutting; 1 = up to 25%; 2 = 26%–50%; 3 = >50%–75%; 4 = >75%; and 5 = cutting totally brown and dead.

Experimental Design and Statistical Analyses

Arrange Experiments 1 and 3 in a randomized block design with a minimum of four replicates for each treatment. Use a split-level design for Experiment 2, with each of the two propagation beds being a treatment. Data from Experiments 1 and 3 are analyzed using an analysis of variance (ANOVA) after the angular transformation of percentage data where required. Compare treatments using the least significant difference test at a probability of 5% ($P = 0.05$). Use a Student's t test to analyze data from Experiment 2.

EXERCISES

Experiment 1. Development and Chemical Control of Gray Mold on *Calluna vulgaris* Cuttings

Gray mold is difficult to control on cuttings because the conventional environment for rooting is conducive to serious outbreaks. The damp and humid environment established for optimum propagation promotes the spore germination of the pathogen and disease spread. A single-control measure is unlikely to be effective, and integrated control is more likely to provide effective and durable disease control. Fungicide sprays, in combination with cultural control measures, such as practicing good crop hygiene and reducing humidity, are recommended for effective control.

Preventative spray programs, as part of an integrated control program, are generally the most effective. Once gray mold is established, spraying often provides very poor disease control. The aim of this experiment is to compare the efficacy of two foliar-applied fungicides, iprodione (or thiram) and azoxystrobin (or pyrimethanil or cyprodinil), against gray mold on *C. vulgaris* cuttings. Iprodione is slightly phytotoxic at recommended rates, and products that contain it are labeled either "general use pesticides" or, under some circumstances, "restricted use pesticides." Azoxystrobin is a "reduced risk" fungicide. The exercise will require about 8 to 10 h in the course of 8 weeks to be complete.

MATERIALS

The following items are needed for each group of three to five students:

- Access to healthy, disease-free stock plants of *C. vulgaris*
- 2 to 3 pairs of scissors
- Peat bark propagation media
- 12 plastic trays (approximately 23 × 18 × 6 cm)
- Labels and permanent marker
- Calculator

The following items are needed for each class of students:

- Fungicide A (iprodione; Rovral and other trade names, such as Chipco 26019, DOP 500F, Kidan, LFA 2043, NRC910, and Verisan)
- Fungicide B (azoxystrobin; Amistar and other trade names, such as Abound, Bankit, Heritage, and Quadras)
- Azo precision sprayer fitted with a medium flat fan nozzle
- Personal protective clothing (overall, boots, gloves, face shield, and mask)
- Propagation beds with sub-irrigation

Caution: It is recommended that fungicide sprays are prepared and applied by a trained competent member of the staff with proper pesticide applicator's licensing and wearing the appropriate personal protective clothing.

Follow the protocol outlined in Procedure 12.1 to complete this experiment.

Procedure 12.1
Development and Chemical Control of Gray Mold on *Calluna vulgaris* Cuttings

Step	Instructions and Comments
1	Fill 12 plastic trays (~23 × 18 × 6 cm) with peat, bark propagation medium. Take cuttings from stock plants, prepare, and insert into each tray (30 per tray) as described under stock plants and cutting preparation.
2	Label four replicate trays "Control—Water" and spray with tap water.
3	Label four replicate trays "Fungicide A—Iprodione" and the remaining four "Fungicide B—Azoxystrobin." Place the trays in a designated spraying area for spraying by a trained member of staff.
4	Preparation and application of fungicides by trained staff only Wearing appropriate personal protective clothing, prepare the following rates of fungicide products per liter of water: iprodione (as Rovral WP; 50% a.i. w.p.), 4 g; and azoxystrobin (as Amistar, 25% a.i. s.c.), 4 mL. Adjust rates if different products of these fungicides are used. Apply fungicides to the appropriate labeled trays of cuttings (~10–15 mL per tray) using an Azo precision sprayer (first spray).
5	After 24 h post-spraying, arrange the trays in a randomized block design under a low-polyethylene tunnel on a sand bed with sub-irrigation and drainage or other suitable high-humidity, stagnant-air environment. Introduce inoculum of *B. cinerea* if necessary as described under inoculum production.
6	After 2 weeks, remove the trays from the low-polyethylene tunnels and place them in a designated spraying area for respraying (second spray) by a trained member of the staff.
7	Wearing gloves, replace the sprayed cuttings into the high-humidity environment.
8	Assess each cutting for incidence and severity of *Botrytis* after 4 and 8 weeks as described under disease assessments. The class should share and analyze the data using ANOVA after the angular transformation of percentage disease incidence data.

Caution: Always wash your hands after handling trays of treated cuttings.

ANTICIPATED RESULTS

After 2 to 3 weeks, cuttings in the untreated trays will show typical symptoms of *Botrytis*. Symptoms should first appear on the leaf tips or the cut ends and will include dieback, browning, and cutting death. Sporulating *B. cinerea* may also be seen as characteristic furry, gray-brown masses of spores on affected cutting parts. After 4 and 8 weeks, disease incidence and severity should be greatest on the untreated cuttings. The number of rooted cuttings is also likely to be reduced. Disease incidence and severity should be less on fungicide-treated cuttings. Of the two fungicides (iprodione and azoxystrobin) tested, disease incidence and severity should be less on cuttings treated with azoxystrobin. A reduction in size may be observed in some varieties of cuttings treated with azoxystrobin, demonstrating slight toxicity of the fungicide.

QUESTIONS

- Several propagators and growers have reported poor control of *Botrytis* using iprodione. Why?
- What strategies should a propagator/grower adopt to minimize the development of fungicide resistance?
- Describe the mode of action of azoxystrobin against *B. cinerea*.
- Why are fungicides rarely incorporated into a propagation medium when rooting cuttings?

Experiment 2. Effect of Ventilating Propagation Beds on the Development and Control of Gray Mold on *Calluna vulgaris* Cuttings

Prolonged high humidity and/or leaf-surface wetness favor infection by *B. cinerea* and the subsequent development of gray mold. Therefore, crop management practices such as ventilation, which reduces prolonged periods of high humidity, could affect disease development (see Chapter 11). Cuttings grown under reduced humidity should be at less risk of infection. This exercise is designed to demonstrate the effect of ventilating propagation beds, by opening low-polythene tunnels periodically, on the development of gray mold on *C. vulgaris* cuttings. The exercise will require about 8 to 10 h in the course of 8 weeks to be complete.

MATERIALS

The following items are needed for each group of three to five students:

- Access to healthy, disease-free stock plants of *C. vulgaris*
- 2 to 3 pairs of scissors
- Peat, bark propagation media
- 8 plastic trays (approximately 23 × 18 × 6 cm)
- Labels and permanent marker
- Calculator

Procedure 12.2
Effect of Ventilating Propagation Beds on Development and Control of Gray Mold on *Calluna vulgaris* Cuttings

Step	Instructions and Comments
1	Fill eight plastic trays (~23 × 18 × 6 cm) with peat, bark propagation medium. Take cuttings from stock plants, prepare, and insert into each tray (30 per tray) as described under stock plants and cutting preparation.
2	Label four replicate trays "Unventilated" and the remaining four "Ventilated."
3	Place the four replicate "No ventilation" trays under a low-polyethylene tunnel on a sand bed with sub-irrigation and drainage, together with the trays of other groups of the class. Label the propagation bed "Unventilated," and do not remove the polyethylene during the course of the exercise. Polyethylene tents over greenhouse benches or tables near windows may also be used provided that there is sub-irrigation.
4	Place the four replicate "Ventilated" trays under a second low-polyethylene tunnel on a sand bed with sub-irrigation and drainage, together with the trays of other groups of the class. Label the propagation bed "Ventilated." During the course of the exercise, ventilate the bed by opening the tunnel on alternate mornings for 4 h.
5	Assess each cutting for incidence and severity of *Botrytis* after 4 and 8 weeks as described under disease assessments. The class should share and analyze the data using a Student's *t* test.

The class requires two propagation beds with sub-irrigation: one ventilated and one unventilated.

Follow the protocols described in Procedure 12.2 to complete this experiment.

ANTICIPATED RESULTS

Cuttings will show typical symptoms of *Botrytis* after 2 to 3 weeks as in Experiment 1. After 4 and 8 weeks, disease incidence and severity should be greatest on the "unventilated" and least on the "ventilated" cuttings.

QUESTIONS

- How may periodic ventilating of propagation beds affect the rate of rooting?
- When is prolonged high humidity in a greenhouse likely to occur?
- Explain why it is important to maintain stock plants free of gray mold and not to take cuttings from diseased plants.

Experiment 3. Biological Control of *Pythium* Damping-Off During Seed Germination

Controlling disease caused by *Pythium*, *Phytophthora*, and *Rhizoctonia* during seed germination and seedling establishment is one of the most important tasks for the propagator. Traditional methods of control have included the integration of cultural practices, such as the manipulation of the environment, and the application of fungicides to the growing media. However, fungicide application is not the most desirable method of controlling damping-off for several reasons. For example, they are expensive, and their registration and use varies between countries. Pathogenic strains of *Pythium* have also developed resistance, or considerable tolerance, to the commonly used fungicides etridiazole, furalaxyl, and propamocarb hydrochloride. Furthermore, there are often problems with chemical treatments causing the stunting and chlorosis of young seedlings. The difficulty in managing damping-off has increased the need for alternative non-chemical methods of disease control. Considerable progress has been made in the development of bacteria and fungi as commercial biocontrol products for the control of plant diseases (Whipps and McQuilken, 2009). The number of products available continues to increase, although there is an ongoing flux as some products appear and others are removed from the market. The majority of the products are sold for the control of seed- or soil-borne pathogens, with fewer for foliar and post-harvest pathogens.

Prospects are good for the control of damping-off by treating the seed, roots, or the propagating medium with a number of commercial biocontrol products. For example, Verdera Oy (Finland) has produced Prestop®, a 32% w/w wettable powder formulation containing mycelium and spores of *Clonostachys rosea* f. *catenulata* (=*Gliocladium catenulatum*) strain J1446. It is effective against *Pythium* and *Rhizoctonia*, which are the major causes of seed decay and pre- and post-emergence damping-off during propagation. Prestop® is incorporated into a seed sowing medium or applied as a drench for the control of damping-off. The recommended rate for the incorporation of the product before sowing is 0.26 to 0.52 g per 1 L of seed sowing medium. It is incorporated evenly as a 2.5% (w/v) aqueous suspension to ensure effective control. The aim of this experiment is to evaluate a seed sowing medium incorporation of Prestop® for the control of pre- and postemergence damping-off in cress (*Lepidium sativum* L.) caused by *Pythium ultimum*. Cress seed is used in this experiment because it is relatively cheap compared to other seeds (e.g., herbaceous bedding plant seed) and is an ideal size for ease of sowing. Suitable alternative seeds to use include pansy, *Lobelia* spp., alyssum, *Salvia* spp., pea, beet, tomato, and cucumber. The exercise will require about 6 to 8 h in the course of 4 weeks to be complete.

MATERIALS

The following items are needed for the technical staff to produce a supply of *P. ultimum* inocula for five groups, each of two to three students, to use in the experiment:

- Pathogenic culture of *P. ultimum*
- Cornmeal agar (CMA; Fisher Scientific, Atlanta, Georgia)
- Petri dishes
- Twenty 250-mL Erlenmeyer flasks
- 1-L distilled water
- 1-L and 50-mL measuring cylinders
- 1-L Campbell's V8 tomato juice
- 200 g of vermiculite (grade DSF)
- Autoclave
- Sterile spatula
- Five 8 × 5 cm containers with lids

The following items are needed for each group of two to three students:

- 1-L distilled water
- 12 g of vermiculite (pathogen free)
- 24 g *P. ultimum* vermiculite inoculum
- 200-mL beaker
- 100-mL measuring cylinder
- 2.5 g cress seed (~350 seeds per gram)
- 2 large plastic bags
- 18-L peat-based propagating media (without sand) designed for seed sowing and general purpose propagation
- 12 plastic seed trays (~20 × 15 × 5 cm)
- Wooden striking-off ruler and presser board

Procedure 12.3
Production of *Pythium ultimum* Inoculum by Technical Staff

Step	Instructions and Comments
1	Prepare CMA according to the manufacturer's instructions printed on the product, label, and autoclave at 120°C and 103 kPa for 15 min. Use aseptic technique to pour plates of CMA in Petri dishes (9 cm diameter).
2	Culture *P. ultimum* by inoculating three to five plates of CMA with a 0.5 cm^2 agar plug of the pathogen, and incubate at 20 to 22°C in the dark for 3 days.
3	Place 10 g samples of vermiculite in each of the twenty 250 mL Erlenmeyer flasks, and moisten with 35 mL V8 juice broth (200 mL V8 tomato juice, 2 g $CaCO_3$, and 800 mL distilled water).
4	Place foam stoppers in each flask and wrap with aluminium foil.
5	Autoclave flasks at 120°C and 103 kPa for 15 min, cool, and inoculate each flask with two to three 0.5-cm diameter discs cut from a three-day-old culture of *P. ultimum*.
6	Incubate flasks for 12 to 14 days at 25°C. Shake flasks vigorously by hand each day to ensure mixing of nutrients.
7	Empty the contents of each flask into a 1 L beaker, and mix thoroughly with a sterile spatula.
8	Add 30 g of *P. ultimum* inocula to each of the five 8 × 5 cm containers. Attach the lids and use inoculum within 1 to 3 days.

Procedure 12.4
Biological Control of *Pythium* Damping-Off During Seed Germination

Step	Instructions and Comments
1	Add 6 L of seed sowing medium and 12 g of vermiculite (pathogen free) to a large plastic bag. Add 120 mL of distilled water and mix thoroughly. Label the bag "Healthy control."
2	Add 6 L of seed sowing medium and 12 g of *P. ultimum* vermiculite inocula (equivalent to 2 g per 1 L) to a large plastic bag. Add 120 mL of distilled water and mix thoroughly. Label the bag "Pathogen control."
3	Repeat step 2 above and label the bag "Pathogen + Prestop®." Prepare a 2.5% (w/v) aqueous suspension of Prestop® in a 200-mL beaker by mixing 3.75 g of the product in 70 mL of distilled water. Agitate sufficiently to disperse the product evenly, and dilute further with 80 mL of distilled water to give a final volume of 150 mL. Add 150 mL of the 2.5% (w/v) aqueous suspension to the pathogen-infested seed sowing medium and mix thoroughly.
4	Fill four plastic seed trays (20 × 15 × 5 cm) with seed sowing medium from the bag labeled "Healthy control" (~1.5 L per tray). Overfill each tray slightly and use a striking-off ruler from the middle of each tray outward to remove excess medium. Press the medium surface with a presser board to create a firm level surface for seed sowing. Label the four replicate trays "Healthy control," and sow each tray with 60 seeds (six rows of 10 seeds each, with each seed being ~2 cm from its neighbor and 5 mm deep).
5	Fill four seed trays with propagating medium from the plastic bag labeled "Pathogen control" as described in step 4. Label four replicate trays "Pathogen control," and sow each tray with 60 seeds as described in step 4.
6	Fill four seed trays with propagating medium from the plastic bag labeled "Pathogen + Prestop®" as described in step 4. Label four replicate trays "Pathogen + Prestop®," and sow each tray with 60 seeds as described in step 4.
7	Arrange the trays in a randomized block design in a controlled environment growth room or glasshouse maintained at 18 to 22°C, with alternating periods of 16 h light and 8 h dark. Water trays regularly with a fine rose to keep the propagating medium moist, but not saturated.
8	Record diseased seedlings every 3 to 5 days for 21 days. Calculate the total percentage of mortality (both pre- and postemergence damping-off) after 21 days for each replicate. The class should share and analyze the data using ANOVA after the angular transformation of percentage data. Caution: Always wear a laboratory coat and waterproof gloves when handling the product and seed trays, and wash your hands afterward.

- Labels and permanent marker pen
- 2 to 3 pairs of small blunted forceps for seed sowing
- Calculator

The following items are needed for each class of students:

- 100 g pack of Prestop® available in the United States from MGS Horticultural U.S.A., Inc., Detroit, Michigan. Details of the product distributors in other countries are listed on the Verdera Oy Web site (http://www.verdera.fi/en/front-page). Prestop® will keep unopened for 1 year if stored in cool, dry conditions below 8°C.

Caution: It is essential to wear a laboratory coat and waterproof gloves when handling the product and wash your hands afterward.

Follow the protocols outlined in Procedures 12.3 and 12.4 to complete this experiment. Procedure 12.3 should be conducted by a technical staff at least 2 weeks before the start of the experiment to produce a supply of *P. ultimum* inoculum for five groups of two to three students to use.

ANTICIPATED RESULTS

Damping-off should not occur in trays not infested with *P. ultimum* ("healthy control"), whereas extensive damping-off will occur in trays infested with *P. ultimum* ("pathogen control"). Preemergence damping-off (i.e., seeds rotted in the growing medium or seedlings failed to emerge) should be evident within 3 to 5 days. Postemergence damping-off (seedlings emerged, but collapsed and died) should be evident within 7 to 10 days and will increase to reach a maximum by 21 days. There should be less damping-off in trays infested with *Pythium* and treated with Prestop® ("pathogen + Prestop®").

QUESTIONS

- Outline how you would isolate *Pythium* from a diseased seedling and undertake Koch's postulates (Mullen, 2007).
- How may the propagator spread *Pythium*? State the main factors that encourage damping-off.
- List the key diseases that Prestop® controls.
- What are the main factors affecting the efficacy of Prestop?
- What are the constraints in using commercial biocontrol products?
- Identify commercial biocontrol products available to the propagator.

LITERATURE CITED AND SUGGESTED READING

Chatterton, S. and Z. K. Punja. 2010. Factors influencing colonization of cucumber roots by *Clonostachys rosea* f. *catenulata*, a biological disease control agent. *Biocontrol Sci Tech* 20:37–35.

International Plant Propagator's Society. http://www.ipps.org/. Accessed on March 10, 2014.

McQuilken, M. P. 2000. Evaluation of novel fungicides and irrigation methods for gray mold control on *Calluna vulgaris*. *Comb Proc Intl Plant Prop Soc* 50:137–141.

McQuilken, M. P. and J. Thomson. 2008. Evaluation of anilinopyrimidine and other fungicides for control of gray mould (*Botrytis cinerea*) in container-grown *Calluna vulgaris*. *Pest Manag Sci* 64:748–754.

McQuilken, M. P., J. Gemmell, and M.-L. Lahdenperä. 2001. *Gliocladium catenulatum* as a potential biological control agent of damping-off in bedding plants. *J Phytopath* 149: 171–178.

MGS Horticultural, Inc. http://www.mgshort.com. Accessed on March 10, 2014.

Mullen, J. M. 2007. *Plant disease diagnosis*. In Plant Pathology Concepts and Laboratory Exercises, 2nd ed., Trigiano, R. N., M. T. Windham, and A. S. Windham (Eds.) CRC Press, Boca Raton, FL. pp. 447–463.

O'Neill, T. M. and M. P. McQuilken. 2000. Influence of irrigation method on development of gray mould in glasshouse crops of *Calluna*, *Cyclamen*, and *Primula*. *Proc BCPC* 2000: 267–272.

Verdera Oy. http://www.verdera.fi/en/front-page. Accessed on March 10, 2014.

Whipps, J. M. and McQuilken, M. P. 2009. *Biological Control Agents in Plant Disease Control*. In: Disease Control in Crops: Biological and Environmentally Friendly Approaches (Ed. D. Walters), Wiley-Blackwell, Oxford, U.K. Chapter 3.

13 Disinfestation of Soil and Planting Media

James J. Stapleton

Infestation of nursery soil and planting media with pests, including pathogens, nematodes, insects, and weed propagules, can be extremely destructive to crops. The pests may be disseminated widely via plant shipments and cause "exotic" pest outbreaks in their new locations. This, in turn, may trigger financially ruinous regulatory quarantines, lawsuits, and/or customer boycotts. To protect nursery industries, national and regional governments usually require specific soil treatments and handling procedures to ensure against infestation by certain soilborne pests of field-, container-, flat-, and frame-grown nursery stock for commercial planting. Ornamentals and plants destined for noncommercial uses, such as in homes and gardens, are sometimes exempted from these regulations. Chemical fumigation and heat treatments have been widely used for many years to disinfest soil and planting media for nursery production.

For the purpose of demonstrating soil disinfestation to students, the use of dangerously toxic chemical fumigants is not recommended. The same eradicative effects may be obtained with heat, using the relationship of (temperature × exposure time) to achieve lethal dosage rather than (active ingredient concentration × exposure time), which governs lethal dosage of chemical toxicants. In addition to the soil or planting media, nursery plant containers for reuse also should be disinfested.

THERMAL DISINFESTATION

Exposure of soil or planting media to excessively high temperature (greater than 80°C [176°F]) may release phytotoxic compounds via thermal decomposition of the organic constituents. For that reason, the use of aerated steam at a lower temperature (e.g., 70°C [158°F]) rather than live steam or autoclaving may be preferable. The introduction of moisture to heated soil or planting media will allow more uniform heating and will also hydrate resistant structures of pests to increase their metabolic rate and render them more susceptible to treatment effects. Certain solarization techniques that heat soil and planting media to 60 to 70°C (140–158°F) may also be useful.

FACILITIES

This experiment may be conducted in the laboratory, greenhouse, or a growth chamber.

EXERCISE

Experiment 1. Effect of High-Temperature Dosages on Soilborne Pest Survival

This procedure describes the use of weed propagules as test organisms for the determination of soil disinfestation because they are the most ubiquitous. However, other pest organisms, such as fungal and bacterial plant pathogens, phytoparasitic nematodes, and soil-dwelling arthropods may be used if present in the experimental soil or planting medium. In this case, appropriate materials and procedures for the determination of numbers should be devised by the instructor. Also, portions of the treated or untreated soil may be used to conduct bioassays of pest survival and activity on susceptible plants. Achievement of lethal (temperature × exposure time) dosage will provide 100% control of unwanted pest propagules. Sublethal dosages will allow survival of varying levels of pest organisms. Students should use appropriate caution when working with hot water to avoid spills or burns.

MATERIALS

The following items are needed for individual students or teams:

- Sufficient soil or planting media infested with known pest organisms (pathogens, nematode parasites, insects, and/or weed propagules). Seeds (50–100 per tube) from weedy plants important to local horticulture can be used from previous collection of wild sources. Alternatively, check with weed scientists for seed supplies and suggestions. The same principles can be illustrated, substituting crop or turfgrass seeds for weed seeds (Figure 13.1).

- Soil or planting media should be uniformly moistened to approximate field capacity prior to experiment initiation.
- Three large glass or plastic culture tubes (e.g., 25 × 150 mm) with fitted lids or other closures
- Two tube racks
- Three 100 × 15 mm Petri dishes for each treatment
- Three discs of 14-mm #1 filter paper for each treatment
- One or more large plastic vegetable crispers
- One squirt bottle for distilled water
- Two constant-temperature water baths
- Thermometer or electronic temperature logger

Follow the instructions provided in Procedure 13.1 to complete this experiment.

ANTICIPATED RESULTS

Numbers of weed propagules and other pest organisms will be maximal in soil or planting media from the non-heated (ambient) treatment. Reduced survival will be noted in the treatment heated for 10 min, whereas

Procedure 13.1
Effect of High-Temperature Dosages on Soilborne Pest Survival

Step	Instructions or Comments
1	Treat the soil or planting medium to be disinfested at a constant temperature of 60°C (140°F). Place sufficient untreated moist soil or planting medium to fill approximately half the volume of each of the tubes when the soil is packed by gently tapping the bottom of the tube on a solid surface. Prepare three tubes for each treatment.
2	Cover tube openings and place in racks that are then immersed in water baths for treatment. The surface of the water bath should reach to at least 25 mm above the level of the soil in the tube. Control (unheated) tubes are maintained in a water bath at room temperature (e.g., 18–24°C [64–75°F]). Use appropriate care with hot water to avoid spills or burns.
3	Remove one tube from the 60°C water bath after 10 min of exposure; remove the other tube after 60-min exposure time (more tubes may be prepared for other intervals of time, e.g., 20, 30, 40, and 50 min if desired). After the tubes are lifted from the water baths, soil is removed by tapping the inverted tubes onto moistened filter paper discs in three labeled 100 × 15 mm Petri dishes containing moistened discs of #1 filter paper. If the soil will not tap out of the tubes, scraping tools may be used, taking care not to contaminate treated soil with less treated soil.
4	Spread soil over filter paper surfaces and cover with lids.
5	Place the Petri dishes in the vegetable crisper to retain moisture and then incubate in a growth chamber, a greenhouse bench, or other location with a light/dark and temperature cycle appropriate for the growth of weed and/or bioassay plants.
6	Numbers of germinated weed propagules should be determined for each dish at least once per week. Germinated weeds may be removed after recording data from each time to avoid overgrowth, but be sure to keep a running total of germinated seeds for each of the treatments. Deionized water can be added to the Petri dishes or crispers as needed to maintain the original moisture level during the weed emergence period. After the final weed emergence determination, the cumulative percent weed emergence from each heated tube is compared with that of the control (unheated) tube. If additional time samples of the high-temperature soil treatment are used (step 3), then plot the means of each treatment and calculate the regression line of the mean number of germinating seeds versus the time in the heating water bath.
7	The number of emergent weeds from each treatment should be determined at least once each week for 5 weeks. If soilborne nematodes or insects, or fungal or bacterial pathogens are present, and recovery or bioassay of these organisms is desired, the appropriate procedures should be done once at the most suitable time for the organism. This may be immediately after heat treatment or after a period of plant growth in the experimental soil, which will allow for sufficient symptom development or pest reproduction.

FIGURE 13.1 Heat (60°C) disinfestation of soilless medium using grass seeds as a substitute for weed seeds. (a) Soilless medium not heated. (b) Soilless medium heated for 10 min. (c) Soilless medium heated for 20 min. (d) Soilless medium heated for 30 min. (e) Soilless medium heated for 40 min. (f) Soilless medium heated for 60 min. Note that the numbers of grass seeds that germinated and developed into seedlings were dramatically reduced by heat treatments of 20 min or greater. (Courtesy of R. N. Trigiano, University of Tennessee.)

complete disinfestation will be observed in the treatment heated for 60 min (Figure 13.1).

QUESTIONS

- What kinds of organisms are soilborne pests?
- Why is pest-free soil or planting media so important to nursery operations?
- What are the parameters for calculating the lethal dosage for chemical disinfestation? For thermal disinfestation?
- Why is heat treatment of moist soil more effective than that of dry soil?
- What is the risk to nursery plants of disinfesting soil at temperatures greater than 80°C (176°F)?

14 Crop Certification Programs and the Production of Specific Pathogen-Free Plants

Danielle J. Donnelly and Adam Dale

CONCEPT BOX 14.1

- Certified plants are produced under all the guideline requirements of the certification program. These plants are tagged accordingly, with the name of the cultivar and the name and address of the certified grower. For potato only, seed tubers designated certified include additional information on the farm and field in which they have been grown.

- Elite plants are grown from cuttings (elite cuttings) taken from nuclear level plants. Elite cuttings are planted into an elite nursery source block by certified growers. Cuttings can be taken from these elite plants and sold as certified plants (elite level) or used to produce a foundation nursery increase block. In the potato industry only, pre-elite potato plants are derived from nuclear level seed tubers and are grown under a system of rigorous inspection in the field. Elite 1 describes the potato plants and seed tubers grown in the field during the next year, and these seed tubers, in turn, are used to produce successive seed generations described as elites 2, 3, or 4.

- Foundation stock plants are grown from cuttings (elite cuttings) removed from elite plants in an elite nursery source block and used to produce a much larger propagation block called a "foundation nursery increase block." Cuttings can be taken from these foundation plants and sold as certified plants (foundation level). In the potato industry only, foundation level seed tubers are derived from elite seed tubers in the field, increased over successive generations from 1 to 4 years. Tolerances for diseases and pests are less stringent at the foundation level compared with elite level seed tubers.

- Isolation distance is the planting distance between blocks. These distances are described in the certification guidelines for each species and are designed to minimize cultivar mixing and disease transmission between each individual nursery source and increase block.

- A meristem tip is an apical dome and a pair of attached leaf primordia with an overall length of less than 0.5 mm. Meristem tip culture is used following thermotherapy for specific virus elimination from virus-infected clones.

- A meristem tip source clone includes one meristem tip explant and all micropropagated propagules derived from it. One, or a limited number of, meristem tip or shoot tip source clones may represent the elite source material for the certified propagation of a cultivar.

- The nuclear level refers to a group of plants held in a germplasm repository as micropropagated plantlets or in a greenhouse or screen house that are true to cultivar, physiologically healthy, and free of specific disease organisms and pests known to affect the industry. The nuclear level is the first (entry) level in the certification program. It is the cuttings from these plants that are designated "elite level." In the potato industry only, nuclear level plantings may include up to two generations of potato plants grown under protected cultivation in a greenhouse or screen house. These propagules are used to plant a pre-elite generation in the field.

- The shoot tip is the growing point of a stem consisting of an apical dome and several pairs of attached leaf primordial with an overall length of more than 0.5 mm. Shoot tips are used as explants for Stage I micropropagation.
- Specific pathogen–tested plants have repeatedly tested negative for specific pathogens (viral, fungal, bacterial, etc.) that are known to affect the industry. This applies to nuclear plants held in germplasm repositories as source material and elite cuttings taken from them for the certification programs.
- Rogueing is the process of removing and destroying poor-quality plants. This is practiced during the certification process to eliminate plants that appear to have symptoms of disease or physiological problems of any kind, or are mixed in with other cultivars.

CROP CERTIFICATION PROGRAMS

The term "certification" describes both a strategy and a process used to restrict or eliminate crop pests and diseases. Through the certification process, a crop is endorsed if it meets specified standards involving pest and disease occurrence, physical appearance, and genetic off-types. These standards are described in published guidelines that are prepared by agricultural authorities and may be supported by legislation. Propagators who participate in the certification process (certified propagators) must follow these guidelines to produce plants that may be distributed both nationally and internationally as "certified plants."

THE NEED FOR CROP CERTIFICATION

Crop certification originated because scientists and growers recognized that when propagules were transported (vegetative propagation materials and true seed), they sometimes spread plant pests and diseases. In agricultural industries that use seeds, bulb, corms, and other relatively hardy structures, diseases can be controlled if the propagules are heated or treated with chemicals to eliminate pathogens or pests. However, this cannot be done easily when herbaceous cuttings or woody plants are the major propagule. For this reason, a system of practices has evolved to ensure that a source of specific pathogen-free planting material is supplied by various governmental and non-governmental agencies to a select group of propagators (certified propagators) who, in turn, supply planting material to commercial growers. In Canada, as in many other countries around the world, certification programs are in place to limit the spread of diseases of vegetatively propagated, commercially important crops. The crops for which certification programs are most developed in Canada and elsewhere, and where governmental agencies are often involved, include potato (*Solanum tuberosum* L.), red raspberry (*Rubus idaeus* L.), strawberry (*Fragaria* × *ananassa* Duch.), and a limited number of other species such as currant and gooseberry (*Ribes* spp.). Within the ornamentals industry, similar programs are used for select species, but with less involvement from government agencies.

REGULATORY AGENCIES

The United Nations Food and Agriculture Organization's International Plant Protection Convention of 1951 was signed by Canada, the United States, and many other countries that recognized the need to regulate the international movement of plants and plant products. Under this agreement, countries can request that imported plant material be accompanied by a phytosanitary certificate and an import permit. They can also prohibit the importation of certain plant species. A phytosanitary certificate is a document attesting that a consignment has been grown and inspected according to the import requirements of the destination country. Under the authority of two parliamentary acts, the Seeds Act and the Plant Protection Act, the Canadian federal government regulates the movement of plants and plant products, both internationally and between provinces; conducts national certification programs; and attempts to control the entry, spread, and export of diseases and pests. In the United States, this authority is granted under the Federal Seed Act and coordinated by the Association of Official Seed Certifying Agencies.

IMPORT AND EXPORT OF PLANTS

To import or export plants, growers and plant propagators may have to fulfill certain conditions. Growers often import plants across national borders, and propagators import and export them. First, they need to know whether they can legally ship the plants to and from the country concerned. Many crop species are prohibited from entering a particular country, and the main regulatory agencies, the Canadian Food Inspection Agency (CFIA) in Canada and the Animal and Plant Health Inspection Service (APHIS) in the United States keep detailed records of these plants. Permission to import plants may require an import permit. In Canada, information on individual species and application forms can be obtained from the CFIA website. Applications are quickly processed for a small fee. The import permit is issued by the national regulatory agency; may have conditions attached to it; and may require an inspection, quarantine period, or testing, after the plants have entered the country concerned. Once plants can be shipped legally, a propagator might find that his/her sales territory for a certain variety is restricted by a plant patent or plant variety right. Oftentimes, these commercial agreements come with restrictions, for example, a sales territory may be restricted to North America. This means that the propagator can only sell plants in Canada and the United States and is prohibited from selling plants to the rest of the world. It also means that a grower may be unable to buy European cultivars to grow in North America. Finally, the propagator will need to provide a phytosanitary certificate from the relevant regulatory agency, the CFIA in Canada or the State Plant or Animal Inspection Services in the United States; APHIS delegates the export inspection to state regulatory agencies.

CERTIFICATION PROGRAMS

Many agencies cooperate in certification programs, including federal (CFIA and APHIS), provincial, and non-governmental agencies, such as universities and grower's associations. These authorities produce guidelines that describe the certification procedures needed for each species, and these are published. The guidelines are designed to direct the agencies that produce specific pathogen–tested (SPT) plants and the certified propagators who produce high-quality planting stock. They detail the conditions needed to minimize virus and mycoplasma infection and control fungal diseases, insects, mites, and nematode pests. The examples used in this chapter are taken primarily from the guidelines distributed by the Ontario Plant Propagation Programs for strawberry, red raspberry, currant, and gooseberry and the New Brunswick Plant Propagation Program for potato.

SPECIFIC PATHOGEN-TESTING

When new cultivars enter the program, they are tested for pest and disease presence. Usually, an organization that enters the cultivar into the program provides material that has been tested and determined to be specific virus negative. This comes either as a cultured plant material from a laboratory (micropropagated shoots or plantlets *in vitro*) or field plants. The material is then examined critically by a pathologist trained to recognize all of the major pests (insects, mites, nematodes, etc.) and disease organisms (bacteria, fungi, viruses, etc.) known to affect the crop. The type of testing done goes well beyond standard microbiological tests for bacterial and fungi. It includes serological assays such as enzyme-linked immunosorbant assay, DNA probe technology, and tests where plant parts are grafted or sap is inoculated onto indicator plants. Most pests and disease organisms are readily eliminated through a combination of routine propagation by cuttings from noninfected parts of the plant or, if these do not exist, the selective use of chemicals such as systemic fungicides. However, if virus is detected, a complex system of virus elimination is necessary. This is done through a combination of heat treatment (thermotherapy) at a relatively high temperature (37–38°C) for a prolonged interval (2–8 weeks, sometimes longer), followed by meristem tip culture. Meristem tip explants are tiny shoot apices less than 0.5 mm in length, which are composed of the meristematic dome and one pair of leaf primordia. Short lengths of stem, with one to several buds, are trimmed off leaves and washed under running water for several minutes. They are then disinfested (cleaned on all their outer surfaces) in a 10% solution of household bleach (5% sodium hypochlorite) and rinsed in sterile water. The meristem tips are removed from shoot buds on these stems, under a dissecting microscope, using sterile technique and placed into tissue culture. This procedure is described in Procedure 14.1 for meristem tip isolation. Within the medium, which may be different for each plant species, antiviral agents such as ribavirin may be included. The preparation of isolation stage (Stage I) culture media for meristem tips is described in Procedure 14.2 for red raspberry, strawberry, and potato.

Each meristem tip, and its clonally derived propagules, is known as a "meristem tip source clone." When sufficient plant material is available, these meristem tip source clones are tested and retested for the virus(es) that were present in the source plant. Testing is done on plantlets both *in vitro* and after they are transferred to a greenhouse or screen house. In the greenhouse or screen house, plants of each clone are kept separate under conditions that are designed to favor virus multiplication within the plants. Under these conditions, the virus, if present, will reach tissue levels that are detectable by serological, molecular, or plant assays. If these meristem tip source clones repeatedly test free of the specific virus in question (specific virus free), they will be used to increase the nuclear stocks of this cultivar. Each cultivar–virus combination has a characteristic small percentage of meristem tip source clones that are likely to be specific virus free following this procedure. Extreme measures may be required to eliminate virus in the case of mixed infections or persistent host–virus combinations. These could include prolonged or repeated episodes of thermotherapy or thermotherapy applied to *in vitro* plantlets, followed by dissection of tiny meristem tips (0.1–0.3 mm). Eventually, each cultivar may be represented by a limited number of specific virus–free meristem tip source clones (one to three is common). Negative test results for a specific disease agent, such as a viral pathogen, do not guarantee freedom from that virus as this is limited by the sensitivity of the test used. Also, other virus(es) may be present for which there is no test. For this reason, the terms "specific pathogen tested" or "specific virus tested" are legally and scientifically preferred to the older terms "pathogen free" or "virus-free."

TRUE TO CULTIVAR

Within each certification program, economics dictates the number of meristem tip source clones or mother plants that are used to initiate the propagation sequence that can last for several years before a new source material is used. A great many propagules will be generated from a relatively small amount of source material. During this process, cultivars can be mixed and genetic mutations can occur, so confirmation is necessary that source material is true to cultivar (true to type). Until recently, the only way that this could be done accurately was by a grow-out test in which horticultural characteristics are assessed through to commercial harvest; for berry crops, these include foliage, flower, and fruit characteristics, and in the case of potato, foliage, and stem tuber characteristics. More recently, cultivar mixtures can be detected using various types of DNA analysis, including restriction fragment length polymorphism (RFLP), polymerase chain reaction (PCR), or similar technology. Unfortunately, DNA analysis has not yet developed to the stage where point mutations can be detected.

BERRY CERTIFICATION GUIDELINES (STRAWBERRY, RED RASPBERRY, CURRANT, AND GOOSEBERRY)

Stock plants of the berry cultivars grown in Canada are maintained by Agriculture and Agri-Food Canada (AAFC) research stations at Agassiz (British Columbia), Kentville (Nova Scotia), and the University of Guelph (Ontario). These stock plants are referred to as "nuclear level stock plants" as they have been tested and determined to be negative for specific viruses and other disease organisms (specific pathogen tested). They also have been examined by experts, who have confirmed that they are true to type and physiologically healthy. These nuclear level plants may be micropropagated plantlets *in vitro* or greenhouse- or screen house–grown plants.

Annually, nuclear stock plants are used to make cuttings that are made available to a small group of select propagators (certified plant propagators or certified propagators) on a contractual basis. These cuttings or plants are referred to as "elite plants" to show that they are derived from nuclear level stock plants. Certified propagators must request elite cuttings from the clonal repository through a formal contractual application process. This is done up to one year prior to planting, depending on the province. Many cultivars that are requested by certified propagators are protected under Canadian Plant Breeder's Rights. When this is the case, additional agreements and costs are the responsibility of the certified propagator who transfers these costs to the fruit grower. The price for elite level material varies with the province; these costs were subsidized for many years, but are now more likely to cover the real cost of production. In some provinces, when the application for nuclear material is made, an application must also be made to register the growing sites where cuttings will be planted.

Once they have received the elite material, the certified propagators must follow a strict set of guidelines that varies for each berry species and, to some extent, among provinces. These elite cuttings have tested negative for all organisms that are known to be detrimental to the industry, including viruses and other serious pathogens, insects, mites, and nematodes. The elite cuttings are used by certified growers to plant an elite nursery source block (stock block) for each cultivar received. This elite nursery source block is used to produce saleable plant material or plant material for a second generation of nursery plants.

The elite plants that are not sold to growers are used by certified propagators to establish another larger planting or nursery increase block. This is called the "foundation level" to show that it is derived from the elite level, and the increase block is known as a "foundation nursery increase block." Plants from this nursery block may be less healthy than the elite block as they have already been exposed to potential infection in the field for 1 year. This foundation nursery increase block may be used to produce saleable material for a limited number of years. The life span of this planting is determined by the industry and specified in the guidelines. This can sometimes be extended for additional years, upon request, if inspection shows that the plants have remained relatively free of disease organisms.

Under less-than-ideal circumstances, elite cuttings or plants may not be available for distribution to certified propagators. In this case, administrative authorities of the program may allow foundation stock derived from elite cuttings to be used to establish foundation nursery increase blocks. The guidelines are similar, but the duration of the nursery increase block will be limited (e.g., to 2 years).

Different systems of propagation are used for the different berry crops. With strawberry, nuclear stock is usually grown in a screen house for 1 year, followed by 1 year of elite and foundation crops. Raspberry may be propagated from the elite block for up to 4 years, or the elite block is dug and replanted as a foundation crop. For currant and gooseberry, an elite planting referred to as a stool bed is used for up to 10 years to take cuttings that are either sold as unrooted shoots or grown for 1 year and sold as rooted plants.

PLANTING AND CULTURAL PRACTICES

Elite and foundation plantings are determinate plantings in the sense that they may be used as a source of propagules for the commercial industry for a limited number of years (typically, 1–4 years) before they must be destroyed.

Prior to planting, the land should not have been used for at least 5 years (this varies with the province and is sometimes reduced if the land is fumigated) to grow related species or species that are alternate hosts for the diseases of the crop species. Soil fumigation for nematode control may be advised following soil testing for nematode numbers. Each nursery block must be planted, respecting minimum isolation distances from wild plants and those that are not being grown under the propagation program, for example, fields where a commercial crop of the same species is being grown. Plants that may host pests and diseases of the crop being propagated may be tested if they are growing near the nursery block. These isolation distances are maintained, and plant testing is done to prevent pests and pathogens from being introduced. Also, minimum distances between nursery blocks of different cultivars are specified to prevent cultivars from being mixed.

A cultural control program is specified in the guidelines to control insects, diseases, and weeds. A spray program to control insects must be maintained and fully documented. For example, log books are copied and included in inspection reports. Weed control is essential. Weeds compete very successfully with the nursery crops and can reduce propagation rates substantially. They can also harbor pests and pathogens. Blossom removal is usually advised. This is done to control pollen-borne viruses and to prevent seed production, which can lead to off-types if the seeds germinate and grow. Any fruit that is produced cannot be consumed because of pesticide residues related to strict pest control. Inspection is carried out at the discretion of the agricultural authorities and is done following planting and at intervals (usually annually) throughout the duration of the planting. The plants or propagules are usually stored before sale; stored stock may also be subject to inspection and testing.

Inspectors are usually, but not always, government agents. They are trained in the identification of pests and plant diseases, and in cultivar recognition. They visually inspect the crop and review planting plans, test reports from soil- and disease-testing agencies, and grower log reports for cultural practices such as fertilizer and spray applications. It is their responsibility to work with the certified grower to follow the certification guidelines. It is up to them to determine if the crop falls within the maximum tolerances permitted by the guidelines for disease and pest incidence, cultivar mixing, and off-types.

MAXIMUM TOLERANCES

Viral, bacterial, and fungal disease tolerances may permit up to five mother plants or their progeny with visible symptoms per 1000 mother plants. All mother plants with visible symptoms and all plants derived from them must be destroyed. For diseases subject to quarantine, a zero tolerance may be set. In this case, if any visible symptoms are found, the whole nursery block of that cultivar will be rejected. Nematode numbers per kilogram of soil are recorded and must not exceed the guidelines for certain problem-causing nematodes. Threshold levels are designated for all potentially problematic organisms. The tolerance for cultivar mixing is zero. Cultivar mixtures and plants from an area that have become mixed must be removed. Any off-type plants or diseased plants spotted in the field are immediately rogued (removed from the field and destroyed). The size (height, stem diameter, number of roots, and aggregate length) and condition of the plants (dormant) at the time of sale are clearly designated in the guidelines.

CERTIFICATION TAG

At the time of sale, certified plants carry a tag to designate that they were produced under all guideline requirements. On the tag, the name of the cultivar and the name and address of the certified propagator are written. Advertisements are carefully worded to avoid confusion regarding the virus status of plants. For example, terms such as "specific virus tested" or "specific pathogen tested" are used.

SEED POTATO CERTIFICATION GUIDELINES

Potato nuclear stock is held *in vitro* in a limited number of facilities in Canada; one is federal (the "node" for potato is at the AAFC research station in Fredericton, New Brunswick), several are provincial, and some are private tissue culture laboratories. All nuclear stocks are comprehensively tested by CFIA and multiplied by tissue culture, greenhouse, or screen house production. When new cultivars enter the program, they are disease tested. If virus is detected, these plants must undergo thermotherapy and meristem tip culture (see Procedure 14.1 for meristem tip isolation). These explants are micropropagated to yield plantlets that can be retested *in vitro* and at intervals after transfer to a greenhouse or screen house. If these meristem tip source clones test free of the specific virus in question, they will be used to increase the nuclear stocks of this cultivar.

Within the potato industry, terminology varies slightly from that used in the berry industry. This is related, in part, to the greater influence of nuclear propagules derived from micropropagation and microtuberization. These plantlets and microtubers (pea-sized tubers formed from plantlets in culture) may be planted under protected cultivation in a greenhouse or screen house to produce a crop of minitubers (golf ball–sized tubers that develop on plants derived from plantlets or microtubers) and described as nuclear level over the course of one to two generations within these protected structures.

Starting with nuclear stock, a field generation (pre-elite) is produced. The number of field generations for propagule increase is limited to no more than seven generations from nuclear stock. These seed tubers advance at least one class every year after this, including elites 1, 2, 3, and 4, foundation, and certified. Fields of each class of seed are inspected from two to three times annually. Rogueing is used to eliminate diseased or off-type plants. Tolerance for cultivar mixing is zero until elite 4 and is less than 0.1% for certified seed tubers. Disease tolerance is zero for certain severe disease-causing organisms such as bacterial ring rot. Disease tolerance for other pathogens remains zero until elite 2 and may increase to 2% for certified seed tubers. Inspection is done at harvest, during storage, and at packing and shipping points. Certification tags are used to identify the cultivar and seed tuber class, province, farm, and field of production.

ECONOMIC AND LEGAL LIABILITIES

Clonally propagated plant material is traded between commercial growers, retail nurseries, propagators, and governmental and non-governmental agencies. Some plants are traded across international borders. This trade is controlled by commercial law and phytosanitary regulations in the countries and legal jurisdictions involved. For the most part, this trade functions smoothly, but problems can occur and disputes need to be settled. Since different countries have different trade and commercial laws, dispute resolution can be difficult.

In Canada, the trade in clonally propagated plant material is based on the following three principles: the buyer is expected to pay for the plants that are received, the propagator is expected to provide material fit for the purpose for which it is designed, and the liability for negligence cannot be limited.

The buyer is expected to pay fair value for the plants, plus any royalties, and applicable taxes. Propagators will usually vary the price that they charge depending on the quantity of plants bought; larger quantities command a lower price. Oftentimes, groups of growers will join together to obtain a lower price for their plants. Propagators have to deal each year with growers defaulting on their payments. In response, propagators demand at least part of the purchase price with the plant order and may not extend credit on the balance of the payment.

Despite the propagators' and the agencies' best efforts, problems do occur in the propagation programs. These usually involve cultivar mixtures, off-types, or the latent (hidden) transmission of economically devastating pathogens. The effects of these problems can be minor but can lead to a shortage of plant material for several years and affect large numbers of growers. The propagators and the agencies usually insert clauses into their sales agreements and guidelines to limit their liabilities. The legal effect of these clauses is uncertain. Propagators routinely prominently display on their order forms and invoices statements that liability will be limited to the replacement costs of the plants sold. Governmental and non-governmental agencies are increasingly inserting clauses that do not guarantee the disease-free or true-to-type status of the plants.

Usually, most cases are settled by negotiation and only occasionally do cases reach the courts, but when they do the, consequences can be far reaching. There have been two cases in Canada that have reached the courts. In the first, Blue Mountain Nurseries successfully sued the province of Ontario over 'Latham' red raspberry plants, which were not true to type. As a direct result of this case, the province abruptly cancelled the Ontario Strawberry and Raspberry Plant Certification Act. This curtailed the supply of SPT strawberry and raspberry plants to Ontario and eastern Canadian growers. It took the concerted efforts of the whole strawberry and raspberry industry, several years of negotiations, and the formation of the Ontario Berry Growers Association to reinstate the program as the Ontario Strawberry and Raspberry Plant Propagation Program, which is not backed by legislation. In the second, the Ministère Agriculture, Pêcheries, et Alimentation du Québec was sued because hundreds of acres of the strawberry cultivar 'Kent,' propagated under the Quebec propagation program, produced an albino-type fruit. As a result, the Quebec government closed the provincial micropropagation laboratories.

THE CONTRIBUTION OF CERTIFICATION PROGRAMS

Certification programs serve the agricultural community by providing a carefully monitored procedure to restrict the spread of crop pathogens and ensure the distribution of the highest-quality planting material to commercial growers. The guidelines are prepared by agricultural authorities and may be supported by legal means. These guidelines outline a set of standards and describe cultural processes to limit pests, diseases, and genetic off-types. Certified propagators who follow these guidelines are able to produce certified plants that meet both national and international standards.

EXERCISES

Experiment 1. Meristem Tip Isolation

The purpose of meristem tip isolation is to obtain specific virus-free plants from infected source plants. Usually, the source plants have been subjected to thermotherapy. In the thermotherapy of temperate species, plants are exposed to elevated temperatures (37–38°C) for extended intervals (2–8 weeks is typical). Heat treatment is stressful to the plant and often results in elongated, thin (spindly)

Procedure 14.1

Meristem Tip Isolation from the Heat-Treated Source Plants of Red Raspberry or Its Relatives (*Rubus* spp.)[a]

Step	Instructions or Comments
(The first five dissection steps are illustrated in Figures 14.1 to 14.4.)	
1	Sever sections of stem (potato, *Rubus* spp.) from potted, heat-treated plants.
2	Remove leaves, leaving a short length of petiole attached (1–2 cm).
3	Shorten stems to sections (two to three nodes long) that will fit into the 1-L container that you will use for disinfestation.
4	Wash stem sections for 30 min under a gentle stream of cool tap water. Use a square of plastic netting held in place with an elastic band to prevent loss of stem sections from the container.
5	Disinfest these stem sections in 10% liquid chlorine bleach containing a wetting agent such as Tween 20 (1 mL/L) for 15 min, with agitation at intervals, and rinse three times with sterile distilled water. Leave a small amount of sterile water in your disinfestation container and place the container of disinfested shoots toward the back of your clean laminar airflow cabinet.
6	Remove one short length of stem with three to four nodes from the container of disinfested shoots and lay it horizontally on the sterile paper toweling on the dissecting platform of your microscope.
(The next four dissection steps are illustrated in Figures 14.6 to 14.9.)	
7	Hold the shoot down with one pair of forceps and pull the petiole down firmly with another pair of forceps. Pull until the petiole strips away from the stem to reveal the axillary bud.
8	Hold the shoot firmly and work quickly to avoid the shoot tip drying. Use the microscope (10–50×) to help you remove the outer pairs of leaf primordia. Do this using the back of the scalpel tip; slide the scalpel down to the base of each leaf primordium and apply outward pressure to break it off at the base or to cut it off. The meristem tip is usually surrounded by two to four pairs of leaf primordia that must be removed.
9	Once exposed, sever the meristem tip (meristematic dome plus one pair of leaf primordia) at the base using a fresh scalpel.
10	Steady the scalpel, holding the tip close to the paper, and measure the length of the meristem tip using your calibrated eyepiece (make a note of this measurement).
11	Quickly transfer the meristem tip to a test tube containing Stage I medium by touching it to the top of the wick; the surface tension will pull on the meristem tip and permit you to orient it vertically using the scalpel tip. Cap the tube quickly and set it into the rack.
12	Pull back the petiole at the next node of your shoot, exposing the axillary bud, and repeat the explantation procedure. Use fresh scalpels each time—one or more to remove the leaf primordia and another to cut off the meristem tip.
13	Once all of the meristem tips have been dissected from the shoot, discard the plant remnants and used paper, clean the dissecting platform using 70% ethanol, and place fresh paper toweling to receive the next shoot.
14	Test tubes should be placed on a lighted bench (16 h cool, white fluorescent light of ~36 μmol $m^{-2}sec^{-1}$) at 23 ± 2°C for at least one month.
Following 1 month in culture, your meristem tip should have developed into a small shoot (see Figure 14.10).	

Note: These steps are illustrated in Figures 14.1 to 14.10.

[a] Potato (*Solanum tuberosum* L.), currant, or gooseberry (*Ribes* L. spp.) can be used instead of red raspberry (*Rubus idaeus* L.) or other *Rubus* species with no change in the dissection protocol. Strawberry (*Fragaria* × *ananassa* Duch.) can also be used. However, in preparing strawberry runners for disinfestation and dissection, nodes are usually handled individually.

FIGURE 14.1 Primocanes of red raspberry cut from potted, heat-treated plants.

FIGURE 14.2 Petiole sections, 2 cm long, of red raspberry are retained when leaves are cut from the canes.

FIGURE 14.3 Stems of red raspberry are cut into two- to three-node lengths and inserted into 1-L Erlenmeyer flasks.

FIGURE 14.4 Stem sections of red raspberry are washed under cool tap water for 30 min and retained by plastic netting held in place with an elastic band.

FIGURE 14.5 Stem sections of red raspberry are disinfested for 15 min in 10% household bleach followed by three rinses with sterile water.

FIGURE 14.6 The axillary bud of red raspberry is located just above the scar remaining after the removal of the petiole.

FIGURE 14.7 The meristem tip of red raspberry is exposed once several pairs of leaf primordia have been removed.

FIGURE 14.8 The meristem tip of red raspberry is removed using a sterile scalpel.

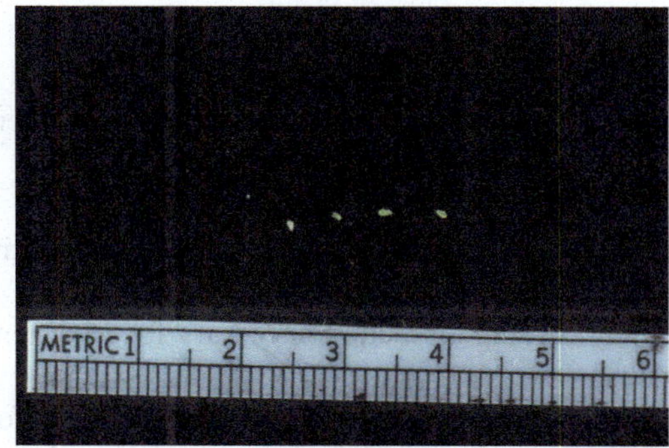

FIGURE 14.9 Meristem tip explants of red raspberry are shown lined up against a metric ruler so that their size can be put into perspective.

branches, and sometimes, plants die. Meristem tip isolation occurs when experience suggests that the virus in question will have greatly reduced titer; by this time, the plant may be severely heat stressed. Procedure 14.1 can be carried out using heat-treated red raspberry or other *Rubus* spp. or simulated on healthy greenhouse-grown plants of a wide range of other species. An experienced horticultural technician can isolate one meristem tip per minute. This technician might dissect 10 to 20 meristem tips per day, at intervals of several days, once source plants have been subjected to the recommended heat treatment. The laminar air flow cabinet is set up in a conventional way, except that a dissecting microscope is cleaned (70% alcohol) and located centrally within the unit.

FIGURE 14.10 After 1 month in culture, the meristem tip of red raspberry has grown into a small shoot (~1 cm long).

MATERIALS

- Heat-treated source plants (can simulate this)
- 1-L Erlenmeyer flask
- Square of plastic or wire netting or cheese cloth
- Elastic band to secure the netting or cloth
- Plastic cover or a square of Parafilm
- Single-sided razor blade or scissors
- 10% Javex bleach (10.8% sodium hypochlorite and contains a wetting agent) or substitute
- 1-L sterile, distilled water (refrigerated)
- 70% ethanol (for cleaning the microscope and tools)
- Kimwipes or paper towel (for cleaning)
- Sterile pack of paper towels or alternate materials (for sterile dissection surface)
- Sterile pack containing several pairs of long- or short-handled forceps
- Sterile pack containing several pairs of scalpels (#7 scalpel handles, #11 blades)
- Dissecting microscope with a calibrated, reticulated eyepiece
- Prepared medium (described in Procedure 14.2)

Follow the instructions in Procedure 14.1 to complete the experiment.

ANTICIPATED RESULTS

The dissection procedure may seem challenging at first. This is especially true for students who do not have much experience at dissecting shoot tips or other explants. However, these difficulties can be quickly overcome by practicing under a dissecting microscope placed on the

Procedure 14.2
Media Preparation for Meristem Tip Explants of Red Raspberry, Strawberry, and Potato (Stage I)[a]

Step	Instructions and Comments
1	Add 700 mL of double distilled water to a 1-L Erlenmeyer flask.
2	Prepare Murashige and Skoog (1962) basal salt medium formulation using your stock solution or commercially prepared powdered medium.
3	Dissolve 30 g sucrose and 100 mg myo-inositol.
4	Add 1.0 mg (4.44 μM) benzylaminopurine (BAP) and 0.1 mg (0.49 μM) indolebutyric acid (IBA) (red raspberry); 0.25 mg (1.11 μM) BAP, 0.025 mg (12.30 μM) IBA, and 0.025 mg (72.17 μM) gibberellic acid (GA_3) (strawberry); or no growth regulators (potato).
5	Make it to 1-L volume by adding double distilled water.
6	Adjust pH to 5.7.
7	Dispense into the test tubes in which you have inserted a filter wick. The wick should be made of Watman's #2 or #3 filter paper, cut chromatographic paper, or fiber cloth such as Miracloth. The wick should project 0.5 to 1.0 cm above the medium meniscus.
8	Cap your tubes and autoclave for 15 min at 15 psi.
9	Cool autoclaved medium in a laminar airflow cabinet.
10	Use fresh or stored in the refrigerator in plastic bags for up to 1 month.

[a] If Stage II culture will follow, growth regulators may have to be adjusted, and the medium is usually solidified with agar.

open bench. Practice in pairs, with one student acting as the "spotter" while the other student works. A brief discussion on tactful, helpful criticism may be in order. All manipulations should be practiced: dissection and the transfer of the isolated meristem tip onto the wick in the test tube. Be sure to touch the meristem tip to the wick without bringing the fingers too close to the lip of the test tube. There are various ways to safely position the test tube cap when you insert the explants. These can be discussed and practiced. When dissection efficiency and test tube manipulations have improved, the procedure can be carried out in the laminar airflow cabinet with full attention to aseptic practice.

Once the meristem tip is positioned upright anywhere near the top of the wick, the tube is capped and carefully set into the test tube rack. Meristem tips that are displaced from the wick in the first few days of incubation rarely survive. These "sinkers" should be aseptically rescued and replaced onto their wick.

The meristem tips, which have little color at first, should start to become green and enlarge over the first 2 to 3 days. These can be measured at intervals and their growth rate plotted. A lag period may precede growth for the tiniest meristem tips, and some may not survive. Growth rate tends to be faster for the larger meristem tips, which may not exhibit an initial lag period. Contaminated tubes may indicate poor surface disinfestation (microorganisms associated with the explant) or poor laminar airflow technique (microorganisms randomly distributed within the test tube). After they have become established, shoots or plantlets can be readily subcultured to fresh, solidified medium. This is usually done one month following excision.

QUESTIONS

1. Why do specific pathogen assays precede thermotherapy treatment?
2. What is a meristem tip source clone?
3. What is the relationship between meristem tip size, survival, and virus elimination?
4. Why is rapid dissection necessary for meristem tips?
5. How could you investigate possible sources for the contamination of meristem tip cultures?

Experiment 2. Media Preparation for Meristem Tip Explants

Meristem tip explants are usually placed onto a wick, made of one of several possible materials. Wicks can be cut from #2 or #3 filter paper, chromatography paper, or Miracloth. The wicks are folded in half and inserted into small test tubes. The medium is usually Stage I medium, but in liquid form, as more explants survive in this medium. It is carefully dispensed to the side of the wick to avoid collapsing the wick. The wick should project just above the level of the medium. A small test tube is preferred so that the scalpel on which the meristem tip is balanced can be safely inserted into the tube without touching the fingers to the lip of the container.

MATERIALS

- Small test tubes (<20 mL)
- Wicks cut to fit tubes and folded in half, inserted into tubes
- 1-L Erlenmeyer flask
- Double distilled water
- Stock solutions of basal salts or prepared, packaged salts for 1 L medium
- 30 g sucrose and 100 mg myo-inositol
- Stock solutions for growth regulators (usually 10 mg growth regulator per 100 mL water)
- Medium dispenser
- pH apparatus and buffers
- Stirring platform
- Autoclave

Follow the instructions in Procedure 14.2 to complete the experiment.

ANTICIPATED RESULTS

Choose test tubes carefully. When the test tubes are too large, explant insertion is harder, so you get more contamination and the wicks tend to collapse against the sides of the tube. Wicks should fit snugly and remain erect and intact during medium dispensing and autoclaving.

QUESTIONS

1. What are the characteristics of an ideal wick, and what are these used for?
2. Why is liquid medium preferred over agar-solidified medium for meristem tip explants?
3. Why is double distilled water preferred over tap water for medium preparation?
4. Why should autoclaved medium be cooled in a laminar airflow hood?
5. Why is fresh medium preferred over medium that has been stored for several months?

LITERATURE CITED AND SUGGESTED READING

Animal and Plant Health Inspection Service. http://www.aphis.usda.gov. Accessed March 10, 2014.

Buonassisi, A. J., H. A. Daubeny, and B. Peters. 1989. The B. C. Raspberry certification program. *Acta Hort* 262:175–185.

Canadian Food Inspection Agency. http://www.inspection.gc.ca. Accessed March 10, 2014.

Dale, A. 1999. *Guidelines of the Ontario Raspberry Plant Production Program*. Ontario, Canada: Department of Plant Agriculture, University of Guelph, 11 pp.

Dale, A. 1999. *Guidelines of the Ontario Ribes Plant Production Program*. Ontario, Canada: Department of Plant Agriculture, University of Guelph, 10 pp.

Dale, A. 1999. *Guidelines of the Ontario Strawberry Plant Production Program*. Ontario, Canada: Department of Plant Agriculture, University of Guelph, 10 pp.

de Boer, S. H. 1994. The role of plant pathology research in the Canadian potato industry. *Can J Plant Path* 16:150–155.

de Boer, S. H., S. A. Slack, G. W. van den Bovenkamp, and I. Mastenbroek. 1996. A role for pathogen indexing procedures in potato certification. In *Advances in Botanical Research*, Volume 23: *Incorporating Advances in Plant Pathology*. London: Academic Press, Ltd., pp. 217–242.

Hartman, H. T., D. E. Kester, F. T. Davies, Jr., and R. L. Geneve. 2002. *Plant Propagation Principles and Practices*, 7th Edition. Upper Saddle River, NJ: Prentice Hall, 880 pp.

McDonald, J. G. 1995. Disease control through crop certification: Herbaceous crops. *Can J Plant Path* 17:267–273.

Murashige, T. and F. Skoog. 1962. A revised medium for rapid growth and bioassays with tobacco tissue cultures. *Physiol Plant* 15:473–497.

15 Building a Plant Propagator's Integrated Pest Management Toolkit

William E. Klingeman

Concept Box 15.1

An integrated pest management (IPM) strategy will allow plant propagators to adapt their site-specific experiences and choices of multiple management options. The propagator's decision-making process can follow six simple steps to help keep pest populations below levels that are expected to cause economic or aesthetic plant loss.

- The first step in successful IPM is to identify the pest.
- Scouting, or routine monitoring of propagated plants, is essential for maintaining healthy plants.
- Propagators have several options for managing pests in production systems, including host plant resistance; cultural, physical, mechanical, biological, and chemical control; as well as doing nothing.
- Records of the IPM actions that were chosen and the outcomes that were observed can be used to identify site-specific "hotspots" and predict seasonal pest activity.
- IPM programs are ineffective unless observations and recommended management actions are recorded. A propagator's records can be used to train new employees and IPM scouts. Well-kept pesticide and action records may limit grower liability in wrongful injury claims.
- IPM is incomplete until the effect of the management action has been assessed.

The objectives of integrated pest management (IPM) for plant propagation include restricting pest entry into production systems and managing pest populations below levels that would be expected, often based on past experience, to cause either economic or aesthetic crop losses. More than ever, resources are available to apply the principles of IPM to nursery production fields, container-grown ornamentals, and plant propagation houses.

IPM relies on the knowledge-based selection and application of one or several pest management tactics to limit pest pressure on target plants. A grower determines which IPM tactic to use based on his or her direct knowledge and expectations of pest-related injury. An experienced grower will also consider historical observations of site-specific pest populations at their production facility. To be successful, this knowledge is combined with an awareness of alternative management actions and their potential consequences.

Growers have occasionally argued that rotation cycles for their propagation stock are too short and that the ornamental plant species that they grow are too diverse to enable IPM as a viable option for their commercial propagation systems. In fact, IPM principles are just as readily applied in propagation operations as in more conventional greenhouses and nurseries, and to landscape management. Indeed, most experienced plant propagators are practicing several elements of IPM already. The rationale for adopting IPM into management strategies for propagation systems can be readily explained: nursery managers and propagators all want solutions to the challenges raised by reliance on pesticides, including worker reentry intervals, employee safety, ongoing elimination and increased restrictions for the use of chemical pesticides, pest resistance and tolerance to select pesticides, and growing consumer demands for environmental sustainability.

From an economic perspective, rooted cuttings or tissue-cultured plantlets offer substantial potential for monetary gain. It is important to remember that, early in propagation, plants represent a minimal investment.

Individual rooted cuttings and liners are the cheapest that they will ever be. Infested and poor-quality plants can be thrown away with the least economic loss. As production continues, however, plants must be transplanted into larger containers and surrounded by amended soilless substrate or media. From that stage on, plants will also receive regular pesticide, fertilizer, and irrigation applications. Plants must be pruned and spaced, repotted and re-fertilized, lifted and stored, and finally, shipped to the market and sold. Because each step adds significant labor and materials costs, competent plant propagators should demand superior plant quality and performance from the onset of their efforts. A critical eye and a willingness to discard inferior plant products help the grower optimize their eventual return on continuous management investments.

DESIGNING PROPAGATION FACILITIES AND MANAGING CROP ENVIRONMENTS WITH IPM IN MIND

A gardening hobbyist can easily propagate many plant species and cultivars with very little investment or need for structural facilities. For example, English ivy (*Hedera helix* L.), pothos, and many willow (*Salix* sp.) species produce functional roots if cut stems are immersed in just water. On the scale of commercial production though, where heavy volume and a diverse plant mix are commonplace, plant propagators will need to provide either greenhouses or polyethylene film (sheet plastic)–covered "polyhouses" (Figure 15.1) that are used to protect and acclimate their plant material. In addition, an isolated quarantine or staging area should be included so that newly purchased and imported plant materials can be held before introduction to the full production environment. All new plant materials, including cuttings and budwood, should be thoroughly inspected for pests and diseases before the crop is transferred into the general greenhouse, polyhouse, or other production area. Infested plants or plant parts can be returned to the supplier, treated, or discarded before they contaminate the main production area. Plants leaving quarantine areas and propagation and overwintering houses can be routed to the nursery to limit transport to more mature crops that are nearing their final dates of sale.

It is important to recognize that the environment within propagation structures, as well as the way that plants are maintained during production, can lead to conditions particularly well suited to the rapid development of pest populations, as well as the buildup of debris, algae, and fungi that often serve as pest food resources (Figure 15.2). Regardless of the time of the year, propagation structures generally have reduced airflow and higher-than-ambient temperatures, particularly early and late in the season. Cuttings and media are often kept moist with mist systems, spray stakes, or overhead irrigation, and humidity in the greenhouse and under plastic frequently exceeds 90% (Figure 15.3). Rooted cuttings and container-grown liners are stored pot tight during winter (Figure 15.4). In spring, temperatures in plastic-covered polyhouses can rise quickly. When overwintered crops are left too long under cover in spring, pests, weed populations, and plant diseases can result in rapid plant decline. To limit these effects, synthetic woven fabrics of different shade densities can be used to reflect heat and reduce the level of direct light that the plants receive.

FIGURE 15.1 Commercial polyethylene film (sheet plastic)–covered propagation and overwintering polyhouses. (Courtesy of W. E. Klingeman.)

FIGURE 15.2 A well-maintained propagation space should be free of leaf litter, algae, and other debris that can harbor pests and plant pathogens. (Courtesy of W. E. Klingeman.)

FIGURE 15.3 Propagation seedling trays may be covered to enhance germination, yielding both higher-than-ambient temperature and humidity. (Courtesy of W. E. Klingeman.)

FIGURE 15.4 Propagation stock and nursery plants are often stacked pot tight during overwintering. (Courtesy of W. E. Klingeman.)

FEW SIMPLE "MUST-HAVE" TOOLS FOR THE IPM SCOUT

It costs very little to prepare a toolkit for monitoring crops for pests and plant diseases (Table 15.1). Every scout should own and carry a small 10× to 15× hand magnifier. Ideally, the lens should not be less than 20 mm (0.75 in) across and should be large enough for comfortable use during extended periods. A pocketknife (or pruners); a marking pen; and small, sealable plastic bags are needed to collect and preserve the root or leaf tissues until they can be more thoroughly examined. Production rows and borders adjacent to crops, as well as blocks of sturdy stock plants, can be sampled quickly using a canvas sweep net. Branches and foliage of more delicate plants can be shaken over a square "beat" cloth, a piece of paper, or a light-colored pan. Alternatively, a clipboard or field notebook with a light surface can be used as a backdrop to count mites or thrips shaken from plant foliage. A light surface is important to allow contrast with darker colors and shadows of small and slow-moving pests.

TABLE 15.1
The IPM Scout's Toolkit

A few simple and inexpensive tools are invaluable to a well-prepared IPM scout. These essential items will fit in a small tackle box and beneath the seat or in the glove box of a car, truck, or golf cart.

1. 10× to 15× hand magnifying lens
2. Pocketknife (or pruners)
3. Small, sealable plastic bags
4. A "beat" cloth, a piece of paper, or a light-colored pan
5. Small, battery-powered aspirator
6. 70% isopropyl (rubbing) alcohol preservative
7. Several 10 to 20 mL (0.4–0.8 fl. oz) watertight vials
8. Notebook with pre-printed scouting sheets
9. Permanent marking pen or pencil
10. Digital camera
11. Yellow (and blue) sticky cards
12. Canvas sweep net (optional)

A scout should also invest in a small aspirator, which is essentially a miniature vacuum that enables small arthropods to be drawn into a collection vial. Battery-powered aspirators suck small volumes of air and the insect pest into a screened intake tube. With manual aspirators, the opening of a second flexible tube is held close to the insect, and the vacuum created by sucking on the first tube pulls the insect into the vial. A small volume of 70% isopropyl (rubbing) alcohol preservative can be added to collection vials, which should be labeled, and the specimen should be returned to the office for subsequent identification. Logically, a scout's portable kit should include 10 to 20 mL (0.4–0.8 fl. oz) watertight vials and a permanent labeling pen or pencil.

Digital cameras are invaluable tools, particularly if they can be mounted onto a dissecting microscope. With digital images and e-mail messages, professionals who are miles away from a grower or a problem can quickly identify the causal agent. Many universities have established "Distance Diagnostics" Extension resources that allow digital images of pests, pest injury, disease symptoms, and signs to be submitted online to specialist experts. Correct identification of pest and disease organisms is often impossible without clear images of small but diagnostic characteristics. This technology also lets experts rapidly recommend action or request additional information that may initially have been overlooked in the field.

BASIC SCOUTING AND MONITORING TECHNIQUES

Consistent plant monitoring is the key to successful IPM for plant propagation. Monitoring allows the pest scout to identify pest outbreaks before economic and aesthetic losses occur. Production areas should be divided into logical units and mapped before scouting. Maps may reflect individual propagation houses and production areas, which can be subdivided by similar crop types, irrigation type or schedule, or crop age. Plugs and easily rooted cuttings are short-cycle crops that may need to be scouted every 2 to 4 days. Slower-to-root cuttings and seedling liners can be monitored at weekly or biweekly intervals. Pests of field-grown tree whips and container-grown liners of woody ornamentals can be detected by scouting biweekly throughout the ornamental growing season.

Scouts do not need to inspect every plant within a production block to be confident that pest populations are not approaching economically or aesthetically damaging levels. Many pest populations develop in patchy distributions rather than evenly throughout production areas. A scout will have to inspect more plants in a production block to detect pests in patchy or aggregated populations. Fewer plants will need to be scouted to detect pests that are distributed evenly throughout the block. To maximize scouting efforts, mark the point of entry into a production block with a colored flag. Vary the point of entry into production blocks and travel in a routine pattern through stock blocks and propagation beds. At regular intervals, select and inspect a plant for signs and symptoms of pest or disease presence. It is appropriate to be "biased," particularly in commercial operations, in selecting which plants are inspected. Chlorotic, wilted, and damaged foliage are often indicators of diseases or arthropod pest activity. Plants with characteristic signs and symptoms should not be ignored. On return visits to the production block, start above or below the last entry point and inspect different plants.

Within about 100 m^2 (1075 ft^2) of production area, inspect 20 or more plants. It takes an experienced scout less than 15 min to scan the upper and the lower leaf surfaces of both new and old foliage, flowers, leaf axils, bark on the main trunk and stems, and roots of plants showing evidence of pests or disease. Inspect whole individual plugs or cuttings. On larger specimens, examine three to five, 10 cm (4 in)–long terminal stem sections per plant. Inspect stock blocks for cuttings at low and high positions on each cardinal direction around the surface of individual trees or shrubs. In propagation and greenhouses, start inspections near the entrance. Pay attention to the bench space adjacent to vents and horizontal-airflow fans. Do not ignore bench middles, but concentrate on the edges and the ends of benches where pest populations develop. Scout beneath benches for pest refuges on weeds. Do not forget to inspect hanging baskets. When pests are encountered in an area, inspect a greater number of plants to provide a more accurate estimate of pest population size. Insert a colored flag into the pot or flat to mark the outbreak location. Make a rough count of the ratio of affected to unaffected plants within the production unit.

FIGURE 15.5 A Dutch door is kept closed to limit the size of the drought shadow that develops adjacent to an open polyhouse doorway. (Courtesy of W. E. Klingeman.)

While most pests develop in patchy distributions, the probable location of outbreaks can often be predicted. "Hotspots" are problem areas that, more or less regularly, contribute to seasonal population outbreaks. These often include shaded corners' drought-prone areas near polyhouse doors (Figure 15.5); areas near greenhouse vents, sidewalls, and along gravel roadways; and low areas that collect standing water and drip zones (e.g., around greenhouse cooling pads and beneath greenhouse purlins and mist irrigation lines).

USE STICKY CARDS, SENTINEL, AND "BANKER" PLANTS TO MONITOR AND MANAGE PESTS

Propagation pests can be attracted to yellow (e.g., whiteflies, aphids, fungus gnats, leafminer flies, and shore and crane flies) or blue (for thrips) sticky cards (Figure 15.6) and rolls of yellow sticky ribbons. Sticky cards and ribbons are used to passively monitor pest presence and gauge adult population densities. Cards are generally not effective at controlling pest populations once they become established. Cards or ribbons should be mounted on bamboo or metal stakes and suspended at canopy height to intercept flying pests. For best effect, sticky cards should be counted weekly. Use a grid layout with at least one card per 100 m^2 (~1075 ft^2).

Plants that are susceptible to pests can be exploited within a propagation IPM program. For example, rye

FIGURE 15.6 Yellow and blue sticky cards can be mounted on stakes at crop canopy height to attract and help monitor populations of many propagation and nursery crop pests. (Courtesy of W. E. Klingeman.)

(*Secale* sp.) and wheat (*Triticum* sp.) seedlings can be grown beside ornamentals as trap crops for fungus gnats. These plants, and also ornamental pepper plants, have also been used as "banker plants," on which an aphid colony sustains a supply of beneficial parasitic wasps contained within aphid "mummies." Garden beans (*Phaseolus vulgaris* L.) and roses (*Rosa* sp.) can be used as a trap crop to monitor spider mite and broad mite populations. Tomatoes, lantana, and poinsettia can become trap crops for whiteflies, whereas aphid outbreaks can often be detected first on a chrysanthemum trap plant. Because pest populations can quickly build to plant-threatening levels on untended sentinel and trap plants, once installed, sentinel and trap plants must be observed regularly. Infested sentinel plants should be removed and destroyed to avoid contaminating the ornamental crop.

ARTHROPOD PROFILES: A ROGUE'S GALLERY OF PROPAGATION PESTS

Ornamental plants have high aesthetic thresholds. In short, plants must look good to keep buyers coming back. For this reason, economic losses to propagated plants can rapidly increase, even when relatively few pests are found. Competent growers can quickly identify the propagation pests that threaten ornamental crops as well as the evidence of signs and symptoms showing that pests may be active. The most efficient propagators also train their employees and sales staff to recognize both pests and plant diseases that are common production problems. Once a problem is found, the cause must be identified. To make an informed management decision, the actual type of pest and its presence must also be confirmed. Next, the scout should thoroughly sample production blocks to determine if pest numbers can be expected to translate into plant damage and economic losses.

COMMON PROPAGATION PESTS

Fungus Gnats

The cosmopolitan distribution of *Bradysia* sp. and *Sciara* sp. fungus gnats has made them the most familiar pests of commercial and hobbyist propagators. Fungus gnats develop quickly during mild winters and in spring (Figure 15.7a). Fungus gnats persist in propagation systems that provide too much irrigation, leading to algal growth. Seedlings, stem cuttings, and young plants grown in rich, organic rooting media are particularly susceptible. Individual adult females may live as long as 7 days, during which they can deposit 100 to 150 eggs. Adult fungus gnats are readily observed on moist media surfaces where they deposit eggs. Larval fungus gnats (Figures 15.7 and 15.8) can be found in groups feeding on foliage, fine roots, and root hairs on or near the surface of the medium, and at the basal wound on cuttings. Larval feeding injury provides entry points for plant pathogenic fungi. Plant pathogens that cause root and crown rots, including *Pythium*, *Verticillium*, and *Fusarium*, are also transferred on the bodies and feet of adult fungus gnats. Subsequent plant wilting may be attributed as much to plant disease as to direct root loss from larval feeding.

Correctly identifying fungus gnats is important. Fungus gnats are often confused with shore flies, which are generally a nuisance pest. In contrast to shore fly larvae that have a light-colored head that is largely soft tissue, fungus gnat larvae have a hardened, black head capsule. Adult fungus gnats are about 2.5 mm (0.1 in) long and have long legs and long antennae. Shore fly adults (Figure 15.7b) are smaller than fungus gnats (~2.0 mm or 0.08 in long) and have shorter legs and stubby antennae.

Adult fungus gnats can be attracted to sentinel pots in which rye, wheat, or other small grains are grown. Eggs hatch within 4 days, larvae mature in about 14 days, and pupation requires about 3 days at 22°C (72°F). Once infested, sentinel pots should be removed and destroyed, and sentinel pots should be replaced at 2-week intervals during propagation cycles. Adult fungus gnats are also attracted to yellow colors and can be monitored with commercially available paper or plastic sticky traps. The presence of fungus gnat larvae can be monitored by using small potato wedges about 25 mm (1 in) in diameter. The potato wedge should be placed flat on the media surface and scouted regularly for the presence of larvae. Insecticides applied as drenches control larval fungus gnats, rather than adults. Results of control efforts may not be immediately apparent.

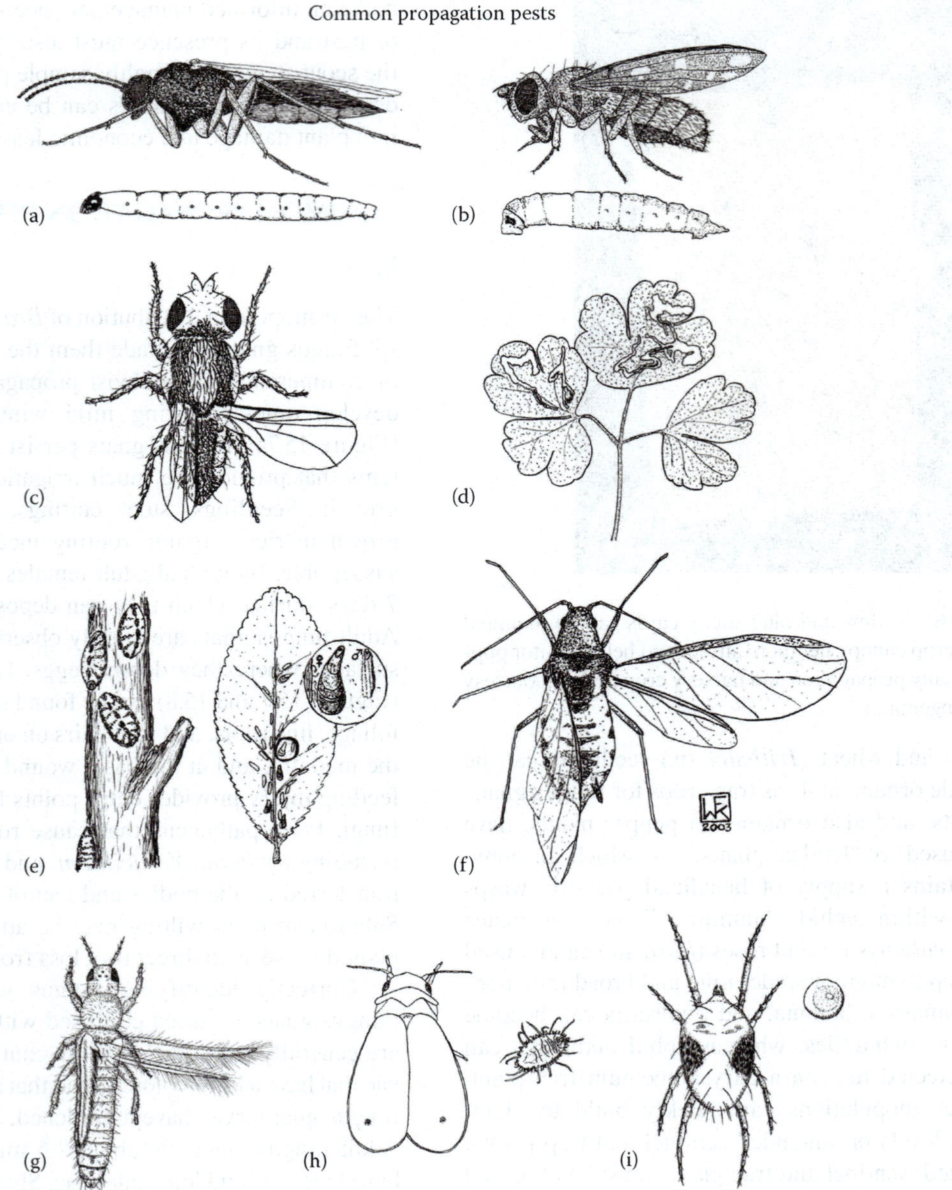

FIGURE 15.7 Common propagation pests. Growers, propagators, and employees should learn to identify the most common arthropod pests that they will encounter in propagation systems. Frequent pests include fungus gnat adults (a, top) and larvae (a, bottom), shore fly adults (b, top) and larvae (b, bottom), serpentine leafminers (c) and leafminer damage (e.g., on columbine leaflets) (d), soft scales (e, left) and armored "oystershell" scales (e, right), wingless and winged (alate) adult aphids (f), thrips (g), whitefly adults (h, left) and a pupa (h, right), and two-spotted spider mite adults and an egg (i). (Drawings by W. E. Klingeman.)

Shore Flies

Shore flies (*Scatella* sp.) are also common inhabitants of propagation houses (Figure 15.7b). Unless propagation areas are adjacent to retail sales areas, shore flies are generally of very little economic concern. While adult shore flies are primarily a nuisance pest, large populations of flies can spread plant pathogens. Populations can build quickly where moisture and humidity have allowed algal food resources to develop. Shore fly larvae also feed on decaying organic matter. Prompt sanitation to reduce these food resources is an important strategy for effective cultural control.

Leafminers

Larvae of several fly and beetle species develop in tunnels cut beneath leaf epidermal cells or within the palisade or spongy mesophyll cell layers of leaves. Chrysanthemum

FIGURE 15.8 Fungus gnat larva will feed on soft and decaying plant tissues, including roots and foliage in contact with the substrate or media surface. (Courtesy of F. A. Hale, University of Tennessee.)

leafminers (*Chromatomyia* [=*Phytomyza*] sp.) and serpentine leafminers (*Liriomyza* sp.) (Figure 15.7c) are pests of many different greenhouse crops. Annual and perennial crops, including sunflowers, zinnias, daisies, columbines (Figure 15.7d), and fruits and vegetables (e.g., tomatoes and onions), are susceptible. Sanitation is essential in infested propagation areas. When humidity is high, chrysanthemum leafminer larvae can successfully develop and pupate within cut or aborted foliage that drops onto greenhouse benches and floors. Adult chrysanthemum leafminers, which are about 2.5 mm (0.1 in) long, first appear in April and early May in mid-Atlantic U.S. states. Within a year, three to four generations can be completed, but flies can persist yearlong in climate-controlled greenhouses.

Scale Insects and Mealybugs

Some 6000 species of scale insects occur worldwide—about 1000 are found in North America. Among the most commonly encountered are mealybugs and armored and soft scales. Adult female mealybugs are small (1–5 mm long), wingless, soft-bodied insects. Tufts of cottony white wax exude from specialized cells and often highlight the body segmentation of these pests. Around the margins of many mealybug species, wax is exuded into short filaments. In species like the longtailed mealybug (*Pseudococcus longispinus*), these waxy secretions extend from the body in filaments up to 4 mm long. Waxy secretions also hide the tiny, yellow eggs of mealybugs, forming a protective "ovisac." Soft scales are generally larger (2–6 mm long) than armored scales (1–3 mm long). Soft scales do not produce the hardened wax covering that armored scales exude to protect the insect and its eggs. Soft scales like *Pulvinaria* sp. (Figure 15.7e, left) can infest a broad range of host plants. Tea scale (*Fiorinia theae* Green) is an armored "oystershell" scale that can develop on many ornamental hosts, including hollies, dogwood, euonymus, and camellia (Figure 15.7e, right).

Feeding activity of scale insects and mealybugs can occur on leaf undersides, along shoots and stems, on roots, and in shoot axils. Throughout their lives, mealybugs remain mobile, moving slowly throughout host plants and inserting piercing–sucking mouthparts to feed on plant sap. Many species of mealybugs inject a salivary toxin into plants while feeding. These toxins can result in plant stunting, contorted leaves and branches, stem dieback, and plant death. Plants infested with scales may look generally unhealthy. Infested leaves may become chlorotic, particularly on the upper surface of leaves where scales are feeding. Leaves may wilt and drop prematurely.

In spring and early summer, newly hatched scale nymphs, called "crawlers," migrate away from the protection of their mother's dead body and water-repellent wax testa to colonize new plant tissue. Unlike adults, the crawler stage is readily controlled with insecticides, horticultural oils, and insecticidal soaps. The emergence of crawlers can be monitored by placing a small section of double-sided sticky tape around a small stem just above a female scale. Many scales are capable of multiple generations a season and require continuous monitoring. Scale populations may develop quickly following broad-spectrum insecticide applications that kill parasitic wasps and other natural enemies of scales.

Blocks of ornamental stock plants, cuttings, and bare-rooted liners should be inspected for scale and mealybug populations before they are introduced to propagation areas. Ants tend scale insects and mealybugs for the sugary secretions that scales exude and can be used to locate pest populations.

Aphids

Several species of aphids, including green peach aphid (*Myzus persicae* [Sulzer]) and melon = (cotton) aphid (*Aphis gossypii* Glover), are common pests in propagation and production systems. Individual female nymphs are about 3.0 mm (0.12 in) long. Female aphids reproduce for 20 to 30 days, during which they can give live birth to 60 to 100 female nymphs. Newborn nymphs become reproductive in 7 to 10 days, depending on temperature. If aphids become crowded or plant nutritional quality declines, winged females (alates) develop and migrate to new hosts (Figure 15.7f). Like many aphid species, green peach aphid produces pink and green color morphs.

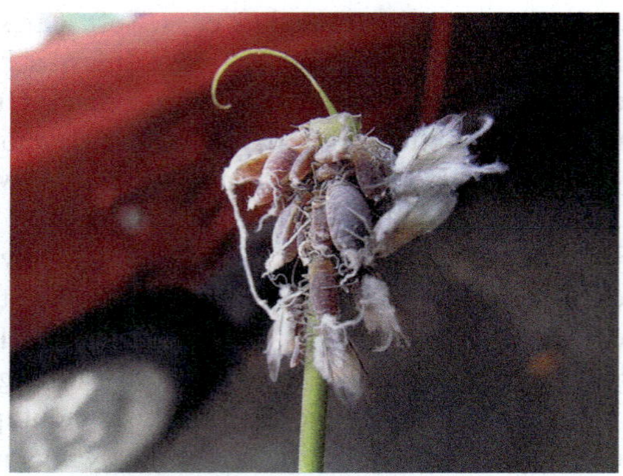

FIGURE 15.9 Wooly apple aphids clustered on the branch tip of a maple tree. (Courtesy of W. E. Klingeman.)

Color dimorphism in aphid species may be controlled by genetic, temperature, and plant nutritional factors.

Feeding aphids extract plant sap through piercing stylet mouthparts that penetrate phloem tissue. Excess ingested fluids are excreted as "honeydew," a sugar-rich solution. Honeydew may be forcibly ejected by the aphid and will thinly coat adjacent foliage. Ants quickly find insects that produce honeydew and protect them from natural enemy predators. Ants can be followed to locate aphid, mealybug, and scale insect outbreaks. In time, a layer of dark fungal hyphae called "sooty mold" will grow on honeydew. While sooty mold is aesthetically unpleasant, it also reduces plant photosynthetic potential and subsequent plant growth. Because sooty mold needs time to develop, it is found more commonly among blocks of propagation stock plants than among rooted cuttings or container-grown liners.

The wooly apple aphid (*Eriosoma lanigerum* [Hausmann]) is another pest in propagation systems (Figure 15.9). Wooly apple aphids use elms (*Ulmus* spp.) as their primary host, but will colonize roots of rosaceous plants as secondary hosts. Wooly apple aphids are frequently overlooked on rootstocks of apples and crabapples (*Malus* spp.), pyracantha, pear (*Pyrus* spp.), and quince (*Chaenomeles* spp.). As a result of aphid feeding, wooly apple aphids cause bark cracks and galling of stem and root tissues. Fibrous roots are reduced on heavily infested plants. Affected trees can become stunted and die.

Thrips

There are numerous species of thrips common in ornamental production and landscape systems (Figure 15.7g). Many species, including *Echinothrips americana* (Morgan), western flower thrips (*Frankliniella occidentalis* [Pergande]), greenhouse thrips (*Heliothrips haemorrhoidalis* [Bouché]), and banded greenhouse thrips (*Hercinothrips femoralis* [O. M. Reuter]) have an extensive host range among ornamental plants. Adult thrips are about 1.5 mm (0.06 in) long. Thrips have rasping–sucking mouthparts used to pierce plant cells and ingest cell contents. Feeding injury from thrips may range in severity from silvery stipples and streaks of chlorosis to contorted flowers and leaves. Affected leaves fail to develop fully and may prematurely senesce. Thrips are also a vector for tomato spotted wilt virus and plant diseases.

Like many insects, reproductive rates of thrips are accelerated at higher temperatures. Thrips can complete 12 to 15 generations during a growing season, but persist yearlong in greenhouse environments. Adult and larval thrips are found in partly opened flower buds, at the base of expanded flowers and on leaf undersides. Slow-moving and light-colored larval thrips can be encouraged to move (and thus be detected) by exhaling warm air onto flowers and buds. Adult thrips are attracted to yellow and blue cards coated with sticky adhesive and placed at plant canopy height.

Whiteflies

Silverleaf whitefly (*Bemisia argentifolii* Bellows and Perring), banded-winged whitefly (*Trialeurodes abutilonea* [Haldeman]), and greenhouse whitefly (*Trialeurodes vaporariorum* [Westwood]) are common propagation pests. Each of these species has an extensive host range among ornamental plants. Adult whiteflies are active fliers when disturbed. Species of whiteflies are difficult to identify from the adults, which look like small (~2 mm long) white moths (Figures 15.7h, left, and 15.10). Pupae or pupal cases are examined to identify species correctly (Figure 15.7h, right). Whiteflies colonize and feed on leaf undersides, reducing plant vigor. Whiteflies also vector

FIGURE 15.10 A whitefly adult and pupa. (Courtesy of F. A. Hale, University of Tennessee.)

plant viruses and diseases. Like aphids and scale insects, whiteflies produce honeydew. Aesthetic loss from sooty mold will reduce the economic value of infested propagation stock.

Mites

Several species of mites challenge commercial and hobbyist propagators. Until population levels become high, mites generally feed on leaf undersides. Feeding injury results in chlorotic stippling of leaves followed by premature leaf drop. Mites are often first active within the protected centers of shrubs, where they can be overlooked. Two-spotted spider mites (*Tetranychus urticae* Koch) (Figures 15.7i and 15.11) are widespread pests of hundreds of plant species, including ornamental hosts and weed species utilized as refuges. Two-spotted spider mite populations develop quickly in hot, dry environments and in the drought shadow near entrances to propagation houses (Figure 15.5). Production areas adjacent to fan vents and dirt or gravel roads are also common hotspots for spider mite outbreaks. Individual adult female spider mites, which are about 0.35 mm (0.015 in) long, lay 100 to 300 eggs during their 2 to 4-week lifespan. As populations build, the silk webbing that two-spotted spider mites produce becomes evident, particularly on the tips of new, young shoots (Figure 15.12).

Other mite pests that feed on a broad range of hosts include European red mites (*Panonychus ulmi*) and southern red mites (*Oligonychus ilicis*). Like European and southern red mites, spruce spider mites (*Oligonychus ungunuis*) are cool-weather pests. Populations of these mites develop rapidly in spring and fall.

FIGURE 15.11 A two-spotted spider mite and eggs. (Courtesy of F. A. Hale, University of Tennessee.)

FIGURE 15.12 Two-spotted spider mite webbing and stippling feeding injury, which makes injured crops appear chlorotic. (Courtesy of F. A. Hale, University of Tennessee.)

Broad mites (*Polyphagotarsonemus latus* [Banks]) and cyclamen mites (*Steneotarsonemus pallidus* [Banks]) feed on a wide range of ornamental and weedy host plants. Broad and cyclamen mite populations can be found in high numbers on buds and small adjacent leaf initials. Broad mites (Figure 15.13) have been collected from legs of whiteflies and other insects. This behavior, called "phoresy," allows broad mites to spread rapidly through production areas. Broad and cyclamen mites can also become passengers on humans and are readily spread on infested cuttings and discarded plant tissues. As broad and cyclamen mites feed on plants and weeds, newly emerging leaves fail to expand fully and become contorted and brittle (Figure 15.14). Studies indicate that salivary toxins cause plant injury when they are excreted into plant tissues as mites feed.

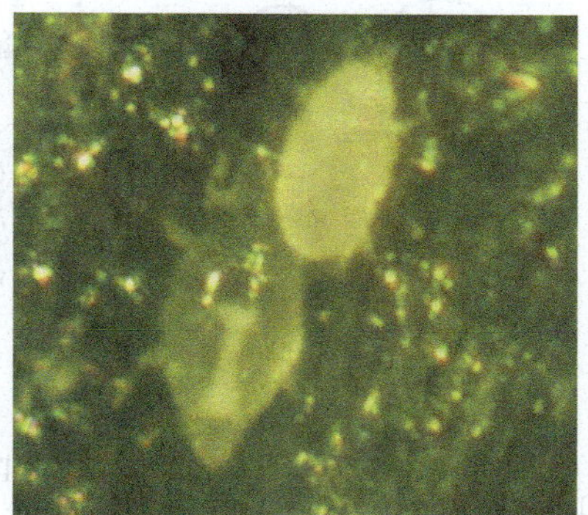

FIGURE 15.13 A female broad mite and nymph. (Courtesy of W. E. Klingeman.)

FIGURE 15.14 Weedy bittercress shows evidence of broad mite feeding injury caused by phytotoxic saliva injected during feeding. (Courtesy of W. E. Klingeman.)

Seed Parasites

Propagators who rely on seeds to generate commercial stock may encounter insect species that parasitize plant seeds. Insect larvae slowly consume seeds while they develop inside the protective seed coat. Rose chalcids and several torymid wasps (e.g., Figure 15.15a) parasitize seeds of rosaceous plants. Seeds of *Rosa*, hawthorn (*Crataegus* sp.), and mountain ash (*Sorbus* sp.) are often infested. Bruchid beetles (Figure 15.15b) parasitize seeds of many legumes, including ornamental Scotch broom (*Cytissus* sp.), honey locust (*Gleditsia triacanthos* L.), Kentucky coffee tree (*Gymnocladus dioicus* [L.] K. Koch), and eastern redbud (*Cercis canadensis* L.). Acorns of oaks (*Quercus* sp.) can be infested with a wide range of beetle, wasp, fly, and moth larvae. Infested seeds of many species can be detected using a simple float test, in which seeds are submerged in tap water. A drop of soap can be added to break the surface tension surrounding small seeds. Infested seeds will float and can be discarded or cut open to confirm the presence of pest larvae.

Occasional Propagation Pests

Cockroaches (Figure 15.15c) and earwigs (Figure 15.15d) inhabit tight, protected spaces. These pests can be transported to greenhouses and nurseries in cardboard boxes and supplies used for shipping and growing ornamental liners. Once established, these pests are difficult to manage. As populations of these pests expand, they shift from dead or dying plant tissue to consume root hairs, roots, and young foliage.

Collembola, or springtails, are also occasional pests of propagation stock (Figure 15.15e). Large springtail populations develop in damp environments and may move into propagation and production areas

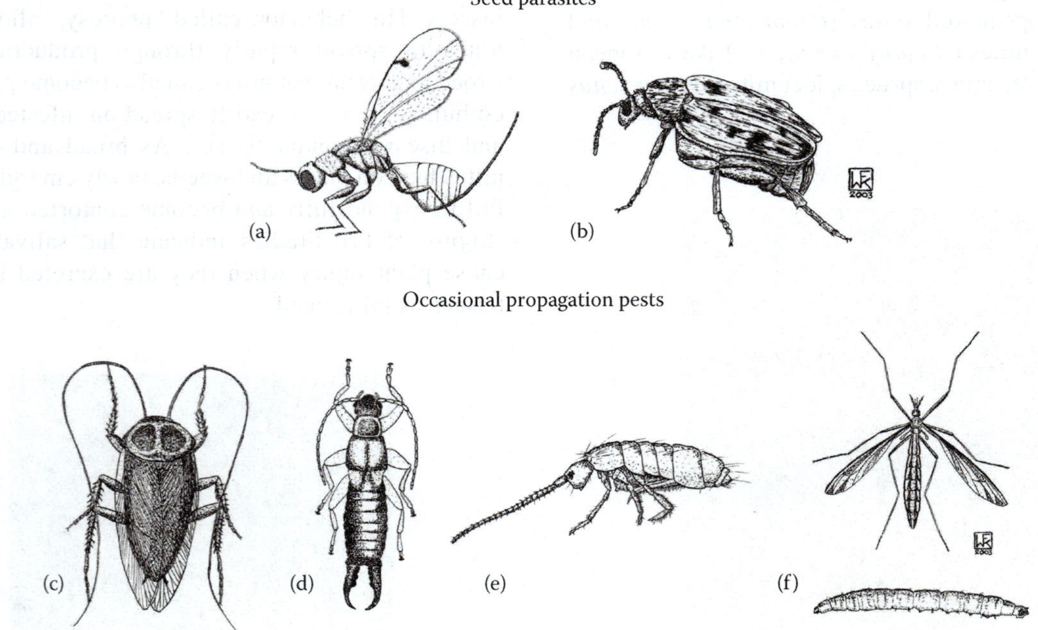

FIGURE 15.15 Seed parasites and occasional propagation pests. Larvae of pests like rose chalcids and torymid wasps (a) and bruchid weevils (b) can feed on the endosperm of seeds while hidden and protected within the hard seed coat. Cockroaches (c), earwigs (d), springtails or collembola (e), and larval crane flies (f, right) are also occasional pests in propagation facilities. An adult crane fly is also illustrated (f, left). Large populations of these pests can develop in mismanaged propagation operations and when temperature and humidity conditions are optimum. (Drawings by W. E. Klingeman.)

during drought. Collembola are opportunists and will consume fine roots in addition to decaying organic matter. Similarly, larvae of winter crane flies and other crane fly species (Figure 15.15f) develop in shaded, wet greenhouse environments. Like springtails, crane fly larvae will feed on plant roots as well as dead plant tissue.

MANAGEMENT STRATEGIES AND OPTIONS

Pest management is expensive. Costs include people-hour labor associated with scouting for pests, mixing and applying pesticides, and cleaning pesticide equipment. Pesticides themselves are expensive. The time and money that a grower invests in liner stock are lost when plants are discarded. For grafted liners of named cultivars ready for sale, losses may exceed several dollars per plant. Earning a reputation for shipping poor-quality plants or infested stock is not only the most costly loss associated with production pests—it is a recipe for failure. Preventing pest populations before their outbreak is not only logical… it makes "cents."

After identifying the pest and determining the size and location of pest populations, the scout must make a recommendation for action. If propagated liners are ready to be sold, pest presence demands immediate control. Other management recommendations may be based on prior experience with the pest and expectations of plant injury. An informed grower has several management options to choose from, including cultural, physical, mechanical, biological, and chemical controls as well as incorporating host plant resistance. In addition, the grower may elect to do nothing. Natural enemies and plant vigor can out-compete many pest population cycles.

CULTURAL CONTROL STRATEGIES

Sanitation

In propagation and nursery production systems, sanitation is consistently the most neglected option for cultural control, which limits pest population introduction and outbreaks. Dead and dying plant tissues, including discarded cutting material, should be regularly removed from propagation areas and destroyed. Nursery cull piles are decomposing mounds of substandard-quality, dead, and disease- or pest-damaged plants. Plants in nursery cull piles should be buried, composted, burned, or hauled offsite (Figure 15.16). Too frequently, unsaleable plants are dumped into a garbage can or cart and left in a greenhouse or polyhouse for several days before being emptied outside the production area. These slow-dying, drought-stressed plants accelerate pest development or become attractants to opportunistic pests that leave cull piles and move into production areas as their life cycles are completed and their food resources decompose.

Pasteurization

Disease and pest populations may develop in soilless and soil-based propagation media mixes exposed to the environment. Media can be pasteurized by injecting steam beneath the surface of the media covered with a tarp. Media temperatures at the center of the pile should reach 60 to 82°C (140–180°F) and be kept stable for 30 min. At this range, temperatures will kill arthropod pests, weed seeds, nematodes, and most harmful bacteria and fungi. Fewer beneficial organisms are killed at lower temperatures of the range. Media pasteurized at lower

FIGURE 15.16 A propagator's cull pile should be discarded away from production areas and burned, composted, or buried. (Courtesy of W. E. Klingeman.)

temperatures will contain populations of surviving beneficial organisms that out-compete pathogens in the event of recontamination.

Irrigation

Unrooted cuttings and young liners in containers and field beds need different volumes of water to thrive. Too much water washes nutrients and minerals from the plant root zone. High humidity and standing water enable fungal and mold growth, and fungus gnats follow. Providing plants with too little water causes plant stress and subjects plants to outbreaks of moisture-intolerant mite pests. An efficient propagator recognizes plant needs and provides the appropriate irrigation volume needed to maintain plant health and sustain optimal plant growth. Morning and early-afternoon irrigation is preferred. Foliage will dry before nightfall, limiting fungal development.

Weed Management

Many weeds and grasses supply pollen and nectar food resources for arthropod natural enemies and refuges for both predatory and prey insects. Yet, well-managed ornamental crops perform the same function. In general, plant propagators and nursery managers cannot tolerate the competitive and aesthetic losses attributed to weed populations. Weedy plants should be excluded from production rows and greenhouses. Aggressive weeds and weeds that produce abundant seeds should be closely controlled throughout production areas, including nursery borders. Grassy weeds and grass borders near propagation areas should be regularly mowed to reduce pest populations.

Weeds that border propagation and production areas are also overlooked as reservoirs of pests and diseases. For example, common ragweed (*Ambrosia artemisiifolia* L.) and spiny sowthistle (*Sonchus asper* [L.] Hill) support banded-wing, greenhouse, and silverleaf whitefly populations. Two-spotted spider mites can persist on wood sorrel (*Oxalis* sp.) and sowthistles (*Sonchus* sp.). Broad mites can feed and cause a characteristic strap-shaped leaf injury on many weeds common among commercial production systems, including bittercress (Figure 15.14). Broad mites overwinter on hairy galinsoga (*Galinsoga ciliata* [Raf.] Blake), prickly sida (*Sida spinosa* L.), and beggar's tick (*Bidens* sp.). Western flower thrips overwinter on spiny sowthistle and common chickweed (*Stellaria media* [L.] Vill.) and find seasonal refuge on Japanese honeysuckle (*Lonicera japonica* Thunb.), multiflora rose (*Rosa multiflora* Thunb.), and dandelion (*Taraxacum officianale* L.). The American serpentine leafminer can be found on spiny sowthistle and hairy galinsoga.

Hot Water Disinfestations

Cuttings, rooted liners, bulbs, or tubers and even entire flats of plants can be immersed in heated water for brief periods to successfully control all life stages of scale insects, mites, mealybugs, and nematodes. Propagators may be challenged to make this control option practical, however. Not all plants respond similarly to hot water immersion. Plant tissues of many species can be easily injured. Plants should be tested individually before entire blocks of plants are treated. Water-bath temperatures should be closely observed and held constant during submersion. Plants can be submerged in heated water kept at 38 to 43°C (100–110°F) for durations ranging from 5 to 20 min, depending on subsequent plant performance. Many plants respond poorly unless preconditioned in 26°C (78°F) water and benefit most by being hosed or rinsed off with cool water following submersion.

Employee Work Patterns and Clothing Colors

Routine and consistent monitoring efforts keep nursery managers and propagators aware of pest hotspots and problem areas. Once identified, plants in these production areas should be the last ones visited during daily work routines. Pest populations can be factored into efficient task scheduling to keep employees from spreading pests to uninfested areas. In general, choose neutral tan- and beige-colored uniforms and restrict clothing colors that attract pests, like red, yellow, light blue, and green.

PHYSICAL CONTROL STRATEGIES

Screening and Fabrics

Commercially available insect screens and spun-bonded fabric barriers exclude arthropod pests from propagation areas. Screen mesh openings as small as 0.15 mm (0.0059 in) prevent thrip entry into greenhouses. As mesh openings get smaller, however, airflow is restricted. Powerful (and expensive) fans are needed to move the same volume of air through production areas. Solutions to this challenge include pleated screens with larger surface areas and screened annexes that are built externally around air intake vents. Spun-bonded and woven fabrics are also available, which are lightweight and reflective. Fabrics shelter field-grown propagation blocks from pests as well as protect plants from light frost damage.

Barriers and Bands

Several pests can be restricted from greenhouse propagation benches by inverting and securing plastic or aluminum pans where the support post meets the bench. The container interior is coated with Pestik® or Tanglefoot®

insect glue. These barriers are effective against flightless pests. Snails, slugs, some beetles, aphid-tending ants, and juvenile cockroaches and earwigs, which are flightless when young, are kept below the barrier.

MECHANICAL CONTROL STRATEGIES

Thrip, aphid, and mite populations decline following some mechanical actions like regular plant handling and brushing or shaking foliage. Because brushed plants are typically smaller than unbrushed plants, mechanical brushing has also been used as a non-chemical alternative to plant growth regulators. Brushing and handling activity physically disrupt pest-feeding activity. Mechanical actions also induce plant stress, which stimulates chemical plant defenses. In limited greenhouse trials, mechanically induced stress has been used as a pest management tactic on some seedling species and rooted cuttings.

BIOLOGICAL CONTROL STRATEGIES

Predators and parasites are often efficient foragers for prey arthropods in open environmental systems. If they can be conserved or released into production areas, populations of natural enemies and entomopathogens are also valuable components of propagation IPM. Many of the most effective natural enemies are mass reared for commercial sale (Tables 15.2 and 15.3). Once established,

TABLE 15.2
Natural Enemy Predators, Parasites, and Entomopathogenic Organisms That Are Commercially Available for Use in Managing Populations of Many Common Pests in Propagation Systems[a]

Arthropod Pest	Common Name	Scientific Name
Fungus gnats	Bt	*Bacillus thuringiensis* ssp. *israelensis*
	Predatory mites	*Hypoaspis miles*
	Nematodes	*Steinernema feltiae*
Shore flies	Bt	*Bacillus thuringiensis* ssp. *israelensis*
	Predatory mites	*Hypoaspis miles*
	Nematodes	*Steinernema feltiae* and spp., *Heterorhabditis* spp.
Leafminers (Diptera)	Parasitic wasps	*Dacnusa* spp., *Diglyphus* spp.
Scales	Predatory ladybeetles	*Chilocorus* sp., *Rhyzobius* spp.
	Parasitic wasps	*Aphytis* spp., *Encarsia* spp., *Metaphycus* spp.
Mealybugs	Lacewing predators	*Chrysopa* and *Chrysoperla* sp. larvae
	Parasitic wasp	*Leptomastix dactylopii*
Aphids	Parasitic fly	*Aphidoletes aphidimyza*
	Parasitic wasps	*Aphelinus semiflavus*, *Aphidius matricariae*, *Lysiphlebus testaceipes*
	Lacewing predators	*Chrysopa* and *Chrysoperla* sp. larvae
	Predatory bug	Orius insidiosus
Thrips	Lacewing predators	*Chrysopa* and *Chrysoperla* sp. larvae
	Minute pirate bug	Orius insidiosus
	Predatory bug	*Eusieus* spp.
	Predatory mite	*Macrotracheliella* sp.
	Predatory thrips	*Franklinothrips vespiformis*, *Scolothrips maculates*
Whiteflies	Predatory ladybeetle	*Delphastus pusillus*
	Parasitic wasps	*Encarsia* spp., *Eretmocerus* spp.
Spider mites	Lacewing predator	*Chrysopa* sp.
	Predatory bug	Orius insidiosus
	Predatory mites	*Amblyseius californicus*, *Phytoseiulus persimilis*, and *P. longipes*
	Predatory thrips	*Franklinothrips vespiformis*, *Scolothrips maculates*
	Spider mite destroyer ladybeetle	*Stethorus punctillium*

[a] Many of the most effective natural enemies of propagation pests are mass reared or cultured for commercial sale. In general, natural enemies sold for pest management have specific prey preferences. Correct pest identification in a propagation system allows IPM practitioners to pair the appropriate parasite, predator, or pathogen with its preferred prey or host.

TABLE 15.3
Commercial Sources of Beneficial Arthropods and Nematodes for Propagation IPM

Becker Underwood 801 Dayton Avenue Ames, Iowa 50010 1-800-232-5907 http://www.beckerunderwood.com	Koppert Biological Systems 1502 Old US-23 Howell, MI 48843 1-800-928-8827 http://www.koppertonline.com
Beneficial Insectary 9664 Tanqueray Court Redding, California 96003 (800) 477-3715 http://www.insectary.com	Rincon-Vitova P.O. Box 1555 Ventura, California 93002-1555 (800) 248-2847 http://www.rinconvitova.com
BioBest Ilse Velden 18 2260 Westerlo, Belgium +32 14 257 980 http://www.biobest.be	Southeastern Insectaries, Inc. 606 Ball Street Perry, Georgia 31069 (877) 967-6777 http://www.southeastern insectaries.com
BioLogic P.O. Box 177 Willow Hill, Pennsylvania 17271 (717) 349-2789 http://www.biologicco.com	Syngenta Bioline P.O. Box 2430 Oxnard, California 93034 (805) 986-8265 http://www.syngenta.com/ global/Bioline
IPM Laboratories 980 Main Street Locke, New York 13092 (315) 497-2063 http://www.ipmlabs.com	

survival of biological control agents is dependent on their successful integration with other management strategies. Long-residual, systemic, and broad-spectrum (general) insecticide and miticide applications should be avoided before and after releasing natural enemies. Fungicidal sprays are toxic to entomopathogenic fungi like *Beuveria bassiana* (Balsamo) Vuillemin and *Metarhizium anisopliae* (Metsch.) Sorokin. Many natural enemies must be supplied with alternative pests, pollen, or nectar resources to survive when their preferred food resources decline.

CHEMICAL CONTROL STRATEGIES

Chemical pesticides are a practical component of successful IPM programs. Within an IPM framework, a pesticide is chosen after considering its expected ability to control the pest, the action of the pesticide on predatory and parasitic natural enemies, and the mode of action of its chemical constituents. Emphasis is placed on making spot treatments to centralized pest populations. Spot sprays are used to manage localized populations rather than relying on calendar-based or cover sprays. Spot treatments are seldom effective against general outbreaks that require "curative" action. Re-treatment intervals are timed to catch pests that emerge from pesticide-resistant eggs and pupae. The reentry interval, during which unprotected workers are excluded from treated areas following a pesticide application, is also an important consideration in propagation systems. Growers prefer to use pesticides that allow workers to handle plants as quickly as possible after treatment.

UNDERSTANDING AND MANAGING PEST RESURGENCE AND PESTICIDE RESISTANCE

Pest resurgence occurs when a broad-spectrum (nonselective) insecticide is used to control a pest. When natural enemies of a pest are also eliminated, pests that survive treatment or that re-colonize the crop develop without pressure from parasites and predators. Subsequent pest populations may be larger than the initial pest population. Efficient pesticide use is accomplished by closely monitoring populations of identified pests. Pesticides that are selective for that pest are applied to the most susceptible life stages of the pest while pest populations are still growing.

Mode of action describes the insect physiological or functional metabolic pathways interrupted by the chemical pesticide. Disrupting these pathways leads to pest or plant mortality. Mode of action is often related to the chemical class that categorizes a pesticide active ingredient. For example, Sevin® is a carbamate class insecticide that, like other carbamates, disrupts the nervous system by allowing continuous axon excitation when arthropod pests contact or ingest the active ingredient carbaryl. In general, pests that develop rapidly and have multiple generations within a season can more readily develop resistance to pesticides. Resistance develops by selection of genetic mutations or physiological shifts that allow recessive alleles to be expressed. Resistance is more likely to occur when the same pesticide or mode of action is applied continually to control pests. Pesticide resistance can also be limited by rotating modes of action after several treatments have been applied. Alternatively, strips or groups of untreated plants allow populations of susceptible gene-dominant pests to survive and interbreed with resistant, populations.

PHYTOTOXICITY

Plants in propagation systems are often susceptible to phytotoxicity or damage from pesticide volatization, drift, and direct contact with pesticides, which would not have affected the parent plant. Rootlets and leaflets on tissue-propagated microplants are not fully functional until they have been acclimated. Microplants and unrooted cuttings are more sensitive to pesticides than parent plants. Exposure to direct sunlight and temperatures in excess of 30°C (86°F) also increase the severity of plant phytotoxic response. Common symptoms of phytotoxicity include the loss of the protective waxy bloom (particularly among conifers treated with horticultural oils and insecticidal soaps), premature anthocyanin production (chlorosis and reddening of foliage), spot chlorosis and necrosis on foliage and flowers, leaf curl, and leaf drop.

Tissue Culture Systems

Tissue culture presents unique challenges for managing pests on plantlets. By volume, very little pesticide product would be purchased by tissue culture propagators. Few, if any, chemical pesticides are labeled for use in tissue culture systems. Thus, propagators have very few chemical options for controlling arthropod pest, mite, thrip, and aphid populations that develop in tissue culture. These pests are most readily managed with sterile techniques and the use of only pest-free stock plant material.

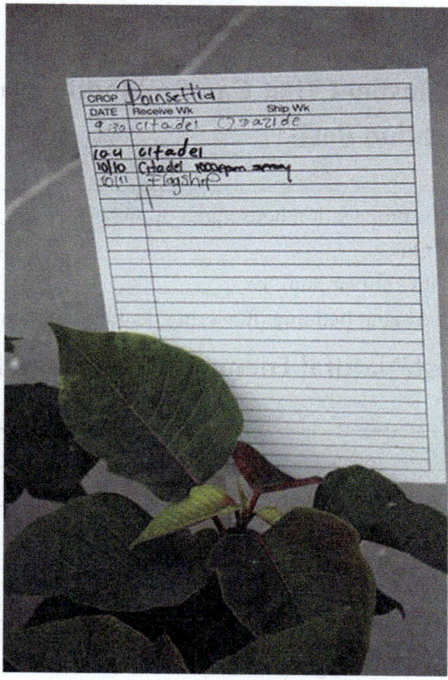

FIGURE 15.17 In-house management records can also be kept right with the crop and may include a propagator or production manager's notes on fertilization and pesticide history records, as well as targeted dates of sale. (Courtesy of W. E. Klingeman.)

HOST PLANT RESISTANCE STRATEGIES

Many chemical, morphological, and physiological resistance mechanisms have evolved in plants. Susceptibility to pests and diseases often varies within genera and among cultivars of ornamental plant species. For example, resistance to pests may be attributed to the density or repellency of trichomes on leaves and stems. Pests may simply be unable to reach the leaf surface of some hairy species or cultivars. Plants may also have glandular trichomes that exude chemical byproducts of cellular metabolism. Terpenoids, essential oils, and tannins in plant tissues offer a measure of protection against many sap-feeding and defoliating pests.

Disease- and pest-resistant plants offer advantages to growers and landscape managers. Fewer chemical, labor, and maintenance expenses are needed to produce these crops. Lower-maintenance plants can be developed as a commercial niche and marketed to homeowners. Landscapes designed using pest- and disease-resistant plants are environmentally sustainable. Yet, even resistant plants can be overcome by pest pressure when subjected to water, nutritional, or other stress. The "right plant" must be grown in the "right place" under appropriate cultural care, whether in the landscape or the production nursery.

THE JOB IS NOT COMPLETE WITHOUT THOROUGH RECORDKEEPING

There are several incentives for growers to maintain thorough and accurate IPM and scouting records. Good records can limit a grower's liability in workplace-related lawsuits. A scout's IPM records provide an invaluable, site-specific training tool for new employees in the event that an IPM scout quits or has his or her employment terminated. Written records should provide an easy method of tracking the type, rate, and volume of pesticides applied, as well as the environmental conditions in place at the time of treatment. These types of data, as well as dates for fertilization records and planned sale readiness, can also be recorded on indelible note cards and placed within each crop (Figure 15.17). Perhaps most importantly, a scout's records can be used to reveal the location of pest and plant disease hotspots where pests routinely first appear. Hotspots can be found in even the best-run operations.

Outdoors and in the field, scouting reports should be entered into a database or made in indelible ink or pencil. The report log can consist of individual pages that are designated by date or production area. Pages should be maintained and filed in a logical order, with the expectation that they will become a useful management resource (Table 15.4).

IT IS NOT ABOUT *WHAT YOU DID*, BUT *HOW IT WORKED*: ASSESSING MANAGEMENT ACTIONS

A propagation IPM program is incomplete unless the management tactics that were implemented are evaluated during follow-up scouting. The scout's decisions should be reevaluated based on pest control results and

TABLE 15.4
Sample Datasheet That Can Be Modified by an IPM Scout for Use in Monitoring Crops and Production Areas for Propagation Pests[a]

Date: ______________________ Name of Scout: ______________________
Stock Block (or propagation house/area of nursery): ______________________
Time In/Time Out: __________/__________
Host Plant (with cultivar name): ______________________ Plant Size: ______________________
Plant Stage (dormant, budding, flowering, leafing out, fruiting, mature): ______________________

Type of Pest/Natural Enemy	Stage (egg/larvae/nymph/adult)	Number Observed	Location	Comment

Damaged Area (percentage of stem, bark, flowers, leaves): ______________________
Number of damaged plants: ______________________
Natural enemies (none, few, common, abundant): ______________________
General Comments: ______________________

Recommended Management Action: ______________________

Pesticide/Date/Rate/Volume Applied: ______________________
Environmental Conditions: (temperature/RH/sunlight): ______________________
Date of Follow-Up Evaluation: ______________________
Success of Treatment/Comments: ______________________

[a] A sample sheet from a scout's notebook should provide space for thorough records. List the pests and natural enemies that were observed, the number and the type of host plants (and what parts) that were affected, and what management actions are recommended. Specify the pesticide used and the date, the rate that was used, and the volume of material applied. Additional space should be provided to allow for follow-up observations and future recommendations about managing the pest.

the aesthetic and economic satisfaction with the crop. A final assessment should be recorded and filed, along with any associated observations (e.g., weather, new plant/pest introductions, changes in staff, new equipment, and others) that could help explain the success or failure of management actions taken. Invariably, these notes will become a valuable resource as the IPM program matures. More importantly, maintenance of accurate and thorough records is prudent insurance against the loss or relocation of employees who have been responsible for implementing the program. A newly hired scout or IPM manager should use past records as a practical training guide.

SUGGESTED READING

Ali, A. D. 1987. Integrated arthropod management in plant tissue culture production. *Comb Proc Intl Plant Prop Soc* 37:104–106.

Daar, S., H. Olkowski, and W. Olkowski. 1992. *IPM Training Manual for Wholesale Nursery Growers*. Berkeley, California: Bio-Integral Resource Center, and Washington, D.C.: Wholesale Nursery Growers of America, 84 pp.

Dreistadt, S. H. 2001. *Integrated Pest Management for Floriculture and Nurseries*. Publication 3402. Berkeley, California: University of California Press, 422 pp.

Flint, M. L. 1992. Resources for establishing an IPM program for ornamental plants. *Comb Proc Intl Plant Prop Soc* 42:251–255.

Flint, M. L. and S. H. Dreistadt. 1998. *Natural Enemies Handbook*. Berkeley, California: University of California Press, 154 pp.

Frank, S. D. 2010. Biological control of arthropod pests using banker plant systems: Past progress and future directions. *Biol Cont* 52:8–16.

James, R. L., R. K. Dumroese, and D. L. Wenny. 1995. *Botrytis cinerea* carried by adult fungus gnats (Diptera:Sciaridae) in container nurseries. *Tree Planters' Notes* 46:48–53.

Johnson, W. T. and H. H. Lyon. 1991. *Insects that Feed on Trees and Shrubs*, 2nd Edition. Ithaca, New York: Cornell University Press, 560 pp.

Latimer, J. G. and R. D. Oetting. 1994. Brushing reduces thrips and aphid populations on some greenhouse-grown vegetable transplants. *HortScience* 29:1579–1581.

LeBude, A. V., S. A. White, A. F. Fulcher, S. Frank, W. E. Klingeman III, J.-H. Chong, M. R. Chappell, A. Windham *et al.* 2012. Assessing the integrated pest management practices of southeastern U.S. ornamental nursery operations. *Pesticide Manage Sci* 68:1278–1288.

Parella, M. P. 1992. An overview of integrated pest management for plant propagation. *Comb Proc Intl Plant Prop Soc* 42:242–245.

Parrella, M. P., K. M. Heinz, and L. Nunney. 1992. Biological control through augmentative releases of natural enemies: A strategy whose time has come. *Amer Entomologist* 38:172–179.

Rosetta, R. 1995. How to use biological control to manage propagation pests. *Comb Proc Intl Plant Prop Soc* 45:272–275.

Part V

Evaluation of Propagated Plants

16 Evaluation of Data from Propagation Experiments

Michael E. Compton

Concept Box 16.1

- The most commonly used form of hypothesis for plant propagation research is the "null hypothesis." This popular hypothesis states that all treatments will elicit a similar response.
- An experimental unit (EU) is the smallest unit that receives a single treatment. The observational unit (OU) is the object being observed or measured. However, a cutting may be considered as both the EU and OU in cases in which one cutting stuck per pot.
- The completely randomized, randomized complete block, incomplete block and split plot designs are most commonly used for plant propagation research. These designs doffer in their randomization schemes and experimental requirements.
- Many researchers find it beneficial to culture multiple explants in culture vessel. This conserves resources and time, but leads to a situation in which multiple measurements are made for each replicate (e.g., culture vessel). Statistically this is referred to as subsampling or repeated measures.
- The nature of the data influences the type of statistical procedures that should be used. Most observations made in plant propagation experiments result in data that can be classified as continuous (shoot length, root length, etc.), counts (number of organs per explant or callus [shoots, embryos or protocorm-like bodies]), binomial (response or no response [e.g., percentages]), or multinomial (shoots, roots, leaves or no response). Therefore, the statistical methods used to evaluate the treatment effects vary for each type of data observation.

Efficient and reliable propagation practices are necessary to supply the millions of plants required annually by the green industry. Practices used by nurseries and greenhouses to propagate plants were developed through research projects aimed at pursuing ideas that optimized propagation and/or solved specific problems associated with propagating plants. From these ideas, planned experiments were developed in which treatments were devised and tested in controlled conditions under which plants were typically propagated. Data gathered from observations made during experimentation were evaluated using statistical analysis to ascertain the effectiveness of each treatment at improving plant propagation rates or correcting specific problems associated with propagation.

The purpose of this chapter is to introduce the basic elements of plant propagation research and data analysis. After reading this chapter, you should understand the basics of the scientific approach and be able to select and apply experimental designs and statistical methods to analyze, interpret, and present plant propagation data.

METHODOLOGIES ASSOCIATED WITH EVALUATING DATA

Organization is an important part of conducting research. Thoughts and materials must be organized into a logical progression of events that can be applied in experimental conditions. This section focuses on the factors and events important for conducting research and provides examples of how these items can be organized to improve research effectiveness.

Establishing a Hypothesis and Experimental Objectives

Ideas aimed at optimizing plant propagation need to be formulated into a hypothesis that can be tested. When formulating a hypothesis, it is important to examine published literature to learn from peers that have conducted similar research. Your hypothesis can be refined, if necessary, and a set of experimental objectives can be developed.

The most commonly used hypothesis for propagation research is the "null hypothesis." This popular hypothesis states that all treatments will elicit a similar response among all propagules. Application of the null hypothesis to plant propagation research can be illustrated by using an example in which treatments consisting of one concentration of four rooting factors (indole-3-butyric acid [IBA], K-IBA, and naphthaleneacetic acid [NAA], and a solvent control with no rooting factor) were tested to evaluate their ability to stimulate adventitious root formation in softwood cuttings of Korean spice viburnum (*Viburnum carlesii* Hemsl.). In this case, the null hypothesis would be stated as follows: Softwood cuttings of Korean spice viburnum will root similarly when treated with the same concentrations of IBA, K-IBA, or NAA, or a solvent (50% ethanol) control. In this case, the control is added to ensure that the results obtained were due to the rooting factor treatments and not to the solvent or an internal factor.

Selecting the Components of the Experiment

The treatments and supportive materials required to conduct the experiment are selected following the development of a hypothesis and experimental objectives. Treatments are usually any condition(s) that characterize the population of interest and whose effect on the population can be measured and tested (Lentner and Bishop, 1986). In asexual propagation experiments, treatments may be rooting compounds, as demonstrated in the previous paragraph, or may include environmental factors, such as misting frequency and duration; light level and photoperiod; temperature and humidity; and rooting media or physiological factors, such as stage of development (softwood vs. semi-hardwood vs. hardwood), developmental phase (mature vs. juvenile), etiolation (banding), degree of wounding, or any other treatment that influences the development of the stock plant or cutting. Researchers may choose to examine one treatment factor at a time or test several factors simultaneously (e.g., stock plant genotype and plant growth regulator [PGR] concentration). Regardless of the number of treatments selected, it is important that they will be chosen according to the experimental objectives and their ability to evoke a wide range of responses so that an optimum level can easily be identified.

Supportive materials are not treatments, but are items found under conditions in which the problem is typically observed. In asexual propagation studies, these materials may be mist irrigation, rooting substrate, growth regulator type and concentration, growth environment (light level and photoperiod), or any aspect important for the satisfactory development of cuttings. Supportive materials should not interfere with the treatment effects.

The experimental unit (EU) and observational unit (OU) are chosen once the treatments and supportive materials are selected. By definition, an EU, or replicate, is the smallest unit that receives a single treatment, whereas an OU is the object being observed or measured (Lentner and Bishop, 1986). A replicate may be a container housing several cuttings to which one treatment is applied or individual cuttings stuck in a container to which all treatments are applied. In the latter situation, each propagule would be considered as both an EU and an OU. During an experiment, each OU is observed, and its reaction to the experimental treatment(s) is recorded.

Replication occurs when more than one EU is evaluated for each treatment. Replication is necessary in research so that an accurate estimate of treatment effects can be obtained. The number of replicates required per treatment is based on the amount of variability expected to occur in the data set and the degree to which the researcher wishes to detect treatment effects (Zar, 1984). At least three replicates are required per treatment to calculate the mean and measure error. Most researchers rely on prior experience, review of literature, and speculation to determine the number of replicates to be used in an experiment. However, insufficient replication can reduce the ability of statistical procedures to detect treatment differences (Lentner and Bishop, 1986). To provide assistance in determining replicate numbers, Kempthorne (1973) provided a formula that uses information from previous studies to estimate the number of replicates required per treatment.

Experimental designs are used to arrange replicates and OUs in a specific randomization scheme to measure the variation associated with experimental treatments and non-treatment factors that might influence the outcome of an experiment. Randomization reduces experimenter bias by assigning treatments in a planned random, fair, and unbiased manner. It is important to examine the supporting materials and the population of interest for possible variation. The experimental design should be chosen after the EUs and OUs have been selected, and should be simple to use and efficient with respect to measuring treatment effects and residuals, and should match the experimental objectives (Compton and Mize, 1999). Typical experimental designs include the completely randomized (CR) and the randomized complete block (RCB) designs. These designs differ in their randomization schemes and experimental requirements.

The CR design is most commonly used for propagation research because it is easy to use and does not use a patterned randomization scheme. This allows researchers to maximize the number of replicates used and employ equal or unequal treatment replication (Little and Hills, 1978). This is important because unequal replicate numbers often occur in propagation studies due to the death of propagules from non-treatment factors, such as mist failure or inaccurate mist coverage, insufficient number of available cuttings, or other similar factors not associated with the treatment.

Randomization of replicates in the CR design can be illustrated using an example in which the effectiveness of four rooting factors on adventitious root formation on the softwood cuttings of Korean spice viburnum was examined. Cuttings were dipped in a 48 mM aqueous solution of one of three rooting factors (IBA, K-IBA, or NAA) for 30 sec before being placed in a mist bed containing sand. Control cuttings were dipped in 50% ethanol (solvent for the auxins) for the same duration. Cuttings were misted with reverse-osmosis water for 10 sec duration on 10 min intervals. Application of mist was controlled by a Gemini 6 control unit with a photoelectric cell. The randomization of replicates for this experiment would appear as shown in Figure 16.1a.

Several techniques can be used to assign propagules to treatments. It is usually best to prepare and sort cuttings into groups of similar size and quality before assigning them to treatments. Sorted cuttings from each size group are selected at random and the treatment is applied. A plan for the randomization of treated cuttings in the propagation area should be established beforehand by drawing numbers representing treatment replicates from a hat and diagramming the results on paper or by using a statistical analysis software package to establish the randomization scheme.

From a statistical perspective, the CR design is most efficient because it allows researchers to maximize degrees of freedom (df) for error, which is important for detecting treatment differences during statistical analysis (Little and Hills, 1978). However, for the CR design to be effective, it is essential that the replicates should be

(a) Completely randomized

①	②	④	②	③	④	④	①	③	④
④	③	②	①	④	①	②	③	①	②
②	①	①	③	④	②	③	④	③	③
③	④	①	②	②	③	①	②	④	①

(b) Randomized complete block

1	2	3	4	5	6	7	8	9	10
①	②	④	③	③	②	①	④	③	②
②	④	③	①	①	④	④	③	①	③
③	①	②	④	②	③	②	①	④	④
④	③	①	②	④	①	③	②	②	①

FIGURE 16.1 Randomization schemes for CR (a) and RCB (b) designs with four treatments. Each treatment was replicated 10 times in each design.

as homogeneous as possible. Situations in which EUs or OUs are highly heterogeneous increase variability in the experiment and decrease the likelihood that treatment differences will be detected (Little and Hills, 1978).

Designs that employ blocking are used when a high degree of heterogeneity exists among the EUs. Unevenness or inconsistency among environmental factors in the greenhouse, differences among stock plants from which propagules are obtained, differences among seasons in which cuttings are harvested, variability among technicians, as well as differences among batches of cuttings are few non-treatment factors that introduce heterogeneity.

When using the RCB design, replicates of each treatment are assembled into blocks. This creates a high level of uniformity among treatments within a block and reduces variation not associated with a treatment (Little and Hills, 1978). When using blocking, it is important that only measurable factors are blocked and that all propagules in a block are as uniform as possible. For example, propagules from the same stock plant often have similar rhizogenic competence and can be used for treatments that appear in the same block.

The viburnum study could be used to illustrate the RCB design by using propagules from one stock plant to establish replicates of each treatment in a block. In this case, 10 stock plants would be required to produce enough blocks to meet the requirement of 10 replicates per treatment (Figure 16.1b). Using the RCB in this situation would maximize uniformity within blocks by utilizing the homogeneity present within each stock plant.

It is impossible to be absolutely sure if blocking will be required when designing experiments. The need for blocking can be determined by testing for blocking when conducting the statistical analysis. Kempthorne (1973) outlined a simple method to determine if blocking is necessary.

Determining How Many Times an Experiment Should Be Conducted

Most statisticians and researchers believe that an experiment must be repeated to validate the results (Lentner and Bishop, 1986). However, repeating experiments does not guarantee a similar outcome even if the same stock plants are used as rhizogenesis is often influenced by the stock plant environment as well as endogenous physiological factors. In addition, repeating experiments may introduce time as a non-treatment factor. For example, the physiology of a stock plant can change during the growing season or as the plant ages. Either situation can change the outcome of an experiment. Researchers generally have two options when differences between runs in an experiment are obtained. The entire experiment can be repeated, which can be costly, or each run may be examined as a block or main plot. The best option would be to first look at the data from the two runs and determine if similar trends exist. If the trends are similar, the experiment need not be repeated, but the circumstance(s) that caused the difference(s) should be examined and explained. In contrast, an experiment should be repeated if the results obtained in the two runs differ significantly. If differences among runs persist, the possibility of seasonal effects on the results should be examined experimentally. As a basic rule, an experimenter should decide how many times the experiment will be conducted and what will be done if differences between runs of the experiment are observed when planning and designing the experiment.

Choosing the Best Data and Planning the Statistical Analysis

The data recorded should help evaluate treatment effects in accordance to the experimental objectives. Recording the wrong data or failing to record important parameters will reduce the likelihood that the researcher will be able to properly assess propagule response. Most researchers review relevant literature and combine that information with their experience when deciding which observations to measure. In the viburnum illustrations used throughout this chapter, the number and percentage of propagules producing roots, the number of roots per propagule, root dry weight, and root length are observations that most researchers would use to evaluate propagule competence. However, to get the most information from the project, the researcher may wish to make additional observations, such as a rating of cutting vigor, shoot weight, plus the number of and the percentage of propagules that produced callus. Nondestructive measurements should be used if evaluation of subsequent survival and establishment of propagules is desired. It is usually a good idea to record as much relevant data as possible and choose the best parameters when presenting results.

CONDUCTING THE EXPERIMENT

Experimentation may begin once planning is completed and thoroughly reviewed. Personal bias must be avoided during all phases of the experiment to avoid swaying the outcome. This can be accomplished by using the structure of the experimental design to set up the experiment (Mize *et al.*, 1999). Do not prepare all propagules for one treatment at a time. Instead, you should prepare the cuttings and sort them into groups of like sizes or vigor ratings before applying the treatments.

It is also important to avoid circumstances such as experimenter fatigue or severe environmental conditions while setting up experiments. These conditions can introduce error into the experiment and alter the results. The time required for each phase of the experiment should be estimated during planning, and tasks that are taxing should be staggered to avoid worker strain. Mistakes may also occur when propagules are stuck into the rooting substrate or labeled. Be careful to label each cutting or replicate with the correct information. If using codes, be sure to record keys to your codes in your laboratory notebook. There are many avenues for introducing operator error into your experiment, so it is important to take the precautions necessary to encourage accurate and safe experimentation.

Recording Data

Observations made during an experiment are recorded on data sheets. This activity can lead to mistakes if the sheets are not well organized. Designing data sheets in the manner that each datum will be entered into the computer helps avoid mistakes during transcription. Many statistical software packages will produce data sheets for you. If not, data sheets should be designed during planning to ensure that the researcher is considering observations that measure treatment effects.

As when starting an experiment, you should use the structure of the experiment when recording data. Observations of replicates in one block should be recorded before moving to the next block. This helps avoid personal bias. Be meticulous and take your time. Sloppy technique introduces errors and inconsistency. Take breaks to avoid fatigue when recording data. When entering data on the computer, it is important to save your work frequently. In addition, be sure to save several electronic copies of your files and distribute them into several secure locations to avoid losing valuable information. One year, a colleague's laboratory was destroyed in a fire, and over 26 years of valuable data were lost.

Data should be checked for errors after they are recorded. Erroneous data entries lead to incorrect treatment evaluation and interpretation (Mize *et al.*, 1999). Make sure that data of each treatment fall within the expected boundaries. Data outside these boundaries are called "outliers" and may result from errors in reading and recording or transcribing data, or may be values obtained from plant tissues with inherent problems or genetic mutations. Outliers are often discarded. However, researchers should exercise caution when deleting data as personal bias may be unintentionally introduced and distort the results. Several statistical procedures can be used to detect outliers and are helpful to reduce bias when editing (Barnett and Lewis, 1984). Treatment codes should be examined while checking data. Do not assume that codes were entered correctly. In the viburnum example, the four treatments were identified by assigning them numbers 1 to 4 (e.g., IBA was assigned treatment number 1; K-IBA, number 2; NAA, number 3; and the control treatment, number 4). Errors associated with entering codes can be identified by calculating the mean for the coded variables (Mize *et al.*, 1999). In this example, obtaining a mean of 2.5 for the rooting factor treatment identification codes indicates that the numbers were entered correctly. Editing data can be facilitated by printing data sheets and thoroughly reviewing them. This can be a time-consuming task, but is worth the effort.

DATA ANALYSIS AND INTERPRETATION

Data must be statistically analyzed, and the results must be interpreted as planned. The nature of the data influences the type of statistical procedure that should be used. Most observations recorded in propagation experiments result in data that can be classified as counts (number of roots per cutting), binomial (response or no response [e.g., percentages]), multinomial (root, callus or no response), or continuous (root length) (Mize *et al.*, 1999). The statistical methods used to evaluate treatment effects vary for each of these types of data. Outlining the sources of variation and df during planning can help identify statistical procedures most appropriate for the type of data recorded as well as experimental objectives. When analyzing data, a general analysis is conducted first to determine if there are any differences between the treatment levels. Further analysis of treatment means is conducted if differences are detected by the general analysis (Compton, 1994; Mize *et al.*, 1999).

The previously described example in which the rooting of *V. carlesii* softwood cuttings was examined following the treatment of cuttings with three rooting factors will be used to discuss the common types of general analyses. To understand how each statistical analysis applies to the experiment, it is important to know the methods and the materials used. Softwood cuttings (~15 cm long) containing five to six nodes were collected on two separate dates, 2 weeks apart, from field-grown plants in July. Each collection date represented a replication of the experiment in time. Since the stock plants themselves were all the same clone, it was decided that cuttings from all plants could be pooled together. All cuttings were sorted, and only the most vigorous propagules were used for the experiment. The stems of all remaining cuttings were trimmed to just below the fourth node from the apex, with leaves from the lower two nodes removed before being dipped into a solution containing 48 mM concentration of one of three rooting factors (IBA, K-IBA, or NAA dissolved in 50% ethanol) or a 50% ethanol control for 30 sec. Following treatment, cuttings were placed in 5 (H) × 27.5 (W) × 55 (L) cm containers filled with coarse sand and incubated in the greenhouse in diffused light (maximum, 250 μM m^{-2} sec^{-1}) and natural photoperiod. Cuttings were arranged within the containers in a CR design, with 10 propagules per treatment. Cuttings were irrigated using mist applied through flora-mist nozzles (15 L/h) for 8 sec every 10 min. Misting commenced 1 h before sunrise and concluded 1 h after sunset. Rooting of treated cuttings was observed for 8 weeks, after which time, the number of cuttings with roots (binomial), the number of roots per cutting (count), and the root length (continuous) were recorded.

Analyzing Response/Percentage Data

Response data, the ability of propagules to survive and produce roots, are important to evaluate the influence of treatments on rhizogenesis. These data have a binomial distribution, which causes treatment variances to be dependent on cutting performance (Lentner and Bishop, 1986; Mize *et al.*, 1999). Percentages calculated from observations are often considered important because researchers are interested in identifying treatments that optimize propagule response and propagation efficiency. This can translate directly to increased profits. Logistic regression is best suited for analyzing response data because the procedure calculates a special coefficient and standard error value that is independent of cutting performance (Mize *et al.*, 1999).

In the viburnum example, data on the percentage of cuttings that produced roots were binomially distributed, with datum being recorded as 0 (no roots/response) or 1 (roots observed). Analyzing these data using logistic regression revealed that the percentage of propagules that produced shoots was influenced by the rooting factor used (Table 16.1). The *P* value (level of significance) was obtained by comparing the calculated statistic with a known set of values. Your computer software will calculate the logistic regression coefficients and generate a *P* value. *P* values equal to or less than 0.05 are considered significant. This level is chosen by most

TABLE 16.1

Logistic Regression Analysis Table Demonstrating the Influence of Three Rooting Factors and a Control on the Percentage of *Viburnum carlesii* Softwood Cuttings that Produced Roots[a]

Predictor Variables	Logistic Regression Coefficient (A)	Standard Error (B)	Logistic Regression Statistic (A/B)	*P*
Constant	–2.579	0.6876	–3.75	0.0002
Rooting factor	1.032	0.2551	4.04	0.0001

[a] Data are hypothetical.

researchers because 0.05 indicates that a similar result would be obtained 95% of the time that the experiment is conducted, signifying a high level of confidence for the outcome. Another reason for selecting 0.05 is that most journals require researchers to test at this level of significance. In our experiment, a significant *P* value of 0.0001 for rooting factor was obtained, indicating that there were significant differences among the treatments and that there is a 99.999% probability that we would observe a similar response the next time that we conduct the experiment. This is important when considering that there may be nurseries or greenhouses throughout the world that will be interested in using our technique.

Before analyzing response data, it is important to determine if there are treatments in which the response did not vary. This usually occurs when all or none of the propagules responded. This occurred in our control treatment in which none of the propagules produced roots. Researchers have two options when this occurs. Data in treatments in which no response or a 100% response was observed can be altered before analysis (Mize *et al.*, 1999). Zeros may be changed to a slightly higher value (0.000001), and 100% values may be reduced slightly (0.999999). On the other hand, data in treatments with an all-or-nothing response may be discarded. It is important to decide during planning how these values will be handled if this situation occurs. In addition, you should also indicate if treatments were dropped or if values were altered when writing reports and manuscripts. In this example, none of the control cuttings responded, and it was decided that zero values obtained in the control treatment be altered to 0.000001 before analyzing the data.

Many researchers wish to use analysis of variance (ANOVA) for binary data because of its popularity and ease of use. This practice may lead to erroneous interpretations because percentage data, or proportions, are generally binomially distributed (Zar, 1984; Finney, 1989; Fernandez, 1992; Compton, 1994). In this situation, treatment variances are dependent on the success of the treatment in stimulating a response (Mize *et al.*, 1999). Treatments with poor success percentages (0–30%) often have small variances, whereas those that stimulate a high success rate (70–100%) have large variances. Treatments in which the response rate is between 30% and 70% generally have more normal variances compared to treatments with greater or smaller success rates. Because ANOVA produces separate estimates for treatments and experimental error, and because the standard errors for treatment parameters are a function of success rate, the use of ANOVA for data sets containing treatment parameters with poor or high success rates often leads to a poor estimate of treatment differences (Mize *et al.*, 1999).

However, ANOVA can be used for analyzing response data provided that the data are transformed before analysis (Compton, 1994). The arcsin transformation is generally used before analyzing response data with ANOVA. The equation $P' = \arcsin (P)^{0.5}$ (*P* equals the original percentage value) is commonly used to transform response data (Bartlett, 1937; Finney, 1989; Zar, 1984). However, when values of 0% or 100% occur, the formula arcsin (1/4n) and arcsin (100 – [1/4n]) should be substituted, respectively. This transformation helps correct some of the distribution problems associated with response data, allowing ANOVA to work more accurately. The arcsin transformation is not required when treatment values fall between 30% and 70% as the distribution of data is usually more normal (Fernandez, 1992) and the results of ANOVA and logistic regression are similar (Mize *et al.*, 1999).

Means of transformed data sets should be expressed in the original units when presented in tables and graphs. In these cases, results of the statistical analysis are applied to the original data. This is done because means of transformed data provide little information to the reader (Compton, 1994). Zar (1984) instructs that data transformed by arcsin can be converted back to the original scale by using the formula $P'' = (\sin P')$, 2 where P' equals the arcsin value and P'' is the new data value. In either case, it is important to explain your statistical methods to the readers.

Count Data

Count data are not normally distributed because treatment variances are equal to the average response to the treatments (Zar, 1984). Because of their distribution, count data (number of roots per cutting) should be analyzed using Poisson regression because the procedure uses a logarithmic value of the mean counts, which normalizes the data (Mize *et al.*, 1999). Poisson regression calculates a coefficient that is divided by the standard error to determine if the model is significant. In the viburnum example, observations on the number of roots per cutting are considered count variables. Because none of the control cuttings responded, there are no data values for the number of roots per cutting among propagules in this treatment. It is not correct to place zero values for the number of roots per cutting in this situation because doing so would skew the data set and influence our interpretation. Therefore, only data for cuttings treated with the rooting factor that responded were analyzed. In our case, Poisson regression indicated that the rooting factors influenced the number of roots per cutting ($P = 0.0001$; Table 16.2).

As stated in the section "Analyzing Response/Percentage Data," many researchers often wish to use ANOVA to analyze count data. Mize *et al.* (1999) stated that ANOVA is unreliable when count values are less than 10. In this case, Poisson regression yields more accurate results than ANOVA. However, ANOVA may be used as

TABLE 16.2
Poisson Regression Summary Table for the Number of Roots per *Viburnum carlesii* Softwood Cuttings Treated with Three Rooting Factors[a]

Predictor Variables	Poisson Regression Coefficient (A)	Standard Error (B)	Poisson Regression Statistic (A/B)	*P*
Constant	0.1247	0.293	0.43	0.6703
Rooting factor	0.3527	0.088	4.00	0.0001

[a] Data are hypothetical.

long as the data are transformed using the square-root transformation prior to analysis (Compton, 1994).

Analysis of Continuous Data

Continuous data are more normally distributed than binomial and count data, having treatments with similar variances that are well suited for statistical analysis using ANOVA (Mize *et al.*, 1999). During ANOVA, the value of each observation is subtracted from the overall mean and the differences between each datum and the overall mean (considered random error or residual) used in the analysis (Mize *et al.*, 1999). A model statement is created and tested, which identifies the dependent variables (observations recorded) and independent variables (treatments, treatment interactions) based on the experimental design used. A summary table is generated during ANOVA, which provides the results of the model tested (Lentner and Bishop, 1986). Information in the summary table includes the df, the sources of variation (sum of squares [SS] and mean square [MS]), the mean square error (MSE), the F-statistic (F), and the *P* value. In ANOVA, the F-statistic is obtained by dividing the MS for the treatment by the MSE. The resulting value is used to calculate a *P* value that is compared to a standard table to determine if the hypothesis is rejected or accepted (Lentner and Bishop, 1986).

In the viburnum example, the root length variable is considered continuous. Because a CR design was used in this study, rooting factor and experimental error were identified as sources of variation (Table 16.3). Because cuttings receiving the control treatment did not produce roots, entering zero values would skew the data set. Therefore, values for this treatment were not included in the analysis. Likewise, values for cuttings that failed to root in the rooting factor treatments were not analyzed. According to the results, root length was influenced by the type of rooting factor used ($P = 0.0001$; Table 16.3). This was determined by looking at the *F* and *P* values for the rooting factor variable. In our ANOVA, the F-statistic

TABLE 16.3
ANOVA Summary Table for the Number of Roots Formed per *Viburnum carlesii* Softwood Cuttings Treated with 48 mM of IBA, K-IBA, or NAA[a]

Source	df	SS	MS	*F*	*P*
Rooting factor	2	3061.4	1530.7	181.46	0.0001
Experimental error	38	320.553	8.43561		
Total	40	3381.95			
		Contrast			
NAA vs. IBA derivatives	1	4040.6	4040.6	478.99	0.00001
IBA vs. K-IBA	1	37.604	37.604	4.458	0.08

[a] Propagules (softwood cuttings) were prepared from plants grown in the nursery. Excised softwood cuttings were trimmed to 15 cm and their basal 3 cm were dipped for 30 sec in a 50% ethanol solution containing 48 mM of IBA, K-IBA, or NAA, or a 50% ethanol control. There were 10 propagules per treatment, and the experiment was completed twice. Cuttings were arranged in a CR design. Data for the controls as well as data for non-responding cuttings in the auxin treatments were deleted before analysis. Data for runs were combined since they were not significantly different. Data are hypothetical.

(181.46) was calculated by dividing the MS for rooting factor (1530.7) by the MSE (8.43561). The MS value for rooting factor reflects the degree of variation associated with the treatments. The MSE measures variation from other non-treatment sources. The *P* value (level of significance for the F-statistic) can be obtained by comparing the calculated *F* value against standardized values for ANOVA found in a statistical textbook.

ANOVA is preferred by most researchers because it is easy to perform, can be used to evaluate data obtained from virtually all experimental designs, and generates an MSE value that is considered to be the best estimate of experimental error (Mize *et al.*, 1999). ANOVA provides a good overall analysis because it accurately measures how treatments relate to each other when used in the proper circumstances.

Methods for Comparing Treatment Means

One of the main objectives of propagation research is to identify treatments that optimize propagule response. To fulfill this objective, it is necessary for researchers to apply a method or methods to determine significant differences among the treatments applied. In experiments in which only two treatments are tested, a significant *P* value indicates that the two treatments differ. The researcher can then calculate the mean for each treatment and declare the treatment with the greater mean to be the best. In this case, the general analysis provides enough information about treatment differences.

In most cases, researchers design experiments that compare more than two treatments. In this situation, the general analysis does not provide enough specific information about how the treatments differ or which treatment or treatments is or are the best. Therefore, further analysis of treatment means is required. The easiest way to compare treatment means is to rank them in ascending or descending order and pick the best treatment. However, this method does not consider within treatment variation and does not account for variation in how the propagules responded to the treatments. Two cuttings within a treatment seldom respond identically to the same treatment, and sometimes, the difference in response can be sizeable. Therefore, it is important to consider variation among cuttings within a treatment as well as cuttings receiving different treatments. Considering these variances greatly decreases the likelihood that the researcher will make a mistake when interpreting the results (Mize *et al.*, 1999).

There are many mean separation procedures that account for within treatment variation that can be used for evaluating data from propagation experiments. It is important to use the mean separation test that best suits the type of observations recorded. Several types of mean separation tests are presented in the following paragraphs, with a discussion on their most appropriate use.

Standard Error of the Mean

The standard error of the mean (SE) is commonly used by researchers to compare treatment means. An SE value is easy to calculate and is derived from the sample standard deviation (Zar, 1984). Most researchers use SE values for mean separation purposes by presenting the treatment means and their respective SE. In these cases, the difference between paired values calculated and treatments are declared different if their collective values do not overlap. Problems with this method occur when standard error terms are used to compare the means of treatments ranked far apart. It is important to remember that treatments with small values have a reduced SE when compared to treatments with larger values. This situation increases the likelihood of detecting differences but violates the assumption of ANOVA that treatment variances are equal, which often leads to erroneous interpretations that do not accurately reflect population variance. A more accurate way to use SE values for mean comparison is to calculate one SE value from the ANOVA MSE or from Poisson or logistic regression (Mize *et al.*, 1999). This value is then used to compare adjacent means. Most statisticians believe that the use of SE to compare treatment means yields poor information compared to mean comparison procedures that use the sample variance when calculating the mean comparison statistic. Some examples of more meaningful mean comparison tests follow.

Multiple Comparison and Multiple Range Tests

Multiple comparison and multiple range tests are statistical procedures that use the population variance to calculate a numerical value for comparing treatment means. Means are ranked, and the difference between the compared means are calculated. The calculated difference between means is compared with the critical value computed by the mean comparison statistic. If the difference between the compared means exceeds the computed statistical value, the treatments are considered statistically different. However, if the difference between treatment means is equal to or less than the computed statistical value, the treatments are considered similar (Mize *et al.*, 1999). When presenting means in a table or graph, means designated as different are assigned different letters, whereas treatments declared similar are assigned the same letter (Table 16.4).

In the viburnum example, Bonferroni's, Tukey's honestly significant difference (HSD), and least significant difference (LSD) mean separation tests were used to compare the effect of different rooting factors on rhizogenesis (Table 16.4). Normally, only one comparison test is used to compare treatment means. However, three are used here to demonstrate differences that can occur due to the procedure used to calculate statistical values. Means of each treatment were ranked from highest to

TABLE 16.4
Comparison of Results Obtained from Analyzing Treatment Means Using LSD, Tukey's HSD, and Bonferroni's Mean Comparison Tests[a]

Auxin Type	Root Length (mm)	Results of Mean Comparison Tests: Bonferroni's[b]	Tukey's[b,c]	LSD[d]
IBA	27.5	a	a	a
K-IBA	25.3	a	a	b
NAA	6.9	b	b	c

[a] Data were analyzed following ANOVA for a single-factor CR design. Data are hypothetical.
[b] Means with the same letter are not significantly different according to Bonferroni's test at 0.05.
[c] Means with the same letter are not significantly different according to Tukey's HSD at 0.05.
[d] Means with the same letter are not significantly different according to the LSD test at 0.05.

lowest and compared. Different results were obtained from the three tests. The results of LSD suggested that the cuttings responded differently to each of the three rooting factor treatments. However, Bonferroni's and Tukey's HSD differed from LSD, indicating that the root length of cuttings treated with IBA and K-IBA was similar and collectively greater than cuttings treated with NAA. When looking at the treatment means, a difference of 1.8 mm in root length between the K-IBA and IBA treatments is probably not practically different and probably would not impact the establishment of rooted cuttings in containers or the field. It is the opinion of many statisticians that the use of LSD for mean comparisons should be limited or not used unless treatments are considered different in a general test (ANOVA, and others). LSD is considered by many statisticians to be a mean comparison procedure most likely to declare means different. This often leads researchers to declare means different that are not practically different, which occurred in this example. Bonferroni's and Tukey's HSD are moderately conservative tests and are considered to yield more accurate results. Results of mean comparison tests should be presented in either tables or bar graphs. If using bar graphs, letters assigned to specific treatments should be positioned above each bar (Figure 16.2).

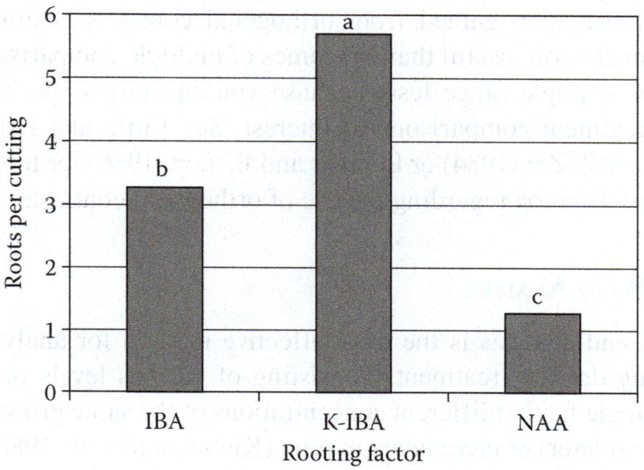

FIGURE 16.2 Bar graph demonstrating the use of lowercase letters to illustrate differences among rooting factor treatments as assigned by Bonferroni's multiple comparison test. Bars with different letters are considered statistically different at the 0.05 level.

Multiple comparison and multiple range tests should only be used when treatments are unrelated (Lentner and Bishop, 1986). In propagation studies, these are treatments in which different species, substrate media, or unrelated rooting factors are tested. These mean comparison procedures should not be used when treatments are related, for example, different concentrations of the same rooting factor (Compton, 1994; Mize *et al.*, 1999). Multiple comparisons and multiple range tests can be used for count and binomial data, but the data must be normalized first using square root (count data) or arc sin (binomial data) transformation procedures (Compton, 1994). Once results of transformed data are calculated, the data are converted back to the original scale for presentation (Zar, 1984). Letter designations obtained using transformed data should be assigned to the treatment means after converting back to the normal scale.

Orthogonal Contrasts

Orthogonal contrasts are used to compare treatments or groups of treatments with similar characteristics (Compton, 1994; Mize *et al.*, 1999). In propagation studies, these may be PGRs or rooting factors with similar activity. Contrast statements are usually performed as part of ANOVA, but the number of comparisons must be restricted to the number of df for the treatment variable (Lentner and Bishop, 1986). A contrast statement is written, which specifies the treatments or group of treatments to be compared, and ANOVA calculates the df, SS, and MS for the comparison. An F-statistic is calculated by dividing the MS for the contrast by the MSE and the level of significance (*P* value) determined. One df is used for each comparison.

In the viburnum example, two contrasts of interest were planned before the experiment (1, IBA and K-IBA vs. NAA, and 2, IBA vs. K-IBA) for the root length variable. Results of orthogonal contrasts are usually presented in a table containing ANOVA computations (Table 16.3). Interpreting the results of the contrasts for root length indicated that cuttings that rooted following treatment with 48 mM IBA or K-IBA produced longer roots than those treated with a similar concentration of NAA ($P = 0.0001$). However, no difference in root length was observed among cuttings treated with IBA or K-IBA ($P = 0.08$).

Orthogonal contrasts can be used to analyze binomial or count data if the data are transformed before analysis (see previous sections for these types of data). Information gained from orthogonal contrasts is often much more useful than outcomes of multiple comparison or multiple range tests because you can target specific treatment comparisons of interest. See Little and Hills (1978), Zar (1984) or Lentner and Bishop (1986) for more information regarding the use of orthogonal contrasts.

Trend Analysis

Trend analysis is the most effective method for analyzing data of treatments consisting of various levels of a single factor (different concentrations of the same growth regulator) or increments in time (Kleinbaum *et al.*, 1988). The primary objective in these types of experiments is to identify a dose (concentration), or time period, that stimulates optimal propagule response. Models identifying specific trends (linear, quadratic, and cubic) are tested in a stepwise fashion from simplest (linear) to most complex (cubic, in most cases) until a nonsignificant trend is identified. At this time, the last significant trend is considered to best describe the data (Kleinbaum *et al.*, 1988). Trend analysis uses SS, *T*, and R^2 values to indicate significant trends. Trends may be tested through regression analysis or polynomial contrast statements in ANOVA (Compton, 1994).

Trend analysis can be illustrated by using an example in which the effect of five concentrations (0, 12, 24, 36, and 48 mM) of K-IBA on the rooting of *V. carlesii* softwood cuttings was examined. In this example, softwood cuttings were prepared as described previously and dipped in one of the K-IBA solutions for 30 sec. Data recorded included the percentage of cuttings with roots, the number of roots per cutting, and root length. Since percentage values are binomial in nature, all data were transformed before analysis using the arc sin transformation. Also, since none of the cuttings that received the control treatment (0 K-IBA) produced roots, all data for this treatment were excluded from the analysis. Results from trend analysis indicated that rooting performance (percentage of cuttings with roots) was dependent on K-IBA concentration (Table 16.5), with the linear model ($y = -22 + 2.23 \times$ K-IBA) giving the best fit ($P < 0.00001$; $R^2 = 0.36$). Higher-order models, such as the quadratic model ($P = 0.07$; $R^2 = 0.389$), did not provide a better explanation of how the cuttings responded to the application of K-IBA. Because the highest K-IBA level yielded the best results, it would be wise to conduct another experiment comparing at least the two highest levels used in this experiment (36 and 48 mM) with even higher doses.

TABLE 16.5
Results Obtained from Simple Linear Regression Analysis of the Percentage of *Viburnum carlesii* Softwood Cuttings that Produced Roots Following Treatment with 0, 12, 24, 36, or 48 mM K-IBA[a]

Predictor Variables	Coefficient	Standard Error	Student's *T*	*P*
Constant	–0.22	0.06913	–3.18	0.0017
Treatment	0.0223	0.00210	10.62	0.0001
R^2	0.3627	Residual MS		0.159
Adjusted R^2	0.3595	Standard deviation		0.399
Source	df	MS	*F*	*P*
Linear trend	1	17.956	115.16	0.0000
Quadratic trend	1	0.5000	3.21	0.0748
Cubic trend	1	0.4840	3.10	0.0797
Residual	196	0.15931		
Total	199			

[a] Data for 0 K-IBA were deleted because none of the cuttings produced roots. Data are hypothetical.

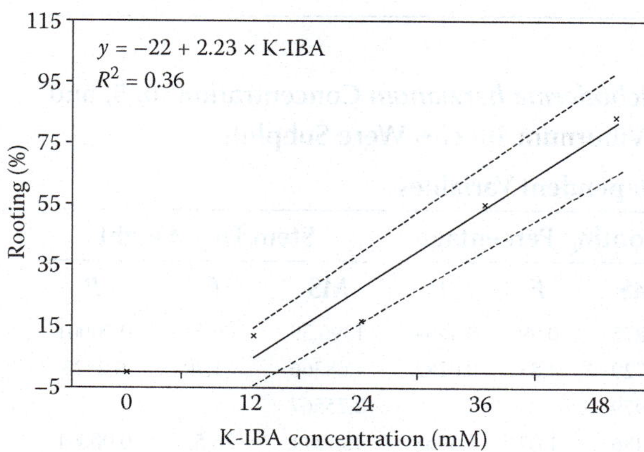

FIGURE 16.3 Line graph demonstrating the use of simple linear regression to illustrate differences in rooting percentage among softwood cuttings treated with 0 to 48 mM K-IBA. Treatment means (×), calculated regression line (—), and 95% confidence intervals (--) are shown.

Means of dosage treatments should be presented in graphs, displaying the regression equation and fitted line, R^2, individual data values, and confidence intervals (Figure 16.3). Similar to mean comparison tests, confidence intervals of treatments that overlap are considered similar. In the viburnum example, none of the confidence intervals for the treatments overlapped. Therefore, the 48 mM treatment would be considered best for promoting the rooting of softwood cuttings. However, as stated before, most researchers would experiment with higher levels to see if better results could be obtained.

Trend analysis and polynomial contrasts may be used for evaluating optimum treatment levels, even if those levels were not directly tested but lie within treatment boundaries (Compton, 1994). Predicted values can be obtained using information generated from the analyses, allowing researchers to estimate the effects of treatments that were not evaluated.

Multifactor Experiments

In the previous examples, single-factor experiments were emphasized (e.g., experiments in which one treatment factor, such as auxin type, is examined during experimentation). However, it is common for researchers to look at the effects of several factors simultaneously (multifactor treatment designs). In some experiments, multifactor treatment designs are established in such a way that all treatment factors are present in each treatment (complete factorial treatment designs). These experiments are popular among researchers because the effects of each factor individually as well as interactions among them can be examined in one experiment (Lentner and Bishop, 1986). Another advantage of multifactor experiments is that they use resources most efficiently (Compton, 1994). It is not a requirement that treatments for multifactor experiments be arranged in a complete factorial. For example, augmented factorial treatment designs are set up so that some treatment interactions are not compared. Another challenge often encountered with multifactor experiments is the large number of propagules required to make all comparisons. Nas *et al.* (2005) describe how a fractional factorial design can be used when there is a shortage of propagules for your experiment. In any case, it is always most important that the treatments selected and the interactions examined make good practical and scientific sense.

The split plot (SP) design is often used for multifactor experiments. In this design, levels of one factor are placed into groups (blocks) called "main plots," which contain all levels of the second factor. Separate error terms are used to test for the significance of each factor. For example, the interaction between the main plot treatment and the block is used as the error for the main plot factor(s) (Table 16.6). In contrast, the block by subplot treatment interaction is used to test for the significance of the subplot treatment and the main plot treatment by block by subplot treatment term was used as error to compare the main plot by subplot treatment interaction.

To illustrate the use of the SP experimental design in factorial treatment experiments, an example in which the influence of two concentrations (5 or 25 g/kg of rooting medium) of *Trichoderma harzianum* isolate T-22, a biological control agent for pathogenic fungi, on the rooting of four viburnum species will be examined. Three of the viburnum species selected were considered easy to root—arrowwood viburnum (*V. dentatum* L.), dwarf American cranberry viburnum (*V. trilobum* Marsh 'Compactum'), and wayfaringtree viburnum (*V. lantana* L.)—whereas Korean spice viburnum (*Viburnum carlesii*) was considered difficult to root (Dirr, 1998). Softwood cuttings of each species were given a 30 sec dip in a 48 mM solution of K-IBA before transfer to rooting medium treated with either 0, 5, or 25 g of *T. harzianum* (isolate T-22) per kilogram of medium. The rooting medium consisted of a mixture containing equal parts (v/v) of sphagnum peat and perlite. The prescribed amount of *T. harzianum* (isolate T-22) was added to moisten the medium before dispensing it into 24 cell 26.25 (W) × 52.5 (L) × 7.5 (D) cm plastic trays. Cuttings were arranged in an SP design, with *T. harzianum* concentration as main plots and viburnum species as subplots. Each tray was considered a main plot and only possessed medium containing one concentration of *T. harzianum* to avoid cross-contamination. There were five main plot blocks per *T. harzianum* concentration, with eight replicates of each species per block, resulting in 40 replicates of each species per treatment (Figure 16.4). Individual trays (blocks) were randomized on greenhouse benches with bottom heat set to maintain a minimum root zone temperature of 25°C. Cuttings were

TABLE 16.6
ANOVA Summary Table for the SP Design in which *Trichoderma harzianum* Concentration (0, 5, and 25 mg/kg of Rooting Medium) Was the Main Plot and Viburnum Species Were Subplots[a]

		Dependent Variables								
		Root Dry Weight			Rooting Percentage			Stem Dry Weight		
Source	df	MS	F	P	MS	F	P	MS	F	P
T. harzianum conc. (T)	2	732.040	290.58	0.00001	0.0875	0.89	0.4464	1396.39	506.74	0.00001
Block (B)	4	5.69179	2.26	0.1518	0.4729	4.83	0.0282	4.68366	1.70	0.2423
T × B	8	2.51924			0.0979			2.75567		
Viburnum species (V)	3	36.2634	9.67	0.0016	0.4486	1.07	0.3965	48.7371	26.59	0.00001
B × V	12	3.74831			0.4174			1.83292		
T × V	6	2.72972	0.65	0.6917	0.0819	0.42	0.8604	5.43448	1.90	0.1223
T × B × V	24	4.21614			0.1965		2.86139			
Residual	898	0.84473			0.06375		0.39606			
Total	957									
***T. harzianum* Contrasts**										
0 vs. 5 and 25 mg	1	988.66	196.22	0.0005	NA	NA	NA	1890.5	343.02	0.0011
5 vs. 25 mg	1	475.43	94.36	0.0005	NA	NA	NA	902.29	163.72	0.0005

[a] Dependent variables examined were root dry weight, rooting percentage and shoot weight. Data are hypothetical.

Trichoderma harzianum	1_4	3_1	2_5	2_2	1_4	3_4	....	1_2	2_1	3_5	3_3	2_4	1_3
	①	④	②	③	③	②		④	②	①	①	②	③
	③	①	④	②	①	①		③	①	④	③	④	①
Viburnum species	④	②	③	①	④	③		①	③	②	④	①	④
	②	③	①	④	②	④		②	④	③	②	③	②

FIGURE 16.4 Randomization scheme for a split plot design in which main plots (*Trichoderma harzianum* concentration) are randomized into blocks and subplots (*Viburnum* species) are randomized within each main plant. *T. harzianum* concentrations (0, 5, and 25 mg/kg) were replicated five times. *Viburnum* species were assigned to subplots (*V. carlesii* [①], *V. dentatum* [②], *V. lantana* [③], and *V. trilobum* 'Compactum' [④]).

irrigated with mist for 10 sec at 10 min intervals. Cutting growth was measured by recording the stem and root weight as well as the percentage of rooting 8 weeks after initiating the experiment. The experiment was conducted twice.

Examining the results of ANOVA, significant differences in root dry weight were observed for *Trichoderma* concentration (0.0001) and viburnum species ($P = 0.0016$; Table 16.6). However, there was no interaction between species and *Trichoderma* concentration for root dry weight ($P = 0.6917$), indicating that the four species reacted similarly to the *Trichoderma* treatments. Two general contrasts (0 *Trichoderma* vs. 5 and 25 mg *Trichoderma* per kilogram of rooting medium, and 5 vs. 25 mg *Trichoderma* per kilogram of rooting medium) were conducted to examine the influence of *Trichoderma* on root growth. Examining the contrasts (Table 16.6) and treatment means (Table 16.7) revealed that adding *Trichoderma* to the rooting medium improved the root growth of viburnum softwood cuttings ($P = 0.0005$). In addition, the second contrast and treatment means indicate that adding 25 mg of *Trichoderma* per kilogram of rooting medium at the time of sticking cuttings improved root growth more than adding 5 mg/kg ($P = 0.0005$). Examination of rooting percentages among species revealed that all viburnum species tested rooted similarly when using 48 mM

TABLE 16.7
Use of Orthogonal Contrast for Determining Significance in Rooting among Viburnum Softwood Cuttings Treated with Three Concentrations of *Trichoderma harzianum*[a]

T. harzianum Inoculation Density[b]	Number of Cuttings	Root Dry Weight (kg)	Rooting Percentage	Shoot Dry Weight (kg)
0	320	1.6	91	2.3
5	320	2.9	91	4.1
25	318	4.6	94	6.5
		Contrasts[c]		
0 vs. 5 and 25		0.0005	NS	0.0011
5 vs. 25		0.0005	NS	0.0005

Note: NS = not significant.
[a] Data are hypothetical.
[b] *Trichoderma harzianum* inoculation density in milligram per kilogram of rooting medium.
[c] *P* values as determined by ANOVA at 1 and 8 df as calculated using an SP design, with *T. harzianum* inoculation density as main plots.

TABLE 16.8
Root and Shoot Dry Weight Plus Rooting Percentage of Softwood Cuttings Collected from Four Viburnum Species[a]

Viburnum Species	Number of Cuttings	Root Dry Weight[b]		Rooting Percentage[c]	Shoot Dry Weight	
V. dentatum	240	3.5	a	95	4.8	a
V. lantana	238	3.2	ab	95	4.4	a
V. trilobum 'Compactum'	240	2.8	bc	92	3.9	b
V. carlesii	240	2.7	c	86	3.9	b

[a] Data are hypothetical.
[b] Means within columns with the same letter are not different according to Bonferroni at 3 and 12 df.
[c] Means not significant according to ANOVA.

K-IBA ($P = 0.3965$) and that *Trichoderma* acted similarly among the species ($P = 0.4464$). Addition of *Trichoderma* did improve the stem growth of all viburnum cuttings ($P = 0.0001$). However, it is unclear if increased stem growth occurred directly as a result of *Trichoderma* treatment or was stimulated by improved root growth. Stem growth also varied among viburnum species, with cuttings of *V. dentatum* and *V. lantana* displaying more growth than those of *V. carlesii* and *V. trilobum* 'Compactum' likely because the former displayed better root growth (Table 16.8).

In this experiment, the SP was the most efficient experimental design to use because the use of separate trays allowed researchers to minimize the cross-contamination of *Trichoderma* between treatments. However, the SP design may also be used in situations in which two treatment factors are expected to have highly different variances. In any case, the experimental setup, data recording, as well as data analysis and interpretation are more complicated for the SP design than the CR and RCB designs.

CONCLUSIONS

This chapter has outlined the steps of experimentation and many of the procedures used in the analysis of plant propagation data. It is important to remember that well-conceived research projects are planned carefully before implementation. This helps ensure that the experiment is conducted properly and will answer the desired questions. Make sure to use the most appropriate experimental design for your conditions and experimental objectives. Doing so can enable you to maximize experimental precision and minimize the influence of non-treatment factors. Be sure to minimize bias, fatigue, and errors during the course of conducting an experiment. These factors lead to untrue results that mislead researchers and growers that may use the information. Use the experimental design while conducting the experiment, and recording and analyzing data. This helps provide an objective, unbiased way to accurately evaluate treatment differences. When used properly, statistics is a valuable tool to the scientific researcher, allowing investigators to declare experimental results with confidence.

LITERATURE CITED

Barnett, V. and T. Lewis. 1984. *Outliers in Statistical Data*. Chichester, United Kingdom: John Wiley & Sons, 463 pp.

Bartlett, M. S. 1937. Some examples of statistical methods of research in agriculture and applied biology. *J Royal Statist Soc Suppl* 4:137–170.

Compton, M. E. 1994. Statistical methods suitable for the analysis of plant tissue culture data. *Plant Cell Tiss Org Cult* 37:217–242.

Compton, M. E. and C. W. Mize. 1999. Statistical considerations for *in vitro* research: I. Birth of an idea to collecting data. *In Vitro Cell & Dev Biol (Plant)* 35:115–121.

Dirr, M. A. 1998. *Manual of Woody Landscape Plants*. Champaign, Illinois: Stipes Publishing, LLC, 1187 pp.

Fernandez, G. C. J. 1992. Residual analysis and data transformations: Important tools in statistical analysis. *HortScience* 27:287–300.

Finney, D. J. 1989. Was this in your statistical textbook? V. Transformation of data. *Expt Agr* 25:165–175.

Kempthorne, O. 1973. *The Design and Analysis of Experiments*. Malabar, Florida: Robert E. Krieger Publishing Co, 631 pp.

Kleinbaum, D. G., L. L. Kupper, and K. E. Muller. 1988. *Applied Regression Analysis and Other Multivariable Methods*, 2nd Edition. Boston, Massachusetts: PWS-Kent Publishing Co, 789 pp.

Lentner, M. and T. Bishop. 1986. *Experimental Design and Analysis*. Blacksburg, Virginia: Valley Book Company, 565 pp.

Little, T. M. and F. J. Hills. 1978. *Agricultural Experimentation: Design and Analysis*. New York: John Wiley & Sons, Inc., 350 pp.

Mize, C. W., K. J. Koehler, and M. E. Compton. 1999. Statistical considerations for *in vitro* research: II. Data to presentation. *In Vitro Cell & Dev Biol (Plant)* 35:122–126.

Nas, M. N., K. M. Eskridge, and P. E. Read. 2005. Experimental designs suitable for testing many factors with limited number of explants in tissue culture. *Plant Cell Tiss Org Cult* 81:213–220.

Zar, J. H. 1984. *Biostatistical Analysis*, Second Edition. Englewood Cliffs, New Jersey: Prentice-Hall, Inc.

Part VI

Propagation by Stem Cuttings

17 Cloning Plants by Rooting Stem Cuttings

John M. Ruter

Concept Box 17.1

- Vegetative propagation of plants by stem cutting is the most commonly used method for producing herbaceous and woody landscape plants in many plants of the world.
- Stem cuttings can be classified as follows: deciduous hardwood cuttings, narrow-leaf evergreen hardwood cuttings, semi-hardwood cuttings, softwood cuttings, and herbaceous cuttings.
- When taking cuttings; plants should be true-to-type, stock plants should be pest free, and cuttings must be taken at the correct time of year.
- Mist systems are critical for propagation of most stem cuttings taken during the growing season.
- Care should be given to the selection of proper propagation substrates, fertility, water quality, and containers.
- Most woody cuttings benefit from the application of rooting hormones.
- Proper sanitation in propagation areas is essential for commercial success.

Vegetative propagation of plants by cuttings is the most commonly used method of propagation for herbaceous and woody landscape plants in many parts of the United States and around the world. A cutting can be described as a detached plant part that, when placed in a suitable environment, will regenerate its missing vegetative parts. These parts may be roots, shoots, or both. The three major advantages of propagation by cuttings are the following: (1) cuttings provide a more successful vegetative method of propagation compared to budding or grafting, (2) the cuttings will have the same genetics as the parent plant, and (3) a number of cuttings may be taken from one stock plant. Since many of our important fruit and landscape species are clonal (same genetic makeup as the parent), cuttings are an important method for propagating such taxa.

PROPAGATION OF PLANTS USING STEM CUTTINGS

Stem cuttings are a section of stem tissue that consists of lateral or terminal buds. Cuttings can be placed into the following categories:

- Hardwood cuttings
  - Deciduous
  - Narrow-leaf evergreen
- Semi-hardwood cuttings
- Softwood cuttings
- Herbaceous cuttings

This type of classification is based on the maturity of the wood used as compared to basing the cutting type on the part of the plant from which the cutting came (e.g., stem cuttings, leaf cuttings, leaf-bud cuttings, or root cuttings.

Deciduous hardwood stem cuttings are taken from the previous season's growth in later winter or early spring while the plant is dormant. In general, the best hardwood cuttings are taken from the midpoint of a stem and may be 15 to 30 cm in length. Cuttings of this type usually have a stem diameter equal to or greater than a pencil and contain at least three vegetative buds. Direct field sticking of deciduous cuttings in raised beds is a common practice. Examples of some plants that are propagated by deciduous hardwood cuttings are figs (*Ficus carica*), grapes (*Vitis* sp.), mulberries (*Morus* sp.), pomegranates (*Punica granatum*), and roses (*Rosa* sp.).

Narrow-leaf evergreen stem cuttings are generally taken in winter, although certain species such as prostrate junipers and certain cypresses may be propagated during periods of summer growth dormancy. Cuttings are usually 10 to 20 cm in length and less than pencil thickness in diameter, and have the bottom branches or leaves removed. Narrow-leaf evergreen cuttings are responsive to bottom heat, wounding, and rooting hormones, yet may still take several months to root. The three different types of cuttings used for the propagation of most conifers are the following: (1) stem cuttings, (2) heel cuttings, and (3) mallet cuttings (Figure 17.1).

Most narrow-leaf evergreen cuttings are taken in late fall or early winter because they generally root best after some exposure to chilling temperatures. Many growers will stick conifer cuttings in late winter so that the plants do not sit in the propagation house for several months before callusing and root formation occurs. Certain types of conifers are considered easy to root. These are *Chamaecyparis*, *Platycladus*, *Podocarpus*, *Taxus*, *Thuja*, and the low-growing forms of *Juniperus*. Taxa considered more difficult to propagate from cuttings include *Cedrus*, *Cupressus*, and *Pinus*.

Semi-hardwood stem cuttings of broadleaf evergreens are taken in late spring to mid-summer when the wood has matured but is not yet woody. The cuttings are often 8 to 16 cm in length, and the basal thickness of the stem will vary. The lower 1/3 of the leaves may be removed, and cuttings with large leaves can be cut back to reduce transpirational water loss. Plants propagated by semi-hardwood cuttings can include boxwood (*Buxus* sp.), camellias (*Camellia* sp.), citrus (*Citrus* sp.), hollies (*Ilex* sp.), and evergreen rhododendrons (*Rhododendron* sp.) (Figure 17.2).

Softwood stem cuttings are usually tip cuttings taken from new growth. The same procedures are followed as

FIGURE 17.1 Variation in the rooting of the narrow-leaf evergreen plant × *Cuprocyparis leylandii*. Note the callused cutting with no roots on the left and the well-rooted cutting on the right.

FIGURE 17.2 Camellias propagated from semi-hardwood cuttings under intermittent mist.

for semi-hardwood cuttings. These cuttings are often quick to root, usually 8 weeks or less. Many plants flower from this new growth, so all flowers and flower buds should be removed. Terminal growth is often removed from the cuttings to prevent wilting or decay under mist. Plants rooted from softwood cuttings include deciduous magnolias (*Magnolia* sp.) and deciduous azaleas (*Rhododendron* sp.).

Herbaceous cuttings are handled in the same way as softwood cuttings. These cuttings can often be taken throughout the year. Some plants propagated by this method are chrysanthemums (*Dendranthema* sp.), dahlias (*Dahlia* sp.), and geraniums (*Pelargonium* sp.).

Some points that need to be considered before taking cuttings are the conditions of the stock plant, the time of the year at which the cuttings are taken, and care of the cuttings once they have been taken. It is important that stock plants are insect- and disease-free. Contaminated plants will not perform well and may infect other plants that are being propagated. Cuttings should be taken from a plant that is in good nutritional health. In general, cuttings taken from non-vigorous or rapidly growing plants do not root well.

The time of the year in which cuttings are taken may vary from year to year because of different environmental conditions. Propagators should keep records from year to year of when certain plants were rooted and their performance (e.g., rooting percentage, vigor, etc.). Although a recommendation may call for sticking cuttings after the first frost of the year has occurred, this does not mean that the cuttings are stuck on the same calendar date each autumn.

Growth phase (juvenile vs. adult; see Chapter 6) and the location of the cutting on the plant (topophysis) have a great influence on cutting success and subsequent growth. In general, the closer the cutting to the root–shoot interface of the plant, the more juvenile the cutting will be. Many plants such as conifers root best from juvenile cuttings versus cuttings taken from the flowering or adult parts of the plant. If juvenility is a problem, once a few cuttings are rooted from a difficult cultivar, those plants can be used as stock plants. After several generations of selection, cuttings should be juvenile and root with better consistency. Conifers are also good examples of plants that show orthotropic (vertical) growth and plagiotropic (lateral) growth. When propagating plants such as *Cunninghamia lanceolata* or *Sequoia sempervirens*, if lateral shoots are used for cuttings, then the resulting plants will grow horizontally for several years. Many conifers prone to plagiotropic growth benefit from having cuttings taken only from vertically growing shoots. Growing stock blocks of hedged plants that produce large quantities of vertical shoots are often used.

PROPAGATION STRUCTURES, BENCHES, AND MIST SYSTEMS

Environmental factors to consider once the cuttings have been stuck are air temperature, humidity, light intensity, and substrates. Not all plants can be propagated outdoors without protection; therefore, it is necessary to provide structures where environmental conditions can be maintained. Structures can be as elaborate as modern greenhouse ranges or as simple as small polyhuts in the field. Items that should be taken into consideration when deciding which type of structure to use include the following: (1) the type of structure, (2) cost, (3) covering to use, (4) structure and bench/bed orientation, and (5) heating and cooling needs.

A greenhouse can be defined as an enclosed structure designed to provide ideal conditions of light, temperature, and humidity for the growth of plants. In general, propagation greenhouses are no more than 11 m in width by 30 m in length to provide optimum environmental conditions (Figure 17.3).

A variety of coverings are available for propagation houses. Many types of plastic are used instead of glass. Many types of rigid plastics are preferred since they will withstand many of the environmental abuses that glass and plastic will not. Plastic is preferred in areas that are prone to hurricane damage because it can be easily removed.

Polyhouses are popular structures for propagating cuttings since they have a lower initial construction cost. The frame is considered permanent, but the covering may be replaced yearly or once every 3 to 5 years, depending on the longevity of the plastic used.

The purpose of a shade house is to protect nursery plants from environmental factors such as wind, temperature, hail, heavy rain, and solar radiation. Shading may be provided using lath or shade cloth. These structures are usually permanent and can be used for the propagation or hardening-off of rooted cuttings or newly potted plants. Saran or shade houses have replaced lath houses in most areas in part due to initial lower costs. Shade cloth can have a useful life of several years. Red shade cloth has become popular in recent years and may increase growth once cuttings are rooted.

Full sunlight works well for many species once rooted because it allows for greater photosynthesis and, thus, greater food reserves and hormone production during the rooting-out process. High temperatures may be a problem under conditions of high light intensity. If this is the case, shading compound can be applied to greenhouse structures or shade cloth may be used to reduce the amount of sunlight by up to 70%.

A cold frame is among the simplest and most economical of all propagation structures. In the past, cold frames were built low to the ground; were covered with glass sashes; and were used for rooting cuttings, germinating seeds, and hardening off tender plants. Simple cold frames covered with white or opaque plastic are used for propagation beds directly in the field. Many growers now use hoop houses covered with plastic and/or shade cloth as propagation structures (Figure 17.4).

FIGURE 17.3 A typical propagation greenhouse covered with shade cloth.

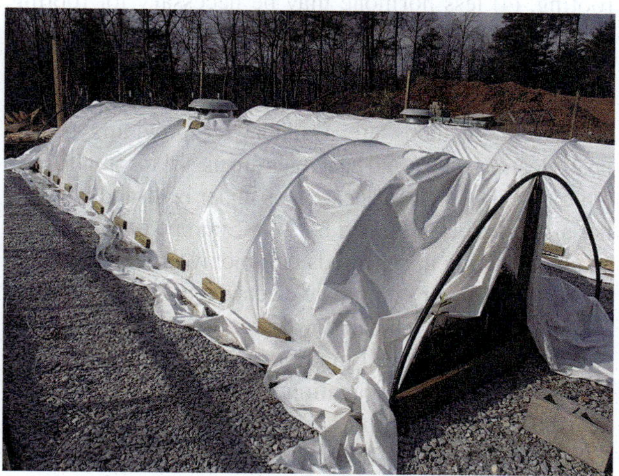

FIGURE 17.4 A simple overwintering structure for rooted cuttings.

Mist tents are commonly used inside greenhouses that have existing mist systems. The purpose of such tents is to increase and/or keep the relative humidity constant while rooting cuttings. Polyethylene sheeting is often draped over a structure constructed above the mist bed. For cuttings that are prone to rot under mist, the same structure can be used as a humidity tent, where mist is applied infrequently, just enough to keep sufficient relative humidity within the tent. Care must be taken to prevent excessive buildup of heat in such tents.

Raised propagation benches are often preferred over ground beds for the propagation of cuttings. Benches allow for greater control of the environment around the cuttings (improved drainage and isolation from soilborne diseases and pests) and also allow work to proceed in a more efficient manner. Bench size is often determined by the type of propagation (sticking in flats or individual containers), the size of the propagation house, the location of benches within the structure, and the number or cuttings to be propagated. When designing a bench system, items to be taken into consideration include the following: (1) longevity of the bench material, (2) ease of assembly, (3) versatility, (4) weight, (5) air circulation, and (6) cost. While propagation benches can be made from many materials, many growers now prefer benches made from galvanized metal.

In general, the ideal air temperature for rooting cuttings is 18 to 24°C (~65–75°F). Ideal substrate temperatures for rooting are between 21°C and 24°C (70°F and 75°F) for most plants. These temperatures may be difficult to maintain, especially air temperature. Remember the rule when it comes to rooting cuttings: Cool Hands, Warm Feet!

Bottom heat provides a number of advantages in propagation, especially for the rooting of cuttings. These advantages include the following: (1) bottom heat is economical, (2) increased rooting percentages, (3) quicker rooting, (4) less hormone may be necessary for rooting, (5) fewer disease problems, and (6) higher-quality rooted cuttings. Bottom heat is especially beneficial for rooting conifers and evergreens during winter months.

Several methods are available for providing bottom heat. Heating cables and heat mats are useful for small areas. On a greenhouse scale, heat pipes under beds and benches, hot air under benches, and hot water pumped through pipe or plastic tubing are methods commonly used. Circulating hot water also helps eliminate electrical hazards as heating cables and mats can break down over time.

A functioning, well-maintained mist or fog system is a necessity during propagation since it prevents excessive transpiration and respiration while allowing the cuttings to photosynthesize. The purpose of a mist system is to maintain cutting turgidity by increasing atmospheric relative humidity at the surface of the leaf and to decrease the temperature of the leaves on the cuttings through evaporation. The two most popular mist systems in use are intermittent mist and fog systems. Fog systems have become popular because they attempt to provide 100% relative humidity while resolving the problems of poor aeration and excessive moisture associated with conventional mist systems. In hot, humid climates, the lack of ventilation with fog systems often leads to excessive heat buildup in propagation structures (Figure 17.5).

A typical mist system consists of the following components: (1) mist line, (2) water filter, (3) solenoid valves, (4) gate valves, (5) time clocks, and (6) mist nozzles. Polyvinyl chloride pipe is the most common material used for mist lines because of its ease of installation, cost, and longevity. Flexible black plastic pipe available in large rolls is also cost effective and is often used for overhead mist systems. The size of pipe necessary for a mist system will depend on the type of nozzle used, the water pressure (psi), the liters per minute of water needed, and the size of the system. For fog systems, copper tubing is often preferred. Mist lines can be placed in the propagation beds or can be placed directly overhead. Mist heads are generally placed a minimum of 30 cm above the cuttings. If mist beds are greater than 90 cm across, it is advisable to install more than one mist line per bench. Many modern mist systems consist of two mist lines that are 6 to 11 m in width in propagation houses.

A filter is a good idea when installing a mist system so that mist heads will not get clogged with particulate matter. Solenoid valves are installed as a means to electronically control the mist system. Plastic solenoid valves work well for low-pressure systems, but brass valves should be used if high-pressure water lines are utilized. Solenoid valves are available as "normally open" and "normally closed" valves. The "normally open" type is considered best for rooting cuttings because it remains open if the electricity goes off; thus, the mist line can continue operating as long as there is water pressure. Always put a gate

FIGURE 17.5 An overhead mist system in action.

valve in the line so that the system can be operated manually should power not be available, and always design mist systems with low voltage to ensure worker safety.

The following three general types of mist nozzles can be utilized for rooting cuttings: (1) whirl or pressure jet types, (2) deflection or baffle type, and (3) fog nozzles. Whirl type nozzles spin when water under pressure exits the openings on the ends of the nozzles. As water is forced through the grooves in the nozzle, it is reduced to smaller water particles. Many whirl-type systems can use low amount of water per hour compared to other systems. Deflection or baffle-type nozzles are the most commonly used nozzles for rooting cuttings. Since deflection nozzles require a larger volume of water, they are less prone to clogging. The orifice size on a fog nozzle is very small and is prone to clogging, especially with hard water. Water droplet sizes of 7 to 10 μm are achieved by running the fog system at pressures greater than 450 psi (Figure 17.6).

Control devices for mist systems can either be preset (time clocks) or variable (environmentally controlled). The preset system usually consists of two time clocks, a 24 h clock and an interval clock. The 24 h clock controls when the system comes on in the morning and when the system shuts down in the evening. The interval time clock controls the frequency and duration of the mist cycle while the system is on. The interval clock is generally set to keep the foliage moist throughout the day, but this system often requires manual adjustment during the day for optimum effectiveness. Many modern electronic mist systems control both the 24 h clock and the interval cycles (Figure 17.7).

Several types of variable or environmentally controlled mist devices are available. The electronic leaf works on the principle of electrical conductivity between two electrodes. When a simulated leaf made from nonconductive material is misted, an electrical current is carried between two electrodes. As the simulated leaf dries out, the current is broken, and the solenoid on the mist system is activated. Once enough moisture collects on the simulated leaf, current flow is reestablished, and the solenoid shuts off.

The electronic leaf is based on the principle of evaporative loss from a real leaf. A simulated leaf is actually a screen that collects water when the mist system is on. Once the screen leaf is weighted down, it shuts off the solenoid. As water evaporates, the screen leaf rises until it again triggers the solenoid to be activated. Calcium and other hard-water mineral deposits on the screen leaf can be a problem with this type of system (Figure 17.8).

A third system is based on a photocell that activates the solenoid once a predetermined quantity of light (photons) has been absorbed. New computer-controlled systems are also available. Such systems are useful for automating mist systems, but the importance of manually checking the propagation systems several times during the day should be stressed.

FIGURE 17.6 A whirl-type mist nozzle used for rooting cuttings. A cutoff valve is often placed on each riser.

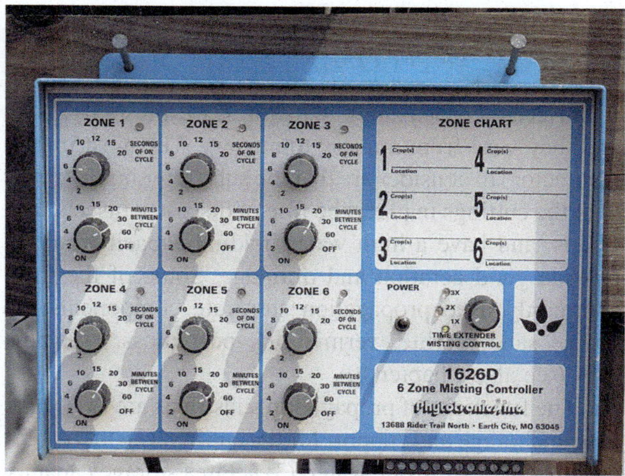

FIGURE 17.7 A six-zone mist controller. The duration of time between cycles and the number of seconds that each cycle comes on can be set for each of the six stations.

FIGURE 17.8 Example of an electronic leaf used for controlling mist frequency.

PROPAGATION SUBSTRATES, FERTILITY, AND CONTAINERS

A variety of propagation substrates are presently used, and they vary from nursery to nursery. The most important factors to consider are that the substrate should firmly hold the cuttings in place; be free of diseases, insects, and weeds; and have a high water holding capacity and yet provide good aeration and drainage. Some of the most commonly used propagation substrate components are coarse sand, perlite, vermiculite, peat moss, and pine bark (see also Chapter 10).

The functions of propagation substrates are to (1) provide anchorage and support for the cutting, (2) provide for the storage of substrate water, (3) act in a storage capacity for nutrients, and (4) supply adequate aeration for good root growth. A good propagation substrate for rooting cuttings should possess the following characteristics:

1. Have large, non-capillary pores for aeration and drainage
2. Have small capillary pores for water holding capacity
3. Have a good, stable structure that will not break down
4. Have adequate porosity to allow for easy root penetration
5. Buffered against rapid changes in chemical properties
6. Able to hold nutrients for plant use
7. Lightweight but able to provide stability for plants

Three main factors that should be taken into consideration when choosing a propagation substrate are the following: (1) economics, (2) physical properties, and (3) chemical properties. Economic considerations to evaluate are the cost of the substrate, availability, and reproducibility. Plant propagators should always use the best substrate available that works for their operation. Commercially prepared propagation substrates are available. The availability and quality of such products are very reliable. Some substrate components can be difficult to find. Harwood bark may be available in some areas, whereas only Douglas fir bark may be available in another parts of the country. Once an acceptable propagation substrate has been found, reproducibility becomes an important factor. Having to change components, grades, or particle sizes can have deleterious effects on the physical and chemical properties of a substrate.

Physical factors that influence the properties of a substrate are the following: (1) water holding capacity, (2) aeration, (3) bulk density, (4) particle size distribution, (5) uniformity, and (6) shrinkage. Nurserymen can perform their own tests to determine water holding capacity and air space at container capacity, or commercial laboratories can determine these physical properties as well. An ideal propagation substrate will have 25 to 35% air space at container capacity. Altering the particle size, uniformity, or using substrate components that will shrink will alter the amount of air space in the substrate. In general, air space decreases and water holding capacity increases as particle size decreases.

The bulk density of a substrate is its weight compared to the volume of space that it occupies. Bulk density is usually expressed in terms of grams per cubic centimeter. The bulk density of a substrate should be considered since the weight of a propagation substrate will influence shipping and handling costs. Uniformity of particle size and resistance to shrinkage are also important considerations. Hardwood bark and sawdust are examples of substrate components that will shrink if they are not composted first. During the course of a crop cycle, such shrinkage will lead to a decrease in the depth of the substrate in the container and a decrease in air space, both of which will adversely influence plant growth.

Chemical factors that contribute to the success or failure of a propagation substrate include the following: (1) cation exchange capacity (CEC), (2) pH, (3) soluble salts, (4) nutrient levels, and (5) sterility. The CEC of a substrate refers to its ability to retain positively charged cations against leaching from rainfall and/or irrigation. Organic matter will generally have a greater CEC than inorganic matter such as sand. Calcined clays or mineral soils have greater CECs than sand and other inorganic components such as perlite.

The pH of a rooting substrate is a measure of the acidity or alkalinity of the substrate solution. Although different plants may have different preferences in regard to pH, the general accepted range of pH for most plants in organic-based substrates is 5.2 to 6.5. Adjustment of pH in the rooting substrate influences the availability of nutrients, the rooting of different species, and the incidence of certain disease organisms. Nutrient concentrations in propagation substrates are influenced by CEC and pH. Most micronutrients are available at a pH of 6.5 or less. Some debate exists over the application of fertilizers to the propagation substrate, but most research clearly shows that, while controlled-release fertilizers do not help rooting percentages, growth after the cuttings root is improved. Considerable research has been conducted with different methods of nutrient application such as (1) liquid feed, (2) top dressing fertilizer, (3) incorporation of fertilizer, and (4) nutrient mist. So far, no one method has proven superior for all plants. A variety of slow- or controlled-release fertilizers have been used successfully on a number of crops. If cuttings are beginning to root or are turning off-color, many growers will apply a combination of potassium nitrate and calcium nitrate or 20–10–20 at a rate of 50 to 75 mg/L of nitrogen as a liquid feed application.

Soluble salt levels and sterility of substrate components can be influenced by the component itself or the irrigation water used. When considering water quality, city and/or deep well water is often preferred over pond or stream water because of fewer disease organisms, algae, and suspended solids. Collection basins may have high soluble salt levels and organisms that can cause damage to plants and clogging of irrigation and mist lines. Table 17.1 is an interpretation of water quality for nursery production and plant propagation.

Some of the problem elements and ions in irrigation water are the following: (1) calcium and magnesium, (2) sodium, (3) boron, (4) chloride, (5) bicarbonate, (6) fluoride, (7) iron, and (8) sulfur. Calcium and magnesium are found in water considered to be hard and, therefore, can cause buildups on mist equipment, which can lead to mist system components not working. These two elements along with bicarbonate are also often responsible for the high pH and/or alkalinity found in irrigation water. Sodium and chloride add to the soluble salt problem and can occur in toxic concentrations. Chloride levels of less than 140 mg/L are not a problem, but levels greater than 350 mg/L are unsuitable. Boron and fluorine can also cause toxicity problems. Boron is not usually a problem if found at levels less than 0.5 mg/L, but at higher levels (>2.0 mg/L), irrigation water is not suitable for use. Fluoride is found in fluorinated city water, superphosphate, and perlite. Many foliage plants and monocots are sensitive to fluorides. If fluorides are a problem (>0.2 mg/L), problems can be avoided by increasing shading, adjusting the pH of the substrate and/or irrigation water, or not using fertilizers or substrate components that contain fluorides. Iron and sulfur can leave residues on leaves and also cause the formation of bacteria that can plug mist nozzles.

The following functions of a container for rooting cuttings are to (1) provide protection for the root system, (2) facilitate transplanting and maximal survival, (3) permit the ease of handling or shipping, and (4) provide consumer appeal for the end user. The criteria for the selection of a good container are the following:

1. Provides adequate drainage
2. Holds sufficient volume of substrate
3. Is lightweight
4. Is easy to handle
5. Is durable
6. Is free of toxic substances
7. Prevents root circling

Containers are often used for rooting cuttings instead of growing plants in ground beds because transplanting losses are often minimized along with other advantages such as lower shipping and handling costs. One of the main problems associated with rooting plants in small containers is root circling and girdling. Different methods have been used in an attempt to solve problems associated with root circling. Some of the successful methods for many species have included air pruning of the roots; the use of bottomless containers, containers with mesh bottoms that prune the roots as they enlarge, and stair-stepped containers; or treatments with root pruning chemicals such as copper hydroxide (Spin Out®) (Figure 17.9).

TABLE 17.1
Estimation of Propagation Water Quality Based on Conductivity

Conductivity Reading (dS/m or mmhos/cm)	Water Quality
0.0–0.25	Very good
0.26–0.75	Good
0.76–1.5	Fair
1.51–2.0	Questionable
Above 2.0	Unsuitable

Growers use several types of containers to root plants. Some examples include the following: (1) flats (wooden, plastic, or galvanized metal), (2) clay pots, (3) plastic pots known as liners, (4) compressed fiber pots, (5) peat pots, (6) plastic cell packs, (7) paper pots, and (8) polybags. Each grower should select a container that works best for their system of propagation and marketing, and produces the highest-quality plants. Plastic liner pots or cell packs are preferred by many large commercial operations. A 32 count or 40 count cell pack may work well for a short-term crop (softwood cuttings, propagation time of 2–3 months), whereas larger individual liner pots are preferred for rooting of certain tree species that are slower to root (Figure 17.10).

FIGURE 17.9 An Accelerator® propagation container. Root circling is prevented by air-root pruning holes in the container.

FIGURE 17.10 Rooted cutting propagated in a compressed fiber container.

USE OF ROOTING HORMONES AND STICKING CUTTINGS

Cuttings are commonly treated with growth-regulating compounds to promote the formation of adventitious roots. The reasons for the application of such root-promoting chemicals are the following: (1) to hasten root formation, (2) to increase the percentage of rooting, (3) to increase the uniformity of rooting, and (4) to increase the quality and number of roots per cutting.

The first root-promoting chemical to be discovered in the 1930s was the naturally occurring auxin indole-3-acetic acid (IAA). IAA is not used for the propagation of plants via cuttings because it is sensitive to environmental factors and is easily broken down by light and loses activity. The two commonly used synthetic auxins are indole-3-butyric acid (IBA) and naphthalene acetic acid (NAA). These two chemicals can be applied to cuttings as either talc (powder) or liquid (dip or spray) formulations.

IBA and NAA can be purchased as reagent grade powders, but these forms are not water soluble and must be dissolved in alcohol or other organic solvents before being diluted with water. A solution containing a final concentration of 50% denatured alcohol is often used. The potassium salt of IBA and the sodium salt of NAA are water soluble but are not registered with Environmental Protection Agency (EPA). Under the Federal Insecticide, Fungicide, and Rodenticide Act, only products that are registered can be utilized by growers. Alternatives to K-IBA include powder or talc formulations, concentrates that can be diluted, or water-soluble salts of IBA.

Concentrations greater than 20,000 mg/L (2%) are generally difficult to keep in solution. Formulation of auxins are sensitive to light and break down; thus, they should be kept in amber or dark containers that limit exposure to light. In general, auxin solutions can be kept for up to 6 months if refrigerated. Precipitation problems can be avoided by using distilled or deionized water instead of tap water. Commercially formulated rooting hormones can also be dissolved in carboxymethylcellulose to form a gel. Lower concentrations of hormone can be used since more gel is present on the cutting compared to a liquid dip.

When making calculations for the preparation of rooting hormones, the most useful values to remember are as follows:

$$1 \text{ mg/L} = 1 \text{ ppm}$$

and

$$1\% = 10{,}000 \text{ ppm or mg/L}$$

To prepare a liquid formulation that contains 3000 ppm or 0.3% IBA, one should dissolve 3000 mg of IBA in 1 L of solvent. Always remember to check the amount of active ingredient in your chemical and compensate when it is less than 100%. For example, if your chemical is only 75% active ingredient, then it will require 1.33 times as much of that chemical compared to a formulation that is 100% active ingredient. Several commercially available formulations of liquids, water-soluble tablets, and powders can be purchased.

Propagation of cuttings can be broken down into several stages including (1) preparation and taking of cuttings, (2) sticking of cuttings, (3) callusing of cuttings, (4) development of roots, and (5) hardening-off of cuttings. Cuttings should be true to type: they should have the same look and genetic makeup. Variegated plants may not be stable, and therefore, uniform cutting selection is important. Also, pay attention to rules regarding the propagation of patented plants. Such plants should be propagated only after a license has been granted from the patent owner or an authorized assignee. All cuttings should be disease and pest free (see Chapter 11), should come from healthy stock plants, and should be similar in size and maturity. Remember that many plants have optimum windows of opportunity for rooting. Good record keeping is essential.

Before the cuttings arrive in the propagation area, make sure that the area where the cuttings will be rooted has been disinfected and is free of pests and weeds (Chapter 15) and that all environmental and mist controllers are functioning properly. Sufficient numbers of propagation containers or flats should be available, selected, and sanitized, if necessary. Fill all containers in advance with appropriate substrate, and check to make sure that enough labor is available to stick the cuttings when they come in.

Cuttings are often harvested the morning that they are to be propagated. If this is not possible, many cuttings can be packed moist and stored overnight in a cooler. Certain cuttings such as *Ilex x attenuata* 'Foster No. 2' often defoliate if stored overnight during the growing season. See Dole and Gibson (2006) for a list of herbaceous taxa and how quickly they should be handled in propagation (Table 17.1). Make sure that stock plants are pest-free. Cuttings being harvested should be uniform in stem diameter and size. Be sure to harvest cuttings at the correct stage of maturity (softwood, semi-hardwood, or hardwood). Once the cuttings are taken, they are often kept in plastic bags and placed in portable ice coolers. Do not leave plastic bags out in the sun; they heat up very quickly, and damage can occur to the cuttings. Keep cuttings cool until they are stuck. This may be several hours from the time that the cuttings are taken.

Oftentimes, the lower leaves are removed before rooting hormones are applied. Some cuttings respond best to wounding before they are treated with hormone and placed in the propagation substrate. Light wounding can be a slice in the low part of the stem or an action as simple as stripping leaves from the stem. Heavy wounding often refers to cutting a 2 to 4 cm slice of the base of the cutting on one side. Many conifers, hollies, magnolias, and evergreen rhododendrons benefit from wounding. Once the cutting is wounded, cells begin to divide, and callus formation begins. As callus formation takes place, certain cells divide, and root development is initiated. The length of time that the cuttings are dipped in rooting hormones and the depth of sticking the cuttings in the hormone solution vary by species. Once cuttings are stuck, misting should be initiated and shade should be provided, if necessary. Some cuttings such as *Magnolia grandiflora* benefit from a combination treatment of a quick dip followed by a talc application before the cutting is stuck.

If substrate temperatures are appropriate—21 to 24°C (70–75°F)—callusing often occurs within a few weeks. High relative humidity must be maintained during this period so cuttings do not wilt. Mist frequency and longevity should be adjusted according to environmental conditions during the course of each day. Mist may be required for softwood or herbaceous cuttings for the first few nights but should then be adjusted so that the cuttings do not have wet foliage through the night. Once the cuttings have developed callus, mist frequency and duration can be reduced. Heating mist water to a temperature of 21°C (70°F) in winter can speed the rooting process since cold water will drop the temperature of the rooting substrate.

As roots are initiated and begin to elongate, mist cycles can again be reduced, and light intensity can be increased. Cuttings often benefit from the application of liquid or controlled-release fertilizers at this point. Check the cuttings for disease problems, and treat as necessary. For some herbaceous species, plant growth regulators can be applied at this stage to reduce stretching. Fertilizers low in phosphorus should be used as well to prevent stretching of cuttings under conditions of low light intensity. If cuttings have been in the propagation areas for several weeks, begin to check the pH and the electrical conductivity of the substrate, and adjust fertility regimes as required.

Finally, cuttings need to be hardened off before they are shipped or potted up and placed in production areas. Cuttings should be well rooted such that the root system remains intact when the plant is removed from the propagation container. Rooted liners are either moved to a different area or mist frequency is reduced to lower the relative humidity in the propagation area. Light intensity is often increased at this point in time. Increasing light intensity decreases internode elongation, increases leaf thickness, and often improves branching of rooted cuttings.

Some cuttings do not survive overwintering unless a new flush of growth is forced during the summer that the cuttings are rooted. Some taxa with this problem include *Acer*, *Betula*, *Cornus*, *Hamamelis*, *Magnolia*, *Quercus*, *Rhododendron*, *Stewartia*, and *Viburnum*. Placing rooted cuttings under long-day conditions often helps plants break bud. Growers often use incandescent light bulbs spaced 1 m^2 and 1 m above the crop. Lights can be turned on for 5 min every hour during the night or for a continuous period from 10:00 P.M. to 2:00 A.M. High concentrations of rooting hormones applied to enhance the rooting process often suppress budbreak, thus reducing winter survival. Using lower concentrations of rooting hormones in conjunction with the antioxidant ascorbic acid (1.5–2.5%) can improve rooting and overwintering survival. Ascorbic acid also helps improve the rooting of cuttings such as camellias that have high levels of phenolic compounds.

SANITATION DURING PROPAGATION

An effective propagation nursery requires a good disease and pest management program. The following advantages can be realized by the grower when a good sanitation program is followed during the rooting of cuttings:

- Improved plant quality
- Decreased plant mortality
- Reduced production costs
- Fewer production delays
- Increased profits
- Returning customers

The pests that nurserymen often have to combat in the nursery are weeds, fungi, insects, bacteria, nematodes, algae, viruses, and mycoplasmas (see Chapters 11 and 15). These pests reach the propagation area in a variety of ways, which may include the following: (1) transmission by insects, (2) entering through openings in propagation structures, (3) splashing of contaminants by water,

(4) hoses dropped on the ground, (5) contaminated water or hormone solutions, (6) infested soil or substrates brought into the propagation area by foot or non-sterile equipment, and (7) the use of previously infected plant material.

The two major nursery areas that must be considered are the stock plant area and the propagation area. The stock plant area may consist of actual blocks of plants being grown in the field that are used exclusively as parent material for cuttings, or may be a production area in the nursery from which cuttings are taken. Since field-grown stock plant areas can be costly to maintain, many nurseries use cuttings taken from container production areas. Some production nurseries allow propagators to take cuttings from container or field-grown stock, and then, later, purchase the rooted cuttings back from the propagator.

The following two sections develop "best practices" ideas for woody plant vegetative propagation and, more generally, propagation of all plant materials.

PREVENTIVE PRACTICES FOR STOCK PLANTS

1. Stock blocks should be physically separated from the propagation area.
2. Field-grown stock should be established in well-drained areas to prevent soilborne root diseases.
3. Container stock should be grown on raised sloping beds for ideal drainage.
4. When preparing a field, as much native plant material and debris should be removed to control root diseases and hosts for foliar diseases and insects.
5. If planting in the ground, controls for noxious weeds, nematodes, and plant diseases should be applied before planting.
6. Allow adequate spacing between rows of plants. This will aid in the application of pesticides, allow for good air circulation and the rapid drying of foliage, and slow the spread of insects and diseases among the plants.
7. Irrigation scheduling should be done as early in the day as possible so that the foliage is not wet during evening hours. Irrigation scheduling should also take into account natural rainfall so that the soil does not become waterlogged, a situation that is ideal for root rot organisms.
8. Weed growth should be controlled to prevent the spread and multiplication of new and existing weeds, either by seed or by vegetative means. Cover crops and turf around the nursery and in roadways should be kept mowed to prevent the movement of pests into stock areas.
9. Container-grown plants should be grown in pest-free substrates.
10. Used containers should be disinfected before reuse if disease- and pest-susceptible species are being grown.
11. Preventive applications of fungicides, insecticides, and preemergent herbicides can be used to keep pest populations low.
12. Proper cultural practices (pruning, fertilization, etc.) will help ensure that stock plants remain healthy.

PREVENTIVE PRACTICES IN THE PROPAGATION AREA

1. Plan propagation cycles for the most rapid rooting period for a given species or selection, thereby limiting exposure time to potential pests.
2. Propagation substrates can be treated with steam or a fumigant to eliminate disease and pest organisms.
3. Beware of potential problem areas such as canals, wells, and collection basins as sources of disease pathogens.
4. Do not use water in the propagation area that is high in soluble salts or has high levels of certain elements such as boron or chlorine. High soluble salt levels may cause necrosis of plant tissue or predispose the cuttings to other plant diseases.
5. If possible, take cuttings from the upper portions of the stock plant away from the soil, where disease organisms cannot be splashed onto the foliage.
6. Always collect cuttings that are free of insects and diseases.
7. Do not mix batches of plant material that have been collected from different locations or were purchased from another nursery.
8. Treat cuttings that are prone to diseases with appropriate chemicals before propagation.
9. Check to make sure that a proper environment is provided for the plant material and not for the pest organism.
10. Select a proper mist cycle that does not saturate the propagation substrate.
11. Check the propagation areas daily for signs of diseases or other pests.
12. Select a propagation substrate that has appropriate chemical and physical properties.
13. Use raised beds or benches whenever possible. Flats or containers placed on the ground should be placed on supports so that they do not remain in standing water.
14. Group plants according to their needs.

Care should be taken to clean all equipment and facilities before they are utilized. Bleach is a good general purpose chemical that can be used to clean equipment and benches, although it can be corrosive to certain metals and paper products such as greenhouse cooling pads. Several copper-based products, as well as others, are available for use in propagation areas. Denatured alcohol can also be used as a good general disinfectant for pruners and other metal tools.

LITERATURE CITED AND SUGGESTED READING

Blythe, E. K., J. L. Sibley, K. M. Tilt, and J. M. Ruter. 2007. Methods of auxin application in cutting propagation: A review of 70 years of scientific discovery and commercial practice. *J Environ Hort* 25:166–185.

Crawford, M. 2006. Improving root initiation of cuttings. *Amer Nurseryman* 201:36–38, 40.

Davis, T. D., B. E. Haissig, and N. Sankhla, eds. 1988. *Adventitious Root Formation in Cuttings.* Portland, Oregon: Dioscorides Press, 315 pp.

Dirr, M. A. and C. W. Heuser. 2006. *The Reference Manual of Woody Plant Propagation: From Seed to Tissue Culture*, 2nd Edition. Cary, North Carolina: Varsity Press, Inc., 424 pp.

Dole, J. M. and J. L. Gibson. 2006. *Cutting Propagation: A Guide to Propagating and Producing Floriculture Crops.* Batavia, Illinois: Ball Publishing, 400 pp.

Hartmann, H. T., D. E. Kester, F. T. Davies, Jr., and R. L. Geneve. 1997. *Plant Propagation: Principles and Practices*, 6th Edition. Upper Saddle River, New Jersey: Prentice Hall, Inc.

Jones, R. K. and D. M. Benson, eds. 2001. *Diseases of Woody Ornamentals and Trees in Nurseries.* St. Paul, Minnesota: The American Phytopathological Society, 482 pp.

McDonald, B. 1986. *Practical Woody Plant Propagation for Nursery Growers*, Vol. 1. Portland, Oregon: Timber Press, 669 pp.

Wilson, P. J. and D. K. Struve. 2004. Overwinter mortality in stem cuttings. *J Hort Sci Biotechnol* 79:842–849.

18 Adventitious Rooting of Woody and Herbaceous Plants

Lori D. Osburn, Zong-Ming Cheng, and Robert N. Trigiano

Plant propagation can be divided into the following two primary categories based on reproduction: sexual and asexual. Any method of multiplying plants other than by seed, or sexual reproduction, is considered asexual reproduction and is generally referred to as "vegetative propagation." Where sexual reproduction results in unique, individual plants from two parents, vegetative propagation results in replicated plants (or clones) from one parent. Some common methods of vegetative propagation include layering (Chapter 24), grafting (Chapters 25–27), budding (Chapters 25–27), division and separation of storage organs or geophytes, (Chapters 28 and 29), micropropagation, (Chapters 30–34), and cuttings. Propagation of plants by rooted cuttings is the most widely used method for producing clones of many herbaceous plants and most woody ornamentals and fruit and nut crops. Whether a plant is propagated by cuttings is usually determined by the ease of which they root or if other propagation methods are not as successful. Cuttings can be taken from roots (Chapters 21 and 23), leaves (Chapters 21 and 22), or stems (Chapters 17 and 19), but because cutting propagation is such a broad topic, and propagation using roots and leaves has its own inherent matters, this chapter will focus on the use of stem cuttings to propagate woody and herbaceous plants.

When using stem cuttings for propagation, the resulting progeny are identical to the parent plant, or stock plant, from which they were taken, and are called "clones." Some of the reasons to use stem cuttings for propagation include the following:

1. To propagate plants that cannot be easily reproduced via seed. This is because some seeds are difficult to germinate or have too strict or exacting germination requirements. Some seeds also have a short storage life.
2. To shorten the juvenile stage of growth and more quickly obtain the desired mature phase of growth that allows for flowering and fruit bearing. This is the case with many fruit, nut, and ornamental trees and shrubs, as well as some herbaceous perennials.
3. To retain the genetic characteristics of a particular plant, allowing the propagator to generate and maintain clones. Many cultivars (cultivated varieties) are desired for their special attributes that might be lost through sexual reproduction. Asexual propagation retains a plant's unique characteristics by maintaining the genotype, whereas sexual reproduction consists of combining DNA from two parents and results in progeny that have new genotypes (Chapter 5). Some unique attributes might include, but are not limited to, special foliage or flower color, leaf variegation, plant size, fruit type, etc.

Leaf and root cuttings are not sufficient to obtain true clones—stem cuttings that contain at least one bud from the stock plant are required. The bud is important because it contains a shoot meristem that will develop into a shoot system. However, leaf and root cuttings must form adventitious buds to produce a new shoot system. These adventitious buds produced by leaf and root cuttings may not retain all of the desired phenotypic characteristics of the stock plant. Thus, some desired mutations (variegation of flowers, leaves, etc.), such as with chimeras (Chapter 7), will not be expressed in the progeny.

Plant chimeras are organisms which, because of a genetic mutation (natural or induced) within the layered shoot meristem, contain two different genotypes within one individual and are phenotypically expressed as variegation or some pattern of distinction, such as thornless blackberries (McPheeters and Skirvin, 1983), depending on the affected layer. A good example of a chimeric plant is *Sansevieria trifasciata.* It typically has a green leaf, but at some point, a mutation within the shoot meristem occurred, resulting in a loss of pigment in the leaf margins, making them appear cream colored. This plant with its mutated, yet attractive, appearance has been widely marketed and is named *S. trifasciata laurenti.* However, when propagated via leaf cuttings, the leaf margins revert back to the original green color because the leaf cuttings produced adventitious buds that did not contain the original mutation (Chapter 7). In this instance, rhizome (modified stem) cuttings are the appropriate starting material to maintain the desired appearance. Buds are also important because they contain an auxin signal that causes rooting to initiate. In many cases, roots will not form on a stem cutting that has had all of the buds removed before rooting is initiated (Went, 1934; Haissig and Davis, 1994).

Rooting requirements can be very species specific, especially in woody plants. Based on the species being propagated, a specific type of cutting is usually required. For example, some species root more easily with hardwood cuttings, whereas others require softwood cuttings. Still, others may root easier when semi-hardwood cuttings are used. Softwood cuttings are taken after the first flush of new spring growth while the wood is green and tender. Semi-hardwood cuttings are taken in summer after a flush of growth when the wood begins to mature but still retains its leaves. Hardwood cuttings are taken during dormancy, after growth has ceased and the leaves have abscised. Of course, there are advantages and disadvantages related to each of these types of cuttings. Softwood and semi-hardwood cuttings are similar to herbaceous cuttings—they require protected structures and a misting or humidity control system to reduce transpiration until they produce their own root system (Chapter 8). Hardwood cuttings are much more versatile and economical because transpiration is not an issue, but not all species will produce roots from a hardwood cutting.

As with other propagation methods, there are many factors to consider when selecting a plant to propagate by stem cuttings, including the health, age, growth stage, and maturity of the stock plant; the time of the year that the cuttings are taken; the type of cutting used (in woody plants); the plant growth regulators (PGRs) to be used; and the availability of an appropriate rooting environment (media, light, temperature, and water requirements). In many cases, after a cutting is taken from a woody stock plant, it is usually treated with an auxin and "stuck" into a suitable rooting medium to produce adventitious roots. Auxins are plant hormones (referred to as "phytohormones" if natural and "PGRs" if synthetic) that induce rooting among other responses (Chapter 4). Although auxin PGRs are used during cutting propagation because they are chemically more stable, auxin phytohormones are sometimes used during micropropagation (Chapter 30) because of the sensitive nature of those plants. During propagation, and especially while under mist, nutrients are leached from the leaves, and the susceptibility to pathogens increases. The longer a cutting remains under mist without roots, the weaker it can become. Thus, auxin PGRs are widely used to increase rooting efficiency and uniformity.

When placing a cutting into the rooting medium, polarity (top versus bottom) must be observed because roots will develop naturally on the end of the cutting (proximal end) that was closest to the crown of the stock plant from which it was taken, even if placed upside down (Bloch, 1943). If placed into the rooting medium upside down, several negative outcomes are possible, including either that rooting will not occur, roots will arise on the distal (stuck) end, or roots will arise on the proximal (unstuck) end. Regardless of outcome, the cuttings will not develop normally (Nanda *et al.*, 1969), which means an economic loss for a propagator. Even with leaf and root cuttings, polarity must be observed or new roots and shoots will not develop.

The following laboratory/greenhouse experiments are designed to provide some experiences with rooting cuttings of various common plants. Each of the experiments requires some common facilities (Table 18.1), materials and supplies (Table 18.2), and procedures (Procedure 18.1).

TABLE 18.1
Facility Requirements for All Cutting Experiments

- *Greenhouse*. Maintain temperatures of about 25°C day/20°C night.
- *Shadecloth* (47–52%). Place over house or individual benches at least until the cuttings acclimate, or longer, if needed.
- *Overhead bench mist system*. Programmed to supply mist or fog at a set duration and interval as needed, depending on the crop and the climate. Approximate settings for leafy cuttings are 10 sec of mist every 15 min, beginning at dawn and ending 1 h pre-dusk. When rooting leafless, hardwood cuttings, the mist interval may be longer and the duration may be shorter than those for herbaceous cuttings. As the cuttings acclimate and develop a root system, the mist duration should be gradually decreased, and the mist interval, gradually increased.
- Bottom heat system is recommended: heating mat, warm water circulation tubing, etc.

TABLE 18.2
Materials Needed for All Cutting Experiments

- Propagation trays (e.g., Dyna-Flat™ F1020 [Hummert International])
- Petri dishes (15 × 100 mm [Fisher Scientific]) or small cups
- Peat-based soilless potting medium (Pro-Mix HP)
- 4-in plastic labels
- Paper towels
- Surface disinfectant for pruners or scissors (10% Clorox®, 70% isopropyl alcohol, Physan 20®, Greenshield®, Triathalon®, etc.)
- Plants: Those listed in each experiment are merely suggestions and may not be suitable for your class, depending on the climate, season, and availability.
- Black Sharpie markers
- Rulers
- Pruners or strong scissors
- Latex gloves
- Dust masks

Procedure 18.1
Common Preparation for All Cutting Experiments

Step	Instructions and Comments
1	For details specific to each experiment, refer to the appropriate section number.
2	Work in small groups, each consisting of four to five members.
3	Prepare one propagation tray per experiment per group: • Wet the potting media, then mix and test the moisture content by squeezing a handful. It should slightly clump. If not, add more water or dry media. Retest. • Fill the tray to within 1/2 in of the rim.
4	Prepare plant labels with the following information for each species and treatment: group number + species + treatment + date, then insert into the prepared propagation tray.
5	Label both the tops and the bottoms of the Petri dishes for water and the various auxins.
6	While wearing gloves and a dust mask, measure 3 tablespoons of each auxin talc from the stock container and place it into its appropriate Petri dish. Never dip the cuttings directly into the stock container.
7	Fill one Petri dish 3/4 deep with water to moisten cut stems prior to auxin talc application.
8	If using Dip'N Grow liquid, a water dip is not needed.
9	Disinfect pruners or scissors by dipping them into one of the recommended disinfectants listed in Table 18.2. Initially soak them for 5 min then use a quick dip in the solution between cuts.
10	Choose two plant species for each experiment. Select nonflowering stems.
11	Referring to the appropriate experiment for the cutting length and the number of nodes needed, obtain uniform cuttings from each plant by making your cuts as described below: • Make a straight cut at the distal end of each cutting. • Make an angled cut at the proximal end of each cutting (Figure 18.1).
12	Refer to the appropriate experiment for instructions on preparing leafy cuttings. Do not remove any leaves unless instructed to do so because each experiment has different requirements.
13	Place your prepared cuttings between sheets of moist paper towels until you are ready to treat and stick them.
14	Prior to treating with auxin talc, dip the proximal end of each stem in water to ensure that the talc sticks to the stem. Lightly tap the cutting on the side of the Petri dish to remove any excess clumping talc (Figure 18.2b and c). Note: In Experiment 2, some treatments will require a distal end dip.
15	Treat all auxin controls with a water dip.
16	Stick 1/3 of the cutting length into the medium within its pre-labeled row (Figure 18.2d).
17	After sticking all of the cuttings for the experiment, place the tray on a mist bench and lightly water. Avoid excess moisture as it cannot only encourage disease organisms to grow, but can also reduce gas exchange in the rooting zone.
18	Clean up: Place laboratory supplies in a designated location, dispose of any trash, and wipe off your now-empty table.
19	Data collection: Each group checks the cuttings for roots at 3 or 4 weeks after sticking, depending on time constraints. Arrange it so that the cuttings are checked only once, instead of by each group member at different times. • Gently tug on the cutting. If it resists, gently push away the surrounding medium and carefully remove the cutting. • If it has any roots, then it is counted as having rooted. Return the cutting to the tray. • Record the data in the appropriate table and calculate the rooting percentage.
20	Once they are rooted, the cuttings may be transplanted into individual pots.

FIGURE 18.1 (a) Cutting the main stem into segments for treating and sticking. Notice the angled cut on the proximal end and that the student is making a straight cut on the distal end. (b) A close-up of the final cuts. All leaves were removed to show the details.

EXERCISES

Experiment 1. The Effect of Various Exogenously Applied Auxins on the Adventitious Rooting of Selected Woody and Herbaceous Cuttings

Question: Does exogenously applied auxin improve the adventitious rooting of selected woody and herbaceous stem cuttings?

Once a cutting is detached from the parent plant, several physiological processes occur, ultimately resulting in the development of adventitious roots. However, the capacity of different plant species to form adventitious roots varies greatly, from very easy to nearly impossible. To optimize rooting during cutting propagation, certain stock plant management practices are utilized (Hartmann *et al*., 2002). Above and beyond those practices, auxins are used to enhance adventitious rooting in several ways. Auxins can increase the uniformity of rooting within a crop, the number of roots formed, and the rate at which rooting occurs (Davis *et al*., 1991). Although auxin plays a role in many plant processes, this experiment will focus only on its function in rooting.

The primary auxin produced by plants is indole-3-indoacetic acid (IAA). However, this natural auxin is not commercially utilized because it is less chemically stable than other available auxins. The two most common auxins found in rooting preparations are indole-3-butyric acid (IBA) and 1-naphthalene acetic acid (NAA) (Chapter 4). While many species readily respond to IBA, some may respond better to NAA. In addition, the specific formulation, combination, and concentration of auxin can also influence adventitious rooting.

The goals of this experiment are to learn the procedures used in cutting propagation and to compare the effect of various exogenously applied auxin formulations on the rooting of selected woody and herbaceous stem cuttings.

MATERIALS

Students will need the following materials to complete this experiment:

- Materials listed in Tables 18.1 and 18.2
- Water
- Hormodin 2 talc (0.3% [3000 ppm] IBA)
- Dip'N Grow liquid (full strength is 1.0% IBA + 0.5% NAA), diluted to half strength to yield 5000 ppm IBA and 2500 ppm NAA
- Rootone talc (0.20% [2000 ppm] NAA)
- Woody plant suggestions: *Hydrangea macrophylla* (florists' hydrangea), *Ficus benjamina* (weeping fig), and *Ficus elastica* (rubber tree)
- Herbaceous plant suggestions: *Solenostemon scutellarioides* (coleus), *Pelargonium* spp. (scented geranium), and *Plectranthus australis* (Swedish ivy)

Follow the instructions outlined in Procedure 18.2 to complete this experiment.

FIGURE 18.2 (a) Removing the lower leaves from an herbaceous stem cutting. (b) Dipping the trimmed cutting into the auxin talc after a water dip. (c) The trimmed cutting is now ready to stick. (d) Sticking the prepared cutting into the propagation tray.

Procedure 18.2
Effect of Various Exogenous Auxins on Adventitious Rooting – Experiment 1

Step	Instructions and Comments
1	See Procedure 18.1 for the preparation of trays, labels, auxins, and cuttings.
2	Choose one herbaceous and one woody plant species.
3	For each species, use five cuttings per treatment. Example: 5 cuttings × 4 treatments = 20 cuttings per species
4	Prepare cuttings 3 to 4 in long with 3+ nodes. Remove the leaves from the lower half of the stem and trim any large leaves on the upper portion of the stem (Figure 18.2a–d).
5	Apply treatments: Dip the proximal end of each stem into one of the following: 1, water only; 2, Dip'N Grow liquid; 3, Hormodin 2 talc; and 4, Rootone talc.
6	Repeat for each species. Stick cuttings into the prepared trays as directed in Procedure 18.1.
7	After all treatments are stuck, place the tray on the mist bench as directed in Procedure 18.1.
8	Collect data in 3 to 4 weeks and record in Table 18.3. See Procedure 18.1 for the complete instructions for collecting data.

TABLE 18.3
Data Collection for Experiment 1 (Effect of Various Exogenous Auxins on Adventitious Rooting)

Plant	Treatment No.	Experimental Treatments	No. Cuttings	No. Rooted	% Rooting[a]
Species 1	1	No auxin (water only)	5		
Species 1	2	Dip'N Grow liquid	5		
Species 1	3	Hormodin 2 talc	5		
Species 1	4	Rootone talc	5		
Species 2	1	No auxin (water only)	5		
Species 2	2	Dip'N Grow liquid	5		
Species 2	3	Hormodin 2 talc	5		
Species 2	4	Rootone talc	5		

[a] Number of rooted cuttings/number of cuttings used.

ANTICIPATED RESULTS

Most cuttings will root in response to auxin application. Therefore, one could expect that the rooting percentage will be increased with the treatment of a higher auxin concentration. However, the higher concentrations required for woody plants may be too strong for the herbaceous plants and could inhibit rooting or cause stem damage. Some of the tender herbaceous cuttings may produce roots without an auxin application. It will be interesting to see which species respond better to NAA than IBA.

QUESTIONS

- Plant propagation by rooted cuttings is an example of what type of reproduction?
- What is the purpose of using intermittent mist while rooting cuttings?
- If you were a large grower and needed to propagate 1000 *Buddleia davidii* liners for Bob's Nursery, which auxin formulation would you use—a liquid or a talc? Think about the advantages and disadvantages. Justify your answer.
- Why do some cuttings root so easily without auxin application? Would the use of an auxin on that species be a justifiable expense?
- Under what situation might auxin inhibit rooting in a plant?

Experiment 2. The Effect of Polarity and Exogenously Applied Auxin on the Adventitious Rooting of Selected Woody and Herbaceous Stem Cuttings

Question: If polarity affects the production of adventitious roots on stem cuttings, can an exogenously applied auxin overcome that effect?

During commercially rooted cutting propagation, stems are typically treated with an auxin that is applied to the proximal (base) end prior to being inserted into a rooting medium. Propagators are careful to ensure that they treat and stick the proper end of the cutting because plants exhibit a phenomenon called "polarity." To distinguish the proximal from the distal ends of the cuttings during propagation, one end of the stem is cut straight, and the other is cut at an angle. Polarity dictates that roots will form on the proximal end of the cutting, that is, the end closest to the crown of the plant from which it was taken, and shoots will form on the distal end of the cutting, that is, the end farthest from the crown (Bloch, 1943; Hartmann *et al.*, 2002). If the cutting is inserted into the medium with the distal end down (inverted), several not-so-desirable outcomes are possible (Nanda *et al.*, 1969). In rare cases, an exogenously applied auxin may be able to overcome reversed polarity, and the cutting will root on the inserted (distal) end but not develop normally (Went, 1941; Cristoferi *et al.*, 1988), and in some cases, polarity cannot be overcome (Sheldrake, 1974).

The goal of this experiment is to observe the effect of polarity on the rooting percentage of selected woody and herbaceous stem cuttings and to determine if an exogenously applied auxin can overcome the influence that polarity exerts on adventitious rooting capability.

MATERIALS

Students will need the following materials to complete this experiment:

- Materials listed in Tables 18.1 and 18.2
- Hormodin 3 talc (0.8% IBA)
- Water
- Woody plant suggestions: *Rosa* spp. (rose), *Salix babylonica* (weeping willow), and *Buddleia davidii* (butterfly bush)
- Herbaceous plant suggestions: *Solenostemon scutellarioides* (coleus), *Schefflera arboricola* (schefflera), and *Dendranthema* × *grandiflorum* (chrysanthemum)

Procedure 18.3
Effects of Polarity and Exogenous Auxin on Adventitious Rooting – Experiment 2

Step	Instructions and Comments
1	See Procedure 18.1 for the preparation of trays, labels, auxins, and cuttings.
2	Choose plants from two woody species.
3	For each species, use five cuttings per treatment. Example: 5 cuttings × 4 treatments = 20 cuttings per species
4	Prepare cuttings 3 to 4 in long with 4+ nodes. For each species: *Remove the lower leaves from only half of the cuttings* and remove the upper leaves from the other half of the cuttings. Example: 20 cuttings of rose—Remove the lower leaves from 10 cuttings and remove the upper leaves from the other 10 cuttings. Repeat for second species.
5	Prepare to apply all treatments (1–4) to both of the species. Refer to Figure 18.3 to observe proximal versus distal sticking.
6	Treatment 1: Dip the proximal end into water only (no auxin). Stick the cuttings into the prepared trays as directed in Procedure 18.1.
7	Treatment 2: Dip the proximal end first into water, then into Hormodin 3. Stick the cuttings into the prepared trays as directed in Procedure 18.1.
8	Treatment 3: Dip the distal end into water only (no auxin). Stick the cuttings into the prepared trays as directed in Procedure 18.1.
9	Treatment 4: Stick the cuttings into the prepared trays as directed in Procedure 18.1. Dip the distal end first into water, then into Hormodin 3.
10	Repeat for each species. After all treatments are stuck, place the tray on the mist bench as directed in Procedure 18.1.
11	Collect data in 3 to 4 weeks and record in Table 18.4. See Procedure 18.1 for the complete instructions.

Follow the instructions outlined in Procedure 18.3 to complete this experiment.

ANTICIPATED RESULTS

Based on polarity alone, it would be expected that those cuttings inserted into the medium distal end down will not root. Some species have a very strong polar influence and will root at the proximal end even if they are inverted. Nevertheless, it may be possible for an auxin to overcome polarity such that, if the auxin-treated distal end of an herbaceous species or an easy-to-root woody species is inserted, roots might be produced (Nanda *et al.*, 1969). However, the addition of auxin would not likely be able to overcome polarity in recalcitrant plants.

QUESTIONS

- Explain the differences between a phytohormone and a PGR described throughout this chapter. How would you utilize each one?
- Why might one plant species be responsive to exogenously applied auxin, whereas another species is unresponsive?
- How might auxin overcome the effect of polarity?
- Describe one way that propagators can ensure that cuttings are not inverted.
- Based on the results of this experiment, which of the conditions tested would you recommend a propagator to use to maximize rooting percentage?

Experiment 3. The Effect of Leaf Removal and Exogenously Applied Auxin on the Adventitious Rooting of Selected Deciduous Woody Stem Cuttings

Question: If the presence of leaves on a stem cutting is needed to initiate adventitious roots, can auxin exogenously applied to defoliated stem cuttings fulfill that need?

Endogenously produced auxin, primarily in the form of IAA, plays a role in many plant functions and might be best known for its role in rhizogenesis (root initiation and formation). Auxin, produced in the leaves, shoot apices, and lateral buds, is transported basipetally (downward) via the phloem. Should those auxin-producing tissues be

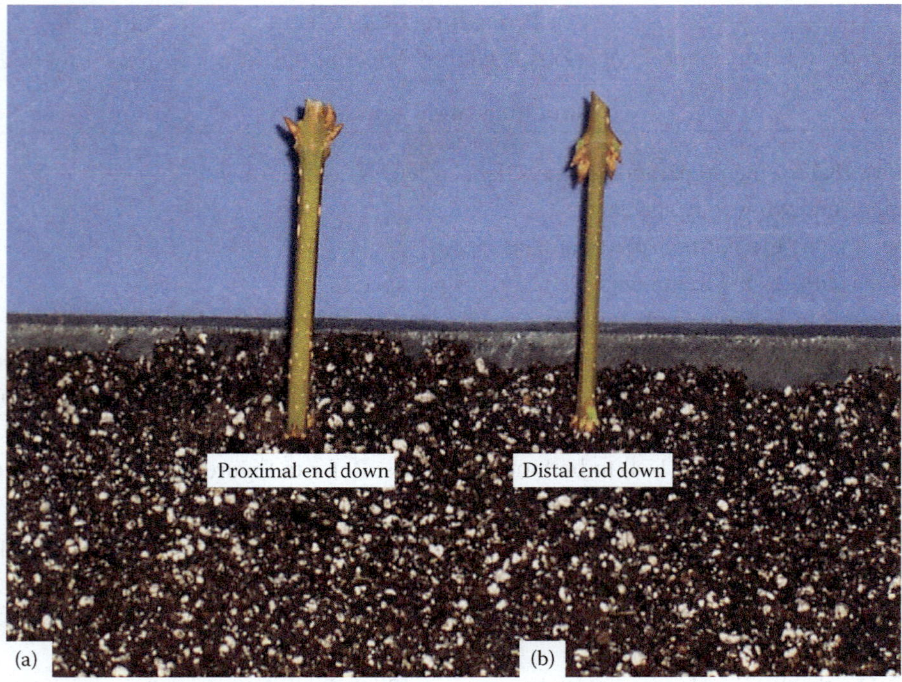

FIGURE 18.3 Orientation of cuttings (leaves have been removed for illustration). (a) Proximal end in medium. (b) Distal end in medium.

removed from a cutting, auxin will not be synthesized, and rooting will not be initiated, even with the application of auxin (Went, 1934; Haissig and Davis, 1994). The presence of leaves on a cutting has also been found to be critical for successful rooting (Reuveni and Raviv, 1980). Leaves that remain are able to photosynthesize and produce carbohydrates that provide energy and are a source of auxin for the development of roots. It is important to note that there should be a balance of leaves on a cutting, especially softwood, such that enough remain to carry out photosynthesis, but not so many that the cutting wilts from too many transpiring leaves.

The objective of this experiment is to observe the influence of an exogenously applied synthetic auxin on the rooting of defoliated deciduous stem cuttings.

MATERIALS

Students will need the following materials to complete this experiment:

- Materials listed in Tables 18.1 and 18.2
- Hormodin 3 talc (0.8% IBA)
- Water
- Sharp knife (disinfected)
- Deciduous plant suggestions: *Salix* spp. (willow tree), *Lagerstroemia indica* (crape myrtle), *Hydrangea quercifolia* (oakleaf hydrangea)

Follow the instructions outlined in Procedure 18.4 to complete this experiment.

TABLE 18.4
Data Collection for Experiment 2 (Effects of Polarity and Exogenous Auxin on Adventitious Rooting)

Plant	Treatment No.	Experimental Treatments	No. Cuttings	No. Rooted	% Rooting
Species 1	1	Proximal end down (no auxin)	5		
Species 1	2	Proximal end down (auxin)	5		
Species 1	3	Distal end down (no auxin)	5		
Species 1	4	Distal end down (auxin)	5		
Species 2	1	Proximal end down (no auxin)	5		
Species 2	2	Proximal end down (auxin)	5		
Species 2	3	Distal end down (no auxin)	5		
Species 2	4	Distal end down (auxin)	5		

Procedure 18.4
Effects of Leaf Removal and Exogenous Auxin on Adventitious Rooting – Experiment 3

Step	Instructions and Comments
1	See Procedure 18.1 for the preparation of trays, labels, auxins, and cuttings.
2	Choose two deciduous plant species.
3	For each species, use five cuttings per treatment. Example: 5 cuttings × 4 treatments = 20 cuttings per species
4	Prepare cuttings 3 to 4 in long with 4+ nodes. See Procedure 18.1 for the complete instructions. Remember to make angled and straight cuts to designate the proper polarity.
5	For each species: In addition to the instructions in Procedure 18.1, carefully remove all of the leaves from half of the cuttings using a sharp, disinfected knife. Example: 20 cuttings of willow—Remove all of the leaves from only 10 cuttings. Repeat for second species. For those cuttings that retain leaves, remove the leaves from the lower 1/3 of each stem for sticking, retaining the upper 2/3 leaves.
6	Treatment 1: Leaves removed, dip the proximal end into water only (no auxin). Stick the cuttings into the prepared trays as directed in Procedure 18.1.
7	Treatment 2: Leaves removed, dip the proximal end into water, then Hormodin 3. Stick the cuttings into the prepared trays as directed in Procedure 18.1.
8	Treatment 3: Leaves intact, dip the proximal end into water only (no auxin). Stick the cuttings into the prepared trays as directed in Procedure 18.1.
9	Treatment 4: Leaves intact, dip the proximal end into water, then Hormodin 3. Stick the cuttings into the prepared trays as directed in Procedure 18.1.
10	Repeat for each species. After all treatments are stuck, place the tray on the mist bench as directed in Procedure 18.1.
11	Collect data in 3 to 4 weeks and enter in Table 18.5. See Procedure 18.1 for the complete instructions.

ANTICIPATED RESULTS

Because leaves photosynthesize to produce carbohydrates needed for energy, the defoliated cuttings without the exogenously applied auxin treatment are least likely to root. It is always possible that application of auxin will provide a sufficient growth regulator to overcome the loss of carbohydrates, but if there is not enough stored energy, defoliated cuttings might still be less likely to root. Those species that tend to have naturally occurring preformed root initials, including, but not limited to, willow, crepe myrtle, and hydrangea, may root regardless of treatment. Alternatively, *de novo* adventitious roots can be triggered to grow in response to being severed from the stock plant, so other species lacking preformed root initials should be included in the study.

TABLE 18.5
Data Collection for Experiment 3 (Effects of Leaf Removal and Exogenous Auxin on Adventitious Rooting)

Plant	Treatment No.	Experimental Treatments	No. Cuttings	No. Rooted	% Rooting
Species 1	1	Leaves removed (no auxin)	5		
Species 1	2	Leaves removed (auxin)	5		
Species 1	3	Leaves remain intact (no auxin)	5		
Species 1	4	Leaves remain intact (auxin)	5		
Species 2	1	Leaves removed (no auxin)	5		
Species 2	2	Leaves removed (auxin)	5		
Species 2	3	Leaves remain intact (no auxin)	5		
Species 2	4	Leaves remain intact (auxin)	5		

QUESTIONS

- In which tissue(s) is endogenous auxin produced, and which direction is it transported throughout the plant?
- If endogenous auxin is produced in the buds and the leaves, how does it promote roots at the proximal end of a stem cutting?
- Describe the difference(s) between an endogenous auxin and an exogenous auxin. Where does each of these come from?
- Describe how the presence of leaves on a cutting can influence the rooting of that cutting.
- Based on the results of this experiment, what conditions tested would you recommend a propagator to use to maximize rooting percentage?

LITERATURE CITED

Bloch, R. 1943. Polarity in plants. *Bot Rev* 9:261–310.

Cristoferi, G., N. Filiti, and F. Rossi. 1988. The effects of reversed polarity and acropetal centrifugation on the rooting of hardwood cuttings of grapevine rootstock Kober 5BB. *Acta Hort (ISHS)* 227:150–154.

Davis, T. D., S. W. George, A. Upadhyaya, and J. Parsons. 1991. Propagation of firebush (*Hamelia patens*) by stem cuttings. *J Environ Hort* 9(2):57–61.

Haissig, B. E. and T. D. Davis. 1994. A historical evaluation of adventitious rooting research to 1993. In Davis, T. D. and B. E. Haissig, eds. *Biology of Adventitious Root Formation*. New York: Plenum Press, 343 pp.

Hartmann, H. T., D. E. Kester, F. T. Davies, and R. L. Geneve. 2002. *Plant Propagation: Principles and Practices*, 7th Edition. Upper Saddle River, New Jersey: Prentice Hall, 880 pp.

McPheeters, K. and R. M. Skirvin. 1983. Histogenic layer manipulation in chimeral 'Thornless Evergreen' trailing blackberry. *Euphytica* 32:351–360.

Nanda, K. K., A. N. Purohit, and V. K. Kochhar. 1969. Effects of auxins and light on rooting stem cuttings of *Populus nigra*, *Salix tetrasperma*, *Ipomea fistulosa* and *Hibiscus notodus* in relation to polarity. *Physiol Plant* 22(6):1113–1120.

Reuveni, O. and M. Raviv. 1980. Importance of leaf retention to rooting of avocado cuttings. *J Amer Soc Hort Sci* 106:127–130.

Sheldrake, A. R. 1974. The polarity of auxin transport in inverted cuttings. *New Phytol* 73:637–642.

Went, F. W. 1934. A test method for rhizocaline, the root-forming substance. *Proc Kon Ned Akad Wet* 37:445–455.

Went, F. W. 1941. Polarity of auxin transport in inverted *Tagetes* cuttings. *Bot Gaz* 103:386–390.

19 Long Cuttings

A Novel Method to Propagate Woody Plants

Wolfgang Spethmann

Each year, millions of cuttings are propagated worldwide, so on the surface, it would seem to be a simple technique. That is true for many easy-to-root species like *Forsythia*, *Ribes*, *Salix*, *Philadelphus*, and many others in the ornamental and nursery industries. For these species, the propagation method can be modified in many ways and can still be successful. Growth regulators can be used or not, substrate mixtures can be changed, cuttings can be stuck in different months, and cuttings can be supported with a mist or fog system. The more difficult a species is to root, the more specific or optimized the method must be. There are many environmental and intrinsic factors influencing the rooting process. In addition to those mentioned above, there are factors such as the age of stock plants, the location on the stock plant that the cutting is taken (harvesting at the base or at the top), the use of the basal or the terminal part of the ramet (a plant that is a clone), fertilizer and the pH of the substrate, the use of bottom heat, growth in pots or in ground beds, and many others. However, what are the most important parameters for rooting cuttings?

Most of the publications evaluating methods for rooting cuttings deal with growth regulators and growth media. After many years of research with cuttings, when comparing combinations, media formulations, and concentrations of growth regulators, results can and do vary from year to year. One combination might be better in one year, and in the next year, another combination, concentration, or formulation would be better. In most years, 0.5% indole-3-butyric-acid (IBA; 5000 ppm) in talc or as a liquid would often be the treatment that stimulated the production of roots. So, for years, 0.5% IBA was used routinely for nearly all species. However, for some rhododendrons, IBA was increased to 1% (10,000 ppm). Similar variable results would occur with rooting substrates or media. We had the best results with the difficult-to-root *Quercus robur* or *Q. petraea* in peat/sand and perlite/bark mixtures as well as pure inorganic gravel 3 to 8 mm in diameter. The efficacy of a specific substrate depends on the method used to increased humidity. If mist systems are used, then the substrate has to have adequate drainage, and the percentage of peat must be reduced. If fog systems are used, the percentage of peat can be higher. In addition, it is beneficial to add 2 kg/m^3 of a slow-release fertilizer such as Osmocote. Another often-neglected parameter is the pH of the substrate. After some years of investigation, a lower pH of 3.5 to 4.5 was found to be best for rooting and further shoot growth.

This variable response suggests that the mysterious mixtures that some growers use may not be so significant for success. So, there are many, often interacting, factors that affect rooting performance, and these are often difficult to measure and control. Experience gained from many years of investigation has allowed the development of a ranking of parameters important for the success of cutting propagation and includes the following from most important to least important:

1. Age of stock plant (cyclophysis/topophysis)
2. Time of setting (phenology)
3. Humidification System (irrigation)
4. Overwintering Survival and further cultivation

The most important of these parameters is the age of the stock plant age combined with the topophysis effect (cuttings from the base of the stock plant are more juvenile than cuttings from higher positions). Another important factor is the time of collecting and sticking of cuttings. Every species has an optimum period to be rooted; broad-leaved species often have their optimum in May/June, whereas conifers perform better in autumn to early spring. For easy-to-root species, a conventional mist irrigation system is often enough for good rooting. Difficult-to-root species, however, need a fog system to maintain 90 to 100% relative humidity during the rooting period.

In addition to all the factors identified here, survival after the first winter and further growth has often been neglected as an important consideration for the propagation of plants via rooted cuttings. For example, cuttings of *Hamamelis* and *Fagus* could be rooted, but survival after the first winter was poor. Cuttings of *Abies* could be rooted and they survived the winter, but they only grew in a horizontal manner or plagiotropically (Spethmann, 1998; Hartmann *et al.*, 2011).

Researching different techniques with a goal of getting a high rooting percentage is only the first step of a successful cutting propagation method. After rooting, the cuttings must survive winter, grow in the field for years, and show the expected growth habit. Most of the literature on cutting propagation deals with the factors described, and optimized cutting propagation systems usually means fine tuning these aspects, but there was an additional factor that we have discovered and used quite successfully. This was the discovery and development of the long cutting method.

THE LONG CUTTING METHOD

Nearly all conventional cuttings propagated worldwide are relatively short, generally only 5 to 20 cm in length. Longer shoots are used only with *Salix* and *Populus*, and these can be rooted successfully, but why is there no research on using cuttings 20 or 30 cm or longer? The answer is that nobody expected that rooting was possible with longer cuttings of most woody plants, particularly with hard-to-root species.

CONVENTIONAL PRODUCTION OF STANDARD ROSES

In Europe, *Rosa canina* 'Pfaenders' or *R. pollmeriana* 'Schreibers' are used as cold hardy rootstocks. Young plants can be harvested in autumn from seeds sown in March after receiving 10 weeks of warm stratification (18–25°C) then 10 weeks of cold stratification (2–8°C). Of those seedlings, only 10 to 20% attain the necessary 8 to 12 mm root collar diameter to be used for rootstock production. Replanted the following March, plants are grown until they produce shoots up to 2.50 m long, which requires up to 2 additional years. In the second and third growing seasons, all side branches are removed, and only a single long shoot remains to form the stem. After replanting in March in the fourth year, budding consists of two or three buds and usually takes place in July or August. At the end of the fifth year, plants are ready for sale. The production of standard roses is very time consuming and expensive.

LONG CUTTING PRODUCTION OF STANDARD ROSES

The potential for using long cuttings is quite exciting in view of the shortened time for the production of salable plants. Very long rose cuttings were used for the production of rootstocks for standard roses from a nursery. On these stock plants, only one shoot was left to form the stem, and all the other branches were removed. Up to 90% of these long shoots (150–250 cm) rooted in just 6 weeks in July, and by September, roots were well developed (Figure 19.1a). Rooted long cuttings were potted in containers in October (Figure 19.1b), and 1 year later, standard roses were ready for sale (Figure 19.2). Root development on long cuttings exceeded that of standard roses grafted on 'Pfaenders' rootstock (Figure 19.3). So,

FIGURE 19.1 Long cuttings of Rosa 'Pfaenders' in September after being collected and stuck in July, showing prolific adventitious root development (a) and either outplanted or transplanted to containers in October (b).

FIGURE 19.2 Standard stem roses ready for sale only 2 years after the propagation of rootstocks by long cuttings.

FIGURE 19.3 Comparison of the root systems of rose conventionally grafted onto 'Pfaenders' rootstock versus the prolific root system of 'Pfaenders' rootstock produced by long cuttings.

by rooting long shoots, unlike conventional production that often takes up to 5 years, standard roses could be produced in only 2.5 years.

LONG CUTTING PROPAGATION OF TREE SPECIES

The same techniques can be used for the propagation of various tree species including fruit tree species such as apple. The long cutting technique has been used successfully with M9 apple rootstock. Long cuttings were placed in a rooting medium in July, and by November, prolific root development had occurred (Figure 19.4).

FIGURE 19.4 Long cuttings of M9 apple rootstock set in July, showing the extent of root development achieved by November of the same year.

In another experiment, long (60–120 cm long) cuttings were taken from six tree species (*Acer pseudoplatanus*, *A. platanoides*, *Carpinus betulus*, *Ulmus* 'Regal,' *Quercus robur*, and *Tilia cordata*) of varying stock plant ages and stuck in June and July 2000. The age of the stock plants ranged from 9 to 30 years. *Quercus* and *Ulmus* stock plants were pruned back in spring. The long cuttings rooted at 74 to 97% (Figure 19.5).

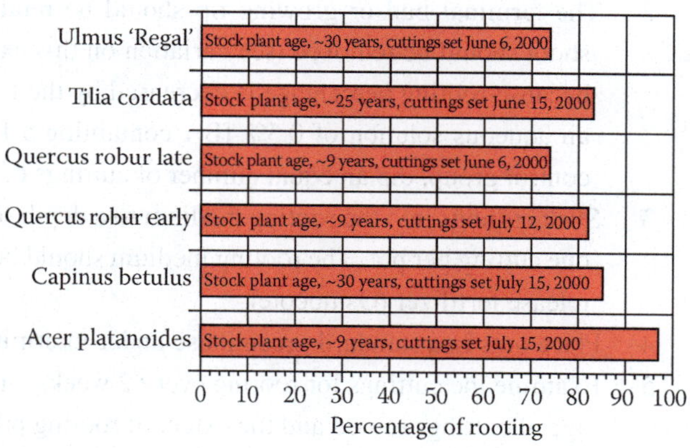

FIGURE 19.5 Percentage rooting of long cuttings of six tree species of stock plant ages ranging from 9 to about 30 years. Cuttings (60–120 cm long) were collected on various dates in June and July of 2000. *Quercus robur* and *Ulmus* 'Regal' stock plants were pruned in spring prior to collecting cuttings.

Although the following experiment applies long cutting propagation techniques to roses, it is readily apparent that the technique could be applied successfully to a great number of species.

EXERCISE

Experiment 1. Comparison of Long and Short Cutting Rootstocks for the Production of Standard Roses

MATERIALS

Each student or groups of students will need the following:

- Stock plants of standard roses sufficient to supply 20 cuttings (five each for four treatments) for each group of students. Note: This experiment can be adapted for other species including *A. pseudoplatanus, A. platanoides, Carpinus betulus, Ulmus* 'Regal', *Q. robur, T. cordata*, and *Pyrus, Malus*, and *Prunus.*
- Long side branches from stock plants of standard roses harvested between July and August. These should be current year's growth and about 150 to 250 cm long. Short side branches from stock plants of standard roses harvested between July and August. These should be current year's growth and about 30 to 60 cm long.
- Pots (2–3 L size) containing peat/sand (3:1, v/v), pH 4.5, with 2.5 kg slow-release fertilizer (Osmocote)
- Greenhouse with high-pressure fog, 10 to 20 sec every 2 to 5 min
- IBA 0.5% plus a fungicide active against powdery mildew
- Clippers for collecting cuttings
- Marking pen and labels

Follow the protocols listed in Procedure 19.1 to complete this experiment.

ANTICIPATED RESULTS

Rooting starts very quickly, and after 6 weeks, 80 to 100% of cuttings should be rooted. Rooted cuttings can be transplanted to the field or to 5 L containers set out in early September in the shade. The cuttings can remain in the rooting bed or in the containers in an unheated greenhouse during winter. Although not possible to complete in one semester, in August of the following year, the rootstocks can be budded in the same way that conventional

Procedure 19.1

Comparison of Long and Short Cutting Rootstocks for the Production of Standard Roses

Step	Instructions and Comments
1	Each group of students should collect five cuttings 150 to 250 cm in length from current year's shoots. Shoots as long as 350 cm in length can be used. Collect an equal number of more conventional cuttings 30 to 60 cm long for rooting and comparison to long shoots.
2	The terminal bud or growing tip should be retained, and only leaves from the basal 10 to 20 cm of the shoot should be removed (see variation on this experiment). Stock plants should be free of powdery mildew because rooting percentage is decreased by the fungus. Dip the bases of both the long and short shoots into an aqueous solution of 0.5% IBA containing a fungicide recommended to control powdery mildew. As a control group, dip an equal number of cuttings of both lengths in water only before sticking.
3	Stick cuttings into the rooting medium to a depth of 8 to 10 cm (several nodes). Pots (2–3 L) can be used, with one cutting per pot. The rooting medium should be a mixture of peat/sand (3:1, v/v), pH 4.5, with 2.5 kg slow-release fertilizer (Osmocote).
4	Place pots in a standard greenhouse under intermittent high-pressure fog, 10 to 20 sec every 2 to 5 min.
5	Examine the cuttings for rooting every 2 weeks, and then after rooting is evident, collect data on the percentage of cuttings rooted and the extent of rooting prior to transplanting the rooted cuttings to 5 L pots containing the same medium.
6	Variation: Remove the apical bud from one-half of the long and short cuttings, and then proceed from step 2 to determine what impact the retention of the apical bud has on rooting success.

rootstocks are, with 2 to 3 buds. After an additional year's growth, the plants are typically ready for sale.

QUESTIONS

- Which shoots root better—long or short?
- Do cuttings with terminal buds root better or worse than those without a terminal bud?
- What effect does IBA have on rooting?

LITERATURE CITED

Hartmann, H. T., D. E. Kester, F. T. Davies, R. L. Geneve. 2011. *Plant Propagation: Principles and Practices*, 8th Edition. Upper Saddle River, NJ: Prentice Hall, 880 pp.

Spethmann, W. 1998. Factors affecting rooting of difficult-to-root plants. *Comb Proc Intl Plant Prop Soc* 48:200–205.

20 Rooting Cuttings of Tropical Plants

Richard A. Criley

While many people think of tropical plants as something grown only in warm climates, many of them are grown as houseplants and for interiorscapes originated in tropical climates. The techniques used to propagate herbaceous and woody tropical plants are much the same as for similar plants of temperate zones. For a plant propagation class, however, the quantities of stock material may be limited in northern areas as a small conservatory may be the sole environment in which to grow the plants year round. Examples of tropical plants that may be used in the laboratory exercises are listed in Table 20.1.

Cutting types used in the propagation of tropical plants include terminal (or tip) cuttings of varying degrees of maturity (herbaceous, softwood, greenwood, intermediate wood, and hardwood), stem cuttings with no terminal bud (often called "cane cuttings"), and stem cuttings with one or two nodes or with opposite buds and the stem split between them. For the most part, cuttings will retain at least some foliage, but some foliage plant propagators will use single- or double-node cuttings with no attached leaf to save space in the propagation flat. Removal of some foliage from cuttings is used for the following reasons: (1) to save space in a flat, (2) to reduce the "umbrella" effect of large leaves over neighboring cuttings, (3) to reduce transpiring surface areas and minimize desiccation, (4) to ease the insertion of the cuttings into the medium, and (5) to eliminate foliage that might rot in the medium or reduce air circulation among densely planted cuttings. Leaves provide the photosynthate and hormones needed to initiate and develop roots.

CALCULATING THE ROOTING INDEX

While rooting can be evaluated by counting the number of roots, cutting them off and weighing them wet or dry, or measuring root lengths, another method is to calculate the rooting index (RI). This is an easy method and basically sorts the rooted cuttings into categories. Arrange a replicate of 10 cuttings from the largest (visually) to the smallest rooted cuttings, including those that are alive but have no roots and those that are dead. When you have done this for several replicates, it should become obvious what constitutes heavy, medium, or light rooting. The cuttings in these categories are each assigned weights: 5 for heavy rooting, 4 for medium rooting, and 3 for light rooting. Alive, but not rooted, cuttings receive a score of 2, and dead cuttings, a 1. The weighted values for the 10 cuttings are added together and divided by 10. This value is the RI for that replication. Since the number falls anywhere between 1 and 5, it can be related to the original category; a 4.3 RI, for example, suggests that most of the cuttings were medium to heavily rooted. If there are several replications of a treatment, the RI values for each are averaged together for a treatment RI. Treatment RI values can be compared to distinguish between treatment effects using statistics or represented graphically. Table 20.2 shows an example of RI calculations. If using a spreadsheet such as Excel, a formula can be entered into the RI column that will automatically calculate the RI. A column for percent rooted is also included.

Several exercises are offered that use different tropical plants, mostly from the foliage plant grouping of tropical plants, as tropical fruits and many woody ornamentals are less available in temperate climates. For each exercise, an objective is suggested that illustrates an important aspect of plant propagation. At a minimum, record the percent rooting, the root quality, and the time from sticking the cuttings to rooting.

EXERCISES

Experiment 1. Determination of the Position on the Stem Where Roots Develop Most Readily

For this exercise, choose plants that produce relatively long, unbranched stems so that you can select wood of differing maturities along the length of the stem (Figure 20.1). If the terminals are still rather soft and succulent and the lower portion of the stem has matured and thickened, you may find differences in both the extent and the quickness to root. It is, of course, problematic to compare a terminal against a stem cutting, but in this case, we are interested in the age of the wood as the principal feature. Rooting hormones can be used but may mask some of the differences in "rootability" among the different portions of the stem.

TABLE 20.1
Tropical Plant Materials for Various Cutting Propagation Exercises

Common Name	Botanical Name	1 or 2 Nodes	Position of Basal Cut	Response to Leaf Area	Maturity of Wood	Cane, Stem Pieces
Flowering maple	*Abutilon X hybridum*				X	
Beefsteak plant, Chenille plant	*Acalypha wilkesiana, A. hispida*		X	X		
Aglaonema	*Aglaonema* species and hybrids					X
Allamanda	*Allamanda cathartica*		X		X	
Anthurium	*Anthurium andraeanum*	X				X
Zebra plant	*Aphelandra squarrosa*	X				
Bougainvillea	*Bougainvillea spectabilis, B. glabra*	X			X	
Yesterday, today, and tomorrow	*Brunfelsia pauciflora*				X	
Grape ivy	*Cissus rhombifolia, C. discolor, C. antarctica*	X				
Glory-bower	*Clerodendrum thomsonae, C. X speciosum*		X		X	
Croton	*Codiaeum variegatum*	X		X		
Ti	*Cordyline fruticosa*					X
Crape ginger	*Costus speciosus*					X
Dumb cane	*Dieffenbachia maculata, D. amoena*					X
	Dischidia major	X				
Madacasgar dragon tree, Corn plant, Striped dracaena	*Dracaena marginata, D. fragrans, D. fragrans* 'Warneckii'					X
Song of India	*Dracaena reflexa (Pleomele reflexa)*	X				X
Pothos	*Epipremnum pinnatum*	X				
Poinsettia	*Euphorbia pulcherrima*		X			
India rubber tree	*Ficus elastic*	X			X	
Creeping fig	*Ficus pumila*				X	
Gardenia	*Gardenia augusta (G. jasminoides)*			X		
Caricature plant	*Graptophyllum pictum*	X	X	X		
Ginger	*Hedychium* species, *Zingiber* species	X				
Hibiscus	*Hibiscus rosa-sinensis, H. schizopetalus, H. tiliaceus*			X	X	
Wax plant	*Hoya carnosa*	X				
Ixora	*Ixora coccinea, I. chinensis*			X		
Arabian jasmine	*Jasminum sambac*		X			
Shrimp plant	*Justicia brandeegeana*		X			
Jacobinia	*Justicia carnea*		X			
Japanese privet	*Ligustrum japonicum*			X		
Mandevilla	*Mandevilla X amabilis, M. splendens*	X				
Prayer plant	*Maranta leuconeura*	X				
Stephanotis	*Marsdenia floribunda*	X			X	
Oleander	*Nerium oleander*		X			
Fire spike	*Odontonema cuspidatum*		X			
Passion flower	*Passiflora* species	X				
Peperomia	*Peperomia obtusifolia, P. crassifolia*	X				
Philodendron	*Philodendron scandens* var. *oxycardium*	X				
Aluminum plant, Creeping Charlie	*Pilea cadierei, P. nummulariifolia*	X				

(continued)

TABLE 20.1 (Continued)
Tropical Plant Materials for Various Cutting Propagation Exercises

Common Name	Botanical Name	1 or 2 Nodes	Position of Basal Cut	Response to Leaf Area	Maturity of Wood	Cane, Stem Pieces
Panax, Ming aralia, Balfour aralia	*Polyscias guilfoylei, P. fruticosa, P. scutellaria*					X
False eranthemum	*Pseuderanthemum carruthersii*	X	X	X		
Shingle plant	*Rhaphidophora celatocaulis*	X				
Vireya rhododendrons	*Rhododendron* species		X			
	Rhoicissus capensis	X				
Purple bignonia	*Saritaea magnifica*		X			
Octopus tree, Umbrella tree	*Schefflera actinophylla*					X
Dwarf brassaia	*Schefflera arboricola*	X	X		X	
False aralia (Dizygotheca)	*Schefflera elegantissima*				X	
Persian shield	*Strobilanthes dyerianus*	X	X			
Arrowhead vine	*Syngonium podophyllum*	X				
Cape honeysuckle	*Tecoma capensis (Tecomaria capensis)*		X			
Yucca	*Yucca elephantipes*					X

TABLE 20.2
Calculating the RI

Treatment	Replication	No. of Cuttings Set	No. of Cuttings: Heavy (×5)	Medium (×4)	Light (×3)	Alive (×2)	Dead (×1)	Sum of Weights	RI	% Rooted
Control	A	10				9	1	19	1.9	0
	B	10				10		20	2.0	0
	C	10				10		20	2.0	0
	Mean								1.97	0
1000 ppm	A	10	2	2	4	2		34	3.4	80
	B	10	2	1	3	2	2	29	2.9	60
	C	10	1	3	2	2	2	29	2.9	60
	Mean								3.07	66.7
3000 ppm	A	10	4	1	1	2	1	38	3.8	60
	B	10	3	2	5			38	3.8	100
	C	10	2	2	3	3		33	3.3	70
	Mean								3.63	76.7

MATERIALS

The instructor may reduce the quantity of material to be used if plant materials are limited. For each student or team of students, provide the following materials:

- 30 stems 2 ft or longer of species such as allamanda, bougainvillea, glory-bower, croton, Indian rubber tree, and hibiscus
- 3 flats of medium (coarse vermiculite or a 1:1 mixture of vermiculite and perlite)
- Pruning shears with which to cut the stems into pieces
- Labels
- Rooting hormone (if used, try 0.1% or 1000 ppm auxin concentration)

FIGURE 20.1 From tip to base, different carbohydrate and auxin gradients occur. Single-node cuttings of *Ficus longifolia* (or similar plant materials) can be rooted to show readiness to root and extent of rooting along such gradients.

Each flat will constitute one replication using similar parts of 10 stems. For example, a row of 10 terminals, a row of 10 greenwood subterminals, one or more rows of 10 intermediate-wood stem pieces, and a row of 10 basal stem pieces.

Follow the instructions in Procedure 20.1 to complete this experiment.

ANTICIPATED RESULTS

Speed of rooting, percent rooting, and the quantity of roots should show a gradient from tip to base, but they may also be optimum for midsections of the stem, tapering off toward the tip and the base.

QUESTIONS

- Where along the original long stem does your best rooting occur?
- Did the different portions of stem begin to root at different times?
- Is there a portion of the original stem that gave you the best results for root initiation and percent rooting?
- What internal factors are responsible for the good rooting that occurs? The poor rooting?
- How might you manage the stock plants to improve the amount of stems from which you can get good rooting?

Procedure 20.1
Position on the Stem Where Roots Develop

Step	Instructions and Comments
1	Place moistened medium in each flat to a depth of 7 to 8 cm and level the surface. Flats will be designated A, B, and C.
2	Divide the 30 stems into three groups of 10 stems so that each group is comparable to the others.
3	Determine how to divide a stem into pieces of differing maturity. Example: green terminal, next 10 cm, next 10 cm, and so on. Remove basal foliage that might be under the medium after insertion.
4	Gather the first 10 terminals and insert the basal 3 to 4 cm into the medium in one row in flat A. (If an auxin treatment is used, apply it to all cuttings.) Do the same for the 10 stem pieces that were just below the terminals; do the same for each set of 10. Repeat for the other two groups of 10 stems using flats B and C.
5	Label the rows in each flat: A-1, A-2, A-3, etc.; B-1, B-2, B-3, etc.; C-1, C-2, C-3, etc. Add the plant name, the propagator's name, and the date. Record the characteristics (e.g., diameter, degree of woodiness, stem or bark color) that distinguish the cuttings from the terminal to the base by their numbers into a notebook.
6	Place flats of prepared cuttings into your propagation area under intermittent mist or fog set to come on for 6 s every 6 min.
7	Evaluate the cuttings for evidence of root initiation about 2 to 3 weeks after planting them. When the earliest signs of rooting are observed, record the date and wait another 2 to 4 weeks before carefully removing them to evaluate the extent of rooting and the percent rooting. For each 10 cuttings, calculate the RI (Table 20.2) and the percent rooted.
8	Average the data for the three replications of each cutting type and represent these graphically.

Experiment 2. The Effect of the Position of Basal Cut on the Rooting from Stems

Some plant materials root readily in the middle of an internode, while others produce more roots when the basal cut is made just below or just above the node (Figure 20.2). For this exercise, it is best to choose plant materials with relatively long internodes so that a clear distinction can be made in the position of the basal cut (Figure 20.3a). Cuts will be made in the middle of an internode, just above a node, and just below a node (Figure 20.3b). The cuts should be squared off rather than slanted. The stem cuttings can either be terminals (tips) or stem pieces of several nodes. A variant of this experiment for plants with long internodes that do root in the internode is to examine the differences among different internode lengths.

FIGURE 20.2 Influence of the position of basal cut on rooting.

FIGURE 20.3 Select materials with long internodes, such as allamanda, to demonstrate the effect of basal cut location. (a) Single-node cuttings are made from a long stem. (b) Cut just above the node. Middle: cut at mid-internode; right: cut just below the node.

MATERIALS

The instructor may reduce the quantity of the materials to be used if plant materials are limited. Each student or group of students will need the following materials to complete this experiment:

- 90 cuttings will be required, 30 of each type (location) of basal cut from plant materials, such as *Acalypha*, allamanda, glory-bower, poinsettia, caricature plant, shrimp plant, oleander, coleus, *Pseuderanthenum*, *Schefflera arboricola*, or cape honeysuckle
- 3 flats of medium (coarse vermiculite or a 1:1 mixture of vermiculite and perlite)
- Pruning shears or sharp knife to make the basal cuts
- Labels

Each flat will have rows of 10 cuttings, each row representing a different position for the basal cut. Where more than one plant material is used, two kinds of plants can share a flat.

Follow Procedure 20.2 to complete this experiment.

ANTICIPATED RESULTS

Many of the suggested species will root anywhere along an internode, but some, such as poinsettia with its hollow stems, will root better with the basal cut made just below a node.

Cape honeysuckle is a plant that roots mainly at the node.

QUESTIONS

- Do the different sites for the basal cuts alter the speed or the extent of rooting?
- Where you find a difference in rooting between different positions of the basal cut, what factors do you think are responsible?
- What practical impact does making the basal cut in a certain position have on the propagation operation?

Procedure 20.2
Effect of the Basal Cut Position on Rooting from Stems

Step	Instructions and Comments
1	Place moistened medium in each flat to a depth of 7 to 8 cm and level the surface. Flats will be designated A, B, and C.
2	Select cutting materials such that similar stem lengths and diameters and leaf areas can be compared. Cuttings can be terminals or stem pieces, but the two types should not be mixed.
3	Make 30 cuttings in which the base of the cutting is made about 8 mm above a node. Make another set of cuttings with the base just below a node. A final set of cuttings is made with the base in the middle of the internode. (If an auxin treatment is used, apply it to all cuttings.) Except in the case of single- or double-node cuttings, your cuttings will be about 7 to 12 cm long. Remove any foliage that might be under the medium after insertion.
4	Insert 10 cuttings of each type in a row in each of the flats (A, B, and C).
5	Label the rows in each flat: A-1, A-2, A-3, etc.; B-1, B-2, B-3, etc.; C-1, C-2, C-3, etc. Add the plant name, the propagator's name, and the date. Record the characteristics (above or below the node cuts, middle of the internode) that distinguish the cuttings by their numbers into a notebook.
6	Place the flats of prepared cuttings into your propagation area under intermittent mist or fog set to come on for 6 sec every 6 min.
7	Evaluate the cuttings for evidence of root initiation about 2 to 3 weeks after planting them. When the earliest signs of rooting are observed, record the date and wait another 2 to 4 weeks before carefully removing them to evaluate the location, the extent of rooting in terms of root number and lengths, and the percent rooting. For each 10 cuttings, calculate the RI (Table 20.2) and the percent rooted.
8	Average the data for the three replications of each cutting type and represent these graphically.

Experiment 3. The Relationship between the Amount of Leaf Surface on the Cutting and the Ability to Root

There are two ways to approach the reduction in leaf surface: cutting leaves to 1/2, 1/3, or 1/4 their leaf areas or to remove the whole leaves. Leaf removal can be the removal of one leaf of each pair on opposite-leaved plants, the removal of upper leaves, or the removal of lower leaves. Cutting leaves may lead to leaf abscission because of injury to the leaf, which causes ethylene production, or from an opportunistic microorganism infection, so this approach should be reserved for large-leaved plant materials. The number or area of the remaining leaves will often influence the speed or the extent of rooting (Figure 20.4). Once plant material is on hand, the student should decide which approach is most appropriate and design an experiment that represents a gradient of leaf areas or counts. There should be 30 cuttings for each choice, with at least three choices representing, for example, all leaves remaining, 2/3 leaves (area) remaining, 1/2 leaves (area) remaining, 1/3 leaves (area) remaining, and no leaves remaining. Other options include the selective removal of top leaves, one each of a pair of opposite leaves, or basal leaves. Note that cuttings with zero leaves may fail to produce roots or rooting will be much delayed and a poor root system may result.

FIGURE 20.4 Influence of leaf area on the rooting of cuttings. Single-node croton cuttings with leaf blade intact or trimmed to 50% or 25% of the original area. Greater root production develops on intact cutting.

MATERIALS

The instructor may reduce the quantity of the materials to be used if plant materials are limited.

- Sufficient plant material that a minimum of 90 cuttings (30 for each treatment) can be produced
- 3 flats of medium (coarse vermiculite or a 1:1 mixture of vermiculite and perlite)
- Pruning shears or knives to cut leaves
- Labels

Procedure 20.3
Amount of Leaf Surface Remaining on the Stem and the Ability to Root

Step	Instructions and Comments
1	Place moistened medium in each flat to a depth of 7 to 8 cm and level the surface. Flats will be designated A, B, and C.
2	In this example using Japanese privet, all cuttings must be very uniform to start. For example, make cuttings with three pairs of leaves and a leafless basal node. Leaf size and area should be equivalent across all the cuttings.
3	Divide the cuttings into groups of 30. Leave all foliage on one set and insert 120 cuttings into each flat. Remove the basal pair of leaves on another 30 cuttings and insert 10 cuttings into each flat. Remove all except the top pair of leaves on another set of 30 cuttings and insert 10 cuttings into each flat. Another group of cuttings with no leaves or with one leaf per cutting could also be prepared and distributed similarly.
4	Label the rows in each flat: A-1, A-2, A-3, etc.; B-1, B-2, B-3, etc.; C-1, C-2, C-3, etc. Add the plant name, the propagator's name, and the date. Record the characteristics (full leaf, 2/3 leaves, 1/3 leaves, etc.) that distinguish the cuttings by their row numbers into a notebook.
5	Place the flats of prepared cuttings into your propagation area under intermittent mist or fog set to come on for 6 sec every 6 min.
6	Evaluate the cuttings for evidence of root initiation about 2 to 3 weeks after sticking them. When the earliest signs of rooting are observed, record the date and wait another 2 to 4 weeks before carefully removing them to evaluate the location, the extent of rooting, and the percent rooting. For each 10 cuttings, calculate the RI (Table 20.2) and the percent rooted.
7	Average the data for the three replications of each cutting type and represent these graphically.

Each flat will have rows of 10 cuttings, each row representing a different leaf area treatment. Where more than one plant material is used, two kinds of plants can share a flat.

Follow the instructions outlined in Procedure 20.3 to complete this exercise.

ANTICIPATED RESULTS

In general, better rooting is obtained with more foliage left on the cutting. However, some cuttings have sufficient reserves in the stem that they will develop roots even if totally defoliated, but this is usually accompanied by lateral budbreak and new shoot and leaf development. The shock of cutting off a portion of the leaf can lead to the rapid senescence of that leaf and the loss of its contributions to the rooting process. Large leaves can provide an "umbrella" over neighboring cuttings, leading to their dehydration. Too much foliage can also reduce air circulation in the cutting flat, leading to *Botrytis* (Chapter 12) or other diseases that attack the cuttings.

QUESTIONS

- Did you observe a difference in the speed or the extent of rooting that could be related to the leaf area?
- Do your results agree with the hypothesis that more rooting occurs with greater foliage retention? Why or why not?
- Would your results be the same if you removed the upper rather than the lower leaves? Why or why not?
- What impact does the practice of reducing the foliage (the number of leaves or the leaf area) have on the propagation operation?
- If you compared similar leaf areas from a leaf removal treatment and a leaf cutting treatment, were the results similar? Why or why not?

Experiment 4. Comparison of the Effectiveness of Single- and Double-Node Cuttings on Rooting

A number of foliage plants with a vining habit of growth are propagated by single- or double-node cuttings (Figure 20.5). The technique is not limited to vines, however, and many woody tropicals can be propagated by single nodes, leaf-bud cuttings, or split-stem cuttings (Figures 20.6 and 20.7). The retention of a leaf also improves rooting, but on double-node cuttings, the basal leaf is usually removed. Recently matured stems are preferred to younger or older cutting materials. The time to produce a saleable plant is often longer with small

FIGURE 20.5 Some tropical vines have preformed roots at their nodes and can be propagated by single- or double-node cuttings. A rooted double-node cutting of *Scindapsus exotica*.

FIGURE 20.6 Two cuttings can be made by splitting the stem at the node of opposite-leaved plant materials. *Sanchezia speciosa* (Acanthaceae) is representative of a tropical plant easily propagated by split-node cuttings.

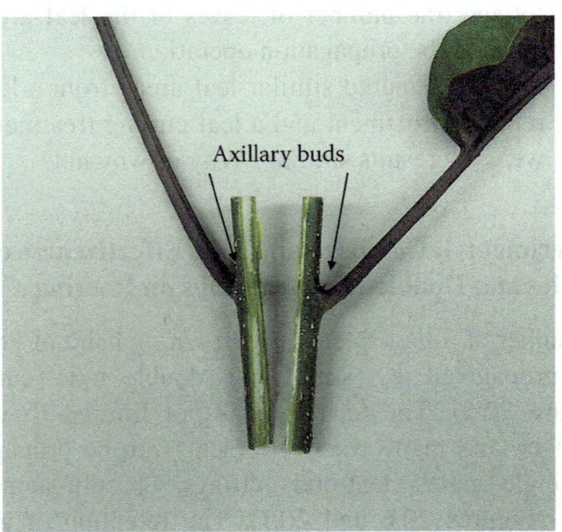

FIGURE 20.7 Opposite-leaved plants can often be propagated by splitting the stem at the node to include the axillary bud and the leaf blade.

starting units. The stem length of the cutting unit is usually 2.5 to 5.0 cm. Cuttings are inserted into the medium so that the bud is just below or at the medium surface as rooting often occurs at the base of the new shoot.

MATERIALS

The instructor may reduce the quantity of the materials to be used if plant materials are limited.

- Sufficient plant material (such as philodendron, pothos, *Syngonium*, grape ivy, Indian rubber tree, caricature plant, *Hoya*, mandevilla, stephanotis, passion fruit, coleus, *Pseuderanthemum*, and *Schefflera arboricola*) is needed to provide 30 uniform cuttings of each type
- 3 flats of medium (coarse vermiculite or a 1:1 mixture of vermiculite and perlite)
- Pruning shears or knives to prepare the cuttings
- Labels
- Rooting compounds (concentration ranges of 0.1–0.3% or 1000–3000-ppm auxin)

Each flat will have rows of 10 cuttings, each row representing a different type of cutting. Three flats are used so that three replications of 10 of each type of cutting can be produced. Where more than one plant material is used, two or more kinds of plants can share a flat.

Follow the instructions in Procedure 20.4 to complete the experiment.

ANTICIPATED RESULTS

In general, nodes with a leaf left on will root faster and with more roots than leafless cuttings. Cuttings with two nodes, the bottom one in the medium, will often root faster than single-node cuttings. Cuttings with a complete piece of stem will retain more reserves to support rooting than leaf-bud cuttings with only a sliver of stem. Split-node (split-stem) cuttings (Figures 20.6 and 20.7) may rot because of the exposed tissue, but where successful, they double the number of plantlets that can be obtained from a stem.

QUESTIONS

- Does rooting occur from the original stem or from the base of the elongating shoot?
- Does node count make a difference in your results? Why or why not?
- Does the presence of a leaf on the cutting make a difference in the results? Why?
- If you used split-stem cuttings (stem split between a pair of opposite leaves), where did the roots develop?

Procedure 20.4

Comparison of the Effectiveness of Single- and Double-Node Cuttings on Rooting

Step	Instructions and Comments
1	Place moistened medium in each flat to a depth of 7 to 8 cm and level the surface. Flats will be designated A, B, and C.
2	**Vine example**: Cut a nephthytis, pothos, or philodendron vine into 30 pieces about 4 cm long, containing a leaf with one node. Prepare a similar set of 30 cuttings, but remove the leaf. Prepare a third set of double-node cuttings and remove the lower leaf. Prepare a fourth set of double-node cuttings from which you remove both leaves. For all cuttings, leave at least 18 mm of stem below the bottom node. This end will be inserted into the medium.
	Stems with opposite leaves: Cut a caricature or coleus plant stem into 30 single-node and 30 double-node units. The length of the stem should be about 1.2 cm above the top node and about 2.5 cm below the bottom one. A third batch of single-node cuttings will be made and split down the middle between the two leaves. It may be necessary to cut the leaves in half transversely for the efficient use of space. Treat the basal ends with auxin.
	Stems with alternating leaves: Cut croton or India rubber tree stems above and below the node such that the axillary bud and leaf have 8 to 12 mm of stem above and below them. Treat the basal end with auxin. (The large leaves of the India rubber tree can be rolled under and held with a rubber band. A wooden skewer or small bamboo stake in the middle can be used to hold the leaf upright when inserted into the medium.)
3	**Vine**: Insert the stem into the medium so that the bud (or a lower bud) is just at or below the surface of the medium.
	Stems: Insert the stem into the medium so that the leaf–stem juncture is just above the medium surface.
4	Label the rows in each flat: A-1, A-2, A-3, etc.; B-1, B-2, B-3, etc.; C-1, C-2, C-3, etc. Add the plant name, the propagator's name, and the date. Record the characteristics that distinguish the cuttings by their row numbers into a notebook.
5	Place the flats of prepared cuttings into your propagation area under intermittent mist or fog set to come on for 6 sec every 6 min.
6	Evaluate the cuttings for evidence of root initiation about 2 to 3 weeks after sticking them. When the earliest signs of rooting are observed, record the date and wait another 2 to 4 weeks before carefully removing them to evaluate the extent of rooting and the percent rooting. For each 10 cuttings, calculate the RI (Table 20.2) and the percent rooted.
7	Average the data for the three replications of each cutting type and represent these graphically.

- What is your opinion of the value of the practice of single-, double-, or split-node cuttings in a nursery operation?

Experiment 5. The Effects of Length and Orientation on the Rooting of Cane Cuttings

A number of foliage plants are propagated from long, leafless stem pieces known as "canes" (Figure 20.8). For the most part, these are straight stems, but some plant materials have been marketed in spirals or other contortions. Cane pieces can be as short as a single node or as long as up to 1.5 m in length. Since the axillary buds have not begun to push, rooting is often delayed until the hormones produced by the buds are available and translocated to the rooting zone. Succulent materials, such as *Dieffenbachia*, require a period of time to suberize and form a protective periderm layer over the cut surface before being placed in a moist medium. The ends of cane pieces of *Yucca* and *Dracaena* from offshore producers are often waxed to prevent desiccation, but this also prevents the uptake of moisture needed for rooting, so recutting the base is necessary. Making vertical cuts above the base permits root development further up the stem to provide better anchorage after potting. Length and diameter influence budbreak and rooting by virtue of the moisture and carbohydrate contents of the cane. Short cane pieces may be laid on their sides, nearly covered in medium with

FIGURE 20.8 *Dracaena marginata* is a classical example of a woody tropical plant propagated by stem pieces called "canes." A slant cut is made at the base to indicate the "down" end of the cutting. Cuttings can be inserted vertically in or laid horizontally on the medium. Roots develop at the basal end, and buds and shoots develop toward the original apical end, no matter which placement orientation is used.

a prominent bud on the upper side at the physiologically apical end. Even if the propagator does not recognize which end of the stem is basal, the cane will always root at the basal end because of polarity. Thus, the orientation of sticking the cutting is important for vertical insertion, but it is of less concern for the horizontal placement of the cane piece. If terminals are available, compare the results of cane piece versus similar-length terminals.

MATERIALS

The instructor may reduce the quantity of the materials to be used if plant materials are limited. The following materials should be provided for each student or team of students:

- Cane material (*Aglaonema*, *Anthurium*, ti, *Dieffenbachia*, *Dracaeana* species, *Polyscias* species, octopus tree, and *Yucca*)
- Flats of medium (coarse vermiculite or a 1:1 mixture of vermiculite and perlite)
- Pruning shears or knives to prepare the cuttings
- Labels
- Rooting compounds (concentration ranges of 0.1–0.3% or 1000–3000-ppm auxin).

Follow the instructions in Procedure 20.5 to complete the experiment.

ANTICIPATED RESULTS

While long, thick cane pieces might be expected to produce the most roots because of the amount of reserve nutrients in the cane, the base can also be hard to root because of little parenchyma. Cane pieces from the middle portion of the stem may root more quickly than those from the base. Whether a cane piece is laid horizontally or placed vertically in the medium, the roots will always

Procedure 20.5

Effects of Length and Orientation on the Rooting of Cane Cuttings

Step	Instructions and Comments
1	Place moistened medium in each flat to a depth of 7 to 8 cm and level the surface. Flats will be designated A, B, and C. (Depending on the availability of materials, three flats may not be needed.)
2	Cut pieces of dracaena or ti cane into 10-, 20- or 30-cm lengths, keeping in mind which end is basal and which is distal. Longer cane pieces can be used if available. (One useful technique uses a sloping cut at the base to indicate the end to be inserted into the medium.)
3	One-half of the cane pieces of each length will be inserted vertically into the medium, and one-half will be pressed horizontally into the medium so the top edge is barely out of the medium. On long cane pieces, make 5- to 7.5 cm longitudinal cuts above the base. If diameters differ among canes, distribute similar diameters equally between the two orientations.
4	Label the rows in each flat: A-1, A-2, A-3, etc.; B-1, B-2, B-3, etc.; C-1, C-2, C-3, etc. Add the plant name, the propagator's name, and the date. Record the characteristics that distinguish the cuttings by their row numbers into a notebook.
5	Place the flats of prepared cuttings into your propagation area under intermittent mist or fog set to come on for 6 sec every 6 min.
6	Evaluate the cuttings for evidence of root initiation about 2 to 3 weeks after planting them. When the earliest signs of rooting are observed, record the date and wait another 2 to 4 weeks before carefully removing them to evaluate the extent of rooting and the percent rooting. For each 10 cuttings, calculate the RI (Table 20.2) and the percent rooted. If the time permits, evaluate shoot emergence, which is often later than rooting.
7	Average the data for the three replications of each cutting type and represent these graphically.

emerge from the basal end, while budbreak and new shoots will develop at the distal end (the original top) of the cane. Terminals of canes usually retain foliage that aids in rooting, but differences can be observed when the terminal tissue is too soft (it usually rots); stem tips of such material need some recently matured woody tissue at the base.

QUESTIONS

- Does cane length or diameter make a difference in the speed or the extent of rooting? Why?
- When you compared the horizontal and the vertical insertion of the cane pieces, what differences in root and shoot development did you observe?
- Does cane size affect the number of buds that develop into shoots?
- What is the practical value of using canes for propagation when terminals often root more readily?
- When you compared cane pieces versus terminals, what were the differences in the speed and the extent of rooting?

Experiment 6. Rooting Cuttings in Water

A common practice among hobbyists is to stick the stem of a plant that they wish to root into a glass of water, adding or changing the water periodically. For plants with which this approach is successful (see cane plants in Exercise 20.5), it represents one of the most simple propagation experiences. However, the root systems that develop usually do not have the fine feeder roots and root hairs that are found in plants growing in a porous solid medium. Indeed, when transplanted to soil, as amateurs often do with their cuttings rooted in water, the water roots may die instead of adapting to the new environment. The roots produced in water are adapted to a different oxygen environment and do not have the stimulus

Procedure 20.6

Rooting Cuttings in Water

Step	Instructions and Comments
1	Prepare cardboard or Styrofoam™ caps large enough to sit on top of each jar and use the knife to cut a hole large enough to support the stem of the cutting that will be inserted into it. You will need enough jars so that each cutting has its own jar. The number of jars will depend on which of the following treatment factors (2A, 2B, or 2C) are evaluated.
2A	To evaluate the difference in rooting ability among several species, have each student or group of students take 10 cutting from each of at least three species of plants listed above. Place each cutting into its own jar and label with the genus and species name of the plant. Note: Adventitious rooting response to water temperature can also be evaluated using the same three species by placing an equal number of cuttings in a water bath with a temperature set 5° above ambient temperature and another one set at 5° below ambient temperature. Evaluate the cuttings for evidence of root initiation about 2 to 3 weeks after placing them in the jars. When the earliest signs of rooting are observed, record the date and wait another 2 to 4 weeks before carefully removing them to evaluate the extent of rooting and the percent rooting. For each 10 cuttings, calculate the RI (Table 20.2) and the percent rooted. Average the data for the three replications of each cutting type and represent these graphically.
2B	To evaluate the effect of additives to the water on the speed or the extent of rooting, choose at least two additives to evaluate in addition to the treatment rooted in the jar containing water only. Use cuttings from one species and place one cutting into each jar. Each student or group of students will replicate this three times. Label each jar with what it contains. Observe the rooting and collect data as in 2A.
2C	Prepare pots with at least three different medium types, for example, potting soil, sand, and perlite. Using cuttings that have been rooted in jars containing water, each student or group of students will transfer cuttings to pots containing the three different media. Three weeks after transfer to pots containing various media, students will collect data on cutting survival, how many are actively growing, and the extent of growth.
3	Students should share data for each of the experimental variations described above, and if there were differences among students or student groups, discuss possible reasons for those differences.

of pushing against solid particles and through pore space. Komissarov (1969) found that 20 out of 30 species rooted as well in water as in sand. In this exercise, several variants are possible. These include allowing new roots to develop to various lengths before transplanting to a solid medium and observing how successful the new plant is to adapting and continuing growth; transplanting the water-rooted cuttings into media of difference porosities and densities (e.g., soil, perlite, mixtures of peat and perlite in different ratios, and others); and adding fertilizers or different stimulating substances to the water during rooting to see the effect on root development. A variant of interest for science fair projects could be to use water of different temperatures or salinities.

MATERIALS

The instructor may reduce the quantity of the materials to be used if plant materials are limited. The following materials should be provided for each student or team of students:

- Cane material (*Aglaonema*, *Anthurium*, ti, *Dieffenbachia*, *Dracaeana* species, *Polyscias* species, octopus tree, and *Yucca*)
- Plastic or glass jars (18-oz jars are a good size) to hold water
- Cardboard or Styrofoam™ with holes cut to support the stem
- Knife or pruning shears to prepare the cuttings
- Labels
- Water (tap, deionized). You may wish to characterize its pH or electrical conductivity
- Pots, appropriately sized to the root system and cutting
- Potting media such as soil, coarse sand, peat, perlite, and coir
- Rooting stimulants: liquid rooting compounds, SuperThrive, algal extracts, soluble fertilizers, or compost tea
- Hot water bath, thermometer—to evaluate the optimum temperature for rooting

Follow the instructions in Procedure 20.6 to complete this experiment.

ANTICIPATED RESULTS

Roots formed in the water environment may not adapt successfully when the rooted cuttings are transplanted to pots containing various media. In some cases, there may be a dieback of the root system, and new roots may form quickly, which are adapted to the new medium. Water containing rooting stimulants and/or nutrient sources may enhance the growth and development of the adventitious root system, resulting in greater length and more branching. If a range of temperatures are compared to determine an optimum for adventitious root formation, the temperatures above and below the optimum at which rooting is negatively impacted can be determined. This will most likely be evidenced by a reduced percentage of rooting and a lack of root extension growth.

QUESTIONS

- How long does it take for roots to appear?
- Where do the roots occur along the stem? Is callus produced?
- Are the water roots initially coarse, stubby or finely divided, branched, or bearing root hairs?
- After a period of time in water, do the roots change their character?
- After potting into solid media, does the plant appear to be retarded in initiating new growth?
- After potting into solid media, how long does it take for roots to reach the pot wall?
- Is this approach to rooting of cuttings likely to be adopted for commercial nursery practice? Why or why not?

LITERATURE CITED AND SUGGESTED READING

Auld, R. E. 1988. The basics of propagating bougainvillea. *Comb Proc Intl Plant Prop Soc* 36:211–213.

Avery, J. D. and C. A. Beyl. 1986. Caliper of semi-hardwood cutting influences footing of kiwi fruit. *The Plant Propagator* 32:5–7.

Calma, V. C. and H. W. Richey. 1930. Influence of amount of foliage on rooting of coleus cuttings. *Proc Amer Soc Hort Sci* 27:457–462.

Chadwick, L. C. 1949. The influence of the position of the basal cut on the rooting and arrangement of roots on deciduous softwood cuttings. *Proc Amer Soc Hort Sci* 53:567–572.

Conover, C. A. 1994. Storage temperature and duration affect propagation of *Dracaena fragrans* 'Massangeana.' *Foliage Digest* 17:1–4 (see also *Proc Fla State Hort Soc* 104:331–333).

Conover, C. A. and R. T. Poole. 1993. Propagation of *Dracaena fragrans* "Massangeana" affected by cane position on stock plants. *Foliage Digest* 16:1–3.

Davies, F. T., Jr., and J. N. Joiner. 1978. Adventitious root formation in three cutting types of *Ficus pumila* L. *Comb Proc Intl Plant Prop Soc* 28:306–313.

Griffith, L. P., Jr. 1998. *Tropical Foliage Plants: A Grower's Guide*. Batavia, Illinois: Ball Publishing, 318 pp.

Henley, R. W. 1979. Tropical foliage plants for propagation. *Comb Proc Intl Plant Prop Soc* 29:454–467.

Higaki, T. 1981. Single-node propagation of *Dracaena goldieana*. *The Plant Propagator* 27:8–10.

Jackson, H. C. 1981. Propagation of *Dieffenbachia* by node cuttings. *Comb Proc Intl Plant Prop Soc* 31:248–249.

Komissarov, D. W. 1969. Biological Basis for the Propagation of Woody Plants by Cuttings. Israel Program for Scientific Translations, Jerusalem, Israel.

Lane, B. C. 1987. The effect of IBA and/or NAA and cutting wood selection on the rooting of *Rhaphiolepis indica* "Jack Evans." *Comb Proc Intl Plant Prop Soc* 37:77–82.

Larkman, B. 1988. Single-node vs. double-node cuttings for the propagation of *Pyrostegia venusta*, *Hardenbergia violacea* "Happy Wanderer" and *Clytostoma callistegiodes*. *Comb Proc Intl Plant Prop Soc* 38:106–109.

Lee, C. W. and D. A. Palzkill. 1984. Propagation of jojoba by single-node cuttings. *HortScience* 19:841–842.

Loach, K. 1985. Rooting of cuttings in relation to the propagation medium. *Comb Proc Intl Plant Prop Soc* 35:472–485.

Marlatt, R. B. 1969. Propagation of *Dieffenbachia*. *Econ Bot* 23:385–388.

McConnell, D. B. 1983. Students learn by evaluating commercial foliage propagation techniques. *Foliage Digest* 6:4–5.

Morgan, J. V. and H. W. Lawlor. 1976. Influence of external factors on the rooting of leaf and bud cuttings of *Ficus*. *Acta Hort* 64:39–46.

Neel, P. L. 1979. Macropropagation of tropical plants as practiced in Florida. *Comb Proc Intl Plant Prop Soc* 29:468–480.

Oka, S. 1979. The effect of IBA and NAA on rooting cuttings of selected *Dracaena* species and cultivars. *The Plant Propagator* 25:11–12.

O'Rourke, F. L. 1944. Wood type and original position on shoot with reference to rooting in hardwood cuttings of blueberry. *Proc Amer Soc Hort Sci* 45:195–197.

Petersen, T. H. 1981. Propagation of philodendrons from node cuttings. *Comb Proc Intl Plant Prop Soc* 31:219–220.

Poole, R. T. and C. A. Conover. 1982. Propagation of spineless yucca. *Foliage Digest* 5:14–15.

Poole, R. T. and C. A. Conover. 1984. Propagation of ornamental *Ficus* by cuttings. *HortScience* 19:120–121.

Poole, R. T. and C. A. Conover. 1987. Vegetative propagation of foliage plants. *Comb Proc Intl Plant Prop Soc* 37:503–507.

Poole, R. T. and C. A. Conover. 1990. Propagation of *Ficus elastica* and *Ficus lyrata* by cuttings. *Foliage Digest* 8:3–5.

Poole, R. T., A. R. Chase, and L. S. Osborne. 1986. *Yucca*. *Foliage Digest* 16:4–6.

Poole, R. T., A. R. Chase, and L. S. Osborne. 1991. *Schefflera*. *Foliage Digest* 16:4–8.

Poole, R. T., A. R. Chase, and L. S. Osborne. 1993. *Dracaena*. *Foliage Digest* 16:1–6.

University of Florida, Mid-Florida Research & Education Center Website for foliage. http://mrec.ifas.ufl.edu/foliage/folnotes.asp.

Wang, Y.-T. 1987. Effect of warm medium, light intensity, BA, and parent leaf on propagation of golden pothos. *HortScience* 22:597–599.

Part VII

Propagation by Leaf and Root Cuttings

21 Adventitious Shoot and Root Formation on Leaf and Root Cuttings

Caula A. Beyl

Concept Box 21.1

Root Cuttings

- Root cuttings are a useful means of propagating species that do not come true from seed or root poorly when stem cuttings are used. They are also useful for obtaining clones of trees that have been selected for propagation based on mature characteristics such as flowering, fruiting, or sex, but have lost the ability to root easily from stem cuttings.
- Many hardwood species that exhibit natural suckering are capable of being propagated by root cuttings.
- In general, thicker cuttings of 5 to 15 cm in length that have adequate carbohydrate reserves are more successful for root cutting propagation, with a greater percentage of rooting and more vigorous shoots.
- Most root cuttings typically have a definite "window of opportunity" for success and should be taken from October to March during the dormant season. Cuttings taken during summer usually do not develop shoots and roots well.
- The polarity of a root cutting is important and which end of the cutting was nearest the crown could easily be confused. This is alleviated by using a straight cut across the cutting at the proximal end (nearest the tree trunk) and a slanted cut across the cutting at the distal end (farthest from the trunk).
- New shoots arise from the proximal end, and new roots, from the distal end of the root cutting or from the base of the new shoot.

Leaf Cuttings

- Some plants can be propagated by cuttings made from detached leaves, without any part of the original stem or the axillary bud being used.
- Plants like African violet (*Saintpaulia ionantha* Wendl.) are propagated using the leaf, with the petiole still attached. Other species, like jade plant (*Crassula ovata* [Miller] Druce) or burro's tail (*Sedum morganianum* E. Walther), have leaves that can be separated from the parent plant and merely inserted as they are into the rooting medium.
- With some species, leaves are cut into pieces, sections, or strips to prepare them as cuttings. Examples are *Sansevieria* (snake plant) and *Streptocarpus* (cape primrose).
- Most plants propagated by leaf cuttings regenerate roots from secondary meristems induced as part of the wounding response. Other species, such as piggyback plant (*Tolmeia menziesii*), mother of thousands (*Bryophyllum daigremontianum* [Raym.-Hamet & E. P. Perrier] Berger), and walking fern (*Asplenium rhizophyllum* L.), develop plantlets from primary meristems that are preformed.
- Cuttings that are only pieces of the original leaf will work for some plants, such as Rex begonia (*Begonia rex* Putz.), snake plant (*Sansevieria trifasciata* Hort. ex. Prain), and streptocarpus (*Streptocarpus formosus* Hilliard & Burt). Begonia leaves can be cut into squares about 3 to 5 cm on the side and containing a vein. *Sansevieria* and *Streptocarpus* leaves can be cut transversely into strips.
- Roots are formed more easily on leaf cuttings than shoots. With some species, roots will form, but shoots may never develop. Examples are *Ficus elastica* Roxb. (rubber tree) and some *Lilium* species.

In conventional vegetative plant propagation, much attention is focused on the use of various types of stem cuttings, with much less attention paid to the use of structures such as roots and leaves. For many species, particularly hardwood trees, some fruit species, and some herbaceous perennials, root cuttings present a viable alternative to seed or stem cutting propagation. Leaf cuttings also receive less attention than stem cuttings for the propagation of tropical foliage plants, but for those plants that root easily from leaves, either with or without petioles, leaf cuttings remain a useful technique for obtaining additional plants. There are many surprising ways in which plants can be successfully propagated.

ROOT CUTTINGS

Species that naturally sucker and form clumps of trees that are clonally related generally make good candidates for propagation from root cuttings. Examples of genera that tend to form large clonal stands are *Acacia*, *Ailanthus*, *Diospyrus*, *Populus*, *Quercus*, *Robinia*, *Salix*, *Sassafrass*, *Ulmus*, and *Rhus*. Clonal stands typically arise when the original seedling tree dies. The term used to describe the original seedling tree is "genet." The cluster of stems that arise asexually from the surviving root system are the "ramets." One male clonal stand of *Populus tremuloides* Michx. (quaking aspen) consisted of more than 47,000 ramets spreading across more than 43 hectares (107 acres) (Figure 21.1). Clearly, these clonal stands arising from suckering at the root crown can become very large. Plant propagators have taken advantage of this natural ability to regenerate new plants from roots or root pieces in these species and others including many woody and herbaceous perennials.

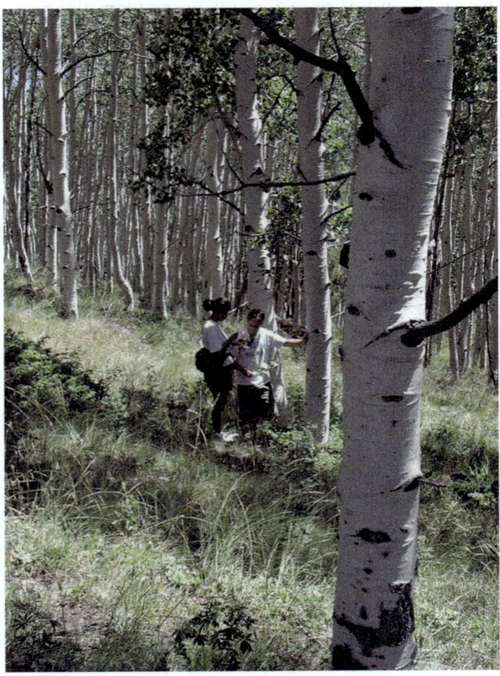

FIGURE 21.1 Male aspen clone in Utah composed of more than 47,000 stems of genetically identical *Populus tremuloides* that developed as a consequence of prolific suckering. The clone was named "Pando," which, when translated, means "I spread" in Latin. (Courtesy of Jeff Mitton.)

ADVANTAGES AND LIMITATIONS OF USING ROOT CUTTINGS

Root cuttings are most advantageous for selections of plants that cannot be produced from seed or are difficult to root using other types of cuttings. Seedlings of many species display considerable variability, making it desirable to select and clonally propagate individual plants with desirable characteristics. A variety of techniques have been discussed in earlier chapters to clonally propagate elite selections, but many times, plants are recalcitrant (uncooperative) to these methods and fail to root successfully. Root cuttings often succeed where other methods have low or nonexistent success rates.

Many species are difficult to root from stem cuttings once they have gone beyond their juvenile period. Native American pawpaw (*Asimina triloba* [L.] Dunal) is a good example. Stem cuttings taken from seedlings up to 2 months old will root, but beyond that time, this species becomes progressively more difficult to almost impossible to root from cuttings from mature trees. Root cuttings, however, generally work well. For other species, shoot cuttings derived from the aerial portion of the tree will not root, but if they are forced from root cuttings, they root more easily. For example, mature silk-tree or mimosa (*Albizia julibrissin* Durz.) is extremely difficult to root from shoot cuttings, but if shoots are harvested from root cuttings, these can be rooted easily. Although these new shoots have the bipinnate leaf form of the mature tree, they will root even without application of indolebutyric acid (IBA). They cannot be said to be "juvenile," but they do exhibit "juvenile characteristics." With other species, such as tree of heaven (*Ailanthus altissima* [P. Mill.] Swingle), only the female tree is a desirable ornamental since the flowers on male trees smell foul. Unfortunately, once the tree has flowered and the sex of the tree can be identified from its flowers, the tree is mature and hard to propagate from stem cuttings. Seeds of *Ailanthus* germinate readily, but since the sex of the seedlings is unknown, it is far better to propagate this species by taking root cuttings from female trees. Other species such as shining sumac (*Rhus copallina* L.), smooth sumac (*R. glabra* L.), and staghorn sumac (*R. typhina* L.) have seeds that germinate slowly or inconsistently, but will propagate successfully from root cuttings. The cutleaf cultivars of *R. typhina* like 'Lacinata' will not root from stem cuttings.

The use of root cuttings offers other advantages in that only simplistic technology is needed for success. Very little expertise is involved in collecting root cuttings compared to performing other propagation techniques such as grafting, so skilled labor is not a requirement. Root cuttings are also easily transported and stored. The requirements of the facility for producing root cuttings successfully are also much less elaborate than those needed for rooting softwood stem cuttings, which require a mist or high-humidity system to prevent desiccation.

One distinct disadvantage is the difficulty of obtaining large numbers of cuttings. Typically, cuttings are taken when plants are lifted from the field or by digging and harvesting roots while they are still in place; however, this is often the time when nursery operations slow down. For unknown reasons, the success rate of root cuttings varies from year to year, particularly with some species such as *Paulownia*. Predicting the numbers of finished liners that will be produced by taking root cuttings from a certain number of stock plants is often difficult, and experience may vary from location to location and year to year.

Root cuttings are not a good technique to propagate plants that are chimeras (Chapter 7) because the origin of shoots is from tissue that may have a different genotype than the one desired. This could be a problem with some plants. Thornless blackberries are an example of a periclinal chimera, where the mutation occurred in the outermost layer of the developing apical meristem. This layer produces the epidermis, which, in thornless blackberry, does not have "thorns," but the shoots produced from root cuttings have the thorny genotype. Root cuttings are also not a good choice for propagating a plant that has been grafted, unless you wish to propagate the rootstock.

OPTIMIZING SUCCESS WITH ROOT PROPAGATION

Stock Plants

Healthy, robust stock plants with adequate carbohydrate reserves stored in root tissues are good sources of root cuttings. In *Paulownia*, secondary roots branching off primary laterals were as good a source for new tissues to develop as primary lateral roots. Roots may be harvested from bare root plants that have been collected and placed in cold storage prior to being sold if care is taken not to take too many cuttings from each plant. Another easily harvested source of root cuttings is containerized plants, particularly if the container is large enough for good growth of the root system, without the contortions associated with being root bound.

In nurseries where some species are routinely propagated by root cuttings, stock plants are lined out specifically for this purpose. They are lifted or dug annually, and their roots are harvested. A piece of equipment that does this job nicely is a spring-trip cultivator with shanks having long, narrow bull-tongue–type shovels. They are then replanted in the same beds or in nearby sites. If the amount of root development is not enough to allow annual collection of roots, a biennial cycle of collection and replanting may be followed. Problems occur when extensive root systems develop, making it difficult to distinguish which roots belong to which plant in the nursery, and sometimes, the root cuttings left behind become a new "weed problem" when they develop shoots and grow.

Although juvenile plants or trees are better sources for obtaining root cuttings, sometimes, the plant to be propagated is a rare or an established mature tree. In that case, taking root cuttings from the juvenile zone closest to the crown (Chapter 6) optimizes the chances for success and allows the propagator to ensure that the roots being taken actually belong to the desired tree.

Size of the Root Cutting

In general, the thicker the cutting, the more likely it is to produce shoots and subsequent roots. The largest cuttings produce the most vigorous root and shoot development. No shoots form on pawpaw root cuttings with diameters less than 5 mm. For royal paulownia (*Paulownia tomentosa* [Thunb.] Sieb. & Zucc. ex Steud.), thicker cuttings with diameters between 10.1 and 15.0 mm produced the most shoots, and those with diameters between 5.1 and 10.0 mm produced the tallest shoots. *Paulownia taiwaniana* Hu & Cheng cuttings with diameters less than 9 mm regenerated poorly no matter what the cutting length. The effect of the diameter of the root cutting on its likelihood for success may be related to its having adequate carbohydrate reserves to support meristem growth and shoot development. The size of the root cutting may also have an impact on how long it takes before shoots appear. In black locust (*Robinia pseudoacacia* L.), shoots appeared on root cuttings with diameters 5 mm and above in as little as 15 days. With fall-flowering *Anemone* × *hybrida*, cutting size had no effect on the initiation of new shoots and roots nor on how much time was required for that regeneration, but it did have an impact on the size of the plant ultimately resulting from the root cutting. Larger cuttings resulted in larger plants.

Polarity

Tissues of root cuttings exhibit an important trait—polarity. Each end of the cutting behaves differently, with new shoots nearly always arising at the proximal end of the cutting (the end nearest the main stem or trunk), and roots, at the distal end (the end farthest from the main stem or trunk) no matter how the cutting is oriented in the rooting medium. This polarity occurs no matter what the

FIGURE 21.2 Root cuttings of oakleaf hygrangea (*Hydrangea quercifolia* Bartr.), illustrating the horizontal cut made at the end nearest the crown and the slanting cut made at the end farthest from the crown of the plant.

original orientation (horizontal, diagonal, or vertical) of the root was on the original root system. Root polarity is very strong in the lateral direction (along the length of the cutting), but there is no corresponding transverse (across the root section) polarity since shoots will arise from all positions around the circumference of the cutting at the proximal end. To keep track of polarity, when root cuttings are taken, the proximal end is cut perpendicular to the root, and the distal end is separated with a slanting cut so that the two ends can be differentiated (Figure 21.2). Auxins are transported from the shoot apices in a polar manner and regulate development, with high concentrations promoting roots and suppressing shoots.

Influence of Juvenility

It has long been recognized that the closer to the trunk that the root cuttings are taken, the more likely they are to regenerate successfully. This is due to a phenomenon called "juvenility," which is discussed more thoroughly in Chapter 6. The early life of some plants, particularly woody plants, is the juvenile phase, characterized by rapid growth, the ability to root easily, and sometimes, even morphological differences such as plant form and leaf shape. Juvenile tissues, although they are the oldest chronologically, are the youngest physiologically. The "cone of juvenility" encompasses the tissues of the tree closest to the crown or the transition zone between shoot and root (Figure 6.3). For the root system, some authors depict the cone of juvenility to be maximal at soil level and decreasing to a certain soil depth. The shape and depth of the juvenile zone may vary from species to species. When the plant becomes capable of reproduction, it is deemed mature. Mature plants have slower growth rates, root more reluctantly or not at all, and may have different morphologies and leaf shapes. Juvenility can be taken into account either by choosing to take cuttings from younger plants or by taking cuttings as close to the crown as possible.

The profound effect of juvenility as a function of the distance from the crown that the root cutting is taken is usually seen in the percentage of the cuttings exhibiting shoot formation. In other species that still produce shoots even if cuttings are taken some distance away from the trunk, the effect of juvenility may be seen in the number of shoots produced per cutting or the growth or vigor of those shoots, or with how extensive a root system is developed. For example, *Paulownia tomentosa* cuttings taken from horizontal roots 101 to 200 mm away from the trunk regenerated more extensive root systems than those from more than 200 mm. On the other hand, in the case of crab apple (*Malus* spp.) trees, it does not seem to matter if cuttings are taken from roots that spread horizontally in the soil or from those that are oriented more vertically or diagonally.

Seasonality

Commercially, those nurserymen who propagate plants by root cuttings typically collect the cuttings from December to March, although some species have smaller windows for success with root cuttings. If cuttings are to be directly stuck in the field, root cuttings can often be taken as early as October and November. This gives them a longer time to develop when soil and temperature conditions are suitable for shoots to form and for roots to become established. The seasonal variation in the suckering capacity of root cuttings of some species may be related to endogenous auxins. For trembling aspen (*P. tremuloides*), the number of suckers produced from root cuttings was inversely related to the quantity of endogenous auxin in the roots, highest in June, a time outside of the best window for success with root cuttings. Treatment with an antiauxin caused an increased number of suckers to be produced on each root. A similar relationship was found between auxin levels in roots and sucker production for European aspen (*P. tremula* L.). This is not surprising since endogenous auxins are known to suppress the growth of lateral buds in shoots.

Responses to Growth Regulators

Root cuttings lack root tips that produce cytokinins, hormones that promote shoot formation but suppress root formation. In tissue culture (Chapters 30–34), cytokinins in the medium stimulate the formation of buds on root segments, and it may be that exogenous application of cytokinins could stimulate the budbreak of preformed buds as well as in root cutting segments. Cytokinins inhibit root formation when applied at the distal end of the cutting or that farthest from the crown. In general, another class of hormones, auxins, suppresses bud development but stimulates root development. In some cases, as with *Anemone*, application of low concentrations of the auxin IBA was helpful in increasing the size of the plant resulting from root cuttings, but in other cases, application of IBA has not resulted in significant improvement or has been detrimental in terms of shoot initiation or growth.

ORIGIN OF THE SHOOTS AND ROOTS THAT EMERGE FROM THE CUTTING

New roots can arise from the distal end of the cutting so that the cutting remains a part of the newly developing plant (*Rhus* spp.), or they can arise from the base of the new shoots emerging from the proximal end of the cutting. In this case, typically, the original root cutting withers.

In roots, two types of buds occur, each of which is capable of giving rise to shoots. They are different in several characteristics, such as when they form, the morphological origin of the bud, and oftentimes, the likelihood of the bud to sprout and grow. "Reparative buds" form in response to damage, injury, or senescence, and their tissue of origin may be phellogen, secondary phloem,

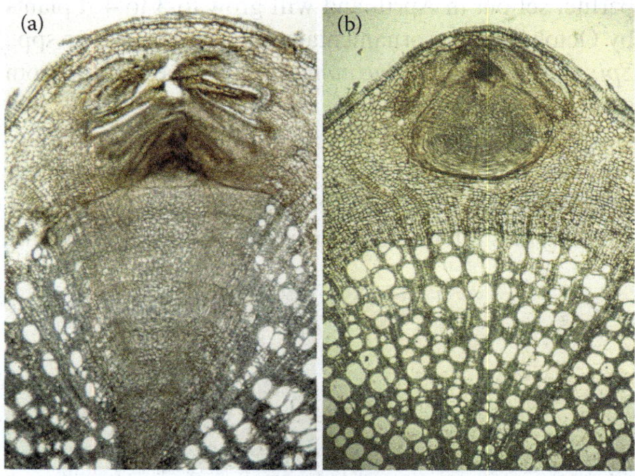

FIGURE 21.3 (a) Cross section of a sassafrass root showing a root bud that is located at the level of the vascular cambium. This "additional bud" formed when the root was young, so as the root grows in diameter, the bud grows outward, with the vascular cambium leaving a clearly visible vascular trace connecting the bud to the center of the root. Also pictured are a "reparative bud" and its associated spheroblast (spherical structure with cambium and xylem). (Courtesy of Michael J. Bosela.) (b) Sassafrass roots also produce buds that are located at more external positions in the bark. These "reparative" buds are found primarily on the roots of older trees, do not appear to require direct root injury for their formation, and show no connection with the vascular tissues of the root, but they do form spherical vascular cambia associated with these buds, resulting in spherical nodules of wood in the bark. In sassafrass, these buds appear incapable of sprouting in field conditions. (Courtesy of Michael J. Bosela.)

or callus. The other type, "additional buds," form very early in the growth of the root, when the plant is young. The bud is visible at the level of the vascular cambium, and it remains connected with the center of the root by a bud trace through the secondary xylem. The trace can be seen in cross section for sassafrass (*Sassafrass albidum* [Nutt.] Nees.) (Figure 21.3a), which has both types of bud, but only the additional buds give rise to root sprouts from root cuttings. The reparative buds have no connection with the interior tissues of the root (Figure 21.3b). In other species, buds that form *de novo* in response to root injury appear to be the primary, if not only, source of root sprouts. For some species such as black locust, additional buds form adjacent to the poles of the protoxylem of the root; this is the reason that shoots sprout from these root cuttings in longitudinal rows.

PREPARATION OF CUTTINGS AND THE ROOTING MEDIUM

Cuttings may be handled in a number of ways once they have been collected: they may be stored, allowed to callus on the cut ends, stuck in containers in a greenhouse or cold frame, or outplanted directly. Aftercare ensures that adequate moisture is available, temperature extremes are avoided, and aeration is sufficient. When cuttings are taken in late fall, winter, or early spring, they are sometimes allowed to dry for several days to remove excess moisture before being placed in the rooting medium. *Paulownia* is a good example of the application of this practice. The rooting medium must be well drained and aerated. In other cases, the root cuttings can be dug, cleaned, trimmed, and after being treated with an appropriate fungicide, stored in moist (not wet) medium until time to plant in the field. The fungicide can be easily applied by placing the root cuttings in a Ziploc® bag containing the fungicide and shaking the bag until the cuttings are well coated.

Root cuttings can be sown either vertically or horizontally. When sown horizontally in flats or in the field in shallow trenches and lightly covered with about 1 to 2 cm of medium or soil, the polarity of the cutting is not as important as when sowing root cuttings vertically. In that case, the end of the cutting that was nearest to the crown or the proximal end with the crosscut should be just below the soil line or just above it. The distal end with the slanted cut should be planted more deeply in the well-aerated medium. The medium that the cuttings are placed in must remain moist to prevent the root cuttings from drying out, but not so wet as to encourage rotting of the cutting. Some propagators recommend waxing the cut ends of the cuttings when they are first taken to minimize loss of moisture.

Root cuttings are sometimes bundled in groups of 10 to 25 for ease of counting and handling and placed in flats containing a well-drained medium such as peat and either sand or perlite for callusing. Callusing is the development of wound tissue at the cut surfaces, and this protects the cuttings when they are later lined out in the field in early spring. Callusing proceeds best at cool temperatures slightly above freezing (~5°C).

Cuttings may also be propagated in a cold frame by direct sticking, the more traditional approach to root cuttings, in a prepared soil bed or by placing root cuttings vertically in deep flats or containers, with their proximal ends just below the surface. The cuttings are then covered with a thin layer of sand or perlite. Root cuttings that are placed in the cold frame during winter or early spring are grown out throughout spring and summer. Those stuck in containers may require fertigation and lifting sooner than those grown directly in the prepared bed. Using the traditional approach, root cuttings would be left in the cold frame to develop until fall of the first growing season, when they would be lifted and potted in containers or lined out in the field.

If cuttings are stuck in flats or containers in the greenhouse, temperature control is extremely important. Initially, the temperature range is held at 2 to 7°C, but as the season progresses, temperatures in the greenhouse will also begin to rise. If this happens too quickly, cuttings will not have developed root systems sufficient to survive water loss and, thus, will die. From March to April, the temperatures can be allowed to range between 7 to 10°C. Plants are typically ready to go to the field by July.

Species Most Suitable for Use of Root Cuttings

A common observation is that any plant that suckers readily in the wild is likely to be successfully propagated with root cuttings. Root cuttings can be used for commercially important hardwood species such as *Paulownia* spp., *Populus* spp., and *Robinia* spp. One of the fastest-growing tree species in the world, *Paulownia* is valued for its strong, lightweight wood. Its ability to produce commercial crops of timber in 6 years makes it advantageous to be able to clone selected individuals with desirable traits. In the United States, most *Paulownia* trees are produced from containerized seedlings, but in other countries, root cuttings are used extensively. *Paulownia* is cross-pollinated, so although it is easily propagated by seeds, there is no guarantee that seedlings from an outstanding tree will share all of its traits. Only clonal propagation can guarantee a tree with an identical genotype, but *Paulownia* does not propagate easily from stem cuttings. Root cuttings, on the other hand, give rise to shoots readily and root relatively easily.

Several *Populus* species are important commercially, including quaking aspen and European aspen. Quaking aspen is valued for its light wood, which is resistant to shrinkage and used for lumber and matches. It is also a fast-growing pioneer species commonly found in pure clonal stands. It does not propagate easily from softwood cuttings, but suckers from root cuttings can be induced to form roots relatively easily. The same is true for most *Populus* species. Root cuttings also provide an inexpensive method to propagate a large number of plants for reforestation and bank stabilization purposes. The beauty of this is that they can be planted in the soil directly at the site where establishment is desired.

Root cuttings can be used for ornamental woody species such as crape myrtle (*Lagerstroemia indica* L.), panicled goldenraintree (*Koelreuteria paniculata* Laxm.), lilac (*Syringa vulgaris* L.), flowering quince (*Chaenomeles* spp.), hypericum (*Hypericum calycinum* L.), European mountainash (*Sorbus aucuparia* L.), roses (*Rosa* spp.), wisteria (*Wisteria floribunda* [Willd.] DC.), and lacebark elm (*Ulmus parvifolia* Jacq.). Only a very few of the many species that can be propagated with root cuttings are listed here. Oakleaf hydrangea (*Hydrangea quercifolia* Bartr.) is another of these, and in the landscape setting, it suckers readily (Figure 21.4). Root cuttings of forsythia (*Forsythia* × *intermedia* Zab.) can be taken in fall, overwintered in nursery flats containing perlite, set out in April, and will grow to 3 to 4 ft plants by October. Some ornamentals, such as *Euonymus* spp., *Spiraea* spp., and *Viburnum* spp., can be grown from summer softwood cuttings taken from the shoots that emerge from the original root cuttings. If an ornamental plant is grafted, cuttings taken from the root system (rootstock) will not be of the same genotype as the ornamentally desirable scion (portion of the plant above the

FIGURE 21.4 Root used for the root cuttings in Figure 21.2a, showing the root sucker origin of the ramet on the right and its root connection to the original oakleaf hydrangea plant (ortet) on the left.

graft). Root cuttings are only useful when the root system is of the same genotype as the shoot (as in own-rooted shrubs and trees). For some ornamentals that are notoriously difficult to propagate asexually, such as Georgia plume (*Elliottia racemosa* Elliott), root cuttings are still the only practical method of propagation.

Root cuttings can also be used to propagate fruit species, such as kiwifruit and its relatives (*Actinidia* spp.), apples (*Malus* spp.), and stone fruits (*Prunus* spp.), but is most commonly used for *Rubus* spp., such as blackberry and raspberry. Blackberry and raspberry plantings have been established easily using root cuttings placed directly in the field, but the ability to regenerate successfully at a high percentage may vary with cultivar. Remember though that, with thornless blackberry, the new shoots that form from root cuttings lack the thornless characteristic of the original shoots because the shoots arise from the tissue of the internal layers. For other fruit species, a specialized application for root cuttings is for obtaining dwarf fruit trees by propagating genetically dwarf selections on their own roots rather than using the more common approach of grafting standard cultivars onto dwarfing rootstocks.

Probably the best known group of plants for which root cuttings are particularly well suited are herbaceous perennials such as windflower (*Anemone* spp.), bellflower (*Campanula* spp.), coneflower (*Echinacea* spp.), blanket flower (*Gaillardia aristata* Pursh), oriental poppy (*Papaver orientale* L.), sages (*Salvia* spp.), thrift (*Phlox subulata* L.), and some *Geranium* species. If the roots are thin as those of yarrow (*Achillea millefolium* L.), flowering spurge (*Euphorbia corollata* L.), gaillardia, thrift, and Stokes' aster (*Stokesia laevis* [J. Hill] Greene), they can be cut into pieces about 2 in long, scattered on the rooting medium, and lightly covered with a thin layer of the medium. If the roots are fleshier like those of peony (*Paeonia* spp.) and oriental poppy, the cuttings can be inserted into the medium, with the distal end pointing downward. Some propagators advocate allowing the top 1/4 in to stick out of the medium, whereas others suggest burying the entire cutting about 1 in under the surface of the rooting medium.

LEAF CUTTINGS

Many different parts of the plant can be used as cuttings for propagation, even roots, as seen in the first part of this chapter. The use of leafy stems has been discussed in Part VI of this book, but for some plants, only the leaf or the leaf with petiole is needed to propagate a new plant that is genetically identical to the original plant. These plants can regenerate new ones, without any portion of the original stem still attached.

ADVANTAGES AND LIMITATIONS OF USING LEAF CUTTINGS

Leaf cuttings are a relatively simple way to propagate some plants. Very little expertise is needed, and leaf cuttings will root without elaborate specialized facilities. The medium should be one that will retain moisture but still drain freely. A typical rooting medium is one consisting of 1/2 peat to retain moisture and 1/2 sand, perlite, or vermiculite to ensure good drainage. A variety of containers can be used as long as some means of restricting moisture loss and ensuring high humidity around the cuttings is established. This can be as simple as the base of a plastic liter soft drink container cut into half and inverted over the pot, a "cloche" donated by the soft drink industry.

In some cases, the way the cutting is taken determines the number of plants that can be obtained. For example, with *Begonia rex*, the leaf can be separated from the parent plant at the petiole and inserted into the rooting medium or even water. New plantlets will form at the base of the petiole (Figure 21.5a). Sometimes, the petiole is trimmed off the leaf, and the leaf blade is rolled and then inserted into the medium (Figure 21.5b). However, if a larger number of plants is desired, the Rex

FIGURE 21.5 Rex begonia, showing the different methods for conducting leaf propagation, including leaf blade and petiole rooted in water (a); leaf with petiole and part of the blade, rolled and inserted into the medium (b); and leaf cut into squares secured by florists' pins to hold them in firm contact with the rooting medium (c).

begonia leaf can be subdivided into squares about 3 to 5 cm on a side, each containing a portion of a vein, and this will result in a much larger number of plants (Figure 21.5c). Another good example is *Streptocarpus*, which can be propagated using an intact leaf cutting resulting in one plant being formed. Cutting the leaf into chevrons (V-shaped pieces) will allow more plantlets to develop from the cut surfaces, but cutting the leaf into two halves with the midrib removed and embedding the cut surface firmly into the rooting medium will result in the largest number of plantlets formed. *Sansieveria* can be propagated by division, which results in fewer plants being obtained compared to being propagated by 2 to 3 in pieces of the leaf (Figure 21.6), but this technique cannot be used with the variegated forms; otherwise, the variegation will be lost.

Leaves of some plants will easily form roots when used as leaf cuttings, but they do not form shoots. *Hoya*, *Clematis*, and *Mahonia* leaves can be rooted, but either fail to form adventitious shoot buds or only rarely do so. Obviously, successful propagation by leaf cuttings is dependent not only on adventitious root initiation, but also on the ability to form adventitious shoots. Although plants may be related as closely as within the same genera, one may have the ability to form adventitious shoots and roots and can be propagated with leaf cuttings, whereas others may not. Sometimes, the ability differs among cultivars of the same species, as has been demonstrated with patchouli (*Pogostemon patchouli* Pelleto), a plant valued for its fragrant oil. Young leaves of sugar beet (*Beta vulgaris* L.) can be rooted with the use of IBA, but adventitious bud formation occurs with only very low frequency.

This peculiarity of rooting without shoot formation has also been used in plant research. Leaves of common bean (*Phaseolus vulgaris* L.) will root, but no shoots form. After adventitious roots form at the end of the petiole, they continue to develop (increasing root-to-shoot ratio), and the original leaf blade photosynthesizes, allowing researchers to investigate the effects of immersion in various growth regulators and the changes in endogenous cytokinins without the complications of shoot buds. The leaves of English ivy (*Hedera helix* L.) root from cuttings consisting of only a leaf blade and a petiole. This feature has been used to investigate root initiation in reciprocal grafting experiments using mature leaf lamina (blades) grafted onto juvenile petioles and vice versa. Root initiation, in response to IBA, was more rapid for combinations with juvenile petioles. The capacity to form adventitious shoots has also been explored as a means to obtain whole (non-chimeral) mutants as a result of ionizing radiation when the adventitious bud is the result of only one cell—in the case of Saintpaulia, one epidermal cell.

Plants that are chimeral may not propagate with the desired appearance, depending on the origin of the newly formed shoots and roots. In African violets (*Saintpaulia ionantha* Wendl.), new shoots arise from newly dedifferentiated cells lying just beneath the epidermis. If those shoots arise from the albino portion of the variegated *Saintpaulia* shown in Figure 21.7, albino plantlets could be formed, which cannot survive on their own.

FIGURE 21.6 Original leaf section cutting of *Sansevieria trifasciata*, showing the new plant originating from the base of the leaf cutting.

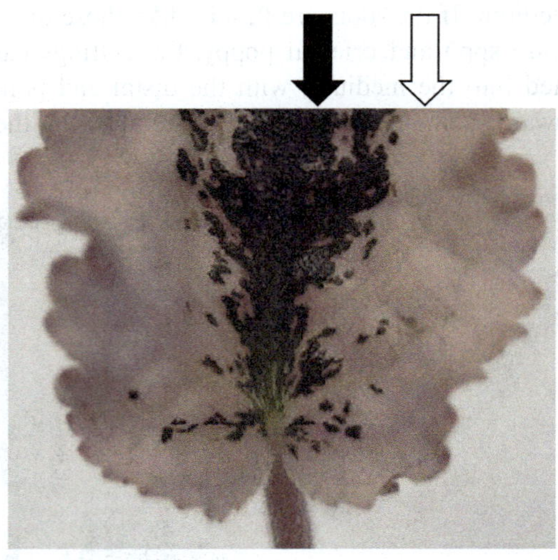

FIGURE 21.7 Cross section of a variegated African violet leaf showing pigmented (solid arrow) and albino (white arrow) tissues. Shoots arising from nonpigmented tissue will be albino and unable to survive on their own.

THE ORIGINS OF NEW SHOOTS AND ROOTS ON LEAF CUTTINGS

Plants propagated by leaf cuttings can be divided into two groups based on the histology of the shoots and roots that form: whether they form from new adventitious meristems or from preformed primary meristems. In the first group, meristematic cells of plants such as *Saintpaulia* or *Begonia* must form from mature differentiated tissue. In the case of new adventitious roots, this means the parenchyma located near the xylem–phloem boundary of the vascular bundle, and in the case of new adventitious shoots, epidermal cells sometimes near a large vein. Leaves of some species, such as *Sedum*, detach from the parent plant very easily, and this stimulus causes them to form secondary meristems that develop from vacuolated, differentiated parenchyma. These secondary meristems then give rise to roots and shoots. The location of where the new meristems form on the leaf may also vary. For *Saintpaulia*, *Peperomia*, and *Streptocarpus*, and the monocots *Sansevieria* and *Lilium*, new meristems form at the base of the petiole or where the leaf was severed. For the carnivorous plant sundew (*Drosera*), new meristems develop randomly on the leaf blade.

In the second group, embryos are preformed early in the development of the leaf, and they retain their meristematic activity. These primary meristems can be located on the periphery of the leaf in notches (*Bryophyllum*), at the ends of the leaf (*Asplenium*), or at the base of the leaf blade where it joins the petiole (*Tolmeia*). This ability to form buds and, oftentimes, roots on leaves that are still attached is not as prevalent as the ability to form roots and shoots on leaves detached from the parent plant. Still, more species can form roots on detached leaves but cannot form shoots.

The Crassulaceae family contains the genus *Bryophyllum*, many of which are well known for the ability to produce plantlets from preformed meristems (sometimes called "foliar embryos") located in the indentations at the leaf margins. These foliar embryos usually develop a couple of leaf primordia first, and then a couple of roots form. This ability is the reason why bryophyllums are sometimes referred to as "maternity plants." Not all *Bryophyllum* species have this ability, and some produce plantlets on the floral stem instead. Plantlets (Figure 21.8a) on this *B. daigremontianum* develop one to two roots (Figure 21.8b) then detach from the leaf margins, falling to the surface of the growing medium becoming established. Oftentimes, older leaves, whose meristems are quiescent, drop from the plant and fall to the growing medium, and as the leaf senesces, new shoots and then roots are formed from the indentations at the margins of the leaf.

For piggyback plant (*Tolmiea menziesii* [Pursh] Torr. & A. Gray), new plantlets form at the juncture where the petiole meets the leaf blade on mature leaves (Figure 21.9), which explains another of its common names, "youth on age." If the leaf is separated from the parent plant and placed on moist medium, the new plantlet can further develop. The original leaf may stay green while the new plantlet begins to grow, but eventually shrivels and turns brown.

FIGURE 21.8 Leaf of mother of thousands showing the plantlet forming from preformed meristems at the periphery of the leaf (a) and the roots that form on the plantlets after the leaves have formed (abaxial surface) (b).

FIGURE 21.9 Piggyback plant with new plantlets forming at the juncture of the leaf blade and the petiole. (Image © 2005 Barry Glick, http://www.sunfarm.com. Used with permission from Barry Glick, Sunshine Farm & Gardens.)

FIGURE 21.10 (a) Walking fern in its normal habitat on limestone outcroppings on Monte Sano Mountain, Huntsville, Alabama, showing the long slender fronds. (Courtesy of Christopher Sutton.) (b) New plantlet forming at the extreme tip of the slender attenuated frond of walking fern. This plantlet will root easily once it has made contact with the ground. (Courtesy of Christopher Sutton.)

Walking fern (*Asplenium rhizophyllum* L.) is a hardy fern found in moist shady limestone or calcareous outcroppings (Figure 21.10a) where there is enough soil to retain moisture and a nearly neutral pH. It has the ability to propagate by forming plantlets at the apex of the leaf (Figure 21.10b). The long fronds arch, and when new plantlets form, the weight of the plant allows the frond to touch the ground and root. This results in a network of new plants spaced around the original plant and still connected to it by the long tapered fronds, giving the impression of the fern "walking" across the ground.

RESPONSES TO GROWTH REGULATORS

Benzyladenine, a cytokinin, increased the number of shoots obtained from the leaf cuttings of *Sedum* but reduced the number of roots formed. Treatment with auxins, such as IBA or naphthaleneacetic acid, promotes root formation. Rieger elatior begonias (*Begonia* × *hiemalis* Fotsch) of the 'Aphrodite' group do not consistently produce adventitious buds at the base of the petiole, but application of 6-(benzylamino)-9-(2-tetrahydropyranyl)-9H-purine, a cytokinin, as either a foliar spray or a petiole dip stimulated bud and shoot development.

SPECIES SUITED FOR PROPAGATION BY LEAF CUTTINGS

Some of the most common examples of species that propagate easily from leaf cuttings are members of the Crassulaceae family, the succulents. Examples include *B. daigremontianum* (mother of thousands), *Crassula ovata* (jade plant), some members of the genera *Echevaria* and *Pachyphytum*, and *Sedum morganianum* (burro's tail). *C. ovata* and *S. morganianum* are both propagated easily from leaves broken off to separate them from the parent plant and placed cut surface downward, in firm contact with a well-drained rooting medium (1/2 peat:1/2 perlite works well).

A surprising example of a species that can be propagated by leaf petiole cuttings is Korean spice viburnum (*Viburnum carlesii* Hemsl.), a woody ornamental shrub valued for its fragrant flowers. The entire process takes several months, but after the petiole has rooted, a new shoot will form. Many of the carnivorous plants can be propagated using leaf cuttings, including *Cephalotus*, *Dionaea*, *Drosera*, and *Sarracenia*. Although each of these species has highly specialized leaves (*Cephalotus* and *Sarracenia* have leaves like a pitcher containing fluid to trap insects, *Drosera* has leaves with sticky trichomes, and *Dionaea* has leaves that can envelope then digest the insect), they can be separated carefully from the plant and induced to form roots. Venus flytrap (*Dionaea muscipulata* Ellis) has highly specialized leaves that trap and digest insects that land on them, stimulate the trigger hairs, and cause leaf closure. If the leaf is pulled off the plant so that it retains some of the original white tissue at

FIGURE 21.11 Specialized leaves of Venus flytrap can be separated from the parent plant and inserted into shredded sphagnum for rooting. Best results are obtained if the lightly pigmented base of the leaf is retained when the cutting is made. (Courtesy of David Webb.)

FIGURE 21.12 *Lachenalia*, a member of the lily family, roots easily from leaf cuttings but only forms new plants after it has formed a bulblet at the base of the rooted leaf. (Courtesy of Mark Mazer.)

the base of the leaf (Figure 21.11) and its base is inserted into a rooting medium, new plants will form several months later.

Some members of the Liliaceae family root from leaf cuttings but, rather than form shoots directly, go through a bulblet formation stage first. Cape cowslip or *Lachenalia* will root from leaf cuttings, but new plants arise only after it has formed a bulblet at the base of the rooted leaf (Figure 21.12). *Lilium longiflorum* Thunb. can also be propagated from leaf cuttings that form bulblets followed by adventitious roots. There are selected examples from other plant families that follow a similar pattern. *Zamioculcas zamiifolia* (Lodd.) Engl., sometimes known as "ZZ plant," is a member of the Araceae family, although it resembles a cycad. When leaflets of this plant fall to the ground, they easily root and form bulblets. This ability is not shared by the other members of the Araceae family, like *Spathiphyllum* and *Aglaonema*.

In summary, propagation by either leaf or root cuttings may seem simplistic and somewhat old fashioned. Advances in tissue culture have presented viable alternatives for the propagation of many difficult-to-root species; however, even with the considerable amount of research devoted to *in vitro* regeneration, not all species respond favorably. This fact argues that there is still a role for root cutting propagation for some of these species. The ease of producing new plants from leaf cuttings and the lack of need for high technology virtually ensure that leaf cutting propagation will continue to be in use for selected species in the future.

LITERATURE CITED AND SUGGESTED READING

Bosela, M. J. and F. W. Ewers. 1997. The mode of origin of root buds and root sprouts in the clonal tree *Sassafrass albidum* (Lauraceae). *Amer J Bot* 84:1466–1481.

Davies, Jr., F. T. and B. C. Moser. 1980. Stimulation of bud and shoot development of Rieger begonia leaf cuttings with cytokinins. *J Amer Soc Hort Sci* 105:27–30.

del Tredici, P. 1995. Shoots from roots: A horticultural review. *Arnoldia* 54:11–19.

Dirr, M. A. and C. W. Heuser. 1987. *The Reference Manual of Woody Plant Propagation: From Seed to Tissue Culture.* Georgia: Varsity Press Inc, 239 pp.

Ede, F. J., M. Auger, and T. G. A. Green. 1997. Optimizing root cutting success in *Paulownia* spp. *J Hort Sci* 72:179–185.

Geneve, L., M. Mokhtari, and W. P. Hackett. 1991. Adventitious root initiation in reciprocally grafted leaf cuttings from the juvenile and mature phase of *Hedera helix* L. *J Exp Bot* 42:65–69.

Hartmann, H. T., D. E. Kester, F. T. Davis, and R. L. Geneve. 1997. *Plant Propagation Principles and Practices*, 6th Edition. Englewood Cliffs, New Jersey: Prentice Hall, 770 pp.

Miedema, P., P. J. Groot, and J. H. M. Zuidgeest. 1980. Vegetative propagation of *Beta vulgaris* by leaf cuttings. *Euphytica* 29:425–432.

Stoutemeyer, V. T. 1968. Root cuttings. *The Plant Propagator* 14:4–6.

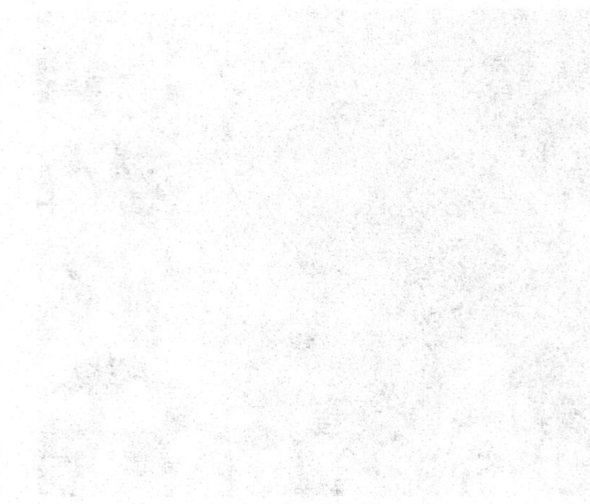

22 Propagation by Leaf Cuttings

John L. Griffis, Jr. and Malcom M. Manners

Successful vegetative propagation from virtually all types of cuttings depends on the ability of the plant to make some organ system adventitiously—stem cuttings must produce adventitious roots; root cuttings must form adventitious shoots. However, with leaf cuttings, there is no root or stem tissue present, so both primary organs must be made adventitiously. Most plant species are not capable of doing this, but in those species that have the ability, the method allows the production of large numbers of new plants from relatively little starting material—at the least, one leaf will produce one new plant, and in some cases, several new plants can be made from a single leaf. Plants commonly propagated from leaf cuttings are listed in Table 22.1.

Leaf cuttings should be made from relatively young, healthy leaves that are fully expanded. Leaves that are damaged or diseased are more likely to rot than they are to form plantlets. Leaves that are not yet fully developed often do not have enough stored nutrients to complete the maturity process and produce new plantlets. In preparing leaf cuttings, an auxin-based growth regulator treatment is generally not used since, while it would promote rooting, it would, at the same time, inhibit the formation of adventitious shoots. Therefore, no pretreatment is generally used.

EXERCISES

Experiment 1. Plantlets That Form from Leaf Cuttings with Petioles

The most reliable method for producing new plantlets from leaf cuttings is to use the entire leaf with the petiole still attached. Far more species have the ability to form both roots and shoots adventitiously from petiole tissue than from leaf blade or vein tissues. If there is a disadvantage to this strategy, it is that relatively few new plants will develop from each leaf. If the stock plant has plenty of leaves that might be removed and used for propagation, then low plantlet yield per leaf hardly matters.

MATERIALS

The following materials are needed for each student or team of students:

- Trays or pots of moistened perlite
- Clippers, knife, or some implement to cut leaves cleanly from the stock plants
- Stock plants of almost any *Peperomia* spp. Ruiz & Pav., African violet (*Saintpaulia ionantha* Wendl.), florist's gloxinia (*Sinningia speciosa* [Lodd.] Hiern), almost any *Bryophyllum* spp. Salisb., and/or piggyback plant (*Tolmeia menzeisii* [Pursh] T. & G.) with mature leaves
- Leaves of *Hoya carnosa* (L. f.) R. Br. (for comparison); other choices for the comparison are the leaves of *Coleus* spp. Lour., *Impatiens* spp. L., and rubber tree (*Ficus elastica* Roxb.)
- Moistened vermiculite or moistened soilless media can be substituted for perlite in any of these experiments, but the water holding capacity of either of these is significantly greater than that of perlite and may lead to a greater loss of leaves to disease unless the watering schedule is closely monitored; as conditions from greenhouse to greenhouse vary, it may also be necessary to apply a fungicidal drench to the medium to discourage disease problems (Chapters 11 and 12)

Follow the protocol outlined in Procedure 22.1 to complete this experiment.

ANTICIPATED RESULTS

Root formation should occur within 3 to 4 weeks, and plantlets should appear visible above the perlite in 4 to 7 weeks (Figure 22.1d, e). After the new plants have a few leaves, they may be transplanted to individual pots of regular potting medium. Note that the florists' gloxinia may take longer to produce visible plantlets as the leaf will first produce a small tuber at the cut end of the petiole, and the new plantlets will sprout mostly from the new tuber rather than arise directly from the petiolar tissue. Note also that, if the piggyback plant is used as the leaf source, it is important to select mature leaves that do not already have plantlets forming on them. The same would be true if *B. pinnata* (Lam.) Kurz, *B. diagremontianum* Hamet & Perr., or *B. fedtschenkoi* Hort. were used as a leaf source. All of these plants actually have certain leaf cells that are predetermined to form asexual embryos within the leaf tissue. A variety of stimuli, including removing a leaf from the plant for propagation purposes, may cause these cells to grow and produce embryonic plantlets. For comparison, it may be noted that the embryonic plantlets will arise from various areas of the leaf blade tissues rather than from the petiolar tissues that

TABLE 22.1
Plants Commonly Propagated from Leaf Cuttings

Family	Scientific Name	Common Name
Asparagacae	*Sansevieria* spp.	Snake plant, mother-in-law's tongue
Araceae	*Zamioculcas zamiifolia*	ZZ
Begoniaceae	*Begonia* spp.	Begonia (rex, angel wing)
Crassulaceae	*Bryophyllum* spp.	Mother of thousands
	Kalanchoe spp.	Kalanchoe
	Sedum morganianum	Burro's tail, donkey's tail
Gesneriaceae	*Saintpaulia ionantha*	African violet
	Sinningia speciosa	Florists' gloxinia
	Streptocarpus spp.	Streptocarpus
Piperaceae	*Peperomia* spp.	Peperomia
Saxifragaceae	*Tolmiea menziesii*	Piggyback plant

Procedure 22.1
Plantlets That Form from Leaf Cuttings with Petioles

Step	Instructions and Comments
1	Each student or team should cut three to five leaves from the stock plants, taking some petiole with each cutting (Figure 22.1a, b).
2	Carefully insert the petiole end of the first cutting into the perlite so that the blade of the leaf is just above the surface of the perlite but not quite touching it. Gently press the perlite around the petiole so that the leaf is held firmly in place.
3	Insert the other cuttings into the perlite in a similar manner, being sure to space the multiple cuttings about 2 to 3 cm apart in the container to allow light to shine on all the cuttings (Figure 22.1c).
4	Set the container of cuttings in a warm, shaded spot (e.g., under a greenhouse bench) and keep the perlite damp at all times.
5	Make observations on the number of new plants obtained from each cutting and the appearance of the new plants. Note when the new plants appeared and the origin of the new plant on the original cutting. Once data have been collected, potting up the new plants into containers of potting medium (with or without the original leaf cutting still attached) will complete this exercise.

will yield the adventitious plantlets formed by African violets or the various peperomias. *Hoya*, *Coleus*, *Impatiens*, or *Ficus* leaves are included in this experiment as an example of species that easily form roots from leaf cuttings but that usually do not form adventitious shoots. So, the result is a rooted leaf that may survive for many months but would not usually produce a plant (Figure 22.1f).

QUESTIONS

- Most plants that are propagated from leaves tend to be rather succulent. What advantage(s) would a succulent leaf have over other leaves in this method?
- Exogenous hormones are not usually used with leaf cuttings. What would likely be the source of auxin for root formation?
- Why do shoots usually form after roots have already formed?

Experiment 2. Plantlets That Form Directly from Leaf Blade Tissues

This method for producing new plantlets from leaf cuttings works well for certain plants that have little or no petiole tissues. If the plants are chimeric, as are some variegated cultivars of snake plant, such as *Sansevieria trifasciata* 'Laurentii' Thunb., then the resulting plantlets may not display the desired, striped pattern like the parent. Therefore, this experiment may also be used to determine whether a particular plant is a chimera (see Chapter 7).

FIGURE 22.1 Propagation of several species from leaf cuttings. (a) Leaf cuttings from three different peperomias and from an African violet. (b) Leaf cutting of *Kalanchoe fedtschenkoi*. (c) Leaf cuttings with petioles inserted in perlite. (d) Peperomia leaf with adventitious roots and plantlets. (e) African violet leaf with adventitious roots and plantlets. (f) Hoya leaf with only adventitious roots.

MATERIALS

The following materials will be required to complete this experiment:

- Trays or pots of moistened perlite
- Clippers, knife, or some implement to cut leaves cleanly from the stock plants (and with certain plants, to cut the leaves into pieces)
- Stock plants of snake plant (*Sansevieria trifasciata*), burro's tail (*Sedum morganianum* Walth.), panda plant (*Kalanchoe tomentosa* Bak.), or cape primrose (*Streptocarpus* spp. Lindl.) that can provide mature leaves. If either *S. morganianum* or *K. tomentosa* is used as the stock plant, the entire leaves will easily break-off of the parent plant, so no cutting is necessary

Follow the instructions listed in Procedure 22.2 to complete this exercise.

ANTICIPATED RESULTS

Roots should form in about 2 to 4 weeks, and shoots, from the leaves by about the seventh week (Figure 22.2d, e). At the end of the experiment, remove the leaf cuttings from the perlite and observe where the plantlets have formed. Note whether the upside-down cuttings made any plantlets and, if so, from what areas the plantlets arose. If a variegated cultivar of burro's tail is selected as the parent plant, whether the parent plant is chimeric can also be determined using this method (Chapter 7).

QUESTIONS

- If a variegated variety of snake plant was used, notice the pattern of variegation on the plantlets (or lack thereof). Does the pattern match that of the parent leaf?
- Is the variegation of the parent plant chimeric?

Experiment 3. Plantlets That Form from Leaf Blade Veins

Some types of begonias, as well as some *Streptocarpus* spp., have the ability to make adventitious plantlets from leaf pieces that contain major or large leaf veins. Since the leaves may be quite large, one can often make several cuttings from a single leaf. Each leaf cutting may produce more than one plantlet, so the main advantage to this type of propagation is that numerous plantlets can be made from a single, mature leaf. Since this method of propagation utilizes small leaf pieces that are prone to rotting, clean tools and equipment, and sterile media should be used at all times. It may or may not be necessary to drench the potting mix with a general purpose fungicide depending on greenhouse conditions.

MATERIALS

The following items are needed for the exercise:

- A number of 10 cm (4-in) pots containing moistened 1 peat:1 perlite (by volume) potting mix
- Stock plants of *Begonia* spp. L. or *Streptocarpus* spp. that can provide mature, fully expanded leaves (*B.* x *rex-cultorum* and other rhizomatous and angel wing cultivars of begonias work very well; tuberous types of begonias and *B. semperflorens* Hort. are not recommended)
- Razor blade, knife, box cutter, scissors, or other cutting device

Follow the instructions outlined in Procedure 22.3 to complete this exercise.

Procedure 22.2
Plantlets That Form Directly from Leaf Blade Tissues

Step	Instructions and Comments
1	When using the dwarf cultivars *S. trifasciata* 'Hahnii,' 'Golden Hahnii,' *S. morganianum*, or *K. tomentosa*, remove entire leaves from the stock plant (Figure 22.2a).
2	When using one of the longer-leafed cultivars of *S. trifasciata*, cut horizontally across the leaf, creating leaf pieces of about 7 to 8 cm long (Figure 22.2b). When using strap-leafed *Streptocarpus* spp., cut horizontally across the leaf, creating leaf pieces of about 5 cm long.
3	Keep track of the original orientation of the cut leaf pieces by writing the word "top" or "up" or making a small arrow on the distal (upper) part of the cutting.
4	Insert the bases of the cuttings into the perlite, leaving about 2 to 3 cm of space between cuttings (Figure 22.2c). Insert most of them upright but also insert one or two of the leaf cuttings upside down for comparison. Entire leaves from the burro's tail or the kalanchoe may also be planted upside down or even left to rest sideways on the medium.
5	Set the container in a warm, shaded spot (e.g., under a greenhouse bench) and keep the perlite damp at all times.
6	Make observations on the number of new plants obtained from each cutting and the appearance of the new plants. Note when the new plants appeared and their point of origin on the original cutting. Once data have been collected, pot up the new plants with the original leaf cutting still attached in a container of potting medium.

FIGURE 22.2 Propagation of plants via leaf cuttings. (a) Stock plant of *Sansevieria trifasciata* 'Hahnii,' with entire leaves removed and inserted in perlite. (b) Leaves of *Sansevieria trifasciata* 'Moonshine,' whole and cut into sections for propagation. (c) Leaf cutting of *Sansevieria trifasciata* 'Moonshine' inserted in perlite. (d) Snake plant leaf with adventitious plantlets. (e) *Kalanchoe fedtschenkoi* leaves with adventitious plantlets.

Procedure 22.3

Plantlets That Form from Leaf Blade Veins

Step	Instructions and Comments
1	Remove a mature leaf from the begonia stock plant (Figure 22.3a).
2	Carefully cut around the petiole to remove it, leaving the palmately arranged major veins of the leaf not quite touching each other (Figure 22.3b).
3	Trim the upper part of the leaf so that the distance from the base of the veins to the top of the leaf is about 3 to 5 cm all around (Figure 22.3c).
4	Cut the remaining leaf piece into multiple cuttings so that each sectional leaf cutting contains a major vein, creating what looks like small, triangular leaves, with a major vein as a midrib (Figure 22.3d).
5	Bury the lower 1/3 to 1/2 of the cuttings in 10 cm (4-in) pots containing medium. If desired, several cuttings may be placed closely together in the same pot; doing so will result in a much fuller plant in less time, making the final product marketable at an earlier date—a commercial advantage (Figure 22.3e).
6	Place the pots in an intermittent mist system for about a week, and then transfer them to a shaded area in a greenhouse. Alternately, they can be placed under a greenhouse bench and hand misted several times a day for the first week.
7	Make observations on the number of new plants obtained from each cutting and the appearance of the new plants. Note when the new plants appeared and their point of origin on the original cutting. Once data have been collected, potting the new plants with the original leaf cutting still attached into containers of potting medium will complete the exercise.

FIGURE 22.3 Propagation of rex and angel-wing begonias. (a) Leaf cuttings from a rex begonia and an angel-wing begonia. (b) Leaf cutting from a rex begonia, with the petiole removed. (c) Leaf cutting from a rex begonia, with the outer leaf edge tissue removed. (d) Leaf cutting from a rex begonia, with the central leaf tissue cut into small, triangular pieces with a major vein running down the center of each. Pieces without large veins usually do not form plantlets. (e) Small, triangular leaf pieces from a rex begonia and an angel-wing begonia inserted into perlite. (f) On the left, a small plantlet develops from the leaf piece of a rex begonia. The slightly older plant on the right developed quickly when provided proper nutrition and growing conditions.

Procedure 22.4
Plantlets That Form from the Leaflets of a Compound Leaf

Step	Instructions and Comments
1	Remove a mature, but healthy, deep green–colored leaf from the ZZ stock plant (Figure 22.4).
2	Using a clipper, knife, or just your fingernails, cut off an individual leaflet from the rachis and allow the cut surface to air dry (Figure 22.5a, b).
3	Bury the lower 1/4 to 1/3 of the leaflet (the end that was attached to the petiole) in the perlite.
4	Repeat the procedure with the other leaflets, lining them up in the tray, leaving 3 cm or more space between them to allow light to shine on all the cuttings (Figure 22.5c).
5	Place the tray in an intermittent mist system or, alternately, in a humid, shady place (e.g., under a greenhouse bench).
6	Make observations on the number of new plants obtained from each cutting and the appearance of the new plants. Note when the new plants appeared and the point of origin on the original cutting. Once data have been collected, potting up the new plants into containers of potting medium will complete the exercise (Figure 22.6).

ANTICIPATED RESULTS

Roots should form in about 2 to 3 weeks, and small plantlets should appear after 6 to 7 weeks. Given proper growing conditions and nutrition, plants can reach saleable size rather quickly (Figure 22.3f). Application of some liquid fertilizer after roots form will dramatically accelerate the formation and development of shoots.

QUESTION

- Do plantlets always form from veins or do some also arise from the leaf blade between the veins?

Experiment 4. Plantlets That Form from the Leaflets of a Compound Leaf

The "ZZ" plant (*Zamioculcas zamiifolia* [Lodd.] Engl.) has recently become quite popular as a shade-tolerant indoor foliage plant. It has long, pinnately compound leaves that resemble those of a palm or cycad and is unusual in that it is propagagted from individiual leaflets cut or brokem from the leaf.

MATERIALS

The following items are needed for each student or team of students:

- Trays or pots of moistened perlite
- Clippers, a knife, or some implement to cut leaflets from the stock plants
- Stock plants of *Z. zamiifolia*

Follow the instructions outlined in Procedure 22.4 to complete this exercise.

ANTICIPATED RESULTS

Root formation should occur within 3 to 4 weeks, and plantlets should appear visible above the medium in 6 to 9 weeks (Figure 22.5d). Like gloxinia, the cuttings will also produce a large corm by the time leaves are produced. After the new plants have a few leaves, they may be transplanted to individual pots of regular potting medium.

FIGURE 22.4 A plant of *Zamioculcas zamiifolia* to be propagated.

FIGURE 22.5 Propagation of *Zamioculcas zamiifolia* with leaflets. (a) Leaflets may be snapped off or pinched off the midrib. (b) Leaflets, ready to be stuck as cuttings. (c) A tray of leaflets, planted with the lower 25% of the leaflets in the medium. (d) Several weeks later, plantlets have formed with roots and corms.

FIGURE 22.6 Plantlets of *Zamioculcas zamiifolia* formed from leaflets are potted up.

QUESTIONS

- Do plantlets always form after the corms or do some also arise directly from the leaflet bases?
- Very few plants can be propagated from leaflets. Why do you think *Zamioculcas* can be propagated this way?

SUGGESTED READING

Browse, P. M. 1979. *Plant Propagation*. New York: Simon and Schuster, 96 pp.

Davis, T. D. and B. E. Haissig, eds. 1994. *Biology of Adventitious Root Formation*. New York: Plenum Publishing Corporation, 358 pp.

Dirr, M. A. and C. W. Heuser. 2006. *Reference Manual of Woody Plant Propagation: From Seed to Tissue Culture*, 2nd Edition. Portland, Oregon: Timber Press, Inc., 424 pp.

Hartmann, H. T., D. E. Kester, F. T. Davies, Jr., and R. L. Geneve, eds. 2010. *Hartmann and Kester's Plant Propagation: Principles and Practices*, 8th Edition. Upper Saddle River, New Jersey: Prentice Hall, 928 pp.

MacDonald, B. 1992. *Practical Woody Plant Propagation for Nursery Growers*, Vol. 1. Portland, Oregon: Timber Press, 660 pp.

23 Propagation by Root Cuttings

Paul E. Read

Many species and cultivars are extremely difficult to propagate from stem cuttings but can be propagated quite successfully from pieces of roots or root cuttings. A few examples of plants propagated commercially by root cuttings are *Albizia julibrissin* Durazz and *Elliottia racemosa* Muhlenb. ex. Elliott, and others are presented in Table 23.1. Plants that sucker readily in nature and produce shoots from roots are generally excellent candidates for propagation by root cuttings. This technique is not commonly used for the propagation of plants because of the time and expense of obtaining the root pieces to be used as cuttings, particularly if the donor plant is in a field setting. Additionally, excising and harvesting of root cuttings may seriously injure or kill the stock or donor plant. You may have already inadvertently experienced the technique, and a familiar example illustrates this point—that of dandelion (*Taraxacum officinalis* Wiggers)—when even a small piece of root is left in a lawn after pulling the dandelion, a new plant will soon grow in its place from the remaining root piece.

STOCK PLANT CHARACTERISTICS

As is true for other cutting types, physiological stage, season of the year, and stored food in the propagules are important factors when propagating by root cuttings. Juvenility is an important consideration, with more successes resulting if cuttings are taken from young plants or from within the 'cone of juvenility' of more mature, older plants such as trees (Chapter 6). The cone of juvenility for the root region is almost a mirror image of the aboveground cone of juvenility for the aerial portion of the plant. The best season of the year for rooting cuttings varies with species. Excellent success with red raspberry (*Rubus idaeus* L.) was achieved when cuttings were taken from fall to spring, but little success was achieved when root cuttings were taken during the summer months (Hudson, 1955). In contrast, root cuttings of oriental poppy (*Papaver orientale* L.) excised during the summer were easy to propagate, whereas horseradish (*Armoracia rusticana* P. Gaertn.) can be propagated throughout the year by root cuttings. With many species, the size of the cutting is important, in part because of the stored food and the amount of surface available for adventitious bud formation. Note that the goal of propagating by root cuttings is to obtain new shoots that arise adventitiously or directly from root tissues—by definition, roots cannot have performed buds. Because these adventitious buds arise from internal root tissue, there may be instances where this form of asexual propagation does not result in a faithful reproduction of the parent plant. For example, root cuttings of some thornless cultivars of blackberry result in the production of new plants that are thorny. This phenomenon occurs when the characteristic, in this case, thornlessness, is a result of the stock plant being a chimera (Chapter 7). Similarly, variegated geraniums, aralias, and bouvardias propagated by root cuttings will produce plants with solid green leaves rather than exhibit the original variegated leaf pattern. Also, root cuttings are an inappropriate way to faithfully reproduce a plant that has been grafted onto a rootstock. Taking a root cutting would result in a new plant resembling the rootstock and not the scion (Chapter 21). Root cutting is only a viable alternative for propagation if the plant is rooted on its own roots (the roots are the same genetically as the aerial portion or scion) and not grafted.

EXERCISES

Experiment 1. Root Cutting Experiment

Propagators may use several aids and techniques, such as bottom heat, high humidity, and chemical treatments, such as cytokinins and auxins, to enhance the success of the process. They also often employ root cuttings to generate a crop of softwood cuttings (adventitious shoots) that can then be propagated by standard softwood cutting protocols (Chapter 17). Forcing shoots from root cuttings is an excellent way to re-introduce juvenility, and shoots harvested from root cuttings will often display juvenile characteristics, including the ability to root more easily. Propagators can take advantage of this when they have an older mature specimen that they would like to propagate, which is impossible or very difficult to root from traditional stem cuttings.

The following experiment will focus on the cutting size and orientation of the root cuttings of staghorn sumac (*Rhus typhina* L.; Figure 23.1). *Rhus typhina* readily develops shoots from roots (Figure 23.2) and is a good candidate for the use of root cuttings. Other readily available species to use as alternatives or in addition to *Rhus typhina* are heavenly bamboo (*Nandina domestica* Thunb.), leatherleaf mahonia (*Mahonia bealei* [Fortune] Carrière), and crape myrtle (*Lagerstroemia indica* L.). Many species are amenable to root cutting, and some alternative choices are noted in Table 23.1.

TABLE 23.1
Examples of Plants That May Be Propagated by Root Cuttings

Species	Common Name
Actinidia deliciosa (Chev.) Liang et Ferg.	Kiwifruit
Albizia julibrissin Durazz	Silk tree
Anemone hupehensis Hort. Lemoine	Japanese anemone
Aralia spinosa L.	Devil's walking stick
Armoracia rusticana P. Gaertn.	Horseradish
Campsis radicans (L.) Seem. ex Bur.[a]	Trumpet vine
Elliottia racemosa Muhlenb. ex Elliott	Georgia plume
Ficus carica L.	Fig
Gymnocladus dioica (L.) C. Koch	Kentucky coffee tree
Gypsophila paniculata L.	Baby's breath
Hydrangea quercifolia Bartr.	Oakleaf hydrangea
Ipomoea batatas (L.) Lam.	Sweet potato
Lagerstroemia indica L.	Crape myrtle
Mahonia bealei (Fortune) Carrière	Leatherleaf mahonia
Nandina domestica Thunb.	Heavenly bamboo
Papaver orientale L.	Oriental poppy
Phlox paniculata L.	Garden phlox
Populus spp.	Aspens, poplars
Pyrus calleryana Decne.	Callery pear
Rhus glabra L.	Smooth sumac
Robinia pseudoacacia L.	Black locust
Rosa spp.	Several rose species
Rubus spp.[a]	Red raspberries, blackberries
Sambucus spp.	Elderberries
Sassafras albidum (Nutt.) Nees[a]	Sassafras
Syringa vulgaris L.	Lilac
Ulmus parvifolia Jacq.	Lacebark elm
Wisteria floribunda (Willd.) DC	Japanese wisteria

[a] These species may be easily substituted for staghorn sumac.

FIGURE 23.1 *Rhus typhina* (sumac). (Courtesy of Dr. Tom Ranney, North Carolina State University.)

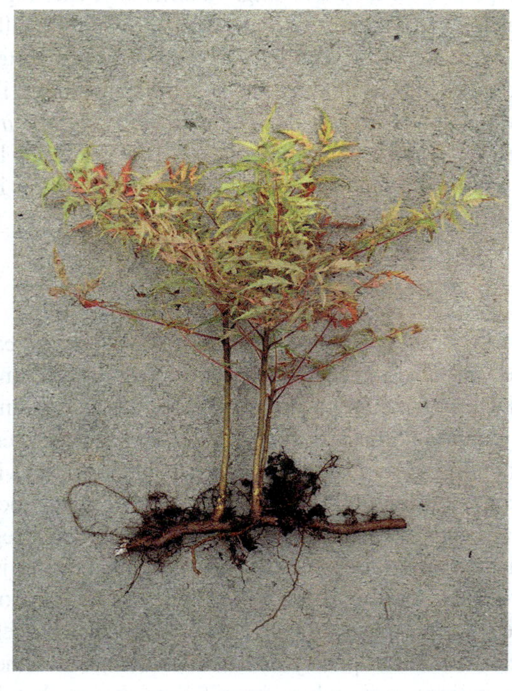

FIGURE 23.2 Adventitious shoots formed from the roots of *Rhus typhina* (Tiger Eyes). (Courtesy of Dr. Tom Ranney, North Carolina State University.)

HARVEST STORAGE

Unless the experiment is to be done immediately following the excision of the cuttings, it may be convenient to collect the cuttings in the fall and store them until they are to be used as propagules. With some species, such as *Robinia* and *Rhus*, fall collection of roots for root cuttings and then cold storage has resulted in greatly reduced success. Care must be taken to store cuttings collected in the fall in a cool place and in a medium that is not too wet; otherwise, rotting can occur.

MATERIALS

Students or student teams will require the following materials to complete this experiment:

- Cuttings from sumac or one or more of the substitute plants listed in Table 23.1. If storing the roots for prolonged periods, dipping the cuttings in captan (see label for rate) is advisable. Always wear protective clothing and follow all safety protocols
- Moist (but not wet) sphagnum peat moss for storing cuttings
- Refrigerator or cold room (5–10°C or 41–50°F) for storing cuttings

- Sharp knife or other utensil used for cutting the stock roots
- Potting medium (vermiculite, heat-treated sand, or commercial soilless medium), flats, labels, and marking pens
- Greenhouse or growth room at 21°C (70°F)

Follow the instructions outlined in Procedure 23.1 to complete this experiment.

Procedure 23.1
Adventitious Shoot Formation from Root Cuttings

Step	Instructions and Comments
1	Dig up the parent (stock) plants in sufficient numbers to provide an adequate number of cuttings based on student numbers and on how many experiments will be performed.
2	Cut roots from the stock plant. It is preferable to obtain roots of 1/4 to 1/2 in (6–12 mm) diameter and 1 to 2 ft in length (30–60 cm).
3	Remove soil adhering to the roots by shaking or washing with a strong spray of water, then allow the roots to dry.
4	If experiments are to be performed immediately, skip to step 7. If roots are to be stored, follow the instructions in steps 5 and 6.
5	Treat the root pieces with an appropriate fungicide such as captan, taking care to follow label directions and to use protective clothing and equipment as may be required.
6	Place the root pieces in moist (not wet) sphagnum moss and store in a cool place at about 5 to 10°C (41–50°F) until needed for the laboratory exercise.
7	When ready to begin the experiment, remove root pieces from storage and cut into pieces of equal length, about 3 in (~7.5 cm). Cut the proximal end perpendicular to the length of the cutting and make a slanting cut on the distal end. This will ensure that the proximal end can be placed up, and the distal end, down, as indicated in steps 12 and 13.
8	Sort cuttings by diameter to obtain 40 cuttings of 1/4-in (6 mm) and 40 cuttings of 1/2 in (12 mm) diameter. The sap of sumac may be irritating to persons with sensitive skin, so wash hands thoroughly after handling the cuttings.
9	Fill four flats with clean, moist propagating medium. A good propagating medium is 1/2 1:1 peat:perlite, giving a medium with good water holding capacity, but with excellent aeration.
10	Place 20 root cuttings of 1/2 in (1.25 cm) diameter horizontally on the medium in the first flat, about 1 in apart, and cover with 1 in of medium. Firm the medium gently to ensure good contact with the root cutting and to eliminate air pockets.
11	Place 20 root cuttings of 1/4 in (0.6 cm) diameter in the second flat and cover as in step 10.
12	Place 20 root cuttings of 1/2 in diameter vertically in the medium, about 1 in apart, and with about 1 in above the soil line. Be sure to push the distal (slanted cut) end into the medium.
13	Place 20 root cuttings of 1/4 in diameter vertically in the medium in the same fashion as in step 12.
14	Label each flat with your name, the date, and the treatment (horizontal 1/2 in, horizontal 1/4 in, vertical 1/2 in, and vertical 1/4 in).
15	Water flats to thoroughly moisten the medium and place them in a greenhouse or room at about 21°C (70°F). Monitor flats carefully to ensure that the medium does not dry out.
16	Record the date when the first shoots are seen. Three weeks later, remove the cuttings from the flats and record the number and length of adventitious shoots per cutting and the number and length of adventitious roots per cutting. Note where along the cutting that the shoots emerged. Present the mean shoot and root length numbers in a table.
17	If sufficient root cuttings are available, alternative experiments could include a treatment with the root cutting placed with the proximal end down into the medium and the distal end sticking up out of the medium against the direction of polarity or a treatment of the root cuttings with various growth regulators.

ANTICIPATED RESULTS

Within 2 to 3 weeks, the ends of the cuttings should exhibit callusing, and shoots will appear as early as 3 to 4 weeks. More shoots will emerge from the proximal ends of the cuttings since the cuttings exhibit polarity. Shoots will continue to emerge from the root cuttings (Figure 23.2) for many months, so a time for data collection should be specified by the group.

QUESTIONS/DISCUSSION

- Did horizontal or vertical cutting orientation result in more shoots? Roots?
- Were there differences between horizontal and vertical orientation in shoot or root length?
- What influence did cutting diameter have on propagation? Did it affect shoot or root numbers or size?
- Explain any differences observed and suggest how information derived from this exercise could benefit a commercial propagator.

LITERATURE CITED AND SUGGESTED READING

Creech, J. L. 1954. Propagating plants by root cuttings. *Proc Intl Plant Prop Soc* 4:164–167.

Cross, R. E. 1981. Propagation and production of *Rhus typhina* 'Laciniata,' cutleaf staghorn sumac. *Proc Intl Plant Prop Soc* 31:524–527.

Flemer III, W. 1961. Propagating woody plants by root cuttings. *Proc Intl Plant Prop Soc* 11:42–47.

Fordham, A. J. 1969. Production of juvenile shoots from root pieces. *Proc Intl Plant Prop Soc* 19:286–287.

Hudson, J. P. 1955. The regeneration of plants from roots. *Proc 14th Intl Hort Congr* 2:1165–1172.

MacMillan-Browse, P. D. A. 1980. The propagation of plants from root cuttings. *The Plantsman* 2:54–62.

Neilson-Jones, W. 1969. *Plant Chimeras*, 2nd Edition. London, United Kingdom: The Camelot Press, 123 pp.

Part VIII

Layering

24 Layering

Brian Maynard

Concept Box 24.1

- Simple layering: covering a bent stem with soil; also known as "tip layering."
- Compound layering: bending a flexible stem in several places and covering with soil; also known as "serpentine layering."
- Continuous layering: laying a shoot in a shallow trench before bud break and allowing new lateral shoots to grow up through the soil; also known as "trench layering."
- Mound layering: forcing numerous new shoots from a cutback plant and covering new shoots with successive layers of sawdust, leaf mold, or soil; also known as "stool layering."
- Air layering: wrapping a girdled shoot with sphagnum moss and plastic.
- Drop layering: replanting a shrub deep in the soil to promote the rooting of shoot tips.
- Basic goals in successful layering include the following:
  - Stimulating root formation by restricting the flow of sugar and auxins in the phloem
  - Excluding light while providing moisture, warmth, and proper aeration in a suitable root growth medium
  - Allowing time for suitable root development

Layering is a method of vegetative plant propagation that encourages adventitious roots to form on stems while they remain attached to the parent or "stock" plant. This ancient propagation method takes advantage of the natural ability of plant stems to initiate roots when they are broken or bent, and to touch the ground—often observed with the stems of rhododendron and the canes of blackberry. Layering can range from simple or tip layering, in which a stem is bent over and buried partially in the ground, to air layering, in which the stem is wounded and wrapped with a moist medium for rooting. By whatever method, plant propagation by layering is a relatively slow process, requiring 1 to 18 months, and increases the number of plants very slowly. Hence, layering does not often come to the mind of a plant propagator as a commercially viable technique. However, to vegetatively propagate difficult-to-root plants, in technologically simple propagation operations, or when only a small number of quality plants are needed, layering can offer a high rate of success without sacrificing the parent plant. Familiarity with the propagation technique of layering is a valuable horticultural skill.

Propagation by layering has a rich history in Europe, particularly in the production of hazels and magnolias, systems that can still be seen in England and Holland, and of fruit tree rootstocks in the states of Oregon, Washington, and the province of British Columbia. In the past, layering was the preferred method of propagating woody plants that needed to be propagated vegetatively (cloned) and on their own root systems, but that were difficult to root from stem cuttings. For certain plant species or cultivars, grafting was not convenient or not desirable (e.g., if there were a high incidence of rootstock suckering). Many of the shrubs that are now rooted under mist and grown in containers were layered in the past. Most growers now consider layer propagation too labor intensive. Yet, the majority of apple and pear rootstocks are still produced by the tens of thousands in extensive stool bedding fields. The mounding and harvesting processes that are considered too laborious for many crops, in this instance, are now highly mechanized.

For noncommercial propagators and hobbyists, layering offers a low-technology method of cloning a favorite plant without risking the loss of the stock plant. While it may take months longer than other propagation methods, success is almost guaranteed, and a larger plant with a better-developed root system can be had for very little cost. Furthermore, there is no risk of delayed graft incompatibility or rootstock suckering or struggling with fragile adventitious roots.

Because layered stems remain attached to the stock plant throughout the rooting period, they are not as sensitive to variations in humidity, temperature, and moisture as are stem cuttings set to root in a humidity tent or under mist. Through its attachment to the stock plant, the layered stem receives a continuous supply of water, nutrients, and carbohydrates. Plant water stress, the bane of many propagation methods, is kept to an absolute minimum.

Layering is particularly useful for plants that produce trailing branches that are able to develop roots where they touch the ground or that produce many suckers or stool shoots when cut back hard (i.e., stooled or coppiced). Indeed, some of the most successful ground cover plants, such as *Hedera helix* L. (English ivy), layer naturally where their stems contact the soil. Even specimen trees such as *Fagus* (beech) and *Carpinus* (hornbeam), which are notoriously difficult to root from stem cuttings, will root, after a time, where their branches touch the ground (Figure 24.1). This can result in a handsome grove of clonal trees. Many of the layering techniques, except for air layering and drop layering, take advantage of flexible stems that can be bent to the ground. This may be accomplished easily or might require some pruning in advance to stimulate vigorous lateral or basal growth, which is also more likely to form adventitious roots. Bear in mind, however, that the number of propagules obtained from each layered stock plant will likely range from one to several, not the hundreds that are available through cutting propagation or tissue culture. On the other hand, layered plants will often be larger and will have better developed root systems than plants started by other vegetative propagation methods, easing transplanting and establishment, and reducing the time needed to produce a salable plant.

FIGURE 24.1 Self-layered European hornbeam, Newport, Rhode Island.

While layering is wonderful in its simplicity, it is important to understand how layering works and what the basic requirements are for successful layering. The first stage of layering is the initiation of adventitious roots. This is accomplished by restricting the flow of sugars and hormones in the phloem, not by removing the stem entirely as with stem cuttings, but by constricting or girdling the stem. Full girdling is usually done only in air layering, where a strip of the living, or green, bark (including the phloem) is removed just below the point where roots are desired. Constricting the stem may be accomplished in various ways, ranging from strongly bending the stem, as in simple layering, to "wrenching" (twisting) the stem to damage the phloem, removing a ring of bark (girdling) (Figure 24.2), or placing a metal ring or wire around the stem so that it girdles the stem as it increases in girth over the growing season. In response to restricting the flow of materials from the buds and leaves, sugars and auxins build up, distal (on the side of the apical meristem) to the constricted or girdled point, and stimulate root initiation. It is common for adventitious roots to emerge first at a stem node, where leaf and bud traces exit the stem, although exceptions to this do occur. Take advantage of this by wounding or constricting the stem just below a node.

Another function of layering is to exclude light from the tissues where roots are expected to form. As you learn about the various layering methods, consider how

FIGURE 24.2 Ringing the stem of fig to initiate an air layer.

each incorporates some aspect of light exclusion from the portion of the stem where roots are expected to form. If light is excluded by shading a previously formed (usually green) stem, it is called "blanching." If a new shoot develops entirely in darkness, the process is known as "etiolation." Generally, etiolation promotes better rooting or successful rooting of a wider variety of plants but is harder to maintain because etiolated shoots are fragile and easily damaged by direct sunlight. Blanching green stems works for many plants, is easier to accomplish, and is more commonly used. In both situations, it is common to see the stem become pale yellow as it loses the green color associated with chlorophyll.

Once roots have initiated, they must elongate (grow out) into a healthy root system. At this stage, good horticultural practices come to bear. Key goals include protecting the roots from light; providing adequate (but not excessive) soil and atmospheric moisture, and suitable warmth; and using a growing medium that is well drained yet high in organic matter. If the soil was prepared properly, all of these goals should be met. Finally, the propagator must be patient. Layering requires time, anywhere from 2 weeks to 18 months, depending on the plant species, season, and size of the plant desired.

METHODS OF LAYERING

Simple Layering

In simple (common) layering (Figures 24.3 and 24.4), the stem to be layered is bent over into a small hollow in the soil, and the tip of the branch is bent back up again at a point 15 to 30 cm (6–12 in) behind the tip. The sharp bend in the branch, usually just below a stem node, restricts the movement of sugar and auxin, and redirects the stem to a nearly vertical orientation. It is also possible to wound the stem lightly by making a diagonal cut up and forward from the bend, no more than one-third of the way through the stem. A matchstick or other sliver of wood may be used to hold the slice open, and rooting powder may be added if deemed necessary. Alternatively, a simple wire twist or hog ring staple may be placed just above the bend and below a node to girdle the stem as it expands. Usually, the stem must be pegged into the soil at the base of the depression, just behind the bend, to prevent the shoot from pulling back out of the soil. The bent stem and peg are then covered with 8 to 15 cm (3–6 in) of soil firmed into place. If necessary, the emerging shoot can be supported with a short stake. The layers should be watered well and irrigated as needed. Once roots have formed, the layer may be dug up, cut free from the parent stem and grown on, or simply cut free behind the buried portion and left in place to develop a larger root system (Figure 24.5). Examples of plant genera (among many) that may be layered by common or tip layering include *Abelia*, *Buddleia*, *Cotoneaster*, *Deutzia*, *Forsythia*, *Ilex*, *Magnolia*, *Rhododendron*, *Rosa*, *Rubus* (tip layering), and *Viburnum*.

FIGURE 24.4 Preparation of a tip layer.

FIGURE 24.3 Simple or common layering.

FIGURE 24.5 A freshly harvested simple layer.

FIGURE 24.6 Compound or serpentine layering.

Compound Layering

Compound (serpentine) layering (Figure 24.6) is identical to simple layering, except that longer and more flexible stems are run in and out of the ground several times with stem nodes buried in subsequent depressions in the soil. Buried stems are bent or wounded as described under simple layering. Rooted layers can then be separated by cutting between each point in the system, closest to the next buried stem section. Examples of plants that are well suited to compound layering include *Campsis* (trumpet creeper), *Clematis*, English ivy, *Hydrangea petiolaris* (climbing hydrangea), *Vitus* (grape), and *Wisteria*.

Trench Layering

Trench or continuous layering (Figure 24.7) is slightly more involved and takes some advanced preparation, but is the method used to start most commercial apple root stock layering operations. One to two years before propagation, a stock plant is established and repeatedly cut back or pinched to develop multiple vigorous basal stems. In one variation, the stock plant is initially planted at an angle of about 30° above the horizontal and allowed to establish for a season. The layering process is started just before budbreak in spring by digging a 15-cm (6-in)–deep trench in which the stems and/or entire plants are laid down and covered with 5 to 13 cm (2–5 in.) of soil. Shoots from the buried stem grow vertically through the soil and are mounded again with soil as they develop. The method is thought to etiolate the base of the layered shoots as they develop, thereby increasing the rooting potential of hard-to-root plants. In apple root stock production systems, these layered shoots are harvested by running a rotating saw blade just above the buried stem; the soil is then swept free from the horizontal stem, and it is ready to be layered again as buds start to grow the following spring. Trench layering is also used commercially in Europe for the propagation of hazels, magnolias, lindens, and many other plants.

Mound Layering

Mound layering of stooled plants (Figure 24.8) takes advantage of the juvenile tissues associated with sucker shoots arising from roots or the lower stem of plants that have been hedged or cut back hard (stooled or coppiced). A plant established especially for stooling is cut back hard to 2.5 to 5 cm (1–2 in) above the soil in early spring. Shortly after new shoots begin to emerge from basal buds, soil, sawdust, wood shavings, or leaf mold is mounded to an 8 to 15 cm (3–6 in) height around the growing stems. If the stem is to be constricted with wire or a hog ring, this stricture would be applied just

FIGURE 24.7 Continuous or trench layering.

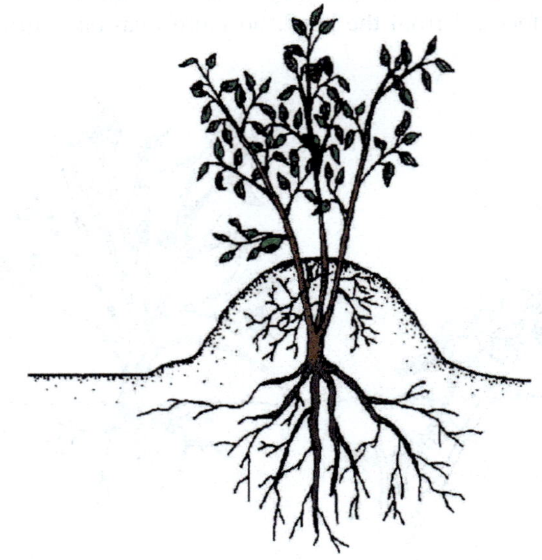

FIGURE 24.8 Mound layering.

before the first mounding. Three to four weeks later, the shoots are mounded again to a final depth of 15 to 23 cm (6–9 in). In commercial application, the stool beds may be enclosed in a 30-cm (12-in)–deep wood frame. Some nurseries have experimented with foliar applications of auxin to stimulate rooting, such as 200 ppm (w/v) indole-3-butyric acid (IBA) in an aqueous solution, sprayed once or twice to drip or run off in the hope that additional auxin taken up by the foliage will move to the rooting zone of the layered stems. Commercial mound layering is used for the production of *Pyrus* (pear) and *Malus* (apple) dwarfing rootstocks, and is used in Europe and the Pacific Northwest for the propagation of fruiting and ornamental cultivars of *Corylus avellana* (Figure 24.9). Mounding is also successful with *Cotinus* spp., *Syringa vulgaris* cultivars, and many species of *Prunus* (almond, cherry, and plum).

Air Layering

Air layering—also known as "pot layering," "Chinese layering," or even by the Chinese name "gootee" (Figure 24.10)—might be the first type of layering that comes to mind. This method is usually reserved for plants that have large or mature stems that cannot be bent to the ground easily or that do not sucker sufficiently to permit stool layering. Because of the intensive nature of this method, only a few branches can be air layered at a time. Air layering is accomplished by ringing or slice wounding the stem; sometimes treating the wound with a powder form of auxin; wrapping with moist sphagnum peat or milled sphagnum; and sealing with waxed paper, saran, or similar plastic wrap. Ringing is usually reserved for dicotyledonous plants that possess a ring of vascular cambium. Monocots are best wounded by slicing upward through the stem about 1/3 to 1/2 the way. Key factors for success include keeping the sphagnum moss moist, but not wet; maintaining a good seal around the stem to prevent the moss from drying out; and allowing sufficient time for new roots to form. The wound must remove a large enough ring to prevent regrowth of the bark over the girdle or must be held open with a toothpick or a wooden match stick to prevent healing if a slice wound is used. A bark girdle 1.3- to 2.5-cm (0.5–1.0-in) wide is usually sufficient. (Note: Students are sometimes tempted to just scratch the bark and must be encouraged to remove the bark down to the cambium.) Sometimes, an additional wrap of aluminum foil is used to exclude light and reduce solar heating of the layer. The term "pot layering" derives from the use of open, or split, bottom containers to contain the rooting medium around the stem, to the same effect as a plastic wrap. When roots have formed, the wrappings are removed, and the shoot is cut free below the root zone and potted up to grow on. Newly potted layers should be placed in a humidity tent or covered by a plastic bag and held in the shade for several weeks before being weaned to ambient light and humidity. Air layering is often recommended for the propagation of

FIGURE 24.9 Freshly harvested mound layer of *Corylus avellana* 'Contorta.'

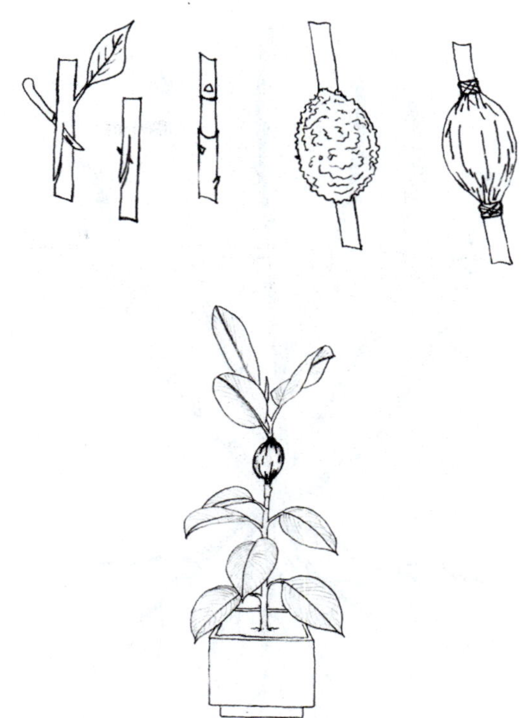

FIGURE 24.10 Air layering.

FIGURE 24.11 Drop layering.

indoor landscape plants such as ornamental figs, *Ficus elastica* (rubber plant), *Croton*, *Dieffenbachia*, and large slow-growing landscape plants such as *Magnolia*, holly, *Camellia*, and azalea.

Drop Layering

Drop layering (Figure 24.11) is a relatively uncommon method of propagation in which an entire stock plant is planted (i.e., "dropped") 15- to 23-cm (6–9-in) deep so that only the shoot tips are exposed. Because of the depth of the receiving hole that is required, this method is limited to small, well-branched shrubs such as heath, heather, and woody herbs. The setup of this method lends itself well to girdling the stem with a wire or a hog ring so that the stem is gradually constricted as the stem increases in girth. Drop layering requires excellent soil drainage to avoid low-oxygen stress or rotting of the plant stems. The stock plant is usually discarded after the procedure. Drop layering is reserved for small, fine-textured shrubs such as *Buxus* (dwarf boxwood), *Calluna* (heather), *Daphne cneorum* (Garland flower), *Erica* (heath), woody herbs (*Lavandula*, *Thymus*), and dwarf rhododendrons.

STEPS IN SUCCESSFUL LAYERING

Stock Plant Health

As with all other methods of vegetative plant propagation, the health of the stock plant is crucial for success. Plants suffering from water or nutrient stress will not form adventitious roots as readily. On the other hand, high levels of nitrogen in the stem tissue may also inhibit rooting. Depending on the method of layering, stems should either be growing actively (for stool mounding) or formed the previous growing season (simple or tip, compound, trench, and drop layering). Air layering may be done with older wood and, accordingly, may take longer to produce roots.

The act of cutting the stock plant back hard and repeatedly also leads to the development, in theory, of more juvenile growth, which usually roots more easily. While the various layering techniques may be attempted on, virtually, any plant (even certain monocot species), the best results will be obtained when well-established stock plants are cultivated for several years, by pinching and cutting back, to produce many vigorous basal shoots. A well-maintained layering bed is able to yield thousands of propagules each year.

Soil

If you will be layering in native soil, the soil should be screened before establishing the stock plants to remove stones that might interfere with digging holes and trenches for layering or with harvesting rooted layers. Improve the drainage and add organic matter, if necessary, to improve the characteristics of moisture retention, soil tilth, and soil warming. A sandy loam soil is a good starting point, but any soil may be suitably prepared as a layering bed so long as the drainage and moisture retention are optimized.

Some layering operations have been compromised by the buildup of soilborne pathogens, such as *Agrobacterium* (crown gall), *Thielaviopsis* (black rot), *Verticillium*, and *Phytophthora* (Chapter 11). This highlights the importance of selecting clean stock plants, pasteurizing or fumigating the soil, rotating the crop, or periodically reestablishing the layering bed.

Timing

There is a general agreement that the best time to start layering stems is late winter to early spring, before growth begins. *Rubus* (blackberry, loganberry) are tip layered in late spring and harvested the following autumn. Air layering of tropical plants may be initiated almost any time. The rooting capacity of many woody plant species begins to decline about 6 to 8 weeks after budbreak as the stem lignifies. Indeed, the rooting capacity of the stems of *Syringa vulgaris* (common lilac) and *Chionanthus* spp. (fringetree) drops off within 2 to 4 weeks of budbreak. Taking advantage of this "window of opportunity" for rooting might be the key to successful layering of some plant species. Similarly, if blanching or etiolation of the stem is an important part of the layering program, consider that the sooner light is excluded from the newly developing stem, the greater the rooting response will be. Blanching fully expanded shoots will have little or no effect on layering success. Light can be excluded from breaking shoots using a shade enclosure to yield truly etiolated stems and leaves. Mounding or air layering could then be applied, and the remaining nonlayered part of the shoot could be acclimatized gradually to higher light levels. Be forewarned that etiolated shoots can be quite brittle and burn readily if exposed to full sunlight, so extreme care must be taken if this method is attempted. Research has shown that heavy shading (60%–80% reduction of light) can be nearly as effective as full etiolation, and the resulting shoots are much less delicate.

Stem Treatments

As stated previously, restricting the flow of auxins and sugars moving from the shoot tip is fundamental for successful layering. Bending, girdling, wire wrapping, hog rings, slicing, or simply twisting (wrenching) the stem are all acceptable means of restricting the transport of materials in the phloem while allowing the free movement of water from the roots in the xylem. Which method works best with each layering method is mainly a matter of convention, for example, removing a strip of bark during air layering, although a wire girdle might work just as well. Experimentation is always encouraged. With any method of stem constriction, an additional auxin rooting compound may be added, either as a powder applied to the girdle or slice wound, as a foliar spray, or even as auxin-soaked toothpicks inserted into the stem or slice wound. One modification of air layering is to use 2.5-cm (1 in.) wide bands of Velcro™ hook-and-loop fabric dipped in rooting powder to exclude light, wound the stem, and apply auxin, all at the same time.

MAINTENANCE OF THE LAYERING PLANT

Part of the attraction of layering relative to other propagation methods is how little attention is needed during the rooting process. Unlike cutting propagation, the layer is supported by the parent plant and is much less likely to experience water or nutrient stress. However, drought, poor stock plant nutrition, or poor drainage can reduce root formation on layered stems. Light mulching, irrigation, and moderate fertilization of the stock plant might promote better layering success. Regular inspection of layered plants is absolutely necessary to evaluate soil moisture (particularly when air layering), to ensure that the rooting area remains covered and secure, and to identify and respond to rodent damage and insect or disease problems.

Monitoring Root Development

Roots should not be expected to form on woody plant layers sooner than 8 to 12 weeks after the layer is initiated. However, certain species and many herbaceous layers can be expected to start forming roots in as soon as 2 to 4 weeks. Root development appears to accelerate during late summer and early fall as carbohydrate storage increases. One of the attractions of air layering is the ability to see roots as they reach the outside of the sphagnum wrap, but resist the temptation to unwrap the rooting stem as the newly formed roots are weakly attached at first and are also susceptible to drying and injury from strong light. Simple layers may be inspected by digging gently in the soil, but take care not to dislodge the stem or fully expose the roots for more than a few minutes as they can dry out quickly.

Harvesting the Layer

As with evaluating root development, it is best to err on the side of caution when harvesting newly layered stems. Heavy root development, particularly if the stems were ringed or girdled and snapped off easily below the root zone, is a sign that the plant is ready to be harvested and potted on or replanted in the field or garden. Newly harvested layers should be potted into a moist soil and shaded for several weeks as new roots develop, depending on the time of the year. Well-rooted layers should be less susceptible to winter damage than newly rooted cuttings, but stem splitting may still be a risk in some climates. It is quite acceptable to re-cover a layering stem or sever the connection to the parent plant and leave the rooted layer in place for an additional 3 to 9 months. The rooted layer will be well established by the end of the following growing season, usually much sooner. Normal precautions for transplanting should be taken when moving rooted layers.

Preparing the Stock Plant for Next Year

Ideally, the layering process will not severely damage the stock plant, and new shoots will emerge the following spring. Soil or sawdust used in the mounding process should be removed so that the plant crown can dry out to prevent bark rot or root suffocation. In mound and trench layering, a number of stems are left un-ringed or just allowed to grow on to produce the next crop of stems for the following year. These should be pruned or supported as necessary to prepare for the next layering event. A light mulch or application of preemergent herbicide will help prevent weed growth in spring, although chemical herbicides should not be used if soil residues might interfere with root development in simple, compound, or trench layers.

EXERCISES

Layering is one of the simplest and lowest-technology methods of vegetative propagation, but requires the most time for rooting to occur and for plants to be ready to be harvested, and yields fewer propagules than most other methods. If the time for preparing the soil and stock plants is included, the entire process can take up to 3 years. Of course, layering may be the only method that will reliably produce the rooted shoots of particularly difficult-to-root or graft plants. The following exercises use easily rooted plants to demonstrate the method of layering. They also allow the student to observe the effects of girdling and other means of wounding, auxin application, and where roots form on layered plants. In concert with the Concepts section of this chapter, the student should be able to devise a protocol for layering more difficult-to-root plants.

Experiment 1. Drop or Mound Layering with Ringing

Both the mound (stooling) and drop methods of layering involve covering growing stems with soil or growing medium (Figures 24.8 and 24.11). Mounding is applied to stock plants cut back to ground level to produce a number of vigorous sucker-like shoots, sometimes called "stool shoots." These are covered by 15 to 23 cm (6–9 in) of soil as they grow in early summer and remain mounded until fall or the following spring. In the commercial mound layer propagation of *Corylus avellana* 'Contorta,' Harry Lauder's walking stick, hog ring staples are placed around the base of the young shoot to girdle the stem gradually as it expands during the growing season. In this exercise, a simple system will be constructed for observing the rooting process in mounded or drop layered plants and the effects of ringing or wounding stems. The steps below are just a guideline—you should be able to identify new methods of wounding or, otherwise, stimulating root formation. (Can you design a method for etiolating the stock plant before setting up the layer?)

MATERIALS

The following items are needed for each student or student team:

- A healthy, rapidly growing potted plant that has been pinched hard or cut back to within 2.5 to 3 cm (1–2 in) of the soil surface and has produced 5 to 10 vigorous, upright growing shoots *Dendranthema* × *grandiflorum* Kitam. (florists' chrysanthemum), coleus, *Mentha* spp., *Lavandula* spp., or similar plant material)
- Similar-sized pot, can, or sleeve of stiff plastic to hold soil over the potted plant. Transparent plastic soda bottles are ideal for this purpose
- Suitable soilless growing medium that has excellent drainage yet holds some moisture, and which can easily be shaken off rooted stems (e.g., sphagnum peat, vermiculite, perlite, growing mix)
- 16 to 20 gauge florists' wire suitable for tying around a stem
- Grafting or similar knife
- Low-strength powdered rooting compound (e.g., Hormodin 1, Hormex 1, Rootone) containing 1000 ppm IBA or 1-naphthalene acetic acid (NAA) or a combination of both auxins
- Plastic or wooden labels and pencil for marking

To complete the following exercise, follow the protocols listed in Procedure 24.1.

ANTICIPATED RESULTS

After 4 to 6 weeks, roots should be visibly emerging above wounded areas and, possibly, at stem nodes. Students should be able to note the location and degree of root production and relate their observations back to the treatments applied. If other groups used different rooting media or stem treatments, more can be learned about the influence of these factors on root initiation and development.

QUESTIONS

- Was girdling necessary for root initiation in this method of layering? Was the rooting powder necessary or helpful?
- Which other plant species that you are familiar with might be able to be propagated in this manner? Would other propagation methods be more efficient for the species you used in this exercise?

Experiment 2. Air Layering

Air layering is an excellent method for propagating large tropical plants that have lost their lower leaves or become leggy. While air layers are easy to set up, many fail because not enough attention is given to maintaining the proper moisture level in the sphagnum moss used for the layer (i.e., the moss dries out). If the plastic wrap is tied well enough and the initial moisture was adequate, the layer should be in good shape; if the wrap is too tight, water may collect in the layer and drown any developing roots; if the wrap is too loose, the moss may dry out. Rooting normally takes 8 to 16 weeks or longer—so be patient. (Note: *Ficus carica*, common fig, usually roots within 2–3 weeks and is one of the best species for this experiment.) The layer has succeeded when new white roots can be seen on the inside of the plastic wrap. Remember that, when harvesting the rooted layer, there might not be enough roots formed to support the plant

Procedure 24.1
Drop or Mound Layering with Ringing

Step	Instructions and Comments
1	Carefully strip the leaves off of the lower 3/4 of the stock plant.
2	Choose stems for control (no treatment), ringing with wire, slice wounding, girdling with a knife, or twisting a single strand of wire about the stem. Draw a diagram showing the treatments or controls in your notebook, making sure to mark the pot to orient the plant to the top of your diagram.
3	Apply treatments to a point midway between the base of the shoot and the foliage. Either use the knife to make a slice wound upward into the stem about 1/3 of the way through or wrap a strand of wire once around the stem and twist the ends together to cinch the wire firmly against the stem. Hold slice wounds open by inserting a small toothpick into the slice. Apply a small dusting of rooting powder to half of the slice wounds as an additional treatment; leave the rest untreated as a control.
4	Remove the bottom from the empty container and cut the container down one side to allow it to be pulled open and slipped around the stock plant. Reduce the height of the container if necessary so that the foliage of the stock plant will be just above the rim. Push the container lightly into the soil surface of the potted plant. Wrap the container with tape or a heavy elastic band to keep it from opening up when soil is added. If necessary, stake the container to prevent it from tipping over.
5	Add growing medium to the container, working the medium down around the shoots with your fingers or a thin stick. Add enough medium to fill the container or to a point just below the foliage. Water from the top until water drains from the bottom of the potted plant.
6	Monitor for insect or disease problems. Maintain adequate, but not excessive, moisture in the upper container. If necessary, enclose the entire system in a large plastic bag tied loosely at the top.
7	Check for rooting every 2 weeks and harvest when ready by undoing the container and shaking the growing medium off the stock plant. Make observations of the root location and the number or extent of rooting for all control and treated stems, referring to your diagram as needed. Remove and pot up rooted stems as desired.

at first, so pot it into a good growing mix, cover with a plastic bag tied loosely at the top, and gradually wean the plant to ambient light and humidity.

MATERIALS

The following items are needed for each student or student team:

- A healthy, rapidly growing, green potted (foliage) plant (e.g., *Ficus*, *Cordyline*, *Dracaena*, etc.)
- Grafting or other similar knife
- Low-strength powdered rooting compound (e.g., Hormodin 1, Hormex 1, Rootone) containing 1000 ppm IBA or NAA or a combination of both auxins
- Moistened whole sphagnum peat, milled sphagnum moss, coarse coir fiber, or shredded newsprint
- Plastic wrap or similar 1-mil (0.0254-mm)–thick clear plastic
- Waterproof tape, waxed string, or heavy elastic bands for tying the wrap
- Aluminum foil
- Plastic or wooden labels and pencil for marking

To complete the following exercise, follow the protocols listed in Procedure 24.2.

ANTICIPATED RESULTS

Roots will form at the upper edge of the girdled or sliced section of the stem. Once roots are visible at the surface of the layer, the plant may be harvested and potted up in a suitable growing medium. It may take 8 weeks or longer before roots become visible through the plastic and the air layer is ready to harvest.

QUESTIONS

- Did the type of wound or auxin treatment affect the type or amount of roots formed? Did an additional wrap of aluminum foil affect rooting? If yes, why?
- Was the air-layered stem ready to be harvested as soon as roots were visible, or should you have waited longer for more roots to form?

Procedure 24.2
Air Layering

Step	Instructions and Comments
1	Remove any leaves on the stem over a 15 to 30 cm (6–12 in) section of stem to leave about 23 to 46 cm (9–18 in) of leafy stem above the air layer.
2	Wound monocot or dicot stems by slicing 1/3 way through in an upward directed cut. Hold the slice wound slightly open with a toothpick or a small amount of moss. Wound dicots, but not monocots, by removing a 1.3- to 2.5-cm (0.5–1.0-in)–wide strip of bark down to the outermost wood.
3	Dust a small amount of rooting powder onto the wounded section of the stem. Alternatively, use toothpicks pre-soaked in a 1000 mg/L auxin solution to hold the slice wound open.
4	Squeeze a fist-sized ball of milled sphagnum moss to drain excess water and place around the wounded section of the stem. Hold this in place while wrapping with plastic.
5	Wrap a sufficiently large piece of plastic wrap around the sphagnum moss, join the edges of the wrap, and roll together to tighten the wrap against the ball of moistened sphagnum. Tape or tie the lower and then the upper edge of the plastic wrap firmly against the stem. If desired, place an additional wrap of aluminum foil around the finished plastic wrap. Twist in place, but do not tape or tie the foil.
6	Label the layered plant and record treatment data in your notebook.
7	Check the layer every 3 weeks by removing the foil wrap and observing for root development. Record data on the time that the first root become visible and describe the roots that are seen.
8	Harvest the layer when ready by removing the wraps and tape. With the moss and roots intact, cut below the layered stem and transplant into a suitable growing mix. Cover the potted layer with a plastic bag loosely tied around the base of the plant. Gradually wean the rooted layer to ambient greenhouse conditions.

SUGGESTED READING

Browse, P. M. 1988. *Plant Propagation*. New York: Simon & Schuster, Inc., 96 pp.

Hartmann, H. T., D. E. Kester, F. T. Davies, Jr., and R. L. Geneve. 2002. *Hartmann and Kester's Plant Propagation: Principles and Practices*, 7th Edition. Upper Saddle River, New Jersey: Pearson Education, 880 pp.

Macdonald, B. 1986. *Practical Woody Plant Propagation for Nursery Growers*, Vol. 1. Portland, Oregon: Timber Press, 669 pp.

Mahlstede, J. P. and E. S. Haber. 1957. *Plant Propagation*. New York: John Wiley & Sons, Inc., 413 pp.

Maynard, B. K. and N. L. Bassuk. 1988. Etiolation and banding effects on adventitious root formation. In Davis, T. D., B. E. Haissig, and N. Sankhla, eds., *Adventitious Root Formation in Cuttings: Advances in Plant Science Series*, Vol. 2. Portland, Oregon: Dioscorides Press, 315 pp.

Maynard, B. K. and N. L. Bassuk. 1992. Stock plant etiolation and stem banding effects on the cutting propagation of greening stems of *Carpinus betulus* L. *J Amer Soc Hort Sci* 117: 740–744.

Thompson, P. 1992. *Creative Propagation: A Grower's Guide*, 3E. Portland, Oregon: Timber Press, 220 pp.

Toogood, A. R. 1981. *Propagation*. Briarcliff Manor, New York: Stein and Day, 320 pp.

Toogood, A. R. 1999. *American Horticulture Society Plant Propagation*. New York: DK Publishing, 320 pp.

Wells, J. S. 1955. *Plant Propagation Practices*. New York: The MacMillan Company, 344 pp.

Part IX

Grafting and Budding

25 Grafting
Theory and Practice

Kenneth W. Mudge

Concept Box 25.1

- "Grafting" refers to the natural or deliberate fusion of plant parts so that vascular continuity is established between them and the resulting composite unit functions as a single plant.
- Grafting involves uniting two (or more) distinct genotypes into a compound genetic system in which the stock and the scion are fused anatomically, but which largely maintain their genetic identity throughout the life of the plant.
- Grafting has its broadest application with woody plants (tree fruit crops and ornamentals), but is also important in forestry and some herbaceous ornamentals and vegetables.
- Grafting is one of the several methods that are commonly used today for asexual (clonal) propagation.
- When grafting is used for clonal multiplication, clonal scions are usually grafted onto seedling understocks, but for some crops (notably fruit trees, that is, apples), genetically improved clonal rootstocks are used. The improved rootstocks are selected for disease resistance, effects on the scion (e.g., dwarfing), and performance in different soil environments.
- If the stock and the scion are sufficiently closely related to form a functional and durable graft union, they are termed "compatible"; if not, they are termed "incompatible." In general, the more closely related the scion and the rootstock are, the more likely the chances of creating a successful graft.
- For a graft to be successful, the vascular cambia of the scion and the rootstock must be brought into intimate contact with each other.

PROPAGATION BY GRAFTING AND BUDDING

Concepts and Definitions

Innovations in propagation have profoundly influenced plant domestication since the beginning of agriculture. Ten to twelve thousand years ago, unintentional selection of self-pollinating species that reproduced true-to-type from seed gave rise to wheat, barley, and other staples. Four or five thousand years later, the discovery of asexual propagation paved the way for the domestication of heterozygous species such as grape, olive, and fig, which rooted easily when, for example, a broken branch was stuck in moist soil. The domestication of more difficult-to-clone pome and stone fruits awaited the relatively recent discovery of grafting. Despite considerable advances in the science and technology of plant propagation since then, particularly over the last 50 years, grafting still plays an important role in the production of horticultural crops, including fruits, ornamentals, and even some vegetables.

"Grafting" refers to the natural or deliberate fusion of plant parts so that vascular continuity is established between them and the resulting composite unit functions as a single plant. Although two adjacent intact plants can become naturally or intentionally grafted together (approach grafting), or different branches of the same plant (autografting), most "horticultural" grafting involves deliberately cutting a shoot from a donor plant and inserting it into an opening in another plant growing on its own root system (detached scion grafting). The shoot piece cut from a donor plant that will grow into the upper portion of the grafted plant is referred to as the "scion," and the plant that receives and fuses with the scion and functions as the root system of the grafted plant is called the "stock" or "understock" (Figure 25.1).

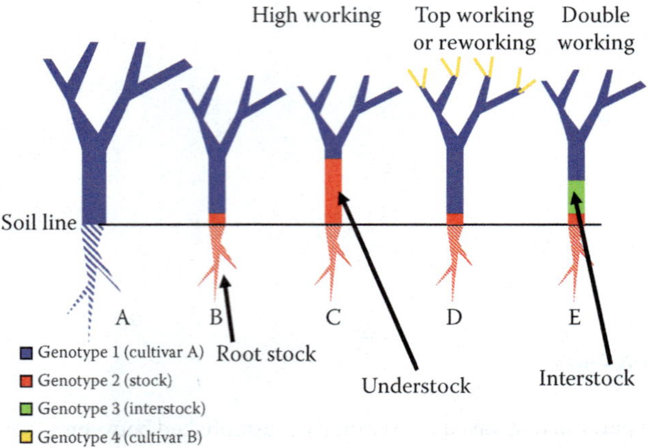

FIGURE 25.1 Components and configurations of grafted plants as compound genetic systems. See text for explanation. (Copyright K. W. Mudge, 2000.)

The term "rootstock" is similar, but not entirely synonymous with stock because "rootstock" implies that only the root system is derived from the stock (Figure 25.1B), whereas the terms "stock" and "understock" are more appropriate when the lower portion of the grafted plant includes not only the root system, but also some portion of the shoot system on which the scion is grafted (Figure 25.1C, D), which may include most of the tree. In the simplest case, grafting consists of one stock and one scion genotype (the genetic identity or DNA sequence of an organism), but in some applications, a third genotype (or more) is involved. "Double working" refers to a genetically compound plant (usually a tree) consisting of three genetically distinct parts, separated in linear sequence by two graft unions, as illustrated in Figure 25.1E.

Grafting involves uniting two (or more) distinct genotypes into a compound genetic system in which the stock and the scion are fused anatomically but otherwise maintain their genetic identity throughout the life of the plant. In sexual reproduction (Chapter 5), the genes from two parent plants recombine in a single cell (zygote) that grows into a plant made up of genetically uniform hybrid cells throughout. A grafted plant is not a hybrid in this sense, but rather more aptly described as a constructed chimera. A natural chimera is a single plant composed of some configuration of genetically different tissues, either arranged in separate layers or sectors (Chapter 7).

Although each part of the new plant resulting from the fusion and subsequent growth of the stock and the scion maintains its original genetic identity (DNA sequence) and much of its original phenotype, the stock and the scion (and the interstock in the case of double working) do have an influence on the growth and development of each other. For example, a dwarfing (size-controlling) rootstock results in a decrease in the vegetative vigor and ultimate size of the scion growing on it, while at the same time, a more vigorous scion accelerates the growth of a less vigorous rootstock, causing it to grow larger than if it were not grafted. The term "stion" (stock + scion) was introduced in 1926 by H. J. Webber to emphasize how this reciprocal influence results in a functionally unique composite organism that is fundamentally different from either of the original donors or any other combination of stock and scion genotypes. More recently, the term "stion" has been replaced by the wordier, but perhaps more descriptive, term "compound genetic system," which stresses the almost universally held notion that the stock and the scion each maintain their own genetic integrity, rather than a mixing of parental genes (hybridization) as what occurs in the offspring of sexual reproduction.

An understanding of this concept that the genetic make-up of a scion is not altered by the genotype of the stock, or vice versa, is important for a practical understanding of grafting. For example, a scion taken from a 'Red Delicious' apple tree and grafted (e.g., top worked) onto an 'Empire' apple tree does not grow into branches that bear 'Empire' apples (stock genotype), nor does it grow apples that are a hybrid between 'Red Delicious' and 'Empire' (i.e., graft "hybridization"). It grows only 'Red Delicious' apples. This is why the term "compound genetic system" is a more accurate descriptor of a grafted plant than "hybrid genetic system."

Nonetheless, throughout the history of grafting, there have been claims that a scion took on the characteristics of the stock. As far back as the 16th century, an Italian named Giambattista della Porta, in his vast and learned tome known as *Natural Magick*, claimed that the fruit of a red mulberry tree (scion) grafted to a poplar tree (stock) produced white mulberries, which might be interpreted as some sort of "graft hybrid." These and similar claims were little more than wishful thinking. In modern times, however, similar claims have been made. For example, Japanese researchers grafted two pepper varieties (*Capsicum annuum*) each known to produce fruit (peppers) of distinctly different morphologies. New growth from the scion of the grafted plant did not produce the fruit shape typical of that variety, but rather an intermediate shape between the fruits of the stock and the scion. Furthermore, this modified morphology fruit of the scion variety was reported to be heritable (passed on to seedlings) (Mudge *et al.*, 2009). Additionally, another researcher claimed that these apparent pepper graft hybrids were brought about by the migration of DNA (chromatin) across the graft union, which, presumably, was incorporated into the scion genome (DNA). Nevertheless, these claims regarding pepper graft hybridization have been viewed with skepticism and have not dispelled the prevailing notion of the genetic integrity of the stock and the scion, that is, the lack of permanent genetic alteration of one by the other. Recently, however, an alternative hypothesis explaining these and similar observations has emerged, involving the epigenetic alteration of gene expression rather than permanent genetic change. This alternative hypothesis involves "gene silencing," which refers to the "switching off" of a gene rather than an alteration of its DNA. For example, in *Nicotiana benthamiana*, grafting of a scion expressing a coat protein (CP) gene (not silenced) onto a stock bearing a silenced version of this gene resulted in the silencing of the CP gene in the scion (Sonoda and Nishiguchi, 2000). Their work showed that some chemical signal was translocated across the stock/scion junction, which resulted in the silencing of the CP gene in the scion. However, the chemical nature of this signal was not determined. More recently, it has been hypothesized that the graft-transmissible factors associated with gene silencing are (at least in some cases) small RNAs (23 base pairs) known as "micro-RNA" (miRNA). These are posttranscriptional regulators of gene expression that block the translation of mRNA into protein.

"Potato tuberization," known to be associated with miR172, is normally induced by short days (SD). Grafting of a scion from an induced plant (exposed to SD) to a stock plant grown under long days (not induced to tuberize) resulted in the tuberization of the uninduced stock as well as an increase in the activity of miR172 in the stock, despite the fact that it was not exposed to SD (Martin *et al.*, 2009). This suggests, but does not prove, that the miR172 was translocated from the scion to the stock, which may have been involved in gene silencing. However, there is clear evidence of miRNA translocation across graft unions. In *Arabidopsis*, for example, Buhtz *et al.* (2010) showed that miRNAs 395 and 399 were translocated from the scion to the stock. These studies suggest that miRNA signaling from the stock to the scion (or vice versa) may have significant gene silencing–related effects on the growth and development of grafted plants. If so, this might account for some of the earlier claims that grafted scions take on some characteristics of the stock that they are grafted to. Gene silencing may be adding a whole new dimension to the concept of grafting as a compound genetic system. As of yet, there is no indication that gene silencing plays an important role in the horticultural applications of grafting, but if it could be "harnessed" by molecular plant breeders, it might play an exciting role in the future.

Grafting qualifies as an asexual or clonal propagation method in so far as the scion-derived portion of the grafted plant is a genetic replica of the scion donor plant. Keep in mind, however, that since the rootstock retains its unique genetic identity, different from the scion, the rootstock is in no way a clone of the scion donor plant. Understocks themselves are either clonal if propagated asexually (by some method other than grafting) from a stock donor plant or seedling if propagated from a normal zygotic (not apomictic) seed. This is an important distinction in terms of rootstock performance as will be seen below and from the standpoint of nursery production.

The term "grafting" or "graftage" can be used either in the broad sense to refer to the process resulting in the fusion of any configuration of stock and scion, which results in a genetically compound plant, or more narrowly to refer to the use of a scion shoot piece that consists of multiple buds, whereas the term "budding" or "bud grafting" refers to the use of a scion that consists of only a single bud. Ultimately, this is a minor distinction because there are no fundamental differences in the post-grafting (or budding) growth or performance of a grafted or budded plant.

Grafting has it broadest application with woody plants, especially tree fruit crops and ornamental trees and shrubs, but it also plays an important role in modern clonal forestry and even in the production of vegetables (Chapter 27) and a few ornamentals (e.g., cactus). In Japan and Korea, for example, much of the greenhouse production of cucurbit vegetables (melons, cucumbers) and tomatoes relies on grafting selected scions onto selected seedling rootstocks that confer resistance to root system–associated diseases such as *Fusarium* and nematode pests, thereby increasing the production period by as much as several months (Lee, 2003).

GRAFT UNION FORMATION

Although the anatomy and physiology of graft union formation varies among different species, methods of grafting, seasons (timing), and the process of graft union formation share certain common features across all these variations. Within hours or days of the placement of a properly cut and aligned stock and scion, cell division occurs near the cut surface, in the vicinity of the vascular cambia, of the stock and the scion (Figure 25.2), eventually filling any reasonable gap that might exist between the two with intermingled and, to some

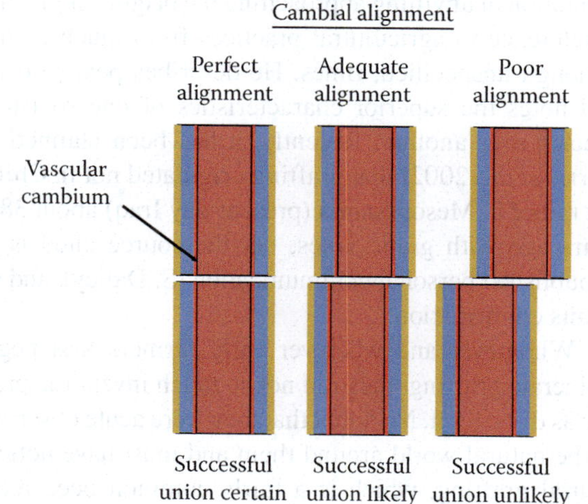

FIGURE 25.2 Alignment of vascular cambia of the stock and the scion. (Copyright K. W. Mudge, 2001.)

extent, interlocking callus cells. Differentiation of new meristematic cells proceeds from the cut edges of the existing cambia of the stock and the scion inward, until a new vascular cambium cylinder forms a continuous bridge between the stock and the scion, across the callus. Through the same process that produces new xylem and phloem from any established vascular cambium, cell division of this new bridging cambium generates cells to the inside, which differentiate into a new xylem, and cells to the outside, which differentiate into a new phloem (inner bark). Along the outer perimeter of these bridging tissues, periderm (bark) formation occurs, and eventually, the junction between the stock and the scion is (more or less) indistinguishable from normal (non-grafted) tissues. All these growth and developmental processes that differ in timing and spatial sequence among species and season are influenced by temperature, cellular water relations, phytohormone gradients, and poorly understood factors related to genetic compatibility. The resemblance of this process to wound healing in non-grafted plants is more than coincidental, as has been pointed out by several authors. It could be argued that the only fundamental difference between graft union formation and wound healing is the absence of periderm (bark) formation that would form a barrier between the stock and the scion.

NATURAL AND HUMAN HISTORY OF GRAFTING

The question of who invented grafting is difficult to answer. It has been claimed by some modern authors that grafting was invented 3000 or 4000 years ago in China, but definitive historical evidence is hard to find. The earliest written account of grafting in China may be no older than the book *Ch'I Min Yao Shu* (Essential Techniques for the Farming Populace) written by Jia Sixie sometime between 500 and 600 A.D. Its only (weak) claim to documentation of anything approaching the origins of grafting is reference to agricultural practices from much earlier, although unspecified, times. He describes pear grafting and notes the superior characteristics of one rootstock species over another. Recently, it has been claimed by Harris *et al.* (2002) that grafting originated not in China, but rather in Mesopotamia (present-day Iraq) about 3800 years ago with grape vines, but the source cited is an unpublished personal communication (S. Dalley), and so, awaits confirmation.

Whomever and wherever early farmers first began deliberate grafting, they did not so much invent the practice as discover it. No doubt that they were acute observers of the natural world around them and must have noticed natural grafting, which is a fairly common occurrence between either the branches or the roots of many species of trees. Natural shoot grafting results in the fusion of two branches within the canopy of a single tree and, even occasionally, between the branches of adjacent trees. Natural root grafting, which is even more common than natural shoot grafting, occurs extensively within the root systems of many species, including white pine (*Pinus strobus*), birch (*Betula alleghaniensis*), and white cedar (*Chamaecyparis thyoides*), and many others. In some cases, natural root grafting results in the interconnection of adjacent trees (Figure 25.3), with potentially important ecological and agricultural consequences. In natural forest or managed landscapes, root graft–interconnected trees may share water, nutrients, and even diseases. For example, Dutch elm disease can spread from tree to tree through root grafts between American elms (*Ulmus americana*), and viral transmission across root grafts has been documented in citrus orchards. Having noticed natural shoot or root grafting, early agriculturists probably first mimicked it using some version of approach grafting in which the shoots of both trees remain attached to their own root systems while the graft union is forming. Eventually, the more challenging practice of detached scion grafting enabled the cloning of pears, plums, and probably, citrus by the Chinese.

In early classical times (~500 B.C.), knowledge of grafting probably migrated along with the earliest domesticated apple trees, from Central Asia to Western civilization via Greece. In 323 B.C., the Greek philosopher Theophrastus (a student of Aristotle), who is sometimes called the "Father of Horticulture," described cleft grafting accurately, but offered this odd description of the process of graft union formation: "The twigs [scion] use the stock as a cutting uses the earth…a kind of planting and not a mere juxtaposition." This would seem to imply that the scion roots (as a cutting) into the stock, which is certainly not the case. In about 29 B.C., the Roman poet Virgil wrote extensively about grafting in his epic poem, the *Georgics*, which functioned as a sort of extension bulletin for gentlemen farmers. The following description

FIGURE 25.3 Natural root grafting of Atlantic white cedar. (Copyright K. W. Mudge, 2001.)

clearly shows his understanding of the essential elements of cleft grafting and is indeed poetic in its expression: "…in knotless trunks is hewn a breach, and deep into the solid grain a path with wedges cloven; then fruitful slips are set herein, and—no long time—behold! To heaven upshot with teeming boughs, the tree strange leaves admires and fruitage not its own." On the matter of graft compatibility, however, Virgil was very much mistaken, suggesting that his information was secondhand and that he had little direct experience: "But the rough arbutus with walnut-fruit is grafted; so have barren planes (*Platanus* sp.) ere now stout apples borne, with chestnut-flower the beech, the mountain-ash with pear-bloom whitened o'er, and swine crunched acorns 'neath the boughs of elms."

REASONS FOR GRAFTING

There are almost as many reasons for grafting as there are ways to graft (see below). Grafting has been used for thousands of years to propagate plants asexually, which cannot readily be cloned by other methods. As advances have been made in other vegetative propagation techniques, grafting merely for the sake of propagation per se has declined in importance while the use of grafting to take advantage of the unique combinations of the characteristics associated with compound genetic systems has increased. These and several other reasons for grafting are discussed below.

GRAFTING FOR CLONAL MULTIPLICATION

Grafting is used as an alternative to cuttage or layerage for the propagation of difficult-to-root species. Along with cuttage, layerage, division, and micropropagation, grafting is one of the several methods that are commonly used today for asexual propagation. For some species, grafting is the only, or at least the most practical, method of cloning despite some dramatic improvements in cutting propagation technology (e.g., misting, synthetic rooting compounds, and polyethylene) and the introduction of micropropagation over the last several decades. Bud grafting onto seedling understocks is the most cost-effective way to propagate many clonal shade tree cultivars, such as Norway maple (*Acer platanoides* L.), red maple (*A. rubrum*), sugar maple (*A. saccharinum* L.), honey locust (*Gleditsia triacanthos* L.), ash (*Fraxinus* spp.), Ohio buckeye (*Aesculus glabra* Willd.), horse chestnut (*A. hippocastanum*), and the few oak (*Quercus* spp.) cultivars in the nursery trade. Clonal selections of unusual growth forms of otherwise common woody ornamentals are often propagated by grafting to seedling understocks. Selections in this category include cultivars of spruce, fir, pine, and other conifers that have been selected for dwarfness or other unusual growth habits, such as weeping, or variegation. Grafting is also used to propagate unusual growth forms of deciduous species, including filberts (hazelnuts) (*Corylus* spp.), flowering cherries (*Prunus subhirtella*), Japanese maple (*A. palmatum*), crabapple (*Malus* spp.) and others selected for weeping, "corkscrew" habit, fasciations, variegation, dissected leaf forms, and others.

Although grafting for the propagation of otherwise difficult-to-clone species usually involves detached scion grafting or budding, approach grafting is occasionally used for especially difficult-to-graft species because graft union formation occurs while the scion is still attached to its own root system (Figure 25.4).

When grafting is used primarily for clonal multiplication, clonal scions are usually grafted onto seedling understocks rather than selected clonal rootstocks. Investment in elaborate clonal rootstock breeding programs is not justified when clonal propagation is the only goal of grafting since the genotype of the understock is relatively unimportant and essentially any seedling will do as long as it is graft compatible with the scion.

In most cases discussed so far, and indeed for most grafted plants, the rootstock becomes a permanent part of the grafted plant for the remainder of its lifespan regardless of the reason for grafting. In the special case of nurse grafting, however, the scion's dependence on the stock is only temporary—long enough for the scion to become self-rooted. Hence, nurse grafting is used to facilitate the asexual propagation of difficult-to-root species, such as lilac, clematis, and avocado. A difficult-to-root scion is grafted to a seedling understock (in many cases, just a piece of root, as illustrated in Figure 25.5) and planted unusually deep so that the graft union is below ground level. Here, the buried base of the scion becomes self-rooted in response to the dark, moist environment. This is really a layering process that is facilitated by grafting and, hence, should be called "nurse graft layering," but it is usually referred to simply as "nurse grafting" or "nurse root grafting" (when the stock is a root piece).

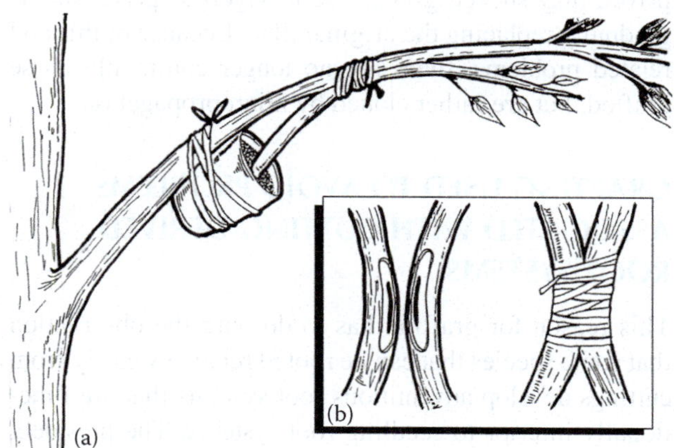

FIGURE 25.4 Approach grafting used for the propagation of difficult-to-root species. (a) Potted understock plant tied to scion donor tree. (b) Splice graft method and wrapping used to attach the scion and the stock. (Drawings by J. Aluko and A. Gachucha.)

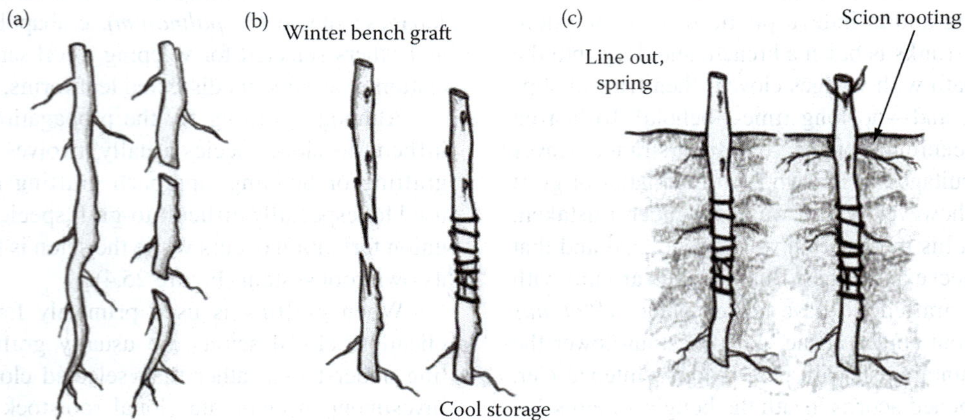

FIGURE 25.5 Nurse root grafting. Roots are cut into 4 to 8 cm sections (a), joined by whip and tongue graft (b), and planted deep, below the graft union, to encourage scion rooting (c). (Illustration by Vanessa Gray, 1999.)

Nurse grafting is a rare case where graft incompatibility (described below) is an advantage rather than a disadvantage since the later stage of the process involves failure of the graft union and death of the stock while the scion lives on, on its own root system. In the case of the nurse root grafting of lilac (*Syringa vulgaris* L.), for example, delayed incompatibility is encouraged by using a stock of a different genus (California privet [*Ligustrum ovalifolium*]) within the same family (Oleaceae). In this case, the stock and the scion are related closely enough (same family or intrafamilial graft) to form a temporary graft union, but distantly enough related (different genera or intergeneric graft) that the graft is not likely to succeed in the long run because of genetic incompatibility. In fact, a problem associated with this propagation strategy is unintended survival of the stock, which, in the case of privet, may sucker, giving rise to vigorous privet shoots gradually replacing the original lilac. Because of this and related problems, lilacs are no longer commonly nurse grafted, but are rather cloned by micropropagation.

GRAFTING USED TO AVOID PROBLEMS ASSOCIATED WITH CUTTING-DERIVED ROOT SYSTEMS

This reason for grafting has to do with the observation that some species that can be rooted relatively easily from cuttings develop adventitious root systems that are functionally inferior to seedling root systems. The president of Princeton Nursery, William Flemer, pointed out several examples of this, including *Cornus florida* 'Rubra,' *Acer palmatum* 'Bloodgood,' and *Pyrus calleryana* 'Bradford,' in his 1989 presentation to the International Plant Propagator's Society entitled "Why we must still graft and bud."

GRAFTING FOR INDEPENDENT OPTIMIZATION OF STOCK AND SCION GENOTYPES

By combining the useful characteristics of scion and stock genotypes, a compound genetic system (stion) can be created, which may be more agriculturally useful than either genotype growing on its own root system (non-grafted). Compound genetic systems can be used to achieve multiple objectives that would be difficult, if not impossible, to find or breed into a single (non-grafted) genotype. When grafting is performed to take advantage of a compound genetic system, the specific genotypes of both the stock and the scion (and the interstock in the case of double working) are important to achieving the desired ends, in contrast to grafting for clonal multiplication in which the genotype of the stock is relatively unimportant and seedling rootstocks generally suffice. Hence, when grafting is used for independent optimization of the stock and the scion, the rootstock genotype is critically important. This can be accomplished in some cases by choosing rootstocks at the species level, that is, rootstocks of a species different from the scion, which confers one or more useful characteristics not present in the scion. Although of different species, obviously, the stock and the scion must be related closely enough (same genus or at least same family) so that the two are compatible. Examples of interspecific grafting for independent optimization of stock and scion genotypes include the use of Northern California black walnut (*Juglans hindsii*) as a rootstock for English walnut (*J. regia*), peach (*Prunus persica*) rootstock for almond (*P. amygdalus*), and horned melon (*Cucumis metuliferus*) as a Fusarium-resistant rootstock for cucumber (*C. sativus*) (Lee, 2003).

In contrast to interspecific compound genetic systems, even greater optimization can be achieved through selection at the intraspecific level, that is, breeding and/or selection of desirable rootstock clones within the same species. The most technologically advanced use of genetically optimized rootstocks has come about through long-term systematic clonal rootstock breeding programs for several of the world's most important temperate fruit crops, including apples (*Malus domestica*), pears (*Pyrus communis*), several stone fruits (*Prunus* spp.), and grapes (*Vitis* spp.).

In fact, rootstock breeding would be a relatively simple task if genetic improvement of plant cultivars could be accomplished by selecting for a single desirable trait, such as rootstock control of scion vigor (size control) or rootstock disease resistance. More realistically, however, a successful rootstock (or scion) genotype is chosen based on a complement of several genetically controlled characteristics that give rise to a desirable set of phenotypic characteristics. Unlike scion cultivars that are selected for characteristics that affect fruit yield and quality and vegetative growth, a desirable rootstock genotype is selected for the following: (1) adaptation to belowground environmental constraints, (2) effect on scion vigor and other scion phenotypic characteristics, and (3) resistance to root system diseases and pests, as well as to shoot (aboveground) diseases and pests during the pre-grafting nursery production phase.

BIOTIC AND ABIOTIC STRESS RESISTANCE/TOLERANCE

In many crops, rootstocks have been chosen because of their natural resistance to or tolerance of various biotic stresses, including bacterial, fungal, and viral diseases; insect and non-insect pests, including nematodes; and even small herbivorous mammals (voles and others), as well as tolerance of adverse environmental (stress) conditions. Examples for each of these categories are shown in Table 25.1. Grapes are a dramatic and historic example of the importance of pest-resistant rootstocks. Wine grape production (primarily, *Vitis vinifera*) in Europe during the 19th century was nearly eliminated because of an aphid-like root-feeding insect pest known as phylloxera (*Daktulosphaira vitifoliae*). The existence of the wine grape industry today in phylloxera-infested regions (most of the world where grapes are grown) is heavily dependent on the use of phylloxera-resistant rootstocks that have been bred from naturally phylloxera-resistant North American grape species (e.g., *Vitis labrusca*). Similarly, the apple industry in New Zealand and Australia and other parts of the world was based on the original East Malling (EM or M) series clonal rootstocks during the early part of the 20th century, but it was seriously compromised because of infestations of wooly apple aphids (*Eriosoma lanigerum*), which feed on the shoots of apple rootstock plants during the pre-grafting phase of nursery production. EM rootstocks, chosen primarily for their effect on the size (vigor) of the scion cultivar, were highly susceptible to wooly aphid damage. The Malling Merton (MM) breeding program was undertaken largely to incorporate wooly aphid resistance by crossing susceptible EM cultivars with the highly resistant Northern Spy apple. The result was the MM series of size-controlling rootstock that are resistant to wooly aphids.

Rootstocks are also chosen for the tolerance of adverse abiotic environmental stresses, including low temperature, and soils that are drought prone or excessively wet or poorly drained. Some examples can be seen in Table 25.1.

ROOTSTOCK EFFECTS ON SCION VIGOR AND VICE VERSA

Rootstocks are selected not only for characteristics that directly affect the root system and/or the shoot system of the pre-grafted rootstock plant in the nursery, but also for rootstock effects on scion growth and development. Rootstock genotype can affect scion phenotype in different ways, but most importantly, in terms of scion vigor. Although less important, rootstock genotype can also affect fruit size and fruit quality.

Vegetative vigor and ultimate plant size can, of course, be influenced by many environmental and genetic factors. Selection for genetically controlled dwarfness is considered a desirable variant from the norm in many species. The centuries-old quest for dwarf fruit trees is motivated by the fact that small trees are easier to harvest and to apply pesticides to, and often have higher fruit yields per acre than full-sized trees. Dwarf selections of ornamental trees and shrubs are valued mainly as specimen plants, that is, novelties. Genetic dwarfness is brought about by a decrease in vegetative vigor. With respect to grafting, it has long been understood that grafting a "normal" (non-dwarf) scion onto a genetically dwarf rootstock will result in dwarfing of the scion. However, there may be other causes of rootstock-associated scion dwarfing, such as a mild incompatibility between the stock and the scion. As has been discussed above, rootstock effect on the vegetative vigor (size control) of scion cultivars was the primary selection criterion for the EM and MM series apple rootstocks that launched the modern era of clonal tree fruit rootstock selection and breeding, and it continues to be an important selection criterion in essentially all apple rootstocks and many other tree fruit breeding programs today.

The terms "dwarf," "semidwarf," "vigorous," and "very vigorous" have been used to describe the range of size-controlling rootstocks in apple, but they are misleading because the range of size control falls along a

TABLE 25.1
Examples of Biotic Stress Resistance and Abiotic Stress Tolerance in Rootstocks

Stress Category	Stressor[a]	Crop	Susceptible Rootstocks	Resistant/Tolerant Rootstocks
Biotic (disease or pest)				
Microbial				
Bacterial	Fireblight (*Erwinia amylovora*)	Apple	Budagovsky (Bud.) 9, Ottawa (O) 3, EM.9, MM.111	Geneva (G) 11, G.65 EM.7, Robusta.5
		Pear	Quince	OH × F (Old Home × Farmingdale) series
	Bacterial canker (*Pseudomonas* spp.)	Cherry	Mazzard	Colt, Mahaleb
	Crown gall (*Agrobacterium tumefaciens*)	Cherry	Colt	Mahaleb
		Walnut	Paradox	Northern California black walnut (*Juglans hindsii*)
Fungal and fungal-like organisms	Crown/root rot (*Phytophthora* sp.)	Apple	EM.26, MM.106	G.65
		Walnut	N. Cal. black walnut	Paradox
	Oak root fungus (*Armillaria* sp.)	Plum	Myrobalan 29-C	Marianna 26-24
Viral	Tomato ringspot virus	Apple	MM.106	
	Tristeza virus	Citrus	Sour orange (*Citrus aurantium*)	Rough lemon (*Citrus jambhiri*), Citrange (*C. sinensis* × *Poncirus trifoliata*)
Pest				
Insects	Phylloxera (*Daktulosphaira vitifoliae*)	Grape	Harmony and Freedom (*Vitis vinifera hybrids*)	*Vitis riparia × rupestris hybrids*
	Wooly aphid (*Eriosoma lanigerum*)	Apple	EM.9, EM.26	MM.111
Nematodes	Root knot nematode (*Meloidogyne* sp.)	Peach	Lovell peach	Nemagard peach
		Walnut	N. Cal. black walnut	Paradox
		Coffee	*Coffea arabica*	*C. canephora*
Mammal	Vole (*Microtus*)	Apple	Bud. 9	Novole
Abiotic				
Low temperature		Apple	MM.111, EM.7	EM.26, Ottawa 3
		Peach	Nemagard	Lovell, Guardian
Wet soils		Peach	Nemagard, Viking	Citation
		Cherry	Mahaleb	Colt
		Apple	EM.26	EM.7, MM.111
Dry soils		Apple	Mark	MM.104

[a] Name of disease (scientific name of disease organism) or common name of pest (scientific name of pest).

continuum rather than in discrete steps. Furthermore, the ultimate size of the grafted tree is the result of the interaction between the vigor of the stock and that of the scion. Each affects the other. For these two reasons, it is more useful to express the size-controlling effect of a given rootstock clone as a percentage of the size of the tree that would have resulted if the scion cultivar were not grafted (on its own roots) or grafted onto a seedling (non-dwarfing) understock. Figure 25.6 categorizes the relative effect of a number of common apple rootstocks on the vegetative vigor of a scion cultivar. Keep in mind that size control is only one of the characteristics associated with common clonal apple rootstocks. Although rootstock effects on scion vigor as an orchard management tool is most developed with apples, size-controlling rootstocks are used for other tree fruit crops as well. For example, among traditional cherry (*Prunus avium* L., *P. cerasus* L.) rootstocks, sweet cherry cultivars grafted on Mahaleb seedling rootstocks grow considerably larger than the same cultivars on Mazzard seedling rootstocks.

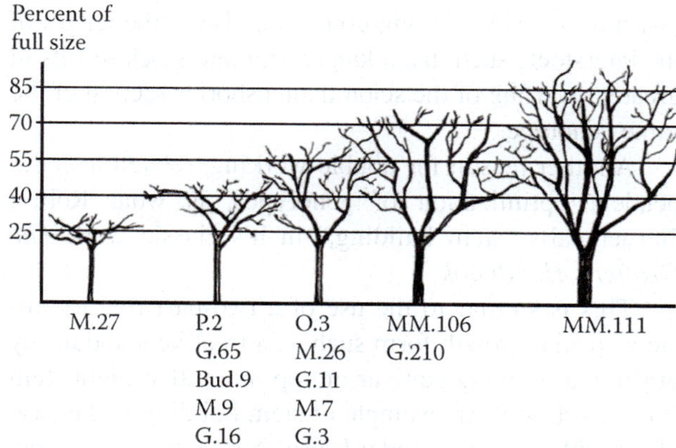

FIGURE 25.6 Effect of common apple rootstock on the relative size of a fruiting scion. Abbreviations for rootstock series (breeding programs): M = East Malling; P = Polish; G = Geneva; Bud = Budagovsky; O = Ottawa; MM = Malling Merton. (Drawing by Vanessa Gray, 1999.)

Newer clonal rootstocks for cherry and other stone fruits have been selected for size control along with other desirable characteristics (disease resistance, soil tolerances, and others). Quince (*Cydonia oblonga*) seedlings are sometimes used as dwarfing rootstocks for pear, and a number of clonal quince rootstocks varying in size control are available. For oranges and other citrus, trifoliate orange (*Poncirus trifoliata*) is an especially cold hardy stock that is used for dwarfing.

Not only does the vigor of the rootstock affect the vigor of the scion, but as noted above, the vigor of the scion affects the vigor of the root system. Nonetheless, size control of the rootstock by the scion is rarely considered a reason for grafting, except in the unusual case of Mukabit grafting of cassava. Although a rather obscure case, it is worth describing here in part because it is a modern confirmation of the point that Liberty Hyde Bailey made in the *Standard Cyclopedia of Horticulture* (Bailey, 1928) quoted at the beginning of the section below on the requirements for successful grafting. He mentioned that, despite grafting's ancient history, anyone with knowledge of the four basic requirements could literally invent a new grafting method to suit his/her needs. In 1952, a peasant farmer named Mukabit on the Indonesian island of Java discovered that the yield of the edible tuberous root crop cassava (*Manihot esculenta*), an important dietary staple in tropical regions, could be increased by at least twofold (100%) or more (5- to 10-fold increases have been claimed) by grafting onto it a scion of the Ceara rubber tree (*Manihot glaziovii*) (de Foresta *et al*., 1994). This is rather extraordinary because the normal (non-grafted) root system *M. glaziovii* produces no tubers at all. In fact, the percentage increase in growth of the cassava stock induced by the Ceara rubber scion far exceeds any claim regarding the effect of the stock on scion yield for apple or any other commonly grafted crop. This profound growth stimulation of cassava root is apparently because of the fact that the Ceara rubber grows later than cassava into the dry season before going dormant, allowing more time for starch accumulation in the cassava tubers.

SPECIALIZED GROWTH FORMS

A special case of grafting to achieve independent optimization of stock and scion genotypes is the practice of top working for the construction of arborescent (treelike) specimens from scions that would grow on their own roots into shrubs. Top working for this purpose is sometimes referred to as "high working." This approach is used to construct such beauties or oddities (depending on your point of view) as tree roses (*Rosa* spp.), tree peonies (*Paeonia suffruticosa*), and weeping forms of flowering cherries.

GRAFTING TO INFLUENCE GROWTH PHASE AND FLOWERING

All plants grown from seed undergo a natural age-related maturation process before they are capable of flowering (Chapter 6). Seedlings begin in the juvenile growth phase, during which they are, by definition, incapable of flowering. In woody plants, this period of juvenility typically lasts several years in fruit trees to several decades in forest species, until the juvenile undergoes a transition to the mature (sometimes called "adult") phase when they become capable of flowering under inductive environmental conditions (appropriate photoperiod, temperature, and others). Even when propagated vegetatively by cuttage, graftage, layerage, and others, a mature-phase plant tends to retain its ability to flower. Grafting a mature scion onto any rootstock will produce a flowering tree several years sooner than growing the same species from seed because, unlike a seedling, the scion has already undergone its juvenile-to-mature phase transition. For example, grafted apple will flower in 2 to 3 years from the time it leaves the nursery, in contrast to a seedling apple, which will not flower until it is approximately 7 years old. Hence, in many fruit crops, this is one, but rarely the only, reason for grafting. The duration of the nonflowering period before even a mature scion commences reliable fruit production is influenced by the growth phase and the genotype of the stock. Some rootstock genotypes (cultivars) result in a shorter delay in the onset of scion flowering than others, and in general, it has been observed that the greater the dwarfing effect of the rootstock, the shorter the delay in fruiting of the scion. Rootstocks that accelerate the onset of flowering and fruit production are said to be precocious.

GRAFTING FOR MULTIPLE (MORE THAN TWO) GENOTYPES

The reasons for grafting described above pertain to two-part (stock/scion) grafted trees, but the same and/or additional advantages can also be conferred by grafting together more than two genotypes, either in "series" or in "parallel," by analogy with the terminology used to describe simple electrical circuitry. Double working is a multi-genotype compound genetic system assembled "in series" with the stock, the interstock, and the scion in a linear sequence (Figure 25.1E). Two or more scion genotypes grafted at different places on a common stock or an interstock could be considered a "parallel" compound genetic system (Figure 25.1D).

DOUBLE WORKING

Each of the three parts of a double worked tree—scion, interstock, and rootstock—generally consists of separate genotypes. Typically, the interstock is several to many centimeters long depending on its function. Double working is performed for any of several reasons. Some are similar to the reasons for grafting a two-part tree ("single working," although the term is not commonly used), that is, to independently optimize the components of a compound genetic system. However, double working involves optimization of three component genotypes rather than just two. Whereas the components of a two-part tree correspond more or less to the root system (stock) and the shoot system (scion) of the grafted tree, the components of a double-worked tree correspond to the root system and, sometimes, the lower portion of the trunk (stock), all or some portions of the trunk (interstock), and more or less of the upper trunk, branches, and leafy canopy (scion). Some desirable characteristics associated with an interstock-derived trunk include cold tolerance or resistance to specific diseases, where the trunk of the nongrafted tree is especially vulnerable. Even the size control function exerted on the scion by the rootstock of a two-part tree, as described above, can instead be assumed by an appropriately selected interstock. For example, an interstock rather than a rootstock is sometimes used to achieve size control in apples grown on wet soils that would be prone to lodging (falling over) under a heavy crop load if grafted onto a small, poorly anchored dwarfing rootstock such as M.9. Although trellising (mechanical support) is the most common solution to this problem, some growers, particularly in the past, resorted to a double-worked tree consisting of a relatively large, well-anchored, vigorous rootstock, such as MM.111 with an M.9 dwarfing interstem, underneath the fruiting cultivar. An advantage of this approach is that the degree of scion dwarfing imparted to the scion variety by the interstock is proportional not only to the genotype of the interstock (M.9, dwarfing vs. MM.111, vigorous), but also to the length of the interstock, such that a longer M.9 interstock results in greater dwarfing of the scion than a shorter section of the same genotype.

Another reason for double working, related to independent optimization of genotypes, is what Robert Garner calls "stem building," in his classic work *The Grafter's Handbook*.

This is similar to the use of a two-part tree to create a special growth form such as a treelike standard by grafting a weeping cultivar on top of a tall straight stem (high working). An example of stem building is the case of a double-worked standard rose. A rootstock genotype, such as Fortuniana, might be selected (in part) for its nematode resistance; Grifferaie, as an interstock, for its tall, straight, thornless trunk; and a scion of a bush-type flowering rose high worked at 36 in.

Double working can also be used to overcome some cases of graft incompatibility by inserting a mutually compatible interstock between otherwise incompatible stock and scion cultivars. An example of this strategy to overcome the incompatibility between quince understock and some pear cultivars is described in the section on compatibility below.

MULTIPLE SCION CULTIVARS

Unlike double working, which involves a linear sequence of three genotypes separated by two graft unions (in "series"), it is possible to graft multiple scion cultivars onto different branches of the same understock (or interstock) in a "parallel" configuration (Figure 25.1D). Two or more scion cultivars grafted onto a single tree can be used to make interesting combinations of various fruit species, such as apple, for example, 'Macintosh', 'Red Delicious', and 'Crispin', on the same tree, or multiple citrus species on the same tree, such as orange, lemon, and grapefruit. Using the methods described in the hibiscus grafting exercise below, it is possible to construct a single hibiscus tree with three or more different branches bearing red, yellow, or orange flowers simultaneously. Such combinations of multiple fruit or flowering cultivars are usually constructed as novelty specimen plants by or for amateur gardeners rather than for commercial crop production. On a larger orchard scale, it is more practical to grow different cultivars as separate trees. Aside from its novelty value, placing more than one scion cultivar on a certain fruit tree species has another advantage related to pollination and fruit set. Many tree fruit crops in the Rosaceae family, including apples, cherries, and pears, are sexually self-incompatible, that is, the pollen of a single tree or even of multiple trees of the same clone cannot fertilize flowers from the same tree or clone. Pollen from another cultivar is required to set fruit. The requirement for a pollinizer cultivar is accomplished in most commercial-scale fruit orchards by planting pollinizer trees of a different genotype from the main crop (e.g., a crabapple) at strategic locations throughout the orchard. For smaller-scale "backyard" production, however, pollination and fruit set can be increased by top working a pollinizer cultivar onto one or more branches of the original tree. It is important to choose a pollinizer that flowers at the same time as the primary cultivar.

GRAFTING FOR DIAGNOSIS OF VIRAL INFECTIONS AND THE PROBLEM OF GRAFT TRANSMISSION OF VIRUSES

Most plant viruses are not seed transmitted, but they are spread by scions from an infected plant to a stock plant. Indeed, viruses are spread by any vegetative propagation method, including cuttage, layering, division, and micropropagation (with the exception of meristem tip culture). Viruses are readily translocated across a graft union from an infected plant to a non-infected one. For this reason, agricultural regulatory agencies in some countries and states attempt to protect regional commercial fruit or other crop industries by maintaining and enforcing virus-indexed bud or scion wood certification programs. These involve rigorous indexing of new germplasm introductions to assure that they are free from specific viral pathogens, maintaining virus-indexed donor trees of stock and scion cultivars in foundation blocks under semi-quarantined conditions to avoid reinfection, and careful regulation of bud/scion wood distribution to growers. Some definitive means of detecting virus in plants suspected of infection is necessary for the success of these clean stock programs because many crop species show no visually apparent viral disease symptoms even when infected but, nevertheless, do respond adversely with a reduced vigor and a declining yield of harvestable crop (Table 25.1). It may seem ironic, but fortunate, that graft transmission of plant viruses can be used to a considerable advantage as a means of screening for and detecting virus-infected plants through a process called "biological indexing" or "graft indexing." Graft indexing involves grafting a scion from an asymptomatic crop species or cultivar suspected of being infected onto a stock of a more sensitive indicator species. If the virus is present and transmitted across the graft union, the indicator stock plant will display characteristic visual symptoms such as streaking or mottling. This technique of viral detection is used not only with normally graft-propagated fruit tree crop species, such as citrus and cherry, but also with non-woody crops, such as sweet potato and strawberry. Although graft indexing is still an important tool of the plant pathologist, it has declined in importance as more specific and, in some cases, more sensitive virus testing methods have been developed, most notably the immunological assay know as enzyme-linked immunosorbent assay.

GRAFTING FOR REPAIR

Just as human organ transplantation (a type of grafting) can be used to replace failing organs and heal sick people, so too is plant grafting occasionally used to repair or prevent future structural problems in trees. Although in modern commercial orchard production, it is usually more cost effective to replace a damaged tree, repair grafting is sometimes used as a last resort to rescue a high-value landscape specimen. Three kinds of damage that can be addressed by grafting are girdling of the trunk, damage to the root system, and splitting of narrow branch crotches. Girdling of young trees is often caused by voles, mice, or rabbits feeding on the inner bark (phloem) when other food sources are in short supply during winter. Additionally, string weed trimmers can damage ornamental trees if the operator allows the string to remove the bark near the soil line. Girdling blocks the downward translocation of photosynthate to the root system, via the phloem, eventually killing the tree. Figure 25.7 shows how bridge grafting can be used to restore a continuous living bark across a girdled area. As a general rule, one bridging "interstock" (although double working terminology is not usually applied to bridge grafting), about the diameter of a pencil, is inserted for each 2.5 cm (1 in) of trunk diameter. Damage or decline of a root system from freezing injury, disease, mechanical damage, or even gradual (delayed) graft incompatibility is another problem that can be addressed by a type of approach grafting known as "inarching." Figure 25.8 shows how inarching (a type of approach grafting) can be used to replace/rescue an existing rootstock that is in decline due to disease and/or pests. English walnuts (*Juglans regia*), grown in California's Central Valley, are propagated by grafting onto seedling or clonal rootstock, each with advantages and disadvantages. Until about the middle of the 20th century, the rootstock of choice for English walnut was the Northern California black walnut (*J. hindsii*) because of its vigor and resistance to oak root fungus. However *J. hindsii* is susceptible to Phytophthora root rot and lesion nematodes. Later in the century, a new rootstock called 'Paradox' (seedling

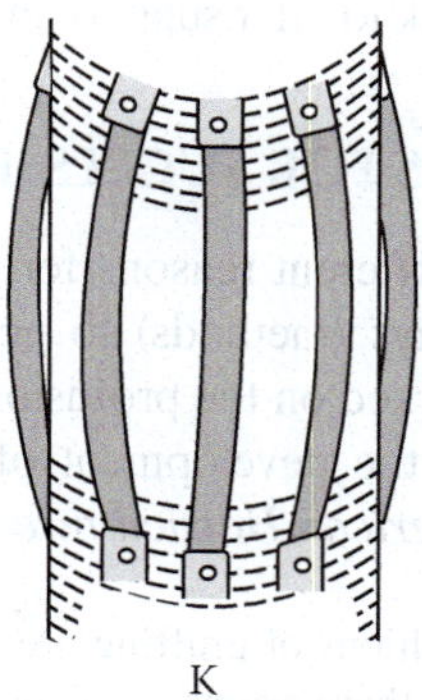

FIGURE 25.7 Bridge grafting using inlay bark graft to attach bridging pieces. (Copyright L. H. Bailey Hortorium, 1978.)

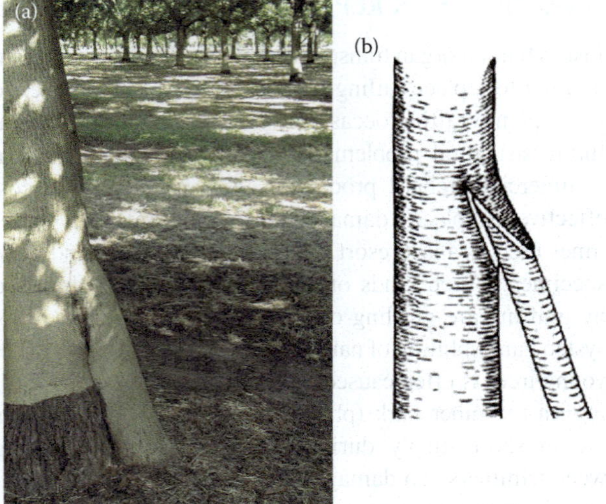

FIGURE 25.8 (a) Inarching for the replacement of a damaged or incompatible root system, using an inlay bark graft to attach new rootstock to established trunk. Note that inarching is a type of approach grafting. (b) Diagramatic representation of an inlay bark graft. (From *Hortus III*, LH Bailey Hortorium. Cornell University. With permission.)

of *J. regia* × *J. hindsii*, as well as more modern 'Paradox' clones) displaced the use of Northern California black walnut because of its tolerance to diseases and pests that plagued this species of tree.

Similar to other types of approach grafting, formation of the inarched graft union occurs while both the stock and the scion are attached to and, hence, sustained by their own root systems. When approach grafting for propagation purposes (Figure 25.4), the potted rootstock along with its newly fused scion is removed for replanting, but in the case of inarching to replace a damaged root system, the stock is planted beside the tree with the damaged root system and left in place permanently to serve as a new, surrogate root system for the remaining life of both trees. Brace grafting is occasionally used to mechanically reinforce a branch with a narrow crotch angle that is likely to split away from the main trunk. In this case, the scion is grafted between both the branch and the trunk to offer support and prevent breakage.

REQUIREMENTS FOR SUCCESSFUL GRAFTING

There are many different reasons for grafting, but there are even more ways (methods) to go about it. Liberty Hyde Bailey remarked on the profusion of existing grafting methods and the development of new ones in his *Cyclopedia of American Horticulture* (1928):

> "The ways or fashions of grafting are legion. There are as many ways as there are ways of whittling. The operator may fashion the union of the stock and the scion to suit himself, if only he apply cambium to cambium, make a close joint, and properly protect the work."

Bailey's advice should be reassuring to the beginner who might otherwise be overwhelmed by all the dozens (at least) of different grafting and budding methods practiced at commercial nurseries, orchards, and backyard gardens, and described in various books, articles, Web sites, and others. His advice is as true today as it was in 1928. If the student of grafting pays careful attention to how the universal requirements for successful grafting are satisfied, while learning just a few basic methods, he/she will find it easier to learn additional methods as the need arises. Bailey's remark emphasizes three of the four universal requirements that can be stated with less oratorical flourish as cambial alignment, pressure, and avoidance of desiccation. The fourth requirement, not mentioned in the quotation from Bailey, is genetic compatibility. In fact, the issue of genetic compatibility is so important that it will be discussed first because if the stock and the scion are not genetically compatible, nothing else matters (except in the case of nurse grafting).

COMPATIBILITY

If the stock and the scion are sufficiently closely related genetically to form a functional graft union, they are said to be compatible. Conversely, a stock/scion combination that is not sufficiently closely related is described as incompatible, and a graft performed between them will display either immediately or eventually incompatibility symptoms such as leaf drop, bulging and cracking of the initial graft union, and death of the scion. Failure of grafting from poor technique or timing is not incompatibility. Delayed incompatibility refers to the initial formation of an apparently normal graft union followed some later time, typically one to several years, by a gradual failure of the graft union accompanied by a decline and, eventually, death of the scion. Incompatibility greatly limits the practice of grafting to closely related species. If the Roman poet Virgil (29 B.C.) had been correct in his "observation" that "…swine crunched acorns 'neath the boughs of elms" (interfamilial oak/elm compatibility), the agricultural possibilities would be endless. For example, most common apple cultivars can be grafted together as can most common citrus species (oranges, lemons, grapefruit, and others), but apples and oranges, which are in different families, cannot be successfully grafted. The more closely that the stock and the scion candidates are related to each other, the more likely it is that they will be compatible, but it is difficult to generalize beyond that point. Interfamilial (between different families) combinations are universally incompatible, except for occasional claims of a few rare, horticulturally unimportant exceptions. At the other end of the spectrum, intraclonal (within a clone) combinations (same species) are universally compatible, but when combinations between different clones, different species, or different genera are considered, the degree of "relatedness" necessary for compatibility, known as the limits of compatibility, varies with different taxa. Generalizations are of limited predictive value. As shown in Table 25.2, intraclonal combinations are universally compatible. Intraclonal grafting is uncommon except for repair grafting of a damaged tree by bridge or brace grafting with a piece of shoot of the same clonal cultivar. Interclonal combinations within the same species are nearly always compatible. Interclonal grafting is often the basis for the nursery production of many difficult-to-root species, such as Norway maple and other shade trees. However, there are economically significant exceptions where interclonal grafting is not ultimately successful, such as the sporadic (unpredictable) delayed incompatibility between red maple (*A. rubrum*) cultivars and red maple seedling rootstocks. Interspecific combinations within the same genus are often compatible and, frequently, horticulturally important. For example, almond (*Prunus amygdalus*) scions are successfully grafted on peach (*P. persica*) rootstock, but many interspecific combinations are not feasible, such as the incompatible combination of almond and apricot (*P. armeniaca*). Note that, while some interspecific/intrageneric combinations of stock and scion are compatible, such as Myrobalin B plum (*P. cerasifera*) scion on Hale's Early peach (*P. persica*) stock, but the reciprocal combination of 'Hale's Early' peach scion on Myrobalin B stock is incompatible. Successful combinations among different genera (intergeneric) within the same family are relatively rare, but several of those that do succeed are horticulturally significant such as some (but not all) pear (*Pyrus communis*) cultivars (e.g., 'Farmingdale') on quince (*Cydonia oblonga*) rootstock. Nonetheless, the closely related pear cultivar Bartlett is incompatible with quince. As described above, double working 'Old Home' interstock between the 'Bartlett' and the quince stock can avoid the incompatibility entirely. Table 25.2 gives only a few examples in each category. Since there are important exceptions to generalizations at the level of interclonal/intraspecific, interspecific/intrageneric, and intergeneric/intrafamilial combinations, any potentially useful combination of unknown compatibility must be tested to have confidence in the outcome. Such empirical testing is often prohibitive in terms of time and money. It would be helpful to have access to a definitive listing of all or nearly all potentially horticulturally useful combinations. A number of worthy attempts have been made to publish the results of empirical tests (such as Andrews and Marquez, 1993), but there is currently no comprehensive guide available. Such a guide would be valuable considering the substantial economic losses that some growers have incurred after investing several years of time, labor, and money in

TABLE 25.2
Compatibility

Relationship	Likelihood of Success[a]	Compatible Example	Incompatible Example[b]
Intraclonal	100% (a sure thing)	Bridge graft (scion from the same tree, for the repair of girdled trunk)	None
Intraspecific	High	'Macintosh' apple (*Malus domestica*)/EM.9 (clonal rootstock cultivar of *M. domestica*)	Red maple clonal cultivars (*Acer rubrum*)[c]/red maple seedling
Intrageneric (interspecific)	Moderate	Sweet orange (*Citrus sinensis*)/rough lemon (*C. jambhiri*) Sweet cherry (*Prunus avium*)/tart cherry (*P. cerasus*)	Almond (*Prunus amygdalus*)/peach (*P. persica*)
Intrafamilial (intergeneric)	Unlikely	'Old Home' pear (*Pyrus communis*)/quince (*Cydonia oblonga*)	'Bartlett' pear (*P. communis*, Rosaceae family)/quince (*C. oblonga*, Rosaceae family)
		Blue spruce (*Picea pungens*)/Norway spruce (*Picea abies*)	Lilac (*Syringa vulgaris*, Oleaceae family)/California privet (*Ligustrum ovalifolium*, Oleaceae family)[d]
Interfamilial	0% (considered impossible, despite a few claims to the contrary)		

[a] Probability of genetic compatibility, assuming that grafting technique and environmental factors are not the cause of failure.
[b] Scion/stock.
[c] Varies with scion cultivar, that is, some clones exhibit essentially no incompatibility, while for others, as high as 25% of grafted individuals develop incompatibility.
[d] Delayed incompatibility is the desired outcome of nurse root grafting used to obtain scion rooting.

growing a crop of grafted trees, only to have some or all of it destroyed by a delayed incompatibility.

What is equally frustrating to propagation physiologists is that there is no consistent explanation for the causes (mechanisms), let alone a cure for incompatibility. Some incompatibilities appear to be related to the production of an organic compound by one member of the pair that is toxic to the other, whereas in other cases, the mechanism appears to be related to differences in the biochemistry of wood formation between the stock and the scion. One of the best understood incompatibility mechanism is that between 'Bartlett' pear (*Pyrus communis*) and quince (*Cydonia oblonga*) rootstock. In this case, a cyanogenic glucoside called "prunasin" present naturally (harmlessly) in the quince understock is broken down chemically when it comes into contact with an enzyme produced by the pear tissue, releasing toxic cyanide that results in the destruction of the pear phloem.

Considering the potential economic loss associated with the delayed incompatibility of a nursery crop or even years later in an orchard of grafted trees, it would be extremely useful to have a reliable short duration assay to test a particular combination for compatibility. Unfortunately, no such quick assay exists despite considerable research directed at developing one.

A final category of incompatibility worth mentioning is the so-called "induced incompatibility," which is related to a more or less asymptomatic virus in the stock or the scion, which triggers an incompatibility reaction at the graft union that would not otherwise occur. Black line of walnut (English walnut [*Juglans regia*] scion/Northern California black walnut [*J. hindsii*] stock) is induced by cherry leaf roll virus, and apple union necrosis and decline is induced by tomato ringspot virus.

CAMBIAL ALIGNMENT

Examples of several different grafting and budding techniques are illustrated in Figure 25.9. The drawings were done by L. H. Bailey's granddaughter, Ethyl Zoe Bailey. The composite figure (with Bailey's original caption) illustrates Bailey's (1928) assertion that "The ways or fashion of grafting are legion. There are as many ways as there are ways of whittling. The operator may fashion the union of the stock and the scion to suit himself, if only he apply cambium to cambium, make a close joint to, and properly protect the work." This statement underscores the importance of cambial alignment for grafting success.

The vascular cambium is a cylinder of living tissue, a single-cell layer between the secondary xylem (wood) to the inside and the secondary phloem (inner bark) to the outside (Figure 25.2). The "carpentry" or knife cuts made while performing any of the various grafting and budding methods, such as those illustrated in this chapter, result in the exposure of a section of the cambial cylinder of both the stock and the scion. The grafter's carpentry must be sufficiently accurate so that the shapes of the cut surface of the stock and the scion complement each other to the point where the cambia of each come into contact, or nearly so, along as much of the perimeter of the cambial cylinders as possible. In fact, some grafting methods, by the nature of their carpentry, expose more cambial surface than others, which allows for more cambial contact between the stock and the scion if the two are cut complementarily to each other. For example, the interlocking "tongues" of the whip and tongue graft (Figure 25.9f) increase the potential zone of cambial contact by almost 100%, compared to a splice graft, which involves the same diagonal cuts joining the stock and the scion without the interlocking tongues. Similarly, the greater zone for cambial contact with a chip bud compared to a T-bud accounts at least in part for the fact that chip budding allows for a quicker, stronger, and more winter-hardy union and an overall better performance than T-budding.

As described above, the process of graft union formation involves the formation of a callus bridge across any gap that exists between the stock and the scion, followed by de-differentiation of a new cambium across the callus bridge to connect the vascular cambia of the stock and the scion. This occurs in an inward direction from the stock and the scion vascular cambium until they connect somewhere in between in the callus tissue. Without the so-called "cambial contact" (misalignment), the two converging ends of the newly forming cambium do not "find" each other, and vascular continuity is never established. Some particularly difficult-to-graft species, such as temperate nut trees like hickories (*Carya* sp.) and walnuts (*Juglans* sp.), produce very little callus, and hence, actual cambium-to-cambium contact (minimal gaps) is critical. In such cases, the carpentry must be especially well executed to assure a near-perfect fit. Figure 25.2a represents perfect alignment that allows more or less direct cambial contact between the stock and the scion when the two are brought together, whereas in Figure 25.2b, the alignment is imperfect but still acceptable (sufficiently close); however, in Figure 25.2c, the gap between the cambia of the stock and the scion is too great for the graft union to develop successfully.

PRESSURE

Once the stock and the scion are fitted together so that their cambia are well aligned and preferably in direct contact, pressure is necessary to maintain this alignment; otherwise, mechanical disruption from handling or wind in the nursery could easily cause the scion to move with respect to the stock. Movement of the scion could create or increase a gap, disrupting the cambial contact and/or the alignment established when the stock and the scion were initially joined by the grafter. When cut surfaces are

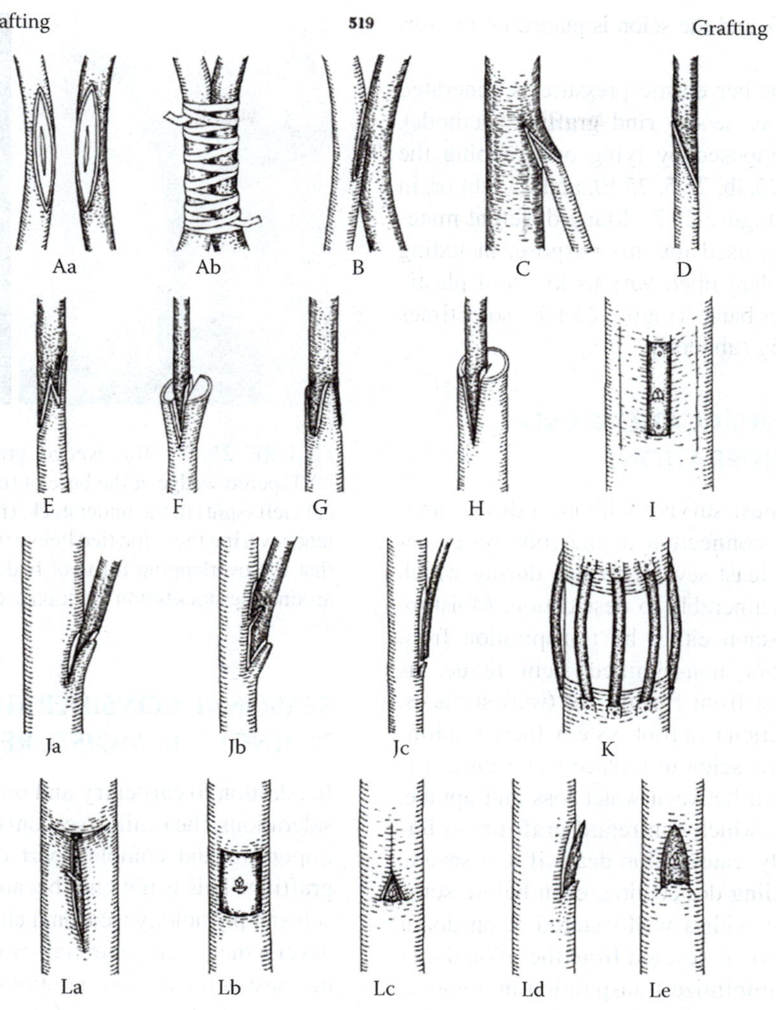

FIGURE 25.9 Grafting. A, simple splice approach graft; Aa, stock and scion showing cut surface; Ab, cut surfaces of stock and scion joined and bound; B, tongued approach graft, tongues of stocks and scion interlocking; C, inarching, stock cut and inserted into scion; D, whip graft, a simple splice of stock and detached scion; E, whip and tongue graft; F, cleft graft; G, saddle graft; H, inlay graft; I, veneer graft; J, side grafts; Ja, side cleft graft; Jb, side tongue graft; Jc, side veneer graft; K, bridge graft; L, budding; La, shield bud; Lb, patch bud; Lc, microbud; Ld, Le, chip bud in side and face view. (Scion with dark shading; stock with light shading.) (Drawings by Ethel Zoe Bailey. LH Bailey Hortorium. With permission.)

exposed to air as occurs when a gap is formed between the stock and the scion, the surfaces tend to dry out more readily, creating a problem related to the next universal requirement—avoidance of desiccation. Furthermore, as mentioned above, the greater the gap between the stock and the scion, the more callus must be generated to fill that gap, and once callus formation has occurred within weeks after placing the graft, further disruption will tear apart the extremely soft (non-lignified) callus cells. Firm pressure between the stock and the scion will minimize the chance of unintended movement between the two and thereby will facilitate graft union formation. In addition to these mechanical considerations, pressure also influences the alignment of the planes of cell division and dedifferentiation in the developing callus during graft union formation, giving rise to the spatially organized (properly lined up) tissues of cambium, xylem, and phloem process (Barnett and Asante, 2000).

In the case of some grafting methods, pressure between the stock and the scion is generated by the carpentry itself (i.e., shape of the cuts) related to the elasticity of the wood of interlocking portions of the stock and the scion. This can be seen for several non-rind grafts illustrated in Figure 25.9, including the whip and tongue graft (E), and the cleft graft (F). Little or no elastic pressure is generated in the case of other non-interlocking, non-rind grafting methods such as the splice approach graft (Figure 25.4b), chip bud graft (Figure 25.9b), or the veneer graft (Figure 25.9c), nor in the cases of rind grafting methods, including T-budding (Figure 25.9a) or bark grafting used for top bark grafting (Figure 25.9e), bridge grafting (Figure 25.7), or inarching (Figure 25.8), in which the bark is peeled back and the scion is placed on the surface of the wood.

Regardless of whether elastic pressure is generated from the carpentry per se (no rind grafting methods), pressure is usually imposed by tying or wrapping the graft union (Figures 25.4b, 25.5, 25.10, and 25.11b) or, in some cases, nailing (Figure 25.7). Many different materials are or have been used for this purpose, including string, tape, natural plant fiber, various kinds of plastic strips, and latex rubber bands (Figure 25.11b), sometimes referred to as "budding rubbers."

AVOIDANCE OF DESICCATION AND SEASONAL CONSIDERATIONS

A freshly cut scion must survive without a direct, anatomically continuous connection to the root system of the stock plant for at least several weeks, during which time it is extremely vulnerable to desiccation. Moisture can be lost from a scion either by transpiration from exposed foliage and/or non-lignified stem tissue, as well as by evaporation from freshly cut (wet) surfaces. Without direct connection to a root system, there is a limited opportunity for the scion to replace water once it is lost. If there is a deficit between water loss and uptake, drought stress results, which can retard graft union formation and, eventually, cause scion death if it is severe. The first step to avoiding desiccation, even before scion collection, is to begin with a well-watered scion donor plant. Once a leafy scion is severed from the scion donor plant, it is critical to minimize transpiration by removal of leaves, the use of moisture barriers after grafting to prevent water vapor loss (waterproofing), and/or by grafting at a time of the year when the scion is dormant and natural transpiration is minimal.

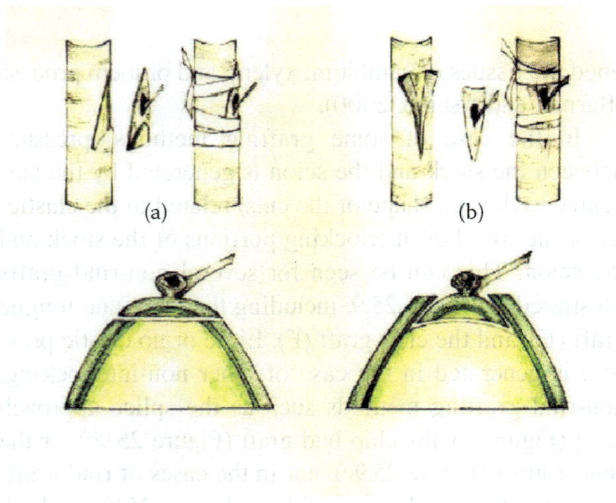

FIGURE 25.10 Two common budding methods. (a) Chip budding. (b) T-budding.

FIGURE 25.11 Top wedge grafting of tropical hibiscus. (a) Tapered wedge at the base of the scion inserted into a vertical cleft (split) in the understock. (b) Polyethylene bag humidity tent covering the scion tied below the stock/scion junction. Note that the overlapping turns of budding rubber wrapped tightly around the stock/scion junction are visible inside the bag.

SEASONAL CONSIDERATIONS (GRAFT TIMING) FOR MOISTURE MANAGEMENT

In addition to carpentry and other technique-related considerations, the timing (season) of grafting is an extremely important and complex part of the art and science of grafting. This is not only because of the influence of season and phenology (seasonal changes in plant growth and development) on potential evapotranspiration, but also because of the obvious relationship between season, temperature, and dormancy. Since the optimum time of the year for performing a particular grafting method for each species is intimately, but not exclusively, related to moisture management and the stage of plant development, the topic of seasonal considerations in grafting will be covered in this section.

For any combination of grafting method and species, there is a critical balance between sufficient heat for growth and potential evaporation. Higher temperature (up to some species-specific optimum) promotes cell division and other temperature-dependent physiological processes involved in graft union formation, while at the same time, transpiration from leafy, and especially non-lignified (succulent), shoots increases with increasing temperature. Hence, as a general rule, grafting is performed when the temperature is adequate, but not higher than necessary for maintaining the physiological processes associated with graft union formation while minimizing transpiration. For many deciduous woody plants, the optimum seasonal combination of these two factors occurs either in early spring when they are still leafless and just emerging from dormancy or late summer/early fall when deciduous plants are still leafy, but shoots are lignified and have already set dormant buds.

DORMANT OR SEMIDORMANT SCION GRAFTING

"Rind grafting" refers to methods that involve peeling back the bark (rind) of the understock and placing the cut surface of the scion beneath the bark (rind), on the surface of, but not into, the underlying wood of the stock. Examples include T-budding (Figure 25.9La), bark grafting (Figure 25.8), patch budding (Figure 25.9Lb), and others. Non-rind grafting methods are those that involve inserting the scion to the bark and into a cleft in the underlying wood. Non-rind grafting methods include cleft grafting (Figure 25.9f), chip budding (Figures 25.9Ld and Le and 25.10a), veneer grafting (Figure 25.9Jc), top wedge grafting (Figure 25.11), and others. The distinction between rind and non-rind grafting is intimately related to seasonal grafting considerations since peeling back the bark is only feasible when the bark is "slipping." This occurs only during periods of active shoot growth, from late spring to early fall/late summer, when the cells of the vascular cambium are actively dividing and their thin cell walls are easily pulled apart when the bark is separated from the underlying wood. Earlier in spring or later in fall, the bark tears rather than slips (peels) away from the wood, and non-rind grafting is the only alternative. Hence, rind grafting is feasible only during periods of active growth when deciduous plants are leafy and potential evapotranspiration is high.

Because potential evapotranspiration is less during periods of dormancy, non-rind, dormant scion grafting methods are more widely practiced in temperate climates than rind grafting methods. For example, dormant scion grafting in early spring is practiced for top-working, established trees for repair, reworking to a new cultivar, inserting a pollinizer branch, and others. The timing is fairly critical, and the "window of opportunity" fairly narrow because the stock plant (established tree in most cases) must be coming out of dormancy and beginning to grow (bud swelling), but must not be fully leafed out. At the same time, it is important that the scion would still be as dormant as possible so that it will not leaf out and transpire excessively before callus formation and vascular (xylem and phloem) interconnection with the stock is well underway. Leafing out of the scion too early would quickly result in the desiccation of the scion since transpiration from the newly expanded, succulent leaf surface would exceed water uptake. This seemingly contradictory requirement for more advanced growth activity in the stock than in the scion is managed by the timing of scion wood collection and handling. Scions are collected while fully dormant in mid- to late winter and stored under refrigeration until the growth activity of the stock resumes in early spring. At that time, the still-dormant (tight-bud) scion can be grafted onto a stock that is already beginning to resume growth (bud swelling). By the time that the scion leafs out several weeks later, it is already well along in the process of forming a functional graft union with the understock.

Whereas moisture management during early spring grafting is accomplished largely by dormant grafting of a non-leafy scion, the strategy for "fall" budding (late summer to early fall), commonly practiced on tree fruit and ornamental nursery stock, is somewhat different. In this case, both the stock plants (field or container nursery stock) as well as the scion bud donor plants are still in leaf, but have set dormant buds in response to decreasing day length and cooler nights of later summer. Excessive transpiration from the bud scion is prevented simply by removing the leaf blade at the time of bud wood collection. A petiole stump is left in place as a "handle" for the grafter. These semidormant, de-bladed bud scions can be grafted onto the still-leafy rootstock either by chip budding (Figure 25.10a) or T-budding (Figure 25.10b) if the bark is still slipping, or somewhat later only by chip budding, for a week or two after the bark has ceased to slip. In fact, it is increasingly common for chip budding to be preferred over T-budding even when the bark is still slipping because chip budding has been shown to result in a better, quicker graft union formation; improved overwintering; and an overall better long-term performance. In either case, transpiration from the dormant scion bud is minimal, even as graft union formation is underway. The bud remains dormant until the following spring when the shoot of the stock plant above the grafted bud is cut back, forcing the new bud to grow. For some nursery crops, especially stone fruits in regions with longer growing seasons (United States Department of Agriculture [USDA] Hardiness zone 6 or greater), "June" (early summer) budding is practiced. Moisture management is critical during June budding because the scion buds resume active growth shortly after insertion. The advantage of June budding is that the nursery tree is "finished" a year earlier than fall-budded plants.

Two additional dormant scion grafting strategies for temperate species include bench grafting of deciduous nursery stock and greenhouse container grafting of narrow-leaved evergreens and some deciduous species. Bench grafting refers to any technique applied to bare-root nursery stock, indoors, "at the bench," during the dormant period. For example, in the production of double-worked apple trees, a whip and tongue bench graft is performed in February or March to join stock and future interstock. Graft union formation occurs before the plant is lined out in the field in spring. New growth from the intended interstock is then semidormant bud grafted, as described above, to a fruiting scion cultivar in fall of the same year. In the case of temperate narrow-leaved evergreens, dormant scion grafting is practiced in a cool greenhouse during winter. Ornamental cultivars (dwarf and other unusual growth forms, and others) of conifers, including pines (*Pinus* spp.), spruce (*Picea* spp.), firs (*Abies* spp.), juniper, and others, are commonly grafted to seedling understocks because they are difficult to root from cuttings. Some deciduous species, such as selections of Japanese maple, are propagated this way. Dormant containerized nursery stock is typically brought into a cool (~50°F) greenhouse during midwinter and allowed for several weeks to begin to emerge from dormancy. Just as the growth of these stock plants is resuming, they are grafted with dormant scions using a side veneer graft and overwintered in a cool greenhouse.

NONDORMANT (GREENWOOD) SCION GRAFTING

In the case of tropical broad-leaved evergreen plants (e.g., citrus, avocado, hibiscus, and others) that do not undergo seasonal defoliation and winter dormancy like deciduous temperate species, a different moisture management strategy is required. Leaves are removed from nondormant scions and top wedge grafted as described in the laboratory exercise below. To minimize transpiration, the actively growing scions and the graft union itself are covered with a polyethylene tent to create a high-humidity environment.

COLLECTION OF SCION OR BUD WOOD

Depending on the seasonal strategy for grafting or budding, scion or bud (scion) wood is usually collected days, or even weeks, before it is used for grafting. Especially when grafting is performed during the growing season, water stress of the scion donor plant and the stock plant should be avoided by keeping them well watered and by collecting the scion wood during the cooler morning hours. Other practices used to minimize the desiccation of scion wood include avoiding exposure to direct sunlight after cutting; removing leaves to minimize transpiration; and storing sticks in a cool, high-humidity environment (e.g., in wet burlap under shade and/or in an insulated cooler). Scions collected for bud grafting are stored as 10 to 20 cm–long "bud sticks" and cut into individual scion buds only as they are needed. If scion wood is to be stored for more than a few hours before grafting, it should be wrapped in moist paper or burlap, placed inside a plastic bag, and kept refrigerated until use. For grafting performed during early spring, before leaf-out of deciduous plants, dormant scions are usually collected weeks earlier, during midwinter, when they are completely dormant, and stored under refrigeration until needed.

WATERPROOFING MATERIALS

The final moisture management strategy to minimize scion desiccation is coating and/or tying the stock/scion junction with some moisture-retardant material that minimizes evaporation from freshly cut surfaces. Wrapping with grafting tape, waxed string, or other semi-porous materials makes a less effective moisture barrier than budding bands (budding rubbers); Parafilm™; and new, specially formulated plastic grafting tapes. Wrapped or tied graft unions are sometimes coated with low–melting-point wax preparations. While Parafilm™ and grafting waxes serve primarily as moisture barriers, waxed string, various tapes, and elastic budding rubbers also serve to create pressure. Some relatively unyielding wrapping materials, such as string and regular polyethylene strips, have the potential for girdling the stem over time as it increases in diameter. Therefore, these must be deliberately removed several weeks after grafting. Traditional latex budding rubbers degrade naturally when exposed to ultraviolet (sun) light. When used outdoors for field budding, they crack and fall off on their own, but when used inside a glass greenhouse where the ultraviolet component of sunlight is filtered out by the glass, budding rubbers must be removed by hand. The use of a polyethylene tent to enclose the entire scion is not commonly practiced except for a few situations, such as the top wedge grafting of tropical fruit tree nursery stock discussed above and in the laboratory exercise below.

METHODS OF GRAFTING

As Bailey said, there are as many different grafting methods as there are "ways to whittle a stick." In his 1672 treatise, *The History of the Propagation and Improvement of Vegetable*," the English cleric Robert Sharrock (1672) listed nine methods, while in 1958, Robert Garner (1958) listed over 40 in *The Grafters Handbook*, and the 7th edition of Hartman *et al.*'s (2002) comprehensive modern reference, *Plant Propagation, Principles, and Practices*, lists 16 grafting and 7 budding methods. Classifying these into several different functional and structural categories helps clarify the relationships among them. Structural classification based on the shape of the stock or the scion cuts (carpentry) is the basis for naming many but not all grafting methods, such as cleft, top wedge, four flap, veneer, and saddle grafting, as well as T-, chip, and patch budding. Names, such as approach grafting, inarching, brace grafting, and bridge grafting, describe the function rather than the structure of the grafts and may not even imply a specific carpentry at all. For example, approach grafting might be performed using the (non-rind) splice approach method (Figure 25.4) or with an inlay bark graft in the case of the inarch approach graft illustrated in Figure 25.8. It is important to note that different authors and practitioners, historically or even currently, sometimes use different names to describe essentially the same method. For example, the most common contemporary name for the type of budding shown in Figures 25.9La and 25.10b is "T-budding," but Hortus III (Bailey, 1976) refers to it as "shield budding," while others call it

"T and shield budding." The non-rind graft referred to as top wedge grafting (Figure 25.11) is quite similar to the cleft graft (Figure 25.9F) except for the relative size of the scion and the understock.

EXERCISE

Experiment 1. Top Wedge Grafting of Hibiscus (*Hibiscus rosa sinensis* L.)

This laboratory exercise will give you an opportunity to learn a non-rind grafting technique that incorporates many different strategies to satisfy three of the four universal requirements for successful grafting described above—cambial alignment, pressure, and avoidance of desiccation (Mudge *et al.*, 2003). Keeping these in mind as you perform this exercise will help you apply what you learn about this method to other grafting and budding methods that you may wish to learn in the future. Tropical hibiscus is recommended for this exercise because it can be grown in a greenhouse or a bright window at any time of the year without the interruption of seasonal dormancy. It is also relatively easy to graft, assuring a positive learning experience. After learning top wedge grafting, you can use the same stock plant to practice other grafting and budding methods. Using different hibiscus cultivars, you can create your own unique compound genetic system that will bear differently colored flowers for years to come.

BACKGROUND

Top wedge grafting is commonly used on tropical fruit tree nursery stock, including avocado, mango, and sometimes citrus, although T-budding is more commonly used with the latter. This and other grafting methods are used to propagate fancy, polyploid cultivars of this species, which tend to be difficult-to-root from cuttings.

MATERIALS

The following materials will be needed to complete the laboratory exercise:

- 3 hibiscus stock plants about 1 m tall (4–5-month-old rooted cuttings) in 4- to 6-in diameter pots. Preferably, each of the three plants should be a different cultivar.
- Paper and string labels
- Sharp grafting knife
- Pruning shears
- Latex rubber budding strips, 5 × 0.25 × 0.016-in gauge (FarmHardware.com, Sacramento, California)
- Spray bottle containing water
- 3 × 6-in polyethylene bags
- Twist ties

Follow the instructions in Procedure 25.1 to complete this exercise.

Procedure 25.1
Top Wedge Grafting of Hibiscus

Step	Instructions and Comments
1	The upper 5 to 7.6 cm (2–3 in) of the shoot of each plant will be removed to serve as the scion for one of the other plants. Each of the three decapitated plants will serve as an understock for a scion from another plant.
2	Collect scion wood from each of the potted hibiscus plants by making a horizontal cut with sharp pruning shears to remove the distal (top) 7.6 cm (3 in) or more of the shoot. Unless it is to be used immediately, the scion should be wrapped in wet paper towels to avoid desiccation. Note: A scion may either be a terminal or a subterminal section from any point along the main shoot. Scions from lower sections of the shoot of the scion donor plant are progressively woodier toward the base and somewhat slower to resume growth when grafted onto another stock plant, but desiccation and wilting during the grafting process is less likely with woodier scions.
3	Use a sharp grafting knife or pruning shears to prepare the scion by trimming off any leaves that are anywhere from half to fully expanded. Regardless of whether it is a terminal section from the top of a scion donor plant or a subterminal section, it should be no longer than about 7 to 10 cm (2.8–4.0 in) and several nodes long. Note: Scions longer than recommended are more easily desiccated and more easily dislodged accidentally. Ideally, the scion diameter should be the same as that of the horizontal cut at the top of the understock. If necessary, it may be up to several millimeters narrower, but it should not be any wider than the understock.
4	With a sharp grafting knife, split the understock vertically, down the center, from the horizontal cut to a depth of approximately 2.5 to 4 cm (1–1.5 in) as shown in Figure 25.11a. Do not cut a V-shaped wedge on the top of the understock; simply split it vertically. Note: The height on the stock plant where a scion from another plant is inserted is largely a matter of preference, as long as it is between 90 cm (36 in) (near the original top) and

about 20 cm (8 in) from the soil line, with at least several leaves remaining on the understock. A lower point of scion insertion will produce a more compact grafted plant, but a higher graft will leave room for the insertion of other scions from the same or other cultivars by side wedge grafting or bud grafting (chip or T-budding). The higher the scion is placed on the stock plant, the more likely it is to be dislodged accidentally.

5 At the basal end of the scion, cut a tapered, V-shaped wedge that comes to a sharp point by making two long, straight, flat cuts on the opposite sides of the scion. While being cut, the scion should be held firmly with the left hand (if right handed) above the end where the wedge is to be cut. The distance from the top to the bottom of the wedge should be no less than 2, but no more than 3 cm long. Note: Terminal non-lignified scions are easier to cut than more basal, partially lignified scions, but it is easier to control the knife to make long, straight cuts that meet in a sharp point on partially lignified scion wood than on those that are softer and non-lignified. Ideally, each side of the V-shaped wedge should be made by a single straight pass with the knife, rather than a concave, convex, and/or uneven "whittled" surface created by multiple passes of the knife.

6 Insert the tapered wedge of the scion into the vertical slit in the understock until the top of the scion V is even with the horizontal cut surface of the stock. If the stock and the scion are of equal diameter, the vascular cambia of the scion should be aligned with that of the stock, but if the scion is slightly smaller in diameter than the stock, the scion should be placed to one side of the stock (rather than centered) so that the cambia align on that side only. Note: Leaving a crescent of the top portion of the scion wedge cut exposed above the horizontal cut of the stock is a common mistake that permits evaporative water loss from the wet cut surface.

7 Wrap the entire stock/scion junction with a budding rubber (inside bag; Figure 25.11b). Begin at the top by holding one end of the band against the scion just above the vertical cut of the stock. Bring the remainder of the band around behind the scion to cross over itself, creating an "X" that prevents the top end of the band from unraveling. Continue to wrap the band downward around the scion, while stretching the band to create pressure and overlapping each turn slightly to prevent gaps that would permit evaporative water loss. Continue wrapping until the entire vertical cut in the stock is covered. Finish off the wrapping by pulling out a loop at the bottom turn of the band, by pinching it between the thumb and the forefinger, and passing the free end of the band through this loop to trap the end and prevent unraveling. Note: Most wrapping of grafts (e.g., chip and T-budding) is made from the bottom up, but in the case of top wedge grafting, the somewhat slippery (wet) cut surface of the scion wedge tends to be squeezed up and pushed out of the cleft in the understock if wrapping pressure is applied from the bottom up. Hence, it is often easier to wrap from the top down, making at least one turn around the scion before continuing to wrap around the top of the stock downward.

8 Spray a minimal amount of water into a plastic bag, and create a humidity tent by placing the inverted bag over the scion, being careful not to crush or dislodge the scion. Tie the bag firmly in place with a twist tie, about 2.5 cm (1 in) below the stock/scion junction (Figure 25.11b).

9 Label each graft with the date, your initials, and the names of the stock and scion varieties. Place the grafted plants in a designated greenhouse under approximately 70% shade cloth.

10 Observe your grafted hibiscus weekly. If water (from transpiration) accumulates at the bottom of the bag where it is tied against the stock, open the bag and drain off the water, then re-tie. Observe the timing and the extent of new scion growth. After 3 weeks, cut a single 1-in–long slit in the upper portion of the plastic bag to begin acclimatization of the scion to ambient relative humidity. One week later (week 4), make another 1-in slit in the bag, and several days later, remove the plastic bag entirely. The budding rubber can be removed at this time to observe the stock/scion junction for callus formation. If the stock and the scion are securely bound together by callus, the band need not be reapplied. Keep the grafted plant under shade for one additional week before moving to full sun. Note: Using the space between the terminal top wedge graft and the base of the stock plant, one or two additional scions may be placed on the opposite sides of the stock plant using any of several different conventional grafting or budding methods (or invent your own). T-budding and chip budding work well (Mudge *et al.*, 2003), but are slightly more challenging than top wedge grafting. Regardless of the method(s), keep in mind the universal requirements for successful grafting, and you will succeed.

ANTICIPATED RESULTS

Within a week, you should notice new scion growth, which is a good sign of anticipated success. After 3 weeks, a careful examination of the junction between the stock and the scion should reveal a white, fluffy callus along the cut surfaces, which is an early stage of graft union formation. By week 4, the graft should be sufficiently strong to remove the wrapping and the plastic bag, entirely assuming that the scion was acclimatized during week 3. One week later, your grafted hibiscus will be able to grow in full sun. With sufficient sunlight and protection from cold winters (avoid temperatures <60°F by moving it indoors), you should get years of enjoyment from your multi–flower-colored tropical hibiscus.

QUESTIONS

- How did your top wedge grafting of hibiscus satisfy each of the four requirements for successful grafting and budding?
- What other methods of grafting would be appropriate for this species if you wanted to end up with a plant with three or more different cultivars (flower colors) grafted to it?

LITERATURE CITED

Andrews, P. K. and C. S. Marquez. 1993. Graft incompatibility. *Hort Rev* 15:183–231.

Bailey, L. H. 1928. *Standard Cyclopedia of American Horticulture*. New York: Macmillan Co., 3639 pp.

Bailey, L. H. 1976. Hortus III. L. H. Bailey Hortorium staff, eds. New York: MacMillan Co.

Barnett, J. R. and A. K. Asante. 2000. The formation of cambium from callus in grafts of woody species. In R. A. Savidge, J. R. Barnett, and R. Napier, eds. *Cell and Molecular Biology of Wood Formation*. Oxford, United Kingdom: BIOS Publishers, pp. 155–168.

Buhtz, A., J. Pieritz, F. Springer, and J. Kehr. 2010. Phloem small RNAs, nutrient stress responses, and systemic mobility. *BMC Plant Biol* 10:64–76.

de Foresta, H., A. Basri, and Wiyono. 1994. A very intimate agroforestry association: Cassava and improved home gardens—The Mukibat technique. *Agroforestry Today* 6:12–14.

Garner, R. J. 1958. *The Grafter's Handbook*. New York: Oxford University Press, 260 pp.

Hartman, H. T., D. E. Kester, F. T. Davies, and R. L. Geneve. 2002. *Plant Propagation: Principles and Practices*, 7th Edition. Upper Saddle River, New Jersey: Prentice Hall, 880 pp.

Harris, S. A., J. P. Robinson, and B. E. Juniper. 2002. Genetic clues to the origins of the apple. *Trends in Genetics* 18:426–430.

Lee, J. M. 2003. Advances in vegetable grafting. *Chronica Hort* 43:13–20.

Martin A., H. Adam, M. Diaz-Mendoza, M. Zurczak, N. D. Gonzalez-Schain, and P. Suarez-Lopez. 2009. Graft-transmissible induction of potato tuberization by the micro-RNA miR172. *Development* 136:2873–2881.

Mudge, K. W., J. Janick, S. Scofield, and E. E. Goldschmidt. 2009. A history of grafting. *Hort Rev* 35:437–493.

Mudge, K. W., K. Hennigan, and P. Podaras. 2003. Use of tropical hibiscus for instruction in grafting. *Hort Technology* 13:723–728.

Sonoda, S. and M. Nishiguchi. 2000. Graft transmission of post-transcriptional gene silencing: Target specificity for RNA degradation is transmissible between silenced and non-silenced plants, but not between silenced plants. *The Plant J* 21:1–8.

Sharrock, R. 1672. The history of the propagation and improvement of vegetables by the concurrence of art and nature: Shewing the several ways for the propagation of plants usually cultivated in England, as they are increased by seed, off-sets suckers, truncheons, cuttings, slips, laying, circumposition, the several ways of graftings and inoculations, as likewise the methods for improvement and best culture of field, orchard, and garden plants, the means used for remedy of annoyances incident to them, with the effect of nature and her manner of working upon the several endeavors and operations of the artist. Written according to observations made from experience and practice. Oxford, United Kingdom: Oxford University, 150 pp.

26 Grafting and Budding with Woody and Herbaceous Species

Garry V. McDonald

Grafting and budding are propagation techniques used when plants cannot be grown from seeds or cuttings. This may be because the plants are sterile or have male flowers only and, therefore, do not set seed. They may also produce nonviable seed. Another reason for grafting and budding may be that the cuttings may only be rooted with great difficulty. Grafting is the process of physically joining together two different plant parts (Chapter 25). The two parts are referred to as the "scion" and the stock or "rootstock." The scion is a stem or branch of the plant that has traits or characteristics of interest. This could be fruit or nut quality, flower size, color, leaf pattern, or some other desirable trait. The root system that the scion will be grafted or budded onto is the rootstock. Rootstocks are usually chosen for vigor, disease resistance, lack of thorns, and ease of rooting. The term "grafting" or "graftage" refers to using a scion that has two or more buds along a stem or a branch. "Budding," on the other hand, refers to a scion that is restricted to one bud. Other than the number of scion buds, the operation is essentially the same. Grafting is both an art and a science. It is important to understand the physiology and anatomy of the plant parts to make sure that there is a proper alignment of cambial tissue so that a graft "takes" (Chapter 25). It is also an art in that it takes much practice and skill to manipulate the plant parts to ensure that a suitable number of grafts are successful. In this chapter, laboratory experiments will detail two grafting techniques and one budding technique commonly used to produce fruit trees, roses, and other ornamental plants.

EXERCISES

Experiment 1. Rose T-Budding

MATERIALS

The following materials will be needed for each student or student team:

- Rooted cuttings of *Rosa multiflora* Thunb. ex J. Murr. thornless cultivars such as 'Brook's 48,' 'Brook's 56,' 'K1,' 'Dan Whiteside,' or 'Ginn 66'
- Bud sticks of any rose cultivar with well-developed buds
- Paraffin film wrap (Parafilm™)
- Grafting or budding knife or sharp pocket knife
- Rubber budding strips or rubber bands cut into a strip about 4 in (10 cm) long
- Bandages and first-aid supplies to treat attempted "finger grafts"

Follow the instructions found in Procedure 26.1 to complete this experiment.

ANTICIPATED RESULTS

After the T-bud is inserted and wrapped, the healing process should begin. The bud should remain green and turgid. The cambium layer around the "T-shaped" cut will begin to grow and cover the edges of the inserted bud. After about 1 week, the bud should still be green and turgid. The healing process should be completed after 2 weeks. At this time, the inserted bud should have formed a union with the surrounding tissue. If the inserted bud has turned brown or has shrunk, the bud probably did not take. After the healing and bud forcing steps have been done, the bud should initiate new growth and produce a new stem with the desired characteristics from the scion. The new shoot should be pruned to force further branching and to reduce the possibility of tearing the bud union before sufficient secondary growth has occurred to make a strong graft union.

QUESTIONS

- Why is it advisable to disbud the rootstock before T-budding?
- Why should the bark be in the "slip" stage when budding?
- What is the purpose of "crippling" the stem above the T-bud?

Procedure 26.1
Rose T-Budding

Step	Instructions and Comments
1	*Rootstock preparation.* Rootstock cuttings should be rooted before needed in 4-in (10-cm) pots and actively growing. The cutting length should be sufficient enough to allow the manipulation of the required cut and the insertion of the vegetative bud (10–12 in [25–30 cm long]). The rootstock should be disbudded (removing lateral vegetative buds), except for the top two or three buds, using a grafting or pocket knife. This will help prevent suckering of the rootstock later after planting out. Rootstock suckers are often more vigorous than the selected scion cultivar. The active growth will ensure that the bark will "slip" or that the cambium is easily cut and manipulated. This may necessitate growing the plants in a greenhouse during the dormant season to force new growth.
2	*Bud stick preparation.* Bud sticks should be about the length and diameter of a #2 pencil. They can be harvested from any rose cultivar of choice. Bud sticks can be prepared beforehand, wrapped in a moistened paper toweling, placed in a plastic re-sealable bag, and stored under normal refrigeration. If the bud sticks have thorns, they may be removed prior to storage to facilitate handling.
3	*Preparing the rootstock to receive the bud.* Take a sharp knife and make a vertical cut about 1.5 in (3.8 cm) long and deep enough to cut through the cambium layer (1/16–1/8 in [0.15–0.30 cm]), but not into the wood. The second cut is a horizontal cut about 1/3 around the diameter of the stem, at the same depth into the cambium as the first cut. Take the tip of the knife blade and insert it into the cut to slightly loosen the flaps of the cambium. The two cuts should resemble a "T" (Figure 26.1).
4	*Preparing the bud to insert into the stock.* Take a bud stick and make an initial cut about 1/2 in (1.25 cm) below the bud. The cut should be deep enough to get under the cambium layer but not so deep as to cut into the wood. The slanting cut should extend about 1 in (2.5 cm) above the bud. The second cut is a horizontal cut about 3/4 in (1.88 cm) above the bud through the cambium and into the wood and excised to yield a shield-shaped bud, with the curved part of the shield toward the bottom of the bud (Figure 26.2).
5	*Inserting the bud into the stock.* Take the bud and push it into the T-shaped cut under the two flaps of the cambium. The horizontal cuts of the bud and the stock should line up (Figure 26.3).
6	*Wrapping the bud and stock.* After the bud has been inserted into the stock, it is necessary to wrap the bud to secure it into place and prevent the bud from drying out. Take a length of the rubber band strip or budding rubber, and leaving the bud exposed, wrap it securely around the two flaps of cambium and extend below and above the cuts. After securing with the rubber strip, take a small square (~2 × 2 in [5 × 5 cm]) piece of paraffin film and wrap it over the entire T-bud section to prevent the bud and the stock from drying out.
7	*Growing on.* After the T-budding operation has taken place, place the potted budded plant in a well-lit place protected from strong winds that might damage or dry out the protected buds. Rose buds will normally "take" in about 2 weeks. If the T-bud is successful, the bud will remain green and turgid during the healing process. There may be some swelling of the bud in size as it starts to grow. After about 2 weeks, remove the paraffin wax film. Leave the rubber strip in place to prevent the bud from being damaged or accidentally dislodged before the cambium completely heals around the new bud.
8	*Forcing the bud.* When the T-bud has callused over and "healed," it is time to force growth in the new bud. This is done by "crippling" or breaking the stem of the stock plant growth above the inserted bud (Figure 26.4). The object is to bend the stem over and snap it, completely breaking the stem off. This will release lateral dormancy in the inserted bud and will allow it to start growing while the semi-attached stem continues to supply photosynthates to the plant. Once the inserted bud has started to grow, the remaining stock growth may be removed right above the new bud union. Care should still be taken to prevent damage to the new shoot. If the new shoot is overly vigorous, it is sometimes desirable to prune it back to force lateral branching and prevent top-heavy growth, which could break off the new shoot before the union is completely healed.

FIGURE 26.1 T-shaped cut, with the cambium loosened to receive the bud.

FIGURE 26.2 Making a shield-shape cut under the bud.

FIGURE 26.3 The T-bud inserted into the rootstock.

FIGURE 26.4 Forcing the bud after healing.

Experiment 2. Splice Grafting Herbaceous Plants

MATERIALS

The following materials will be needed to complete this experiment:

- 4-in (10-cm) potted plants of coleus (*Solenostemon scutellarioides* [L.] L. E. W. Codd) to use as rootstocks
- Stock plants of desirable coleus cultivars to use as scion material
- Plastic straws (soda straws work well) similar to the diameter of the stems being used
- Paraffin film
- Bandages and first-aid supplies to treat cuts

Follow the instructions provided in Procedure 26.2 to complete this experiment.

ANTICIPATED RESULTS

If the scion and the rootstock are properly aligned, the healing process should begin in a few days. The scion will probably wilt or lose turgor for the first few days, but should regain turgor once the healing process begins. This loss of turgor can be compensated by placing the plant under an intermittent mist system or by covering the whole plant with a plastic bag. When the scion or top piece begins to show new growth, it is safe to remove the straw piece used to join the two pieces. This is best done by using a sharp knife or a razor blade to slit the straw. Care should be taken not to damage the new graft union.

Procedure 26.2
Splice Grafting Using Coleus

Step	Instructions and Comments
1	*Stock plant preparation.* Take a 4-in (10-cm) potted plant to use as a rootstock and make a diagonal 45° complete cut through the stem about 2 in (5 cm) above the soil line. Take a section of plastic straw tubing about 1 in (2.5 cm) long and slide it down the length of the stem of the rootstock (Figure 26.5).
2	*Scion preparation.* Take a scion stock plant and find a stem that is about the same diameter as the cut stem section of the rootstock. Cut the scion stem section into a length that has at least two buds (~2–4 in [5–10 cm] long). Make a diagonal 45° cut to match the one made on the stock plant.
3	*Splice graft.* While holding the scion carefully in place, pull the section of tubing up the stem of the rootstock over the inserted scion, making sure that the 45° angles match up. The tubing should be snug enough to ensure that the two stem sections are in intimate contact (Figure 26.6). The straw and the exposed ends of the scion and the rootstock should be wrapped with paraffin film to prevent drying out and give added support while the healing process occurs.
4	*Growing on.* If the stems are soft and succulent, it may be necessary to place the whole grafted plant under an intermittent mist system or enclose it in a plastic bag for a few days until healing begins. After the graft has healed, the plastic tube may be removed by using a razor blade to slit the tubing lengthwise. Care should be taken not to damage the graft union.

QUESTIONS

- When making the cuts on the scion and the rootstock, what is the importance of making identical angled cuts?
- Why is paraffin film used to cover the straw and spliced stem pieces?
- Why is the whole plant either placed under a mist system or in a plastic bag during the healing process?

Experiment 3: Whip and Tongue Grafting

The whip and tongue graft is useful when the plant material is small in size, with stem diameters of 1/4 to 1/2 in (0.63–1.25 cm). The success rate is high if correctly done since the area of cambial contact is great. If aligned properly, the wound heals quickly and a strong union is formed. Ideally, the scion and the rootstock should be of equal diameter, with the scion having two buds.

FIGURE 26.5 Rootstock, with angled cut and plastic straw to support the graft.

FIGURE 26.6 The scion and the rootstock joined by a length of plastic straw.

MATERIALS

The following materials will be required to complete this experiment:

- 4-in (10-cm) potted plants of boxwood (*Buxus sempervirens* L.) or other woody plants with a stem 1/4 to 1/2 in (0.63–1.25 cm in diameter and a length >4 in [10 cm]) for use as rootstock
- Boxwood or other woody plant material using the same species as the rootstock for scion wood
- Grafting or budding knife or sharp pocket knife
- Rubber budding strips or rubber bands cut into a strip about 4 in long
- Bandages and first-aid supplies to treat attempted "finger grafts"

Follow the instructions in Procedure 26.3 to complete this experiment.

ANTICIPATED RESULTS

If the scion is properly matched up with the rootstock, the healing process will begin. Wrapping the two sections with paraffin film will prevent excess drying out of the scion.

Procedure 26.3
Whip and Tongue Grafting Using Boxwood

Step	Instructions and Comments
1	*Rootstock preparation.* A slanting cut from 1 to 2 1/2 in (2.5–6.25 cm) is made at the top of the stock 2 to 4 in (5–10 cm) above the soil line of the potted plant. With a sharp knife, a cut is made in one stroke to make an even, straight cut. A straight cut will ensure better cambial contact between the scion and the rootstock. A second cut in the reverse direction is made about 1/3 down from the top of where the first cut ended and should extend about halfway down the original cut. The second cut should parallel the first cut to ensure a close fit. When completed, the cut should resemble a sideways letter "Z."
2	*Scion preparation.* Choose scion wood with the same diameter as the rootstock. Remove any attached leaves. The scion wood should be long enough to have two to three dormant buds. Two identical cuts as in the rootstock are made (Figure 26.7).
3	*Joining the scion and the rootstock.* The scion is inserted into the rootstock so that the two "Z-shaped" cuts interlock. Care should be taken to ensure that the cambium of the two pieces line up and that there are no gaps between the scion and the rootstock. After the two pieces have been joined, they should be securely tied in place with a length of rubber stripping and then wrapped in paraffin film to prevent drying out of the scion wood while healing (Figure 26.8). If the graft takes, new growth will initiate from the scion buds. Any growth from the buds below the graft union should be removed.

FIGURE 26.7 Making identical initial cuts on the rootstock and the scion.

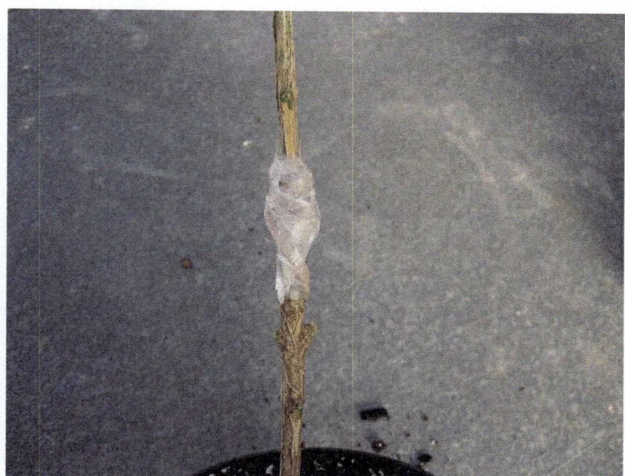

FIGURE 26.8 Graft area wrapped with paraffin film, ready for healing process.

Once the healing process is complete and the graft union has formed, the lateral buds along the scion should begin to develop and grow. The rubber budding strip and paraffin film may be removed after new growth has started. The time required for the healing process will depend on the vigor of the plant, the temperature, and the time of year. This period could range from 3 to 6 weeks, so patience is needed when attempting whip and tongue grafting.

QUESTIONS

- Why is a large amount of cambial surface area important in whip and tongue grafting?
- Why should the scion wood and the rootstock be of the same diameter?
- Why is it important to remove any shoot growth occurring below the graft union?

SUGGESTED READING

Bailey, L. H. 1911. Graftage. In *The Nursery Book*, 15th Edition. New York, New York: The Macmillan Company, pp. 94–108.

Hartmann, H., D. Kester, F. T. Davies, and R. Geneve. 2010. *Plant Propagation: Principles and Practices*, 8th Edition. Englewood Cliffs, New Jersey: Prentice Hall, Inc., 926 pp.

27 Grafting with Vegetable Plants

Chieri Kubota

Concept Box 27.1

- Vegetable grafting is a pest and disease management technique developed many years ago.
- More countries are learning how to graft and use grafted vegetable seedlings effectively in different cultivation systems including greenhouse and open fields.
- Different vegetable crop species have different recommended grafting methods.
- Controlled environment is a key for success in healing grafted seedlings.

Grafting of vegetable seedlings is a unique horticultural technology to overcome soil-borne diseases and pests and to add extra vigor to plants under various environmental stress conditions. Similar to the benefits observed in grafting other crops (Chapters 25 and 26), by selecting an appropriate rootstock, vegetable grafting can accomplish the following: (1) achieve soilborne disease and pest resistances/tolerances, (2) acquire tolerance to abiotic stresses such as chilling or salt stress, and (3) increase vigor and yields. The technique was practiced for many years in East Asia, and more recently, it has become a common pest management technique worldwide. As an example of the intensive use of vegetable grafting in Asia, 94%, 93%, 79%, and 58% of cultivation areas in Japan (including open fields and greenhouses) used grafting for producing watermelon, cucumber, eggplant, and tomato, respectively (2009 survey data by NARO 2011). Over 500 million grafted seedlings are used in Japan (Kobayashi, 2005). Similar levels of grafted seedling production can be seen in Korea, another leading country for this technology. The grafting technology gradually "migrated" to other countries, including the rest of Asia, Middle East, Africa, Europe, and the Americas (Figures 27.1 and 27.2). Today, in Spain, for example, grafting is used for almost all watermelon cultivated in the Almeria region, mainly for controlling Fusarium diseases. Recent rapid expansion of the use of vegetable grafting is also reported for China, Italy, Turkey, and Greece in order to mitigate the impact of soil-borne diseases and pests associated with intensive cultivation and increasing regulations of available chemicals for soil fumigation. Therefore, vegetable grafting is viewed as an environmentally friendly plant propagation technique to reduce the input of soil fumigation.

HISTORY OF VEGETABLE GRAFTING

Vegetable grafting was reportedly practiced in ancient China (Lee and Oda, 2003) for producing large gourd fruits. This specific method was reportedly a self-grafting method in which one scion was grafted with multiple rootstocks of the same species/cultivar to create a large root-to-shoot ratio. Vegetable grafting was not used as a pest management tool until the early 20th century. According to Tateishi (1927), the history of vegetable grafting as a pest management technique began with a venture done by an innovative watermelon farmer in Hyogo prefecture, Japan. The farmer, Ukichi Takenaka, who had struggled with Fusarium diseases of watermelon, had successfully grafted his watermelon plants on pumpkin as a rootstock without major difficulties. Koshiro Tateishi, a researcher in Nara prefecture, Japan, reported the farmer's experiences together with his follow-up research in the *Journal of Japanese Horticulture* in 1927. Tateishi described, "While it has been long known that it is possible to graft watermelon onto pumpkin, it is hard to believe that nobody actually practiced it. If these species are highly compatible in grafting, it may be a technological breakthrough (in watermelon), with potential impact similar to the discovery of rootstock resistant to Phylloxera in viticulture." Ukichi's grafted watermelon plants "continued to be healthy until harvest," but "fruit set was largely delayed," a typical issue we observe today when grafted on strong rootstock. Fruit produced on grafted watermelon had similar flavor and sweetness as non-grafted plants. The effect of grafting on fruit flesh color (redness) was investigated, but the result was not conclusive (small number of samples). Tateishi, as an agricultural researcher and educator, hoped "to obtain more interest and support

FIGURE 27.1 Tomato production greenhouse in Mexico in which plants are grafted onto an interspecific hybrid tomato rootstock to increase resistance from disease and pests.

from the readers of this journal toward possible commercial applications and implications of this technology in the future." Today, almost 90 years later, the technology has become a common practice of sustainable vegetable production in many countries, and some late adopters, including the United States, are trying to catch up by going through somewhat similar efforts and processes as described previously in the old literature.

In North America, introduction of vegetable grafting began very recently, and grafted plants are used mainly for increasing yields of greenhouse tomato (including hydroponics) (Kubota *et al.*, 2008). Because of the limited domestic grafting capacity, large greenhouse growers are importing millions of grafted tomato seedlings from Canada every year. Today, the use of grafting in soil-based cultivation (greenhouse and open fields) is more advanced in Mexico (Figure 27.1) than in Canada or the United States, partly because of the rapid increase of their greenhouse crop production area over the past 20 years. Our survey in 2011 indicated that there were several grafting propagators in operation to produce grafted seedlings not only for tomato, but also for watermelon in Mexico.

Another notable event in the vegetable grafting history in North America is the introduction of grafted vegetable plants to the home gardening market. A U.S. nursery is now producing millions of grafted heirloom vegetable seedlings (tomato, eggplant, and melon) to meet the needs and interests of home gardeners. The use of grafted plants in home gardens has also been practiced in many countries. As a unique application of vegetable grafting, Oda (1990) developed methods to graft multiple

FIGURE 27.2 Grafted tomato and cucumber seedlings in nurseries in (a) Spain and in (b) Japan.

FIGURE 27.3 Grafting two different cultivars of tomato onto one rootstock. (Courtesy of Hishtil, Israel.)

species of scions into one rootstock (tomato, eggplant, and pepino onto tomato or eggplant rootstock) and promoted the technique as an educational demonstration of vegetable grafting at an international science expo held in 1990. Today, "two or more varieties in one" has been seen in the home gardening grafting market in Europe and the United States (Figure 27.3).

ROOTSTOCK SELECTION

Guidelines for selecting rootstocks are summarized in Table 27.1. For grafting tomato, rootstocks resistant or tolerant to Fusarium, bacterial wilt, Verticillium wilt, tomato corky root, and root knot nematode are commercially available from various seed companies (Lee and Oda, 2003). Tomato rootstocks are hybrids either within the same species (*Solanum lycopersicum*), called "intraspecific hybrid," or between *S. lycopersicum* and its relatives (such as *S. habrochaites*), called "interspecific hybrids." Interspecific hybrids are generally more vigorous but less uniform in germination and early growth.

For grafting cucurbits (watermelon, muskmelon, and cucumber), rootstocks resistant or tolerant to Fusarium and melon necrotic spot virus are available from various seed companies. Watermelon is grafted on either squash (*Cucurbita spp.*) or bottle gourd (*Lagenaria siceraria*). Muskmelon is grafted on squash or another type of muskmelon. Cucumber is grafted on squash. Those rootstocks are either intraspecific or interspecific hybrids. Some squash rootstocks are also tolerant to low temperatures.

Although the benefits of grafting discussed previously are expected, the actual benefits and impacts are often "site specific," affected by the environmental conditions, cropping system, management methods, and soil type. Therefore, testing several rootstocks (with different pedigrees or seed companies) is always recommended before deciding to introduce grafting on a large scale.

GRAFTING METHODS FOR TOMATO AND CUCURBITS

Various grafting methods were developed in the past to increase success rate as well as grafting speed. The oldest grafting method is called "approach grafting" (Chapters 25 and 26). This grafting method can achieve high rates of success even for beginners as half of the vascular systems of scion and rootstock are attached through the grafting and healing process. However, the drawbacks of this most reliable grafting method are the slow grafting speed and an additional step of cutting away the scion roots and the rootstock shoot in finishing the grafted seedlings. As a result, this method is practiced to a lesser extent compared to other methods such as tube grafting (tomato, eggplant, and pepper), single-cotyledon grafting (cucurbits), and hole insertion grafting (cucurbits), which are introduced in subsequent sections. Other grafting methods not included in this chapter are

TABLE 27.1
Guidelines for Selecting Grafting Rootstocks (for Tomato and Cucurbits)

Type	Resistance	Other Traits
	Tomato	
Interspecific hybrid (hybrid between different tomato species) (e.g., 'Maxifort') (*Solanum lycopersicum* × *S. habrochaitaes*)	Different for different roostock varieties, but generally include Fusarium and Verticillium wilts, and root knot nematodes. Some include bacterial wilt and race 3 of *Fusarium*.	They are generally vigorous. Some rootstocks have chilling tolerance. However, less uniformity in plant growth at the seedling stage (germination and emergency).
Intraspecific hybrid (hybrid within the same cultivated tomato species) (e.g., 'Aloha') (*Solanum lycopersicum*)	Different for different roostock varieties, but generally include Fusarium and Verticillium wilts, and root knot nematodes. Some include bacterial wilt and a higher race (race 3) of *Fusarium*.	Very uniform growth. Less vigorous.
	Cucurbits	
Interspecific hybrid squash (hybrid between different squash species) (e.g., 'Tetsukabuto') (*Cucurbita maxima* × *C. moschata*)	Fusarium. Some also have vine decline, Verticillium wilt, and anthracnose.	For all cucurbits. Traits varied among different rootstock varieties (vigor, chilling heat or drought tolerance, and others).
Bottle gourd (*Lagenaria siceraria*)	Fusarium. Some also have vine decline, Verticillium wilt, and anthracnose.	For watermelon. Chilling tolerance.

Note: Although these resistances and benefits are expected, the actual benefits and impacts are often "site specific," affected by environmental conditions, cropping system, management methods, and soil type. Therefore, it is always recommended to test several rootstocks (with different pedigrees or seed companies) before deciding to introduce grafting on a large scale.

"cleft grafting," in which scion and rootstock seedlings are cut to create a sharp "V" shape. This method is used for eggplant in Asia and, to a limited extent, in Europe.

EXERCISES

Experiment 1. Tube Grafting

Tomato seedlings are almost always grafted using the "tube grafting" method in commercial propagation. It is a very easy and simple method to learn. Important keys for success are the following:

1. Selecting similar sizes of scion and rootstock hypocotyl diameter is the most important in this grafting procedure. Instructors who want to use this method in their classroom or growers who want to graft a particular combination of scion and rootstock for the first time are encouraged to first grow small trial lots of scion and rootstock seedlings to determine the best seeding schedule to obtain similar sizes of scion and rootstock (Figure 27.4).
2. Selecting grafting clips of appropriate size—clips that are too large cannot hold the grafted union in place, and clips too small impart too much pressure and may deform the union.
3. Limiting irrigation for 1 to 2 days before grafting, but avoiding excessive desiccation during and after grafting.

MATERIALS

Each student or team of students should have the following:

- Scion seedlings at a two true-leaf stage. Target diameter of the grafting position (hypocotyl or epicotyl) is 1.5 mm. Depending on the growing conditions and the germination rate, it takes about 2.5 to 3 weeks for the seed to reach this stage.
- Rootstock seedlings at the same leaf stage. Target diameter of grafting position (hypocotyl or epicotyl) is 1.5 mm (the same as that for the scion). Depending on the growing conditions and the germination rate, it takes about 2.5 to 3 weeks for the seed to reach this stage.
- Cutting tool. Razor blades work fine for this grafting method.
- Grafting tubes (Figure 27.5). Size needs to be selected to match the seedlings. Having two different sizes is a good idea to allow flexibility.
- Disinfectant such as 70% ethanol to clean the tools
- Water spray bottle for humidification
- Paper towel

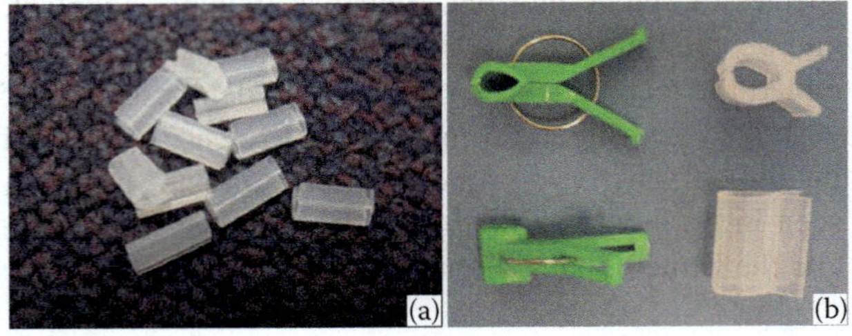

FIGURE 27.5 Grafting tubes for tomato (a) and clips for cucurbits (b).

Follow Procedure 27.1 to graft tomato seedlings using the tube grafting method. For anticipated results, see this section accompanying Experiment 4.

Experiment 2. Single-Cotyledon Grafting for Cucurbits (Melon and Cucumber)

This grafting method was originally developed as one suitable for automation (grafting robot) by Japanese

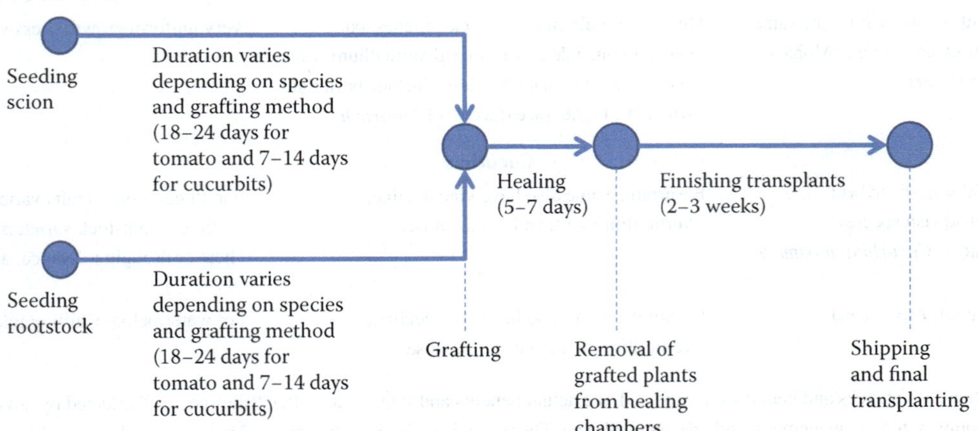

FIGURE 27.4 Sample timetable for producing grafted tomato and cucurbit seedlings.

Procedure 27.1

Tube Grafting for Tomato (Figure 27.6)

Step	Instructions and Comments
1	Cut a rootstock seedling below the cotyledons at a 45° or sharper angle. Cutting rootstock seedlings above the cotyledons requires cautionary consideration as axillary buds may grow out when vigorous rootstocks are used, requiring the removal of those axillary shoots. Interspecific hybrid rootstocks (hybrids between different species) are generally more vigorous than intraspecific hybrid rootstocks (hybrids within the same species) or open pollinated cultivars.
2	Place one tube on top of the cut end of the rootstock hypocotyl, half the length of the tube. Orient the tube so that you can see the angle of the cut through the opening.
3	Prepare the cut for the scion seedling (with matching hypocotyl width) at the same angle at about 5 to 10 mm below the cotyledons.
4	Insert the scion shoot into the grafting tube so that the cut surface aligns perfectly with that of the rootstock. (You need good eyes!)
5	Mist grafted plants regularly with water to prevent desiccation. Damaging water stress occurs before visible wilting.
6	Move the tray filled with grafted plants to the healing chamber for up to 7 days of healing. Please see the section on healing methods for more information.
7	No need to take the tube off as it will fall off as the stem grows.

engineers due to the difficulty to mechanically replicate the conventional grafting methods (hole insertion and approach methods) applied in cucurbits. However, due to its simplicity, it also became a widely accepted manual grafting method. This method can be done with or without cutting the roots from the rootstock. Important keys for success include the following:

1. Selecting similar sizes of scion and rootstock is very critical in this grafting procedure.

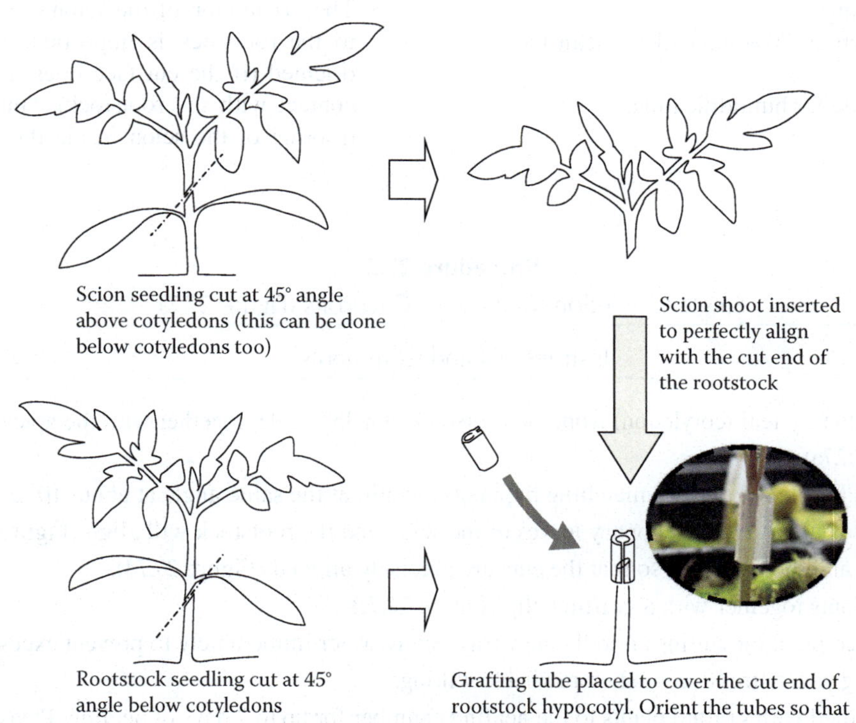

FIGURE 27.6 Tube grafting for tomato.

2. Selecting grafting clips (spring type; Figure 27.5) of the appropriate size. Clips too large cannot hold the grafted union together, and clips too small clip impart too much pressure and may deform the union.
3. Making sure that the prepared rootstocks do not have the axillary bud at the base of the remaining cotyledonary leaf. Rootstock grow-outs in the field are a "must-avoid" situation.

MATERIALS

Each student or team of students should have the following items:

- Scion seedlings at the beginning of first true-leaf stage. Depending on the growing conditions and the germination rate, it takes about 10 to 14 days for the seed to reach this stage.
- Rootstock seedlings at a first true-leaf stage. Long hypocotyls (7–9 cm) are desirable when grafted without roots. Rootstock may be grown at a higher density to make the hypocotyl width smaller to match the scion. Depending on the growing conditions and the germination rate, it takes about 10 to 14 days for the seed to reach this stage.
- Razor blades work best for this grafting method.
- Grafting clips. Size needs to be selected to match the seedlings.
- Disinfectant such as 70% ethanol to clean the tools.
- Water spray bottle for humidification.
- Paper towel.

Follow Procedure 27.2 to graft melon, watermelon, or cucumber seedlings using the single-cotyledon grafting method. For anticipated results, see section accompanying Experiment 4.

Experiment 3. Hole-Insertion Grafting for Cucurbits

This grafting method does not require grafting clips but can establish a very strong graft union. Also, the scion size does not have to match the rootstock size. Scion size can be flexible as long as it is smaller than the rootstock. Grafts using this method are almost always intended to be unrooted grafted cuttings, and so, adventitious rootstock roots will develop during healing of the grafted union; therefore, the healing process after grafting requires extra care (to achieve optimum humidity, temperature, and light). Important keys for success include the following:

1. When rootstocks are too turgid, it is difficult to graft without causing cracking of the hypocotyls. Timing between harvesting rootstock cuttings and grafting must be determined. Rootstock cuttings can be harvested a few hours earlier and kept in a plastic bag if necessary. In contrast, scion cuttings must be fully turgid, and so, they are used for grafting immediately after harvesting them.
2. All axillary buds must be removed from the rootstock. Good vision is necessary. Rootstock grow-outs in the field are a "must-avoid" occurrence.
3. The orientation of the scion's cut faces relative to the rootstock is important. If not properly oriented (if the cut face does not have a good contact with the rootstock), "internal rooting" (rooting of the scion inside the hollow pith of

Procedure 27.2
Single Cotyledon Method for Cucurbits (Figure 27.7)

Step	Instructions and Comments
1	Cut one cotyledonary leaf (cotyledon) from the rootstock at a 45° angle, together with the apical and axillary buds (Figure 27.7a).
2	Prepare the cut for the scion (with matching hypocotyl width) at the same angle at about 10 to 15 mm below the cotyledons so that the cotyledonary leaves of the scion and the rootstock will align (Figure 27.7c).
3	Graft the scion and the rootstock so that the cuts are precisely aligned (Figure 27.7d).
4	Hold the two plants together with a grafting clip (Figure 27.7e).
5	Insert the grafted plant (or cutting) into the new tray. Spray water immediately to prevent excessive dehydration. Damaging water stress occurs before visible wilting.
6	Move the tray filled with grafted plants to the healing chamber for up to 7 days of healing. Please see the section on healing methods for more information.
7	Remove the grafting clip after the union is healed.

FIGURE 27.7 Single-cotyledon grafting for cucurbits (melon and cucumber). (a and b) Cutting off the axillary bud of the remaining cotyledon, the shoot tip, and the other cotyledon of a squash rootstock. (c) Preparing the muskmelon scion by cutting 10 to 15 mm below the cotyledon at an angle matching the rootstock cut. (d and e) Aligning scion and rootstock cuts together to hold with a clip. (f) A grafted muskmelon seedling using the single-cotyledon method.

the rootstock hypocotyl) may occur (Figure 27.8). Other poor grafting procedures could also induce internal rooting. Beginners are encouraged to examine the status of internal rooting after healing by dissecting the hypocotyl below the union within a sample of plants.

MATERIALS

Each student or team of students should have the following materials:

- Scion seedlings at the beginning of first true-leaf stage. The true leaf size is 2 to 3 mm. Depending on the growing conditions and the germination rate, it takes about 10 to 14 days for the seed to reach this stage.
- Rootstock seedlings at a first true-leaf stage. Long hypocotyls (7–9 cm) are desirable. The true leaf blade size is about 2 cm. Depending on the growing conditions and the germination rate, it takes about 10 to 14 days for the seed to reach this stage.
- Scalpels with handles work best for this grafting method.
- Perforating tool. A plastic soldering tool works well. Alternatively, you can create a tool by sharpening the tip of bamboo chopsticks (a pencil sharpener works well).
- New trays filled with well-moistened substrate.
- Disinfectant such as 70% ethanol to clean the tools.
- Water spray bottle for humidification.
- Paper towels.

FIGURE 27.8 Dissected rootstock hypocotyl with (a) and without (b) incidence of internal rooting of scion. Internal rooting can be an issue in the hole-insertion method of cucurbit grafting. Squash rootstock has a hollow pith, and when the grafting method is not done appropriately, internal rooting may be observed at a high rate. By dissecting the hypocotyl below the grafted union, beginners can examine their grafting skill. The picture on (a) shows a squash rootstock, with the hollow filled with scion roots (note the root-free hollow in [b]).

Follow Procedure 27.3 to graft melon, watermelon, or cucumber seedlings using the hole-insertion grafting method. See anticipated results under Experiment 4.

HEALING METHODS

Building Your Own Healing Box

Healing is the process during which the grafted union grows together and the two vascular systems (scion and rootstock) become connected. This permits water and nutrients to be transported from the rootstock to the scion and photoassimilates and other nutrients to be translocated from the scion to the rootstock. Failure to follow all applicable procedures for proper healing can cause grafting failure. Therefore, you should understand the

Procedure 27.3

Hole Insertion Method for Cucurbits (Figure 27.9)

Step	Instructions and Comments
1	Harvest rootstock cuttings 30 to 60 min before starting grafting. Rootstock tissue needs to be less turgid so that cracking can be avoided (when perforated). Harvested rootstock cuttings can be kept for a few hours in a plastic bag.
2	Remove the first true leaf, and the apical and lateral (axillary) meristems from the rootstock with a scalpel. Axillary meristems appear as white dots near the base of the petiole. Failure to remove them will result in rootstock shoot growth.
3	Perforate the tip of the stem at an angle so that the tip of the tool comes out through the hypocotyl below the opposite cotyledon. Leave the perforating tool in so that you do not need to "look for" the hole.
4	Prepare the scion cutting with two cuts, creating an about 5-mm–long sharp edge on the hypocotyl at 5 to 7 mm below the cotyledons. A second sharp cut of similar length on the scion is done after rotating the scion by 90°. Scion cuttings need to be fully turgid.
5	Insert the scion into the hole that you created on the rootstock. Note the orientation of the scion cut surface. The cut surfaces must achieve maximum contact with the rootstock tissue. Therefore, orient the scion with its cut surfaces facing downward.
6	Place the grafted cuttings in the new tray and spray with water to avoid excessive dehydration. Damaging water stress occurs before visible wilting.
7	After filling the tray, move the tray to the healing chamber for up to 7 days of healing and rooting. Please see the section on healing methods for more information.

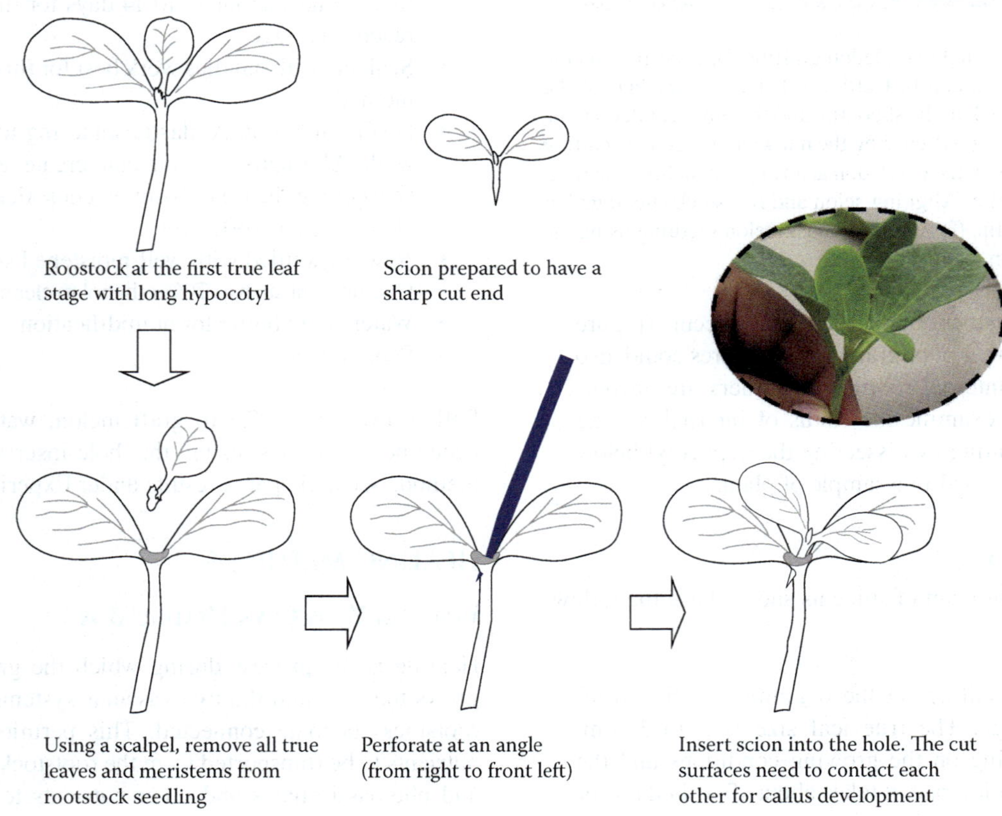

FIGURE 27.9 Hole-insertion grafting for cucurbits.

required conditions as well as practice the protocols recommended here. Duration of healing is generally 6 to 7 days.

Optimum Conditions to Heal the Grafted Plants

The most important environmental factors during healing are temperature, relative humidity, and light. The optimum healing temperature is 28 to 29°C, which is, in fact, slightly warmer than the optimum temperatures for seedling growth (daily average of 20–25°C for most fruiting vegetables). Under this warmer temperature, cell division will be accelerated to form callus that will bridge the cut surfaces of the scion and the rootstock. The optimum relative humidity for healing is 95% or greater during the first 3 to 4 days to prevent excess water loss from freshly grafted plants. Preventing water loss from grafted plants is critical while the vascular systems are not connected, and therefore, water cannot be transported from the rootstock to the scion as easily as intact plants. Once a callus bridge is developed at the union, plants can take up water more easily than those without a callus bridge, but hydraulic conductivity is still less than that for fully healed plants with connected vascular systems. Relative humidity is usually reduced gradually toward the end of the healing process to acclimatize the plants to lower humidity. This higher-humidity environment is often created by enclosing the grafted plants inside a box or a tunnel made with a glazing material that transmits light. Because of light penetration, the internal temperature tends to be warmer than the outside of the healing chamber. Therefore, you must measure the temperature inside the enclosure where your plants are and make sure that the temperature is maintained around the optimum (28–29°C) throughout the healing process.

Light is required to keep plants alive during the healing process, although complete darkness is necessary for the first 24 h. Only a small amount of light is required, compared with the amount of light that plants receive under normal growing conditions. Our recommended light intensity to achieve inside the enclosure is 50 to 100 μmol m^{-2} sec^{-1} photosynthetic photon flux (PPF) (3700–7400 lux for fluorescent lamps or 2700–5400 lux for sunlight). For a healing chamber illuminated with artificial lighting, two fluorescent lamps of the same length as the chamber(s) mounted above at a distance of 15 to 30 cm above the plants in the chambers will create a similar level of light intensity over the plant canopy surface. This light intensity is many times lower than mid-day natural sunlight intensity (1000–2000 μmol m^{-2} sec^{-1}) and is closer to the light compensation point, which provides the minimum energy needed to maintain plant metabolism during the healing process.

Types of Healing Systems

There are generally two types of healing systems: (1) a temperature- and humidity-controlled walk-in compartment (growth chamber) with artificial lighting, and (2) a greenhouse-based healing system with extra humidification, heating, and shading. The former system can create a uniform climate independent from outside environmental conditions and is more suitable for grafts that are difficult to heal. In particular, cucurbit grafting is more sensitive to fluctuating healing environments, and high success rates can be achieved using perfectly controlled environments during healing.

Healing Chambers with Artificial Lighting

A commonly used design of healing chamber with artificial lighting uses fluorescent lamps installed in multilayered shelving units (Figure 27.10). The chamber needs to have a good humidity control to create near-saturating humidity, and shelving units and lamps in such a chamber need to be compatible with wet conditions. Alternatively, small tunnels or compartments can be fabricated to create highly humid microclimates directly around the grafted plants (Figure 27.11). Such a setup may be suitable for healing small numbers of grafted plants.

Healing Chambers in a Greenhouse with Natural Light

Developing healing units inside a greenhouse or high tunnel has been commercially practiced for many years. However, maintaining air temperature, light intensity, and humidity at optimum levels is more challenging especially in warm climate regions or under high solar radiation environments in summer. Key equipment/

FIGURE 27.10 A typical healing system with fluorescent lamps and multi-shelving units, used in commercial operations.

FIGURE 27.11 A small healing setup using boxes for maintaining high humidity around the seedlings.

FIGURE 27.12 Covered benches inside a greenhouse used for healing grafted plants in Spain. Note that the black cloth is used for the first day or two after grafting. Benches are equipped with a bottom heating system and a misting system (under the bench).

materials for such healing systems in greenhouse include shade cloth to reduce light intensity, plastic cover to maintain high humidity, fogging or under-bench misting system, and heating system (during nighttime or cloudy conditions).

Figure 27.12 shows a healing setup used in Spain. Chambers are created over individual benches. Black cloth (in the background) maintains darkness for the first day or two after grafting. Overhead shade cloth is for reducing the light intensity. In this particular system, a misting system is installed under each bench to humidify the air yet avoid wetting the plants. It also has a hot water heating system underneath the seedling trays.

Experiment 4. Healing Grafted Seedlings

The following is a setup for and the methods of healing grafted seedlings, applicable to small amounts of seedlings in research or classroom application.

MATERIALS

Each student or group of students should have the following materials:

- A clear box large enough to hold a tray of seedlings
- A matching transparent lid (if a lid is not available, a transparent plastic film with elastic band)
- Self-recording sensor for air temperature and relative humidity
- Water for humidification
- A temperature control chamber with lights
- Quantum sensor (or lux sensor)

Follow Procedure 27.4 to heal your grafted seedlings.

ANTICIPATED RESULTS (FOR ALL EXPERIMENTS)

Grafted unions will start to heal after about 3 or 4 days when healing chamber conditions are optimum. During this time, a callus bridge is developed to connect the tissues of the scion and the rootstock. Callus development rate is temperature dependent and declines rapidly under either super or suboptimum temperatures (Hartmann *et al.*, 1997). Some resistance to separation of the grafted union should be present when pulling on the scion after 3 or 4 days, but if it can be easily detached from the rootstock, then plants will require a couple more days to heal completely. After complete healing, one more day will be necessary for acclimatization to lower humidity (by opening the lid or cover used in healing). If you still see wilted plants at this point of acclimatization, those plants are likely not, and will not be, healed. Even if they recover from wilting, they may be internally rooted or with limited vascular systems connected. Therefore, those plants may need to be removed without growing on to finish.

A typical error in tube grafting and single-cotyledon grafting is that cut surfaces are not perfectly aligned, and therefore, insufficient vascular bundles are connected, limiting overall growth. Another error is either too low or too high healing temperature. Rooting from aerial portions of the stem and graft union may be seen both for tomato and cucurbits. Better control of relative humidity inside a healing system can eliminate scion rooting. Another issue sometimes reported as problematic is fungal growth over the surface of the substrate. Commercial growers keep the healing chambers and systems as clean as possible to avoid it.

Procedure 27.4
Healing Grafted Plants

Step	Instructions and Comments
1	Measure the light intensity inside the box in the chamber to make sure that it is in the target range (50–100 μmol m^{-2} sec^{-1} PPF).
2	Add a small amount of water (~0.5 cm depth), place a temperature/RH sensor inside the box, and put the box under the light in the chamber set at 26°C air temperature. After running for 24 to 48 h, take the sensor out to download the recorded temperature. If the inside temperature is lower/higher than the target temperature (28–29°C), increase/decrease the set point accordingly. This process requires a few days to finalize the appropriate settings for your chamber.
3	Place your grafted plants inside a clean box. Close the lid tightly.
4	Place the box inside a dark chamber maintained at 28 to 29°C air temperature.
5	After 24 h, turn the lights on and reduce the temperature setting to the one you determined earlier in step 2 that should achieve the 28 to 29°C air temperature inside the box.
6	Check the temperature and plants everyday to determine their status.
7	After 3 to 4 days of healing, slide the lid to create a small vent (or poke several holes) to gradually reduce the humidity inside the enclosure.
8	By day 7, completely open and then move the plants to the greenhouse, preferably in the evening to avoid acute light and humidity stress.
9	Acclimatize your plants for the first 48 h under 50% or greater shade in the greenhouse.
10	Remove the shade cloth.

QUESTIONS (FOR ALL EXPERIMENTS)

- What is the percent success of your grafting?
- What is the strength of your grafted union? Can you pull on the grafted union to see how strongly it is attached? Would the strength be noticeably different among different grafting methods?
- Did any seedlings wilt during the grafting and healing process?
- Examine different healing temperatures to observe differences in the healing process as affected by the temperature.
- Think about the protocols of grafting two or more different scions (cultivars or species) onto one rootstock. What would be the possible approaches to make such grafting successful?

LITERATURE CITED

Hartmann, H. T., D. E. Kester, F. T. Davies, Jr., and R. L. Geneve. 1997. The biology of grafting. In *Plant Propagation: Principles and Practices*, 6th Edition. New Jersey: Prentice Hall, pp. 392–436.

Kobayashi, K. 2005. Vegetable grafting robot. *Res Food and Agric* 28:15–20.

Kubota, C., M. A. McClure, N. Kokalis-Burelle, M. G. Bausher, and E. N. Rosskopf. 2008. Vegetable grafting: History, use, and current technology status in North America. *HortScience* 43:1664–1669.

Lee, J.-M. and M. Oda. 2003. Grafting of herbaceous vegetable and ornamental crops. In Janick J., ed. *Horticultural Reviews*, Vol. 28. New York: John Wiley & Sons, pp. 61–124.

NARO. 2011. *Current Use and Issues in Grafted Vegetable Seedling Production*. Report No. 7. Japan: Institute of Vegetable and Tea Science, National Agriculture and Food Research Organization, 142 pp.

Oda, M. 1990. *Grafted Plant "Tomapena" Produces Three Different Kinds of Fruit in One Plant*. Tochigi, Japan: Japan Tobacco Plant Research Institute, 108 pp.

Tateishi, K. 1927. Grafting watermelon onto pumpkin. *Hortic (Nihon-Engei Zasshi)* 39:5–8.

OTHER SOURCES OF INFORMATION

Vegetable Grafting Information. http://cals.arizona.edu/grafting. Accessed March 13, 2014.

Part X

Bulbs and Plants with Special Structures

28 Storage Organs

Jeffrey A. Adkins and William B. Miller

Concept Box 28.1

- Geophytes are plants that form specialized storage organs that serve as overwinter survival structures.
- The term "bulb" is often used to represent all specialized plant storage organs.
- "True bulbs" have a compressed stem (basal plate) and modified fleshy leaves that serve as the primary storage tissue.
- Corms, tubers, rhizomes, and enlarged hypocotyls derive from stem tissue, but differ in their sites of origin, presence or absence of scales, and orientation.
- Tuberous roots develop from enlarging root radicles and secondary roots.
- Propagation of various storage organs may be accomplished by scaling, scooping or scoring of bulbs, division, stem cuttings, or adventitious shoot production.

Plant species evolve over time in response to changing environmental conditions to ensure survival and continued reproduction. These evolutionary changes are what provide us with great diversity of plant forms and functions. Geophytes represent a unique group of plants that have developed adaptations, allowing them to survive as underground storage organs until environmental conditions are favorable for growth. There are many types of underground storage organs; however, bulbs, corms, tubers, tuberous roots, tuberous stems (enlarged hypocotyls), and rhizomes are the most common in horticulture. Each type of storage organ is defined based on the origin of development on the plant.

THE INDUSTRY

Flower bulbs are grown for the following two major purposes: (1) for use in commercial forcing of cut flowers and pot plants, and (2) as decoration in gardens, landscapes, and parks. The relative proportion of each of these uses varies by country, but worldwide, a significant majority of bulbs are used for forcing. Flower bulb production is a specialized activity and requires unique equipment, industry infrastructure, farming practices, horticultural skill, and proper climactic conditions for consistent success. While flower bulbs are produced worldwide, the center of production is in The Netherlands. Across a wide spectrum of the range of bulbous plants grown, the Dutch industry enjoys specific competitive advantages in each of the areas listed above. The major ornamental bulbous crops grown in The Netherlands include tulip (*Tulipa*), hyacinth (*Hyacinthus*), daffodil (*Narcissus*), hybrid lilies (*Lilium*), and a range of others that include *Crocus*, *Muscari*, *Gladiolus*, *Iris*, and *Allium*.

The path of a bulb from a field in The Netherlands to a market in North America is a long and complicated one. In most cases, bulbs are grown in fields (there are more than 25,000 hectares of bulbs produced in Holland annually) and harvested at or during leaf senescence, after current photosynthesis has "refilled" the bulb. Depending on the species, the basic procedure is that bulbs are cleaned, "peeled" (the process or removing side bulbs, often used for planting stock), sized (generally by circumference), counted, and placed into sales packages (for the retail or commercial landscape market) or into "export crates" for transport to professional forcing markets around the world.

Bulbs are creatures of temperature; thus, the industry is highly advanced in its understanding and use of controlled-temperature environments for bulb storage and transportation. Bulbs exported from Europe are, with very few exceptions, transported via ocean vessel, in temperature-controlled shipping containers. In principle, from the moment that the bulbs are loaded into the container until delivery to the final customer (this takes from 2–5 weeks for shipments to North America, depending on the route and destination), temperature and ventilation can be controlled within the container.

Other than The Netherlands, there are significant production regions that emphasize specialty bulbs or bulbs with unique climactic requirements that give competitive advantages to production in those locations. To cite just a few examples, Israel is a major and specialized producer of paperwhite narcissus (*Narcissus tazetta*) and amaryllis (*Hippeastrum*). Significant numbers of amaryllis are also produced in South Africa and Brazil. The Pacific Northwest region of the United States is known for Easter lily bulb production (*Lilium longiflorum*), and tulips, daffodils, and irises. The development of the southern hemisphere as a production source has revolutionized the bulb industry as it allows the production of bulbs that are 6 months "off-cycle" from the northern hemisphere. This has important implications for scheduling and forcing bulbs in seasons that are traditionally difficult or impossible (e.g., allowing tulip flowers to be available in summer and fall). The next 25 years will increasingly see bulb production activities move away from the physical borders of The Netherlands as pressures of a burgeoning population, stricter environmental and social regulations, and prohibitive costs are encountered.

BULBS

The term "bulb" is often used in general to describe several specialized storage organs. However, a "true bulb" has several unique characteristics that distinguish it from other storage organs. Bulbs are compressed stems with modified fleshy leaves, called "scales," which serve as the primary storage site for carbohydrates, nutrients, organic compounds, and water (Figure 28.1). The compressed stem forms the basal plant and contains a growing point and adventitious roots. Some bulbs have thick, fleshy "contractile roots" that serve to pull the bulb to the proper depth in the soil. Bulbs can be classified as either tunicate or non-tunicate (scaly). Tunicate bulbs such as tulips and daffodils have a dry, papery outer membrane that serves to protect the fleshy inner scales. The scales are attached to the basal plate and arranged in concentric layers. Non-tunicate bulbs have no dry, papery outer membrane, making them more susceptible to mechanical damage and drying out. Like tunicate bulbs, non-tunicate bulbs have scales attached to the basal plant; however, the scales overlap, giving a bumpy scale-like appearance. Lilies are an example of non-tunicate bulbs.

BULB PROPAGATION

Offsets are miniature bulbs (bulblets) formed naturally underground on bulbs and can be used to efficiently propagate many bulbous plants. Offsets will produce vegetative growth and adventitious roots, following removal from the mother plant and planting in the soil. The most important ornamental bulbous plant in the world, the tulip, is propagated by bulb offsets. Unfortunately, some important bulbous plants such as hyacinth (*Hyacinthus*) and lily (*Lilium*) do not produce offsets fast enough for commercial production and, so, must be propagated using different methods.

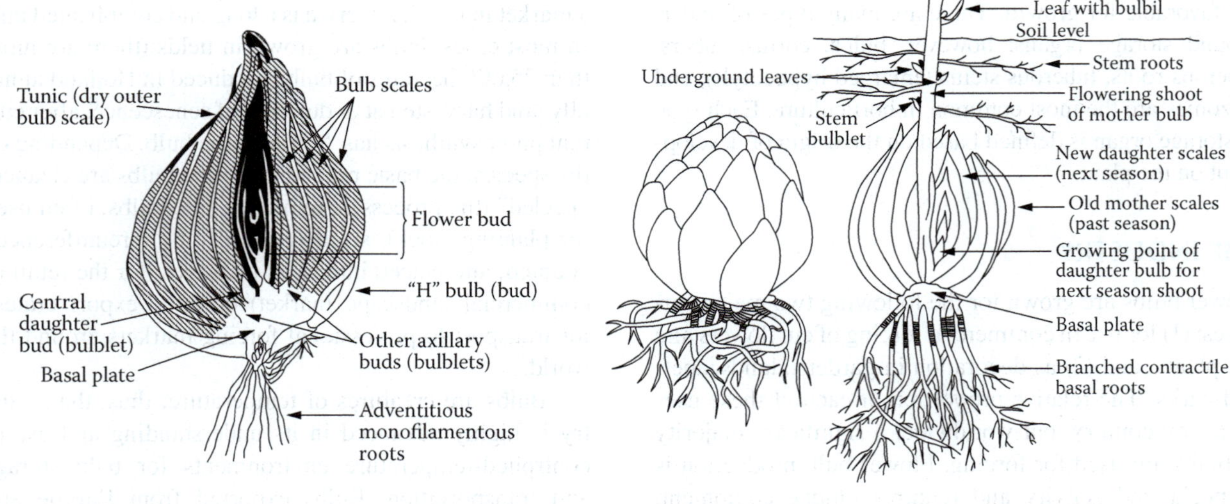

FIGURE 28.1 A true bulb is composed of a compressed stem, or basal plate, and modified leaves called "scales" serve as the primary storage tissue. Examples include tulip, lily, allium (e.g., onion), Dutch iris, hyacinth, daffodil, and others. A bulb is referred to as "tunicated" if the outermost scale is dried out and papery, serving as protection for the fleshy scales beneath. Tulip bulbs are tunicated. Bulbs that lack the papery outer covering, such as lily bulbs, are described as "non-tunicated." (Reprinted from *The Physiology of Flower Bulbs*, de Hertogh, A. A. and M. le Nard, eds., 812 pp., Copyright 1993, with permission from Elsevier.)

Scooping (see also Chapter 29) is a method of bulb propagation that involves the removal of the basal plate and the growing point of a mature bulb and is commonly used to propagate hyacinths. It is important to treat the scooped bulb with a fungicide to prevent rotting or decay. Following scooping and fungicide treatment, bulbs are placed in a dark, dry environment at 21°C (70°F) for a few weeks for wound tissue (callus) to form on the cut surface. After callus has formed, bulbs are exposed to an increased temperature of 30°C (85°F) and a high relative humidity (~85%) to facilitate bulblet formation on the callus tissue. A scooped bulb might produce 30 to 60 bulblets, depending on the species and the condition of the mother bulb.

Scoring (see also Chapter 29) is another bulb propagation method similar to scooping. For scoring, three to six vertical cuts are made through the basal plate and the growing point. The cuts may either be straight or made wedge shaped by removing sections of the bulb to expose greater wounded surface area. Scored bulbs are thereafter treated as with scooped bulbs, including a fungicide treatment. Scored bulbs typically produce 15 to 25 bulblets.

Coring and sectioning are two additional propagation methods that are similar to scooping and scoring. Coring involves the removal of a cone-shaped section of the bulb through the center of the basal plate and the entire growing point. Sectioning differs from scoring only in that the cuts are made vertically through the entire bulb, resulting in six to twelve or more individual wedge-shaped sections. Both methods follow the same protocol described for scooping and scoring but typically result in relatively fewer bulblets formed. With sectioning, there is a trade-off in terms of the number of bulblets produced (oftentimes, a greater total number of bulblets are produced as a bulb is further sectioned) versus the weight of individual bulblets (larger sections yield fewer, but larger, bulblets).

Bulb cuttings begin by first sectioning the bulb as described above; however, each section is then divided by bulb scales into groups of two to four, each containing a portion of the basal plate. When two scale pieces are used, the resulting cutting is commonly referred to as a "twin scale." The bulb cuttings are placed vertically in sand or vermiculite in a warm, moist environment. Bulblets form between the scales from the basal plate in 2 to 3 weeks.

Individual scales of non-tunicate bulbs can be easily separated and used as propagules for new bulb formation. Bulbs to be used for scaling are harvested following flowering, and care must be taken to prevent the bulbs from drying out. Although all the scales present could be used, inner scales are not as productive, and the mother bulb will be diminished or destroyed. Easter lily is an important bulbous crop that is commercially propagated by scaling. Following the removal of scales, they are treated with a fungicide and placed in a rooting medium either in trays basal side down or in sealed plastic bags. Three to five new bulblets will form at the base of each scale within 2 to 3 months.

CORMS

A corm, like a bulb, is a compressed stem, with the basal plate formed from an underground swollen stem surrounded by dry leaves or scales (Figure 28.2). Corms

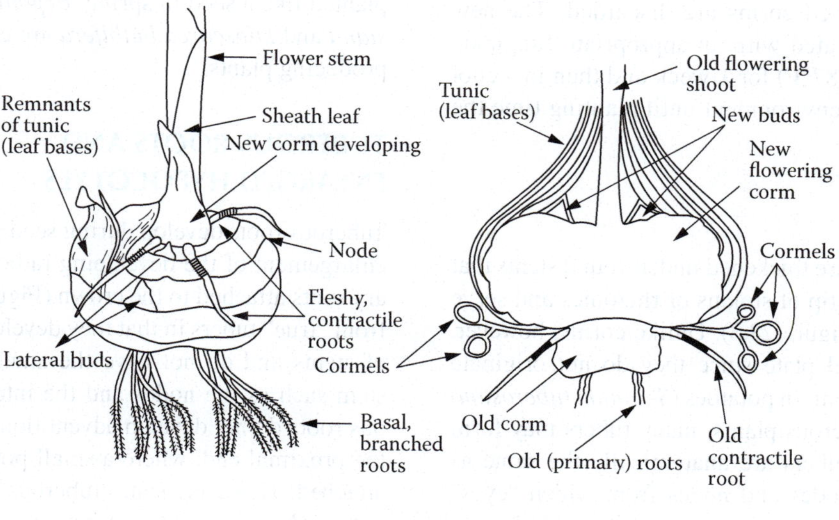

FIGURE 28.2 Like a true bulb, the corm is a modified stem with a basal plate, but the primary storage tissue is the stem tissue itself, rather than the leaf tissue; so, corms are frequently described as "solid bulbs." These organs also may be tunicated or non-tunicated, and they have nodes from which meristems originate. Gladiolus, freesia, crocus, and ixia are some examples. (Reprinted from *The Physiology of Flower Bulbs*, de Hertogh, A. A. and M. le Nard, eds., 812 pp., Copyright 1993, with permission from Elsevier.)

can be tunicate or non-tunicate; persistent dry leaf bases protect the fleshy scales by forming a tunic (cover) often over the entire corm. Unlike true bulbs, the thickened stem serves as the primary storage site, and scales are attached at the nodes, which, along with the internodes, are distinguishable.

Most corms undergo annual replacement, where a new corm is formed each year. One or more new corms usually begin development following flowering on top of the old corm at the base of the current season's stem. The new corm develops fleshy contractile roots that will move the new corm down to the depth of the current season's corm. The mother corm has a fibrous root system at the base. In addition to new corms, miniature corms or cormels often form between the new and old corms.

CORM PROPAGATION

Propagation of corms is usually accomplished by harvesting new corms and cormels. Fortunately, modifying environmental conditions can increase the number of new corms and cormels produced. Corm propagation usually begins in the field. Mother corms are planted and allowed to develop vegetatively and eventually flower. Proper fertilization and moisture control will encourage the development of large corms and an increased number of new corms and cormels. Corms are lifted only after the tops have withered or have been killed by frost. Lifted corms are first cleaned of dirt, placed in slatted trays or on screens for air circulation, and stored for 12 to 24 h at 30°C (87°F). This curing treatment eases the separation of new corms and cormels from the mother corm. New corms and cormels are graded for size, and any diseased corms are discarded. The new corms should be treated with an appropriate fungicide and stored at 30°C (87°F) for 1 week and then in a cool (5°C [40°F]), moist environment until planting time the following season.

TUBERS

Tubers, like corms, are thickened underground stems that often develop at the tip of stolons or rhizomes and serve as storage organs (Figure 28.3). Unlike corms, however, tubers have no basal plate since they do not originate from the base of a stem. In potatoes (*Solanum tuberosum*) and many other tuberous plants, many tubers may form on a single plant. Tubers are anatomically the same as stems having internodes and nodes from which "eyes" develop, containing one or more shoot buds. Tuberous plants produce tubers each season. The tubers then serve as an overwintering storage site producing new roots and shoots during the following season. The new shoots use the reserves from the tuber for initial growth and produce new tubers for the following year.

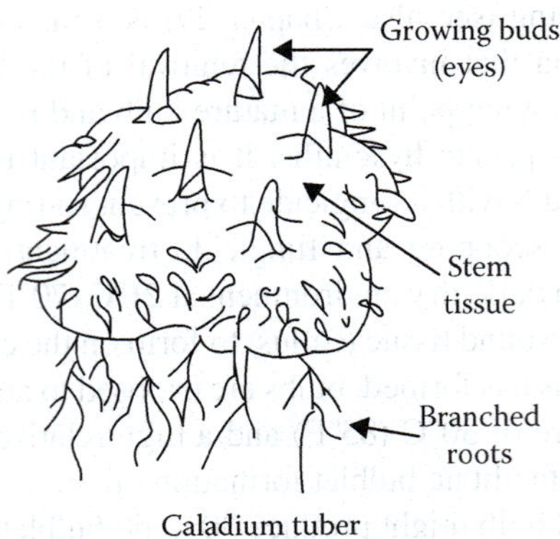

FIGURE 28.3 A tuber is a thickened underground stem, and the stem tissue serves as the primary storage tissue; but unlike a corm, it has no basal plate. Meristems occur on the tuber and are commonly called "eyes" on a potato. Caladium is the most common floricultural tuber. (Reprinted from *The Physiology of Flower Bulbs*, de Hertogh, A. A. and M. le Nard, eds., 812 pp., Copyright 1993, with permission from Elsevier.)

TUBER PROPAGATION

Many tuberous plants are propagated asexually by division of the tuber. Tubers are divided into sections each containing one or more eyes. Tubers can be planted directly; however, the freshly cut surface will be highly susceptible to disease organisms. To avoid this problem, cut sections are often stored in a warm (20°C [68°F]), moist environment to allow the cut areas to heal. A few tuberous plants will produce tubercles that form from buds in the leaf axils aboveground. Tubercles are easily removed from the stems and can be stored overwinter and planted like a seed in spring. *Begonia grandis* ssp. *evansiana* and *Dioscorea bulbifera* are examples of tubercle-producing plants.

TUBEROUS ROOTS AND ENLARGED HYPOCOTYLS

Tuberous roots develop during seed germination with the enlargement of the developing radicle and from secondary roots attached to the crown (Figure 28.4). They differ from "true" tubers in that they develop from roots instead of stems and do not have the anatomical features of a stem such as the nodes and the internodes. Most tuberous roots do not develop adventitious shoots except from the proximal end, where a small portion of the crown is attached. However, some tuberous roots such as sweet potato (*Ipomoea batatas*) have the capacity to produce adventitious shoots under right environmental conditions. Tuberous roots serve as the overwintering portion of the plant and provide energy for growth the following season. Tuberous root declines as reserves are used up and new tuberous roots develop. After a few years of growth,

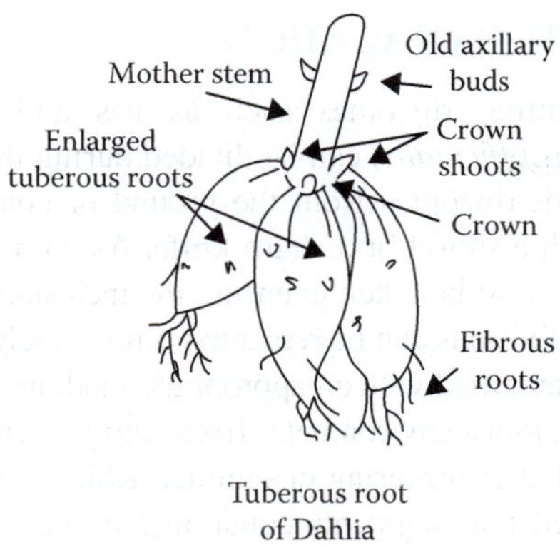

FIGURE 28.4 A tuberous root consists of enlarged fleshy root tissue. Roots are fleshy because they are the primary storage tissue. Growth arises from buds at the top (crown) of the root mass. Examples include dahlia, anemone, and ranunculus. (Reprinted from *The Physiology of Flower Bulbs*, de Hertogh, A. A. and M. le Nard, eds., 812 pp., Copyright 1993, with permission from Elsevier.)

plants may have numerous tuberous roots attached to the same crown.

The hypocotyl is the area of a germinating seedling between the cotyledons and the root radicle. In plants such as cyclamen (*Cyclamen persicum*) and gloxinia (*Sinningia speciosa*), the hypocotyl enlarges to become a storage organ called the "enlarged hypocotyl" or the "tuberous stem" (Figure 28.5). Tuberous roots are usually biennial, whereas enlarged hypocotyls are perennial. Leaves and stems emerge from the crown of enlarged hypocotyls and roots from the base. Cool night temperatures promote carbohydrate storage in the hypocotyl,

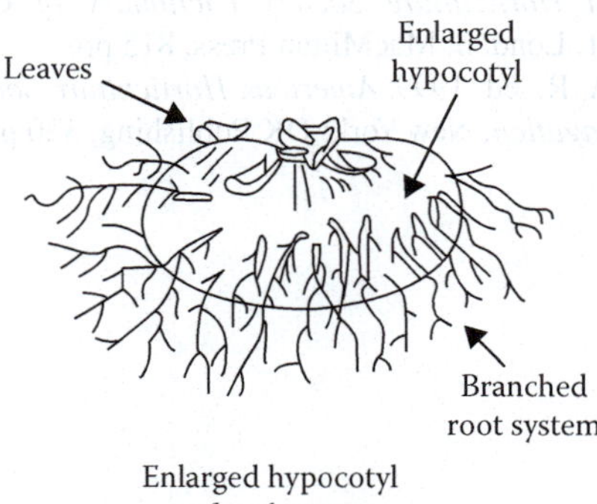

FIGURE 28.5 The hypocotyl of a seedling is the portion of the stem below the cotyledon and above the roots. In some plants, the hypocotyl enlarges, becoming a fleshy storage site as the plant develops. Cyclamen and gloxinia are examples of plants with enlarged hypocotyls. Also, "tuberous" begonias are not actually tuberous but have enlarged hypocotyls. (Reprinted from *The Physiology of Flower Bulbs*, de Hertogh, A. A. and M. le Nard, eds., 812 pp., Copyright 1993, with permission from Elsevier.)

increasing overwintering capacity and vegetative and reproductive growth. This is particularly important in food crops that develop enlarged hypocotyls, such as beets (*Beta vulgaris*), where both the enlarged hypocotyls and the leaves are consumed.

TUBEROUS ROOTS AND ENLARGED HYPOCOTYL PROPAGATION

Tuberous roots and enlarged hypocotyls can be propagated by division. Propagation of tuberous roots by division is accomplished by separating individual tuberous roots from the crown. Care must be taken to ensure that each tuberous root has a portion of the attached crown and has at least one bud present. To divide enlarged hypocotyls, the hypocotyl is sectioned much like a tuber, with each section having at least one bud. The cut sections of the hypocotyl are susceptible to disease organisms and should be allowed to heal before planting. Fungicide treatment may also be beneficial.

Shoot cuttings from plants having enlarged hypocotyls can be rooted similarly to those from other herbaceous and woody plants. Some evidence suggests that including a portion of the hypocotyl may enhance propagation success. Leaf and bud cuttings can also be accomplished by taking a single leaf and a corresponding bud from the hypocotyl. The leaf and bud cutting is placed in a suitable rooting medium and in a moist environment to encourage budbreak and, eventually, adventitious root development. Opportunities for shoot cuttings from tuberous roots are limited. However, plants such as sweet potatoes can be propagated by taking stem cuttings from the current season's vines and treating as with other cuttings. Unlike most tuberous roots, sweet potato tubers will produce adventitious shoots that can be rooted. To promote adventitious shoot development in sweet potato (slips), tuberous roots are placed in an open bed and covered with sand or other suitable medium. If outdoor, beds should be covered to maintain a warm, moist environment. The beds should be ventilated and kept moist after adventitious shoots emerge. Adding medium at the base of the new shoots (i.e., heeling) will promote adventitious root development. When new shoots are about 8 in tall and have three to five nodes, they can be removed from the tuber by cutting below the new root system.

RHIZOMES

Rhizomes are another type of specialized stem that grows horizontally and serves as a storage organ (Figure 28.6). Rhizomes may grow at the soil surface as with irises (*Iris* sp.) or belowground as with ginger (*Zingiber officinale*). Rhizomes that grow on the surface tend to be large and thick, whereas many underground rhizomes

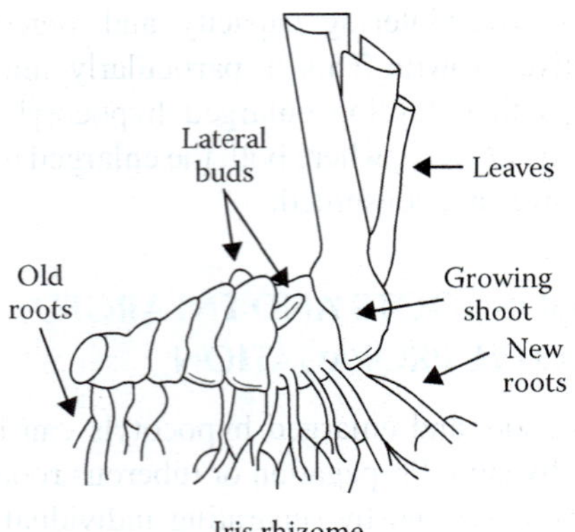

FIGURE 28.6 A rhizome is a modified stem, and the stem tissue itself is the primary storage tissue. The rhizome, however, is unique in that it grows horizontally through the soil. Some irises have rhizomes, as well as lily of the valley, calla lily, oxalis, and achimenes. (Reprinted from *The Physiology of Flower Bulbs*, de Hertogh, A. A. and M. le Nard, eds., 812 pp., Copyright 1993, with permission from Elsevier.)

are more slender and spread rapidly. Many garden weeds such as Johnsongrass (*Sorghum halepense*) and quackgrass (*Elytrigia repens*) have very fast-growing, slender rhizomes, making them difficult to control using mechanical cultivation. When these underground rhizomes are severed with a shovel or hoe, new adventitious shoots and roots develop, and growth continues. Many rhizomes are segmented in appearance, indicating the nodes and the internodes present on the stem. Also, most plants having large rhizomes form crowns and spread slowly, whereas those with slender rhizomes are crownless. Most, but not all, plants with rhizomes are monocots, and new leaves develop from the nodes that initially sheath the rhizome and continue growing vertically away from the rhizome.

RHIZOME PROPAGATION

Crown-forming rhizomes such as iris and asparagus (*Asparagus officinalis*) can be divided during the dormant season. Lift rhizomes from the ground or container and divide with a shovel or a sharp knife. As with other divisions, care must be taken to ensure the inclusion of at least one bud. Divisions can be replanted immediately or placed in containers filled with an appropriate medium and stored in a warm, moist environment. Irises and gingers may also be divided after flowering in summer. Rhizomes are lifted and divided into segments containing at least one culm. Culms are then cut back, and the rhizomes are replanted.

Some rhizomatous plants, such as bamboo, can be propagated by culm cuttings. Culms with lateral branches present are cut from the rhizome and either sectioned (single or double node with branch) or used whole. The cuttings are placed in sand beds, and the culm, but not the culm branch, is covered. Auxin treatment may hasten adventitious root development.

LITERATURE CITED AND SUGGESTED READING

Bryan, J. E. 1989. *Bulbs*. Portland, Oregon: Timber Press, 218 pp.

de Hertogh, A. A. and M. le Nard, eds. 1993. *The Physiology of Flower Bulbs*. Amsterdam, The Netherlands: Elsevier, 812 pp.

Hamrick, D., ed. 2003. *Ball Redbook: Crop Production*, 17E, Vol. 2. Batavia, Illinois: Ball Publishing, 724 pp.

Hartmann, H. T., D. E. Kester, F. T. Davies, Jr., and R. L. Geneve. 2002. *Hartmann and Kester's Plant Propagation: Principles and Practices*, 7E. Upper Saddle River, New Jersey: Pearson Education, 880 pp.

Huxley, A., M. Griffiths, and M. Levy, eds. 1992. *The New Royal Horticulture Society Dictionary of Gardening*, Vol. 1. London: MacMillan Press, 815 pp.

Toogood, A. R., ed. 1999. *American Horticulture Society Plant Propagation*. New York: DK Publishing, 320 pp.

29 Propagating Selected Flower Bulb Species

William B. Miller, Jeffrey A. Adkins, and John E. Preece

Most bulbous plants can be propagated asexually through a variety of means, including offset division, basal cuttage (scooping or scoring), scaling, or division (Chapter 28). Such asexual methods are used to maintain clonal characteristics, which are important since most ornamental plants that are bulbs, tubers, corms, or rhizomes are sold with a cultivar designation. Unless the objective is breeding of new cultivars, where the sexual phase and seed production are necessary, with rare exceptions, almost all ornamental bulbous plants are propagated vegetatively.

Ornamental bulbs are colorful additions to the spring or summer garden. Spring-flowering bulbs, such as tulip, daffodil, hyacinth, and others, are available and planted in fall, whereas summer-flowering bulbs, such as lily, begonia, gladiolus, and others, are typically planted in spring. By definition, bulbs planted in fall and surviving local winter are "hardy." In the case of summer-flowering bulbs, many are, in fact, substantially winter hardy (e.g., most lilies), whereas others are very non-cold hardy and must be planted after all frost is past or even started indoors (e.g., tuberous begonia).

The following laboratory exercises assume a fall course offering as the hyacinths would be commercially available from bulb suppliers or local garden centers. The exercises on lily, dahlia, or potato can, in principle, be done at any time of the year. The lily exercise shows adventitious formation of bulblets at the base of lily bulb scales in relation to different environmental conditions. Recall that a bulb is a compressed stem and that the bulb's scales are actually leaves modified to store large quantities of starch and other carbohydrates. Bulblets that form from the scales have a small basal plate of stem tissue from which new scales grow at the nodes. Adventitious roots are attached to each basal plate, not the mother scale. The influence of new leaf growth from the bulblets on their subsequent growth will also become evident between the two treatments. The hyacinth, potato, dahlia, and garden experiments are mainly descriptive and manipulative in nature and allow students to practice with different forms of asexual propagation of flower bulbs (Figures 29.1–29.5).

EXERCISES

Experiment 1. Effect of the Propagation Environment on Lily Bulblet Formation and Growth

MATERIALS

The following items will be sufficient for individual students or teams:

- One healthy lily bulb (Figure 29.1a)
- Clean potting medium
- Sphagnum peat:perlite:vermiculite (2:1:1, by volume) with nutrients and wetting agent provides a suitable environment for bulblet formation. Commercially available mixes or handmade blends work well if there is a good wettability of the medium.
- One 4 to 6 in (10–15 cm) azalea or standard pot or a 3 × 5 × 8-in market pack, or a similar container. The container should be new or properly disinfested of fungal pathogens. Drainage holes are essential.
- One new 1-qt, clear plastic bag. This can have a zippered top or can be closed with a twist tie.

Follow the instructions presented in Procedure 29.1 to complete this experiment.

ANTICIPATED RESULTS

Almost all lily bulb scales (Figure 29.1b) will form bulblets, regardless of whether they are in the plastic bag or pot or a market pack. The bulblets will sprout and grow faster in the pot or the market pack in full sunlight. The bulblets in the plastic bag will grow slowly and will sprout leaves that become etiolated because of low light. Initially, callus (looks like small bumps on the surface) will form within the first 2 weeks. The bumps grow into the bulblets and can be counted. Root growth will be greater on bulblets in full sunlight. The scales should have well-formed, rooted, and sprouted bulblets within 5 weeks and, certainly, within 6 to 8 weeks.

(a)

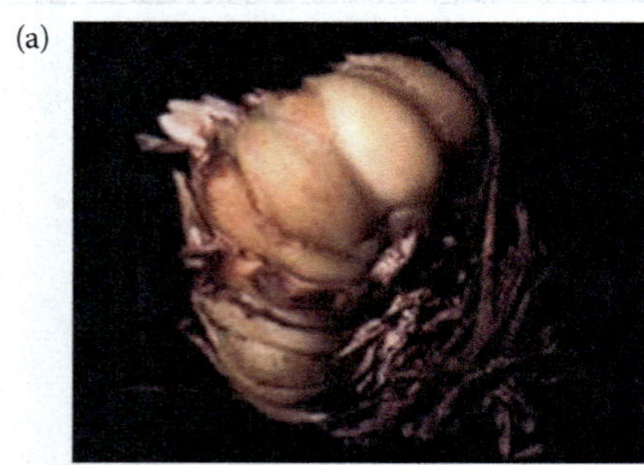

(b)

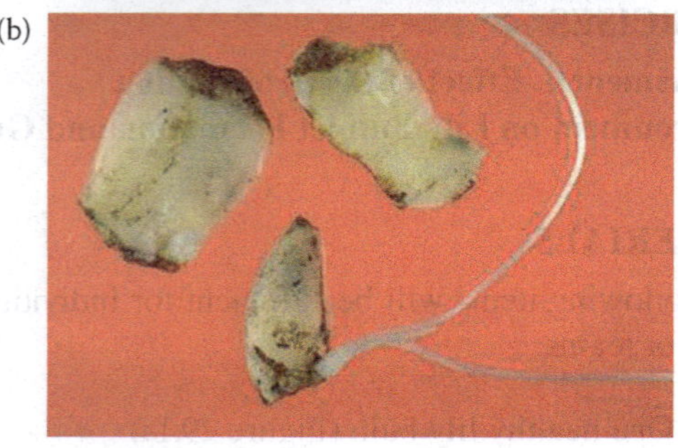

FIGURE 29.1 Nontunicate bulbs, such as lily (a), have scales that can be broken off, placed on moist sphagnum under warm temperatures, and then planted like seeds later. Note the root development (b), at the base of the scale. (a: Courtesy of Rosie Lerner, Purdue University; b: Courtesy of R. Skirvin and K. McPheeters. Available at http://courses.nres.uiuc.edu/hort100/tutorials/stems_roots/55.htm.)

QUESTIONS

- Why does the upper half of the bulb scales that are exposed to full sunlight become green, often with pink, red, or purple?
- If you tried different colored cultivars, is the scale color in the light-exposed area similar to the flower color? Why or why not?
- What are the implications of chlorophyll formation in the scale portions exposed to sunlight on the propagation of bulblets?
- What are the facilities requirements, labor requirements, and overall expenses for each method of bulblet propagation using scales? Which method would you use if propagating lilies commercially?
- Other than passing direct genetic information to the bulblets, mother scale pieces may also transmit pathogens (disease-causing agents). How can this be reduced or avoided?
- What other methods can be used to propagate lily?

Experiment 2. The Effects of Planting Method on the Propagation of Hyacinths by Scaling

Hyacinths (*Hyacinthus orientalis*) produce beautiful, fragrant flowers and are used in gardens and forced in containers. Commercially, hyacinths are propagated

Procedure 29.1

Propagating Lilies from Scale Leaves

Step	Instructions and Comments
1	Fill the pot or market pack with the potting medium, then water. Add enough moist (not wet) potting medium to the bag so that the medium is "three fingers" deep (measure by holding three fingers up to the back of the bag).
2	Select one healthy lily bulb. Unlike onion or hyacinth, there is no papery covering on this bulb. Discard the dried and damaged scales from the outside of the non-tunicate bulb. Keep removing the scales until only white, plump scales remain. Break off at least 10 of the largest, whitest scales and divide them into two uniform groups.
3	Plant all of the scales in one of these groups in the medium in the pot or in the market pack. Plant them so that the pointed end is up and the end where the scale was attached to the mother bulb is within the medium—the lower halves of the scales should be within the medium.
4	In the plastic bag, mix all of the scales from the other group into the medium either with your fingers or by shaking the bag. Close the bag using the zipper or twist tie. Water the scales in the pot or market pack thoroughly, and then place in full sunlight. This is typically done on a greenhouse beach, but a bright window or another protected environment is suitable. The container must be watered when it dries.
5	Place the plastic bag on a bench or table in the classroom. Add water only if necessary. Some bags may not require water for a year. Room temperature and light that is on when the students are present are sufficient for the growth and formation of bulblets.
6	Data should be collected once each week on the number of bulblets, bulblet diameter, rooting, and sprouting.

by removal or by cutting through the basal plate of the stem tissue at the base of the bulb. The basal plate can be scored or scooped out (Figure 29.2), or bulbs can be cored vertically to produce bulblets (Chapter 28). To be successful with these techniques, the proper facilities must be available to allow for proper humidity, temperature, and ventilation control. These "basal cuttage" operations are of prime importance since experience in The Netherlands has shown that each hyacinth cultivar requires a slightly different treatment (e.g., depth or width of scooping, or post-cuttage incubation temperature). Commercially, scooping or scoring is done in early to mid-July.

For the purposes of these exercises, the best-quality hyacinth bulbs (≥18 cm in circumference) are available in late summer and autumn. Sometimes, they may be available in spring but are of lower quality, and cultivar selection will be limited. The cultivar "L'Innocence" will produce bulblets reliably. If the experiments are conducted in spring, it will be necessary to store the bulbs. Hyacinth bulbs can be difficult to store for 5 to 6 months without rotting. If the bulbs are to be stored for more than 2 months, they should be soaked for 1 to 2 h in 0.63 g/L (6 g/gal) Truban 30% wettable powder (wp) fungicide (30% etridiazole, a.i.). Be sure to take appropriate safety measures when working with fungicides. They should then be placed in open containers of "bone-dry" sphagnum moss at 4°C.

To be successful in scaling hyacinths, cuts must be made in the proper area at the base of the scale. Scales are modified leaves that store food and nutrients for the plant during its dormant season. Each hyacinth scale is more-or-less globe shaped, and each completely surrounds the inner scales. The outermost scales have dried into the papery tunic or covering. Each scale is attached to the basal plate (stem tissue) at a node, and the cells at the base of each scale leaf have the competence to dedifferentiate and form bulblets. This ability decreases and then is lost with increasing distance from the basal plate. The basal plate also cannot form adventitious bulblets. Therefore, the cut at the bottom of the scale should be within 5 mm of the basal plate and should include no basal plate tissue. This will ensure that the scale piece contains competent cells for producing callus and adventitious bulblets.

FIGURE 29.2 Hyacinth bulb with bulblets forming at the base where the basal plate has been removed by scooping. (Courtesy of R. Skirvin and K. McPheeters. Available at http://courses.nres.uiuc.edu/hort100/tutorials/stems_roots/81.htm.)

This laboratory exercise will give students experience with the vegetative propagation of a tunicate bulb. Although the focus of the experiment is on scaling, other propagation techniques such as scoring and scooping can be practiced on the same bulb. This will still allow for all of the bulb scale tissue that is necessary for the scaling exercise.

MATERIALS

The following items will be sufficient for individual students or teams:

- Three healthy hyacinth bulbs
- Pasteurized potting medium. A medium containing sphagnum peat, perlite, and vermiculite (2:1:1, by volume) is recommended. Commercially available or handmade can yield good results, if they have good wettability
- One 6 in (15 cm) azalea or standard pot, 3 × 5 × 8-in market pack, or other similar container will work well. Whatever is used, it must have drainage holes and should either be new or properly disinfested
- A tray with approximately 1/2 in (1.25 cm) of sand or dry sphagnum moss in it to help "hold" bulbs upright after scooping or scoring
- Sharp knife
- Greenhouse or bright, sunny windows, or other protected, sunny location
- A temperature of 25/20 ± 3°C should be maintained under a natural photoperiod

Follow the instructions outlined in Procedure 29.2 to complete the primary and optional experiments.

ANTICIPATED RESULTS

Almost all hyacinth scales will form bulblets. There may be slightly more on the horizontal pieces. The bulblets will require at least 6 to 8 weeks to form roots and perhaps longer for green photosynthetic leaves to appear. Initially, cream- to light brown-colored callus at the cut base of the scale pieces will be seen. The callus will gradually form "bumps" that will grow into bulblets. There are similarities and differences between the bulblet formation from hyacinth bulb scale pieces and that

Procedure 29.2

Propagating Hyacinth Bulbs (Including Some Optional Experiment)

Step	Instructions and Comments
	Sectioning
1	Cut from the top down (longitudinally), quarter the bulb. Flake apart the scale pieces with your fingers. Select at least 10 uniform, plump, large, and healthy scale pieces.
2	Plant at least five scale pieces vertically so that the upper half of the scale piece is above the surface of the medium. Plant the remaining scale pieces horizontally so that the curved (convex) side is up and the cut bottom is within the surface 2 to 3 mm of the medium.
3	Dust the horizontally planted scale pieces lightly with the medium, leaving the curved portion exposed to the sunlight.
4	Water and place in a greenhouse, sunny window, or full sunlight in another protected environment.
5	Data should be collected once each week on callus formation, the number of bulblets, bulblet diameter, rooting, and sprouting.
	Scoring
1	Holding the bulb so that the basal plate is facing you, cut completely across and through the basal plate. The basal plate is shaped conically within the bulb, so cut deeper in the center than toward the edges of the basal plate.
2	Angle the knife slightly and make a second parallel cut about 2 mm away to remove a thin wedge. Turn the bulb 90° and make a second wedge cut so that the cuts on the basal plate looks like a "+" sign. The cuts must be sufficiently deep to wound the bottom 5 mm of the scales above the cut. If the cut bottoms of scales are not visible, continue cutting until they are. It is the wounded and exposed portion of the bases of the bulb scales that will produce bulblets. Follow the procedures in step 2 of the "Scooping" section.
	Scooping
1	Hold the scored basal plate toward you and remove the entire basal plate. This can be done by angling the knife and cutting so that the concentric rings of the bases of the scales are visible. A sharpened spoon or melon baller could also be used for scooping. Care must be taken to remove all of the basal plate tissue. Keep cutting if you cannot see the concentric rings at the base of the bulb. If the cuts are too deep, all competent cells will be removed, and the remaining scale pieces either will not produce bulblets or will do so poorly. Scooped bulbs are processed like scored bulbs for bulblet production.
2	For either scooped or scored bulbs, place them vertically, with the cut base plates up, into trays with clean sand (this helps hold them upright). Exposure to 70°F is needed to heal the cuts; then, if possible, gradually increase the temperature to approximately 85°F. If this is not possible, use room temperature, although the process will go more slowly.
3	Make observations on time to see visible callus and on the appearance of any bulblets. This takes a long time (2.5–3 months) and may not be possible within the length of a typical propagation class. In commercial practice, hyacinth propagation is done in July, soon after the bulbs are lifted. Intact, scooped, or scored bulbs, with numerous bulblets attached, are planted intact in fields in October for overwintering and growth the following year.

from lily bulb scales. This makes an interesting comparison if done at the same time as the propagating lily bulbs experiment.

QUESTIONS

- How does hyacinth propagate itself in nature? Is that method more or less efficient than scoring, scooping, or scaling?
- Which technique—scoring, scooping, or scaling—results in the most bulblets per bulb?
- How many years does it take from the formation of a new bulblet to the time that it is 18 to 19 cm in circumference and ready to be stored?
- Why does the mother scale piece change color when exposed to sunlight?
- Is the change in color good or bad for production?

Experiment 3. Division of Dahlia Roots (Figure 29.3) and Cutting of a Potato Tuber (Figure 29.4)

MATERIALS

The following items will be sufficient for individual students or teams:

- One or two healthy tuberous root clumps of dahlia. These may be purchased in spring or may be lifted from a garden in fall before the first frost
- A medium-sized, healthy potato
- Clean, good-quality potting mix
- Two 6 in (15 cm)–diameter plastic pots
- Sharp knife

Follow the instructions outlined in Procedure 29.3 to complete the primary and optional experiments.

Procedure 29.3

Division of Dahlia and Potato (Figures 29.3 and 29.4)

Step	Instructions
	Dahlia
1	Observe the dahlia clump, and identify the plump storage roots and also the stem tissue to which they are joined. Usually, the remnants of the prior growing season's stem are also prominent. Closer examination should reveal several small, pointed buds associated with the stem tissue.
2	Carefully cut two to four roots from the clump, without any stem or bud tissue being present. Cut two to four other roots in such a manner that stem tissue, including at least one bud, is present per root.
3	Plant the roots or each treatment into separate pots. Orient such that the stem (proximal) end is about 1 cm above the surface of the mix. Label the pots.
4	Make observations at weekly or biweekly intervals. Make notes on the presence or the absence of shoots and on shoot length. Continue observations for 6 to 8 weeks.
5	At the conclusion of the experiment, remove the plants and remove soil from them. Observe and make notes on root growth on all the roots.
	Potato
1	Take the potato and cut it in half lengthwise. Save one of the halves. Take the remaining half and cut it into two, making two quarters. Save one quarter and take the other quarter, subdividing it into two. Continue saving one piece and subdividing the other piece until you no longer have at least one eye per piece. The smaller the pieces, the more important this is. The result will be a range of potato pieces, from halves to perhaps sixteenths of a whole tuber.
2	Store the pieces at about 70°F for 2 to 3 days to allow wound healing.
3	Plant such that the eyes are at or slightly under the soil surface.
4	Make observations at weekly or biweekly intervals. Make notes on the presence or the absence of shoots and on shoot length. Continue observations for 6 to 8 weeks.
5	At the conclusion of the experiment, remove the plants and the soil from them. Observe and make notes on root growth and on the condition of the originally planted potato pieces.

FIGURE 29.3 Dahlia is a tuberous root. The eyes on the root are axillary buds that become new shoots after pieces of the root are cut from the stem. (Courtesy of R. Skirvin and K. McPheeters. Available at http://courses.nres.uiuc.edu/hort100/tutorials/stems_roots/55.htm.)

FIGURE 29.4 Shoots emerging from the eyes of the potato tuber. (Courtesy of R. Skirvin and K. McPheeters. Available at http://courses.nres.uiuc.edu/hort100/tutorials/stems_roots/55.htm.)

ANTICIPATED RESULTS

The underground portion of a dahlia is a clump of tuberous roots, which are attached to the stem tissue at their proximal ends. When the roots are separated from the stem tissue (the "crown"), the absence of a viable bud will result in no growth and a failed propagation effort. When one or more buds are present on the removed root, a shoot will grow, and eventually, adventitious roots will form. Roots will also form on the original tuberous root. A potato is a tuber, and the entire volume of the potato is stem tissue. The "eyes" of a potato are buds, arranged in a spiral around the tuber. When cutting the potato, one invariably obtains one to several buds per chunk of tissue, and successful propagation results.

FIGURE 29.5 Corms of gladiolus that have formed cormels at the base, which can be harvested and then planted. (Courtesy of R. Skirvin and K. McPheeters. Available at http://courses.nres.uiuc.edu/hort100/tutorials/stems_roots/55.htm.)

QUESTIONS

- Why is more care needed in dividing dahlia tuberous roots than in cutting a potato?
- With dahlia, is there a difference in visual root growth at the end of the experiment? Why or why not?

Experiment 4. Observation of Natural Propagation Structures in Bulbous Plants

The following exercise assumes the availability of a garden containing examples of bulbous plants such as tulip, daffodil, hyacinth, lily, gladiolus (Figure 29.5), tuberous begonia, canna, iris, etc. Permission to dig plants and observe their underground structures is also assumed. While not an experiment per se, this is a great observational exercise that allows students to gain significant experience with the underground storage organs of many ornamental bulbous plants.

MATERIALS

The following items will be sufficient for a team of students or the entire class:

- 1 garden containing bulbous plants
- 1 shovel

Follow the instructions outlined in Procedure 29.4 to complete the exercise.

Procedure 29.4
Observation of the Natural Propagation of Storage Organs

Step	Instructions
1	In the garden area where you have permission to dig, work the shovel into the ground and lift up the soil clump containing the bulbous plant of interest. Shake off the majority of the soil to expose the bulb/tuber and other clump. Wash in a steam of water, if desired. Try to lift a variety of species. Of particular interest are daffodil (usually, daffodils prominently show offset bulbs, which may be detached from the parent bulb), tulip (may also have offset bulbs), lily (Oftentimes, established bulbs will have multiple growing points [or "noses"], which can be separated. Lilies also commonly have stem bulblets that form on the stem under the soil line.), gladiolus (in fall, right before frost, the new corm can be clearly seen growing on top of last year's corm, and also, you can usually see numerous cormels attached between the old and the new corm), canna (a good example of a rhizomatous plant), and others.
2	Observations can be made about the overall structure of belowground parts. If there are sufficient quantities available, the parts may be sectioned or dissected, and internal components may be identified.

Part XI

Micropropagation

30 Micropropagation

Michael E. Kane, Philip J. Kauth, and Scott L. Stewart

Concept Box 30.1

- Shoot meristem cells retain the embryonic capacity for unlimited division.
- Isolated smaller meristem explants require more complex culture media for survival.
- Meristem culture and meristem tip culture are methods for disease eradication.
- Shoot culture provides a means to multiply periclinal chimeras.
- Cytokinins disrupt apical dominance and enhance axillary shoot production.
- Increased auxin concentration increased percent rooting and root number, but decreased root elongation.
- Negative carry-over effects of auxins used for Stage III rooting may effect *ex vitro* survival and growth of plantlets.
- The application of shoot organogenesis or non-zygotic embryogenesis for commercial micropropagation has been limited in many species.
- For non-zygotic embryogenesis to be a viable propagation method, delivery systems must be developed through which non-zygotic embryos are processed, function as synthetic seeds that survive handling and planting, and further develop into vigorous plants.

Micropropagation is defined as aseptic asexual plant propagation on a defined culture medium in culture vessels under controlled conditions of light and temperature. Since plants are propagated in culture vessels, the procedure is often referred to as *in vitro* (Latin: "in glass") propagation. In reality, numerous types of culture vessels constructed of plastic are often used. The commercial application of micropropagation became established in France during the mid-1960s with orchids. Since this time, commercial micropropagation has developed into a worldwide industry, producing more than 500 million plants yearly. By 2001, there were approximately 90 commercial micropropagation laboratories in the United States. Requirements for specialized equipment, highly-trained technicians, and high labor costs make the technology very expensive. Consequently, commercial micropropagation is limited to crops generating high unit prices, including ornamental plants and food crops such as potato.

Depending on the species and cultural conditions, micropropagation can be achieved by the following methods:

1. Node culture
2. Enhanced axillary shoot proliferation
3. *De novo* formation of adventitious shoots through shoot organogenesis
4. Non-zygotic embryogenesis

Currently, the most frequently used micropropagation method for commercial production, shoot culture, relies on enhanced axillary shoot proliferation from cultured meristems (Figures 30.1 and 30.2a). This method often provides acceptable levels of genetic stability and is more readily attainable for many plant species. Given the importance of shoot culture, our review of micropropagation concepts will focus on this procedure.

SHOOT APICAL MERISTEMS

Given that all four micropropagation methods rely on the formation of shoot apical meristems, it is important to briefly review the general structure of shoot meristems. Shoot growth in mature plants is restricted to specialized regions, which exhibit little differentiation and in which the

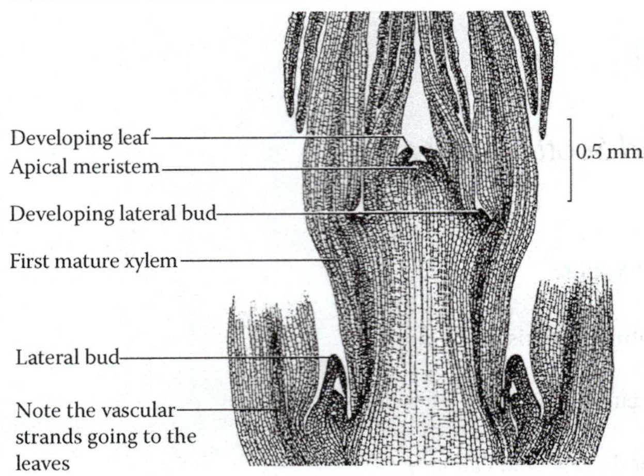

FIGURE 30.1 Diagrammatic representation of a dicotyledonous shoot tip. The shoot tip consists of the apical meristem, subtending leaf primordia, and lateral buds.

cells retain the embryonic capacity for unlimited division. These regions, called "apical meristems," are located in the apices of the main and lateral buds of the plant (Figure 30.1). Cells derived from these apical meristems subsequently undergo differentiation to form the mature tissues of the plant body. Due to their highly organized structure, apical meristems tend to be genetically stable (see also Chapter 2).

Significant differences exist in the shape and size of shoot apices and the associated layers between different taxonomic plant groups. A typical dicotyledon shoot apical meristem consists of a layered dome of actively dividing cells located at the extreme tip of a shoot and measures about 0.1 to 0.2 mm in diameter and 0.2 to 0.3 mm in length. The apical meristem has no connection to the vascular system of the stem. Below the apical meristem, localized areas of cell division and elongation represent sites of newly developing leaf primordia (Figure 30.1; see also Figure 2.4). Lateral buds, each containing an apical meristem, develop

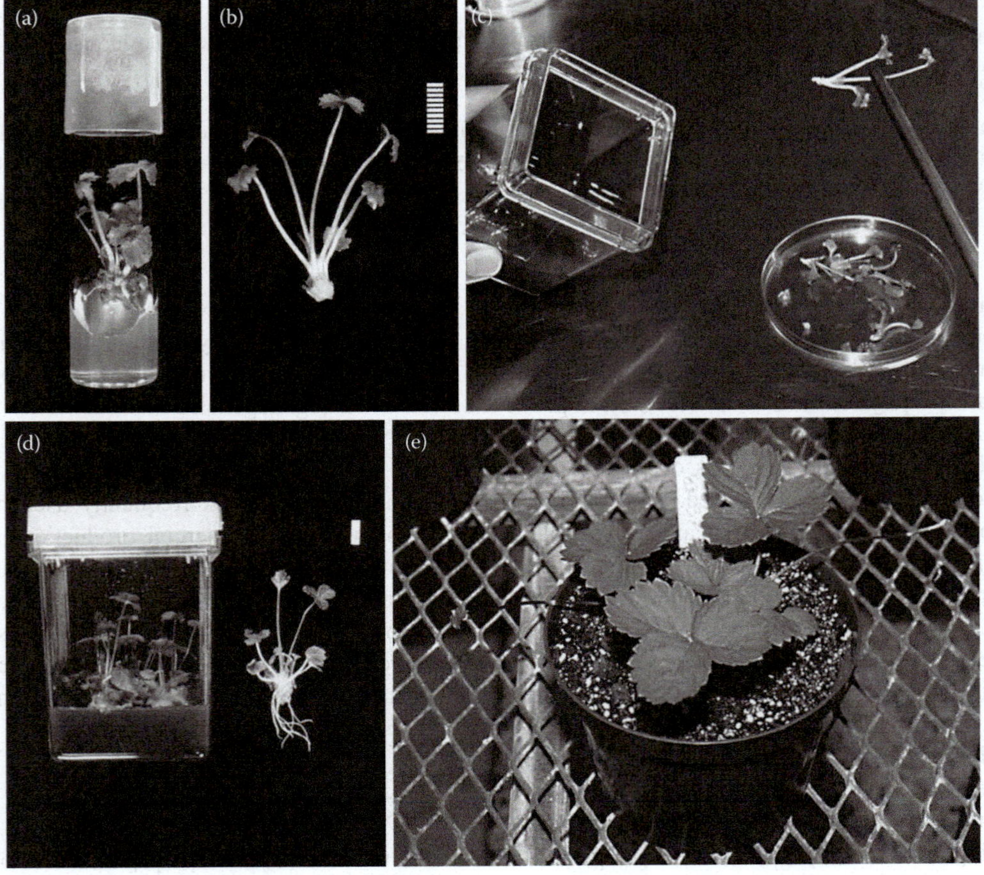

FIGURE 30.2 Shoot culture of strawberry (*Fragaria* × *ananassa*). (a) Stage II culture consisting of multiple axillary shoots produced from a single shoot after 4-week culture. The Stage II medium typically contains a cytokinin to promote axillary shoot production. (b) Stage II shoot clusters are separated into single unrooted microcuttings. Scale bar = 1 cm. (c) Microcuttings are inoculated onto a Stage III rooting medium usually containing an auxin to promote adventitious rooting. (d) Microcutting rooting after 4 weeks. (e) Rooted microcuttings are acclimatized *ex vitro* to lower humidity and higher light conditions in a greenhouse.

within the axils of the subtending leaves. In the intact plant, outgrowth of the lateral buds is usually inhibited by an apical dominance of the terminal shoot tip. Organized shoot growth from plant apical meristems is potentially unlimited and said to be indeterminate. However, shoot apical meristems may become committed to the formation of determinate organs such as flowers.

IN VITRO CULTURE OF SHOOT MERISTEMS

The recognized potential for unlimited shoot growth prompted early and largely unsuccessful attempts to aseptically culture isolated shoot meristems in the 1920s. By the mid-1940s, sustained growth and maintenance of cultured shoot meristems through repeated subculture was achieved for several species. However, in 1946, Dr. Ernest Ball provided the first detailed procedures for the isolation and production of plants from cultured shoot meristem tips and the successful transfer of rooted plantlets into the soil. Ball is often called the "Father of Micropropagation" because his shoot tip culture procedure is the one most commonly used by commercial micropropagation laboratories today. Although these studies demonstrated the feasibility of regenerating shoots from cultured shoot tips, the procedures typically yielded unbranched shoots.

Several important discoveries facilitated the application of *in vitro* culture techniques for large-scale clonal propagation from shoot meristems. The discovery that virus-eradicated plants could be generated from cultured excised meristems led to the widespread application of the procedure also for routine fungal and bacterial pathogen eradication. Demonstration of the rapid production of orchids from cultured shoot tips in 1960 supported the possibility of rapid clonal propagation in other crops. It should be noted that *in vitro* propagation in many orchids does not occur via axillary shoot proliferation, but rather the cultured meristems become disorganized and form spheroid protocorm-like bodies that are actually non-zygotic embryos.

The final discovery critical for the production of plants from cultured shoot meristems was the elucidation of the role of cytokinins in disrupting apical dominance by Wickson and Thimann in 1958. This finding was eventually applied to enhance axillary shoot production *in vitro*. Application of this method was further expedited by the development of improved culture media, such as the medium by Murashige and Skoog (1962), which supported the propagation of a large number of plant species.

MERISTEM AND MERISTEM TIP CULTURE

Although not directly used for propagation, meristem and meristem tip culture will be briefly described since these procedures are used to generate pathogen-eradicated shoots that subsequently may serve as propagules for *in vitro* propagation. Culture of the apical meristematic dome alone (Figure 30.3) from either terminal or lateral buds, for the purpose of pathogen elimination, is termed "meristem culture." Culture of larger explants consisting of the apical meristematic dome plus several subtending leaf primordia is termed "meristem tip culture" (Figure 30.3). Disease eradication by meristem culture is possible because the pathogens do not infect the meristematic cells due to the lack of vascular development in the extreme shoot tip. In reality, true meristem culture is rarely used because isolated apical meristems of many species exhibit both low survival rates and increased chances of genetic variability following callus formation and adventitious shoot formation. Caution should be taken when interpreting much of the early published literature of successful "meristem" culture since, in many instances, meristem tip or even larger shoot tip explants were actually used. The terms "meristemming" and "mericloning" commonly used in the orchid literature are equally ambiguous.

SHOOT AND NODE CULTURE

Although not the most efficient procedure, propagation from axillary shoots has proven to be a reliable method for the micropropagation of a large number of species. Depending on the species, two methods are used: shoot culture and node culture. Both methods rely on the stimulation of lateral shoot growth following the disruption of apical dominance. Shoot culture refers to the *in vitro* propagation by repeated enhanced formation of axillary shoots from shoot tip or lateral bud explants cultured on a medium supplemented with growth regulators, typically a cytokinin. The axillary shoots produced are either subdivided into shoot tips and nodal segments that serve as secondary explants for further proliferation or treated as microcuttings for rooting. In some species, modified storage organs such as miniaturized tubers or corms (Figure 30.4) develop from axillary shoots or rhizomes under inductive culture conditions and may serve as the propagules for either direct planting or long-term storage.

When specific pathogen–eradicated donor plants are used or when pathogen elimination is not a concern, relatively larger (1–20 mm long) shoot tip or lateral bud primary explants (Figure 30.3) can be used for culture establishment and subsequent shoot culture. Advantages of using larger shoot tips include greater survival, more rapid growth responses, and the presence of more axillary buds. However, these larger explants are more difficult to completely surface sterilize and can potentially harbor undetected latent or systemic microbial infection. Compared to other micropropagation methods, shoot cultures (1) provide reliable rates and consistency of multiplication following culture stabilization, (2) are less

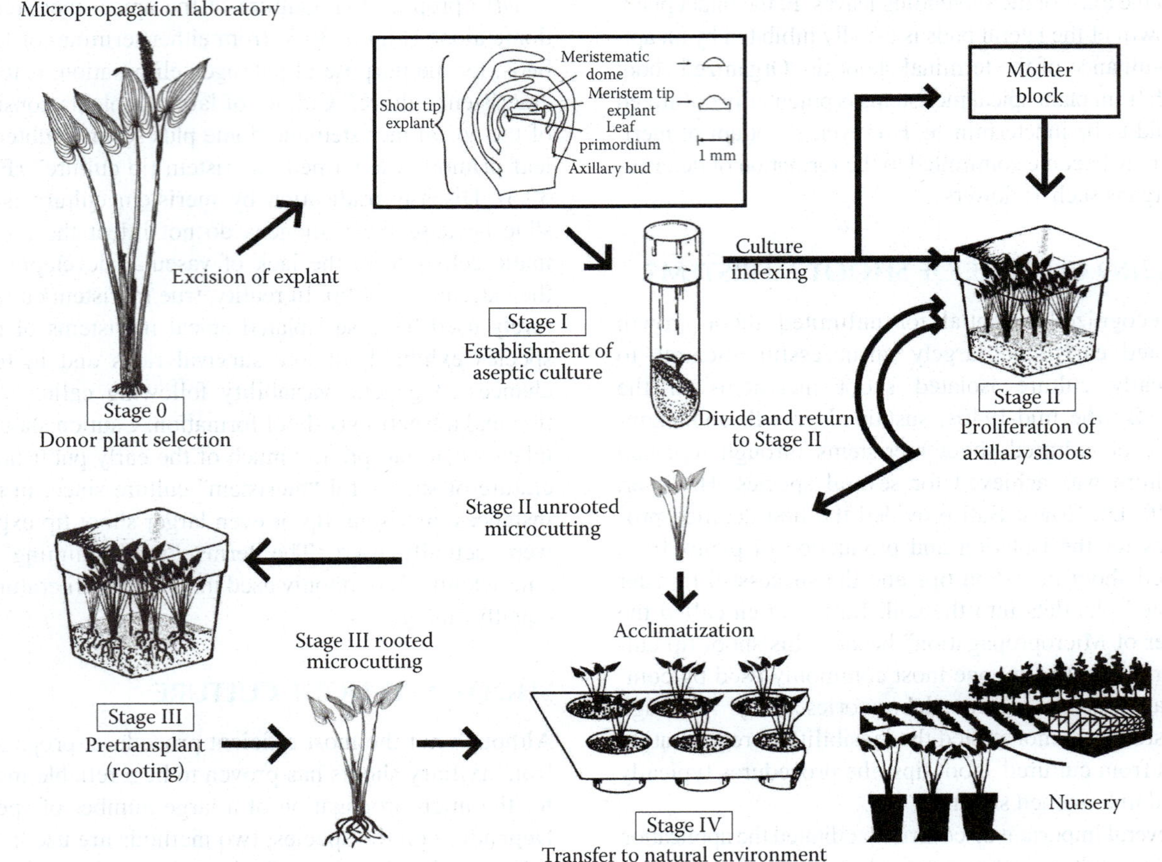

FIGURE 30.3 Micropropagation stages for production by shoot culture.

susceptible to genetic variation, and (3) may provide for the clonal propagation of periclinal chimeras (Chapter 7).

Node culture, a simplified form of shoot culture, is another method for production from preexisting meristems. Numerous plants, such as potato (*Solanum tuberosum* L.), do not respond well to the cytokinin stimulation of axillary shoot proliferation observed in the micropropagation of other crops. Axillary shoot growth is promoted by the culture of either intact shoots positioned horizontally on the medium (*in vitro* layering) or single or multiple node segments. Typically, single elongated unbranched shoots, comprised of multiple nodes, are rapidly produced. These shoots (microcuttings) are either rooted or acclimatized to *ex vitro* conditions or repeatedly subdivided into nodal cuttings to initiate additional cultures. Node culture is the simplest method and it is associated with the least genetic variation between the plants produced.

FIGURE 30.4 In some species, such as *Sagittaria latifolia* L., *in vitro* multiplication may occur through the production of shoots (left) or corms (right), depending on culture conditions of light and temperature. Scale bar = 1 cm.

MICROPROPAGATION STAGES

In 1974, Dr. Toshio Murashige described three basic stages (I–III) for successful micropropagation. Recognition of contamination problems often observed after the inoculation of primary explants prompted the inclusion of a Stage 0. This additional stage describes specific cultural practices that maintained the hygiene of stock plants and decreased the contamination frequency during the initial establishment of primary explants. As a result of our increased awareness of the requirements for successful micropropagation, five stages (Stages 0–IV) are currently recognized. These stages not only describe the procedural steps in the micropropagation process, but also represent points at which the cultural environment is altered. This system has been adopted by most commercial and research laboratories as it simplifies production scheduling, accounting, and cost analysis. Requirements for the completion of each stage will depend on the plant material and the specific method used. A diagrammatic representation of the micropropagation stages for propagation by shoot culture is provided in Figure 30.3.

STAGE 0: DONOR PLANT SELECTION AND PREPARATION

Explant quality and subsequent responsiveness *in vitro* is significantly influenced by the phytosanitary and physiological conditions of the donor plant. Prior to culture establishment, careful attention is given to the selection and maintenance of the stock plants used as the source of explants. Stock plants are maintained in clean controlled conditions that allow active growth but reduce the probability of disease. The maintenance of specific pathogen–tested stock plants under conditions of relatively lower humidity and the use of drip irrigation and antibiotic sprays have proven effective in reducing the contamination potential of candidate explants. Such practices also allow the excision of relatively larger and more responsive explants often without increased risks of contamination.

Numerous practices are also employed to increase explant responsiveness *in vitro* by modifying the physiological status of the stock plant. These include the following: (1) trimming to stimulate lateral shoot growth, (2) pretreatment sprays containing cytokinins or gibberellic acid, and (3) use of forcing solutions containing 2% sucrose and 200 mg/L 8-hydroxyquinoline citrate for the induction of budbreak and the delivery of growth regulators to target explant tissues. Currently, information on the effects of other factors such as stock plant nutrition, light, and temperature treatments on the subsequent *in vitro* performance of meristem explants is limited.

STAGE I: ESTABLISHMENT OF ASEPTIC CULTURES

Initiation and aseptic establishment of pathogen-eradicated and responsive terminal or lateral shoot meristem explants is the goal of this stage. The primary explants obtained from the stock plants may consist of surface-sterilized shoot apical meristems or meristem tips for pathogen elimination or shoot tips from terminal or lateral buds (Figure 30.3). This can be demonstrated by using strawberry (*Fragaria* × *ananassa* Duch.) as an example (Figure 30.5a).

Rapid clonal production of specific pathogen–eradicated plants is a fundamental goal of the micropropagation

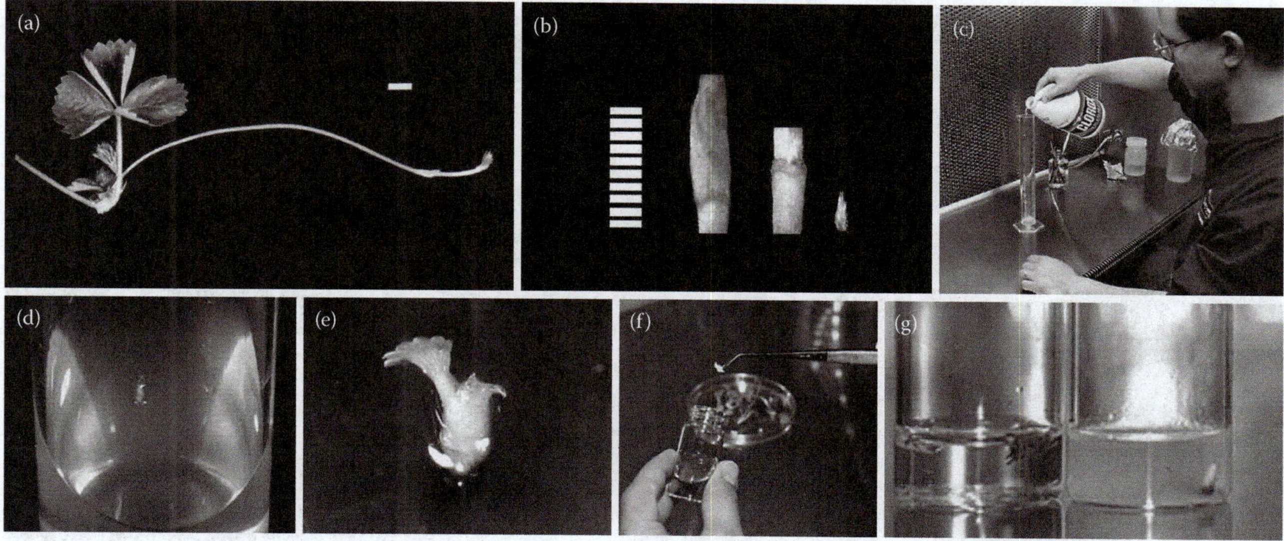

FIGURE 30.5 Establishment of strawberry (*Fragaria* × *ananassa*) shoot cultures. (a) Strawberry stolons serve as the source of axillary and apical shoot meristems for the establishment of Stage I cultures. Scale bar = 1 cm. (b) Excised stolon nodes with bracts covering lateral buds (left) are surface sterilized and rinsed in sterile water. The nodal bract is then removed to expose the lateral bud (center). The lateral shoot tip (right) is removed using a sterile scalpel and forceps, usually with the aid of a dissecting microscope placed in a laminar-flow transfer hood. Scale bar = 1 cm. (c) Diluted bleach (sodium hypochlorite) is often used as the surface-sterilizing agent. (d) Excised shoot tip explants are inoculated onto a Stage I medium that promotes establishment. (e) Enlargement of strawberry shoot tip after 4-week culture. (f) Following establishment, small tissue samples are cut from the established shoot culture and inoculated into an indexing medium to screen for the presence of microbial contaminants. The shoot culture is transferred to a fresh Stage I medium. (g) Clouding of inoculated indexing medium (right) is a positive indication of the presence of cultivable contaminants in the tissue sample.

process. However, it is desirable to establish and maintain plant cultures that are also free of nonpathogenic microbial contaminants. The surfaces of plants are naturally populated with a diverse microflora consisting of bacteria, fungi, yeast, and other organisms. A primary objective of Stage I is the elimination of this microflora and the subsequent physiological adjustment of the explants to *in vitro* culture. This is usually accomplished through surface disinfecting explants with alcohol and/or sodium hypochlorite prior to culture inoculation (Figure 30.5c). The transfer of surface-sterilized explants into culture vessels onto sterile medium is performed in a laminar airflow transfer hood (Figure 30.6a), which provides for a particulate-free work area. Tissues are cut and handled using scalpels and forceps (Figure 30.5f) sterilized by flame or dry heating such as that achieved using glass bead sterilizer (Figure 30.6a).

Once explants become established and grow *in vitro* (Figure 30.5e), it is essential that cultures be indexed (screened) for the presence of common microbial contaminants. Bacteria and fungal contaminants often persist within cultured tissues that appear visually contaminant-free. Contaminated cultures may exhibit no symptoms, variable growth, regeneration, reduced shoot proliferation, rooting, or poor survival. Many culture contaminants are not pathogenic to plants under natural conditions, but become pathogenic *in vitro* due to the release of toxic secondary metabolites into the medium. Consequently, it is essential that Stage I cultures be indexed for the presence of internal microbial contaminants prior to serving as sources of shoot tip or nodal explants for Stage II multiplication. Because secondary culture contamination can occur as a result of poor aseptic technique or contaminant vectors such as mites, cultures should be routinely re-indexed. Indexing for microbial contaminants is usually accomplished by inoculating tissue sections or intact shoots into an enriched nutrient medium that will promote the visible growth of bacteria, filamentous fungi, yeast, or other contaminants (Figure 30.5f, g).

FACTORS AFFECTING CULTURE ESTABLISHMENT

Many factors may affect the successful Stage I establishment of meristem explants: explantation time, position of the explant on the stem, explant size, and polyphenol oxidation. The time of explantation can significantly affect explant response *in vitro*. In deciduous woody perennials, shoot tip explants collected at various times during the spring growth flush of the donor plants may vary in their ability for shoot proliferation. Shoot tips collected during or at the end of the period of most rapid

FIGURE 30.6 Commercial micropropagation laboratory. (a) To maintain sterility, culture transfers are performed in a laminar-flow clean bench that provides a particulate-free work area. (b) Culture medium preparation room with autoclave for medium sterilization. (c) Mother block room, where indexed stabilized cultures are maintained to ensure a source of genetically identical culture. (d) Production culture room maintained under controlled light and temperature conditions. (e) Stage II unrooted or Stage III rooted microcuttings are transferred to growing medium in plug trays before being placed under greenhouse conditions. (f) Robotic plug transfer equipment is used to increase efficiency and decrease labor costs. (g) Stage IV plants acclimatized in a large production greenhouse with computer-controlled environmental and irrigation/fertilization systems. (h) Acclimatized micropropagated plantlets.

shoot elongation on the donor plants exhibit weak proliferation potential. Explants collected before or after this period are capable of strong shoot proliferation *in vitro*. Conversely, the best results are obtained with herbaceous perennials that form storage organs, such as tubers or corms, when explants are excised at the end of dormancy and after sprouting.

Explants also exhibit different capacities for establishment *in vitro*, depending on their location on the donor plant. For example, survival and growth of terminal bud explants are typically greater than lateral bud explants. Often similar lateral meristem explants from the top and the bottom of a single shoot may respond differently *in vitro*. In woody plants exhibiting phasic development, juvenile explants are typically more responsive than those obtained from the often unresponsive mature tissues of the same plant. Sources of juvenile explants include root suckers, basal parts of mature trees, stump sprouts, and lateral shoots produced on heavily pruned plants.

The excision of primary explants from donor plants often promotes the release of polyphenols and stimulates polyphenol oxidase activity within the damaged tissues. The polyphenol oxidation products often blacken the explant tissue and medium. Accumulation of these (poly)phenol oxidation products can eventually kill the explants. Procedures used to decrease tissue browning include the use of liquid medium with frequent transfer; adding antioxidants such as ascorbic or citric acids, polyvinylpyrrolidone, and activated charcoal; or culture in reduced light or darkness.

Clearly, there is no one universal culture medium for the establishment of all species; however, modifications to the Murashige and Skoog medium formulation are most often used. Cytokinins and auxins are most frequently added to Stage I media to enhance explant survival and shoot development. The types and levels of growth regulators used in Stage I media are dependent on the species, genotype, and explant size.

Knowledge of the specific sites of hormone biosynthesis in intact plants provides insight into the relationship between explant size and dependence on exogenous growth regulators in the medium. Endogenous cytokinins and auxins are synthesized primarily in root tips and leaf primordia, respectively. Consequently, smaller explants, especially cultured apical meristem domes, exhibit greater dependence on medium supplementation with exogenous cytokinin and auxin for maximum shoot survival and development. Larger shoot tip explants usually do not require the addition of auxin in Stage I medium for establishment. Rapid adventitious rooting of shoot tip explants often provides a primary endogenous cytokinin source. Most Stage I media consist of mineral salts, sucrose, and vitamins, supplemented with at least a cytokinin and solidified with a gelling agent such as agar or gellan gum (Figure 30.5d). The most frequently used cytokinins are N^6-benzyladenine (BA), kinetin (Kin), and N^6-(2-isopentenyl)-adenine (2-iP). Due to its low cost and high effectiveness, BA is the most widely used cytokinin. Substituted urea compounds, such as thidiazuron, exhibit strong cytokinin-like activity and have been used to facilitate the shoot culture of recalcitrant woody species like maples (*Acer* spp.).

Several types of auxins are also used in Stage I media. The naturally occurring auxin indole-3-acetic acid is the least active, whereas the stronger and more stable auxins α-naphthalene acetic acid (NAA), and indole-3-butyric acid (IBA), are more often used. Stage I medium plant growth regulator (PGR) levels and combinations that promote explant establishment and shoot growth but limit the formation of callus and adventitious shoot formation should be selected.

A commonly held misconception is that primary explants exhibit immediate and predictable growth responses following inoculation. For many species, particularly herbaceous and woody perennials, consistency in growth rate and shoot multiplication is achieved only after multiple subculture on Stage I medium. Physiological stabilization may require 3 to 24 months and four to six subcultures. Failure to allow culture stabilization, before transfer to a Stage II medium containing higher cytokinin levels, may result in diminished shoot multiplication rates or the production of undesirable basal callus and adventitious shoots. With some species, the time required for stabilization can be reduced by initial culture in liquid medium.

In many commercial laboratories, stabilized cultures, verified as being specific pathogen tested and free of cultivable contaminants, are often maintained on media that limit shoot production to maintain genetic stability. These cultures, called "mother blocks," serve as sources of shoot tips or nodal segments for the initiation of new Stage II cultures (Figure 30.6c).

STAGE II: PROLIFERATION OF AXILLARY SHOOTS

Stage II propagation is characterized by repeated enhanced formation of axillary shoots from shoot tips or lateral buds cultured on a medium supplemented with a relatively higher cytokinin level to disrupt the apical dominance of the shoot tip (Figure 30.5a). A subculture interval of 4 weeks with a threefold to eightfold increase in shoot number is common for many crops propagated by shoot culture (Figure 30.2a). Given these multiplication rates, conservatively, more than 4.3×10^7 shoots could be produced yearly from a single starting explant.

Stage II cultures are routinely subdivided into smaller clusters, individual shoot tips, or nodal segments that serve as propagules for further proliferation (Figure 30.5a). Additionally, axillary shoot clusters may be harvested as individual unrooted Stage II microcuttings or multiple shoot clusters for *ex vitro* rooting and acclimatization (Figures 30.3). Clearly, Stage II represents one of the most costly stages in the production process.

Both the source and orientation of explants can affect Stage II axillary shoot proliferation. Subcultures inoculated with explants that had been shoot apices in the previous subculture often exhibit higher multiplication rates than lateral bud explants. In some species, inverting shoot explants in the medium can double or triple the number of axillary shoots produced on vertically oriented explants per culture period.

The number of Stage II subcultures possible before new cultures from the mother block are required depends on the species or cultivar and its inherent ability to maintain acceptable multiplication rates while exhibiting minimal genetic variation. Some species can be maintained with monthly subculture from 8 to 48 months in Stage II. In contrast, as few as three subcultures may only be possible before the frequency of off-types increases to unacceptable levels, such as in the fern genus *Nephrolepis*.

Increased production of off-types is often attributed to the production of adventitious shoots, particularly through an intermediary callus stage. Stage II cultures, originally regenerating from axillary shoots, often begin producing adventitious shoots (see Figure 30.7) at the base of axillary shoot clusters after a number of subcultures on the same medium. These so-called "mixed cultures" can develop without any morphological differences being apparent. Selecting only terminal shoots of axillary origin for subculture, instead of shoot bases, decreases the frequency of off-types, including the segregation of periclinal chimeras (Chapter 7).

Selection of Stage II cytokinin type and concentration is made on the basis of shoot multiplication rate, shoot length, and the frequency of genetic variation. Although shoot proliferation is enhanced at higher cytokinin concentrations, the shoots produced are usually smaller and may exhibit symptoms of hyperhydricity (water soaked and glassy appearing). Depending on the species, exogenous auxins may or may not enhance cytokinin-induced axillary shoot proliferation. This is exemplified with the woody shrub *Aronia arbutifolia* L. (Figure 30.8). Addition of auxin in the medium often mitigates the inhibitory effect of cytokinin on shoot elongation, thus increasing

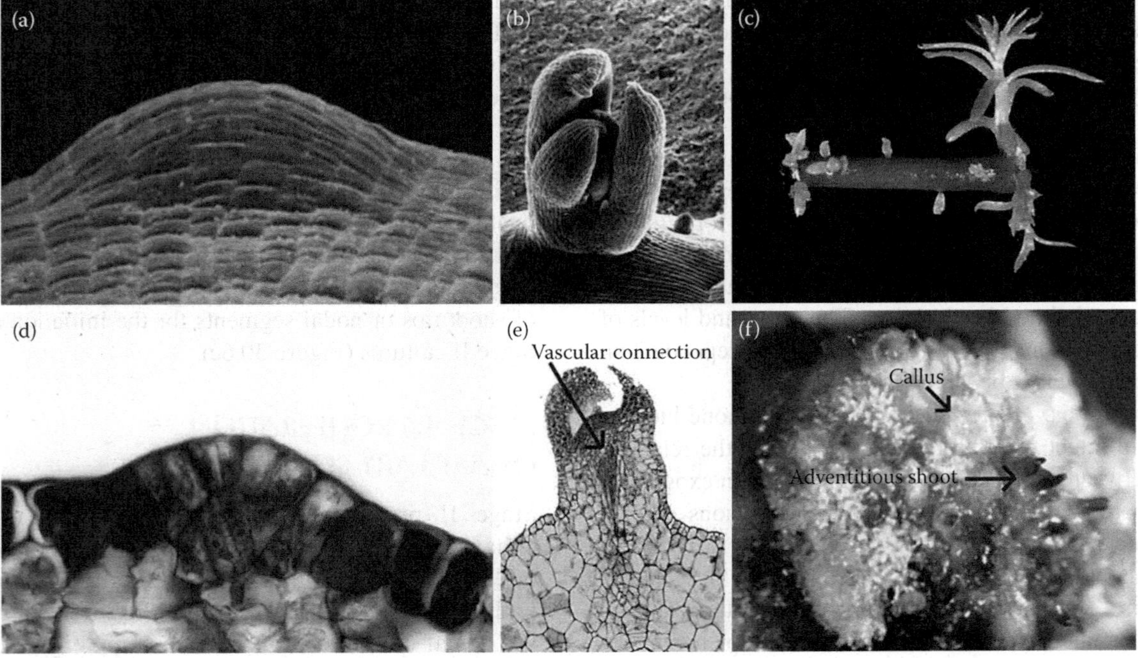

FIGURE 30.7 Micropropagation by shoot organogenesis. (a and d) Epidermal cells of explants divide to form meristematic domes when cultured on a medium containing cytokinin and auxin. Each meristem typically arises from a single cell and may form directly on the explant. (b and e) Meristems develop into adventitious shoots consisting of an apical meristem and leaves. The vascular system of the shoot may connect with the vascular system of the explant. (e) This is an example of direct shoot organogenesis. (c) Adventitious shoots develop to a size sufficient for rooting. (f) Indirect shoot organogenesis occurs when adventitious shoots form from cells on an intermediary callus that forms on the explant. The frequency of genetic variability is often greater in plants that are produced from callus.

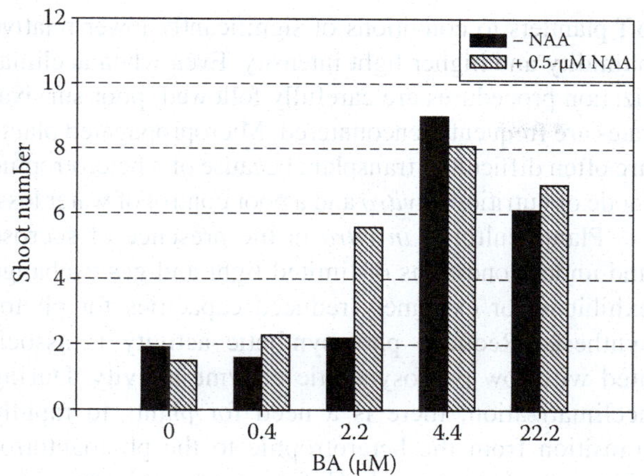

FIGURE 30.8 Effect of BA concentration on Stage II axillary shoot proliferation from two-node explants of *Aronia arbutifolia* L., a woody plant, after 28-day culture in the presence and the absence of 0.5 μM NAA.

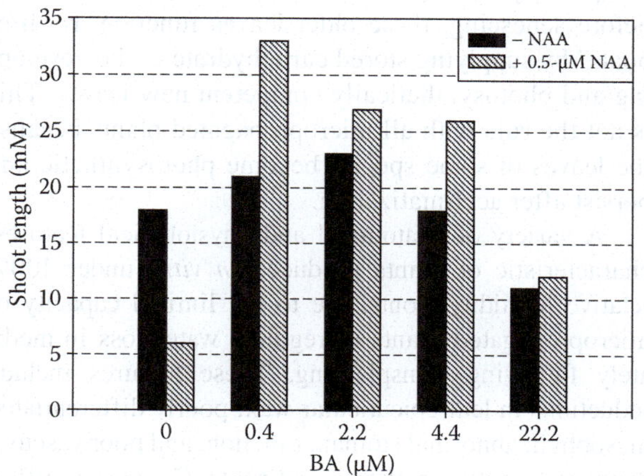

FIGURE 30.9 Inclusion of auxin may reduce the inhibitory effect of BA on axillary shoot elongation. Data shown for shoots are generated from two-node explants of *Aronia arbutifolia* after 28-day culture in the presence and the absence of 0.5 μM NAA.

the number of usable shoots of sufficient length for rooting (Figure 30.9). This benefit must be weighed against the increase chance of callus formation. Similarly, shoot elongation in Stage II cultures may be achieved by adding gibberellic acid to the medium.

The possibility of adverse carry-over effects on the rooting of plantlets and survivability in Stage IV should be evaluated when selecting a Stage II cytokinin. For example, with some tropical foliage plant species, the use of BA in Stage II can significantly reduce Stage IV plantlet survival and rooting to as low as 10%. The use of Kin or 2-iP instead of BA yields survival rates in excess of 90%. In some species, the adverse effect of BA on Stage IV survival and rooting has been attributed to the production of an inhibitory BA metabolite, which can be reduced by substituting with meta-topolin, a BA analog.

STAGE III: PRETRANSPLANT (ROOTING)

This step is characterized by the preparation of Stage II shoots (microcuttings) or shoot clusters for successful transfer to the soil (Figures 30.2b and 30.3). The process may involve the following:

1. Elongation of shoots prior to rooting
2. Rooting of individual shoots (microcuttings) or shoot clumps
3. Fulfilling the dormancy requirements of storage organs *in vitro* by cold treatment
4. Pre-hardening cultures *in vitro* to increase survival

Where possible, commercial laboratories have developed procedures to transfer Stage II microcuttings to the soil, thus bypassing Stage III rooting (Figure 30.3). There are several reasons for eliminating Stage III rooting in commercial production. Estimated costs for Stage III range from 35 to 75% of the total production costs. This reflects the significant input of labor and supplies required to complete Stage III rooting. Considerable cost savings can be achieved if Stage III is eliminated. Furthermore, *in vitro*–formed root systems are often nonfunctional and die following transplanting. This results in a delay in transplant growth prior to the production of new adventitious roots.

For various reasons, however, it may not always be feasible to transplant Stage II microcuttings directly to the soil. Given the aforementioned limitations of Stage III rooting, sometimes, Stage III culture is used solely to elongate Stage II shoot clusters prior to separation and rooting *ex vitro*. Elongated shoots may be further pretreated in an aqueous auxin solution prior to transplanting. Oftentimes, Stage III rooting of herbaceous plants can be achieved on the medium in the absence of auxins (Figure 30.2d). However, with many woody species, the addition of an auxin (IBA or NAA) in Stage III medium is required to enhance adventitious rooting (Figure 30.10) and plant performance *ex vitro*. Optimum auxin concentration is determined based on percent rooting, root number, and length (Figure 30.10). Roots should not be allowed to elongate excessively to help prevent root damage during transplanting. Care must be taken when selecting an auxin. For example, compared to IBA, the use of NAA for Stage III rooting decreases *ex vitro* survival rates or suppress posttransplant growth (Figure 30.11).

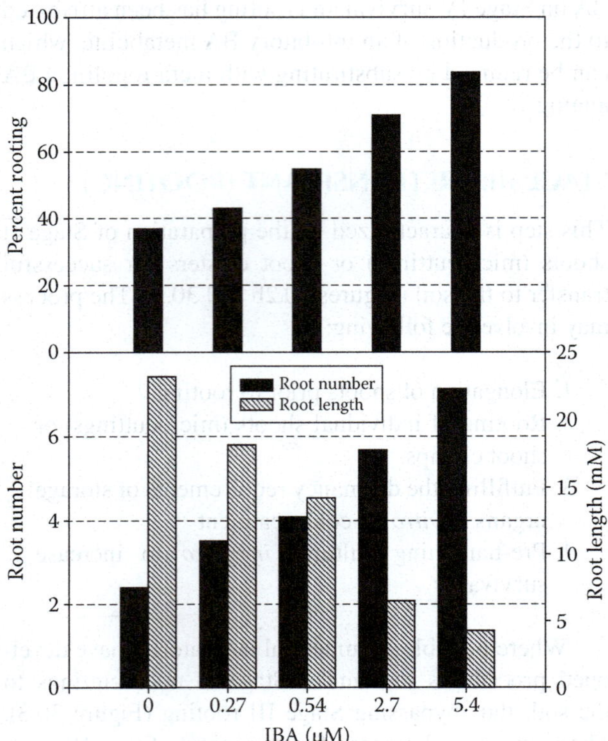

FIGURE 30.10 Effects of IBA on the Stage III rooting of 10 mm *Aronia arbutifolia* microcuttings after 28-day culture. Increased IBA concentrations enhance percent rooting but inhibit root elongation.

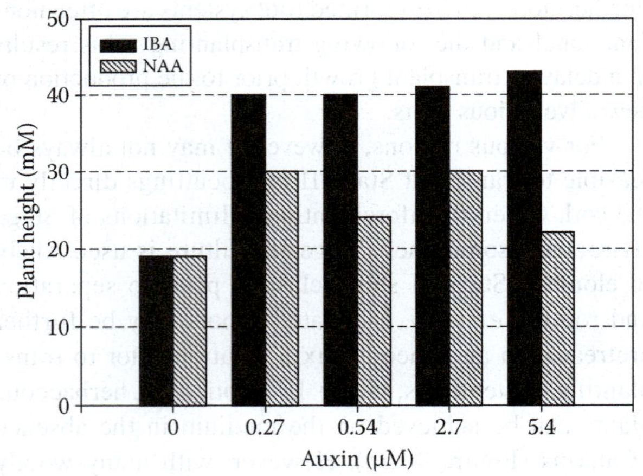

FIGURE 30.11 Comparative effects of the Stage III rooting of *Aronia arbutifolia* microcuttings with IBA or NAA on post-transplant growth *ex vitro* after 28 days.

STAGE IV: TRANSFER TO NATURAL ENVIRONMENT

The ultimate success of shoot or node culture depends on the ability to transfer and reestablish vigorously growing plants from *in vitro* to greenhouse conditions (Figures 30.2e, h; 30.3). This involves acclimatizing or hardening off plantlets to conditions of significantly lower relative humidity and higher light intensity. Even when acclimatization procedures are carefully followed, poor survival rates are frequently encountered. Micropropagated plants are often difficult to transplant because of a heterotrophic mode of nutrition *in vitro* and a poor control of water loss.

Plants cultured *in vitro* in the presence of sucrose and under conditions of limited light and gas exchange exhibit no or extremely reduced capacities for photosynthesis. Reduced photosynthetic activity is associated with low photosynthetic enzyme activity. During acclimatization, there is a need for plants to rapidly transition from the heterotrophic to the photoautotrophic state for survival. Unfortunately, this transition is not immediate. For example, in cauliflower, no net increase in CO_2 uptake is achieved until 14 days after transplantation. This occurs only following the development of new leaves since leaves produced *in vitro*, in the presence of sucrose, never develop photosynthetic competency under greenhouse conditions. Interestingly, before senescing, these older leaves function as "lifeboats" by supplying stored carbohydrate to the developing and photosynthetically competent new leaves. This is not the rule with all micropropagated plants because the leaves of some species become photosynthetic and persist after acclimatization.

A variety of anatomical and physiological features, characteristic of plants produced *in vitro* under 100% relative humidity, contribute to the limited capacity of micropropagated plants to regulate water loss immediately following transplanting. These features include reductions in leaf epicuticular wax, poorly differentiated mesophyll, abnormal stomata function, and poor vascular connection between shoots and roots. Genotypic differences in capacity for acclimatization have been reported in some species.

To overcome these limitations, plantlets are transplanted into a well-drained "sterile" growing medium and maintained initially at high relative humidity and reduced light (40–160 μmol m^{-2} sec^{-1}) at 20 to 27°C. Relative humidity may be maintained with humidity tents, single-tray propagation domes, intermittent misting, or fog systems. However, the use of intermittent mist often results in slow plantlet growth following water logging of the medium and excessive leaching of nutrients. Transplants are acclimatized by gradually lowering the relative humidity over a 1 to 4 week period. Plants are gradually moved to higher light intensities to promote vigorous growth. During this stage, container size and growing medium type can have a profound effect on the quality of the plants produced. Commercial laboratories labor costs are reduced through the adaptation of automation, including robotics for planting (Figure 30.6f), greenhouse environmental control, and fertilization and pesticide application (Figure 30.6g).

OTHER MICROPROPAGATION METHODS

Besides propagation by proliferation of preexisting shoot meristems through shoot culture, plants can be produced *in vitro* via two additional development pathways: shoot organogenesis, the formation of adventitious shoots from tissues without preexisting meristem, or non-zygotic embryogenesis, the formation of embryos from cells other than zygotes. When formed, adventitious shoot buds consist of an apical shoot, but no root meristem (Figure 30.7c). Adventitious shoots may develop vascular connections with the vasculature of the explant from which it arises or produce adventitious roots at the base (Figure 30.7e). Non-zygotic embryos, being bipolar, possess both shoot and root meristems, and develop from non-zygotic cells without any vascular connection to the original explant (Figure 30.12a). Non-zygotic embryos develop through similar developmental stages as zygotic embryos. Most reports of non-zygotic embryogenesis involve development from somatic (vegetative) cells. This method is often termed "somatic embryogenesis." Adventitious shoots may form directly on the explants (direct shoot organogenesis; Figure 30.7b) or indirectly on an intermediary callus stage (indirect shoot organogenesis; Figure 30.7f). Similarly, non-zygotic embryos may develop directly or indirectly from the explant (Figure 30.12a, d). Both adventitious shoots and non-zygotic embryos typically arise from single cells.

Given the single-cell origin of adventitious shoots and non-zygotic embryogenesis, there is great potential for high plant multiplication rates from a single explant. As many as 2600 shoot meristems have been observed on a single leaf explant cultured for 4 weeks. Depending on the species, genotype, and culture conditions, indirect shoot organogenesis and non-zygotic embryogenesis can result in an increased frequency of genetic variability in the plants produced. This variation, called "somaclonal variation," can be a limitation to the use of shoot organogenesis or non-zygotic embryogenesis for the commercial micropropagation of some species.

Many potential advantages exist for producing plants via non-zygotic embryogenesis. These include high rates of clonal multiplication, the presence of both shoot and root meristems, and the adaptation of automated liquid culture/bioreactor technology for production using embryogenic cell suspension cultures (Figure 30.12e). One limitation has been the absence of a protective seed coat, which makes the processing and handling of non-zygotic embryos problematic (Figure 30.12b). For

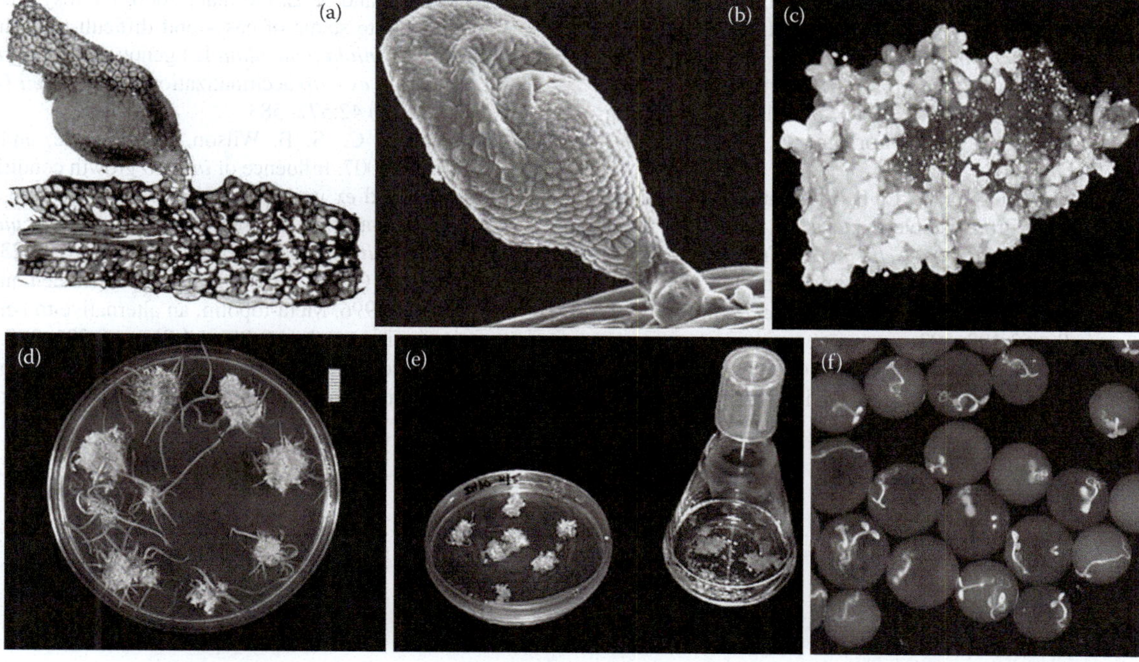

FIGURE 30.12 Micropropagation by non-zygotic embryogenesis in orchardgrass (*Dactylis glomerata* L.) (a) Longitudinal section of a non-zygotic embryo. Note the absence of a vascular connection between the non-zygotic embryo and the explant. (Courtesy of D. J. Gray.) (b) Scanning electron micrograph depicting direct non-zygotic embryogenesis. The absence of a seed coat can make the handling of these embryos problematic. (Courtesy of D. J. Gray.) (c) Formation of multiple non-zygotic embryos on a leaf explant. (Courtesy of D. J. Gray.) (d) Indirect formation of "germinated" non-zygotic embryos on embryogenic callus. (e) Embryogenic callus cultures (left) can be used to establish highly efficient embryogenic cell suspensions (right). The use of bioreactors to maintain cell cultures offers a significant potential for increased propagation efficiency. For example, more than 60,000 carrot non-zygotic embryos per liter can be generated in this manner. (f) Synthetic seeds are produced by encapsulating non-zygotic embryos in alginate beads. Each bead functions as an artificial seed coat that increases survival and improves handling and storage.

non-zygotic embryogenesis to be a viable propagation method, delivery systems must be developed through which non-zygotic embryos are processed, function as synthetic seeds that survive handling and planting, and further develop into vigorous plants. One technique involves the encapsulation of non-zygotic embryos in an alginate gel to create an artificial seed coat (Figure 30.12f). To date, the application of synthetic seed technology as a commercially viable plant propagation method has been limited by high costs.

LITERATURE CITED AND SUGGESTED READING

Aitken-Christie, J., T. Kozai, and M. A. L. Smith. 1994. *Automation and Environmental Control in Plant Tissue Culture*. Dordrecht, The Netherlands: Kluwer Academic Publishers, 574 pp.

de Klerk, G. J. 2002. Rooting of microcuttings: Theory and practice. *In Vitro Cell Dev Biol (Plant)* 38:415–422.

Gaspar, T., C. Kevers, C. Penel, H. Greppin, D. M. Reid, and T. Thorpe. 1996. Plant hormones and plant growth regulators in plant tissue culture. *In Vitro Cell Dev Biol (Plant)* 32:272–289.

George, E. F. 1993. *Plant Propagation by Tissue Culture: Part 1. The Technology*. London: Exegetics, Ltd., 514 pp.

Gray, D. 2005. Propagation from non-meristematic tissues: Non- embryogenesis. In Trigiano, R. N. and D. J. Gray, eds. *Plant Development and Biotechnology*. Boca Raton, Florida: CRC Press, 358 pp.

Jain, S. M. and G. J. de Klerk. 1998. Somaclonal variation in breeding and propagation of ornamental crops. *Plant Tiss Cult Biotech* 4:63–75.

Jenks, M. A., M. E. Kane, and D. B. McConnell. 2000. Shoot organogenesis from petiole explants in the aquatic plant *Nymphoides indica*. *Plant Cell Tiss Org Cult* 63:1–8.

Kane, M. E., G. L. Davis, D. B. McConnell, and J. A. Gargiulo. 1999. *In vitro* propagation of *Cryptocoryne wendtii*. *Aquatic Bot* 63:197–202.

Kane, M. E. 2000. Micropropagation of potato by node culture and microtuber production. In Trigiano, R. N. and D. J. Gray, eds. *Plant Tissue Culture Concepts and Laboratory Exercises*, 2nd Edition. Boca Raton, Florida: CRC Press, pp. 103–109.

Kane, M. E. 2000. Culture indexing for bacterial and fungal contaminants. In Trigiano, R. N. and D. J. Gray, eds. *Plant Tissue Culture Concepts and Laboratory Exercises*, 2nd Edition. Boca Raton, Florida: CRC Press, pp. 427–430.

Lloyd, G. and B. McCown. 1980. Commercially feasible micropropagation of mountain laurel, *Kalmia latifolia*, by use of shoot-tip culture. *Intl Plant Prop Soc Proc* 30:421–427.

Murashige, T. and F. Skoog. 1962. A revised medium for rapid growth and bioassays with tobacco tissue cultures. *Physiol Plant* 15:473–497.

Preece, J. E. and E. G. Sutter. 1991. Acclimatization of micropropagated plants to greenhouse and field. In Debergh P. C. and R. H. Zimmerman, eds. *Micropropagation Technology and Application*. Boston, Massachusetts Kluwer Academic Publishers, pp. 71–93.

Schwartz, O. J., A. R. Sharma, and R. M. Beaty. 2005. Propagation from non-meristematic tissues: Organogenesis. In Trigiano, R. N. and D. J. Gray, eds. Plant Development and Biotechnology. Boca Raton, Florida: CRC Press, pp. 159–172.

Styer, D. J. and C. K. Chin. 1984. Meristem and shoot-tip culture for propagation, pathogen elimination, and germplasm preservation. *Hort Rev* 5:221–277.

Valero Aracama, C., M. E. Kane, S. B. Wilson, J. C. Vu, J. Anderson, and N. L. Philman. 2006. Photosynthetic and carbohydrate status of easy- and difficult-to-acclimatize sea oats (*Uniola paniculata* L.) genotypes during *in vitro* culture and *ex vitro* acclimatization. *In Vitro Cell Develop Biol (Plant)* 42:572–583.

Valero Aracama, C., S. B. Wilson, M. E. Kane, and N. L. Philman. 2007. Influence of *in vitro* growth conditions on *in vitro* and *ex vitro* photosynthetic rates of easy- and difficult-to-acclimatize sea oats (*Uniola paniculata* L.) genotypes. *In Vitro Cell Develop Biol (Plant)* 43:237–246.

Werbrouck, S. P. O., M. Strnad, H. A. van Onckelen and P. C. Debergh. 1996. Meta-topolin, an alternative to benzyladenine in tissue culture? *Physiol Plant* 98:291–297.

Wickson, M. and K. V. Thimann. 1958. The antagonism of auxin and kinetin in apical dominance. *Physiol Plant* 11:62–74.

31 Getting Started with Tissue Culture
Media Preparation, Sterile Technique, and Laboratory Equipment

Caula A. Beyl

Concept Box 31.1

- A tissue laboratory needs adequate physical space for work and storage and equipment such as an autoclave, a distilled water source, balances, refrigerators, various laboratory instruments, culture vessels, and flow hoods, to name a few items.
- There are many growth media available and the type of basal culture medium selected depends on the species to be cultured. The growth of the plant in culture is also affected by the selection of plant growth regulators (PGRs) and environmental (cultural) conditions.
- There are about 20 different components in tissue culture medium, which include inorganic mineral elements, various organic compounds, PGR, and support substances (e.g., agar or filter paper).
- PGRs are typically expressed in media as milligram per liter or micromole (μM). When comparing the effects of several PGRs on tissues in culture, prepare media using micromolar concentrations since an equal number of molecules of the various PGRs will be present in each of the media.

A plant tissue culture laboratory has several functional areas, whether it is designed for teaching or research and no matter what its size or how elaborate it is. It has some elements similar to a well-run kitchen and other elements that more closely resemble an operating room. There are areas devoted to the preliminary handling of a plant tissue destined for culture, media preparation, and sterilization of media and tools; a sterile transfer hood or "clean room" for aseptic manipulations; a culture growth room; and an area devoted to washing and cleaning glassware and tools. The following chapter will serve as an introduction to what goes into setting up a tissue culture laboratory; what supplies and equipment are necessary; some basics concerning making stock solutions; calculating molar concentrations; making tissue culture media; preparing a transfer (sterile) hood; and culturing various cells, tissues, and organs.

EQUIPMENT AND SUPPLIES FOR A TISSUE CULTURE TEACHING LABORATORY

Ideally, there should be enough bench area to allow for both preparation of media and storage space for chemicals and glassware. In addition to the usual glassware and instrumentation found in laboratories, a tissue culture laboratory needs an assortment of glassware, which may include graduated measuring cylinders, wide-necked Erlenmeyer flasks, medium bottles, test tubes with caps, Petri dishes, volumetric flasks, beakers, and a range of pipettes. In general, glassware should be able to withstand repeated autoclaving. Baby food jars are inexpensive alternative tissue culture containers well suited for teaching. Ample quantities can be obtained by preceding the recycling truck on its pick-up day (provided that you are not embarrassed by the practice). Some tissue culture laboratories find presterilized disposable culture containers and plastic Petri dishes to be convenient, but the cost may be prohibitive for others on a tight budget. There are also reusable plastic containers available, but their longevity and resistance to wear, heat, and chemicals vary considerably.

It is also good to stock metal or wooden racks to support culture tubes both for cooling and later during their time in the culture room, metal trays (such as cafeteria trays) and carts for transport of cultures, stoppers and various closures, non-absorbent cotton, cheesecloth, foam plugs, metal or plastic caps, aluminum foil, Parafilm™, and plastic wrap.

To teach tissue culture effectively, some equipment is necessary, such as a pH meter; balances (one analytical to four decimal places, and one, to two decimal places); Bunsen burners, alcohol lamps, or electric sterilizing devices; several hot plates with magnetic stirrers; a microwave oven for rapid melting of large volumes of agar medium; a compound microscope and a hemocytometer for cell counting; a low-speed centrifuge; stereomicroscopes (ideally with fiber optic light sources); large (10- or 25-L) plastic carboys to store high-quality (purity) water; a fume hood; an autoclave (or, at the very least, a pressure cooker); and a refrigerator to store media, stock solutions, plant growth regulators (PGRs), and others. A dishwasher is useful, but a large sink with drying racks, pipette and acid baths, and a forced air oven for drying glassware will also work. Also, deionized distilled water for the final rinsing of glassware is needed. Aseptic manipulations and transfers are done in multi-station, laminar-flow hoods (one for each pair of students).

Equipment used in the sterile transfer hood usually includes a spray bottle containing 70% ethanol, spatulas (useful for transferring callus clumps), forceps (short, long, and fine tipped), scalpel handles (#3), disposable scalpel blades (#10 and #11), a rack for holding sterile tools, a pipette bulb or pump, Bunsen burner, alcohol lamp or other sterilizing device, and a sterile surface for cutting explants (see below). If necessary for the experimental design, uniform-sized leaf explants can be obtained using a sterile cork borer.

There are a number of options for providing a sterile surface for cutting explants. A stack of autoclaved paper towels wrapped in aluminum foil is effective, and as each layer becomes messy, it can be peeled off, and the next layer beneath it can be used. Others prefer reusable surfaces such as ceramic tiles (local tile retailers are quite generous and will donate samples; Figure 31.1), metal commercial ashtrays, or glass Petri dishes (100 × 15 mm). Sterile plastic Petri dishes can also be used, but the cost may outweigh the advantages. A container is needed to hold the alcohol used for flaming instruments, if flame sterilization is used. An ideal container for this purpose is a slide-staining Coplin jar with a small wad of cheesecloth at the bottom to prevent breakage of the glass when tools are dropped in. It has the advantage that it is a heavier glass, and since the base is flared, it is not prone to tipping over. Other containers can also serve the same purpose, such as test tubes in a rack or placed in a flask or a beaker to prevent them from spilling. Plastic containers, which can catch fire and melt, should never be used to hold alcohol.

Instead of using alcohol and flame system for sterilizing equipment, various alternatives are available. Sterilization can be achieved within 60 sec by inserting scalpels or forceps into glass bead sterilizers. After the utensils are removed from the sterilizer, they require a 30–60 sec cool-down period before use. Another option is the use of a Bacti-Cinerator, which used a ceramic heating element to achieve sterility.

FIGURE 31.1 A typical layout of materials in the hood, showing the placement of the sterile tile work surface, an alcohol lamp, a spray bottle containing 70% ethanol, a cloth for wiping down the hood, and two different kinds of tool holders—a glass-staining (Coplin) jar and a metal rack for holding test tubes.

WATER

High-quality water is a required ingredient of plant tissue culture media. Ordinary tap water contains cations, anions, particulates of various kinds, microorganisms, and gases that make it unsuitable for use in tissue culture media. Various methods are used to treat water, including filtration through activated carbon to remove organics and chlorine; deionization or demineralization by passing water through exchange resins to remove dissolved ionized impurities; and distillation that eliminates most ionic and particulate impurities, volatile and nonvolatile chemicals, organic matter, and microorganisms. The process of reverse osmosis, which removes 99% of dissolved ionized impurities, uses a semipermeable membrane through which a portion of the water is forced under pressure and the remainder containing the concentrated impurities is rejected. The most universally reliable method of water purification for tissue culture use is a deionization treatment followed by one or two glass distillations, although simple deionization alone is sometimes successfully used. In some cases, newer reverse-osmosis purifying equipment (Milli-RO™, Millipore™, RO pure™, Barnstead™, Bion™, and Pierce™), combined with cartridge ion exchange, adsorption, and membrane filtering equipment, has replaced the traditional glass distillation of water.

THE CULTURE ROOM

After the explants are placed on the tissue culture medium under the sterile transfer hood, they are moved to the culture room. It can be as simple as a room with shelves equipped with lights or as complex as a room with intricate climate control. Most culture rooms tend to be rather simple, consisting of cool white fluorescent lights mounted to shine on each shelf. Adjustable shelves are an asset and allow for differently sized tissue culture containers and for moving the light closer to the containers to achieve higher light intensities. Putting the lights on timers allows for photoperiod manipulation. Some cultures grow equally well in dark or light. Temperatures of 26 to 28°C are usually optimum. Heat buildup can be a problem if the room is small, so adequate air conditioning is required. Good airflow also helps reduce condensation occurring inside Petri dishes or other vessels. Some laboratories purchase incubators designed for plant tissue culture. If a liquid medium is used, the culture room should be equipped with a rotary or reciprocal shaker to provide sufficient oxygenation. The optimum temperature, light, and shaker conditions vary depending on the plant species being cultured.

CHARACTERISTICS OF SOME OF THE MORE COMMON TISSUE CULTURE MEDIA

The type of tissue culture medium selected depends on the species to be cultured. Some species are sensitive to high salts or have different requirements for PGRs. The age of the plant also has an effect. For example, juvenile tissues generally regenerate roots more readily than adult tissues. The type of organ cultured is important; for example, many roots require thiamine. Each desired cultural effect has its own unique requirements such as auxin (see below) for induction of adventitious roots and altering the cytokinin:auxin ratio for the initiation and development of adventitious shoots.

Development of culture medium formulations was a result of systematic trial and experimentation. Table 31.1 provides a comparison of the composition of several of the most commonly used plant tissue culture media with respect to their components in milligram per liter and molar units. Murashige and Skoog (1962) (MS) medium is the most suitable and most commonly used basic tissue culture medium for plant regeneration from tissues and calluses. It was developed for tobacco based primarily on the mineral analysis of tobacco tissue. This is a "high-salt" medium due to its content of potassium and nitrogen salts. Linsmaier and Skoog (1965) medium is basically Murashige and Skoog (1962) medium with respect to its inorganic portion, but only inositol and thiamine HCl are retained among the organic components. To counteract the salt sensitivity of some woody species, Lloyd and McCown (1980) developed the woody plant medium (WPM).

Gamborg's B5 medium (Gamborg *et al.*, 1968) was devised for soybean callus culture and has lesser amounts of nitrate and, particularly, ammonium salts than the MS medium. Although B5 was originally developed for the purpose of obtaining callus or for use with suspension culture, it also works well as a basal medium for whole plant regeneration. A medium for callus culture of monocots and dicots was developed by Schenk and Hildebrandt (SH medium) 1972. White's medium (1963), which was designed for the tissue culture of tomato roots, has a lower concentration of salts than the MS medium. Nitsch's medium (Nitsch and Nitsch, 1969) was developed for anther culture and contains a salt concentration intermediate to those of the MS and White's media.

Many companies sell packaged prepared mixtures of better-known media recipes. These are easy to make because they merely involve dissolving the packaged mix in a specified volume of water. These can be purchased as salts, vitamins, or the entire mix with or without PGRs, agar, and sucrose. These are convenient and less prone to individual error, and make keeping stock solutions unnecessary. However, they are more expensive than making media from scratch.

COMPONENTS OF THE TISSUE CULTURE MEDIUM

Growth and development of explants *in vitro* is a product of its genetics, surrounding the environment and components of the tissue culture medium, the last of which is easiest to manipulate to our own ends. Tissue culture medium consists of 95% water, macro- and micronutrients, PGRs, vitamins, sugars (because plants *in vitro* are often not photosynthetically competent), and sometimes, various other simple to complex organic materials. All in all, about 20 different components are usually needed.

Inorganic Mineral Elements

Just as a plant growing *in vivo* requires many different elements from either soil or fertilizers, the plant tissue growing *in vitro* requires a combination of macro- and micronutrients. The choice of macro- and micro-salts and their concentrations is species dependent. The MS medium is very popular because most plants react to it favorably; however, it may not necessary result in the optimum growth and development for every species because the salt content is so high.

The macronutrients are required in millimolar (mM) quantities in most plant basal media. Nitrogen is usually supplied in the form of ammonium $\left(NH_4^+\right)$ and nitrate $\left(NO_3^-\right)$ ions, although sometimes, more complex organic sources, such as urea and amino acids like glutamine or casein hydrolysate, which is a complex mixture of amino acids and ammonium, are used too. Although most plants prefer NO_3^- to NH_4^+, the right balance of the two ions for optimum *in vitro* growth and development for the selected species may differ.

TABLE 31.1
Composition of Five Commonly Used Tissue Culture Media in Milligrams per Liter and Molar Concentrations

Compounds	Murashige and Skoog	Gamborg B-5	WPM	Nitsch and Nitsch	Schenk and Hildebrandt	White
Macronutrients in mg/L (mM)						
NH_4NO_3	1650 (20.6)	–	400 (5.0)	720	–	–
$NH_4H_2PO_4$	–	–	–	–	300 (2.6)	–
NH_4SO_4	–	134 (1.0)	–	–	–	–
$CaCl_2 \cdot 2H_2O$	440 (3.0)	150 (1.0)	96 (0.7)	166 (1.1)	151 (1.0)	–
$Ca(NO_3)_2 \cdot 4H_2O$	–	–	556 (2.4)	–	–	288 (1.2)
$MgSO_4 \cdot 7H_2O$	370 (1.5)	250 (1.0)	370 (1.5)	185 (0.75)	400 (1.6)	737 (3.0)
KCl	–	–	–	–	–	65 (0.9)
KNO_3	1900 (18.8)	2500 (24.8)	–	950 (9.4)	2500 (24.8)	80 (0.8)
K_2SO_4	–	–	990	–	–	–
KH_2PO_4	170 (1.3)	–	170 (1.3)	68 (0.5)	–	–
NaH_2PO_4	–	130.5 (0.9)	–	–	–	16.5 (0.12)
Na_2SO_4	–	–	–	–	–	200 (1.4)
Micronutrients in mg/L (μM)						
H_3BO_3	6.2 (100)	3.0 (49)	6.2 (100)	10 (162)	5 (80)	1.5 (25)
$CoCl_2 \cdot 6H_2O$	0.025 (0.1)	0.025 (0.1)	–	–	0.1 (0.4)	–
$CuSO_4 \cdot 5H_2O$	0.025 (0.1)	0.025 (0.1)	0.25 (1)	0.025 (0.1)	0.2 (0.08)	0.01 (0.04)
Na_2EDTA	37.3 (100)	37.3 (100)	37.3 (100)	37.3 (100)	20.1 (54)	–
$Fe_2(SO_4)_3$	–	–	–	–	–	2.5 (6.2)
$FeSO_4 \cdot 7H_2O$	27.8 (100)	27.8 (100)	27.8 (100)	27.8 (100)	15 (54)	–
$MnSO_4 \cdot H_2O$	16.9 (100)	10.0 (59)	22.3 (132)	18.9 (112)	10.0 (59)	5.04 (30)
KI	0.83 (5)	0.75 (5)	–	–	0.1 (0.6)	0.75 (5)
$NaMoO_3$	–	–	–	–	–	0.001 (0.001)
$Na_2MoO_4 \cdot 2H_2O$	0.25 (1)	0.25 (1)	0.25 (1)	0.25 (1)	0.1 (0.4)	–
$ZnSO_4 \cdot 7H_2O$	8.6 (30)	2.0 (7.0)	8.6 (30)	10 (35)	1 (3)	2.67 (9)
Organics in mg/L (μM)						
Myo-inositol	100 (550)	100 (550)	100 (550)	100 (550)	1000 (5500)	–
Glycine	2.0 (26.6)	–	2.0 (26.6)	2.0 (26.6)	–	3.0 (40)
Nicotinic acid	0.5 (4.1)	1.0 (8.2)	0.5 (4.1)	5 (40.6)	5.0 (41)	0.5 (4.1)
Pyridoxine HCl	0.5 (2.4)	0.1 (0.45)	0.5 (2.4)	0.5 (2.4)	0.5 (2.4)	0.1 (0.45)
Thiamine HCl	0.1 (0.3)	10.0 (30)	1.0 (3.0)	0.5 (1.5)	5.0 (14.8)	0.1 (0.3)
Biotin	–	–	–	0.05 (0.13)	–	–
Folic acid	–	–	0.50	–	–	–

In addition to nitrogen, potassium, magnesium, calcium, phosphorus, and sulfur are provided in the medium as different components referred to as the macro-salts. $MgSO_4$ provides both magnesium and sulfur; $NH_4 H_2PO_4$, KH_2PO_4, or NaH_2PO_4 provides phosphorus; $CaCl_2 \cdot 2H_2O$ or $Ca(NO_3)_2 \cdot 4H_2O$ provides calcium; and KCl, KNO_3, or KH_2PO_4 provides potassium. Chloride is provided by KCl and/or $CaCl_2 \cdot 2H_2O$.

Micro-salts typically include boron (H_3BO_3), cobalt ($CoCl_2 \cdot 6H_2O$), iron (complex of $FeSO_4 \cdot 7H_2O$ and Na_2EDTA or rarely as $Fe_2(SO_4)_3$), manganese ($MnSO_4 \cdot H_2O$), molybdenum ($NaMoO_3$), copper ($CuSO_4 \cdot 5H_2O$), and zinc ($ZnSO_4 \cdot 7H_2O$). Micro-salts are needed in much lower (micromolar [μM]) concentrations than macronutrients. Some media may contain very small amounts of iodide (KI), but sufficient quantities of many of the trace elements may inadvertently be provided because reagent-grade chemicals contain inorganic contaminants.

Organic Compounds

Sugar is a very important part of any nutrient medium, and its addition is essential for the *in vitro* growth and development of the culture. Most plant cultures are unable to photosynthesize effectively for a variety of reasons, including insufficiently organized cellular and tissue development; lack of chlorophyll; limited gas exchange and CO_2 in tissue culture vessels; and less-than-optimum environmental conditions, such as low light. A concentration of 20 to 60 g/L sucrose (a disaccharide made up of glucose and fructose) is the most often used carbon or energy source since this sugar is also synthesized and transported naturally by the plant. Other mono- or disaccharides and sugar alcohols such as glucose, fructose, and maltose may be used. The sugar concentration chosen is dependent on the type and the age of the explant in culture. For example, very young embryos require a relatively high sugar concentration (>3%). For mulberry buds *in vitro*, fructose was found to be better than sucrose, glucose, maltose, raffinose or lactose (Coffin *et al*., 1976). For apple, sorbitol and sucrose supported callus initiation and growth equally well, but sorbitol was better for peach after the fourth subculture (Oka and Ohyama, 1982).

Sugar (sucrose) that is bought from the supermarket is usually adequate, but be careful to get pure cane sugar as corn sugar is primarily fructose. Raw cane sugar is purified and, according to the manufacturer's analysis, consists of 99.94% sucrose, 0.02% water, and 0.04% other materials (inorganic elements and also raffinose, fructose, and glucose). Nutrient salts contribute approximately 20 to 50% to the osmotic potential of the medium, and sucrose is responsible for the remainder. The contribution of sucrose to the osmotic potential increases as it is hydrolyzed into glucose and fructose during autoclaving. This may be an important consideration when performing osmotic sensitive procedures such as protoplast isolation and culture.

Vitamins are organic substances that are parts of enzymes or cofactors for essential metabolic functions. Of the vitamins, only thiamine (vitamin B_1 at 0.1–5.0 mg/L) is essential in culture as it is involved in carbohydrate metabolism and the biosynthesis of some amino acids. It is usually added to tissue culture media as thiamine hydrochloride. Nicotinic acid, also known as niacin or vitamin B_3, forms part of a respiratory coenzyme and is used at concentrations between 0.1 and 5 mg/L. The MS medium contains thiamine HCl as well as two other vitamins, nicotinic acid, and pyridoxine (vitamin B_6) in the HCl form. Pyridoxine is an important coenzyme in many metabolic reactions and is used in media at concentrations of 0.1 to 1.0 mg/L. Biotin (vitamin H) is commonly added to tissue culture media at 0.01 to 1.0 mg/L. Other vitamins that are sometimes used are folic acid (vitamin M; 0.1–0.5 mg/L), riboflavin (vitamin B_2; 0.1–10 mg/L), ascorbic acid (vitamin C; 1–100 mg/L), pantothenic acid (vitamin B_5; 0.5–2.5 mg/L), tocopherol (vitamin E; 1–50 mg/L), and para-aminobenzoic acid (0.5–1.0 mg/L).

Inositol is sometimes characterized as one of the B complex vitamin group, but it is really a sugar alcohol involved in the synthesis of phospholipids, cell wall pectins, and membrane systems in cell cytoplasm. It is added to tissue culture media at a concentration of about 0.1 to 1.0 g/L and has been demonstrated to be necessary for some monocots, dicots, and gymnosperms.

In addition, other amino acids are sometimes used in tissue culture media. These include L-glutamine, asparagine, serine, and proline, which are used as sources of reduced organic nitrogen, especially for inducing and maintaining non-zygotic embryogenesis. Glycine, the simplest amino acid, is a common additive since it is essential in purine synthesis and is a part of the porphyrin ring structure of chlorophyll.

Complex organic compounds are a group of undefined supplements such as casein hydrolysate, coconut milk (the liquid endosperm of the coconut), orange juice, tomato juice, grape juice, pineapple juice, sap from birch, banana puree, and others. These compounds are often used when no other combination of known defined components produces the desired growth or development. However, the composition of these supplements is basically unknown and may also vary from lot to lot, causing variable responses. For example, the composition of coconut milk (used at a dilution of 50–150 mL/L), a natural source of the PGR zeatin (see below), differs not only between young and old coconuts, but also between coconuts of the same age.

Some complex organic compounds are used as organic sources of nitrogen such as casein hydrolysate, a mixture of about 20 different amino acids and ammonium (0.1–1.0 g/L); peptone (0.25–3.0 g/L); tryptone (0.25–2.0 g/L); and malt extract (0.5–1.0 g/L). These mixtures are very complex and contain vitamins as well as amino acids. Yeast extract (0.25–2.0 g/L) is used because of the high concentration and quality of B vitamins.

Polyamines, particularly putrescine and spermidine, are sometimes beneficial for non-zygotic embryogenesis. Polyamines are also cofactors for adventitious root formation. Putrescine is capable of synchronizing the embryogenic process of carrot.

Activated charcoal is useful for the absorption of brown or black pigments and oxidized phenolic compounds. It is incorporated into the medium at concentrations of 0.2 to 3.0% (w/v). It is also useful for absorbing other organic compounds, including PGRs, such as auxins and cytokinins, and vitamins, iron, and zinc chelates (Nissen and Sutter, 1990). Carry-over effects of PGRs are minimized by adding activated charcoal when transferring explants to media without PGRs. Another feature of using activated charcoal is that it changes the light environment by darkening the medium so that it can help with root formation and growth. It may also promote non-zygotic embryogenesis and enhance the growth and organogenesis of woody species.

Leached pigments and oxidized polyphenolic compounds and tannins can greatly inhibit growth and development. These are formed by some explants as a result of wounding. If charcoal does not reduce the inhibitory effects of polyphenols, addition of polyvinylpyrrolidone (250–1000 mg/L) or antioxidants, such as citric acid, ascorbic acid, or thiourea, can be tested.

PLANT GROWTH REGULATORS (PGRs)

PGRs exert dramatic effects at low concentrations (0.001–10 μM). They regulate the initiation and development of shoots and roots on explants and embryos on semisolid or in liquid medium cultures. They also stimulate cell division and expansion. Sometimes, a tissue or an explant is autotrophic and can produce its own supply of PGRs. Usually, PGRs must be supplied in the medium for the growth and development of the culture.

The most important classes of the PGRs used in tissue culture are auxins and cytokinins. The relative effects of auxin and cytokinin ratio on the morphogenesis of cultured tissues were demonstrated by Skoog and Miller (1957), and their results still serve as the basis for plant tissue culture manipulations today. Some of the PGRs used are hormones (naturally synthesized by higher plants) and others are synthetic compounds. PGRs exert dramatic effects depending on the concentration used, the target tissue, and their inherent activity, even though they are used in very low concentrations in the media (from 0.1 to 100 μM). To prepare a μM solation of a PGR, follow the calculation shown in Procedure 31.1. The concentrations of PGRs are typically reported in milligram per liter or in micromolar units of concentration. Comparisons of PGRs based on their molar concentrations are more useful because the molar concentration is a reflection of the

Procedure 31.1

Converting Molar Solutions to Milligram per Liter and Milligram per Liter to Molar Solutions Using Conversion Factors

- How to determine how many milligram per liter are needed for a 1.0 molar concentration:
  - First, look up the molecular weight of the PGR. In this example, we will use kinetin. The molecular weight is 215.2, so a 1 M solution will consist of 215.2 g/L of solution.
  - By using conversion factors and crossing out terms, you cannot go wrong!

$$1 \text{ mM solution} = \frac{\cancel{1\text{ M}}}{\text{liter solution}} \times \frac{215.2\text{ g}}{\cancel{1\text{ M}}} = 215.2\text{ g/L}$$

  - To see what a 1.0 mM solution of kinetin would consist of, multiply the grams necessary for a 1 M solution by 10^{-3}, which would give you 215.2 × 10^{-3} g/L or 215.2 mg/L.

$$1 \text{ mM solution} = \frac{\cancel{1\text{ mM}}}{\text{liter solution}} \times \frac{215.2\text{ mg}}{\cancel{1\text{ mM}}} = 215.2\text{ mg/L}$$

  - To see what a 1.0 μM solution of kinetin would consist of, multiply the grams necessary for a 1 M solution by 10^{-6}, which would give you 215.2 × 10^{-6} g/L or 215.2 μg/L.

$$1\ \mu\text{M solution} = \frac{\cancel{1\ \mu\text{M}}}{\text{liter solution}} \times \frac{215.2\ \mu\text{g}}{\cancel{1\ \mu\text{M}}} = 215.2\ \mu\text{g/L or } 0.215\text{ mg/L}$$

- How to determine the molar concentration of a solution that is in milligram per liter:
  - Let us assume that you are given a 10 mg/L solution of IBA and you wish to know its molarity. First, look up the molecular weight of the PGR. In this example, using IBA, the molecular weight of IBA is 203.2.
  - Again, using conversion factors and crossing out terms:

$$\frac{10\text{ mg IBA}}{\text{liter of solution}} \times \frac{\cancel{1\text{ g}}}{1000\ \cancel{\text{mg}}} \times \frac{1\text{ M}}{203.2\ \cancel{\text{g}}} = \frac{0.0000492\text{ M}}{\text{liter of solution}} = 49.2\ \mu\text{M IBA}$$

Note: Remember 1 M = 10^3 mM = 10^6 μM.

actual number of molecules of the PGR per unit volume (Table 31.2).

Auxins play a role in many developmental processes, including cell elongation and swelling of tissue, apical dominance, adventitious root formation, and somatic embryogenesis. Generally, when the concentration of auxin is low, root initiation is favored, and when the concentration is high, callus formation occurs. The most common synthetic auxins used are 1-naphthaleneacetic acid (NAA), 2,4-dichlorophenoxyacetic acid (2,4-D), and 4-amino-3,5,6-trichloro-2-pyridinecarboxylic acid (picloram). Naturally occurring indoleacetic acid (IAA) and indolebutyric acid (IBA) are also used. IBA was once considered synthetic, but has also been found to occur naturally in many plants, including olive and tobacco (Epstein *et al.*, 1989). Both IBA and IAA are photosensitive, so stock solutions must be stored in the dark. IAA is also easily broken down by enzymes (peroxidases and IAA oxidase). IAA is the weakest auxin and is typically used at concentrations between 0.01 to 10 mg/L. The relatively more active auxins such as IBA, NAA, 2,4-D, and picloram are used at concentrations ranging from 0.001 to 10 mg/L. 2,4-D and picloram are examples of auxins used primarily to induce and regulate somatic embryogenesis.

Cytokinins promote cell division and stimulate the initiation and growth of shoots *in vitro*. The cytokinins most commonly used are zeatin, dihydrozeatin, kinetin, benzyladenine (BA), thidiazuron, and 2iP. In higher concentrations (1–10 mg/L), they induce adventitious shoot formation but inhibit root formation. They promote axillary shoot formation by opposing the apical dominance regulated by auxins. Benzyladenine has significantly stronger cytokinin activity than the naturally occurring zeatin. However, a concentration of 0.05 to 0.1 μM thidiazuron, a diphenyl-substituted urea, is more active than 4 to 10 μM BA, but thidiazuron may inhibit root formation, causing difficulties in plant regeneration. Adenine (used at concentrations of 2–120 mg/L) is occasionally added to tissue culture media and acts as a weak cytokinin by promoting shoot formation.

Gibberellins are less commonly used in plant tissue culture. Of the many gibberellins thus far described, GA_3 is the most often used, but it is very heat sensitive (after autoclaving, 90% of the biological activity is lost). Typically, it is filter sterilized and added to autoclaved medium after it has cooled. Gibberellins help stimulate the elongation of internodes and have proved to be necessary for meristem growth for some species.

Abscisic acid is not normally considered an important PGR for tissue culture except for somatic embryogenesis and in the culture of some woody plants. For example, it promotes the maturation and germination of the somatic embryos of caraway (Ammirato, 1974) and spruce (Roberts *et al.*, 1990).

Organ and callus cultures are able to produce the gaseous PGR ethylene. Since culture vessels are almost entirely closed, ethylene can sometimes accumulate. Many plastic containers also contribute to the ethylene content in the vessel. There are contrasting reports in the

TABLE 31.2

PGRs, Their Molecular Weights, Conversions of Milligram per Liter Concentrations into Micromolar Equivalents, and Conversion of Micromolar Concentrations into Milligram per Liter Equivalents

Plant Growth Regulator	Abbreviation	Molecular Weight	mg/L Equivalents for these μM Concentrations				μM Equivalents for these mg/L Concentrations			
			0.1	1.0	10.0	100.0	0.1	0.5	1.0	10.0
Abscisic acid	ABA	264.3	0.0264	0.264	2.64	26.4	0.38	1.89	3.78	37.8
Benzyladenine	BA	225.2	0.0225	0.225	2.25	22.5	0.44	2.22	4.44	44.4
Dihydrozeatin	2hZ	220.3	0.0220	0.220	2.20	22.0	0.45	2.27	4.53	45.3
Gibberellic acid	GA3	346.4	0.0346	0.346	3.46	34.6	0.29	1.44	2.89	28.9
Indoleacetic acid	IAA	175.2	0.0175	0.175	1.75	17.5	0.57	2.85	5.71	57.1
Indolebutyric acid	IBA	203.2	0.0203	0.203	2.03	20.3	0.49	2.46	4.90	49.0
Potassium salt of IBA	K-IBA	241.3	0.0241	0.241	2.41	24.1	0.41	2.07	4.14	41.4
Kinetin	Kin	215.2	0.0215	0.215	2.15	21.5	0.46	2.32	4.65	46.7
Naphthaleneacetic acid	NAA	186.2	0.0186	0.186	1.86	18.6	0.54	2.69	5.37	53.7
Picloram	Pic	241.5	0.0242	0.242	2.42	24.2	0.41	2.07	4.14	41.4
Thidiazuron	TDZ	220.3	0.0220	0.220	2.20	22.0	0.45	2.27	4.54	45.4
Zeatin	Zea	219.2	0.0219	0.219	2.19	21.9	0.46	2.28	4.56	45.6
2-isopentenyl adenin	2-iP	203.3	0.0203	0.203	2.03	20.3	0.49	2.46	4.92	49.2
2,4-dichlorophenoxyacetic acid	2,4-D	221.04	0.0221	0.221	2.21	22.1	0.45	2.26	4.52	45.2

literature concerning the role played by ethylene *in vitro*. It appears to influence embryogenesis and organ formation in some gymnosperms. Sometimes, *in vitro* growth can be promoted by ethylene. At other times, addition of ethylene inhibitors results in better initiation or growth. For example, ethylene inhibitors, particularly silver nitrate, are used to enhance embryogenic culture initiation in corn. High levels of 2,4-D can induce ethylene formation.

AGAR AND ALTERNATIVE CULTURE SUPPORT SYSTEMS

Agar is used to solidify tissue culture media into a gel. It enables the explant to be placed in precise contact with the medium (e.g., on the surface or embedded) but remain aerated. Agar is a high–molecular weight polysaccharide that can bind water and is derived from seaweed. It is added to the medium in concentrations ranging from 0.5 to 1.0% (w/v). High concentrations of agar result in a harder medium. If a lower concentration of agar is used (0.4%) or if the pH is low, the medium will be too soft and will not gel properly. The consistency of the agar can also influence growth. If it is too hard, plant growth is reduced. If it is too soft, hyperhydric plants may be the result (Singha, 1982). To gel properly, a medium with 0.6% agar must have a pH above 4.8. Sometimes, activated charcoal in the medium will interfere with gelling. Typical tissue culture agar melts easily at approximately 65°C and solidifies at approximately 45°C.

Agar also contains organic and inorganic contaminants, the amount of which varies between brands. Organic acids, phenolic compounds, and long-chain fatty acids are common contaminants. A manufacturer's analysis shows that Difco Bacto agar also contains (amounts in parts per million): 0.0 to 0.5 cadmium, 0.0 to 0.1 chromium, 0.5 to 1.5 copper, 1.5 to 5.0 iron, 0.0 to 0.5 lead, 210.0 to 430.0 magnesium, 0.1 to 0.5 manganese, and 5.0 to 10.0 zinc. Generally, relatively pure plant tissue culture–tested types of agar should be used. Poor-quality agar can interfere or inhibit the growth of cultures.

Agarose is often used when culturing protoplasts or single cells. Agarose is a purified extract of agar from which agaropectin and sulfate groups have been eliminated. Since its gel strength is greater, less is used to create a suitable support or suspending medium.

Gellan gums like Gelrite™ and Phytagel™ are alternative gelling agents. They are made from a polysaccharide produced by a bacterium. Rather than being translucent (like agar), they are clear, so it is much easier to detect contamination, but they cannot be re-liquified by heating and gelled again, and the concentration of divalent cations like calcium and magnesium must be within a restricted range or gelling will not occur.

Mechanical supports such as filter paper bridges or polyethylene rafts do not rely on a gelling agent. They can be used with liquid media, which then circulates better, but keeps the explant at the medium surface so that it remains oxygenated. The types of support systems that have been used are as varied as the imagination and include rock wool, cheesecloth, pieces of foam, and glass beads.

STEPS IN THE PREPARATION OF TISSUE CULTURE MEDIUM

The first step in making tissue culture medium is to assemble needed the glassware, for example, a 1 L beaker, 1 L volumetric flask, stirring bar, balance, pipettes, and various stock solutions (Table 31.3 and Procedure 31.2). The plethora of units used to measure concentration may be confusing when first encountered, so a description of the most common units and what they mean is given below. Once familiar with these, you can confidently proceed to the section on making stock solutions.

TABLE 31.3
Macro- and Micronutrient 100× Stock Solutions for MS Medium (1962)[a]

Stock (100×)	Component	Amount
Nitrate	NH_4NO_3	165.0 g
	KNO_3	190.0 g
Sulfate	$MgSO_4{\cdot}7H_2O$	37.0 g
	$MnSO_4{\cdot}H_2O$	1.7 g
	$ZnSO_4{\cdot}7H_2O$	0.86 g
	$CuSO_4{\cdot}5H_2O$	2.50 mg[b]
Halide	$CaCl_2{\cdot}2H_2O$	44.0 g
	KI	83.0 mg
	$CoCl_2{\cdot}6H_2O$	5.0 mg[c]
PBMo	KH_2PO_4	17.0 g
	H_3BO_3	620.0 mg
	Na_2MoO_4	25.0 mg
NaFeEDTA[d]	$FeSO_4{\cdot}7H_2O$	2.78 g
	NaEDTA	3.74 g

[a] The number of grams or milligrams indicated in the "Amount" column should be added to 1000 mL of deionized distilled water to make 1 L of the appropriate stock solution. For each liter of the medium made, 10 mL of each stock solution will be used. To make 100× stock solutions for any of the media listed in Table 31.1, multiply the amount of chemical listed in the table by 100 and dissolve in 1 L of deionized, distilled water.

[b] Because this amount is too small to weigh conveniently, dissolve 25 mg of $CuSO_4{\cdot}5H_2O$ in 100 mL of deionized distilled water, then add 10 mL of this solution to the sulfate stock.

[c] Because this amount is too small to weigh conveniently, dissolve 25 mg of $CoCl_2{\cdot}6H_2O$ in 100 mL of deionized distilled water, then add 10 mL of this solution to the halide stock.

[d] Mix the $FeSO_4{\cdot}7H_2O$ and NaEDTA together and heat gently until the solution becomes orange. Store in an amber bottle or protect from light.

Procedure 31.2
Making Stock Solutions for Vitamins and PGRs

Step	Instructions and Comments
1	To make a stock solution for nicotinic acid, look at how much is required for 1 L of medium. In this example, we are going to assume that we are making MS medium, so we will need 0.5 mg.
2	It would be convenient to be able to add 0.5 mg of nicotinic acid by adding a volume of the nicotinic acid stock that corresponded to 1 mL. If 0.5 mg of nicotinic acid must be in 1 mL of the stock solution, then (multiplying by 100) use 50 mg for the 100 mL of stock solution. One mL may be dispensed into 1.5 mL Eppendorf tubes and frozen (–20°C).
3	Now, prepare a PGR stock—IBA for a rooting medium. If 1.0 mg/L of IBA is require for the rooting medium, then our solution must contain 1 mg in each milliliter of the IBA stock. We must weigh 100 mg of IBA for the 100 mL of IBA stock solution. IBA is not very soluble in water, so first, dissolve it in a small amount (~1 mL) of a solvent such as 95% ethanol, 100% propanol, or 1 N KOH. Swirl it to dissolve and then slowly add the remainder of the water to a final volume of 100 mL. Label the stock solution and add 1 mL for every liter of medium to be made. Store in a brown bottle in the refrigerator (4°C). Note: Many plant growth regulators need special handling to dissolve. IAA; IBA; NAA; 2,4-D; benzyladenine; 2-iP; and zeatin can be dissolved in either 95% ethanol, 100% propanol, or 1 N KOH. Kinetin and ABA are best dissolved in 1 N KOH, but thidiazuron will not dissolve in either alcohol or base, so a small amount of dimethylsulfoxide (DMSO) must be used. By using only 1 mL of each stock, only very small amounts of the solvent are added to the medium and will minimize toxic effects.
4	Now, let us make a stock solution of IBA, but this time, we want to have a 5 μM solution of IBA in the medium. Look up the molecular weight of IBA in Table 31.2 and find that it is 203.2. Using the same rationale in the two examples above, we wish to have 5 μM delivered to the medium by using 1 mL of IBA stock solution. So, to make 100 mL of stock solution, you must have 500 μM or 0.5 mM in the 100 mL of stock solution. If 1 M is 203.2 g and 1 mM is 203.2 mg, so 0.5 mM IBA × 203.2 mg/1 mM = 101.6 mg of IBA is needed for 100 mL of IBA stock solution. To prepare the solution, weigh out the IBA, dissolve it in 1 N KOH as described above, label the stock solution, and store in the refrigerator (4°C) in a brown bottle. Add 1 mL of the IBA stock solution to deliver 5 μM for every liter of the medium.

UNITS OF CONCENTRATION CLARIFIED

Concentrations of any substance can be given in several ways. The following list gives some of the methods of indicating the concentration commonly found in literature on tissue culture:

- Percentage based on volume % (v/v): Used for coconut milk, tomato juice, and orange juice. For example, if 100 mL of a 5% (v/v) coconut milk was desired, 5 mL of coconut milk would be diluted to 100 mL with water.
- Percentage based on weight % (w/v): This is often used to express concentrations of agar or sugar. For example, to make a 1% (w/v) agar solution, dissolve 10 g of agar in 1 L of nutrient medium.
- Molar solution: A mole has the same number of grams as molecular weight (Avogadro's number of molecules), so a 1 molar solution represents 1 M of the substance in 1 L of solution, and 0.01 M represents 0.01 times the molecular weight in 1 L. A millimolar (mM) solution is 0.001 M/L and a micromolar (μM) solution is 0.000001 M/L. Substances like PGRs are active at micromolar concentrations. Molar concentration is used to accurately compare relative reactivity among different compounds. For example, a 1 μM concentration of IAA would contain the same number of molecules as a 1 μM concentration of kinetin, although the same could not be said for units based on weight.
- Milligrams per liter (mg/L): Although not an accurate means of comparing substances molecularly, this is simpler to calculate and use since it is a direct weight. Such direct measurement is commonly used for macronutrients and, sometimes, with PGRs. One milligram per liter means placing 1 mg of the desired substance in

a final volume of 1 L of solution. One milligram is 10^{-3} g.

- Microgram per liter (μg/L): This is used with micronutrients and also, sometimes, with growth regulators. It means placing 1 μg of substance in 1 L of solution. 1 μg = 0.001 or 10^{-3} mg = 0.000001 or 10^{-6} g.
- Parts per million (ppm): Sometimes, media components are expressed in parts per million, which means 1 part per million or 1 mg/L.

Instructions for making media can be found in Procedure 31.3. These instructions describe MS medium preparation, but will work just as effectively for any other media that you choose. Merely follow the same steps and substitute the macro- and micronutrient stocks that you have made for the desired medium. Omit the agar to produce a liquid medium for use in suspension culture. PGRs can be customized for the medium of your choice whether it is intended for the initiation of callus, shoots, or roots, or some other purpose.

Adjusting the pH of the medium is an essential step. Plant cells in culture prefer a slightly acidic pH, generally between 5.3 and 5.8. When pH values are lower than 4.5 or higher than 7.0, growth and development *in vitro* is generally greatly inhibited. This is probably due to several factors, including PGRs, such as IAA and gibberellic acid, becoming less stable; phosphate and ion salts precipitating; vitamin B_1 and pantothenic acid becoming less stable; a reduced uptake of ammonium ions; and changes in the consistency of agar (the agar becomes liquified at a lower pH). Adjusting the pH is the last step before adding and dissolving the agar and distributing it into the culture vessels and autoclaving. If the pH is not what it should be, it can be adjusted using KOH to raise the pH or HCl to lower the pH (0.1–1.0 N), depending on if the pH is too low or too high. While NaOH can be used, it can lead to an undesirable increase in sodium ions. The pH of a culture medium generally drops by 0.3 to 0.5 units after it is autoclaved and then changes throughout the period of culture both due to oxidation and the differential uptake and secretion of substances by growing tissue.

Procedure 31.3
Step-by-Step Media Making

Step	Instructions and Comments
1	If you are making 1 L of media, put about half the volume of deionized, distilled water into a beaker and add a stirring bar. Place the beaker on a stirrer so that, when you add different components, they will be thoroughly mixed.
2	Add 10 mL of the following stock solutions (details on how to make them are provided in Table 31.3): nitrate, sulfate, halide, PBMo, and NaFeEDTA.
3	Add the appropriate amount of the vitamin stocks. How to make the vitamin and the PGR stocks is found in Procedure 31.2. At this time, also add the appropriate amounts of inositol (usually 100 mg) and sucrose (usually 20–30 g/L).
4	Add the appropriate volume of the PGR stocks that you plan to use (details on how to make them is also found in Procedure 31.2).
5	Adjust the pH using 0.1 to 1.0 N NaOH or HCl, depending on how high or low it is. For most media, the pH should be 5.4 to 5.8, and adjust the final volume (1 L) with deionized, distilled water.
6	If you are going to use liquid medium, you can distribute the media into the tissue culture vessels that you plan to use and autoclave. If you plan to use solid medium, weigh out and add agar or other gelling agent to the liquid. If you are going to put it into tissue culture vessels (tubes or boxes), melt the agar by putting it on a heat plate while stirring or use a microwave to melt the agar. You may have to experiment with settings since microwaves differ in output. Once the media is distributed into the tissue culture vessels, these are capped and placed in the autoclave for sterilization. Autoclave for 15 min at 121°C. If a large quantity is being autoclaved, longer times may be necessary. If you plan to distribute the medium into tissue culture vessels after autoclaving (such as into sterile, disposable Petri dishes), cap the vessel with a cotton or foam plug and cover the plug with aluminum foil. When the medium is cool enough to handle (~60°C), it can be moved to the sterile transfer hood, uncapped, and poured into dishes.

STOCK SOLUTIONS OF MINERAL SALTS

Mineral salts can be prepared as stock solutions 10× to 1000× the concentration specified in the medium. Mineral salts are often grouped into two stock solutions, one for macroelements and one for microelements, but unless these are kept relatively dilute (10×), precipitation can occur. To produce more concentrated solutions, the preferred method is to group the compounds by the ions that they contain, such as nitrate, sulfate, halide, phosphorus, boron, molybdenum, and iron and make them up as 100× stocks. Table 31.3 lists the stock solutions for the MS medium made at 100× final concentration, which means that 10 mL of each stock is used to make 1 L of medium. Some of the stock solutions require extra steps to get the components into the solution (e.g., NaFeEDTA stock) or may require making a serial dilution to obtain the amount of a trace component for the stock (sulfate and halide stocks).

Sometimes, the amount of a particular component needed for a tissue culture medium is extremely small, so it is difficult to weigh out the amount even for the 100× stock solution. Because such small quantities of a substance cannot be weighed accurately, a serial dilution technique is used. The following example illustrates how serial dilutions can be used to obtain the correct amount of a component (in this case, $CuSO_4{\cdot}5H_2O$) of the medium for its appropriate stock solution.

The stock solution calls for 2.5 mg of $CuSO_4{\cdot}5H_2O$ as a part of the sulfate stock of the MS medium. Make an initial stock solution by placing 25 mg of $CuSO_4{\cdot}5H_2O$ in 100 mL of deionized/distilled water. After mixing thoroughly, use 10 mL of this solution, which will contain the desired 2.5 mg, and place it into the sulfate stock. This procedure requires only one serial dilution, but any component can be subjected to one or more dilutions to obtain the desired amount. Once a stock solution is made, it should be labeled as in Figure 31.2 to avoid error and inadvertently keeping a stock solution for too long.

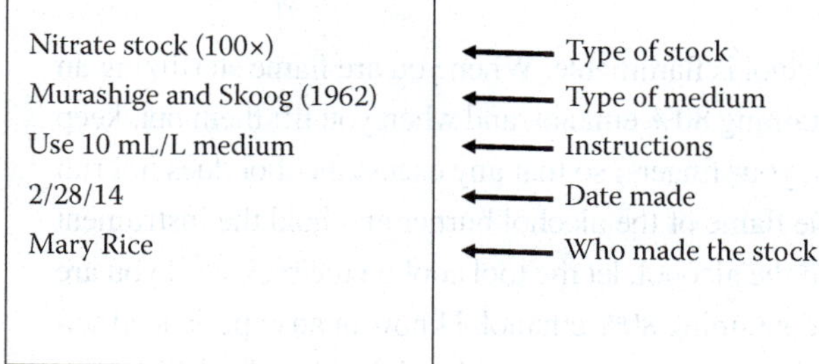

FIGURE 31.2 Information that should be placed on a stock solution label.

MAKING STOCK SOLUTIONS OF THE PGRS AND VITAMINS

Vitamins and PGR stock solutions can be made up in concentrations 100× to 1000× of that required in the medium. Determine the desired amount for 1 L of medium, the volume of stock solution needed to deliver that dosage of vitamin or PGR, and the volume of stock that you wish to make. Procedure 31.2 gives examples of how to make vitamin and PGR stock solutions. Many of the PGRs require special handling to get them into the solution. You will also find this information in Procedure 31.2.

STERILIZING EQUIPMENT AND MEDIA

Tissue culture media, in addition to providing an ideal medium for the growth of plant cells, are also an ideal substratum for the growth of bacteria and fungi. So, it is necessary to sterilize the media, culture vessels, tools, and instruments, and surface disinfest the explants as well. The most commonly used means of sterilizing equipment and media is by autoclaving at 121°C with a pressure of 15 psi for 15 min or longer for large volumes. Glassware and instruments are usually wrapped in heavy-duty aluminum foil or put in autoclave bags. Media (even those that contain agar) are in a liquid form in the autoclave, requiring a slow exhaust cycle to prevent the media from boiling over when pressure is reduced. Larger volumes of media require longer autoclave times. Media should be sterilized in tissue culture vessels with some kind of closure such as caps or plugs made of non-absorbent cotton covered by aluminum foil. This way, they do not become contaminated when they are removed from the autoclave for cooling. The use of racks that tilt the tissue culture tubes during cooling can give a slanted surface to the agar medium. These can be purchased or made from scratch using a little ingenuity.

Some components of the medium may be heat labile or altered by the heat so that they become inactive. These are usually added to the media after it has been autoclaved but before the media has solidified. It is filtered through a bacteria-proof membrane (0.22 μM) filter and added to the sterilized medium after it has cooled, enough not to harm the heat-labile compound (<60°C), and then thoroughly mixed before distributing it into the culture vessels. A rule of thumb is to add the filtered material at a point when the culture flask is just cool enough to be handled without burning one's hands. Some filters are available pre-sterilized and fit on the end of a syringe for volumes ranging from 1 to 200 mL. More elaborate disposable assemblies are also available.

PREPARING THE STERILE TRANSFER HOOD

Successfully transferring explants to the sterile tissue culture medium is done in a laminar-flow transfer hood. A transfer hood is equipped with positive-pressure ventilation and a bacteria-proof high-efficiency particulate air (HEPA) filter. Laminar-flow hoods come in two basic types. Generally, air is forced into the cabinet through a dust filter and a HEPA filter, and then it is directed either downward (vertical flow unit) or outward (horizontal flow unit) over the working surface at a uniform rate. The constant flow of bacteria- and fungal spore–free filtered air prevents unfiltered air and particles from settling on the working area that must be kept clean and disinfected. The simplest transfer cabinet is an enclosed plastic box or shield with an ultraviolet light and no airflow. A glove box can also be used, but both of these low-cost, low-technology options are not convenient for large numbers of transfers.

In the transfer hood, you should have ready some standard tools for use, such as a scalpel (with a sharp blade), long-handled forceps, and sometimes, a spatula. Occasionally, fine-pointed forceps, scissors, razor blades, or cork borers are needed for preparing explants.

Many people prefer doing sterile manipulations on the surface of a pre-sterilized disposable Petri dish. Other

Procedure 31.4

Getting under the Hood

Step	Instructions and Comments
1	Turn on the transfer hood so that positive air pressure is maintained. This ensures that all of the air passing over the work surface is sterile. You should feel air flowing against your face at the opening. Make sure that there are no drafts such as open windows or air-conditioning vents that may interfere with the airflow coming out of the hood.
2	Use a spray bottle filled with 70% (v/v) ethanol and spray down to the interior of the hood. Do not spray the HEPA filter! This is more effective than absolute alcohol for sterilizing surfaces perhaps because 70% (v/v) ethanol denatures DNA. You can also use a piece of cheesecloth saturated with 70% (v/v) ethanol to help distribute the ethanol more uniformly. Allow it to dry. To maintain the cleanliness of the interior, anything that is now placed inside the hood must be sprayed with 70% (v/v). This includes the alcohol lamp, the slide-staining jar filled with 80% (v/v) ethanol for flaming the instruments, a rack for the tools, racks of the tissue culture vessels containing medium, and all of the pre-sterilized wrapped bundles containing your working surface (tiles, paper towels, ashtrays, and others) and your tools (forceps, spatula, scalpel, and others).
3	When you are ready to begin, remove your jewelry and wash your hands thoroughly with soap and water, and then just before placing them in the hood, spray them down with 70% ethanol.
4	You may now open your work surface by peeling back the heavy-duty aluminum foil exposing the surface of the tile (or other alternative). Never block the airflow across the surface coming from the filter unit. Also, do not pass your hands across the surface of the tile. Talking while you are in the unit also compromises sterility. If you must talk, turn your head to one side. If you have long hair, tie it so that it does not dangle onto your work surface.
5	Keep any open sterile containers as far back in the hood as you can. When you open containers that have been sterilized, keep your fingers away from the opening. If you open a glass container or vessel, as a general rule, pass the opening through the flame. This creates a warm updraft from the vessel, helping to prevent contamination from entering it.
6	Flaming instruments can be hazardous if you forget that ethanol is flammable. When you are flame sterilizing an instrument, for example, forceps, dip them into the jar containing 80% ethanol, and when you lift them out, keep the tip of the instrument at an angle downward (away from your fingers) so that any excess alcohol does not run onto your hand (Figure 31.1). Then, pass the tip through the flame of the alcohol burner and hold the instrument parallel to the work surface. When the flame has consumed the alcohol, let the tool cool on the rack until you are ready to use it. Never ever place a hot tool back into the jar containing 80% ethanol. I know of an experienced scientist who momentarily forgot this simple rule and the resulting fire singed his hand and the hair off his forearm.
7	Alternatively, if you wish to avoid open flame, you can use a Bacti-Cinerator or a glass bead sterilizer. Both of these sterilize utensils using heat, so you must allow the utensils to cool down before using them on the cultured plant material.

alternatives that have worked well are stacks of standard laboratory-grade paper towels. These can be wrapped in aluminum foil and autoclaved. When the top sheet of the stack is used, it can be peeled off and discarded, leaving the clean one beneath it exposed to act as the next working surface. Another alternative is to use ceramic tiles. These can also be wrapped in aluminum foil and autoclaved, but tiles with very slick or very rough surfaces should be avoided.

In Procedure 31.4 is a suggested protocol to follow for preparing the sterile transfer hood. It also contains tips for keeping your work surface clean, eliminating contamination and avoiding burns when flaming your instruments.

STORAGE OF CULTURE MEDIA

Once culture media has been made, distributed into tissue culture vessels, and sterilized by autoclaving, it can be stored for up to 1 month, provided that it is kept sealed to prevent the excessive evaporation of water from the medium. It should also be placed in a dark, cool place to minimize the degradation of light labile components such as IAA. Storage at 4°C prolongs the time that the media can be stored, but condensate may form in the container and encourage contamination. By making media 5 to 7 days in advance, you allow time to check for any unwanted microbial contamination before explants are transferred onto the medium. However, media for certain sensitive species or operations or that contains particularly unstable ingredients must be used fresh and cannot be stored.

SURFACE DISINFESTING PLANT TISSUES

Just as the media, instruments, and tools must be sterilized, so must the plant tissue be disinfested before it is placed on a culture medium. Many different materials have been used to surface disinfest explants, but the most commonly used are 0.5 to 1% (v/v) sodium hypochlorite, 70% alcohol, or 10% hydrogen peroxide. Ultra Clorox® Germicidal Bleach Contains 6.15% hypochlorite. Check the active ingredient concentration before using any commercial bleach source. Others include using a 7% saturated solution of calcium hypochlorite, 1% solution of bromine water, 0.2% mercuric chloride solution, and 1% silver nitrate solution. If these more rigorous techniques are used, precautions should be taken to minimize health and safety risks, especially with the heavy metal–containing solutions.

The type of disinfectant used, the concentration, and the amount of exposure time (1–30 min) vary depending on the sensitivity of the tissue and how difficult it is to disinfest. Woody or field-grown plants are sometimes very difficult to disinfest and may benefit from being placed in a beaker with cheesecloth over the top and being placed under running water overnight. In some cases, employing a two-step protocol (70% ethanol followed by bleach) or adding a wetting agent such as Tween 20 or detergent helps increase the effectiveness. In any case, the final step before trimming the explant and placing it onto sterile medium is to rinse it several times in sterile, distilled water to eliminate the residue of the disinfesting agent. Laboratory exercises presented throughout the book will contain detailed directions for various procedures to disinfest tissue.

FINAL WORDS

Tissue culture is much like good cooking. There are simple recipes and then there is "haute cuisine." Cooking is also a very rewarding activity particularly when the end result is delicious. By following the procedures outlined in this chapter and in the succeeding chapters of the book, you should find that, with care and attention to detail, you will be a "chef extraordinaire," and your tissue culture ventures will be successful.

LITERATURE CITED

Ammirato, P. V. 1974. The effects of abscisic acid on the development of somatic embryos from cells of caraway (*Carum carvi* L.). *Bot Gaz* 135:328–337.

Coffin, R., C. D. Taper, and C. Chong. 1976. Sorbitol and sucrose as carbon source for callus culture of some species of the Rosaceae. *Can J Bot* 54:547–551.

Epstein, E., K.-H. Chen, and J. D. Cohen. 1989. Identification of indole-3-butyric acid as an endogenous constituent of maize kernels and leaves. *Plant Growth Regul* 8:215–223.

Gamborg, O. L., R. A. Miller, and K. Ojima. 1968. Nutrient requirements of suspension cultures of soybean root cells. *Exp Cell Res* 50:151–158.

Linsmaier, E. M. and F. Skoog. 1965. Organic growth factors. Requirements of Tobacco Tissue cultures. *Physiol Plant* 18:100–127.

Lloyd, G. and B. McCown. 1980. Commercially feasible micropropagation of mountain laurel, *Kalmia latifolia*, by use of shoot tip culture. *Intl Plant Prop Soc Proc* 30:421–427.

Murashige, T. and F. Skoog. 1962. A revised medium for rapid growth and bioassays with tobacco tissue cultures. *Physiol Plant* 15:473–497.

Nissen, S. J. and E. G. Sutter. 1990. Stability of IAA and IBA in nutrient medium to several tissue culture procedures. *HortScience* 25:800–802.

Nitsch, J. P. and C. Nitsch. 1969. Haploid plants from pollen grains. *Science* 163:85–87.

Oka, S. and K. Ohyama. 1982. Sugar utilization in mulberry (*Morus alba* L.) bud culture. In A. Fujiwara, ed. *Plant Tissue Culture*. Tokyo, Japan: Japan Association of Plant Tissue Culture, pp. 67–68.

Roberts, D. R., B. S. Flinn, D. T. Webb, F. B. Webster, and B. C. S. Sulton. 1990. Abscisic acid and indole-3-butyric acid regulation of maturation and accumulation of storage proteins in somatic embryos of interior spruce. *Physiol Plant* 78:355–360.

Schenk, R. V. and A. C. Hildebrandt. 1972. Medium and techniques for induction and growth of monocotyledonous plant cell cultures. *Can J Bot* 50:199–204.

Singha, S. 1982. Influence of agar concentration on *in vitro* shoot proliferation of *Malus* sp. "Almey" and *Pyrus communis* "Seckel." *J Amer Soc Hort Sci* 107:657–660.

Skoog, F. and C. O. Miller. 1957. Chemical regulation of growth and organ formation in plant tissues cultured *in vitro*. *Symp Soc Exp Biol* 11:118–131.

White, P. R. 1963. *The Cultivation of Plant and Animal Cells*, 2nd Edition. New York: Ronald Press Co., 228 pp.

32 Micropropagation of Mint

Sherry L. Kitto

Mint oils, from both peppermint (*Mentha* × *piperita* L.) and spearmints (*M. spicata* L. and *M.* × *gracilis Sole*), are important flavoring agents for pharmaceuticals, foods, and cosmetics. In 2011, peppermint oil, harvested from roughly 74,000 acres, was valued in the United States at more than $151 million. Infestations with pests such as weeds, insects, or pathogens greatly impact yields of this vegetatively propagated and, therefore, clonal crop. While there is a great deal of interest in improving mints that would be tolerant of such pests, crop improvement through breeding is challenging mostly due to sterility. However, improvement of commercial mint cultivars may be realized in the near future as a result of genetic engineering or transformation. The release of a new, improved mint cultivar for commercial distribution would require the development of a quick economical, vegetative propagation protocol.

The purpose of micropropagation is to produce the highest number of clonally identical, quality plants in the shortest time frame. The objective of this laboratory exercise is to examine some variables that impact the rapidity with which mint can be micropropagated. This exercise will examine the following classical stages of a micropropagation system: establishment of sterile cultures, proliferation of shoots, rooting of microcuttings, and acclimation/reestablishment of plants outside the laboratory environment. Each experiment of this multi-week laboratory is described in detail; however, feel free to experiment with the protocols to discover the "best" system for micropropagating mint under your physical and environmental conditions. Where applicable, the exercise introductions that follow will generally describe some variables that can be explored in each section; pick no more than four variables to examine prior to beginning each week's work. It is a good idea to decide on the table format to be used to report the results so that the correct data can be collected. If possible, use at least 10 explants for each treatment.

EXERCISES

Experiment 1. Establishment of Clean/Sterile Cultures

Sterile cultures can be established using seeds or vegetative material. Although commercial mints are not propagated using seeds, for the purposes of this exercise, purchasing seeds may be the easiest way to obtain mint germplasm. The contact information for sellers of mint seeds can be found on the Internet. Vegetative material may be obtained from a friend's yard or, commonly, in the United States, mint shoots can be found in the produce section of a local grocery store. If you decide to use vegetative material, be sure to select healthy, actively growing, above-ground shoots; actively growing plant material is usually less contaminated by wind-borne pests. If you are unable to find actively growing shoots, a rootstock can be dug, potted, and placed in a greenhouse or on a window ledge to encourage new shoot growth for this exercise. There are advantages in using each initial explant: seeds are more resilient to the surface sterilization protocols and, therefore, are easier to surface sterilize and establish as sterile cultures; however, starting with vegetative material more accurately mimics what would actually occur in a commercial tissue culture facility working with a new cultivar of mint.

Aseptic cultures need to be established for use in the proliferation phase of the laboratory exercise. Protocols on seed germination (Procedure 32.1) and surface disinfestation of the vegetative shoots (Procedure 32.2) can be started on the same day. Within a couple of weeks, seedling shoots will be available for subculture, and the initial identification of clean vegetative cultures should be possible. Keep in mind that there might be "clean" cultures that are false negatives; be sure to carefully observe the cultures for contamination. Holding cultures up to the light and looking through the bottom of the container and the gelled medium helps visualize some bacterial contaminants. Try to avoid turning the cultures upside down. Contaminated cultures should be discarded (disposable containers) or autoclaved (reusable containers); do not open contaminated cultures.

MATERIALS

The following materials will be required for the experiment:

- Seeds and/or vegetative shoots of mint
- Sterile glass Petri dishes containing moist filter paper
- Culture tubes (25 × 150 mm) containing sterile water

Procedure 32.1
Seed Germination

Step	Instructions and Comments
1	Scarify 20 seeds using fine-grit sandpaper.
2	Place 20 scarified and 20 unscarified seeds in separate 5 × 5-cm (2 × 2-in) squares of muslin or cheesecloth (place paper-and-pencil treatment identification labels within each group of seeds). Secure with a rubber band and surface disinfest by dipping in 70% ethanol for 1 min, followed by 20% bleach + 0.1% Tween 20 solution for 20 min. Rinse three times with sterile water, 10 min each.
3	Culture five seeds on presterilized, moist filter paper in glass Petri dishes. It is important that the filter paper be moist. Seal the dishes with Parafilm™ or plastic wrap and label with the date and your name, and the plant ID, that is, *Genus species* = ____________.
4	Place seeds in the dark for germination. Seeds should germinate in a week or two. Examine seeds using a microscope for obvious contamination. Do not open dishes that contain contamination.
5	Keep clear records on the number of seeds germinated, the number of contaminated seeds, and the type of contamination (i.e., bacterium or fungus). Use Table 32.1 as a model for reporting results. Using the results presented in Table 32.1, write a brief summary of this exercise.
6	Go to the protocol for the establishment of proliferating shoot cultures (Procedure 32.3) to continue.

Procedure 32.2
Surface Disinfestation of Vegetative Shoots

Step	Instructions and Comments
1	Collect shoots that visually appear to be free from contaminants. Remove larger leaves from stems and cut 30 nodal/tip explants that are approximately 2.5 to 40 cm (1–1.5 in).
2	Leave enough plant material, petioles, and stem so that the bleach-killed edges can be trimmed later.
3	Surface disinfest 1/2 of mint shoots for 20 min in 20% bleach + 0.1% Tween 20; rinse three times with sterile water, 10 min each.
4	Trim the bleach-damaged ends off the cuttings and culture on the establishment medium. Be sure to label the containers with the treatment, date, your name, and the plant ID, that is, *Genus species* = ____________.
5	Surface disinfest 1/2 of mint shoots for 10 min in 10% bleach + 0.1% Tween 20 solution; rinse three times with sterile water, 10 min each.
6	Culture the shoots in liquid establishment medium and place on a shaker (100 rpm) overnight. Surface disinfest for 10 min by adding an equal volume of 20% bleach + 0.1% Tween 20 solution and shaking. Rinse three times with sterile water, 10 min each.
7	Treat cuttings as described in step 4.
8	Collect data weekly and, after 1 month, compare the two surface disinfestation protocols for the percentage of clean cultures. Count the number of shoots and the number of nodes for each culture/treatment. Use Table 32.1 as a model for reporting results. Using the results presented in Table 32.1, write a brief summary of this exercise.
9	Go to the protocol for the establishment of proliferating shoot cultures (Procedure 32.3) to continue.

TABLE 32.1
Disinfestation Protocols

Explant	Disinfestation Protocol	N[a]	Contamination (%)	Mean No. of Axillary Shoots[b]	Mean No. of Nodes/Shoots[b]	No. of Roots[b]
Seeds	20% bleach					
Vegetative	20% bleach					
Vegetative	10% bleach					

[a] Initial number cultured.
[b] Data based on the number of clean cultures.

- Establishment medium: Murashige and Skoog (1962) (MS) medium supplemented with 3% sucrose and 0.6% Phytoblend (25 mL per culture tube [25 × 150 mm])
- Liquid establishment medium: MS medium supplemented with 3% sucrose (10 mL per culture tube)
- Muslin or cheesecloth, rubber bands, labels, ethanol, commercial bleach, Tween 20 or some other surfactant, and Parafilm™ or plastic wrap cut into strips

Follow the protocols listed in Procedure 32.1 and Procedure 32.2 to complete the experiments.

ANTICIPATED RESULTS

Do not be surprised if many of the seeds do not germinate or if browned-out leaves appear on the vegetative shoots. As long as a shoot's growing point is green, you should be able to establish a viable culture. There is a fine line between "cleaning up" and killing vegetative plant tissue (bleach is phytotoxic), so some damage (browning) is to be expected. The amount of browning, resulting from the sensitivity of the vegetative tissues to the bleach, is dependent on how hardened the initial material is. Mints are fairly malleable, and successful establishment of clean, vigorous shoots should be fairly easy.

QUESTIONS

- What was the most common type (bacterium or fungus) of contamination?
- Why are seeds easier to clean up compared to vegetative shoots?
- Why do some cultures initially appear to be clean but later turn up contaminated?
- What are some alternative procedures that might be tried to establish clean cultures?

Experiment 2. Establishment of Proliferating Shoot Cultures

The objective of shoot proliferation is to maximize axillary shoot (Figure 32.1) production between successive subcultures. Aseptic, proliferating shoot cultures need to

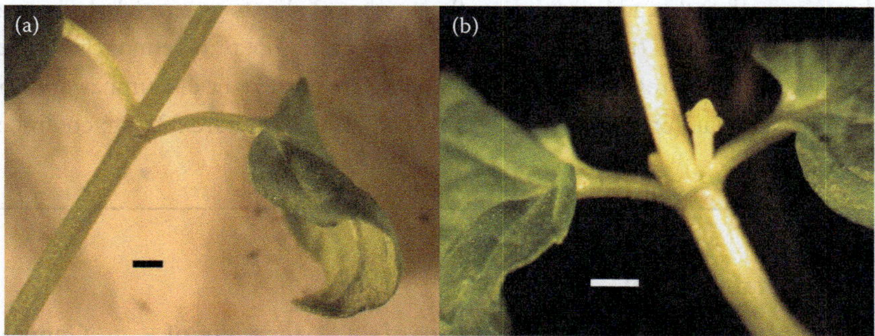

FIGURE 32.1 Leaf axils just after subculture (a) and after a couple of weeks (b); note the axillary shoots in the leaf axils on the right. (Bar = 1 mm.)

be established to generate shoots for use in the maintenance of proliferating cultures experiment.

MATERIALS

- Aseptically grown mint seedlings
- Aseptically initiated vegetative shoot cultures
- Shoot proliferation medium: MS medium supplemented with 3% sucrose, 5 μM benzyladenine (BA), and 0.6% Phytoblend (15–25 mL per 25 × 150-mm culture tube)

Follow the protocol listed in Procedure 32.3 to complete this phase of the experiment.

ANTICIPATED RESULTS

If you initiated shoot cultures from both seeds and vegetative plant material, you may notice that the seedling-derived shoot cultures are less vigorous. Remember that one reason why the vegetative material from the grocery store is being clonally maintained and cultivated is because it is vigorous. Response variation among explants that appeared to be initially identical is normal. A few cultures may become contaminated due to false negatives in the sterile-culture establishment stage.

QUESTIONS

- With respect to efficiency, what is the best size for an explant for shoot proliferation—one node or two nodes?
- Do seedling shoots proliferate differently than vegetative shoots?

Experiment 3. Maintenance and Periodic Subculture of Proliferating Cultures

Once aseptic cultures are established, shoot cultures must be maintained. It is important to develop a protocol to maximize the production of microcuttings or axillary shoots. Mint, having opposite leaves, has two axillary buds at each node; however, some plants can be massed up very effectively from single-node explants. In this experiment, it will be important to know how many axillary shoots are generated as well as the number of nodes present on each shoot. Experiments that can be attempted here include comparing proliferation from shoot-tip

Procedure 32.3
Establishment of Proliferating Shoot Cultures

Step	Instructions and Comments
1	Excise six shoots (three with one node and three with two nodes) each from the surface disinfestation cultures containing the seedlings and/or the vegetative shoots and culture on the shoot proliferation medium. Be sure to label the cultures with the explant source, date, your name, and plant ID, that is, *Genus species* = ________________.
2	Observe weekly and keep clear records for each explant source on the number of contaminated cultures and the type of contamination (i.e., bacterium or fungus).
3	At the end of 4 weeks, collect this additional data: the number of shoots generated, the presence of callus, the general appearance of the shoots (i.e., green, yellow, brown), and root production. Use Table 32.2 as a model for reporting results. Using the results presented in Table 32.2, write a brief summary of this exercise.
4	Go to the protocol for the maintenance of proliferating cultures (Procedure 32.4) to continue.

TABLE 32.2
Contamination and Establishment

Shoot Source	N[a]	Contamination (%) Bacterial/Fungal	Mean No. of Axillary Shoots[b]	Mean No. of Nodes/Shoots[b]	General Appearance[b]	No. of Roots[b]
Seedling						
Vegetative						

[a] Initial number of shoots cultured.
[b] Data based on the number of clean cultures.

cuttings, nodal cuttings (two axillary buds) (Figure 32.2), and split-node cuttings (one axillary bud) or determining if there is a gradient of proliferation based on node position on a shoot (how do nodal explants collected nearer the tip proliferate compared to nodal explants located nearer the base?). Orientation of the explants can also be explored; stem cuttings may be placed in the medium vertically, horizontally, or even upside down. Shoots that result from the establishment of proliferating shoot cultures can either be recultured to maintain shoot proliferation cultures or be rooted.

Methods: Select the treatments that you are interested in comparing. Set up at least 10 containers for each treatment, with one explant per container. After 4 weeks, using sterile technique, collect shoot and/or node production data for each shoot. Repeat the experiment once and collect the data. Present the numerical and observational data in table format (Table 32.3).

FIGURE 32.2 Shoot-tip cutting (left) and nodal cutting (right) may be used to maintain proliferation cultures or may be rooted. (Bar = 1 cm.)

TABLE 32.3
Data for Shoot Proliferation

Explant	Orientation	N[a]	Mean No. of Shoots	Mean No. of Nodes/Shoots	Callus (±)
Apical	Vertical				
	Horizontal				
Nodal	Vertical				
	Horizontal				

[a] Initial number of shoots cultured.

MATERIALS

The following items are needed for the experiment:

- Clean shoot cultures
- Shoot proliferation medium: MS medium supplemented with 3% sucrose, 5 μM BA, and 0.6% Phytoblend (15–25 mL per 25 × 150-mm culture tube)

Follow the protocol listed in Procedure 32.4 to complete the experiment.

ANTICIPATED RESULTS

Seedling-derived cultures may remain less vigorous after repeated subcultures. Shoot-tip cuttings may grow taller, while nodal cuttings may produce more axillary shoots. Commercially, the goal is to produce the highest number of explants, whether for subculture or rooting. Split-node explants may be less responsive due to damage while cutting (the sharper the blade, the more practiced the operator, and the quicker the cut all result in less damage.)

Procedure 32.4
Maintenance of Proliferating Cultures

Step	Instructions and Comments
1	Choose two treatments from those listed in the methods. Excise shoots from proliferating shoot cultures and culture 10 shoots per treatment.
2	Be sure to label the cultures with the explant treatment, date, your name, and plant ID, that is, *Genus species* = ________________.
3	After 4 weeks, count the number of shoots and nodes/shoots generated per original explant for each treatment; score for the presence of callus and the general appearance of the shoots. Use Table 32.3 as a model for reporting results. Using the results presented in Table 32.3, write a brief summary of this exercise.
4	Go to the protocol for rooting (Procedure 32.5) to continue.

QUESTIONS

- Why is it important to look for callus production?
- What are the differences between axillary and adventitious shoots? Is one type of shoot better than the other?
- Which treatment was the best for the production of shoots?
- Which treatment "looked" the best (greenest, least amount of browning or yellowing, least amount of callus, and others)?

Experiment 4. Rooting and Acclimatization

Mint cuttings root fairly rapidly and easily without the use of rooting hormones (Figure 32.3); therefore, the rooting environment, the use of mist and/or humidity domes, will be examined. It is critical that the microcuttings do not dry out at any point. They need to be kept in a high-humidity environment until they have had the chance to initiate roots and acclimate to the *in situ* environment. Microcuttings will be placed into a soilless greenhouse medium and under mist for 1 week and then transferred to the greenhouse bench. Remove the humidity domes prior to placement on the greenhouse bench. Microcuttings may be placed into individual cells or as a group into a larger container; be sure to label them correctly. Select a container and a soilless greenhouse medium that will allow for good drainage.

FIGURE 32.3 Microcutting that has rooted *in vitro* on growth regulator–free medium. (Bar = 1 cm.)

It is not necessary to plant a node in the rooting medium for mint microshoots to root. Place the microcuttings in the moist soilless greenhouse medium. Half of the containers will be placed in a flat with a humidity dome. The other half of the containers will be placed in a flat that is set under mist. You will need to work fairly quickly so that the microcuttings do not dry out; use the water bottle mister frequently.

MATERIALS

The following materials will be needed for this experiment:

- Microcuttings from proliferating shoot cultures
- Water bottle with mist attachment
- Pencil, scalpel, or sharp knife
- Greenhouse containers, flats, and humidity domes
- Greenhouse mist system
- Soilless greenhouse medium such as Promix BX or Redi-Earth®

Follow the protocol listed in Procedure 32.5 to complete the experiment.

ANTICIPATED RESULTS

Mint microcuttings are fairly easy to root and reestablish.

QUESTIONS

- How would you determine which protocols are cost effective?
- What is the challenge between guaranteeing clonal integrity and producing microshoots as fast as you can?
- If mint plants are planted in 50 to 75 cm (20 to 30 in) rows with plants 15 cm (4–6 in) apart in each row, how many plants will you need to plant in each hectare (acre)? How long will it take you, using your protocol, to produce the required number of plants required for an acre?
- If 10 stock cultures, each with 10 shoots, are maintained and some shoots have to be held back for the "mother" stock plants, how far in advance would you have to start production to guarantee spring delivery of 10,000 plants?

Experiment 5. Alternative Species

Now that you understand the basic protocols for micropropagating (establishing, maintaining, and reestablishing) a plant species, a species that can be used in place of mint is chrysanthemum (use the mint protocols as

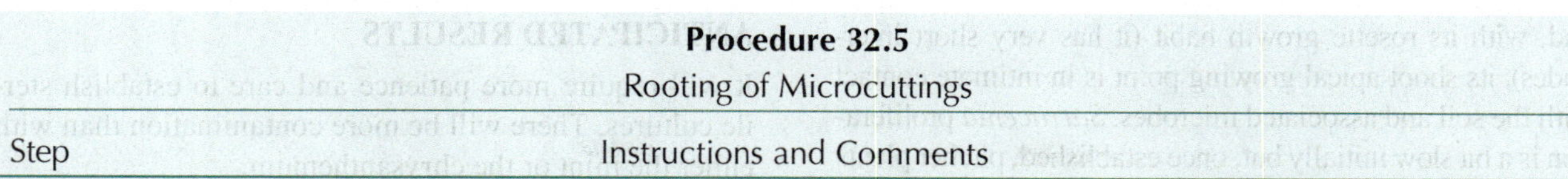

Procedure 32.5
Rooting of Microcuttings

Step	Instructions and Comments
1	In the laboratory, using forceps, grab the base of a proliferating shoot clump and remove it from the container. Working with one container at a time, wash the agar away from the shoot clump using warm tap water and gentle rubbing. Place the washed shoot clumps between a moist paper toweling. It is critical that the shoot clumps do not dry out. Repeat until all the shoot clumps have been cleaned.
2	Again, working with one clump at a time, use a scalpel to cut individual shoots away from a clump. Be careful not to cut yourself. Place these microcuttings between a moist paper toweling. It is helpful once you get to the greenhouse if you have oriented all of the cuttings in the same direction (e.g., all the cut ends together). It is critical that the microcuttings do not dry out. Collect 40 microcuttings.
3	Wrap the moist towel–wrapped microcuttings in plastic wrap or aluminum foil for travel to the greenhouse.
4	In the greenhouse, fill four containers with soilless greenhouse medium and thoroughly wet.
5	Using a small dibble—a pencil works fine—make a shallow hole. Remove one microcutting from the toweling (be sure to keep the remaining microcuttings covered with the moist toweling), place it in the hole, and using the dibble, carefully backfill with the medium and tamp around the microcutting. The microcutting should remain upright when stuck properly. Mist the microcutting.
6	Repeat step 5 until 10 of the microcuttings are stuck. Remember to mist all of the stuck microcuttings with water repeatedly until the flats have been placed under mist.
7	Repeat steps 5 and 6 for the other three containers. Label each container with the treatment (plus or minus dome), date, your name, and plant ID, that is, *Genus species* = ________________. Place each container in a flat. Mist the inside of the humidity domes and place over the two appropriate flats. Put all four flats, two with domes and two without domes, under mist.
8	At 2-week intervals, remove 1/2 of the cuttings from each container, and count and measure the length of the roots. Keep a record of survival.
9	Using the collected data, calculate for each 2-week interval the percentage of survival, the average number of roots, and the average length of roots. Use Table 32.4 as a model for reporting results. Using the results presented in Table 32.4, write a brief summary of this exercise.

TABLE 32.4
Data for Rooting and Acclimatization

		Week 2			Week 4		
Treatment	**N**[a]	**Survival (%)**	**Mean No. of Roots**[b]	**Mean Length of Roots**[b]	**Survival (%)**	**Mean No. of Roots**[b]	**Mean Length of Roots**[b]
+ Humidity dome							
– Humidity dome							

[a] Initial number of shoots stuck.
[b] Data based on five shoots.

described previously). Chrysanthemum shoots are very easy to surface disinfest and establish *in vitro*. As with mint, tip, nodal, or bisected nodes may be subcultured. Once in culture, tip cuttings will, without the addition of growth regulators, more than triple in length about every 4 weeks. Microcuttings will root quickly. After 1 week, cuttings can have roots longer than 2 cm (1 in).

Another fun species to work with *in vitro* is *Sarracenia*, commonly known as the "pitcher plant." It is a bit more challenging to establish because it grows in a boggy environment and, with its rosette growth habit (it has very short internodes), its shoot apical growing point is in intimate contact with the soil and associated microbes. *Sarracenia* proliferation is a bit slow initially but, once established, pitcher plants are very proliferative and easy to root.

MATERIALS

The following materials will be required for the experiment:

- *Sarracenia* plants
- Establishment/shoot proliferation media: 1/2 the salts of MS medium supplemented with 2% sucrose, 5-μM BA, and 0.6% Phytoblend (15–25 mL per 25 × 150-mm culture tube)

Follow the protocol listed in Procedure 32.6 to complete the experiment.

ANTICIPATED RESULTS

It will require more patience and care to establish sterile cultures. There will be more contamination than with either the mint or the chrysanthemum.

QUESTIONS

- What could be done in the greenhouse, prior to collection for surface sterilization, to improve your ability to establish clean cultures of pitcher plant?
- Since there are no insects for the pitchers to degrade, where is their energy derived from?
- What is the purpose of a humidity dome, especially considering that the microcuttings are placed under mist?
- What other plants would be interesting or economically sensible to micropropagate? Why?

Procedure 32.6

Establishment of *Sarracenia* spp.

Step	Instructions and Comments
1	To maximize establishing clean cultures, trim leaves to about 5 cm (2 in), place plants in a container, and place under slowly running cool tap water for 2 h.
2	Remove as much of the leaf as possible prior to surface disinfestation (Procedure 32.2, steps 5 and 6). Any culture that becomes contaminated can be re-surface disinfested (Procedure 32.2, steps 5 and 6).
3	Be sure to label the cultures with the date, your name, and plant ID, that is, *Genus species* = ____________ ____________.
4	Every 4 weeks, subculture the shoots to a fresh medium. Use Table 32.5 as a model for reporting results. Using the results presented in Table 32.5, write a brief summary of this exercise.
5	Follow the protocols listed in Procedures 32.2, 32.3, 32.4, and 32.5 to complete the experiment.

TABLE 32.5
Data for *Sarracenia* establishment

		Aseptic Culture Establishment		
Treatment	**N[a]**	**% Clean**	**Mean No. of Times Surface Disinfested**	**Time to New Growth**
Surface disinfestation				

[a] Initial number of shoots cultured

SUMMARY

Prepare a laboratory report in a scientific manuscript format, which includes the following sections: Abstract, Introduction, Materials and Methods, Results, and Discussion.

LITERATURE CITED AND SUGGESTED READING

Lee, J. 2001. Making a mint of money: Essential oils export success. *AgExporter* 13:20.

Murashige, T. and F. Skoog. 1962. A revised medium for rapid growth and bioassays with tobacco tissue cultures. *Physiol Plant* 15:473–497.

Northcutt, C., D. Davies, R. Gagliardo, K. Bucalo, R. O. Determann, J. M. Cruse-Sanders, and G. S. Pullman. 2012. Germination *in vitro*, micropropagation, and cryogenic storage for three rare pitcher plants: *Sarracenia oreophila* (Kearney) Wherry (federally endangered), *S. leucophylla* Raf., and *S. purpurea* spp. *venosa* (Raf.) Wherry. *HortScience* 47:74–80.

Tucker, A. O. 1992. The truth about mints. *The Herb Companion*. August/September, pp. 51–52.

Tucker, A. O. and S. L. Kitto. 2012. Mint. In Singh, R., ed. *Genetic Resources, Chromosome Engineering, and Crop Improvement: Medicinal Plants*, Vol. 6. Boca Raton, Florida: CRC Press, LLC, pp. 703–761.

USDA National Agricultural Statistics Service. 2012. http://www.nass.usda.gov. Accessed April 25, 2012.

van Eck, J. M. and S. L. Kitto. 1990. Callus initiation and regeneration in *Mentha*. *HortScience* 25:804–806.

Veronese, P., X. Li, X. Niu, S. C. Weller, R. A. Bressan, and P. M. Hasegawa. 2001. Bioengineering mint crop improvement. *Plant Cell Tiss Org Cult* 64:133–144.

33 Micropropagation of Tropical Root and Tuber Crops

Leopold M. Nyochembeng

Concept Box 33.1

- A tuberous root is an enlarged, fleshy secondary root containing carbohydrate reserves (often starch).
- Examples of tropical root and tuber crops are cassava, yam, cocoyam, potato, and sweet potato.
- Cassava is a woody perennial shrub with palmate leaves, which bears multiple nodes containing dormant buds on the stem and branches and produces an edible root.
- A tuber is a swollen subapical portion of a modified stem that develops underground as a storage organ. Tubers contain large amounts of starch reserves serving as the major sink for the products of photosynthesis.
- Edible tubers such as yam, cocoyam, and potato are important sources of carbohydrates, vitamins, and minerals.
- Yam is a tuber-producing annual or perennial climbing monocotyledonous vine. It produces a stem tuber. Yams are a different species than sweet potatoes, and yam tubers have white flesh.
- Cocoyam is an Aroid that produces underground starchy, edible tubers known as "cormels" from a swollen cormous stem base.
- A yam minisett is a piece of yam derived from the main tuber and weighs 30 to 40 g, which is used as a "seed" for yam production.
- Microtuberization is the development of miniature tuberous propagules from shoots in tissue culture.

Root and tubers crops are essentially geophytic plants whose survival rests on the development of fleshy subterranean organs with or without dormant buds. These food storage organs allow the plant to survive during adverse environmental conditions and support renewed growth when environmental conditions become favorable again. For these important food crops adapted to the wet/dry cycle of tropical and subtropical regions (cassava, yam, cocoyam, sweet potato, potato, and taro) and the warm/cold cycle of temperate climates (sweet potato and potato), survival means stashing away large amounts of carbohydrates and water during their annual growth cycle. These root and tuber crops are tropical perennials, cultivated mainly as annuals. The subterranean environment is ideal for storing carbohydrates because its moisture content is relatively stable and temperature fluctuations are not as extreme as in the above-ground environment. After attaining physiological maturity, the tuberous structures undergo variable durations of dormancy prior to sprouting.

Botanically, a tuber is a subapical swelling of a genetically programmed modification of a stem that serves as a subterranean storage structure. Examples of tubers include yam, cocoyam, taro, and potato. Tuberous roots are enlarged swellings of fleshy secondary roots that contain significant carbohydrate reserves, such as cassava and sweet potato. A distinguishing feature of tubers and tuberous roots is that their storage organs are generally bulky due to high starch and moisture content. They also store poorly after harvest if not properly conditioned; thus, some form of postharvest processing is required to increase their shelf life. Root and tuber crops are conventionally vegetatively propagated either by stem cuttings (cassava and sweet potato), whole root or root fragments (sweet potato), and tubers (yam, cocoyam, taro, and potato). This method of propagation offers some advantages to root and tuber crops in that it makes them easy and convenient to propagate and also helps maintain clonal fidelity in the propagated populations with little or no genetic variability. Seed propagation is rare for several reasons, including unpredictable flowering, poor seed set, genetic heterozygosity, and variable germination rate and vigor.

Root and tuber crops are important food resources, serving as staple food crops in the humid regions of many developing countries. The foliage may also be used for food or feed. Many roots and tubers serve as sources of important raw materials for various industries in developed economies. Cassava starch, for example, is useful in the paper and textile industries. It is also used for alcohol production in South America, and in Africa, cassava flour is an important substitute for wheat flour.

CASSAVA

Cassava (*Manihot esculenta* Crantz) is a perennial woody shrub (Figure 33.1) with an edible root, which grows in tropical and subtropical areas of the world. It is also known in other regions as "yuca," "manioc," "manihot," "tapioca," and "mandioca." A unique feature of cassava is its good drought tolerance and the ability to thrive in soils with low fertility. Cassava roots (Figure 33.2) can be stored in the ground for up to 2 years or more depending on the cultivar, which provides much flexibility in harvesting the tuberous roots and cuttings for propagation.

Cassava Plant Propagation

Cassava is propagated conventionally from hardwood stem cuttings consisting of two to six nodes (10–20 cm long), depending on the portion of the stem and the cultivar used. Although cassava plant produces seed, seed propagation is not usually used. A propagation technique that employs ministems was developed at the International Institute of Tropical Agriculture (IITA), Nigeria. This technique involves cutting healthy cassava stems into several short pieces, each with at least one node. If cuttings are taken from hardwood stems, one to two nodes are used; from semi-hardwood, four to six nodes; and from the soft terminal portion of shoots, six to ten nodes. Cuttings are then dipped into a fungicide suspension and then placed loosely in perforated transparent polyethylene bags under shade. The cuttings sprout 3 to 5 days later and are then transplanted into the field. Ministem propagation increases the in-field multiplication ratio of cassava several-fold compared to the traditional hardwood cutting propagation.

FIGURE 33.1 Cassava plant growing from a potted stem cutting. Note the prominent nodes.

FIGURE 33.2 Fresh cassava roots harvested at maturity. (Courtesy of Dr. K. N. Konan, Alabama A&M University, AL.)

In Vitro Propagation

In vitro procedures have been applied extensively to cassava for rapid micropropagation, meristem culture for the generation of pathogen-free plants for international germplasm distribution, embryo culture, non-zygotic (somatic) embryogenesis, and plant transformation for the development of transgenic cassava. Several kinds of explants and regeneration systems exist for micropropagating cassava. Most commonly used systems include apical meristems in meristem-tip culture, somatic embryogenesis from cotyledonary explants, organogenesis and shoot production from stem callus, and rapid micropropagation via multiple shoot production from nodal explants and axillary bud–derived meristems.

Meristem Culture

Shoot meristem culture involves excising shoot tips consisting of the meristematic dome plus two or more subtending leaf primordia and culturing these under well-defined *in vitro* conditions. Introduced over 3 decades ago, meristem culture has been used to eliminate several viral pathogens in important crop plants. For example, African mosaic virus–free cassava has been obtained via meristem culture (Chapter 14). The technique gained prominence in micropropagation with the observation that plants regenerated via meristem culture maintained clonal fidelity in addition to eliminating viral disease symptoms that were present in parental clones.

The following two experiments will demonstrate cassava shoot meristem culture and rapid shoot multiplication in cassava using axillary shoot buds.

Experiment 1. Shoot Meristem Culture

MATERIALS

The following materials will be needed for each team of students:

- Scissors
- Liquid chlorine bleach (containing 5.25% NaClO)
- Whatman #1 filter paper
- Stereo dissecting microscope
- Scalpel and forceps
- Growth chamber
- pH meter
- Stir plate and stir bar
- Balance
- Laminar-airflow bench
- Murashige and Skoog (1962) (MS) basal salt medium plus B5 vitamins (Gamborg *et al.*, 1968) and glassware

Follow the protocols listed in Procedure 33.1 to complete the experiment.

ANTICIPATED RESULTS

The growth and morphogenic response of cassava meristems in culture is influenced by the growth factors in the medium as well as the cultivar. Some combinations of growth regulators should work better than others for inducing callus, shoots, and roots.

QUESTIONS

- Why is meristem culture used for the production of disease-free plants for germplasm distribution? Is meristem culture a viable option for

Procedure 33.1
Shoot Meristem Culture of Cassava

Step	Instructions and Comments
1	Prepare two to three node cuttings per group, place in moist 1 sphagnum peat:1 perlite by volume or sawdust in the greenhouse to force shoots. Shoots will be ready in about 2 weeks.
2	Detach (24 per group) shoot tips about 2 cm long, place in a beaker, and rinse under running tap water for 1 h, followed by two rinses in distilled water.
3	Within the sterile transfer hood, disinfest with 100 mL 10% liquid chlorine bleach (v/v; 0.525% NaClO) containing 1 drop of Tween 20 (a wetting agent) for 15 min. Rinse four times in sterile distilled water.
4	Place shoot tips under a dissecting microscope, and using a sterile scalpel and forceps, dissect meristems approximately 0.5 to 1 mm long at a magnification of 4×. Place the meristems in fresh Petri dishes containing a moist sterile Whatman #1 filter paper and cover. Students should work in groups of two to three and take turns in dissecting the meristem tips. Care should be taken to minimize the time that meristems are exposed on the dissecting scope to avoid dehydration and microbial contamination.
5	Transfer the meristem tips onto agar-solidified MS basal medium in 60 × 15-mm Petri dishes containing 100 mg/L inositol, 80 mg/L adenine, 0.58 μM GA_3, B_5 vitamins, and 0.7% agar, supplemented with BA (0, 0.44, and 1.32 μM) and NAA (0 and 0.05 μM). Culture two meristem tips per Petri dish, seal with Parafilm™, and label the dishes with your name, date, and the concentrations of the growth regulators used. Make four replications per growth regulator treatment combination.
6	Incubate cultures at 26°C to 28°C in a 16 h photoperiod at 90 $\mu mol{\cdot}m^{-2}{\cdot}sec^{-1}$ and observe weekly for up to 6 weeks. Record your observations on viability, callus formation, root formation, the number and length of roots, shoot development, and the number and length of shoots.
7	Analyze your data using analysis of variance (ANOVA), and compare treatment means where necessary using Tukey's Honestly Significant Difference (HSD) at 5% probability (Chapter 16).

the production of virus-free plants from virus-infected ones?

- What factors should be considered in the use of meristem culture as a tool for micropropagation?

Experiment 2. Rapid Multiplication from Cassava Axillary Shoot Buds

MATERIALS

The following materials will be needed for each team of students:

- Tissue cultured cassava shoots
- Scalpel and forceps
- Tissue culture glassware
- Growth chamber, tissue culture room, or laboratory bench equipped with lights
- pH meter
- Stir plate and stir bar
- Balance
- Laminar-airflow bench
- Magenta GA7 containers or baby food jars
- Ziploc® bags
- Marker pens and plastic labels
- Vermiculite
- Perlite
- Plastic pots (10 × 10 cm)
- MS basal salt medium

Follow the instructions outlined in Procedure 33.2 to complete this experiment.

ANTICIPATED RESULTS

Shoot and plantlet development will depend on the morphogenic potential of the cassava cultivar used and its response to Benzyladenine (BA) treatment. Rapid shoot multiplication can be achieved from shoot- or axillary bud–derived explants of cassava; however, the rates of shoot multiplication may be quite variable with different cultivars or clones.

QUESTIONS

- How would you improve the efficiency of the *in vitro* rapid shoot multiplication of cassava?

Procedure 33.2
Rapid Multiplication from Cassava Axillary Shoot Buds

Step	Instructions and Comments
1	Prepare single-node explants from the shoots and plantlets of preexisting axenic tissue cultures of cassava under aseptic conditions. Axenic cultures are advantageous for use since additional disinfestation is not required.
2	Place two explants in MS medium containing 0.7% agar and 100 mg/L inositol, supplemented with 0, 22.2, 44.4, and 66.6 μM BA in Magenta GA7 containers. For each treatment, make 10 replications.
3	Incubate cultures at 26 to 28°C in a 16-h photoperiod at 90 $\mu mol \cdot m^{-2} \cdot sec^{-1}$ for 5 days.
4	Excise axillary shoot buds from each stock culture, reculture in the same formulation, but with fresh medium (first subculture), and incubate as above. Monitor shoot development in each treatment.
5	Transfer axillary bud–derived regenerated shoots (second subculture) onto fresh MS basal medium containing 3% sucrose and incubate as above to root and develop as plantlets.
6	Remove well-rooted plantlets, and wash the agar off the roots with distilled water. **Caution**: *In vitro*–derived roots of cassava plantlets are very brittle so care must be exercised in rinsing the agar off the roots. Place the plantlets in pots (10 × 10 cm) containing a mixture of 1 vermiculite:1 perlite (by volume) and cover the newly potted plantlet with an inverted Ziploc® bag to retain high levels of humidity. Systematically widen the opening of the bag (once every 5 days) until the cover is completely removed in about 2 weeks. This process is called "acclimatization" and prepares the plantlet for survival in the harsher *ex vitro* environment. The Ziploc® bag is placed over the plantlets in the laboratory and then continued in the shaded greenhouse.
7	Record the number of shoots, roots, and plantlets produced per nodal explant for different concentrations of BA. Determine the survival rate for the acclimatized plantlets.
8	Analyze the BA concentration effect on shoot, root, and plantlet formation, including the number and lengths of roots and shoots formed using ANOVA. Share the results with other groups of students.

- What is the relative advantage in using axillary bud–derived meristems for the rapid multiplication of cassava compared to meristem tip culture?
- What characteristics does a tissue-cultured plantlet have that makes acclimatization necessary?

YAMS

Yams (*Dioscorea* spp.) are tuber-producing annual or perennial climbing monocotyledonous plants that should not be confused with sweet potato (*Ipomoea batatas* [L.] Lam.). The stem is a vine, which may bear aerial tubers (tubercles) in some species (*D. bulbifera* L.). In most edible species, the tubers (Figure 33.3) are subterranean and can grow over 40 cm long and 15 cm wide. Over 95% of the world's edible yams (36 million tons) are cultivated in sub-Saharan Africa, within the region commonly known as "the yam belt," which stretches across West Africa. Other minor producers are found in the West Indies, South and Central America, and parts of Asia. Of the 600 yam species grown worldwide, the following three species are most popular: white yam (*D. rotundata* Poir), yellow yam (*D. cayenensis* Lam.), and water yam (*D. alata* L.), with white yam being the most predominant species. Yam is an important cash crop that competes with cassava as the preferred staple food crop and plays an important sociocultural role within some indigenous communities in West Africa.

Propagation of Yams

Traditionally, yams are propagated from the seed yams, which are made up of relatively smaller tubers (200–1000 g) selected from the current harvest for use in the next planting season. With this method, a significant portion of the harvest has to be set aside as planting material. In the late 1970s, a new technology was introduced, called the "minisett technique," for seed yam production.

FIGURE 33.3 Dormant tubers of white yam.

Yam Propagation by Minisett Production

The minisett technology for yam propagation is a semi-rapid multiplication method for seed yam production in major yam species. The technology was developed at the National Root Crops Research Institute, Nigeria, tested and improved at IITA, and adopted by yam growers in West Africa. Minisetts are obtained by sectioning a yam tuber weighing 1.2 to 1.5 kg into 45 to 50 pieces, each weighing about 25 to 30 g. The sections can be dipped in a fungicide solution and cured (air dried) under shade for a day and then planted in a sprouting medium, such as garden soil or moist sawdust. Sprouting occurs 5 weeks after planting, and the sprouted minisetts are planted in a nursery bed for seed yam production. Seed yams weighing an average of 100 g are produced. Under proper management and production practices, seed yams weighing up to 300 g have been obtained. Still, there is a great potential for higher yields. Larger seed yams are desirable for large yam tuber production. Therefore, the goal is to optimize those factors (e.g., mulching, sanitation, and others) that influence good seed yam production.

In Vitro Propagation

Tissue culture regeneration experiments in yam have been geared toward shoot production and clonal multiplication, germplasm preservation and distribution, microtuber induction, and the subsequent field production of seed yams. Micropropagation of yams can be achieved through the development of multiple axillary shoots or microtuber induction. Microtuberization is a useful method for rapidly propagating yams for seed yam production. It also offers tremendous opportunity for storage and safe international movement and distribution of pathogen-free germplasm and clonal yam propagules. Microtuberization, similar to other tissue culture procedures in yam, is influenced by several growth factors, including all the major classes of endogenous plant growth regulators, sucrose concentration, and environmental factors such as photoperiod.

The following experiments will illustrate sprouting in minisetts and disinfestation procedures for the establishment of the aseptic cultures of nodal vine cuttings.

Experiment 3. Effect of a Portion of Tuber (Head, Middle, Tail, or Mixed) on the Sprouting Frequency of Minisetts

Minisetts obtained from various portions of the yam tuber may sprout at different rates. The purpose of this experiment is to evaluate the response of minisetts taken from various sections of the yam tuber on sprouting frequency.

MATERIALS

The following materials will be needed for each student or team of students:

- Two large ware yam tubers each weighing 1.5 to 2 kg (the large yams used for eating are "ware" yams, and the smaller ones less than 1 kg used for planting are "seed" yams)
- Sharp nursery knife
- Planting trays
- Sawdust or other porous potting mix (1:1 perlite:vermiculite)
- Overhead misting system and bench space in greenhouse
- Labels and permanent marker

Follow the instructions in Procedure 33.3 to complete this experiment.

ANTICIPATED RESULTS

Unequal sprouting among minisetts taken from the different sections of yam tuber will occur. The most sprouting will be obtained in minisetts derived from the middle sections of the yam (treatment B) and minisetts taken from all sections of the tuber (treatment D) that served as control.

QUESTIONS

- What accounts for the differences in sprouting of minisetts taken from various sections of the tuber?
- Is it possible to achieve uniform sprouting with the minisett technology? Why or why not?
- How would you obtain uniform sprouting in minisetts?
- How would you enhance the rate of sprouting in minisetts?

Experiment 4. Effects of Disinfesting Agents on Contamination in Yam Tissue Cultures

Yam tubers sprouted in potting soil or other non-sterilized commercial propagation mixes in the greenhouse or nursery are often difficult to disinfest because of the presence of microorganisms in the shoots and axillary buds. The objective of this experiment is to compare two methods of explant disinfestation, including a post-rinse dip of explants in dilute, aqueous solutions (0.1% or 0.2% [w/v])

Procedure 33.3
Effect of the Position of the Tuber (Head, Middle, Tail, or Mixed) on the Sprouting Frequency of Minisetts

Step	Instructions and Comments
1	Cut a healthy yam tuber weighing about 2 kg into three equal cylindrical sections and label the sections as head (A), middle (B), and tail (C). The head is the section closest (proximal end) to the stem, and the tail comes from the distal end. Cut each section into similar-sized disks from which small wedge-like pieces (30–40 g) can be obtained by cutting longitudinally. These pieces constitute the minisetts. Use another tuber to obtain minisetts for the control treatment. For the control, blend equal numbers of minisetts obtained from all three sections (D). About 50 minisetts can be derived from one 2 kg tuber.
2	Spread the minisetts on a shaded flat surface and allow to air dry for 1 to 2 h.
3	Plant the minisetts, with the "skin" portion down, in trays (50 × 36 × 10 cm) containing sawdust in four rows, with four minisetts per row. The rows should be approximately 10 cm apart. Make sure that the minisetts are completely covered by the sawdust.
4	Prepare enough minisetts for a randomized complete block design with four replications.
5	Place the minisetts in a greenhouse or nursery at 26 to 28°C and irrigate periodically using an overhead fine mist sprinkler to avoid drying.
6	Monitor the sprouting of minisetts that begins in 3 to 4 weeks and continues up to 7 weeks after planting. Sprouting involving root and shoot emergence takes place on the "skin" of the minisett after dormancy is broken.
7	Count and record data on the number of sprouted minisetts, including the number of roots or shoots produced per minisett in each treatment and the root or shoot length.
8	Analyze your data statistically using ANOVA and run mean separation where necessary. Compare your results with those obtained by other student groups (Chapter 16).

of calcium hypochlorite, to ensure success in the aseptic culture establishment.

MATERIALS

The following items will be required for each group of three to four students:

- Sprouted shoots of yam (*D. rotundata* or *D. cayenensis*)
- Scissors
- Cheesecloth
- Liquid chlorine bleach
- Calcium hypochlorite—To prepare 8% (w/v) solution, add 8 g in 100 mL distilled water and stir for 10–15 min. Filter through Whatman No. 1 using funnel and gravity.
- 95% ethanol
- Baby food jars
- Whatman No. 1 filter paper
- Large funnel
- Growth chamber, tissue culture room, or laboratory bench with lights
- pH meter
- Stir plate and stir bar
- Balance
- Laminar-airflow bench
- MS basal salt medium and glassware

Follow the instructions in Procedure 33.4 to complete this experiment.

ANTICIPATED RESULTS

Treatments A and B should result in a low contamination rate. Sub-treatments I, II, or III may or may not enhance this result significantly.

QUESTIONS

- Which agent was more effective for disinfesting nodal yam cuttings? Why?
- What factors would you consider in selecting a chemical agent for disinfesting your explants?
- What was the predominant contaminant—fungi or bacteria?

Procedure 33.4
Effects of Disinfesting Agents on Contamination in Yam Tissue Cultures

Step	Instructions and Comments
1	Collect nodal yam cuttings (3 cm tall), divide into three groups (A, B, and C), and place in distilled water. Trim off the lamina, leaving the petioles attached, and rinse twice with distilled water containing a few drops of Tween 20.
2	Surface disinfest the explants of the two groups (A and B) of cuttings in 8% aqueous calcium hypochlorite solution with one drop of Tween 20 for 10 min. After one rinse in sterile distilled water, place explants in 6% calcium hypochlorite solution plus one drop of Tween 20 for 5 min and then rinse three times with sterile distilled water. Disinfest the third group (C) of explants in a 1 Clorox (5.25% NaOCl):1 ethanol (v/v) solution for 7 min and rinse three times with sterile distilled water.
3	Cut off the bleached "burnt" ends of the nodal stem cutting and the petioles to expose healthy tissue.
4	Immediately dip the explants of group B in either 0.1% (I) or 0.2% (II) calcium hypochlorite (0.1%—0.1-g $Ca(OCl)_2$ in 100 mL distilled water; 0.2%—0.2 g in 100 mL distilled water), or sterile distilled water (III) prior to culture. Sterile distilled water dips serve as controls.
5	Culture all explants in baby food jars containing agar-solidified MS medium supplemented with 10 μM Indoleacetic acid (IAA) and 10 μM kinetin, with one explant per jar. Make seven replications (jars) per treatment.
6	Incubate cultures at 26 to 27°C with 60-$\mu mol \cdot m^{-2} \cdot sec^{-1}$ of fluorescent light for up to 3 weeks.
7	After 3 weeks, determine explant viability and the number of contaminated explants in each treatment, and compare treatments A and C using a *t* test.
8	Analyze the results of treatments I, II, and III using the parameters stated in step 7. Share and compare your results with those of other student groups (Chapter 16).

FIGURE 33.4 Tissue culture–derived cocoyam plants in a greenhouse.

COCOYAM

Cocoyam (*Xanthosoma sagittifolium* L. Schott) (Figure 33.4) is a monoecious (male and female flower parts on the same plant), herbaceous, perennial monocot in the Araceae, which grows in the humid tropics and subtropics. It is known under a variety of names, including "macabo," "tannia," "yautia," "malanga," or "maffafa," depending on the region where it is cultivated. It is mainly grown as an annual over an 8- to 12-month production cycle. The enlarged underground storage organ extending from the base of the crown is a modified stem with compressed internodes referred to as a "corm" (Chapter 28). Toward the end of the first annual growth cycle, it becomes enveloped in senesced or shreddy vestiges of the petioles, which, when removed, reveal rings of small dormant axillary buds. When maintained in the ground for up to 2 or 3 years without cultivation, this structure may grow longer and larger, becoming a prominent stem with several rings of small dormant buds interspersed with larger secondary buds. The primary, small dormant shoot buds need special treatments, such as tissue culture conditions, to grow into shoots. The corms produce small tubers called "cormels" (Figure 33.5) that are detached at maturity and used for food or propagation. Cocoyams are conventionally propagated by fragmenting the corms or cormels into 30 to 40 g pieces with large buds and planted. Propagation by seed is not practical because of the absence or scarcity of flowering.

FIGURE 33.5 Cocoyam cormels.

In Vitro Propagation

Conventional methods of propagating cocoyams are not adequate to meet the demand for good-quality seed stock and also contribute to the dissemination of pathogens. *In vitro* techniques have been considered and applied successfully for rapid micropropagation via shoot-tip culture and axillary bud enhancement. Regenerating cocoyams adventitiously has contributed to a tremendous increase in the multiplication rate of cocoyam and provided opportunities for biotechnological applications.

Shoot-Tip Culture and Axillary Bud Multiplication

The following experiment will demonstrate the rapid multiplication of cocoyam via shoot-tip culture and axillary bud enhancement. This experiment is to be conducted by groups of four to five students.

Experiment 5. The Effects of Kinetin and the Auxin Napthalene acetic acid (NAA) on Shoot Multiplication in Cocoyam

MATERIALS

The following materials are required for each group of students:

- Cocoyam shoots
- Liquid chlorine bleach
- 95% ethanol
- Balance
- Growth chamber, tissue culture room, or laboratory bench with lights
- Petri dishes
- pH meter
- Stir plate and stir bar
- Laminar-airflow bench
- Kinetin and NAA stocks
- Test tubes with caps (150 × 20 mm)
- B5 basal salt medium (Gamborg *et al.*, 1968) and glassware

Follow the instructions in Procedure 33.5 to complete this experiment.

Procedure 33.5

The Effects of Kinetin and NAA on Shoot Multiplication in Cocoyam

Step	Instructions and Comments
1	Collect 1 to 2 week-old cocoyam shoot buds from sprouted tuber fragments in the greenhouse and wash under running tap water to remove growing medium. Three weeks before the laboratory exercise is planned, cocoyam tubers should be cut into 40-g fragments, bearing dormant buds, allowed to air dry, and placed in a mixture of 1 vermiculite:1 perlite (by volume) in plastic flats (50 × 36 × 10 cm) to encourage sprouting. The trays are misted for 30 sec every 2 h.
2	With the aid of a scalpel blade, remove several whorls of leaves to obtain shoot tips 5 to 6 mm long, leaving a thin layer of tuber tissue at the base. Place in a beaker containing distilled water.
3	Disinfest the shoot tips in 15% liquid chlorine bleach (0.79% NaClO) solution containing one drop of Tween 20 for 20 min with constant agitation, followed by four rinses in sterile distilled water under the laminar-airflow hood. Because of their small size, the shoot tips should be disinfested in a small container such as a Petri dish.
4	Transfer explants into a B5 basal liquid medium (Gamborg *et al.*, 1968) containing kinetin (0, 0.46, 4.65, and 13.9 μM) and NAA (0, 0.54, and 1.62 μM) in 150 × 20-mm culture tubes. Place one explant in each culture tube, cap, and seal with Parafilm™. A completely randomized design with five replications should be used.
5	Incubate and monitor cultures at 25° to 27°C and at a photon flux of 50 μmol m^{-2} sec^{-1}, 16-h photoperiod, for up to 6 weeks. Note and remove contaminated cultures from the growth chamber.
6	Collect data on contamination, viability, the number and length of shoots and roots, and plantlet formation, and perform the ANOVA procedure. Perform treatment means separation using least significant difference (LSD) or Tukey's HSD at 5% probability (Chapter 16). Share your data with other student groups and report your findings.

ANTICIPATED RESULTS

Properly dissected shoot tips should float on the surface of the liquid medium. Explants with large amounts of tissue from the original corm at the base may sink in the medium and, thus, fail to grow. Bulging and expansion of explants should occur in 7 to 10 days in the medium containing the highest concentration of kinetin. Multiple shoots may be formed in media containing relatively high concentrations of kinetin in combination with NAA.

QUESTIONS

- How would you improve explant development and morphogenesis when using liquid media?
- How would you increase the frequency of shoot production and plantlet regeneration?

LITERATURE CITED AND SUGGESTED READING

Alizadeh, S., S. H. Mantell, and A. M. Viana. 1998. *In vitro* shoot culture and microtuber induction in the steroid yam *Dioscorea compisita* Hemsl. *Plant Cell Tiss Org Cult* 53:107–112.

Gamborg, O. L., R. A. Miller, and K. Ojima. 1968. Nutrient requirement of suspension cultures of soybean root cells. *Exp Cell Res* 50:157–158.

Hartmann, H. T., D. E. Kester, F. T. Davis, and R. L. Geneve. 2002. *Plant Propagation: Principles and Practices*, 7th Edition. New Jersey: Prentice Hall, 880 pp.

Konan, N. K., C. Schöpke., R. Cárcamo., R. N. Beachy, and C. Fauquet. 1997. An efficient mass propagation system for cassava (*Manihot esculenta* Crantz) based on nodal explants and axillary bud–derived meristems. *Plant Cell Rep* 16:444–449.

Murashige, T. and F. Skoog. 1962. A revised medium for rapid growth and bioassays with tobacco tissue cultures. *Physiol Plant* 15:473–497.

Nyochembeng, L. M. and S. Garton. 1998. Plant regeneration from cocoyam callus derived from shoot tips and petioles. *Plant Cell Tiss Org Cult* 53:127–134.

Otoo, J. A. 1992. Substitutes for chemicals, sawdust, and plastic mulch in improved seed yam production. *Proc 4th Symposium Intl Soc Trop Root Crops—Africa Branch (ISTRC-AB)* pp. 281–284.

Okoli, O. O., A. F. K. Kissiedu, and J. A. Otoo. 1992. Production of seed yams by the minisett technique: Effect of mulch, stakes, and plant population. *Proc 4th Symposium Intl Soc Trop Root Crops—Africa Branch (ISTRC-AB)* pp. 277–279.

Watanabe, K. N. 2002. Challenges in biotechnology for abiotic stress tolerance on roots and tubers. *JIRCAS Working Report* 23:75–83.

Zok, S., Sama A., and L. Nyochembeng. 1992. Elimination of yam culture contamination using double surface sterilization. *Proc 4th Symposium Intl Soc Trop Root Crops—Africa Branch (ISTRC-AB)* pp. 291–293.

34 Micropropagation of Woody Plants

Robert R. Tripepi

Micropropagation is the production of plants from plant parts cultured *in vitro*. Many *in vitro* techniques can be used to produce large numbers of plants. Although micropropagation refers to many tissue culture techniques, axillary shoot proliferation is the procedure most often used during commercial micropropagation to reproduce true-to-type plants. Axillary shoot production is preferred over shoot organogenesis (adventitious shoot production) since the former technique reduces the chances for producing off-type plants. As you read this chapter and complete the laboratory exercises, keep in mind that micropropagation and axillary shoot proliferation are used synonymously.

Micropropagation (axillary shoot proliferation) involves several phases during the production of plants; these phases are referred to as "stages of micropropagation" (Chapter 30). Stage 0 is the growth and care of stock plants from which explants are harvested. Stage I of axillary shoot proliferation involves the establishment of shoot tip or nodal explants in culture. Stage II entails multiplication (proliferation) of axillary shoots from the established explants (propagules). Stage III involves preparing the plant parts (usually shoots) for transfer to the outside (*ex vitro*) environment. This stage usually involves rooting microshoots. Finally, Stage IV involves transferring and acclimatizing plantlets from *in vitro* conditions to a greenhouse or other controlled environment facility.

Many species of woody plants, including temperate hardwoods and conifers as well as tropical plants, can be micropropagated. Some woody species are relatively easy to establish in micropropagation, whereas other species may survive for only several subcultures before all microshoots die. For this reason, micropropagation is a commercially useful tissue culture technique for only amenable species.

Explant establishment is the first objective of this laboratory exercise. Establishment depends on the physiological state of the stock plant, the components in the culture medium, and the environmental conditions under which the explants are grown. European birch (*Betula pendula* Roth) is a good candidate for micropropagation because it is an important landscape species in North America and a prominent timber species in northern European countries. This species can be micropropagated for commercial and research purposes. It also multiplies easily (Stage II of micropropagation) and microshoots easily form roots *in vitro* (Stage III). Finally, a modest amount of care is needed to acclimatize rooted microshoots to the environment outside the tissue culture vessel (Stage IV). Alternatively, birch microcuttings can be taken from Stage II shoot cultures and rooted directly in a potting mix held in a protected environment. This laboratory exercise will provide experiences in all stages of the micropropagation of European birch, and the procedures can be completed in 16 weeks. Although, European birch is easy to micropropagate, the use of two other woody species will be highlighted for comparative purposes.

The objectives of this part of the exercise are the following:

1. To demonstrate the methods used to make and disinfect the nodal explants of a deciduous tree species; and
2. To provide experience in establishing nodal explants, subculturing and multiplying microshoots, rooting microshoots *in vitro*, and acclimatizing the resulting plantlets to the ambient environment.

PROCEDURES

The culture medium used throughout this laboratory exercise is Woody Plant Medium (WPM) basal salts (Lloyd and McCown, 1980) supplemented with various organic components (e.g., vitamins, sugar alcohol, and others; Table 34.1). The pH of the medium is adjusted to 5.2 before sterilization, and about 25 mL of medium are poured into each 250 mL baby food jar.

Stage I: Explant Establishment

Stem nodes used as explants (Figure 34.1) in this stage of micropropagation must be freed from microbial contamination or explant establishment on the medium will be impossible. The concentrations of growth regulators used in the establishment medium can also have significant effects on explant establishment. During Stage I of micropropagation, the effects of different concentrations of auxin and cytokinin growth regulators on explant axillary budbreak and subsequent growth will be determined.

TABLE 34.1
WPM Used as the Culture Medium for Three Stages of European Birch Micropropagation[a]

Ingredient	Stage I Establishment Medium	Stage II Proliferation Medium	Stage III Rooting Medium
WPM salts	1.15 g	1.15 g	1.15 g
Sucrose	10 g	10 g	10 g
WPM organics stock solution[b]	10 mL	10 mL	10 mL
Plant growth regulators	See Table 34.2	See Table 34.2	NAA stock solution,[c] 0.5 mL
Adjust volume	Bring up to 500 mL	Bring up to 500 mL	Bring up to 500 mL
Adjust pH of medium	5.2	5.2	5.2
Agar	3.25 g	3.25 g	3.25 g

[a] Note that each separate column is used to make 500 mL of medium, which is enough for 20 jars of medium. To make more or less culture medium, the amounts of each ingredient need to be adjusted accordingly.

[b] WPM organic stock solution is made by mixing (on a gram-per-liter basis) the following organic compounds: myo-inositol, 5; glycine, 0.1; nicotinic acid, 0.025; pyridoxine HCl, 0.025; and thiamin HCl, 0.05. Dissolve each component, starting with myo-inositol, before adding the next one. These compounds can be purchased from various vendors (Sigma-Aldrich Company, Caisson Laboratories, and PhytoTechnology Laboratories).

[c] A 1 mM NAA stock solution is made by dissolving 0.0186 g of NAA in 10 to 15 drops of 1 N NaOH (40 g NaOH per liter of water) and then bringing the solution up to 100 mL.

FIGURE 34.1 Single-node explants of birch stems being established during Stage I of micropropagation.

MATERIALS

The following materials are needed to complete this stage of micropropagation by individuals or groups of students:

- 10% bleach (0.6% w/v NaOCl—commercial product) plus two drops of Tween 20 per 200 mL of bleach solution
- 70% (v/v) ethanol solution (aqueous)
- Spray bottle with 250 mL capacity
- Three autoclaved baby food jars filled 2/3 full with sterile distilled water
- Several European birch stems, enough to contain 50 nodes when cut into explants
- Razor blades or pruning shears
- Scalpels and forceps (two each per workstation or hood)
- 16 baby food jars, four of each containing a medium with different combinations of cytokinin (benzyladenine) and auxin (naphthaleneacetic acid [NAA])—see Table 34.2
- Sterile paper towels (wrapped in aluminum foil and sterilized)

Follow the protocol provided in Procedure 34.1 to complete this experiment.

ANTICIPATED RESULTS

The axillary buds will begin to grow within 4 to 7 days after being placed on the medium if stems were taken at the proper stage of development. Contamination will be present on the medium if either the explant surface sterilization procedure (using the bleach solution) was inadequate or aseptic techniques (handling procedures) were poor. The buds should grow from nodal explants on all four combinations of plant growth regulators, but the combinations of cytokinin and auxin will affect shoot growth differently. These differences should be noted.

QUESTIONS

- Why must you avoid contaminating any of the sterile water rinses used in Stage I?
- Why were you asked to remove (cut off) 2 mm of tissue from the ends of the single node explants after completing the last sterile rinse in Stage I?

TABLE 34.2
Plant Growth Regulators Used during Stage I and II for European Birch Micropropagation Treatments[a]

Treatment	Stage I Benzyladenine stock solution[b] (mL)	Stage I NAA stock solution[c] (mL)	Stage II Benzyladenine stock solution (mL)	Stage II NAA stock solution (mL)
A	0	0	2.2	0
B	1.1	0	4.4	0
C	2.2	0	6.6	0
D	2.2	0.5	–	–

[a] The stated volumes of stock solutions are added to 500 mL of medium, which is enough to for 20 jars of medium. To make more or less culture medium, the amounts of each stock solution need to be adjusted accordingly.
[b] A 1 mM benzyladenine stock solution is made by dissolving 0.0225 g of benzyladenine in 10 to 15 drops of 1 N NaOH (40 g NaOH per liter of water) and then bringing the solution up to 100 mL.
[c] A 1 mM NAA stock solution is made by dissolving 0.0186 g of NAA in 10 to 15 drops of 1 N NaOH and then bringing the solution up to 100 mL.

Procedure 34.1
Establishing Single-Node Birch Stem Explants in Stage I of Micropropagation

Step	Instructions and Comments
1	Make a 10% (v/v) bleach solution (20 mL liquid chlorine bleach + 180 mL water + 2 drops of Tween 20).
2	Be sure to have three baby food jars filled about 2/3 full with distilled water, and sterilize these jars in an autoclave or pressure cooker for 30 min.
3	European birch stock plants will be used for micropropagation. Please be careful as you cut branches off the plants with razor blades, scalpels, or pruning shears. Note the condition of the leaves and stems (stem appearance: woody, semi-woody, and others; leaf color; disease or tissue damage present). Succulent small stems as well as older woody branches should be avoided. The best stems to use are those that are still green and pliable (softwood to semi-woody).
4	Make stem sections from the branches that you cut, with each section containing one node (axillary bud). Cut the stem about 1 cm above the bud and 2 cm below the bud if possible. Some adjustments for these cutting distances from the bud may be necessary so that as many explants as possible can be made from the plant material available. These stem sections are called “single-node explants.”
5	Remove the leaf blade and the petiole, being careful to avoid tearing or damaging the bud or stem. Cut the petiole if necessary. Make 50 nodal sections for each group (typically 2–3 students).
6	Place the stem sections into the 10% bleach solution (in a baby food jar) for 20 min. The baby food jar used to hold the bleach solution does not need to be sterile (the bleach solution will surface sterilize it). Agitate every few minutes. Spray the jar with 70% ethanol before moving it into the laminar-flow (clean) hood.
7	Spray 70% ethanol on the outside of each capped baby food jar that contains sterile water before placing the jars in the laminar-flow hood. In the hood, rinse the stem sections in three changes of sterile distilled water in baby food jars for 1 to 2 min per rinse. Use sterile forceps to transfer stem sections. Forceps and scalpels can be sterilized in a Bacti-Cinerator® or other appropriate equipment that will heat these tools to high temperatures to kill microbes on tool surfaces. Forceps and scalpels can be surface sterilized using 70% ethanol and an alcohol lamp or gas burner to flame these tools, but a fire can easily be started when flaming tools and using alcohol. Therefore, use equipment that lacks flames (e.g., Bacti-Cinerator®) to maintain laboratory safety.

8 Sterilize forceps between rinses. Leave the stem sections in the rinse water 1 to 2 min before transferring them to another rinse jar. Rinse the stem sections three times.

9 After the third rinse, remove three stem sections at a time and place them on a sterile paper towel. Make two cuts on each stem section to remove about 2 to 3 mm from each end of the stem section. With sterile forceps, place the stem sections in an upright position into one of the jars of culture medium. The base of the explant should be submerged 3 to 5 mm deep into the agar. Be sure to keep the proper orientation of the stem sections—proximal end down (right side up). Put three stem sections in each jar, and use four jars of each medium (treatment). The four treatments are listed in Table 34.2.

10 After the three stem sections are in place, cap the jar before removing more stem sections from the last rinse of sterile distilled water. Label all jars and put these jars on a lighted shelf (25–40 $\mu mol \cdot m^{-2} \cdot sec^{-1}$) in a culture room. The jars can be placed about 34 cm (14 in) under fluorescent shop lights fitted with cool white lamps to provide a 16-h photoperiod in a room at 25°C. These Stage I explants will be allowed to grow 4 weeks before subculturing.

11 Observe the explants once every few days for contamination, and then observe at weekly intervals. Record the following data: the percentage of contaminated stem explants at Stage I, the percentage of stem sections with expanded/elongated buds per treatment, the number of shoots that grew or formed from each single-node explant per treatment, and the length of each elongated shoot (when subculturing it). Calculate the means for the number of shoots formed per explant and the lengths of elongated shoots for each treatment. Measure elongated shoots under aseptic conditions.

- How many days passed before the buds started to expand?
- Was cytokinin necessary for bud elongation?
- What were the effects, if any, of the changes in cytokinin concentrations on the explants?
- Was auxin necessary for bud expansion or did this growth regulator have any effect on the explants?

Stage II: Shoot Multiplication

Axillary buds that have grown to form stems are removed from the original stem (single-node explant) and placed on a multiplication medium to induce more axillary shoots to grow. Therefore, the new growth from the original explant now becomes a propagule that is divided and put back into culture (i.e., subcultured) to produce additional axillary shoots (for axillary shoot proliferation) (Figure 34.2). The concentrations of growth regulators in the medium during this stage can also have significant effects on the number of axillary buds that grow (proliferate) (Figure 34.3). During Stage II of micropropagation, the effects of different concentrations of a cytokinin growth regulator on the number of axillary shoots that grow from the subcultured propagules will be determined.

FIGURE 34.2 European birch shoot culture with proliferating microshoots on multiplication medium during Stage II.

FIGURE 34.3 Axillary shoot proliferation from two stem propagules of European birch during Stage II of micropropagation. Note the number of axillary shoots that have formed.

MATERIALS

The following materials will be needed to complete this stage of micropropagation by individuals or groups of students:

- 70% (v/v) ethanol solution (aqueous)
- Spray bottle with a 250 mL capacity
- Scalpels and forceps
- 12 baby food jars, three of each different combination of cytokinin (benzyladenine) and auxin (NAA)—see Table 34.2
- 15-cm–long ruler that is surface disinfested with 70% ethanol (a flat ruler without raised letters or markings works best)
- Sterile paper towels

Follow the outline provided in Procedure 34.2 to complete this experiment.

ANTICIPATED RESULTS

The shoots will continue to elongate, but the cytokinin concentrations will affect the length of shoot growth and the number of axillary buds that grow on each stem (propagule). These differences should be noted since the results will provide an idea of how much cytokinin should be used in the proliferation medium if this birch species were to be micropropagated commercially.

QUESTIONS

- How did the cytokinin concentration affect the number of axillary buds that grew (proliferated)?
- Did the original establishment (Stage I) medium have any effects on the number of new shoots formed? How do you know?
- How did the cytokinin concentration affect the increase in stem length of the propagules (stems removed from Stage I explants)?

Procedure 34.2

Multiplying Birch Stems Obtained from Stage I Nodal Explants

Step	Instructions and Comments
1	About 4 weeks after making the Stage I explants, prepare to subculture the expanded birch shoots. Be sure to record the data requested in step 11 from Procedure 34.1.
2	Remove the jars containing the microshoots from the culture room shelf. Spray 70% ethanol on the outside of these jars and those that contain the multiplication media before transferring the jars to the laminar-flow hood.
3	Open one jar that contains the Stage I explants, remove one stem explant with sterile forceps, and place it on a sterile paper towel. All of the following procedures should be completed with only the sterile forceps and the sterile scalpels touching the plant tissues. Cut off the expanded shoot at the point where it arises from the old (original explant) stem, and place the new shoot beside the surface-sterilized ruler on a sterile paper towel. Avoid touching the shoot with the ruler to reduce chances of contamination. Be sure to measure the length of the new (expanded) shoot while in the hood.
4	Place the shoot on a new medium (A, B, or C). This shoot and all others should be oriented vertically when placed on a new medium. Repeat this procedure with the next two shoots in the jar and then go on to the next jars to subculture the other shoots. Put four shoots in each jar of new medium. Since 12 shoots are on each Stage I medium (A, B, C, or D), put four shoots into each jar of each multiplication (Stage II) medium (new A, B, or C). See Figure 34.4 for a diagram that shows how the stems should be transferred from Stage I to Stage II media. Be sure to note the original establishment medium on the new jar. You will need to use four jars of each multiplication medium.
5	After completing this work, the jars containing the shoots should be placed in the culture room under fluorescent lights. Let the shoots grow for about 4 weeks before transferring the microshoots to a rooting medium (Stage III).
6	Observe the explants at least twice a week, looking for contamination, axillary budbreak, and stem growth. Record the following data: the number of shoots formed on each propagule per Stage II medium, the mean number of shoots that formed from each propagule per medium, and the length of the original propagule (stem) after 4 weeks on the multiplication medium. Determine this stem length when preparing the shoots for Stage III, and calculate a mean length for each treatment. Make measurements under aseptic conditions within the laminar-flow hood. Also, be sure to note the relative amount of callus formed on stems in each treatment.

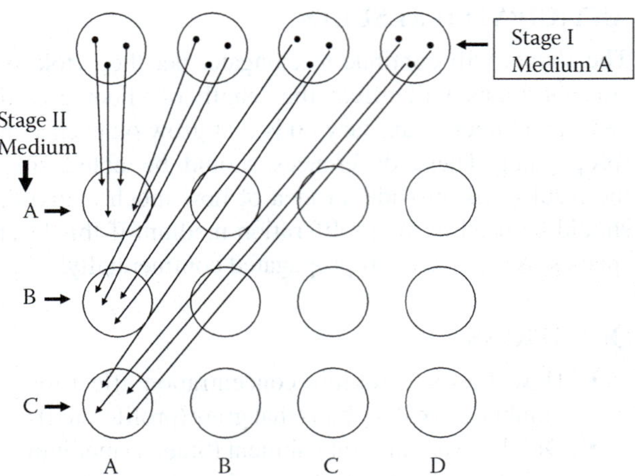

FIGURE 34.4 Schematic diagram for moving stems from one jar of Stage I medium (A) to three different media used in Stage II of micropropagation. Note that the original treatment should also be listed on the Stage II treatment jar.

- Would the number of axillary buds that proliferated increase or decrease if the Stage I propagules were oriented horizontally (placed on their sides) rather than vertically? Why?

Stage III: Transfer Preparation and Rooting the Microshoots

A critical part of micropropagation is to induce roots to form on microshoots. Root formation on microshoots is typically the limiting factor in micropropagation rather than shoot proliferation from axillary buds. Microshoots can be rooted either *in vitro* or in the regular environment (in which case microshoots are now called "microcuttings"). *In vitro*–rooted shoots can be placed on a medium supplemented with auxin or one that lacks cytokinin. Basal mineral concentration is sometimes reduced to half strength to promote rooting. Addition of activated charcoal is sometimes helpful for root formation on microshoots.

In this experiment, axillary buds that have broken (swollen or expanding) on birch microshoots and subsequently grown to form stems are counted and removed from the original explant (propagule). Stems that are longer than 1.5 cm can be rooted to form plantlets (Figure 34.5), which will later be transferred to the *ex vitro* environment. The concentrations of cytokinin used during Stage II may have an effect on the ability of the microshoots to form roots. During Stage III of micropropagation, proliferated microshoots will be cut from Stage II shoot cultures and rooted individually on a culture medium that contains an auxin.

FIGURE 34.5 European birch microshoots rooted during Stage III of micropropagation.

MATERIALS

The following materials will be needed to complete this stage of micropropagation by individuals or groups of students:

- 70% (v/v) ethanol solution (aqueous)
- Spray bottle with a 250 mL capacity
- Scalpels and forceps
- 12 baby food jars of rooting medium containing 1 μM NAA, but lacking a cytokinin—see Table 34.1
- 15-cm–long ruler that is surface sterilized with 70% ethanol (a flat ruler without raised letters or markings works best)
- Sterile paper towels

Follow the procedures outlined in Procedure 34.3 to complete this exercise.

ANTICIPATED RESULTS

Many or most of the microshoots should form roots depending on their health and vigor. The cytokinin concentrations used in Stage II may have an effect on the percentage of shoots that form roots. Proliferated microshoots on Stage II medium that contained higher cytokinin concentrations may also grow to lengths similar to those of proliferated microshoots on lower cytokinin concentrations. Since Stage II European birch microshoots often root readily when transferred to a potting mix, non-rooted shoots can also be used in Stage IV of this experiment.

ROOTING EXPERIMENT FOR BIRCH MICROSHOOTS

European birch microshoots readily respond to different concentrations of auxin in the culture medium.

Procedure 34.3

Rooting Birch Microshoots Obtained from Stage II Shoot Cultures

Step	Instructions and Comments
1	About 4 weeks after making the Stage II explants, prepare to excise the birch microshoots. Be sure to record the data requested in step 6 from Procedure 34.2.
2	Remove the jars containing the microshoots from the culture room shelf. Spray 70% ethanol on the outside of these jars and those that contain the rooting medium before placing the jars in the laminar-flow hood.
3	Open one jar that contains the Stage II explants; remove one microshoot, which should contain one or several shoots, with sterile forceps; and place it on a sterile paper towel. All of the following procedures should be completed with only the sterile forceps and the sterile scalpels touching the plant tissues. Measure the shoot length before cutting off any callus that has formed at the base of the shoot. Make the measurements while under the hood, and be sure to maintain aseptic techniques.
4	Use a scalpel to remove any callus at the base of the shoot. Cut off proliferated axillary shoots at their points of attachment on the original propagules (stems), and place the shoots (also referred to as "microshoots") vertically in a jar that contains rooting medium. This medium contains 1 μM NAA, but lacks a cytokinin. Although any shoot about 1 cm long may root, use shoots longer than 1.5 cm in this part of the laboratory exercise for ease of handling. When placing the shoot in the jar, 3 to 5 mm of the basal end of the stem should be inserted into the medium. Repeat steps 3 and 4 until all jars contain at least five or six microshoots from the same Stage II proliferation treatment. If extra shoots are available, transfer only the most vigorous shoots to the Stage III medium. Be sure to note the Stage II proliferation medium on the jar that contains the rooting medium. You will need to use four jars of rooting medium for each Stage II multiplication medium.
5	After completing this work, the jars containing the shoots should be placed in the culture room under fluorescent lights. Let the shoots grow for 2 to 4 weeks before transferring the microshoots to the *ex vitro* environment (Stage IV).
6	Observe the explants at least twice a week, looking for contamination, adventitious root initiation, and stem growth. Record the following data: the percentage of shoots that formed at least one root; the number of roots per shoot, only for shoots that formed roots; and the number of root initials per shoot. All these data should be grouped for each Stage II medium. To determine if the cytokinin concentrations in the multiplication media affected rooting, calculate a mean number of roots per shoot (ONLY for responding shoots) for each Stage II treatment. A mean for the number of root initials per responding stem can also be calculated. When counting roots, rooted stems can be handled under non-aseptic conditions, but they should be held in a capped vessel since they can readily dry out. Take these data before transferring the rooted microshoots and those that lack roots to the Stage IV potting mix.

Higher concentrations may induce more roots to form, but root and shoot growth can be affected adversely by auxin concentrations that are too high. The question is what concentration is too high? An alternative experiment to try is to use WPM supplemented with different levels of NAA to determine the effects of this plant growth regulator on the root and shoot growth of Stage III plantlets.

For this alternative experiment, the culture media should be supplemented with 0, 0.1, 1.0, or 5.0 μM NAA (0, 0.05, 0.5, or 2.5 mL of 1 mM NAA stock solution per 500 mL of medium, respectively). Follow Procedure 34.3 to prepare Stage II microshoots for Stage III. Cut all microshoots to a consistent length (e.g., 2.5 cm), if possible. Place equal numbers of microshoots in each jar containing one of the four rooting media. The microshoots should form roots in about 2 weeks, but count roots after 3 to 4 weeks to allow time for root initiation and growth to be completed. Note the number of roots formed and their thickness as well as overall root lengths. In addition, measure the length of the plantlet stems (rooted microshoots).

QUESTIONS

- Did the multiplication treatment used in Stage II affect the percentage of rooting and the number of roots formed per shoot? If so, how?
- Did roots form directly on the stem of each shoot or on callus? From what tissue(s) did the root appear to originate?
- What conclusions can you draw about the effects of NAA concentrations on root and shoot growth of the plantlets?
- How would you be able to tell if the NAA concentration used in Stage III was too high for the microshoots? What type of plant symptoms would you expect to see?
- What concentration of NAA would you recommend to use when rooting European birch microshoots?

Stage IV: Transferring and Acclimatizing Birch Plantlets and Microshoots

Rooted microshoots (plantlets) and non-rooted shoots can be transferred to a simple vessel or container to continue rooting and promote acclimatization of the shoots. Although a greenhouse mist bench covered with shade cloth can be used to acclimatize the shoots, other less elaborate environments can be used to help the birch plantlets and non-rooted shoots successfully make the transition from *in vitro* to *ex vitro* conditions. Since European birch shoots require only a modest amount of care during the acclimatization phase, healthy plantlets can make the transition to an environment with higher light and lower humidity in about 2 weeks. The amount of new growth formed in a short period of time depends on the vigor of the plantlets or non-rooted microshoots when transplanted as well as the acclimatization conditions.

MATERIALS

The following materials will be needed to complete this stage of micropropagation by individuals or groups of students:

- Pasteurized potting mix
- Clear polystyrene (plastic) clamshell container (typically, a donut, cookie, or salad container)
- Small-diameter cork borer or dissecting probe

Follow the protocol provided in Procedure 34.4 to complete this experiment.

ANTICIPATED RESULTS

Most of the birch plantlets should acclimatize to the conditions in the clamshell container. Even microshoots that lacked roots may start rooting in the potting mix. Rooted plantlets that have successfully made the transition will form new leaves and roots that are fully functional. In fact, rooted plantlets may grow about 2 to 3 cm within 3 weeks of being transferred out of the baby food jars. Once the plantlets have formed new leaves and roots (1–3 weeks) and have been acclimatized to lower humidity, the plants can be transplanted into pots and grown in a greenhouse or outdoors.

QUESTIONS

- Did the multiplication treatment affect the ability of rooted microshoots to be transplanted successfully?
- Did non-rooted shoots die? Were any starting to form roots?
- Did the new leaves that formed after the plantlets started to grow in the clamshell container appear different than those formed under *in vitro* conditions? If so, how did they differ?

ALTERNATIVE SPECIES FOR MICROPROPAGATION

European birch plants are often easy to establish *in vitro* and they usually form only one shoot per explant. Other plant species, however, respond differently to the plant growth regulators used in the establishment medium, which will affect the number of shoots that form on Stage I explants. Alternative species to micropropagate are 'Kardinal' rose (*Rosa* × *hybrida*) or American elm (*Ulmus americana* L.).

Stems from 'Kardinal' rose and American elm plants growing outdoors or in a greenhouse can be used for explants. Greenhouse plants are preferred since they should have fewer microorganisms on their surfaces, making them easier to surface sterilize compared to outdoor-grown plants. Procedure 34.1 can be used for making single-node explants from rose or elm stems. The differences for these two species will be the tissue culture media and the plant growth regulator concentrations used (Tables 34.3 and 34.4).

Surface-sterilized stem explants from 'Kardinal' rose plants should be placed on Murashige and Skoog (MS) medium (Figure 34.1), whereas American elm stem explants should be cultured on Driver–Kuniyuki Walnut (DKW) medium (Table 34.3). Higher concentrations of benzyladenine can be used for establishing rose stem explants compared to elm stem explants (Table 34.4). After 4 weeks, be sure to record the data described in step 11 of Procedure 34.1. Pay particular attention to the number of buds that expanded per explant and the length of the new shoots.

Stages II, III, and IV for rose and elm microshoots can be completed as described in Procedures 34.2,

Procedure 34.4

Transferring Stage III Plantlets to *Ex Vitro* Conditions

Step	Instructions and Comments
1	Between 2 to 4 weeks after beginning Stage III, roots will form on the birch microshoots. Be sure to record the data specified in step 6 from Procedure 34.3.
2	Obtain a clear polystyrene (plastic) clamshell container. The type of container suggested for this stage of micropropagation has a hinged lid and is often used to hold donuts or cookies. Used food containers can be washed and used rather than purchasing new containers. Use a small-diameter cork borer (≤3 mm) or a dissecting probe to punch several holes in the bottom of the container. These holes will provide a drainage for the vessel. Next, moisten enough pasteurized potting mix (any good brand should work) to fill the bottom of the food container to a depth of 4 cm. The potting mix should be moist NOT wet; if the mix is too wet, the rooted plantlets can become severely stressed and die.
3	Remove the jars containing the rooted microshoots from the culture room shelf and place them on a classroom table or other suitable area for working with the potting mix. Avoid transferring the shoots in the laminar-flow hood or culture room.
4	Carefully remove the plantlets from the jar. You may find it easier to remove the plantlets and agar all at once and separate the plantlets outside the jar. Remove each plantlet from the agar and gently remove any agar adhering to the roots by pulling or rinsing off the material with tap water. Be sure to record the data specified in step 6 from Procedure 34.2 before planting the rooted plantlets. Use your finger or a pencil to poke a small hole in the moistened potting mix, place the plantlet base and roots in the hole, and gently cover the base with the mix. Repeat this procedure for all shoots, noting the multiplication (Stage II) treatment used on the shoots. To keep track of the multiplication treatment used, keep the plantlets receiving the same treatment in the same row in the container. Also transplant non-rooted shoots into the container.
5	The clamshell container with the microshoots can be placed back in a culture room shelf, but insects and the potting mix may contaminate the room. An alternative choice would be to place the container about 30 cm below a 1.3 m (4-ft) fluorescent shop light. The container can also be placed on a shaded greenhouse bench to keep the potting mix out of the culture room. If the plantlets are acclimatized on a greenhouse bench, make sure that the container is kept under a shade cloth rather than direct sunlight. After 1 or 2 weeks, check the plantlets and count those that appear sick, wilted, or dead. Also note if the plantlets have resumed shoot and root growth. Root growth can be checked by placing a finger under the plantlet and gently lifting the shoot out of the potting mix.
6	To help the plants acclimatize to the environment, the lid of the clamshell container can be opened for one to several hours a day. The lid should be held open for longer periods of time each day over a 4-day period. After the plants become acclimatized to ambient humidity conditions, they can be transplanted to pots and grown in a greenhouse or outdoors. Bright light should be avoided for the first week after transplanting.

34.3, and 34.4, respectively. The media formulations described in Table 34.3 can be used for these species for Stages II and III, but the plant growth regulators need to be changed. BA will be the only plant growth regulator needed for Stage II for rose and elm multiplication (Table 34.5). Rose microshoots taken from Stage II can be placed on half-strength MS medium supplemented with 0.5 μM NAA (0.25 mL of 1 mM NAA stock solution per 500 mL of medium) to promote rooting during Stage III. Elm microshoots are somewhat more difficult to root *in vitro* compared to rose and birch microshoots. An NAA concentration of 1 μM (0.5 mL of 1 mM NAA stock solution per 500 mL of medium) in DKW medium should help promote root initiation. Elm microshoots taken from Stage II can also be rooted directly in a clear polystyrene (plastic) clamshell container, bypassing Stage III, but this process can take a long time, and the percentage of successfully rooted microshoots may be low. Follow the directions in Procedure 34.4 for the acclimatization of rooted rose (Figure 34.6) and elm microshoots.

TABLE 34.3

Tissue Culture Media Used for Stage I of the Micropropagation of 'Kardinal' Rose and American Elm Plants[a]

Ingredient	Rose Micropropagation Medium	American Elm Micropropagation Medium
Salt formulation[b]	MS, 2.15 g	DKW, 2.6 g
Sucrose	15 g	15 g
Organic stock solution	MS organics stock,[c] 5 mL	DKW organic stock,[d] 5 mL
Plant growth regulators	See Table 34.4	See Table 34.4
Adjust volume	Bring up to 500 mL	Bring up to 500 mL
Adjust pH of medium	5.6	5.7
Agar	3.25 g	3.25 g

[a] Note that each separate column is used to make 500 mL of medium. To make more or less culture medium, the amounts of each ingredient need to be adjusted accordingly.

[b] Abbreviations for salt formulations: MS is Murashige and Skoog (1962) medium, whereas DKW is Driver–Kuniyuki Walnut (McGranahan *et al.*, 1987) medium. These can be purchased from vendors such as Sigma-Aldrich Company, Caisson Laboratories, and PhytoTechnology Laboratories.

[c] MS organic stock solution is made by mixing (on a gram-per-liter basis) the following organic compounds: myo-inositol, 10; glycine, 0.2; nicotinic acid, 0.05; pyridoxine HCl, 0.05; and thiamin, HCl 0.01. Dissolve each component, starting with myo-inositol, before adding the next one.

[d] DKW organic stock solution is made by mixing (on a gram-per-liter basis) the following organic compounds: myo-inositol, 10; glycine, 0.2; nicotinic acid, 0.1; and thiamin HCl, 0.2. (Note: Pyridoxine HCl is excluded from this medium.) Dissolve each component, starting with myo-inositol, before adding the next one.

TABLE 34.4

Plant Growth Regulators Used during Stage I for 'Kardinal' Rose and American Elm Micropropagation Treatments[a]

	'Kardinal' Rose Stage I	'Kardinal' Rose Stage I	American Elm Stage I	American Elm Stage I
Treatment	Benzyladenine Stock Solution[b] (mL)	NAA Stock Solution[c] (mL)	Benzyladenine Stock Solution (mL)	NAA Stock Solution (mL)
A	0	0	0	0
B	1.8	0	0.8	0
C	1.8	0.5	1.5	0
D	3.5	1.0	1.5	0.5

[a] The stated volumes of stock solutions are added to 500 mL of medium, which is enough for 20 jars of medium. To make more or less culture medium, the amounts of each stock solution need to be adjusted accordingly.

[b] A 1 mM benzyladenine stock solution is made by dissolving 0.0225 g of benzyladenine in 10 to 15 drops of 1 N NaOH (40 g NaOH per liter of water) and then bringing the solution up to 100 mL.

[c] A 1 mM NAA stock solution is made by dissolving 0.0186 g of NAA in 10 to 15 drops of 1 N NaOH and then bringing the solution up to 100 mL.

TABLE 34.5
Amount of Benzyladenine Stock Solution Used during Stage II for Rose and Elm Micropropagation Treatments[a]

Treatment	Rose Micropropagation Stage II Benzyladenine Stock Solution[b] (mL)	Elm Micropropagation Stage II Benzyladenine Stock Solution (mL)
A	1.8	0.8
B	3.5	1.5
C	7.0	3.0

[a] The stated volumes of BA stock solution are added to 500 mL of medium, which is enough to for 20 jars of medium. To make more or less culture medium, the amount of stock solution needs to be adjusted accordingly.

[b] A 1 mM benzyladenine stock solution is made by dissolving 0.0225 g of benzyladenine in 10 to 15 drops of 1 N NaOH (40 g NaOH per liter of water) and then bringing the solution up to 100 mL.

FIGURE 34.6 Rose plantlets that have successfully made the transition from Stage IV plantlets to the ambient environment. Note the new (glossy) leaves on the rooted microshoots, indicating that the plantlets are actively growing.

LITERATURE CITED AND SUGGESTED READING

Lloyd, G. and B. McCown. 1980. Commercially feasible micropropagation of mountain laurel, *Kalmia latifolia*, by shoot-tip culture. *Comb Proc Intl Plant Prop Soc* 30:421–427.

McGranahan, G. H., J. A. Criver, and W. Tulecke. 1987. Tissue culture of *Juglans*. In Bonga, J. M. and D. J. Durzan, eds. *Cell and Tissue Culture in Forestry*, Vol. 3. Boston: Martinus Nijoff Publishers, pp. 261–271.

Murashige, T. and F. Skoog. 1962. A revised medium for rapid growth and bioassays with tobacco tissue cultures. *Physiol Plant* 15:473–497.

Pierik, R. L. M. 1997. *In Vitro Culture of Higher Plants*, 4th Edition. Dordrecht, The Netherlands: Kluwer Academic Publishers, 348 pp.

Preece, J. E. 1997. Axillary shoot proliferation. In Geneve, R. L., J. E. Preece, and S. A. Merkle, eds. *Biotechnology of Ornamental Plants*. Wallingford, United Kingdom: CAB International, pp. 34–43.

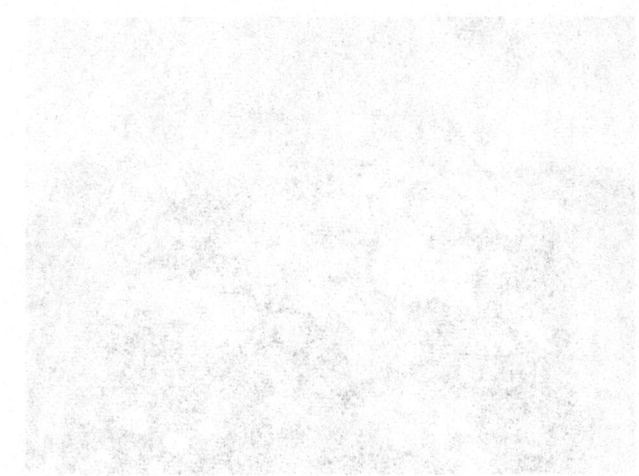

Part XII

Seed Production and Propagation

35 Seed Production, Processing, and Analysis

J. Kim Pittcock

CONCEPT BOX 35.1

- Pure seed is the percentage of actual seeds in the seed lot.
- Inert matter is the percentage of nonviable material in a seed package. These may include broken seeds, rocks, sand, trash, and any material that is not a complete seed.
- Other crop seed is the percentage of seeds found in the seed package that are not the species or variety on the label.
- Weed seed is the percentage of non-crop seeds found in the seed package. Most labels will include the name and the percentage of each type of weed seed found.
- Noxious weed seed is the percentage of weed seeds found that are on the state or federal noxious weed list. Noxious weeds are weeds that a state has determined to be objectionable.

Modern-day farming and horticulture is dependent on seeds and seed production. Seeds are self-contained units, which have the ability to regenerate the species. For the continued use of superior or pure seed lines, they must be produced in sufficient quantities to meet the needs of the market. After the plants have produced seeds, they must also be properly handled and stored to continue this production cycle.

SEED PRODUCTION

Seed production can occur either in regions with commercially produced crops or may be produced in isolated areas away from production centers. For seed crops that are self-pollinated or when pollination can be controlled, the crops are not isolated.

Seed production in isolated regions is necessary when the plants are open-pollinated or cross-pollinated, or when hybrid seed is being produced. Climatic factors, such as low relative humidity or lack of rainfall during harvest, are critical site location factors for high-quality seeds, especially when disease is an issue. Isolated areas are also important when the crop encounters extreme pest problems in the production regions.

Normal production practices are required for quality seed production. Initial field selection is critical, and field history needs to be considered. Seed production of a cereal grain should not be done in a field that was previously planted with a similar grain as volunteer grains could contaminate the seed lot at harvest. Optimum soil fertility and proper moisture levels must be maintained. Additionally, proper insect, disease, and weed control needs to be performed to produce highest quality seeds. Production fields and storage facilities are monitored by the certifying agency to guarantee that the crop meets all the certification standards.

Harvesting of agronomic crops is done mainly after the seeds are mature and have reached the specific moisture content required for the species. Basic harvesting methods are combining/threshing, stripping, and windrowing. Combining/threshing involves removing the seed heads and any plant material in close proximity. Stripping is the process of removing only the mature seed from the plant, without injury to the plant and any future seed. Windrowing involves harvesting the upper portion of the plant before the seed has reached the proper moisture content. The seed and plant are allowed to dry, and then the seed is removed for cleaning and storage.

After the seed is harvested, it may need to be conditioned either by hulling (cereal crops), de-bearding (grasses), or being scalped and rough cleaned. Air-cleaning machines are used to remove trash and to grade the seeds. Additional conditioning is required on grain seeds. They are finished by sizing for length and thickness and then density graded.

If the seed is not dried in the field before harvest, then it must be mechanically dried or naturally air dried after harvest. Seeds are normally dried to 7% moisture content or below and then stored under low temperatures (0–4°C) and low relative humidity (<25%). If the temperature and relative humidity are elevated, the longevity of the seeds will be reduced. To maintain the maximum life of the seed, seed moisture, temperature, and relative humidity must be carefully regulated.

Harvesting of vegetable and flower seeds is more labor intensive than most agronomic crops. Vegetable seed production can be divided into the following two categories: those that are wet seeded (tomato, melons, cucurbits) and those that are dry seeded (onion, legumes, cole crops).

Wet-seeded crops are either hand harvested or machine harvested by removing the entire plant. The fruit may be sorted to select only the superior ones. Fruits are crushed or macerated to pulverize the tissue so that the seeds may be extracted. This gelatinous material is fermented, which allows the mature seeds to sink to the bottom of the tanks. After extraction, the seeds are immediately washed in water troughs. Drying is accomplished in batch dryers, which expose the seed to heated air. A few crops such as cucumber can be air dried in the sun. The seeds are separated from inert matter by an air-screen cleaner. Immature seeds are separated from mature seeds by gravity tables. Seeds are stored under cool, dry conditions at approximately 6% moisture content.

Dry-seeded vegetables have the potential for seed loss due to shattering when the seed is ripe. Immediately after the seed is ready, the entire green plant is harvested and placed in windrows. The plant and fruit are allowed to air dry. The seeds are hand harvested or combined before they shatter. Dry-seeded vegetable seed lots must be prescreened because they normally contain more inert material and undesirable appendages. The appendages are removed by processes such as de-tailing or de-bearding. Conditioning and storage conditions are similar to wet-seeded vegetables.

SEED PROCESSING

Seed Treatments

To assist the grower, many enhancements have been added to the seed and/or coat to make seeding easier or to ensure higher seedling survival rates. Many bedding plants and vegetable seeds are extremely small; therefore, they are pelleted with inert materials, such as clay, that more than double the size of the seed. Larger seed allows for easier mechanical and hand seeding.

Another seed treatment is the addition of improper products that will improve the health or assist the seedling in growth. These include fungicides, insecticides, and micronutrients, such as zinc. Chemical pretreatments are common on cotton (*Gossypium hirsutum* L.), corn (*Zea mays* L.), and rice (*Oryza sativa* L.). A polymer-based or plastic film coating can also be applied to the seed. This allows chemicals to adhere better and also lessens the chance that the pretreatment will rub off during handling. The coating also greatly decreases possible chemical exposure to workers.

Priming seeds is a pretreatment process that has gained popularity, mainly in the vegetable industry. The seeds are hydrated to allow seed germination to begin, but the germination process is stopped before radicle emergence. The seed can be planted or redried and stored. Primed seeds will develop into plants faster when planted and are less disease prone.

Mechanized treatments remove any unwanted part of the reproductive structure, such as wings, barbs, fuzz, or chaff, from the seed. These treatments include de-bearding, bobtailing, defuzzing, and de-tailing. This allows the seed to more easily be separated for mechanical or hand planting.

SEED CLASSIFICATION

Certified seed is an important aspect of the production of genetically pure varieties. Many agronomic crops, except hybrid varieties such as corn, have been certified. New interest has occurred in the horticulture industry to participate in certification programs for sod production and tree seed breeding programs. The use of a certified seed ensures that producers are obtaining varietally pure seeds that is of the highest quality. The four seed classes are breeder seed, foundation seed, registered seed, and certified seed. Breeder seed is directly controlled by the originator or plant breeder and is used for seed increases. It is designated by a white tag and is labeled "breeder seed." Foundation seed is the first generation of breeder seed and is controlled by the foundation producer. White tags are also used for the foundation class. Registered seed is the progeny of foundation seed and is controlled by registered producers. It is also used for seed increases and carries purple tags. Certified seed is commonly available to the farmer and in large quantities. Blue tags represent certified seed. See also Chapter 14.

SEED ANALYSIS

The quality of seed produced from a plant is the determining factor in a crop's future success or failure. Seed quality involves genetic factors, physical purity, germination, and health. By law, seed analyses (germination percentage and purity information) must be conducted, with the results displayed on the tag or seed packet. Numerous seed tests can be performed and normally fall into three basic categories: germination testing, vigor testing, and biochemical testing. The most common laboratory analyses performed are germination test, tetrazolium (TTZ) test (viability), and purity test.

The germination test is a basic seed analysis test and determines the actual percentage of normal seedlings that develop. Controlled environmental conditions of moisture, humidity, and light are combined with a specific time frame for each species of seed tested. Cereal crops are tested for 12 to 14 days, grasses require 21 days, and many woody plant seeds and some wildflowers must be evaluated for 6 weeks. Seeds are observed at regular intervals with newly germinating seedlings removed. A weakness to this test is that all seeds are given the opportunity to germinate under optimum conditions during the time frame, but there is no distinction between strong seedlings and weak seedlings. The weak seeds may never germinate under normal growing conditions, so this measure of germination percentage under typical environmental conditions may be overestimated.

The TTZ test is a biochemical vigor test that gives a rapid estimation of the potential germination percentage by estimating viability. Seeds are soaked in a TTZ chloride (2,3,5-triphenyl TTZ chloride) solution, and staining of various tissues occurs. Respirating cells will absorb the solution at varying amounts and will reveal various staining intensity patterns. Trained and experienced technicians evaluate the staining patterns to determine if the seed is viable.

Purity tests are conducted to establish the percentages of the following: pure seed, inert matter, other crop seed, weed seed, and noxious weed seed. Pure live seed (PLS) is becoming a more commonly used category in purity testing. PLS is a combination of the pure seed percentage and the germination percentage. This calculation gives the percentage of usable seeds in the lot. Many other tests can be included under purity tests depending on individual state laws or for a specific plant species. For most horticultural crops, the seed lot will be cleaned, and the seed package will contain only pure seed.

Vigor testing is more specific to individual species or categories of plants. These tests include the warm germination test, the cold test, and the accelerated aging (AA) stress test. The warm germination test uses sand as the growing medium instead of germination paper and is used by large seed companies because it provides a more realistic germination percentage due to the seed being grown in a natural environment. The cold test is a stress vigor test and is used to determine the germination percentage of seeds planted at low temperatures and high moisture contents. This test mimics environmental field conditions in early spring and is primarily used for corn, soybean (*Glycine max* Merr.), and sorghum (*Sorghum bicolor* [L.] Moench.). The AA test is designed to simulate the aging (harvest problems and storage) of a seed lot. It also provides an estimate of the germination percentage that will be obtained under less-than-ideal conditions. Seeds are exposed to high temperatures and high humidity prior to a standard germination test. AA testing is commonly performed on soybeans and corn.

Another beneficial test that can be conducted in the field or laboratory is the mechanical seed damage test or bleach test. Damaged seeds will appear swollen after about 3 min of soaking in bleach. The percentage of damaged seeds from harvesting or processing can be determined quickly, and adjustments can be made to the equipment to minimize future damage.

EXERCISES

Experiment 1. Germination Test

This exercise is designed to provide students the necessary tools to determine the actual germination percentage of a seed lot. To determine the germination percentage, each species of the seed chosen may need to be placed at different temperatures, although most seeds germinate between 21°C and 23°C. The germination paper towels (commonly called "rag dolls" in the industry) must be kept moist throughout the experiment. After a seed has germinated and has been recorded, it is normally removed. After 10 to 14 days, the majority of agronomic species will have reached maximum germination percentage. Recorded data will be both the number of seeds that germinate to produce healthy, normal seedlings and the number of ungerminated seeds.

MATERIALS

The items needed for each student or student team include the following:

- 10 of each seed provided (corn, soybean, bean, garden pea, tomato, and marigold, and others)
- Distilled water
- Germination paper
- Rubber bands
- Markers
- Sealable plastic container
- Germination or growth chamber

Follow the methods in Procedure 35.1 to complete this experiment.

ANTICIPATED RESULTS

Germination should have begun on the majority of agronomic crop seeds by the first observation date. Vegetable and flower seedlings may not begin germinating until the second or third observation date. Most seeds from herbaceous plant material should be concluded in 21 days or less.

Procedure 35.1
Germination Test

Step	Instructions and Comments
1	Select 10 seeds randomly from a seed lot.
2	Place the seeds along the centerline of the germination paper towel.
3	Roll the paper towel into a cylindrical tube. Place a rubber band on each end and label each towel.
4	Thoroughly soak the germination paper in distilled water, and place the towels in a sealed container in the germination/growth chamber at the appropriate temperature (as indicated by the instructor).
5	Check the seeds at 3-day intervals until maximum germination has been recorded (Figure 35.1a–c). This may take 10 to 14 days or longer (if determined by the instructor). Remoisten the towels if needed.
6	Calculate the final germination percentage ([number of germinated seeds/total number of seeds] × 100).

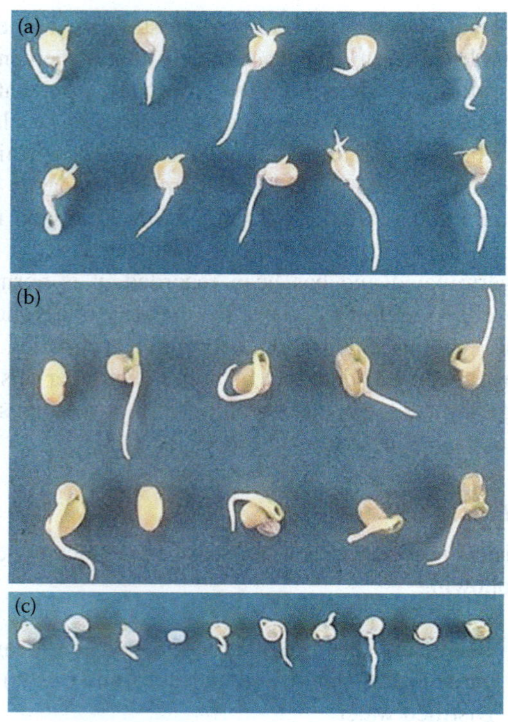

FIGURE 35.1 Germination using the paper towel method illustrating 100% germination for corn (*Zea mays*) (a), 80% germination for soybean (*Glycine max*) (b), and 90% germination for garden pea (*Pisum sativum*) (c). (Courtesy of Jeneen Abrams.)

Most fresh commercial seeds should have germination percentages above 85%. A variation that can be added to the experiment is to conduct germination tests on old seed lots and fresh seed lots to illustrate the reduction in germination percentages as seeds age.

QUESTIONS

- If a seed lot has a published germination percentage of 95%, why would a grower obtain only 80%?
- Why would a grower need to know the germination percentage for a seed lot?

Experiment 2. TTZ Tests

This exercise is designed to provide students the necessary tools to determine the potential for seed germination percentage within a seed lot. This is a more rapid test in comparison to the standard germination test. This test depends on TTZ chloride (2,3,5-triphenyl TTZ chloride), which produces a chemical reaction when applied to living tissue. A red coloration indicates cellular respiratory activity. The intensity of the red staining is correlated with the rates of respiration of different tissue types. Dark-red staining indicates damaged or deteriorated tissues, and no staining is indicative of non-respiring or dead tissue. The potential seed germination percentage is calculated by the proportion of seeds with stained embryos and essential germination structures. Potential germination percentage may be higher than the percentage obtained from a standard germination test. Dormant and diseased seeds may stain, but may not germinate and/or produce healthy seedlings due to additional internal or external factors affecting their successful germination.

MATERIALS

The items needed for each student or student team include the following:

- 10 of each seed provided (bean, corn, garden pea, wheat, and others)
- Distilled water and pH meter
- 10 cm–diameter Petri dishes (3 per group)
- Scalpel or razor blade
- 0.1% TTZ chloride solution (To prepare 0.1% TTZ solution, mix 0.1 mg of 2,3,5-triphenyl TTZ chloride in 99 mL of distilled water. Adjust the pH to 7.0.)
- Marker
- Paper towels
- Dark chamber (box or inside a cabinet)
- Microscope or hand lens

Procedure 35.2
TTZ Test

Step	Instructions and Comments
1	Select 10 seeds (of each species provided) that have been soaked overnight in distilled water. There will be a total of 30 seeds.
2	Cut the seeds longitudinally through the embryo with a scalpel or a razor blade (Figure 35.2a).
3	In a Petri dish, pour 0.1% TTZ solution (2,3,5-triphenyl TTZ chloride) to cover the bottom.
4	Place the seeds, cut side down in the TTZ solution. Soak in a dark location for 1 to 1.5 h.
5	Remove the seeds from the TTZ solution when they are light pink. Rinse and cover the seeds with distilled water until each seed can be evaluated.
6	Place the seeds on paper towels for evaluation of the red staining. Categorize each seed's germination potential by observing the embryo and other essential germination structures (Figure 35.2b). Small seed or seed structures may need to be observed with a hand lens or under a microscope. (Variations of red-stained areas are normally respiring tissue, dead tissue will not be stained, and dark red–stained areas indicate damaged or deteriorated tissue.)
7	Determine the potential for seed germination percentage of the seed lot ([number of seeds with viable germination structures/total number of seeds] × 100).

Follow the methods in Procedure 35.2 to complete this experiment.

Note: Most seeds require 1 to 1.5 h soaking in the TTZ solution, but soybeans require 3 to 4 h at room temperature to complete the test. The soaking time may be decreased by placing the seeds at higher temperatures (~35°C).

ANTICIPATED RESULTS

Students may have a difficult time evaluating the plant structures and staining patterns unless examples are provided as a guide. The Association of Official Seed Analysts and the International Seed Testing Association have good examples on their websites (http://www.aosaseed.com/). The embryo structures will stain a darker red than cotyledon tissues. Radicle tips will normally have a redder staining intensity than other embryo structures.

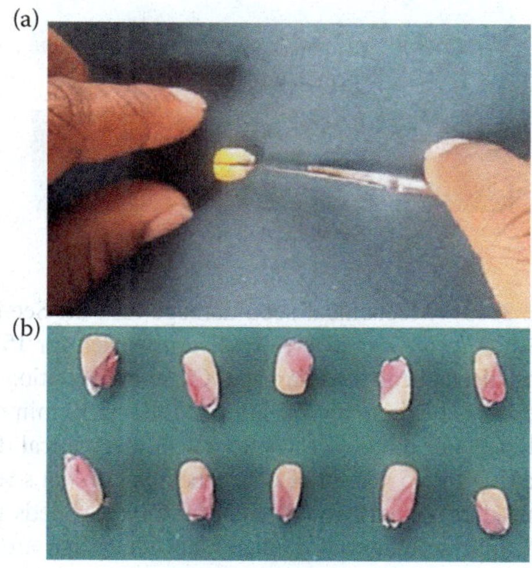

FIGURE 35.2 TTZ test. (a) Cutting seeds of corn (*Zea mays*) in half longitudinally to expose the embryo prior to immersion in 0.1% TTZ solution. (b) Results of the TTZ test showing the pinkish-red stain, indicating the respiratory activity of the viable embryo. (Courtesy of Jeneen Abrams.)

QUESTIONS

- Why is the TTZ test only a measure of the potential seed germination percentage and not the actual germination percentage?
- Why do different structures of an embryo have varying stain intensities?

Experiment 3. Mechanical Seed Damage Test (Bleach Test)

This exercise provides a quick and easy method of determining the percentage of damage that seed lots suffered from harvesting, threshing, or seed conditioning. This test is specifically used for soybean and dry edible beans. An advantage of this test is that it can be conducted quickly under field conditions or in the laboratory. Each seed crop has an optimum moisture content for harvesting, and if seeds are allowed to drop below this level before harvesting, the seed coats are susceptible to cracking. Combines and threshers are also a potential threat that cause damage to the seed when they are not in proper adjustment. If damage is noted by this test conducted while in the field, harvesting equipment can immediately be adjusted. Damaged seeds are a major concern when the seed is sold for "whole seed markets," such as for seed production.

Procedure 35.3
Mechanical Seed Damage Test (Bleach Test)

Step	Instructions and Comments
1	Randomly select 100 seeds from the seed lot.
2	Place the seeds in a Petri dish (Figure 35.3a) containing a bleach solution. (Caution: Bleach may damage clothes.)
3	Soak the seeds for 15 min. (If blistering and swollen seeds are not observed, continue soaking for a total of 30 min.)
4	Pour off the bleach solution and place the seeds on a paper towel for evaluation (Figure 35.3b).
5	Count the number of swollen seeds.
6	Calculate the percentage of damaged seeds ([number of swollen seeds/total number of seeds] × 100).

MATERIALS

The items needed for each student or student team include the following:

- 100 soybean or edible bean seeds (eliminate the severely damaged)
- Household bleach solution (sodium hypochlorite—mix one part bleach to five parts water)
- 10 cm–diameter Petri dishes
- Paper towels

Follow the methods in Procedure 35.3 to complete this experiment.

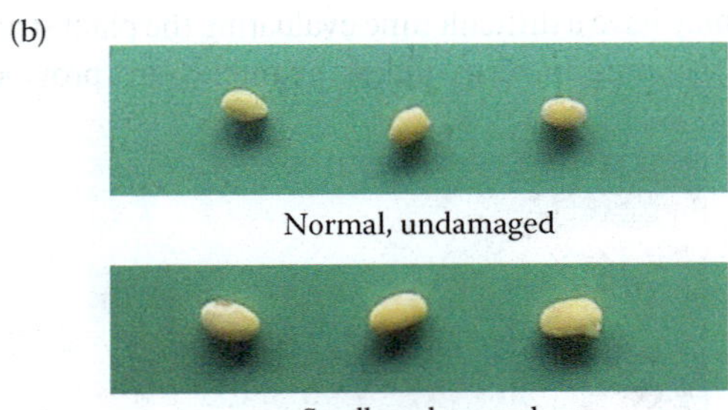

FIGURE 35.3 Mechanical seed damage test. (a) Seeds (100) of soybean (*Glycine max* Merr.) soaked in bleach for 15 min to determine mechanical seed damage. Close examination reveals that differences begin to appear after as little as 10 min of soaking. (b) Results of the bleach test to detect mechanical damage. After soaking for 15 min in bleach, seeds that do not swell have little to no damage and would likely germinate. Seeds that are wrinkled indicate marginal damage, and seeds that are clearly swollen have significant mechanical damage and would probably not germinate. (Courtesy of Jeneen Abrams.)

ANTICIPATED RESULTS

Seeds that do not swell have sustained very little or no damage. Wrinkled seeds (seed coat appears blistered) are only moderately damaged and will normally germinate the subsequent season. Swollen seeds that have absorbed large quantities of bleach are considered damaged and normally do not have the ability to germinate. When the swollen seed percentage is 10% or above, harvest or handling equipment adjustments must be made to lower the quantity of damage seeds.

From commercial bags of soybeans, our results have averaged approximately 50% to 60% of the seeds not swollen; 30% to 40% with blistering; and the remainder, swollen.

QUESTION

- If a mechanical damage test is performed on a trailer load of soybean seeds and the percentage of swollen seeds is 35%, for what can this load of seeds be used?

LITERATURE CITED AND SUGGESTED READING

Acquaah, G. 2002. *Horticulture: Principles and Practices*. New Jersey: Prentice Hall, 760 pp.

Association of Official Seed Analysts. 1999. *Rules for Testing Seeds*. Lincoln, Nevada: Association of Official Seed Analysts, pp. 445–481.

Baskin, C. *Seed Facts*. Information Sheet 1062. Mississippi: Mississippi State Cooperative Extension Service, Mississippi State University, 2 pp.

Bassett, M. J. 1986. *Breeding Vegetable Crops*. Westport, Connecticut: Avi Publishing, 584 pp.

George, R. A. T. 1985. *Vegetable Seed Production*. Essex, United Kingdom: Longman Press, 320 pp.

International Seed Testing Association. 1999. International rules for seed resting. *Seed Sci Technol (Suppl)* 27:1–333.

McDonald, M. B. and L. O. Copeland. 1989. *Seed Science and Technology Laboratory*. Ames, Iowa: Iowa State University Press, 231 pp.

McDonald, M. B. and L. O. Copeland. 1997. *Seed Production: Principles and Practices*. New York: Chapman and Hall, 749 pp.

Thompson, J. R. 1979. *An Introduction to Seed Technology*. New York: Halstead Press, 252 pp.

36 Practices to Promote Seed Germination

Caula A. Beyl

Concept Box 36.1

- Many seeds only require adequate temperature and moisture to imbibe and germinate. Others require more complex conditions to be satiesfied before germinating.
- Seeds with hard seed coats that act as a barrier for gas for gas exchange, imbibition of water or leaching of inhibitors can be encouraged to germinate by processes that break down or weaken the seed coat, such as microbial action, passing through digestive tracts of birds or other animals, insect activity, temperature fluctuations, or even fire. To encourage seed coats with sandpaper, or even use a hammer to crack the impervious seed coat. These processes are called scarification.
- Some seeds require exposure to chilling temperatures for varying lengths of time before they will germinate. Layering the seeds in a moist environment at chilling temperatures for the required duration of time is called stratification and can overcome the need for the need for the seeds to go through the winter before germinating.
- Many species of plants have requirements for light, darkness, fire, smoke, or even a period of time to allow the embryo to complete development (afterripening) to germinate.

In nature, seeds have a varied menu of mechanisms that either prevent them from germinating when there is a little chance of them surviving or delay germination to allow as wide dispersal as possible. As a propagator, an understanding of these mechanisms and how to satisfy them is essential to obtain a high percentage of both germination and seedling establishment. To do this, various seed treatments have been developed to overcome these fail-safe mechanisms, whether they reside in the embryo itself or in the seed coat or fruit that surrounds the embryo.

Seed dormancy occurs when a seed will not germinate even if placed under ideal conditions. This may be a simple dormancy due to the seed coat physically preventing the seed from imbibing water or due to factors in the embryo of the seed or the flesh of the fruit that inhibit germination. If the seed is capable of germinating, but environmental conditions prevent germination, we call these seeds "quiescent." When there are two or more factors that prevent seeds from germinating, such as a hard seed coat and the need for a certain number of days of cold temperatures, this is termed "double dormancy." A secondary dormancy can occur when the seed is capable of germinating, but external conditions, such as high temperature, prevent it from doing so. Lettuce seeds, similar to many small-seeded plants, require light to germinate, but they are also sensitive to high temperatures (26–35°C) and undergo a heat-induced thermodormancy that prevents them from germinating even if later placed in the light and in an optimum-temperature environment.

Some seeds have a hard seed coat that acts as a barrier for gas exchange, water imbibition, or leaching of inhibitors of germination. Many legumes have a thick waxy cuticle and thick-walled macrosclereids in the epidermis that contribute to the hardness of the seed coat. Seeds that have tough, impervious seed coats do not germinate readily until the seed coat is weakened or removed. The seed coat acts as a barrier to water and oxygen movement into the seed. In nature, hard seed coats are broken down by different routes, such as soil microbial action, passing through the digestive tracts of birds or animals, insect activity, fluctuations of temperature, and even fire. "Scarification" is the technical term for this process of weakening or removing the seed coat, allowing the seed to take up water or oxygen or leach out inhibitors. In the laboratory, the natural processes of scarification can be simulated by using concentrated sulfuric acid, sandpaper, files, hot water, or even a hammer. On a large commercial scale, this can be accomplished using a rotating drum lined with an abrasive substance or sand paper (Figure 36.1). Sometimes, seeds are mixed with sand and then put into the rotating drum.

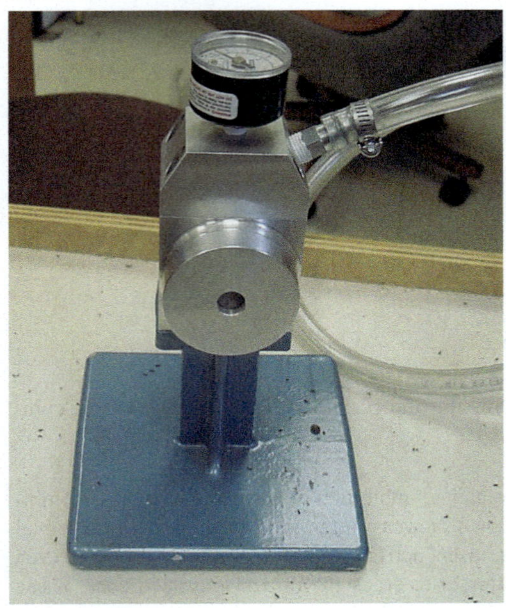

FIGURE 36.1 Hoffman Model SC-50 commercial seed scarifier capable of administering scarification treatment to abrade the seed coats of large lots of seeds.

Immersion in hot water will often soften the seed coat sufficiently so that imbibition or water uptake can occur. Several examples of seeds that require scarification for successful germination are listed in Table 36.1. Note that many of these species are members of the legume family.

Sometimes, seeds are capable of germinating immediately after maturation, but others require after-ripening or a period of time under specific conditions to germinate. This could consist of a period of storage under dry or moist, warm conditions to allow the embryo to finish its physiological and morphological development and/or until the embryo is competent to germinate. Some seeds can retain their viability after an extraordinarily long time. A date seed taken from several seeds found in an excavation in Masada, Israel, and carbon dated at about 2000 years old was germinated successfully after being soaked in hot water, then in a solution of nutrients followed by an enzymatic fertilizer made from seaweed (Roach, 2005). This is extraordinary as most seeds lose viability long before that amount of time has passed.

When a seed needs a prolonged moist, cold treatment to satisfy an inherent physiological dormancy of the embryo and develop into a seedling, propagators

TABLE 36.1
Species Whose Seeds Benefit from Scarification or Stratification

Species	Common Name	Family	Treatment
Albizia julibrissin Durazz.	Mimosa	Fabaceae	Acid scarification, 30 min
Cedrus atlantica (Endl.) Manetti ex Carr.	Atlas cedar	Pinaceae	Cold moist stratification, 14 days
Cercis canadensis L.	Eastern redbud	Fabaceae	Scarification then stratification, 30 days
Chamaecyparis obtusa (Sieb. & Zucc.) Endl.	Hinoki cypress	Cupressaceae	Cold moist stratification, 21 days
Cornus florida L. *C. kousa* Hance	Flowering dogwood	Cornaceae	Remove pulp and stratify, 90 to 150 days
Cotinus coggygria Scop.	Smoke tree	Anacardiaceae	Acid scarify, 30 to 60 min, then stratify, 3 months
Gleditsia triacanthos L.	Honey locust	Fabaceae	Acid scarification, 1 to 2 h
Gynocladus dioica (L.) K. Koch	Kentucky coffee tree	Fabaceae	Acid scarification, 4 to 6 h
Hibiscus syriacus L.	Rose-of-Sharon	Malvaceae	Mechanically scarify
Koelreuteria paniculata Laxm.	Golden rain tree	Sapindaceae	Scarify, 60 min, then cold stratify, 90 days
Magnolia grandiflora L.	Southern magnolia	Magnoliaceae	Stratify, 90 to 120 days
Olea europa L.	Olive	Oleaceae	Stratify, 60 days
Pinus strobus L.	Eastern white pine	Pinaceae	Stratify, 60 days
Pinus taeda L.	Loblolly pine	Pinaceae	No treatment needed
Prunus serrulata Lindl.	Oriental cherry	Rosaceae	Warm stratification, 20 to 60 days, then cold stratification, 90 to 120 days
Rhus glabra L.	Smooth sumac	Anacardiaceae	Mechanically scarify
Robinia pseudoacacia L.	Black locust	Fabaceae	Acid scarification, 1 to 2 h
Zelkova sinica Schneid.	Chinese zelkova	Ulmaceae	Hot water scarify then cold stratify, 60 days
Ziziphus jujuba Mill.	Chinese date	Rhamnaceae	Hot water scarify then cold stratify, 40 days

mimic this by using a process called "cold stratification." Stratification is also used in the case of epicotyl dormancy. In this case, the radicle or embryonic primary root will emerge without special treatments, but the seed must undergo a cold treatment before the epicotyl or the embryonic shoot will emerge. The term "stratification" arose because seeds were arranged (or stratified) between layers of sand or soil and then subjected to chilling at temperatures ranging from 35 to 45°F (1–7°C). The amount of time and the effective temperature under which a seed must be stratified depends on the species and its inherent requirement for chilling (often similar to the cold requirement needed by the dormant shoot buds before they will grow in spring) and the provenance or geographical source of the seeds. Seeds from individuals in northern latitudes may have a longer chilling requirement at a lower temperature than those at lower (and warmer) latitudes. The time period can be as little as 3 weeks for tree of heaven (*Ailanthus altissima* [Mill.] Swingle) and Norway spruce (*Picea abies* [L.] Karst.) or up to 20 weeks for Japanese maple (*Acer palmatum* Thunb.) or tuliptree (*Liriodendron tulipifera* L.). Propagators satisfy this chilling requirement of seeds naturally by layering seed in sand or other medium in pits, bins, frames, or boxes and allowing ambient winter temperatures and rainfall to provide the moist, cold treatment. In refrigerated stratification, a variety of containers can be used as long as they do not impede aeration, including simple Ziploc® bags.

Some seeds require light to germinate, most likely a mechanism to ensure that they are not buried too deeply to reach the surface of the soil successfully. These light-requiring seeds include lettuce cultivars like grand rapids and Waldmann's dark green, and species of *Alyssum*, *Petunia*, and *Coleus*. Others must have an absence of light. Examples of these include species of *Delphinium*, *Nemesia*, and *Gazania*. Whether light or darkness is required, the mechanism involves phytochrome-mediated responses (Chapter 3). With lettuce seed, treatment with gibberellic acid can substitute for the light requirement. When the embryo is the dormant part of the seed, the constraint on germination can be due to the lack of growth promoters, such as gibberellic acid, the presence of inhibitors such as abscisic acid, or the influence of a specific part of the embryo (cotyledons in the case of peach). Sometimes, this type of dormancy can be broken by the application of specific growth regulators such as gibberellic acid, cytokinins, or ethylene (Chapter 4).

No matter what the treatment, in the nursery or greenhouse, obtaining a high percentage of germination is not enough. It is also desirable that the seedlings emerge uniformly and grow vigorously as well. Many times, it is desirable to be able to hasten seed germination and seedling establishment, particularly if there is a chance of the environment not being favorable for germination. This is true even in areas of the world where agriculture is not as advanced as in the more industrialized nations. For example, if there is a narrow time window for planting a seed of a crop that requires it to become well established before a period of drought sets in, then every day spent getting the seed to germinate and the seedling to grow is crucial. Seed pregermination and seed priming are two techniques that can advance seed germination and subsequent seedling establishment.

Pregermination involves allowing the seeds to imbibe water and then germinate only until incipient radicle emergence (when the radicle has just appeared). The radicle or embryonic root is the first organ that emerges from the newly germinating seed. The seed is then dried down to nearly its original dry weight. When it is sown and allowed to regerminate, the germination events leading to seedling emergence from the soil are accelerated. Sometimes, the seeds are not dried but are placed in a gel matrix and fluid drilled using equipment designed to sow seeds suspended in a fluid gel.

Priming involves placing seeds in an environment where either osmotic or matric forces limit how far germination can progress. Water potential is a reflection of the free energy of water. The greater the concentration of water, the greater the free energy and the higher the water potential. Osmotic forces are caused by solutes that decrease the free energy and, thus, the water potential. Since the water potential of pure water is zero, decreasing the water potential causes it to be negative. In the case of priming, the osmotic forces are usually generated by either salt or polyethylene glycol added to the priming solution, with osmotic water potential values in the ranges of −0.5 to −1.5 MPa. These are the units of water potential. Some older scientific literature expresses the water potential in bars (1 bar = 0.1 MPa). The more negative the value, the greater the ability to hold or attract water. In the case of matrix priming, seeds are imbibed in moist vermiculite, silica, or clay. The free energy of water is decreased by adsorption onto the matrix particles. There are three phases of germination. In Phase I, the imbibitional or log phase, water enters the dry seed very rapidly in response to matric forces. A seed does not need to be viable to take up water in Phase I.

During priming, seeds take up water in Phase I, the imbibitional phase (Figure 36.2), because the driving forces for water entry into the dry seed are much stronger than those for the surrounding solution or matrix. Typical water potential values for dry seeds can range from −50 to −250 MPa. The Phase II of germination is the activation phase or the lag phase, so named because the rapid uptake of water slows dramatically, but even though it does not look like much is happening, this is a very busy phase for the seed. In this phase, which may last for several days, respiration starts, membranes reorganize, enzymes become active, and protein synthesis begins. At the same time, the stored food reserves of the seed are being metabolized and mobilized so that the seed can continue germination. Because

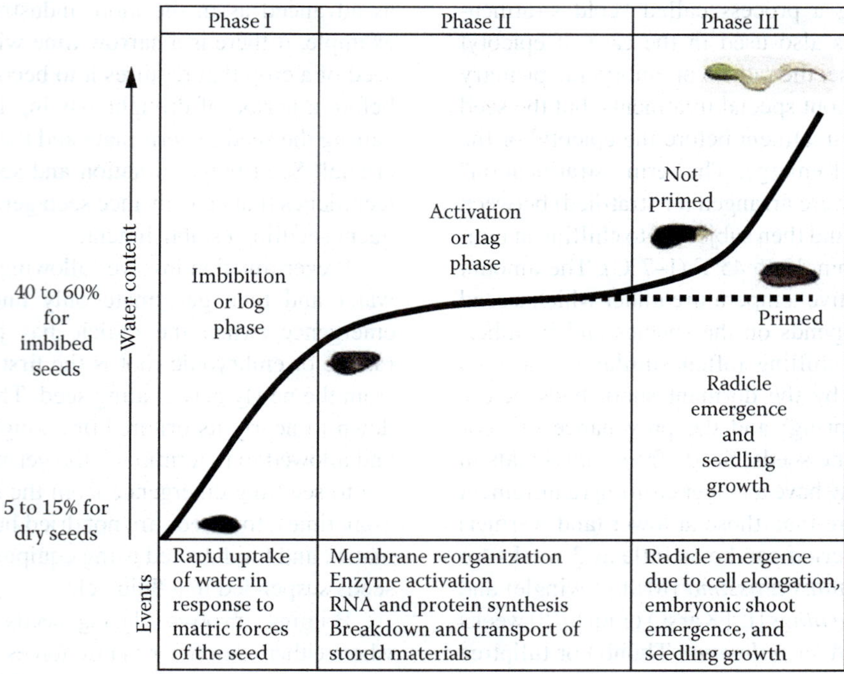

FIGURE 36.2 Water uptake over time, illustrating the three phases of germination and the effect of priming to hold germination in the activation or lag phase (Phase II).

they have taken up water, imbibed seeds have a much less negative water potential than dry seeds and cannot compete with the surrounding priming solution for water. In Phase III, cells in the radicle elongate, and the radicle protrudes from the germination seed. Radicle growth continues with cell division occurring, and the embryonic shoot emerges soon afterward. During priming, seeds are held in the Phase II of germination and cannot proceed to Phase III or radicle emergence. Holding seeds in the lag phase allows all of the important activation processes (membrane reconstitution, enzyme activation, protein synthesis, and others) to occur at optimum temperatures. Primed seeds are then redried. When the seeds are later sown and rehydrated, germination occurs at a faster, more uniform rate than what would occur for seeds that have never been primed. Priming allows seeds to germinate faster under less-than-optimum temperature conditions and overcomes the problems inherent with crops that undergo thermodormancy when germinated under high temperatures, such as lettuce and celery.

The following experiments will allow students to gain a hands-on experience on the use of scarification, stratification, and priming, and the use of liquid smoke to promote germination. Each student or a small group of students can represent one replication of the experiment.

EVALUATING GERMINATION

Students should discuss and agree on the criteria for the measurement of germination success for this experiment. What is germination? Is it when the seed imbibes water, when the radicle appears, when the embryonic shoot appears, or when the seedling emerges above the soil line? Once the criteria for germination have been discussed and developed and the students have decided what type of data should be collected, they can devise a common data sheet for recording the results of the following experiments. Germination or seedling emergence should be checked frequently so that the pattern of seedling emergence over time can be documented. This is a measure of the uniformity of germination, which is also an important consideration.

Experiment 1. Effectiveness of Various Scarification Techniques on Two Species in the Fabaceae (Legumes)

There are several species of trees whose seeds have seed coats that require scarification. *Gleditsia triacanthos* L. var. *inermis* (L.) Zab. (thornless honey locust) is a member of the legume family (Fabaceae) and, like many of its relatives, has a hard impervious seed coat. Honey locust is a medium-sized tree valued for its lacy-textured foliage. Although there are seedless cultivars available, many older trees are of seedling origin and have not been selected for either the thornless or the seedless characteristics of many of the new cultivars. *Albizia julibrissin* Durazz. (mimosa or silk tree), a member of the same family, is a medium-sized spreading tree with showy flowers common throughout the southern United States. Two other species that could serve as alternate choices for this exercise are *Robinia pseudoacacia* L. (black locust), a fragrant-flowered tree known for its ability to withstand harsh conditions and sites, and *Gymnocladus dioicus* (L.) K. Koch (Kentucky coffee tree), used in parks and large-scale landscape settings. If alternate species are chosen, acid scarification times must also be adjusted. It is also wise to check on the progress of scarification when using acid because the thickness of the seed coat, and thus, the time required to break it down, may vary from year to year. All of the species mentioned above (Figure 36.3) have the hard seed coat type of dormancy that acts as a barrier to germination.

The following five treatments will be compared in this exercise, which illustrate various approaches to scarification:

1. Control, with no scarification
2. Hot water soak
3. Cold water soak
4. Sandpaper abrasion
5. Concentrated sulfuric acid soak

FIGURE 36.3 Seeds of *Gleditsia triacanthos* (honey locust), *Albizia julibrissin* (mimosa), *Robinia pseudoacacia* (black locust), *Gymnocladus dioicus* (Kentucky coffee tree), *Koelreuteria paniculata* (Panicled golden rain tree) and *Cercis canadensis* (Eastern Redbud) prior to scarification treatment.

MATERIALS

The following items are needed for each student or group of 3 to 5 students:

- Seeds of *Glditsia tricanthos* var. *inermis* (honey locust) and *Albizia julibrissin* (mimosa). Since there are five treatments, each student or group will need at least 50 seeds of each species. Each student or group may act as one replication of the treatment.
- Thermometer
- Large beaker and hot plate
- Cheesecloth and string
- Labels and permanent marker
- 6 pack flats for each student or group of students
- Seed germination medium, such as Jiffy Mix
- Medium-grit sandpaper and a block of wood (4 in [10 cm]—a piece of 2 × 4 in works well)
- Concentrated sulfuric acid
- Face shield or eye protection and acid-resistant gloves
- Acid-impervious container, such as a large beaker, and a glass stirring rod

CAUTION: Handling concentrated sulfuric acid can be hazardous. This portion of the exercise can be done by the instructor as a demonstration. Adding water to concentrated acid will cause a sudden exothermic (heat-releasing) reaction, resulting in acid splashing up onto hands, face, and clothing. Never add water to concentrated sulfuric acid. Care must also be taken to make sure that seeds are dry when put into the acid.

Follow the experimental protocols outlined in Procedure 36.1 to complete this experiment.

Procedure 36.1

Comparison of Different Scarification Techniques to Enhance the Germination of Honey Locust and Mimosa

Step	Instructions and Comments
1	Each student or group should count out five lots with 10 seeds in each lot for mimosa and honey locust.
2	They should then prepare two flats by filling them with a seed germination medium, such as Jiffy Mix. One flat will be used for the mimosa seeds, and one flat, for the honey locust seeds.
3	Using a label, score five trenches approximately 1/2 in deep lengthwise along the surface of the germination medium in the flats. Each trench will hold 10 seeds from one treatment of one species. Prepare labels for each of the five treatments and place them at the head of their respective trenches.

4 Since control seeds receive no treatment, they can be sown directly into their trenches in the appropriate flat for each species.

5 Place a large beaker on a hot plate, fill with water, and bring to a boil. Once the water has boiled, remove the beaker carefully from the hot plate and allow the water to begin cooling. Place the seeds for the hot water treatment in a square of cheesecloth cut large enough to hold the seeds easily and tie with a string. If the string is long enough to hang over the side of the beaker, it can be used to retrieve the cheesecloth seed bag. Check the temperature of the water, and when it has cooled to 180°F (82°C), immerse the cheesecloth bag containing the seeds (Figure 36.4). If the beaker is large enough, more than one bag of seeds can be accommodated. The seeds should be left to soak in the hot water as it cools. After 24 h, the bags are removed and opened, and the seeds are sown in the flats.

6 At the same time that the seeds are placed in the hot water, place a second set of seeds in a corresponding beaker containing cold water. After the seeds are soaked for 24 h, they can be removed and sown in the flats in their respective trenches.

7 Seeds to be mechanically scarified are placed on one sheet of medium-grit sandpaper. Use another piece of sandpaper wrapped around a piece of 2 × 4 in wood (Figure 36.5), and use a circular motion to abrade the seeds between the two sheets of sandpaper. Check the seeds periodically to determine if the seed coat has been abraded enough to just begin to see the interior tissue, and place them in the flats in their appropriate trench.

8 Acid scarification requires care. The instructor may wish to conduct this portion of the exercise as a demonstration. Enough concentrated sulfuric acid should be placed in a dry beaker to constitute twice the volume of the seeds. Dry seeds are carefully placed in the concentrated sulfuric acid and allowed to remain for the recommended duration of time for the type of seed being treated (Figure 36.6). If this duration is uncertain, seeds may be checked periodically. Seeds that have received enough acid scarification are dull and may be pitted on the surface. Decant the sulfuric acid off by slowly pouring it, a small portion at a time, into a much larger volume of water. Never add water to concentrated sulfuric acid (see previous cautionary note). The seeds can then be rinsed by scraping them into a beaker containing a large volume of rinse water, decanting the water and blotting the seeds dry. Seeds should then be sown in the flats in their appropriate place and labeled with the species name and treatment.

9 Check the flats every 2 to 3 days and count the number of seedlings to emerge each time. Graph the number of germinated seeds over time. Note not only the final percentage of germination, but also the uniformity of the emergence of seeds from each treatment. After 2 months, finalize the data and prepare the report.

FIGURE 36.4 Cheesecloth bag containing seeds of golden rain tree being immersed in 82°C (180°F) hot water for scarification treatment.

FIGURE 36.5 Seeds of the golden rain tree being scarified by abrasion between two pieces of sandpaper—one serves as the base, and the other is wrapped around a piece of 2 × 4 in wood to make handling easier.

FIGURE 36.6 Appearance of golden rain tree seeds only 30 min after being immersed in concentrated sulfuric acid for scarification treatment.

ANTICIPATED RESULTS

Scarification will not only enhance the percentage of germination, but will also result in more uniformity of seedling emergence. Results will vary from year to year due to differences in the thickness of the seed coats. Hot water is more effective than soaking in cold water. The control seeds should have a very low percentage of germination.

QUESTIONS

- What type of equipment is available for large-scale scarification of seeds?
- In a situation where the seed has been scarified, but the resultant germination rate is still poor, what are some of the possible reasons?
- Why is the uniformity of germination important as well as the percentage of germination?
- What happens if the seed is kept too long in the concentrated sulfuric acid?
- If you are working with a species that emerges after fire, what types of treatments could you devise to encourage germination?

Experiment 2. Response of Golden Rain Tree and Eastern Redbud to Stratification with and without Scarification

Koelreuteria paniculata Laxm. (golden rain tree) is an ornamental tree valued for its yellow flowers and brown seed panicles, which make a sound like rain in the wind. *Cercis canadensis* L. (eastern redbud) is a small tree native to the United States and valued for its flowers that are purplish red in bud, later opening into a rosy pink with a purplish hue. There are also cultivars that have white flowers. Seeds of both of these species require scarification prior to stratification. Various techniques are used to stratify seeds, including stratification pits, bins, and frames; sowing in fall and letting stratification occur naturally; or controlled stratification under refrigeration. For this experiment, seeds will be stratified in moist sphagnum peat. Other media may be used as well, including vermiculite, hydrogels, or even sand.

Procedure 36.2

Response of Two Species to Scarification and Stratification

Step	Instructions and Comments
1	Two species, honey locust and golden rain tree, are used in this experiment; however, any other species that require scarification followed by stratification can be used as well. Each student or group of students should take two Ziploc® bags and label them with the species name and the treatment—control (not scarified and not stratified), scarified then stratified, and lastly, stratified only. A handful of peat (or other stratification medium) that has been moistened and squeezed with hand pressure to remove excess water should be placed in the Ziploc® bag. Place 20 scarified seeds of each species in the Ziploc® bag and mixed thoroughly with the moist peat medium. Label the bag and then place it in the refrigerator for cool moist stratification.
2	The same procedure as in step 1 should be followed for the seeds that have not been scarified to determine if scarification is necessary prior to stratification for it to be effective.
3	The stratification treatments should be observed weekly for evidence of germination or to check for fungal contamination.
4	After 4 weeks of stratification treatment, the seeds can be removed from the bags and sown in a flat containing a 1:1 mixture of peat and perlite. At the same time, plant seeds that have received neither scarification nor stratification treatment as controls. Water the seeds thoroughly.
5	Observe for germination weekly and record the cumulative germination over time.

MATERIALS

The following items are needed for each student or team of 3 to 5 students:

- Seeds of golden rain tree (*Koelreuteria paniculata*) and eastern redbud (*Cercis canadensis*). These should be divided into three lots, with only one lot of seeds subjected to scarification (see Experiment 1). This could be done ahead of time by the instructor if time is a constraint. For each student or team of students, 60 seeds of each species will be needed.
- Quart-sized Ziploc® bags
- Sphagnum peat moss
- Standard flats, 21 × 11 in, filled with an appropriate medium (a 1:1 mixture of peat and perlite will work well)
- Labels and permanent marker
- Refrigerator

Follow the experimental protocols outlined in Procedure 36.2 to complete this experiment.

ANTICIPATED RESULTS

Seeds receiving neither scarification nor stratification will exhibit the poorest percentage of germination. Both species of seeds will exhibit better germination when scarified prior to receiving stratification treatment. Eastern redbud may germinate at a greater percentage rate and faster than golden rain tree because it only requires 30 days of stratification, but golden rain tree germinates better with a 60 day period. Fungal contamination may occur in the treatment containing sphagnum peat beyond moisture holding capacity.

QUESTIONS

- Sometimes, seeds stratified in peat moss or other organic materials mold. What can be done to minimize the likelihood of losing seeds in this manner?
- If stratification at 5°C is effective for satisfying the cold requirement of the seed for germination, why do temperatures below freezing not work better?
- Does the source of the seeds for some species have an impact on the required length of the stratification period?
- What other materials would be appropriate choices as stratification media?

Experiment 3. Effect of the Priming of Lettuce and the Terra-Sorb® Pregermination of Sweet Corn on Germination under Temperature Stress

All species have a temperature optima for seed germination. When the temperature of the germination environment goes too much above or below these levels, germination suffers and, in some cases, is prevented altogether. The supersweet cultivars of sweet corn (*Zea mays* L.) have genes (e.g., *su1*, *sh2*) that cause them to accumulate sugars at the expense of starches in the endosperm of the seed, making the seeds appear shrunken. Seeds of sweet corn, particularly the supersweet cultivars, are susceptible to chilling injury if germinated at temperatures below 10°C. Many crops, including tomato, cotton, and okra, whose origins were in tropical regions, have this same susceptibility. To gain market advantages, many farmers plant sweet corn early and, thus, risk exposure to chilling temperatures during germination. Terra-Sorb is a superabsorbent starch polymer that can absorb 3 to 400 times its weight in water, forming a hydrogel that has been used for germinating seeds, improving growing media, and as a root coating on bare root trees prior to planting. In the experiment below, it is being used as a pregermination medium for sweet corn.

Lettuce (*Lactuca sativa* L.) exposed to temperatures at or above 30°C during germination will develop thermodormancy. Even if temperatures are then reduced to more optimum levels, the seeds will not germinate. Priming with an osmoticum, such as KNO_3 or polyethylene glycol, overcomes thermodormancy, allowing the lettuce to germinate normally.

In this experiment, a solution of KNO_3 is used to prime the lettuce seeds. Other salts have been used effectively as well as polyethylene glycol solutions.

MATERIALS

The following items are needed for each group of 3 to 5 students:

- Seeds of lettuce and a 'supersweet' sweet corn, such as 'Silver Queen'
- 0.2% solution of potassium nitrate (KNO_3) for priming (alternatively, 1% potassium phosphate [K_3PO_4] or polyethylene glycol solutions of various molecular weights can be used)
- 25-g/L Terra-Sorb®, a superabsorbent starch polymer (Industrial Services International, Bradenton, Florida)
- 8 Petri dishes containing Whatman #1 filter paper for each student or group
- Labels and permanent marker pen
- 1 125 mL-Erlenmeyer flask per group
- 1 plastic pipette connected by tubing to a source of air to provide aeration in the priming solution for each flask
- 1 shallow tray or large weighing boat per group
- Fine basket strainer for rinsing seeds
- Paper towels for blotting seeds dry

- Refrigerator (to impose chilling stress on sweet corn)
- Seed germinator capable of attaining a temperature of 30°C for lettuce germination

Follow the experimental protocols outlined in Procedure 36.3 to complete the experiment.

ANTICIPATED RESULTS

Priming will allow lettuce seeds to germinate even in temperatures that would normally induce thermodormancy. Alternative treatments could be performed comparing shorter and longer times in the priming solution before the seed is dried and sown. Seeds may be held in the priming solution for as long as 2 weeks, without

Procedure 36.3

Evaluation of the Pregermination of Sweet Corn and the Priming of Lettuce for Germination Outside the Optimal Temperature Range

Step	Instructions and Comments
1	Each student or group should count out 20 lettuce seeds each for the control and the priming treatments at two temperatures (25°C and 30°C). Each student or group should also count out 10 sweet corn seeds each for the control and the Terra-Sorb® pregerminated treatments at two temperatures (5°C and 25°C).
2	The lettuce seeds designated for the priming treatments should be placed into an Erlenmeyer flask containing the priming solution (0.2% KNO_3). A plastic pipette inserted into the flask and connected to a source of air (Figure 36.7) will aerate the priming solution. Anoxia or lack of oxygen in the priming solution reduces the viability of the seeds. Lettuce seeds should be left in the priming solution for 48 h, after which they are removed, rinsed, and allowed to redry for 24 h. These seeds are now designated as "primed."
3	Two Petri dishes lined with Whatman #1 filter paper should be prepared to receive the primed lettuce seeds, one for the 25°C and the other for the 30°C environment. Place 20 primed lettuce seeds in each and moisten the surface with water until it glistens. Prepare two more Petri dishes and follow the procedure above, but using dry lettuce seeds. Place one primed and one control treatment in the 30°C incubator. The other primed and control treatment may be left out at room temperature (25°C) until germination.
4	Terra-Sorb® should be hydrated by placing 25 g in a liter of water. Make up a sufficient quantity to create a volume 2 to 3 times that of the seeds. Mix the Terra-Sorb® matrix with the sweet corn seeds and spread the mixture in a shallow tray to allow the seeds to pregerminate for 24 h (Figure 36.8).
5	After 24 h, the Terra-Sorb pregerminated seeds of sweet corn should be rinsed using the strainer, and then the seeds should be blotted to remove free water. Spread the seeds onto paper towels and allow them to redry for 24 h.
6	Place 10 sweet corn seeds that had been pregerminated in the Terra-Sorb® in a Petri dish lined with Whatman #1 filter paper and moisten the surface of the paper until it glistens. As a control, prepare 10 sweet corn seeds that have received no pregermination treatment in the same way. Label both Petri dishes with the treatment and place in the refrigerator (5°C) to simulate exposure to chilling temperatures that sweet corn planted too early in spring would experience. Now, prepare two more Petri dishes just like those above and place on the bench top in the laboratory to germinate at room temperature (25°C).
7	Place 20 primed lettuce seeds in a Petri dish lined with Whatman #1 filter paper and moisten the surface of the paper until it glistens. Do the same for 20 lettuce seeds that were pregerminated in Terra-Sorb®. As a control, prepare 20 lettuce seeds that had received no pretreatment. Label the Petri dishes with the treatment and place in the seed germinator set to 30°C to simulate thermodormancy conditions. Now, prepare three more Petri dishes just like those above and place on the bench top in the laboratory to germinate at room temperature (25°C).
8	Examine the seeds in the Petri dishes daily for 10 days or until no additional germination has occurred. Collect data on cumulative germination percentage. Count a seed as germinated once the radicle has emerged. Graph the performance of the two seed treatments under temperature stress conditions (either chilling or thermodormancy) and compare with the performance of the two treatments conducted at room temperature. Calculate the T_{50} for germination for each treatment. T_{50} is the time required for 50% germination of the total number of germinated seeds (not the total number of seeds).

FIGURE 36.7 Seeds of lettuce undergoing priming with 0.2% KNO_3. Note how the solution is being aerated to prevent anoxia.

negatively impacting germination. Sweet corn seeds pregerminated in Terra-Sorb® will germinate faster and at a greater percentage than those not pregerminated.

An additional exercise that would demonstrate the rapidity of water uptake in Phase I relative to Phase II would be to have students measure the increase in fresh weight in the early hours of imbibition and then at a set time daily for several days thereafter. Because they are large enough to handle easily and increases in the seed fresh weight can be measured, sweet corn seeds work well for this.

QUESTIONS

- What types of materials could be used for osmotic priming? For matric priming?
- If a seed is nonviable, will it still imbibe water in the Phase I of germination?
- What does calculating the T_{50} of the treatments above reveal about how effective they are?
- What other species have seeds susceptible to chilling injury? To thermodormancy?
- Sweet corn in the field takes proportionally longer to germinate as the soil temperature declines. What are other risks to a seed's survival the longer that it is in the ground? What events occur during chilling that exacerbate the risk to the seed?

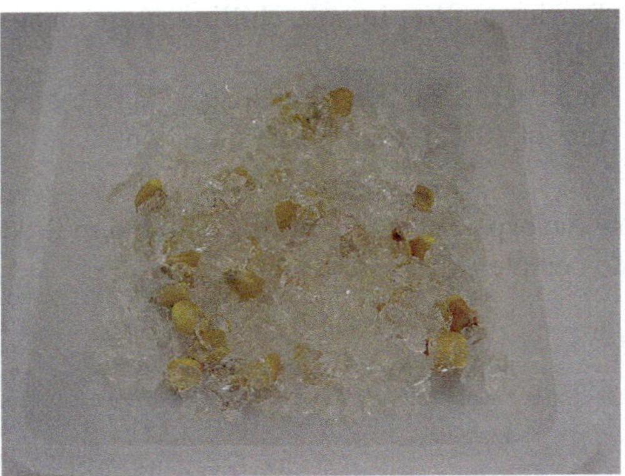

FIGURE 36.8 Seeds of standard sweet corn after being pregerminated in Terra-Sorb superabsorbent gel in a large weighing boat for 24 h.

Experiment 4. Evaluation of Liquid Smoke as a Stimulant of Germination

Some species in Mediterranean climates require exposure to fire or smoke to germinate. Plants have several mechanisms for regenerating after fire events, including the release of viable seeds stored in cones on the tree, breaking of seed dormancies by fire, and seed germination stimulated by smoke (Nelson *et al.*, 2012). Smoke can contain as many as 5000 compounds (Smith *et al.*, 2003), and one class of these has been named "karrikins," from the aboriginal word for smoke, which is *karrik*. Another type of germination stimulant found in smoke is cyanohydrin glyceronitrile (Nelson *et al.*, 2012). In aqueous solutions, this class of compounds releases cyanide (a known stimulant of germination for many species). Smoke water or liquid smoke stimulates the germination of a number of species, not just those requiring fire or smoke to germinate, including tomato, maize, rice, lettuce, some gymnosperms, and some grasses. This laboratory exercise evaluates the use of liquid smoke to promote germination in a wide range of species.

MATERIALS

The following items are needed for each group of 3 to 5 students:

- Seeds of 10 species (20 seeds of each species per group of students). Various species can be chosen for this exercise and can include vegetables like tomato, maize, rice, and lettuce; Arabidopsis; gymnosperms like pine, spruce, and fir; ornamental and turf grasses; and an assortment of weed species.
- Commercial liquid smoke containing no additives (use a 10% v/v aqueous solution of the commercial product)

Procedure 36.4
Evaluation of Liquid Smoke as a Stimulant of Germination

Step	Instructions and Comments
1	Each student or group should count out two lots with 10 seeds in each of the 10 species being evaluated.
2	Two Petri dishes lined with Whatman #1 filter paper should be prepared for each species, with one Petri dish labeled "liquid smoke" and the other labeled "control."
3	Prepare a 10% liquid smoke solution and use this solution to moisten the filter paper in Petri dishes designated for the liquid smoke treatment. Use water to moisten the filter paper in control dishes. Moisten the filter paper until it glistens but is not overly saturated.
4	For each of the 10 species, students should place 10 seeds in the liquid smoke Petri dish and 10 seeds in the control Petri dish and label them with the name of the species.
5	Examine the seeds in the Petri dishes daily for the next 10 days or until there is no further increase in germination. If the filter paper becomes dry, additional liquid of the appropriate type can be added to the Petri dishes. Collect data on cumulative germination percentage. Count a seed as germinated once the radicle emerges. Graph the performance of the two seed treatments for each of the 10 species.

- 20 Petri dishes containing Whatman #1 filter paper for each student or group
- Labels and permanent marker pen
- Pipettes

Follow the experimental protocols outlined in Procedure 36.4 to complete the experiment.

ANTICIPATED RESULTS

Many species respond to liquid smoke even though they are not normally thought of as fire- or smoke-responsive species. Depending on the species and inherent dormancies, seeds may or may not respond to the liquid smoke treatment. Those that do may exhibit an increased germination percentage, a faster rate of germination, and even an enhanced seedling growth rate.

QUESTIONS

- Why is it important to choose a natural brand of liquid smoke with no additives?
- What advantage is conferred to species that respond to smoke or fire by germinating?
- Describe how you could make your own "liquid smoke."

LITERATURE CITED AND SUGGESTED READING

Dirr, M. A. 1998. *Manual of Woody Landscape Plants: Their Identification, Ornamental Characteristics, Culture, Propagation, and Uses*. Champaign, Illinois: Stipes Publishing, 1187 pp.

Hartman, H., D. Kester, F. Davies, Jr., and R. Geneve. 2002. *Plant Propagation: Principles and Practices*, 7th Edition. Upper Saddle River, New Jersey: Prentice Hall, 880 pp.

MacDonald, B. 1986. *Practical Woody Plant Propagation for Nursery Growers*. Portland, Oregon: Timber Press, 669 pp.

Nelson, D. C., G. R. Flematti, E. L. Ghisalberti, K. W. Dixon, and S. M. Smith. 2012. Regulation of seed germination and seedling growth by chemical signals from burning vegetation. *Annua Rev Plant Biol* 63:107–130.

Roach, J. 2000-Year-Old Seed Sprouts, Sapling is Thriving. National Geographic News. November 22, 2005. http://www.news.nationalgeographic.com/news/2005/11/1122_051122_old. Accessed April 8, 2014.

Sabota, C., C. Beyl, and J. Biedermann. 1987. Acceleration of sweet corn germination at low temperatures with Terra-Sorb or water presoaks. *HortScience* 22:431–434.

Smith, C. J., T. A. Perfetti, R. Garg, and C. Hansch. 2003. IARC carcinogens reported in cigarette mainstream smoke and their calculated log *P* values. *Food Chem Toxicol* 41:807–817.

Young, J. A. and C. G. Young. 1994. *Seeds of Woody Plants in North America*. Portland, Oregon: Dioscorides Press, 416 pp.

37 Producing Seedlings and Bedding Plants

Holly L. Scoggins

CONCEPT BOX 37.1

- The following four stages of plug production, which are based on seedling/plant growth and development, are: (1) radicle emergence, (2) hypocotyl and cotyledon emergence, (3) the presence of true leaves, and (4) root and shoot growth until sufficient for transplanting.
- Plugs are seedlings that grow in their own tiny container—a cell in a plug tray.
- Once a plug is transplanted into the final container, the time spent in the greenhouse or nursery until shipping or sale is growing on or finishing.
- One way to grow bedding plants is in flats—often grown and sold in cell packs; groups of four to six plants comprise an individual, separable unit. Several cell packs are joined together and supported by a plastic tray to form a sales unit called a "flat." Larger pots (up to 6 in [15 cm]) are becoming much more popular though.
- The number of crop cycles—one group of bedding plants that occupy the same greenhouse space for a several-week period—is a "turn." The limited window of opportunity to successfully market spring bedding plants is the season.

ABOUT BEDDING PLANTS

Bedding plants are a big business. The United States Department of Agriculture (USDA) National Agriculture Statistics Service survey of the 15 largest producer states noted that there are 1537 businesses producing a wholesale value of $1.96 billion of bedding plants in 2012 (USDA, 2013). This category includes both seed and vegetatively propagated herbaceous plants for the landscape, along with herb and vegetable transplants. Bedding plants make up nearly half of all floriculture industry sales in the United States. Although vegetatively propagated annuals (most are tropical perennials) are rapidly gaining a share of this market, seed-grown annuals and perennials will probably always comprise the majority of sales. In fact, four of the top six bedding plants—garden impatiens (*Impatiens walleriana* Hook.f.), wax begonias (*Begonia* × *semperflorens-cultorum* hort.), marigold (*Tagetes patula* L. and *T. erecta* L.), and pansy/viola (*Viola* × *wittrockiana* Gams.) (Figure 37.1)—are produced from seed. The other two are propagated by both seed and vegetative cuttings—petunias (*Petunia* × *hybrida* hort. Vilm.-Andr.) and the number one bedding and landscape plant, geranium (*Pelargonium* × *hortorum* hort.). Vegetable transplants, another large crop category that is statistically lumped in with bedding plants, are also produced from seed.

TIMING IS EVERYTHING

There is a limited window of sales in the bedding plant industry—most greenhouse growers that specialize in bedding plants will make 75% or more of their gross income during spring months (March–June in most parts of the United States and Canada). The market is expanding for fall-planted, cool-season annuals such as pansies and violas as well as ornamental cabbages and kale, although it is still a relatively small fraction of the total bedding plant production (Figure 37.2).

Having plants ready at the appropriate time requires a clear understanding of what conditions are required for germination, growing on as plugs, and time needed til marketable size (Table 37.1).

CHANGES IN PRODUCTS

The primary product for sale to consumers has been flats—the traditional 11 × 22 in (28 × 56 cm) trays holding plastic cell packs in various configurations, such as

FIGURE 37.1 Impatiens are the number one bedding plant sold in the United States (wholesale value).

FIGURE 37.2 Interest is increasing in "cool-season" ornamental and edible annuals such as violas, pansies, kale and cabbages. (Courtesy of H. L. Scoggins.)

six packs with six cells each, for a total of 32 plants per tray. However, the public has shown an increased interest in larger bedding plants, including herb and vegetable transplants, and now, 3.5 in (8.8 cm), 4 in (10 cm), and 6 in (15 cm) pots make up well over half of the bedding and landscape plant wholesale value. Also gaining in popularity are mixed hanging baskets (more than one species per basket) and "color" or "edibles" mixed containers—larger planters ready for the porch or sundeck. An excellent and ongoing source of information for new selections and cultivars of bedding and vegetable plants is the nonprofit organization All-America Selections (AAS) (Figure 37.3). Their mission is to trial and promote superior seed-grown garden plants (both ornamental and edible) to the gardening public. The AAS website, http://www.all-americaselections.org, lists the annual winners and archives of those selected since 1933. Another window to the newest and best cultivars is also an annual springtime event up and down the coast of California. Greenhouse growers flock to the "pack trials" hosted by major seed companies and brands. These displays feature hot, new bedding plant varieties along with growing and marketing tips and techniques. University gardens, seed companies, and large growers may also feature bedding plant trials; the best of these trials offer plant performance data to both growers and the general public (Figure 37.4).

STARTING BEDDING PLANTS AND SEEDLINGS

In gardening parlance, direct seeding usually implies planting the seed directly into the landscape. This technique is useful for species that either resent transplanting, such as annual poppies (*Papaver rhoeas* L.) or those that have large, easy-to-handle seeds such as sunflower

TABLE 37.1

Producing Seedlings of Some Commonly Grown Ornamental, Herb, and Vegetable Bedding Plants

Name	Germination[a]	Growing on as Plugs	Scheduling	Comments
Ageratum houstonianum Floss flower	Requires light to germinate. Do not cover seeds. Germinates in 8 to 10 days at 24 to 25°C.	Reduce temperature to 18 to 20°C for stage 4.	Ready to transplant in 3 to 4 weeks. Crop can then be finished in 6 to 8 weeks, depending on the container size.	Sensitive to high salts—keep fertilizer levels low.
Begonia semperflorens (Cultorum hybrids) Wax begonia	Do not cover seeds. Requires high humidity and warm temperatures (26 to 27°C).	Supplemental lighting in plug stages 3 and 4 can hasten development. Optimal day/night temperatures of 22/18°C.	Later stages of plug production are relatively slow—up to 8 to 9 weeks. However, once transplanted, the crop can be finished in 4 to 5 weeks.	Tiny seeds—90,000/g. Pelleted seeds are much easier to handle.

(continued)

TABLE 37.1 (Continued)

Producing Seedlings of Some Commonly Grown Ornamental, Herb, and Vegetable Bedding Plants

Name	Germination[a]	Growing on as Plugs	Scheduling	Comments
Calendula officinalis Pot marigold	Requires darkness to germinate. Cover seeds. Germinates at relatively cool temperatures (20°C).	Keep temperatures (16–21°C) moderately cool up to stage 4.	Transplantable in 4 weeks, with 8 to 14 weeks to finish, depending on the time of the year—faster in late spring to summer (long-day plant).	Relatively large seeds (105/g) can be sown directly into packs or pots.
Capsicum annuum Bell, cayenne, and ornamental peppers	Germinates in 6 to 12 days at 21 to 24°C. Seeds can be covered or can be left uncovered.	Maintain relatively warm temperatures up to stage 4 (20–22°C).	Transplantable in 2 to 3 weeks; grow on for 3 to 5 weeks for green plant sales (no flowers).	Transplant no later than 3 weeks after germination for fastest establishment.
Catharanthus roseus Vinca, Madagascar periwinkle	Cover seeds with media. Germinate in very warm temperatures (24–26°C) and in darkness if possible (stage 1 only).	Drop temperature after germination to 23 to 24°C. Maintain relatively dry media—do not overwater.	A relatively slow bedding crop, vinca is ready to transplant in 4 weeks, but requires 9 or 10 weeks more to reach saleable size.	A good crop for late-spring production—grows poorly at cool temperatures.
Celosia argentea var. *plumosa* and var. *cristata* Plumed and crested celosia	Cover seeds. Germinates in 4 to 8 days at 24°C.	Grow up to stage 4 at 18 to 20°C. Reaches transplantable size fairly rapidly—2 weeks or so.	Grow on at moderately warm temperatures (19–21°C); should finish 5 to 6 weeks after transplanting.	Celosia plugs often come into flower before transplanting. Pinching the bloom/bud off will hasten root establishment.
Impatiens walleriana Bedding impatiens	Rapid germination at 21 to 26°C. High light intensity hastens germination time to 1 to 3 days.	Grow at 21 to 24°C with moderate fertilizer levels (100 mg/L of a complete fertilizer).	Can be transplanted to pots/flats in 4 to 5 weeks. Impatiens are one of the fastest bedding crops, with 10 to 11 weeks from seed to sale.	Although often grown in shady sites in the landscape, greenhouse performance is best at moderate to high light levels.
Lavandula angustifolia Lavender	Do not cover. Seeds require light to germinate. Temperatures of 18 to 24°C. Expect relatively low germination percentage (50%–70%).	Germinates over 2 to 3 weeks; transplantable at 4 to 5 weeks. Do not overwater, even at the early plug stages.	Not a fast crop. Requires 7 to 13 weeks to finish, depending on the size of the pot.	Choose newer cultivars for faster germination and crop time. Perennial in USDA zones 6 to 8.
Lobelia erinus Lobelia	Sow multiple seeds per plug cell. Do not cover seeds. Germinates in 5 to 15 days at 21 to 27°C, depending on the cultivar.	For stage 4 and beyond, lobelia grows best under moderately cool temperatures (16–19°C, day and night).	Another relatively slow bedding crop. Expect 7 to 12 weeks from transplant to finish.	Very tiny seeds (up to 45,000/g). Pelleted seed containing multiple seeds per pellet is much easier to handle.
Lycopersicum esculentum Tomato	Cover seeds and germinate at 21 to 24°C. Usually germinates quickly—3 to 9 days.	Reduced temperatures (18–19°C) and high light in stage 4 reduces stretching of seedlings.	Another quick crop—transplant 2 to 3 weeks after sowing, and finish as green plants in 2 to 3 more weeks.	The top-selling vegetable transplant. Watch for purplish cast to foliage—indicates phosphorus deficiency.
Ocimum basilicum Basil	Cover seeds and germinate at 21°C. Germinates in 5 to 10 days, depending on the variety or cultivar.	Grow on fairly warm temperatures—19–20°C. Do not overwater in stage 4.	The Italian types finish 4 to 5 weeks after transplanting— "flavored" varieties like lemon basil or the purple-leaf varieties take a week or longer.	Requires warm temperatures and abundant light for optimum growth—do not start seeds too early in spring.

(continued)

TABLE 37.1 (Continued)

Producing Seedlings of Some Commonly Grown Ornamental, Herb, and Vegetable Bedding Plants

Name	Germination[a]	Growing on as Plugs	Scheduling	Comments
Pelargonium × *hortorum* Garden geranium	Cover very lightly with vermiculite—light aids germination. Germinates in 5 to 10 days at 21 to 24°C.	Spends a long time in stage 4—up to 5 weeks from sowing to transplant. Supplemental light can be beneficial in early spring.	Crop time varies as to the cultivar and the container size, but averages 10 to 12 weeks after transplant to a 10 cm pot. High light levels produce best-quality plants.	Choose deep containers for optimum root development. Refer to the resources listed for in-depth cultural information (growth regulators, photoperiod, and others).
Petunia × *hybrid* Petunia	Most cultivars germinate uncovered. Many benefit from light and warm temperatures (22–24°C) to hasten germination, usually in 4 to 8 days.	Reduce temperature to 17–19°C. Most varieties are ready for transplanting within 5 to 6 weeks. Petunias benefit from the application of a weak fertilizer (75-mg/L N) as early as stage 2.	High light levels at a longer duration speeds flowering. Grows best in mid- to late spring. Plants are usually marketable from 6 to 10 weeks after transplanting, depending on the season and the latitude.	Small seeds (9000/g). Many companies offer pelleted seeds for ease of handling.
Salvia splendens Annual salvia	Traditionally a bit tricky to germinate. Best uncovered and not too wet. Germinates in 10 to 15 days at 24 to 25°C.	Reduce temperature to 19–21°C for stage 4. Transplanted fairly rapidly (1 or 2 weeks after germination).	One of the faster crops—usually 4 to 5 weeks from transplant to market for cell packs.	Salvia plugs are very sensitive to high salts—use weak, low-ammonium fertilizer, and use only in stage 4.
Tagetes erecta African marigold *Tagetes patula* French marigold	Cover seeds lightly. Rapid germination (2–3 days) at 24 to 27°C. Small growers may consider seeding directly to cell packs.	Lower temperatures after stage 1 to 20–21°C. Plugs are ready for transplanting in 4 to 5 weeks for *T. erecta* and 5 to 6 weeks for *T. patula*.	Short-day length (<12 h) hastens flower initiation in *T. erecta*. Depending on the species and the cultivar, crop times range from 7 to 11 weeks from sowing.	Long seeds. Use a 200-cell tray or a larger one. Purchase de-tailed and coated seeds for automated planting.
Viola × *wittrockiana* Pansy and viola	Darkness is necessary for germination—cover seeds. Start at 17 to 20°C with high moisture levels. Germination occurs in 5 to 10 days.	Reduce temperature to 16–17°C for stage 3 (2 weeks) and 13–16°C for stage 4 until transplant (2–3 weeks). Time from sowing to transplant-ready plugs is 6 to 7 weeks.	Supplemental lighting in both stage 4 and after transplant speeds the time to flower. Crop finishes 6 to 8 weeks after transplanting. Consider finishing outside in temperate climates.	One of the challenges of growing pansies for the fall market is starting them in the late summer—warm temperatures produce lanky, weak seedlings. Consider ordering plugs from a northerly source.

Source: Adapted from Nau, J. 1999. Ball Culture Guide: *The Encyclopedia of Seed Germination*, 3rd Edition. Batavia, Illinois: Ball Publishing, 243 pp.; Dole, J. M. and H. F. Wilkins. 2005. *Floriculture: Principles and Species*, 2nd Edition. Upper Saddle River, New Jersey: Prentice Hall, 1023 pp.; and Styer, R. C. and D. S. Koranski. 1997. *Plug and Transplant Production. A Grower's Guide*. Batavia, Illinois: Ball Publishing, 374 pp.

[a] Temperatures given in the "Germination" column refer to growing media temperature.

(*Helianthus annuus* L.). However, the majority of bedding plants find their way into the landscape by the two-step process of indirect seeding, where they are germinated and grown in the first step and transplanted into their permanent location in the second step. This can also be accomplished by the home gardener. However, this chapter focuses on the commercial production of bedding plants, so we will examine the methods of indirect seeding as they apply to bringing the bedding plants to the retail market, finished and ready for purchase by the home gardener.

Most seed-grown bedding plants are started in trays full of tiny cells (known as "plugs"). These can be bought from another grower that specializes in propagation or

FIGURE 37.3 AAS is an organization that trials and promotes the best seed-grown bedding plants and vegetables. (Courtesy of AAS.)

the finishing grower can start from scratch by germinating the seeds and growing the transplants themselves. In reality, most growers do some of both. The ratio is usually determined by a combination of factors—greenhouse space, labor availability, and difficulty/speed of plug production; some species are much trickier than others to grow. With the efficiency and decreasing cost of ground and air freight and the advances in packaging to reduce damage and losses in shipment, many growers are opting out of the seedling business altogether, happy to buy in the propagules, ready to transplant (Figure 37.5). This, in turn, has produced a relatively new category of greenhouse businesses—those that specialize in propagative material, grown for other greenhouse operations. An example is Speedling, Inc., based in Sun City, Florida, which produces over 1 billion ornamental and vegetable transplants annually, with production facilities across the United States and China.

There is another option for growers who start their own seeds—growers will sometimes plant seeds directly into the pot or cell pack that they will be finished in. This is called "direct seeding" (similar to direct seeding into the landscape). This method is useful for large or irregularly shaped seeds such as marigold (*Tagetes* spp.), but is very labor intensive and not done by larger, highly automated greenhouse operations. Another disadvantage is the bench space taken by the flat for the entire production time.

An "old-timey" way of starting bedding plant seedlings is to line the seeds out in furrows in open flats (trays of growing media) (Figure 37.6). Once the seedlings reach a certain size, it is necessary to "tease" the tangled, but delicate roots apart to transplant them. This process is tedious, often injurious to the seedling, and it must be done by hand. The term "transplant shock" refers to the stress that any seedling can undergo when uprooted from one situation to another. The stressed seedling is slow to recover, and transplant shock can add weeks onto a production schedule. Keeping the growing media intact as a unit around the roots is an important benefit of plug production.

Stages of Plug Production

Floriculture researchers have divided herbaceous seedling plug production into four stages, each with its own

FIGURE 37.4 Comparative (and colorful) performance trials of *Zinnia elegans* at C. Raker & Sons, Litchfield, Michigan. (Courtesy of H. L. Scoggins.)

FIGURE 37.5 Bedding plant plugs of various species, ready to transplant. (Courtesy of H. L. Scoggins.)

FIGURE 37.6 Although some growers still "row out" seeds to germinate in open trays, one seed per cell in a cell tray makes for happier transplants. (Courtesy of H. L. Scoggins.)

set of optimum cultural conditions. Stage 1 is the very beginning of germination as the root (radicle) emerges from the seed. In stage 2, the germination process continues as the root grows out into the soil and the shoot emerges, consisting of the hypocotyl (stem) and cotyledons (seed leaves) (Figure 37.7). True leaves begin to grow during stage 3. By the end of stage 4, the plug is well-rooted and ready for shipping or transplanting.

FIGURE 37.7 Stage 2 seedlings. (Courtesy of H. L. Scoggins.)

PLUG SCHEDULING

How long does it take to grow bedding plant plugs to transplant size? Of course, it depends on the species. Some are very fast—impatiens plugs can be ready to ship or ready to transplant within 4 weeks of seeding. Others take a while longer—wax begonias are slow to germinate and are slow growers throughout stages 3 and 4 (although once transplanted, can be finished out nearly as fast as impatiens). The key to profitability for a bedding plant grower is the number of transplant-to-market crop cycles, called "turns." The larger the plug cell size, the bigger the transplant and the faster that the crop can be finished or "turned." For example, it takes up to 6 weeks from transplanting to finish a 48-cell pack flat of impatiens from 512 plugs. By using 288s, the time to finish is reduced to 4 weeks. Although larger plugs cost more per cell to produce or buy in, the reduction in the time spent on the greenhouse bench usually more than makes up for it. Also, growers maximize profits by increasing the number of turns over a growing season. Choosing a tray size may seem like "six of one, half-dozen of the other," but the size of the plug greatly affects how long it takes to finish a crop. Using larger plugs as transplants can dramatically shorten the bench time, allowing for more turns in the same time period. For really quick turns, a relatively new concept in bedding plant production is the prefinished plant. The grower buys the plants from another grower, already transplanted into their final flats. The crop can then be finished in 2 or 3 weeks and resold. Obviously, this is an expensive way to increase turns, with the cost of the flats and the shipping cutting well into profit margins. However, sometimes, just having the product that your customer wants makes up for the smaller profit with greater goodwill.

Sizes and Shapes of Plugs

Plug trays hold a number of cells—tiny chambers where a seed can germinate and produce a self-contained root system that easily pops out of the tray at transplant time. Plug trays are rectangular in shape, with a fairly standard overall dimension of 11 × 22 in (28 × 56 cm); dimensions come in a variety of cell sizes, indicated by the number of cells per tray. Cell counts per tray can range from 36 up to 512 (Figure 37.8). Individual cells may be round, square, or octagonal, and all are perforated for bottom drainage. Cell depth also varies. The greater the cell volume, the more media that it can hold, providing a greater buffer for the seedling to help protect against drying out. Currently, 128s and 288s seem to be the most frequently used for seed-produced bedding plants. Trays with larger cells such as 78s and 36s are usually reserved for vegetatively propagated annuals, perennials, and woody plant liners.

FIGURE 37.8 Plug trays come in a wide variety of cell counts and shapes. (Courtesy of D. Steiner, Blackmore Co.)

Seeding and Transplant Technology

Plug production lends itself to mechanization and has revolutionized the bedding plant industry. Most commercial greenhouses that use plug trays also have an automated seeder, transplanter, or both. However, with the wide variety of plug tray configurations listed above, it is difficult to match the tray to an automated seeder or transplanter. A boon to the industry has been the formation of a "common element" or CE standard, cooperatively established by the major plug tray and equipment manufacturers. These standards will achieve uniformity among the most common sizes of plug trays and will ensure compatibility between trays, automatic seeders, and transplanting equipment.

FIGURE 37.9 Table-top needle seeder with dibbler. (Courtesy of D. Steiner, Blackmore Co.)

Seeding

Spend an hour or so placing barely visible petunia seeds by hand into a 512-cell plug tray, and you will see the need for automation. A wide range of technology is available. Suitable for even the smallest greenhouse operation, hand-held "wands" pick up seeds with a vacuum and deposit them one row at a time. The next step is plate seeding—an entire tray is sown at once. A variety of plates are necessary to match the seed size and the tray configuration. Semiautomatic and reciprocating seeders use needles to pick up seeds of all sizes (Figure 37.9). Semiautomatic seeders are manually moved into place over the tray, and reciprocating seeders are motorized. Drum or cylinder seeders use a revolving drum to pick up and deposit seed onto a passing plug tray (Figure 37.10). The fastest of the lot, a newer-model cylinder seeder, like the one pictured, can sow up to 1200 plug trays per hour.

FIGURE 37.10 Cylinder seeder system. (Courtesy of D. Steiner, Blackmore Co.)

FIGURE 37.11 Automatic transplanter in action. (Courtesy of H. L. Scoggins.)

FIGURE 37.12 A freshly coated batch of seeds at Ball Seed, Inc. (Courtesy of H. L. Scoggins.)

Transplanting

Automated transplanter machines can transplant from 3600 to 50,000+ plugs per hour. Depending on the model, plugs are plucked, tweezed, blown, poked, or cork-screwed from their cells and planted into packs or pots (Figure 37.11). Transplanters can be "manual" (mechanical) or motorized and range in price from \$3500 to \$200,000+. However, even moderately slow transplanters require another piece of automation for maximum efficiency—the flat or pot filler. Filling growing containers with soilless media by hand will simply not keep up with the transplanter.

Having a full plug flat with almost 100% of the cells containing a viable seedling is essential to the successful automation of the transplant process. This is where "patching" comes in. Most growers employ a manual "patch line" to replace dead seedlings or empty cells with living plugs. There is new (although very expensive) technology available involving computer scanning of plug trays, robotic removal of "duds," and refilling of empty cells with viable plugs.

BEDDING PLANT SEED IMPROVEMENTS

With the ubiquity of automated seeding in commercial greenhouses, seed companies have made great strides in improving seed uniformity, germination percentage, and ease of handling. Pelleting adds much-needed mass to ridiculously tiny seeds such as begonia. A clay coating, usually brightly colored, surrounds the seed, increasing its diameter (Figure 37.12). Several seeds can be combined into each pellet for species for which multiple seeds per plug are desired, such as lobelia. Even large but irregularly shaped seeds benefit from the coating to increase uniformity. The author passes along a caveat from experience: do not handle coated seeds with wet hands—an ounce of pelleted (and expensive) basil seed quickly turned to a sticky, unmanageable blob. Other improvements to germination and uniformity include refining and cleaning—simply removing the papery seed coat or other parts that may clog seeders and slow germination. "De-tailing" of marigold seeds is an example.

CULTURAL AND ENVIRONMENTAL FACTORS

Previous chapters (35 and 36) and exercises have touched on the cultural and environmental requirements necessary for the successful germination and growth of seedlings. We will elaborate further on these concepts, especially in terms of the commercial greenhouse production of seed-grown bedding crops.

Temperature

The vast majority of bedding plant species germinate the fastest and with the highest percentage at relatively warm media temperatures of 23 to 25°C. After germination, the temperature should be reduced a bit but should remain fairly warm for most species throughout stages 2 and 3, from 20 to 23°C. Heating of the root zone instead of just the air is the most efficient way to accomplish this. Heat mats with thermostats are useful for starting a few flats but are not suitable or cost effective for large-scale propagation. Some greenhouse systems grow on concrete floors with ebb and flood irrigation. The heating pipes or lines are imbedded in or under the concrete. Other growers run heating cables or pipes under benches. Polyethylene convection tubes that normally run the length of the greenhouse near the roof can also run under benches and direct warm air under the plant. Infrared heating is another option as it warms objects rather than air.

Moisture Management

High humidity aids germination. There are a number of ways to provide and conserve this humid environment. Germination chambers can provide a highly specialized environment—temperature and humidity are optimized for plug stages 1 and 2. Since the bright light necessary for quality stage 3 and 4 growth is not usually necessary for stages 1 and 2, plug trays can be arranged on shelves to maximize the use of space (Figure 37.13). Polytunnels or tents over the bench can be used to maintain high humidity levels. Boom irrigation is ideal for larger areas (Figure 37.14). Regardless of the method of overhead irrigation, choose mist heads that emit a very small droplet size—the finer the better. Monitor regularly as clogged or leaky mist heads can cause serious issues with under- or overwatering. Once the plug reaches stages 3 and 4, both the frequency and the duration of irrigation events should decrease. The media should remain moist but not saturated. Perpetually wet media prevents roots from growing to fill the plug cell and provides an ideal environment for water molds such as *Phytophthora* (Chapter 11). However, if the media completely dries out, the tender seedling will likely become wilted beyond saving. Irrigation management for such a tiny container as a plug cell is closer to an art than a science—it takes an experienced grower to manage the needs of several species and cell sizes at once.

FIGURE 37.13 Germination chambers with multiple shelves are an efficient use of space. (Courtesy of H. L. Scoggins.)

Nutrition

At the time of germination, seeds really require no supplemental fertilizer as they are "packing their own." However, as early as plug stage 2, as soon as the seedling is visible, the seedlings benefit from additional nutrients. This may be accomplished if the media has a starter charge or can be supplemented with a diluted water-soluble fertilizer. Plug-specific fertilizer formulations are available, but any complete fertilizer (macro- and micronutrients) with relatively low ammonium-N (60% or more nitrate-N) will suffice. Reducing relative amounts of phosphorus (instead of the usual 2:1:2 nitrogen phosphorus potassium [NPK] ratio) can slow down stem elongation and can produce a more compact seedling. Begin with 50 mg/L N every few days. Depending on the species, environment, and rate of growth, this can be increased to 100 or 150 mg/L throughout stage 4.

FIGURE 37.14 Boom irrigation system at Battlefield Farms, Rapidan, Virginia. (Courtesy of H. L. Scoggins.)

Media Testing

One of the most common problems associated with plug production is stunting of roots or death of the seedling from high salts. Many plant species grown as plugs are sensitive to high or low pH or high soluble salts. Soluble salts from fertilizer and/or irrigation water can build up quickly in plug cells as there is so little media to act as a buffer. Conversely, the small volume of media results in limited nutrient reserves. Frequent monitoring via measurement of the electrical conductivity (EC) and the pH of the seedling's root zone environment can detect both these problems before damage to the crop is done. Current media tests used by growers and analytical laboratories for greenhouse crops are the 2 water:1 substrate or 5 water:1 substrate (v/v), the saturated media extract, and the pour-through. These tests are used for pH and EC determinations and nutrient analysis. These testing methods each have their own interpretive ranges for substrate pH and EC, which, especially for soluble salt measurements, are not interchangeable. A simpler technique for plug testing is the press extraction method (PEM or press) developed at the North Carolina State University (Scoggins, 2002). One hour after irrigation/fertilizer application, a plug tray is selected, and the grower gently presses on the top of a plug, collecting the root-zone solution (leachate) from the bottom of the plug in a container. This is repeated for four or five plugs until a sufficient volume is collected, usually 10 to 50 mL, depending on the instrument used for testing. Leachate EC and pH can then be measured. Ideally, the grower would select several trays for testing within a certain species and fertilization/irrigation regime and average the results. Testing should be done every 2 weeks for most crops and weekly (or even more frequently) for particularly tricky (prone to nutrient disorders) or high-value crops. The recommended pH range for the plugs of most bedding crops using the press method for testing is 5.4 to 6.1. Media soluble salt levels should fall within an EC of 0.5 to 1.2 dS/m at sowing and germination and between 0.9 and 2.5 dS/m during stage 4, depending on the species.

Height Control

Compact, well-rooted, and well-branched seedlings make better transplants. Tall, lanky seedlings with long internodes are more likely to be damaged in the shipping or transplanting process. Plug height can be manipulated through cultural or chemical means.

Ideal growing conditions for most species of bedding plant seedlings include high light levels; cooler temperatures; adequate, but not excessive, moisture levels; and appropriate fertilizer regimens. As noted in the fertilizer section, research has shown that an imposed phosphorus deficiency may help plugs of some species to stay compact, but this must be managed carefully. An interesting method of height control, mechanical brushing, has been used successfully for years, especially in the production of tomato transplants. A boom, often the same one used for overhead irrigations, runs the length of the greenhouse several times per day. With the boom set low enough to significantly bend the plants, this action causes a release of ethylene, which inhibits apical growth. Chemical plant growth regulators are commonly used in the growing-out phase of the bedding plant crop cycle, but they can also be used for plug production. Extreme care must be taken as too high a concentration or amount applied can continue to limit plant growth for weeks after transplanting, greatly delaying the transplanted crop.

FIGURE 37.15 African marigold (*Tagetes erecta* L.) seedlings, unfortunately a bit past their prime. (Courtesy of H. L. Scoggins.)

Finally, after all the effort necessary to sow and produce high-quality, healthy seedlings, scheduling again comes into play. Most bedding plants are bred to grow rapidly and flower quickly; the size and stage of development optimal for transplanting cannot be maintained for long (Figure 37.15). Overgrown, root-bound, and/or heavily flowering plugs are never desirable; the quality of the finished plant is likely to be compromised. If received from another grower, consider rejecting them or ask for a credit. If grown in-house, the best option is to start over.

LITERATURE CITED AND SUGGESTED READING

Beytes, C., ed. 2011. *Ball Redbook*, Vol. 1: *Greenhouses and Equipment*. Batavia, Illinois: Ball Publishing, 276 pp.

Dole, J. M. and H. F. Wilkins. 2005. *Floriculture: Principles and Species*, 2nd Edition. Upper Saddle River, New Jersey: Prentice Hall, 1023 pp.

Greenhouse Grower. Meister Media Worldwide. http://www.greenhousegrower.com. Accessed March 14, 2014.

Greenhouse Product News (GPN). Scranton Gillete Communications. http://www.gpnmag.com. Accessed March 14, 2014.

GrowerTalks. Ball Publishing. http://www.growertalks.com. Accessed March 14, 2014.

Nau, J. 1999. *Ball Culture Guide: The Encyclopedia of Seed Germination*, 3rd Edition. Batavia, Illinois: Ball Publishing, 243 pp.

Nau, J., ed. 2011. *Ball Redbook*, Vol. 2: *Crop Production*. Batavia, Illinois: Ball Publishing, 768 pp.

Scoggins, H. L., D. A. Bailey, and P. V. Nelson. 2002. Efficacy of the press extraction method for bedding plant plug nutrient monitoring. *HortScience* 37:108–112.

Styer, R. C. and D. S. Koranski. 1997. *Plug and Transplant Production. A Grower's Guide*. Batavia, Illinois: Ball Publishing, 374 pp.

United States Department of Agriculture (USDA), National Agriculture Statistics Service. 2013. Floriculture Crops 2012 Summary. http://www.nass.usda.gov/Publications/Todays_Reports/reports/floran13.pdf. Accessed July 8, 2014.

Part XIII

In Conclusion: Special Topics

38 Myths of Plant Propagation

Jeffrey H. Gillman

CONCEPT BOX 38.1

- Over time, a variety of techniques have been used to improve propagation success, but not all of these techniques are actually beneficial.
- In any research, especially in propagation research, a control is needed to ensure that observed benefits are real and not imagined.
- It is important to try new propagation techniques. Only by experimenting can we discover the best techniques to multiply the plant that we want to produce.

Up until now, most of the propagation techniques that you have been exposed to in this book work to one extent or another. They are based on research, and most have been tested extensively by commercial propagators. This chapter is not about these established and well-tested techniques; it is about propagation techniques and ideas that are not in line with what we currently think of as "mainstream."

One of the great things about propagation is the fact that we have to experiment. Throughout the course of this book, you have had the chance to try your own experiments investigating various methods and techniques. If this chapter does nothing else, hopefully, it will encourage you to try new things when you propagate. New propagation methods are constantly introduced as propagators work to improve germination rate, cutting success, and speed of rooting. Some of these methods have a positive impact and so have become common practice, but a number have found a place at one propagator's bench or another due not to their usefulness, but rather to superstition and rhetoric. How could this happen? Simple! In the world of propagation, it is not unusual for a propagator to try a new technique without simultaneously testing untreated plants, cuttings, or seeds. Without testing these techniques against a control, any success that is seen may be attributed to the applied treatment instead of to a more mundane explanation such as a healthy, well-watered stock plant. In the next paragraphs, a number of techniques and treatments will be introduced, some of which you may want to try. If you do, be sure to test them against untreated controls so that you can be confident with your results.

APPLYING GRAVEL TO THE BASE OF SEED TRAYS TO IMPROVE DRAINAGE

A well-drained medium is important for seed propagators to prevent newly formed roots from suffocating and to control soil-borne diseases. Over the years, a variety of methods have been used to ensure that propagation media drain appropriately, the most common of which is to add coarse materials such as vermiculite and perlite to the propagation medium. However, in the late 19th and early 20th centuries, a technique was implemented to improve drainage, which included adding coarse materials such as oyster shells or gravel to the base of containers rather than dispersing these materials throughout the medium. Even today, it is not unheard of for propagators and gardeners to apply gravel to the base of their containers to ensure that water moves rapidly through the medium and enough air reaches the germinating seeds and roots.

DOES IT WORK?

There is no doubt at all that a container full of gravel will be better drained than a container full of soil or a commercial potting medium. The question is whether filling the bottom half of a container with gravel and then filling the rest of the way with a more typical medium will provide a better drainage to that top section. The answer to this question is no. But why? Well, the answer is that water does not move freely between media with different pore sizes, so there is not a continuous column of water between the growing medium and the gravel. Without this continuous column, water will not be drawn out of

FIGURE 38.1 Placing gravel in the bottom of a container leads to a perched water table where the water sits higher than it otherwise would.

the growing medium and into the gravel, but will instead sit in the media above the gravel, forming something called a "perched water table." In other words, by applying gravel to the base of the container, we end up making what amounts to a shorter, squatter container. Shorter containers have a worse drainage than taller containers with a similar diameter.

Should You Try It?

At this point, you realize that a good drainage is important for seed propagation; however, a good drainage cannot be achieved by placing a more poorly drained medium on top of a well-drained medium. It just does not work. However, if you are a skeptic, there is a quick way to test it out. Take two clear plastic cups. Punch holes into the bottom of both and then fill one with a well-drained medium and the other halfway with gravel and then the rest of the way with the same well-drained medium. Now, thoroughly water both cups. You should be able to see the water level in the media through the sides of the containers. What you will observe is that, in the cup with gravel, the water level will be higher than that in the cup without gravel, and it will drain more slowly over time (Figure 38.1).

LIQUID SMOKE HELPS STIMULATE SEED GERMINATION

You may have heard that the seeds of some plants, such as lodgepole pine, require fire to germinate, but might there be a way for a propagator to simulate fire without actually striking a match or flicking a lighter? Liquid smoke is a commercial product for flavoring food that is made by running smoke through water. Is it possible that seeds of species responsive to fire might also respond to liquid smoke?

Does it Work?

Researchers at the University of Hawaii have tested liquid smoke on pili grass, a native Hawaiian grass that is adapted to living with fire. Seeds that were soaked for 15 min in 1% liquid smoke had a greater germination rate than seeds that were not soaked. Other plants, such as red rice, have also been shown to germinate more successfully when treated with liquid smoke.

Should You Try It?

Soaking seeds of species that tend to germinate better when exposed to fire, or even species that have not shown that they respond to fire, in a dilute solution of liquid smoke and testing their germination against seeds that have not been soaked makes a very interesting experiment that you can try at home while cooking some burgers. In fact, putting a few seeds on the grill for different periods of time could be one of your treatments (be sure to wear eye protection as some seeds will "pop"). Liquid smoke is definitely worth trying out.

YOU CANNOT GRAFT AN APPLE ONTO AN ORANGE

Over the years, there have been all kinds of tall tales about how one propagator or another is so good at the job that he or she can graft an apple onto an orange! Obviously, when we hear something like that, we think that the person doing the talking is making some sort of a joke or exaggerating for the sake of emphasis. But is this really such a far-fetched idea? After all, apples and oranges both have cambial layers that do essentially the same thing. Why can't you just line up the layers and bang! There you go—an apple onto an orange or a fig onto a pear or a walnut onto a plum.

Does It Work?

People have tried to graft different plants onto each other with some very interesting results, although what we usually see is incompatibility. What this means is that the two plants cannot work together, and the graft fails either right away or at a later time, perhaps years in the future. When the graft lasts for a while before it fails, we say that we have delayed incompatibility. Getting around incompatibility is not easily accomplished. In fact, it is not usually accomplished at all.

Grafting an apple onto an orange will not work, but that does not mean that there are not plenty of unrelated plants that can be attached to each other successfully. It is true that grafting is typically most successful when it is accomplished using plants that are of the same species. However, it is not unusual for plants of closely related species, and even genera, to be able to be grafted together. For example, it is common practice for pears to be grafted onto quince rootstock to dwarf the produced tree. Lilac (*Syringa vulgaris*) was once commonly grafted onto privet (*Ligustrum ovalifolium*) because of the difficulty of rooting lilac cuttings, and trifoliate orange (*Poncirus trifoliata*) may be used as a rootstock for *Citrus* spp. to confer cold tolerance. Many maples, including the full-moon maple and Japanese maples, can be grafted onto each other, as can many different apples, ashes, and others.

Should You Try It?

Yes, definitely. It is a lot of fun to try grafting different plants onto each other. You will certainly end up with a lot of disappointments, but you will probably discover a few surprises too. Plus, by trying to graft different plants onto each other, you will have the opportunity to gain a familiarity with the plants and with grafting in general that you might not otherwise have.

GROWING PLANTS IN THE DARK TO IMPROVE ROOTING

The term "etiolation" refers to a plant's response to growing in darkness. This response is characterized by very little chlorophyll in the leaf tissue, resulting in bleached-out leaves and/or flowers, and very narrow elongated stems. Etiolation is an old process that has been used for many years to alter the shape and even the flavor of foods. Most people are familiar with cauliflower that is commonly etiolated when it is grown to retain its white color. There are people out there who swear that growing plants in the dark will increase the likelihood of cuttings from these plants developing roots. But why would it? Intuition says that these people are wasting their time. After all, plants need nice green leaves to photosynthesize and create sugars for new and growing roots. If we etiolate the plant, we will not have these nice green leaves, so how could this process possibly work?

Does It Work?

Believe it or not, plants that are grown in the dark do tend to develop roots more readily than plants that are grown in plenty of sunlight. In fact, many plants, such as lilacs, apples, pears, and others respond well to this technique. But why? Currently, there is not a single definitive answer to this question, but there are a few characteristics of stem cuttings from plants that were etiolated that could potentially make them more likely to root than non-etiolated

FIGURE 38.2 Wrapping a light excluding material around a branch to prepare it for rooting (banding).

cuttings. First, etiolated cuttings respond to auxin much more readily than non-etiolated cuttings. Second, etiolated cuttings may have chemicals in them that can act as co-factors, aiding the activity of the rooting hormone. Third and finally, etiolated cuttings have less lignin, a structural compound, in them. It is possible that reduced lignin allows roots to penetrate out through tissues that would inhibit their extension under normal conditions.

Should You Try It?

Absolutely! There are a few ways to grow plants without light. Most propagators do not keep the plant that they are planning on reproducing in complete darkness the entire time that they are growing them, but rather cover them with a heavy shade, often in the form of a box, for a few days or up to 10 weeks before cuttings are taken. Another easy-to-implement etiolation technique is called "banding" and involves wrapping a band of heavy cloth around the piece of stem where you plan on taking cuttings and leaving it there for a few weeks (Figure 38.2). This region of the stem should be cut so that the banded portion constitutes the bottom of the cutting that is taken. The cutting, and, specifically, the banded region, can be treated with a rooting hormone after it is removed from the plant. If you try etiolation, be sure to compare the rooting success of cuttings from plants that were etiolated with that of cuttings from similar plants that were not. Remember, you will never know how well a treatment works unless you compare it to an untreated control.

USING HERBICIDES TO PROMOTE ROOT FORMATION

Once upon a time, plant propagators used herbicides to help them produce roots on cuttings. It is like a magic trick. Take a cutting and stick the base into a pan with a small amount of herbicide in it, place it in a well-drained rooting media, and put it onto a mist bench, and in a few weeks, poof! Instead of having a dead cutting, you have got a plant with roots. What is going on here?

Does It Work?

The herbicide of choice for rooting cuttings is 2,4-D, also known as "2,4-dichlorophenoxyacetic acid." This is the same stuff that you spray on your yard to kill dandelions while preserving the grass. This compound was first created when researchers, including Dr. Franklin Jones who received the first patent for 2,4-D, tried to create synthetic auxins during the World War II to act as herbicides, perhaps against Japan's rice crop. Indoleacetic acid (IAA) was too chemically unstable to make an effective herbicide, and so something else was needed. Before the end of the war, 2,4-D was synthesized, but was not registered for use until 1946 because of its status as a war secret. This chemical is a strong auxin that does not break down rapidly. It tends to be more effective on what are called "broad-leaved weeds" or, in other words, most plants other than grasses. At high-enough concentrations, this strong auxin causes broad-leaved weeds to have abnormal growth and die. Grasses are not affected to the same extent and tend to be able to tolerate a heavy dose of 2,4-D before they succumb to its effects. At lower doses, 2,4-D may not be lethal, but might, instead, provide the stimulation that a cutting needs for it to develop root initials. Because of its longevity, and the sensitivity of many plants to it, 2,4-D is not anymore the compound that is commonly used to stimulate roots in cuttings; however, it is still sometimes used as an auxin for micropropagation.

Should You Try It?

This is a technique that is better off not tried by the casual propagator. The synthetic auxin 2,4-D is regulated as a pesticide and is not currently labeled for use as a root-promoting compound. If you want to root a cutting by using a hormone, then you are much better off purchasing a product that contains the indole butyric acid (IBA) or naphthalene acetic acid (NAA) that you read about in earlier chapters. If you are interested in trying to use 2,4-D as a root promoter, then do so only under the direction of a qualified instructor who is trained with the use of hazardous materials.

PLANTS THAT ROOT EASILY CAN BE USED TO PROMOTE ROOT FORMATION IN OTHER PLANTS

There is a theory that, if a tree is easy to root, then that tree must contain some sort of chemical that makes the tree easy to root. Furthermore, this chemical should be able to be extracted by soaking the stems or other portions of the plant in water. After soaking, the liquid portion of the concoction can be separated from the stems and used to treat cuttings from other difficult-to-root plants to make them more likely to develop roots (Figure 38.3).

FIGURE 38.3 Cutting willow branches to prepare a willow diffusate.

Does It Work?

Over the years, a number of people have tried using extracts from a variety of easy-to-root plants to promote root formation in cuttings. These extracts are created by soaking the stems of the easy-to-root plant in water for some period of time and then using the liquid that the stems were soaked in to treat cuttings from plants that they want to root (Figure 38.3). The extracts used are usually called "diffusates," the theory being that important rooting compounds will diffuse from the easy-to-root plant into the water. The first person to research diffusates as a means of root promotion was C. E. Hess in the 1950s. Willow stems are the plant most commonly used for root promotion in other plants, although extracts from plants such as ivy and black locust have also been implicated as root promoters. In some studies, these extracts have worked, but in most, they have not. There are people who will swear up and down that these extracts can coax roots from the stems of hard-to-root plants, but hard evidence is currently lacking. Indeed, most researchers who try these compounds (such as the author of this chapter) end up unimpressed with their performance.

One argument for the use of diffusate from willow trees is the fact that willows contain the compound salicylic acid (similar to aspirin), which has been shown to increase the rooting of certain plants, including faba and mung beans. Unfortunately, since these beans are extremely easy to root, showing that salicylic acid helps promote root formation in them is not an appropriate test of whether they will promote root formation in other, more difficult-to-root species.

Should You Try It?

It is fun and easy to test whether various diffusates will help cuttings from different plants develop roots. Simply take stems from plants whose root-promoting abilities you would like to test and steep them in water for a certain amount of time. You may use the amount of time that you steep them as a variable if you would like (the number of stems soaking in the water could also be a variable), perhaps allowing some stems to sit in water for an hour; some, a day; and some, a week. Next, take the stems out and filter the whole mess through a cheesecloth, keeping the filtered water. This filtered water is the "diffusate."

Place the stems from the plants that you would like to promote roots on into the filtered water and let them sit for some period of time. Twenty-four hours is often noted as appropriate and should provide enough time for the cutting to take up the compounds that they need to for root formation, but certainly, this can be treated as a variable, and stems can be allowed to soak in the diffusate for any length of time that you see fit. When the plants are done soaking, they should be placed onto an ordinary mist bench and allowed to sit for a few weeks as with any stem cuttings. Be sure to test untreated cuttings along with the diffusate-treated cuttings so that you can compare the rooting success of both. You may also choose to treat some cuttings with IBA so that you can compare the success of diffusates with a synthetic auxin.

CUTTING A RING AROUND A STEM WILL BENEFIT ROOT FORMATION

The practice of cutting a ring of bark from around a stem 6 or so weeks before cuttings are taken is an uncommon practice, but one which some people claim as a way to encourage roots from ordinarily difficult-to-root cuttings (Figure 38.4). This cut is called a "girdling cut" and is one that is considered detrimental to the tree. In fact, there are certain beetles called "stem girdlers" who cause this type of damage and are considered pests of the trees that they feed on. So, why would girdling the bark around a healthy stem increase rooting? Wouldn't this just damage the plant and, particularly, the stems from which the cuttings are taken?

FIGURE 38.4 Girdling a stem with a budding knife.

Does It Work?

Cutting a ring around a stem will disrupt the physiology of the stem—no doubt about it—but this disrupted physiology is not necessarily bad for propagation. When a stem is girdled, the tissue that is removed is the phloem. When the phloem is damaged, carbohydrates manufactured in the leaves do not have a path to follow to exit the stem; hence, the base of the stem, where the girdle was made, develops an abnormally high content of carbohydrates that may be useful to a cutting when it is taken. Additionally, other compounds that are normally exported from the stem through the phloem will build up, including, potentially, auxins, and auxin cofactors that may aid auxin in stimulating roots.

Should You Try It?

If you girdle every stem on a plant, you could damage that plant, perhaps even severely if the plant is small to begin with. However, if you just want to girdle a few branches and then take cuttings from them, then this is a great idea for an experiment. Make sure to wait a few weeks between the time that you girdle the stems and the time that you take the cuttings so that all of the potentially important compounds can build up in the stem, and be sure to compare the success of these cuttings with that of cuttings from the same plant or plants that have not been girdled so that you can make a reasonable assessment as to whether the girdling worked.

PLACING GRAIN AT THE BASE OF A CUTTING WILL BENEFIT ROOT FORMATION

When people in the Middle East and Europe propagated plants using cuttings centuries ago, they would often place grain into the base of a cutting to promote root formation. In fact, in some regions of Macedonia, it is still a common practice to embed a germinating grain of wheat in the base of a cutting to stimulate root development. Is the use of grain in this way just a superstitious waste of time or is there some practical reason why it could stimulate root development in cuttings?

Does It Work?

There are few studies that have investigated the usefulness of placing or embedding a germinating grain of wheat into a cutting. However, there is a reason to believe that there is more to this practice than superstition. Germinating seeds tend to be higher in auxin content than other parts of the plant. Germinating grains of wheat are packed with auxin (relatively speaking) and so might act as a promoter of roots for a cutting. Another seed that has particularly high auxin content is corn, and it might also serve to encourage roots to form on a cutting. Of course, the act of slicing the base of a cutting can also promote roots in and of itself, and so it might not always be clear whether the benefits of putting grain into the base of a cutting that has been sliced open is actually greater than just making the cut, unless experiments are carefully controlled.

Should You Try It?

Why not give this one a shot? All you need are some cuttings and some grain. To test whether germinating grain benefits root development, all you need to do is to slice open the base of the cutting and insert the grain, which should be soaked for 24 to 48 h before insertion to stimulate germination. Be sure to test this against other cuttings that are also slit open but which have something else inserted besides grain, perhaps a grain-sized piece of sand. It is very important when conducting any experiment to have appropriate controls, especially in a situation like this. Cuttings from certain plants respond favorably when their base is cut open; hence, if cuttings without a slice in them are used as controls, the simple act of slicing the cutting with the grain in it may stimulate roots and lead you to the erroneous assumption that the grain promoted rooting when, in actuality, it was the cut that was the culprit. Alternatively, you could also test whether the slit is even necessary by just placing the grain next to the cut surface of the cutting as it is inserted into your propagation medium just before it goes onto the mist bench. The control for this would simply be to stick cuttings into the mist bench without the grain.

SUGGESTED READING

Liquid Smoke Helps Stimulate Seed Germination

Baldos, O. C., J. DeFrank, and G. Sakamoto. Improving germination of pili grass (*Heteropogon contortus*) seeds using liquid smoke flavoring (poster presentation). *Amer Soc Horti Sci Annual Meeting*, Hilton Waikoloa Village, Waikoloa, Hawaii, September 25–28, 2011.

Doherty, L. C. and M. A. Cohn. 2000. Seed dormancy in red rice (*Oryza sativa*): XI. Commercial liquid smoke elicits germination. *Seed Sci Res* 10:415–381.

You Cannot Graft an Apple onto an Orange

Yelenosky, G. and J. C. V. Vu. 1992. Ability of "Valencia" sweet orange to cold acclimate on cold-sensitive citron rootstock. *HortScience* 27:1201–1203.

Growing Plants in the Dark to Improve Rooting

Koukourikou-Petridou, M. A. 1998. Etiolation of stock plants affects adventitious root formation and hormone content of pea stem cuttings. *Plant Growth Regul* 25:17–21.

Maynard, B. K. and N. L. Bassuk. 1991. Stock plant etiolation and stem banding effect on the auxin dose response of rooting in stem cuttings of *Carpinus betulus* L. "Fastigiata." *Plant Growth Regul* 10:305–311.

Using Herbicides to Promote Roots

Delargy, J. A. and C. E. Wright. 1978. Root formation in cuttings of apple in relation to auxin application and to etiolation. *New Phytol* 82:341–347.

Plants that Root Easily Can be Used to Promote Root Formation in Other Plants

Girourd, R. M. and C. E. Hess. 1964. The diffusion of root-promoting substances from stems of *Hedera helix*. *Comb Proc Intl Plant Prop Soc* 14:162–166.

Hess, C. E. 1959. A study of plant growth substances in easy- and difficult-to-root cuttings. *Comb Proc Intl Plant Prop Soc* 9:39–43.

Sharma, A. R. and R. N. Trigiano. 1999. Rooting flowering dogwood (*Cornus florida*) microshoots. *Proc SNA Res Conf* 44:376–377.

Cutting a Ring around a Stem Will Benefit Root Formation

Cooper, W. C. 1936. Transport of root-forming hormone in woody cuttings. *Plant Physiol* 11:779–793.

Delargy, J. A. and C. E. Wright. 1978. Root formation in cuttings of apple (cv. Bramley's seedling) in relation to ring barking and to etiolation. *New Phytol* 81:117–127.

Placing Grain at the Base of a Cutting Will Benefit Root Formation

Sheldrake, A. R. 1973. Do coleoptile tips produce auxin? *New Phytol* 72:433–447.

39 Intellectual Property Protection for Plants

Christopher Eisenschenk

CONCEPT BOX 39.1

- Plant patents, utility patents, Plant Variety Protection Act (PVPA) certificates and trademarks are forms of federal intellectual property protection that are available for plants and plant materials
- Unauthorized sexual production of plants can be controlled by utility patents and/or PVPA certificates.
- Unauthorized asexual reproduction of plants can be controlled by utility patents and/or plant patents.
- Of all the forms of federal intellectual property protection available for plants, a utility patent provides the strongest intellectual property protection and is the most difficult to obtain.
- A plant patent provides the right to exclude others from asexually reproducing the new and distinct plant variety and parts thereof for a 20-year period starting from the earliest effective filing date of the patent application.
- A plant patent application has only one claim (the plant), whereas a utility patent application may have one to many claims.

This chapter will seek to address the various types of federal intellectual property protection that are available for plants and plant materials. While some forms of state intellectual property rights exist, a discussion of these rights is beyond the scope of this chapter. Additionally, this chapter is not intended to be an exhaustive discussion of the various forms of intellectual property protection available for plants and is not to be construed as a legal advice regarding any form of intellectual property discussed herein. Readers are strongly encouraged to seek legal counsel for advice relevant to any intellectual property issue encountered.

The types of federal intellectual property protection available for plants and plant materials include patents (utility patents and/or plant patents), the Plant Variety Protection Act (PVPA) certificates, and trademark protection. On September 16, 2011, the America Invents Act (AIA) was enacted. AIA substantially changes certain aspects of federal patent law. Namely, the United States has harmonized its law with that of the rest of the world by shifting from a "first-to-invent" system to a "first-to file" system of patent law. Certain aspects of AIA and its impact on intellectual property rights will be addressed later in this chapter.

Most commonly, utility and plant patents are used to protect intellectual property rights; however, PVPA provides a form of intellectual property protection that is analogous to patent rights, albeit with a few exceptions that may impact one's decision to pursue PVPA protection for plants.

COPYRIGHT PROTECTION AND PLANTS

Federal copyright protection exists for "original works of authorship fixed in any tangible medium of expression, now known or later developed, from which they can be perceived, reproduced, or otherwise communicated, either directly or with the aid of a machine or a device" (see 17 U.S.C. §102[a], 1988). Thus, where an author fixes an image of a plant or depicts the image of a plant in some form of a tangible medium (an "original work"), a copyright in that original work is obtained at the moment of its fixation within the tangible medium. However, it is important to note that the scope of protection for the copyrightable material (e.g., an "original work" depicting a plant) only extends to the "original work" and not to the plant itself.

17 U.S.C. §102[a] (1988) provides examples of "original works." This non-limiting list includes the

following: literary works, musical works (including lyrics), dramatic works (including accompanying music), pantomimes and choreographic works, pictorial, graphic and sculptural works, motion pictures and other audiovisual works, sound recordings, and architectural works (see 17 U.S.C. §102[a], 1988). However, specifically excluded from copyright protection is any idea, procedure, process, system, method of operation, concept, principle, or discovery, regardless of the form in which it is described, explained, illustrated, or embodied in such work (see 17 U.S.C. §102[b], 1988). Additionally, the term "tangible medium" refers to any physical medium, non-limiting examples of which include paper, video/audio tapes, or electronic storage medium (e.g., computer disks).

The owner of the copyrighted "original work" has the exclusive right, subject to certain limitations, to do and/or to authorize any of the following: (1) to reproduce the copyrighted work in copies; (2) to prepare derivative works based on the copyrighted work; (3) to distribute copies of the copyrighted work to the public by sale or other transfer of ownership, or by rental, lease, or lending; (4) in the case of literary, musical, dramatic, and choreographic works; pantomimes; and motion pictures and other audiovisual works, to perform the copyrighted work publicly; (5) in the case of literary, musical, dramatic, and choreographic works; pantomimes; and pictorial, graphic, or sculptural works, including the individual images of a motion picture or other audiovisual work, to display the copyrighted work publicly; and (6) in the case of sound recordings, to perform the copyrighted work publicly by means of digital audio transmission (see 17 U.S.C. §106, 1988).

However, the author of an "original work" depicting a plant should also be aware of the "fair use doctrine" that allows for the limited reproduction of the copyrighted work without the consent of the author. The "fair use doctrine" allows for the reproduction of a copyrighted work for purposes including, but not necessarily limited to, criticism, comment, news reporting, teaching, scholarship, or research. Whether the reproduction of a copyrighted work is a "fair use" is determined on a case-to-case basis and includes an analysis of the following factors: (1) the purpose and character of the use, including whether such use is of a commercial nature or is for nonprofit educational purposes; (2) the nature of the copyrighted work; (3) the amount and substantiality of the portion used in relation to the copyrighted work as a whole; and (4) the effect of the use on the potential market for or the value of the copyrighted work (see 17 U.S.C. §107, 1988).

Finally, federal copyright protection exists as of the moment that the "original work" is fixed in a tangible medium, and there is no longer any requirement for marking the first publication of a work, published as of March 1, 1998, with a copyright symbol "©," the term "Copyright" (or the abbreviation "Copr."), and the author's name and the date of publication. Additionally, one should note that there is no requirement for copyright registration with the Copyright Office in the Library of Congress to establish one's copyright in an "original work." However, registration of one's copyright with the Copyright Office and proper marking of the "original work" with a copyright notice provides a number of advantages in case of infringement or unauthorized reproduction of the work. For example, no lawsuit for infringement of the copyright in any U.S. work shall be instituted until registration of the copyright claim has been made (see 17 U.S.C. §411). Additionally, statutory damages and attorneys' fees are not available to a copyright owner unless registration of the copyright has preceded the infringement, although a 3-month grace period exists in this regard (see 17 U.S.C. §§412 and 504–505). Finally, the copyright statutes provide that no weight shall be given to such a defendant's assertion of a defense based on innocent infringement if a copyright notice as specified in 17 U.S.C. §401(b–c) has been placed on the published copy or copies to which a defendant in a copyright infringement suit had access.

Thus, one should recognize that Web sites, photographs, marketing materials, sketches, artwork, or other forms of expression that depict a plant produced by a grower (and are fixed in a tangible medium) are a protectable form of intellectual property associated with the plant. One should also be aware of the rights that are associated with this aspect of intellectual property protection and consider enforcing these rights if infringers are identified. Additional information regarding copyrights can be accessed on the Internet at http://www.copyright.gov.

TRADEMARKS AND THEIR USE WITH PLANTS

The term "trademark" includes any word, name, symbol, or device, or any combination thereof: (1) used by a person or (2) which a person has a *bona fide* intention to use in commerce and who applies to register the trademark on the principal trademark register to identify and distinguish his or her goods, including a unique product, from those manufactured or sold by others and to indicate the source of the goods, even if that source is unknown (see 15 U.S.C. §1127). Thus, a trademark serves to identify the source or point of origin of a particular product sold in the stream of commerce.

As would, perhaps, be expected, the U.S. Patent and Trademark Office (USPTO) has promulgated guidelines addressing the use of varietal or cultivar names as trademarks in the *Trademark Manual of Examination Procedures*, 8th Edition (October 2012). As indicated in §1202.12, *Varietal and Cultivar Names (Examination of Applications for Seeds and Plants)*:

Varietal or cultivar names are designations given to cultivated varieties or subspecies of live plants or agricultural seeds. They amount to the generic name of the plant or seed by which such variety is known to the public. These names can consist of a numeric or alphanumeric code or can be a "fancy" (arbitrary) name. The terms "varietal" and "cultivar" may have slight semantic differences but pose indistinguishable issues and are treated identically for trademark purposes.

Subspecies are types of a particular species of plant or seed that are members of a particular genus. For example, all maple trees are in the genus *Acer.* The sugar maple species is known as *Acer saccharum*, while the red maple species is called *Acer rubrum.* In turn, these species have been subdivided into various cultivated varieties that are developed commercially and given varietal or cultivar names that are known to the public.

A varietal or cultivar name is used in a plant patent to identify the variety. Thus, even if the name was originally arbitrary, it "describe[s] to the public a [plant] of a particular sort, not a [plant] from a particular [source]." *Dixie Rose Nursery v. Coe*, 131 F.2d 446, 447 55 USPQ 315, 316 (D.C. Cir. 1942), cert. denied 318 U.S. 782, 57 USPQ 568 (1943). It is against public policy for any one supplier to retain exclusivity in a patented variety of plant, or the name of a variety, once its patent expires. *Ibid.*

Market realities and lack of laws concerning the registration of varietal and cultivar names have created a number of problems in this area. Some varietal names are not attractive or easy to remember by the public. As a result, many arbitrary terms are used as varietal names. Problems arise when trademark registration is sought for varietal names, when arbitrary varietal names are thought of as being trademarks by the public, and when terms intended as trademarks by plant breeders become generic through public use. These problems make this a difficult area for the examining attorney in terms of gathering credible evidence and knowing when to make refusals.

Whenever an application is filed to register a mark containing wording for live plants, agricultural seeds, fresh fruits, or fresh vegetables, the examining attorney must inquire of the applicant whether the term has ever been used as a varietal name and whether such name has been used in connection with a plant patent, a utility patent, or a certificate for plant variety protection (see 37 C.F.R. §2.61[b]). The examining attorney must also undertake an independent investigation of any evidence that would support a refusal to register, using sources of evidence that are appropriate for the particular goods specified in the application (e.g., laboratories and repositories of the U.S. Department of Agriculture, plant patent information from the USPTO, and a variety name search of plants certified under the PVPA listed at http://www.ars-grin.gov/npgs/searchgrin.html).

If the examining attorney determines that the wording sought to be registered as a mark for live plants, agricultural seeds, fresh fruits, or fresh vegetables comprises a varietal or cultivar name, then the examining attorney must refuse registration, or require a disclaimer, on the ground that the matter is the varietal name of the goods and does not function as a trademark under §§1, 2, and 45 of the Trademark Act, 15 U.S.C. §§1051, 1052, and 1127. See *In re Pennington Seed, Inc.*, 466 F.3d 1053, 1057–1058, 80 USPQ2d 1758, 1761-1762 (Fed. Cir. 2006) (upholding the USPTO's long-standing precedent and policy of treating varietal names as generic and affirming refusal to register REBEL for grass seed because it is the varietal name for the grass seed as evidenced by its designation as the varietal name in the applicant's plant variety protection certificate); *Dixie Rose Nursery v. Coe*, 131 F.2d 446, 447 55 USPQ 315, 316 (D.C. Cir. 1942), cert. denied 318 U.S. 782, 57 USPQ 568 (1943) (holding TEXAS CENTENNIAL, although originally arbitrary, has become the varietal name for a type of rose; *In re Hilltop Orchards & Nurseries, Inc.*, 206 USPQ 1034, 1035 (TTAB 1979) (affirming the refusal to register COMMANDER YORK for apple trees because it is the varietal name for the trees as evidenced by use in the applicant's catalog); *In re Farmer Seed & Nursery Co.*, 137 USPQ 231, 231–232 (TTAB 1963) (upholding the refusal to register CHIEF BEMIDJI as a trademark because it is the varietal name for a strawberry plant and noting that large expenditures of money does not elevate a mark to a trademark; and *In re Cohn Bodger & Sons Co.*, 122 USPQ 345, 346 (TTAB 1959) (holding BLUE LUSTRE as merely a varietal name for petunia seeds as evidenced by the applicant's catalogs).

Likewise, if the mark identifies the prominent portion of a varietal name, it must be refused. *In re Delta & Pine Land Co.*, 26 USPQ2d 1157 (TTAB 1993) (affirming the refusal to register DELTAPINE, which was a portion of the varietal names Deltapine 50, Deltapine 20, Deltapine 105, and Deltapine 506).

In trademark law, a descriptive trademark conveys information regarding an ingredient, quality, characteristic, function, feature, purpose, or use of the product or service. Descriptive trademarks are, generally, not eligible for registration because the mark must become associated with the product or service among consumers (e.g., acquire secondary meaning). With respect to a view that the cultivar name is generic, trademark law holds that generic terms or common words for products or services cannot function as a trademark because it would prevent others from rightfully using the common name for the product or service that they make. Thus, in view of the guidance provided to trademark examining attorneys by the Trademark Manual of Procedure (TMEP), it is clear that cultivar names will be considered generic or, at best, descriptive and ineligible for trademark protection.

Thus, while one may not be able to trademark a particular cultivar name, it should be understood that one can still identify a particular cultivar as originating with a particular source or point of origin. In this context, a nursery or grower of a particular cultivar may trademark a different name and use it in conjunction with the cultivar name to identify the source or point of origin of the cultivar produced by the nursery or grower. This combination of a trademark and the cultivar name would then allow the consumers of the plant to identify the source or point of origin of the cultivar. With respect to marking requirements under trademark law, that portion of the combined trademark–cultivar name that constitutes the trademarked portion should be identified with a "®" or "™" symbol on the tag or label to designate a registered or common law trademark, respectively. The portion of the combined trademark–cultivar name that identifies the actual cultivar name on the tag or label should identify the cultivar using single quotation marks according to the industry norm.

THE PLANT VARIETY PROTECTION ACT

Plant Variety Protection Act (PVPA) is a form of federal intellectual property that is available to growers of new sexually reproduced plants. The PVPA is administered through the Plant Variety Protection Office of the U.S. Department of Agriculture and provides a breeder a somewhat patent-like protection for any sexually reproduced or tuber-propagated plant variety (other than fungi or bacteria).

Thus, if a plant variety is new, distinct, uniform, and stable as defined in 7 U.S.C. §2402 of the PVPA,* it will be eligible for a certificate under the act. Growers should also be aware of the various statutory bars that would prevent one from obtaining a PVPA certificate. These include making the plant variety available to persons, for purposes of exploitation, within the United States more than 1 year prior to the filing date of the PVPA application; making the plant variety available to persons outside the United States more than 4 years prior to the filing date of the PVPA application (although certain exceptions may exist for tubers); or making a tree or vine available to persons outside the United States more than 6 years prior to the filing date of the PVPA application.

Turning now to the scope of protection for plants for which PVPA certificates have been issued, courts have held that PVPA protects plants and seeds that differ "from either of the parent plants as well as from other plants produced from other seeds resulting from the cross-pollination" and that "[p]lants true to type, although different in a strict genetic sense, are protectable under PVPA" (see *Imazio Nursery, Inc. v. Dania Greenhouses*, 69 F3d. 1560 [Fed. Cir. 1996]). Thus, the PVPA allows for some genetic variation within the plants and/or seeds of a variety protected by a PVPA certificate. As discussed below, such genetic variation is not permitted for plants protected under the Plant Patent Act (PPA).

With respect to the rights conferred by the owner of a protected variety, 7 U.S.C. §2541 indicates that:

> … it shall be an infringement of the rights of the owner of a protected variety to perform without authority any of the following acts in the United States or in commerce, which can be regulated by the Congress, or affecting such commerce, prior to the expiration of the right to the plant variety protection but after either the issue of the certificate or the distribution of a protected plant variety with the notice under Section 2567 of this title:
>
> (1) sell or market the protected variety, or offer it or expose it for sale, deliver it, ship it, consign it, exchange it, or solicit an offer to buy it, or any other transfer of title or possession of it;
> (2) import the variety into, or export it from, the United States;
> (3) sexually multiply, or propagate by a tuber or a part of a tuber, the variety as a step in marketing (for growing purposes) the variety;
> (4) use the variety in producing (as distinguished from developing) a hybrid or a different variety therefrom;
> (5) use a seed that had been marked "Unauthorized Propagation Prohibited" or "Unauthorized Seed Multiplication Prohibited" or progeny thereof to propagate the variety;
> (6) dispense the variety to another, in a form which can be propagated, without notice as to being a protected variety under which it was received;
> (7) condition the variety for the purpose of propagation, except to the extent that the conditioning is related to the activities permitted under §2543 of this title;

* §2402. Right to plant variety protection; plant varieties protectable.
(1) In general
The breeder of any sexually reproduced or tuber-propagated plant variety (other than fungi or bacteria) who has so reproduced the variety, or the successor in interest of the breeder, shall be entitled to the plant variety protection for the variety, subject to the conditions and requirements of this chapter, if the variety is
(a) new, in the sense that, on the date of filing the application for plant variety protection, the propagating or harvested material of the variety has not been sold or otherwise disposed of to other persons, by or with the consent of the breeder, or the successor in interest of the breeder, for purposes of exploitation of the variety
(A) in the United States, more than 1 year prior to the date of filing, or
(B) in any area outside of the United States
(i) more than 4 years prior to the date of filing, except that, in the case of a tuber-propagated plant variety, the secretary may waive the 4-year limitation for a period ending 1 year after April 4, 1996, or
(ii) in the case of a tree or vine, more than 6 years prior to the date of filing;
(b) distinct, in the sense that the variety is clearly distinguishable from any other variety, the existence of which is publicly known or a matter of common knowledge at the time of the filing of the application;
(c) uniform, in the sense that any variations are describable, predictable, and commercially acceptable; and
(d) stable, in the sense that the variety, when reproduced, will remain unchanged with regard to the essential and distinctive characteristics of the variety, with a reasonable degree of reliability commensurate with that of varieties of the same category in which the same breeding method is employed.

(8) stock the variety for any of the purposes referred to in paragraphs 1 to 7;
(9) perform any of the foregoing acts even in instances in which the variety is multiplied other than sexually, except in pursuance of a valid U.S. plant patent; or
(10) instigate or actively induce the performance of any of the foregoing acts.

Damages that one can obtain for the infringement of a PVPA certificate shall be adequate to compensate for the infringement, but in no event less than a reasonable royalty for the use made of the variety by the infringer, together with interest and costs as fixed by the court (see 7 U.S.C. §2564). It is also possible to obtain an injunction prohibiting an infringer from continuing to make and use the seeds or plants protected under PVPA.

It is also important to recognize that PVPA provides several exceptions/exemptions for acts of infringement. For example, it is not considered an act of infringement for a farmer to save and replant a seed protected by a PVPA certificate.* Additionally, a *bona fide* sale of a seed produced on a farm (either from a seed lawfully obtained for seeding purposes or from a seed produced by descent from the lawfully obtained seed and produced on the farm) for other than reproductive purposes, made in channels usual for such other purposes, shall not constitute an infringement.† A third, and potentially important, exception/exemption from infringement is a research use exception. 7 U.S.C. §2544 states that the use and reproduction of a protected variety for plant breeding or other *bona fide* research shall not constitute an infringement. Thus, for both growers and academics, the use of PVPA-protected plants and seeds for plant breeding or other true research purposes do not constitute an act of infringement. As we shall see, this is not the case of plants protected by either a utility or a plant patent.

PLANT PATENTS

Plant patents are granted to those who invent or discover, and asexually reproduce, any new and distinct variety of plant (see 35 U.S.C. §161‡). Whereas PVPA can be used to protect a sexually reproduced plant, a plant patent provides the right to exclude others from asexually reproducing the new and distinct plant variety and parts thereof for a 20-year period starting from the earliest effective filing date of a plant patent application (see 35 U.S.C. §163§). However, asexually reproduced plants can also be protected by utility patents as 35 U.S.C. §161 is not an exclusive form of protection that conflicts with the granting of utility patents to plants (see *Ex Parte Hibberd*, 227 U.S.P.Q. 443 [Bd. Pat. App. & Int. 1985]). Additionally, plants capable of sexual reproduction (i.e., from seed) can be protected under the PPA provided that the plant has been asexually reproduced.

As set forth in 35 U.S.C. §161, potentially patentable plants include cultivated sports, mutants, hybrids, and newly found seedlings, and only plants identified in cultivated areas are eligible for protection under the PPA. It should also be noted that plants identified in the wild are not eligible for plant patent protection. Fungi and bacteria are not considered "plants" under this aspect of the patents statutes; however, utility patents can be used to protect isolated fungi and bacteria. Also specifically excluded from protection under the PPA are tuber-propagated plants. Examples of tuber-propagated plants include Irish potato and Jerusalem artichoke. Because these plants are asexually propagated by the same part as is sold for food, the plants are considered ineligible for protection under a plant patent (see *Manual of Patent Examination Procedure* §1601, August 2012, 8th Edition, Revision 9).

Additionally, the single claim of a plant patent only protects the entirety of the plant, not subparts of the plant (e.g., seeds, flowers, fruit, etc.).¶

Among the requirements that must be fulfilled to obtain a plant patent are that (1) the plant is capable of stable asexual reproduction; (2) the plant was invented or discovered and, if discovered, that the discovery was made in

* 7 U.S.C. §2543. Right to save seed; crop exemption.
Except to the extent that such action may constitute an infringement under subsections 3 and 4 of §111 (7 U.S.C. §2541[3] and [4]), it shall not infringe any right hereunder for a person to save seeds produced by the person from seeds obtained, or descended from seeds obtained, by authority of the owner of the variety for seeding purposes and use such saved seeds in the production of a crop for use on the farm of the person or for sale as provided in this section. A *bona fide* sale for other than reproductive purposes, made in channels usual for such other purposes, of seeds produced on a farm either from seeds obtained by authority of the owner for seeding purposes or from seeds produced by descent on such farm from seeds obtained by authority of the owner for seeding purposes shall not constitute an infringement. A purchaser who diverts seeds from such channels to seeding purposes shall be deemed to have notice under §127 (7 U.S.C. §2567) that the actions of the purchaser constitute an infringement.

† *Ibid.*

‡ 35 U.S.C. §161. Patents for plants.
Whoever invents or discovers and asexually reproduces any distinct and new variety of plant, including cultivated sports, mutants, hybrids, and newly found seedlings, other than a tuber-propagated plant or a plant found in an uncultivated state, may obtain a patent therefor, subject to the conditions and requirements of this title.

§ 35 U.S.C. §163. Grant.
In the case of a plant patent, the grant shall include the right to exclude others from asexually reproducing the plant and from using, offering for sale, or selling the plant so reproduced, or any of its parts, throughout the United States, or from importing the plant so reproduced, or any parts thereof, into the United States.

¶ 37 C.F.R. §1.164. Claim.
The claim shall be in formal terms to the new and distinct variety of the specified plant as described and illustrated and may also recite the principal distinguishing characteristics. More than one claim is not permitted.

a cultivated area; (3) the plant is not a plant that is excluded by statute (e.g., the part of the plant used for asexual reproduction cannot be a tuber food part); (4) the person or persons filing the application are those who actually invented the claimed plant, that is, discovered or developed and identified or isolated the plant, and asexually reproduced the plant; (5) the plant has not been sold or released in the United States more than 1 year prior to the effective filing date of the application*; (6) the plant has not been enabled to the public, that is, by description in a printed publication in this country more than 1 year before the application for patent with an offer to sale† or by release or sale of the plant more than 1 year prior to the application for patent‡; (7) the plant be shown to differ from known, related plants by at least one distinguishing characteristic, which is more than a difference caused by growing conditions or fertility levels, etc.; or (8) the invention would not have been obvious to one skilled in the art at the time of invention by applicant.§ Interestingly, the person who identified the plant claimed in a plant patent application (i.e., the applicant for the plant patent) need not own the cultivated area in which the plant is discovered or the plant itself (see *Ex parte Moore*, 115 U.S.P.Q. 145 [Pat. Off. Bd. App. 1957]).

Infringement of a plant patent results whenever someone asexually reproduces, by any means, a plant that is protected by a plant patent. It should also be noted that "experimental use" defenses are exceptionally narrow and would, likely, not be available should one be involved in litigation involving an allegedly infringed plant patent.¶

Should one be successful in proving infringement, the patent statutes provide for a variety of remedies. These remedies include enjoining the infringer from producing any additional plants and requiring the payment of a monetary award to the patent holder. Monetary damages can be calculated in a variety of ways, including actual damages (e.g., lost sales) or a determination of what a reasonable royalty would have been for the plants sold by the infringer. In case of willful infringement, the damages can be trebled and it may also be possible to recover attorney's fees in instances where a court finds exceptional circumstances.

* AIA has created uncertainty with respect to on-sale bars to patentability as discussed below. The new law contains no explicit grace period for inventions that are first sold and for which a patent application is subsequently filed.

† *Ibid.*

‡ *Ibid.*

§ See U.S. Patent and Trademark Office publication "General Information About 35 U.S.C. 161 Plant Patents," accessible at http://www.uspto.gov/web/offices/pac/plant/.

¶ *Madey v. Duke University*, 307 F.3d 1351, 1362 (Fed. Cir. 2002) (stating that, regardless of whether a particular institution or entity is engaged in an endeavor for commercial gain, so long as the act is in furtherance of the alleged infringer's legitimate business and is not solely for amusement, to satisfy idle curiosity, or for strictly philosophical inquiry, the act does not qualify for the very narrow and strictly limited experimental use defense.

UTILITY PATENTS

As noted above, utility patents are also a form of federal intellectual property protection that is available for plants and plant materials. For plants, utility patents are most often directed to transgenic plants (e.g., plants containing herbicide-resistance genes or genes that confer resistance of insect pests). It is far less common for utility patents to be sought for non-transgenic plants, although a utility and plant patent can be obtained for a given plant if desired.

As discussed above, the United States has moved from a "first-to-invent" to a "first-to-file" system of patent law. This change is effective as of March 16, 2013, and creates a situation where two different sets of patent laws exist for patent applications. For those patent applications filed prior to March 16, 2013, "first-to-invent" principles of law apply as do the statutes mentioned in the following paragraphs. For patent applications filed on or after March 16, 2013, "first-to-file" principles of law will be applied to the patent applications.

To be eligible for patent protection, an invention must be one of the statutory classes of invention,** and the specification that describes the inventions must be written such that it teaches those skilled in the art how to make and use the invention that is claimed (this is commonly referred to as the "enablement requirement").†† In some cases, the deposit of seed will also be required to "enable" a patent application. The specification must also provide what is referred to as the best mode of carrying out the invention that is known to the inventors at the time that the application is filed, and it must also contain an adequate written description of that which the inventor considers to be the invention.‡‡ Additional requirements for patentability include a requirement for novelty§§ and

** 35 U.S.C. §101. Inventions patentable.
Whoever invents or discovers any new and useful process, machine, manufacture, or composition of matter, or any new and useful improvement thereof, may obtain a patent therefor, subject to the conditions and requirements of this title.

†† 35 U.S.C. §112. Specification.
The specification shall contain a written description of the invention and of the manner and process of making and using it, in such full, clear, concise, and exact terms as to enable any person skilled in the art to which it pertains, or with which it is most nearly connected, to make and use the same, and shall set forth the best mode contemplated by the inventor of carrying out his invention …

‡‡ *Ibid.*

§§ 35 U.S.C. §102. Conditions for patentability; novelty and loss of right to patent.
A person shall be entitled to a patent unless

(a) the invention was known or used by others in this country, or patented or described in a printed publication in this or a foreign country, before the invention thereof by the applicant for patent, or
(b) the invention was patented or described in a printed publication in this or a foreign country or in public use or on sale in this country, more than 1 year prior to the date of the application for patent in the United States …

non-obviousness.* For patent applications filed prior to March 16, 2013, it is possible to obtain a utility patent for a plant that has been disclosed or made available to the public provided that a patent application is filed within 1 year of the date of disclosure (see 35 U.S.C. §§102[a]† and 102[b]‡). For patent applications filed on or after March 16, 2013, new statutes relating to novelty§ and non-obviousness¶ apply.

As noted above, the AIA has changed the U.S. patent law from a "first-to-invent" to a "first-to-file" system. In this regard, it is important to note that inventors retain a certain "grace period" under the new law. Particularly, disclosures by the inventor or disclosure by those who have derived information from an inventor should not be considered prior art provided that the patent application is filed within 12 months of the disclosure. However, a significant difference between prior law and the new law effective for patent applications filed after March 16, 2013, is that any disclosure by a third party who arrived at the invention without the benefit of information derived from the inventor will bar the patentability of an invention. Thus, it is imperative that one should consider filing a patent application as soon as possible upon the identification of a new plant variety to avoid potential bars to patentability arising from third-party disclosures.

Another potential issue that may create challenges for a grower is the potential for "on-sale" bars to the patentability of a new plant variety. Under the previous law**, a 1-year grace period existed for filing a patent application after the sale of a product corresponding to a patented invention. The new law (in its current form††) contains no explicit grace period for inventions that are first sold and for which a patent application is subsequently filed. Furthermore, the legislative history related to the definition of a "disclosure" is unclear (e.g., is the sale of a product by an inventor a "disclosure" by the inventor?). Thus, the conservative view suggests that plant growers file patent applications (plant or utility applications) prior to selling (or offering to sell) a new plant variety until such time as the law is clarified by judicial interpretation or the Congress clarifies this aspect of the law.

Upon the grant of a patent, the patent owner has the right to exclude others from making, using, selling, importing, or offering to sell the patented invention. Thus, it would be an act of infringement to make, use, import,

---

* 35 U.S.C. §103. Conditions for patentability; non-obvious subject matter.
(a) A patent may not be obtained, although the invention is not identically disclosed or described as set forth in §102 of this title, if the differences between the subject matter sought to be patented and the prior art are such that the subject matter as a whole would have been obvious at the time that the invention was made to a person having ordinary skill in the art to which the said subject matter pertains. Patentability shall not be negatived by the manner in which the invention was made ...

† 35 U.S.C. §102. Conditions for patentability; novelty and loss of right to patent.
A person shall be entitled to a patent unless
(a) the invention was known or used by others in this country, or patented or described in a printed publication in this or a foreign country, before the invention thereof by the applicant for patent, or
(b) the invention was patented or described in a printed publication in this or a foreign country or in public use or on sale in this country, more than 1 year prior to the date of the application for patent in the United States ...

‡ *Ibid.*

§ 35 U.S.C. §102. Conditions for patentability; novelty and loss of right to patent (as amended by AIA).
(a) NOVELTY; PRIOR ART. A person shall be entitled to a patent unless
  (1) the claimed invention was patented, described in a printed publication, or in public use, on sale, or otherwise available to the public before the effective filing date of the claimed invention, or
  (2) the claimed invention was described in a patent issued under §151, or in an application for patent published or deemed published under §122(b), in which the patent or application, as the case maybe, names another inventor and was effectively filed before the effective filing date of the claimed invention.
(b) EXCEPTIONS. (1) DISCLOSURES MADE 1 YEAR OR LESS BEFORE THE EFFECTIVE FILING DATE OF THE CLAIMED INVENTION. A disclosure made 1 year or less before the effective filing date of a claimed invention shall not be prior art to the claimed invention under subsection (a)(1) if
  (A) the disclosure was made by the inventor or a joint inventor or by another who obtained the subject matter disclosed directly or indirectly from the inventor or a joint inventor, or
  (B) the subject matter disclosed had, before such disclosure, been publicly disclosed by the inventor or a joint inventor or another who obtained the subject matter disclosed directly or indirectly from the inventor or a joint inventor ...

¶ 35 U.S.C. §103. Conditions for patentability; non-obvious subject matter.
A patent for a claimed invention may not be obtained, notwithstanding that the claimed invention is not identically disclosed as set forth in §102, if the differences between the claimed invention and the prior art are such that the claimed invention as a whole would have been obvious before the effective filing date of the claimed invention to a person having ordinary skill in the art to which the claimed invention pertains. Patentability shall not be negated by the manner in which the invention was made.

** 35 U.S.C. §102. Conditions for patentability; novelty and loss of right to patent.
A person shall be entitled to a patent unless
(a) the invention was known or used by others in this country, or patented or described in a printed publication in this or a foreign country, before the invention thereof by the applicant for patent, or
(b) the invention was patented or described in a printed publication in this or a foreign country or in public use or on sale in this country, more than 1 year prior to the date of the application for patent in the United States ...

†† 35 U.S.C. §103. Conditions for patentability; non-obvious subject matter.
(a) A patent may not be obtained, although the invention is not identically disclosed or described as set forth in §102 of this title, if the differences between the subject matter sought to be patented and the prior art are such that the subject matter as a whole would have been obvious at the time that the invention was made to a person having ordinary skill in the art to which the said subject matter pertains. Patentability shall not be negatived by the manner in which the invention was made ...

offer to sell, or sell a patented plant in the United States (see 35 U.S.C. §271). Should such an act occur, the patent owner has a right to recover damages as discussed previously with respect to the infringement of plant patents. Namely, one can obtain injunctive relief that precludes the infringer from making, using, selling, offering to sell, or importing the plant. One can also obtain a monetary award of actual damages (e.g., lost sales) or a reasonable royalty for the plants sold by the infringer. As discussed previously, treble damages can be obtained in some cases as can attorney's fees (in instances where a court finds exceptional circumstances).

CONCLUSION

The Supreme Court has compared and contrasted rights conferred by PVPA, plant patents, and utility patents in a decision entitled *J. E. M. Ag Supply, Inc. v. Pioneer Hi-Bred Intl, Inc.*, 534 U.S. 124, 140, 142 (2001). In this decision, the Supreme Court has written:

> The PVPA also contains exemptions for saving seeds and for research. A farmer who legally purchases and plants a protected variety can save the seed from these plants for replanting on his own farm. See §2543 ("[I]t shall not infringe any right hereunder for a person to save seeds produced by the person from seeds obtained, or descended from seeds obtained, by authority of the owner of the variety for seeding purposes and use such saved seed in the production of a crop for use on the farm of the person ..."); see also *Asgrow Seed Co. v. Winterboer*, 513 U.S. 179, 115 S.Ct. 788, 130 L.Ed.2d 682 (1995). In addition, a protected variety may be used for research. See 7 U.S.C. §2544 ("The use and reproduction of a protected variety for plant breeding or other *bona fide* research shall not constitute an infringement of the protection provided under this chapter"). The utility patent statute does not contain similar exemptions.

The Supreme Court continues, at pages 142 to 143,

> To be sure, there are differences in the requirements for, and coverage of, utility patents and PVP certificates issued pursuant to PVPA. These differences, however, do not present irreconcilable conflicts because the requirements for obtaining a utility patent under §101 are more stringent than those for obtaining a PVP certificate, and the protections afforded by a utility patent are greater than those afforded by a PVP certificate. Thus, there is a parallel relationship between the obligations and the level of protection under each statute.
>
> It is much more difficult to obtain a utility patent for a plant than to obtain a PVP certificate because a utility patentable plant must be new, useful, and non-obvious, 35 U.S.C. §§101–103. In addition, to obtain a utility patent, a breeder must describe the plant with sufficient specificity to enable others to "make and use" the invention after the patent term expires (§112). The disclosure required by the Patent Act is "the *quid pro quo* of the right to exclude." *Kewanee Oil Co. v. Bicron Corp.*, 416 U.S. 470, 484, 94 S.Ct. 1879, 40 L.Ed.2d 315 (1974). The description requirement for plants includes a deposit of biological material, for example, seeds, and mandates that such material be accessible to the public. See 37 CFR §§1.801–1.809 (2001); see also App. 39 (seed deposits for U.S. Patent No. 5,491,295).
>
> By contrast, a plant variety may receive a PVP certificate without a showing of usefulness or non-obviousness. See 7 U.S.C. §2402(a) (requiring that the variety be only new, distinct, uniform, and stable). Nor does PVPA require a description and disclosure as extensive as those required under §101. PVPA requires a "description of the variety setting forth its distinctiveness, uniformity, and stability, and a description of the genealogy and breeding procedure, when known." 7 U.S.C. §2422(2). It also requires a deposit of seed in a public depository (§2422(4)), but neither the statute nor the applicable regulation mandates that such material be accessible to the general public during the term of the PVP certificate. See 7 CFR §97.6 (2001).
>
> Because of the more stringent requirements, utility patent holders receive greater rights of exclusion than holders of a PVP certificate. Most notably, there are no exemptions for research or saving seeds under a utility patent. Additionally, although the Congress increased the level of protection under PVPA in 1994, a PVP certificate still does not grant the full range of protections afforded by a utility patent. For instance, a utility patent on an inbred plant line protects that line as well as all hybrids produced by crossing that inbred with another plant line. Similarly, PVPA now protects "any variety whose production requires the repeated use of a protected variety." 7 U.S.C. §2541(c)(3). Thus, one cannot use a protected plant variety to produce a hybrid for commercial sale. PVPA protection still falls short of a utility patent, however, because a breeder can use a plant that is protected by a PVP certificate to "develop" a new inbred line while he cannot use a plant patented under §101 for such a purpose. See 7 U.S.C. §2541(a)(4) (infringement includes the "use [of] the variety in producing (as distinguished from developing) a hybrid or a different variety therefrom"). See also H. R. Rep. No. 91-1605, p. 11 (1970), U. S. Code Cong. & Admin. News 1970, pp. 5082, 5093; 1 D. Chisum, Patents §1.05(2)(d) and (i), p. 549 (2001).

Thus, it is apparent that the exclusionary rights conferred by various forms of intellectual property protection for a plant can have significant ramifications for both a grower and those who seek to infringe that grower's rights to the plant that has been developed. For example, a grower may wish to rely on a plant patent and a PVPA certificate to protect the plant and its progeny. However, as noted above, certain limitations exist with respect to the intellectual property rights conferred by plant patents and PVPA. To retain complete control of the intellectual property rights associated with a plant, a utility patent may provide the most appropriate vehicle to accomplish this goal because it provides one with the right to exclude others from making, using, offering for sale, or selling the invention in the United States or importing the invention into the United States regardless of whether the plant was asexually or sexually reproduced.

40 Molecular Biology and the Future of Plant Propagation

Govind C. Sharma, Manan Sharma, and Caula A. Beyl

CONCEPT BOX 40.1

- Information is transcribed from DNA to mRNA and then translated to proteins.
- Each eukaryotic cell is potentially endowed to develop into a multicellular organism (totipotency).
- Advances in sequence analysis techniques have driven discovery in molecular genetics and its potential to elucidate processes essential for plant propagation success.
- RNAi modifies or silences a gene and its function by transcriptional and post-transcriptional regulation.
- Epigenetic changes (somaclonal variation) are often observed in tissue-cultured plants and can occur with or without changes in gene sequences or chromosomal structure.
- Several specific genes (mostly auxin related) have been identified to be involved in embryogenesis and rooting.

In this chapter, we attempt to describe some molecular biology/genetic concepts that are at the forefront of understanding plant biology and, thus, plant propagation. Our descriptions are intentionally very general and do not speak to specific cases. The ultimate purpose of this chapter is not only to acquaint and provide readers with this knowledge, but also to stimulate students to read and explore these concepts more by themselves or to enroll in courses specifically designed for advanced studies.

THE CENTRAL DOGMA AND RELATED CONCEPTS

Plant propagation follows the strict codes that govern overall plant life. A new root initiated on a cutting must go through cell division and cell enlargement just like a young embryo or a whole plant. At the cellular level, the central dogma of life reminds us that information flows from DNA (genes) to messenger RNA (mRNA; via transcription) and from RNA to protein (via translation). In a given cell, mRNA is the transcript of a gene. Periodically, information can also flow from RNA to DNA through reverse transcription and replication of RNA. Although certain proteins can, in turn, influence the rate of activity of a gene through interactions with specific genes or promoters, the information contained in the protein cannot flow back to code RNA or DNA. Since genes are the initiator of a specific protein/enzyme, an understanding of the function of genes is critical to understanding complex processes such as somatic embryogenesis, rooting, or the inability to germinate or to initiate roots (recalcitrance), and to the overall processes of plant propagation. It is this understanding that may ultimately overcome recalcitrance that still occurs with some species with respect to adventitious root formation.

Although hormones have reigned supreme in understanding the physiology of plant propagation, the role of specific genes in the functioning of tissues important to propagation, such as meristems, is increasingly being understood. A number of genes have been identified that are expressed mainly in shoot meristems and not in leaves and, thus, provide insight into the process of meristem function (Barton, 2010). A few genes that functionally regulate cell proliferation, differentiation, and patterning of growth include *Knotted1* of maize (Ramirez *et al.*, 2009) and the *AGAMOUS* of Arabidopsis (Weigel and Meyerowitz, 1993). More plant propagation–related genes are described later in this chapter.

Traits such as root initiation, adventitious bud formation, and rapid rooting of a cutting are all under genetic control (either as complex or quantitative genetic traits) and, therefore, are regulated by several genes. Better fundamental understanding of these processes will be obtained by more work on the molecular biology of plant propagation. For example, the rooting ability of larch cuttings is under strong genetic control. Rooting ability as a trait has been examined in controlled crosses or hybrids made for this purpose. For example, hybrids of two species of larch *Larix leptolepis* × *Larix olgensis* have enabled the identification of a specific genetic marker associated with high rooting ability. Among the 328 primers that were studied, only one primer (S356) could amplify a specific region in the DNA sequence associated with high rooting, and this primer can now be utilized to screen larch varieties or genotypes with high rooting ability (Li *et al*., 2008). There are also proteins that act to alter gene expression, thus affecting growth patterns, and some of these groups are called "transcription factors."

Transcription Factors

Genes are transcribed through RNA, and this step is often regulated by one (or more) transcription factors or sequence-specific DNA-binding factors. These generally consist of a protein that binds to specific DNA sequences, thereby controlling the flow (or transcription) of genetic information from DNA to mRNA. A defining feature of transcription factors is that they contain one or more DNA-binding domains that attach to specific sequences of DNA adjacent to the genes that they regulate. Additional transcription factors are being identified that are important to propagation and, thus, to growth. For example, they may influence or regulate morphogenesis or the external shape or form of a plant or play a developmental role in vascular differentiation, such as in the formation of xylem cells (Lau *et al*., 2012).

Totipotency

Vegetative propagation has benefitted by the development of *in vitro* tissue culture techniques, which use tiny to microscopic explants that are exponentially multiplied for many plant species. *In vitro* somatic cell culture has underscored the concept of totipotency, which means that (1) each cell is potentially endowed with structural and functional information essential to develop into a multicellular organism and (2) an undifferentiated cell (like a stem cell) can divide and differentiate into different cells, specialized tissues and organs, and ultimately, a complete multicellular organism. Based on the principle of totipotency, entire plants have been regenerated from a single cell of tobacco or carrot root, among many other species.

ADVANCES IN MOLECULAR BIOLOGY TECHNIQUES

Gene and Genome Sequencing

Much attention has been paid to the human genome, but humans are just one species, whereas there are over 250,000 higher plant species known to science. The capacity to perform relatively rapid and increasingly cost-effective genome sequencing has been a strong driver of the improved understanding of plants and plant processes, such as embryogenesis and adventitious root regeneration. At the same time, improvements and innovation in research tools, such as real-time polymerase chain reaction (PCR), comparative transcriptomes, and microarrays, all provide greater insight into the complexities of meristem growth and rooting, root patterning, and root development.

Polymerase Chain Reaction (PCR)

PCR is common to many molecular biology/genetics techniques and combines the use of a heat-stable DNA polymerase for rapid and exponential amplification (thousands to millions) of copies of a template or a desired DNA sequence so that the sequence can then be studied. PCR has been used in investigating human, animal, and plant diseases, and in relation to plant propagation, in identifying genes associated with rooting efficiency, in genetic testing, in tissue typing, and for genetic fingerprinting a distinct plant variety or an individual (genotype) of a unique plant. For example, mRNA signals associated with root induction in walnut differed in the basal part of the microcutting between auxin-treated and control treatments (Caboni *et al*., 1997).

Complementary DNAs

Messenger RNA sequence can be reverse transcribed to reveal the original DNA sequence. It provides a snapshot of which proteins are being made at a given time in the life of a cell, tissue, or organism. There are several methods for isolating complementary DNA (cDNA). One type of PCR is reverse transcription PCR (RT-PCR). RT-PCR uses mRNA as a template and the enzyme reverse transcriptase to convert mRNA to cDNA, which can then be amplified using primers and DNA polymerase enzyme.

Microarray

Before appreciating the importance of microarray techniques, it is important to understand what a primer is. A primer is a short DNA fragment that complements a targeted larger fragment or piece of DNA and "tells" the DNA polymerase where to start copying the DNA strand. Microarray is a process during which RNA is exposed to primers of numerous genes, and the fluorescent signal emanating during such hybridization is coded to indicate the upregulation or the downregulation of a gene when a comparison is made between the two contrasting treatments or growth conditions. For example, the differences in gene expression between the juvenile and the mature cuttings of *Eucalyptus grandis* were studied in a microarray experiment. Just before root induction, mRNA was extracted from both cuttings and hybridized with a DNA microarray of *E. grandis*. The mRNA from juvenile cuttings showed 363 transcripts (~genes) that were upregulated as opposed to only 245 transcripts that were upregulated for mature cuttings. For example, the genes coding for nitrate reductase were upregulated in juvenile cuttings, but were not found in mature cuttings (Wang *et al.*, 2003; Abu-Abied *et al.*, 2012). Such studies provide an approximation of the number of genes participating in a process and their role through increased or decreased activity.

DNA–RNA Sequencing Advances

Gene and genome sequencing is in high demand now, and it will continue to be used to understand the physiology, pathology, and genetics of organisms. The Sanger method (Sanger *et al.*, 1977) was the first generation of nucleic acid sequencing that ushered us into the genomic era. It is the classical chain-terminating method that requires a single-stranded DNA template, a DNA primer, a DNA polymerase, and normal deoxynucleoside triphosphates (dNTPs). The "N" in dNTPs represents the four nucleotides adenine (A), thymine (T), guanine (G), and cytosine (C) in DNA or other modified nucleotides. The DNA is replicated, and the DNA primer is elongated one base at a time, which is being recorded. The second-generation sequencing methods have used a sequencing-by-synthesis approach that has increased the throughput and enabled many applications previously unfeasible (Metzker, 2005). These methods reduced sequencing costs to approximately $100K per genome. Third-generation sequencing includes advances such as single-molecule real-time DNA sequencing (manufactured by Pacific Biosciences of Menlo Park, CA; http://www.pacificbiosciences.com), whereas Ion Torrent's technology uses ion-sensitive field effect transistors (Niedringhaus *et al.*, 2011). Accelerated technology changes will be the norm in gene sequencing for the decades to come, and such advances will also reduce the sequencing cost. This will allow scientists to sequence genes never considered worth the cost to investigate. After sequencing is achieved, then plant scientists are faced with the more intensive task of identifying the gene functions.

RNA Interference

Gene-silencing signals move between cells, often in the phloem tissue, and are caused by small RNAs (sRNAs) that are only 21, 22, and 24 nucleotides long, generated by a dicer-like protein. Plant biologists along with other life scientists are improving applications of RNA interference (RNAi) to promote productivity, solve food- and fiber-quality issues, and better manage plant biotic and abiotic stresses. RNAi is the process by which foreign, double-stranded RNA is recognized and degraded by specialized protein complexes within eukaryotic cells. There is a tremendous potential for the use of RNAi to confer resistance to pathogens and transmission across graft unions. This provides an additional advantage in that, for food crops (where there has been a resistance to the use of genetically modified organisms), a transgenic rootstock can be grafted to a non-transgenic scion. Small interfering RNAs (siRNAs) can be transmitted across the graft union, conferring resistance to a particular pathogen without the scion (the source for the edible portion of the grafted plant) having to be transformed. In some cases, where the rootstock is the source of the harvested portion of the plant (e.g., cassavas, potatoes, and others), the scion can be the transgenic portion conferring resistance across the graft union. A relevant example of this occurs with cassava. RNAi-mediated resistance to cassava brown streak Uganda virus (CBSUV) in a scion was expressed in the non-transgenic rootstock after grafting (Figure 40.1; Yadav *et al.*, 2011). Another example of the role of RNAi involves not pathogen resistance, but desiccation tolerance in the resurrection plant *Craterostigma plantagineum* (Hilbricht *et al.*, 2008). As we learn more about RNAi-mediated effects, we may see more pairing of transgenic plants with non-transgenic food crops via grafting.

These examples illustrate that, although we undertake macropropagation at the tissue level, the success or failure of propagation efforts is being determined by the interplay between hormones, genes, and the environment at the cellular level. Here, our focus is more on the function of genes. Gene responses are affected by response factors, transcription factors, transposons, and silencing efforts such as siRNAs. The microRNA390 participates in targeting multiple auxin response factors (ARFs) ARF3 and ARF4 transcripts (Table 40.1) (Overvoorde *et al.*, 2010).

Gene Silencing

One may assume that, once the DNA has been synthesized, every trait or character of a plant is now set forever. However, this is not always the case in plants where genes can be turned on or off even after they have become part of the overall genome. Several changes can and do occur even after gene transcription (production of mRNA from DNA). A gene can be inactivated or "silenced" post-transcriptionally. These silencing signals travel to meristematic regions, causing epigenetic changes (see Chapter 6 on juvenility) or changes by mechanisms other than those induced by modifications in the DNA sequence. These changes may influence growth and development, and may influence a phenotype (Melnyk *et al.*, 2011).

FIGURE 40.1 RNAi mediated resistance to cassava brown streak Uganda virus (CBSUV) transmitted via graft union. Nontransgenic cassava scion infected with CBSUV grafted onto transgenic cassava cv. 60444 rootstock (a) free of CBSUV. In (b), orange arrows indicate old leaves showing CBSUV symptoms, and white arrows indicate old leaves of transgenic stock free of CBSUV symptoms 60 days after grafting. Cassava brown streak disease (CBSD) necrosis symptoms on tuberous roots of symptomatic nontransgenic cassava cv. 60444 (c) and absence of necrosis in tuberous roots of non-symptomatic transgenic cassava cv. 60444 (d), 120 days after grafting with CBSUV-infected cassava scion. (From Yadav, J. S. *et al.* 2011. *Molec Plant Path* 12:677–687.)

TABLE 40.1
Rooting Induction and Root Growth–Related Genes

Genes/Regulators	Gene Function or Roles	Reference
AUX1/LAX	Lateral root initiation and lateral root emergence	Overvoorde *et al.* (2010)
ALF4	Maintains the pericycle in a mitosis-competent state for lateral root formation	Hodge *et al.* (2009)
PLT	*De novo* root formation from shoot structures	Hodge *et al.* (2009)
Auxin f-box proteins (*AFBs*)	Members of the Aux/IAA protein family that are encoded by 29 genes	Overvoorde *et al.* (2010)
ARFs	Transcription factors, coded by 23-member gene family active in auxin signaling pathway	Overvoorde *et al.* (2010)
ARF10 and *ARF16*	Defective root tip by overexpression of *MIR160* and repression of *arf10* and *arf16* double mutants	Hodge *et al.* (2009)
ARF7 and *ARF19*	Regulation of auxin-responsive genes	Overvoorde *et al.* (2010)
slr-1/iaa14 and *iaa28*	Root growth–related gain-of-function mutants that show reduced lateral root formation	Overvoorde *et al.* (2010)
SLR/IAA14 promoter	Drives expression in both the root epidermis and the stele	Overvoorde *et al.* (2010)
Brevis radix (*BRX*)	Transcription factor regulating cell proliferation and elongation proximal to the root tip	Hodge *et al.* (2009)
mir164a and *mir164b*	*Arabidopsis* mutants with microRNA (miRNA) in promoting lateral root growth	Guo *et al.* (2005) in Hodge *et al.* (2009)
Salt overly sensitive 4 (*SOS4*), *rhd6*	Pyridoxal kinase gene that functions in root hair development	Shi and Zhu (2002)
TTG and *GL2*	Functions in both the roots and shoots as epidermal developmental regulators	Shi and Zhu (2002)
WER	Specification of root hair	Shi and Zhu (2002)

Transposable elements (TEs), also known as "jumping genes" or "transposons," are sequences of DNA that move (or jump) from one location in the genome to another. Please do not laugh at the concept of jumping genes because they truly can regulate gene expression and some are also responsible for giving rise to different cell types. In plants, gene silencing can inactivate TEs. A *Nicotiana benthamiana* transgenic plant containing a green fluorescent protein (GFP) expressed a gene-silencing signal linked to this visually observable GFP that started its journey near a leaf vein in the scion and was observed as far away as in the meristem of a lateral root (Figure 40.2). This transmissible signal survived through tissue culture and was subsequently inherited by the clonal progeny (Bai *et al.*, 2011). These results suggest that the potential application of mobile promoter-targeting siRNA in horticulture is considerable for the improvement of plant cultivars by grafting. For example, a plant that has acquired the ability to transmit siRNA could be grafted onto an appropriate graft-compatible stock or to a scion, and consequently, the siRNA would induce gene silencing and, thus, improve a graft partner and can also result in transmitting flowering stimulus.

SOMACLONAL VARIATION

Oftentimes, clonal (somatic) plant propagation progresses as planned, and a genotype reproduces exactly by

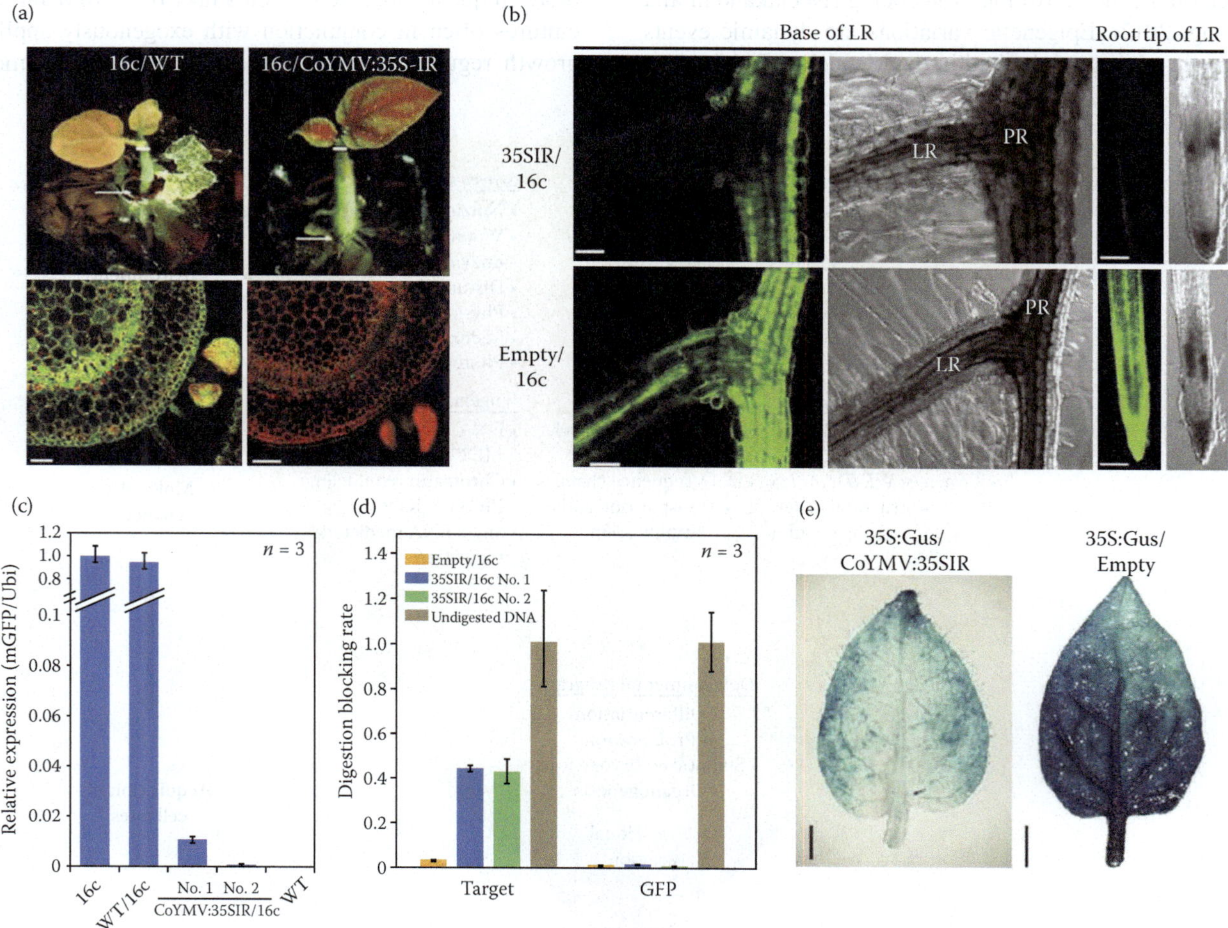

FIGURE 40.2 Evidence of transcriptional gene silencing (TGS; loss of red color especially close to leaf veins) by an siRNA as demonstrated for a transgenic plant of *Nicotiana benthamiana*. (a) The left upper pair of plants shows the loss of GFP in the cross-section of a 16c plant grafted onto CoYMV:3SIR rootstock due to gene silencing and equivalent silencing of GFP in a cross section of the stem shown just below it. (b) TGS is being induced across the graft union by a signal from the scion to the 16c rootstock, resulting in complete suppression of signal in the lateral root including the root tip in all micrographs in the upper row, but not in the lower row. Wild type (WT), lateral root (LR), primary root (PR). (c) Quantitative PCR analysis clearly showing the loss of modified green fluorescent protein (mGFP) mRNA signal in two roots (numbers 1 and 2) compared to the non-grafted mGFP signal containing the stock and the scion. (d) Similar data that analyze DNA methylation status. (e) The loss of β-glucuronidase (GUS) stain in a newly developed leaf (see the left one as compared to the right leaf, especially around the leaf veins). (From Bai, S. *et al.*, *J Exp Botany* 62:4561–4570, 2011.)

following its genetic and epigenetic program. However, sometimes, in tissue culture work, dedifferentiation and the *in vitro* environment, including the media components, may affect the developmental processes occurring at the molecular level, and off-types or variants (e.g., larger or smaller leaf, taller or shorter, and (rarely) even a sterile plant) emerge among these clonally propagated progenies. *In vitro* plant cell and tissue culture methodologies are important to plant propagation and breeding programs. Oftentimes, these changes are deleterious (not improvements on the original), but occasionally, they are desirable and, thus, selected precisely for the changes induced. Somaclonal variation in clonally propagated plants results in a heritable alteration in gene expression, that is, they may or may not be due to gene sequence modification, including gene activation or silencing (Neelakandan and Wang, 2012). Epigenetic variations are dynamic events that are often not inherited in the Mendelian fashion (see Chapter 5). Examples of aberrant phenotypes in regenerated plants include abnormal leaf structures and variant floral morphology. One example of a variant phenotype that has occurred in many primary maize plants regenerated from tissue culture is seed development on the tassel, the male inflorescence (Kaeppler *et al.*, 2000).

The factors that contribute to somaclonal variation can be grouped as physiological, genetic (e.g., gene silencing or gene inactivation), and biochemical. Plant breeders and geneticists have often welcomed them as a source of variation in developing new cultivars with desirable agronomic and horticultural, including nutritionally important, traits. Genetic variation seems to occur at relatively modest frequencies; however, variation in DNA methylation patterns seems to be much more frequent. Epigenetic events take place in plant cell cultures often in conjunction with exogenously applied growth regulators under artificially provided chemical

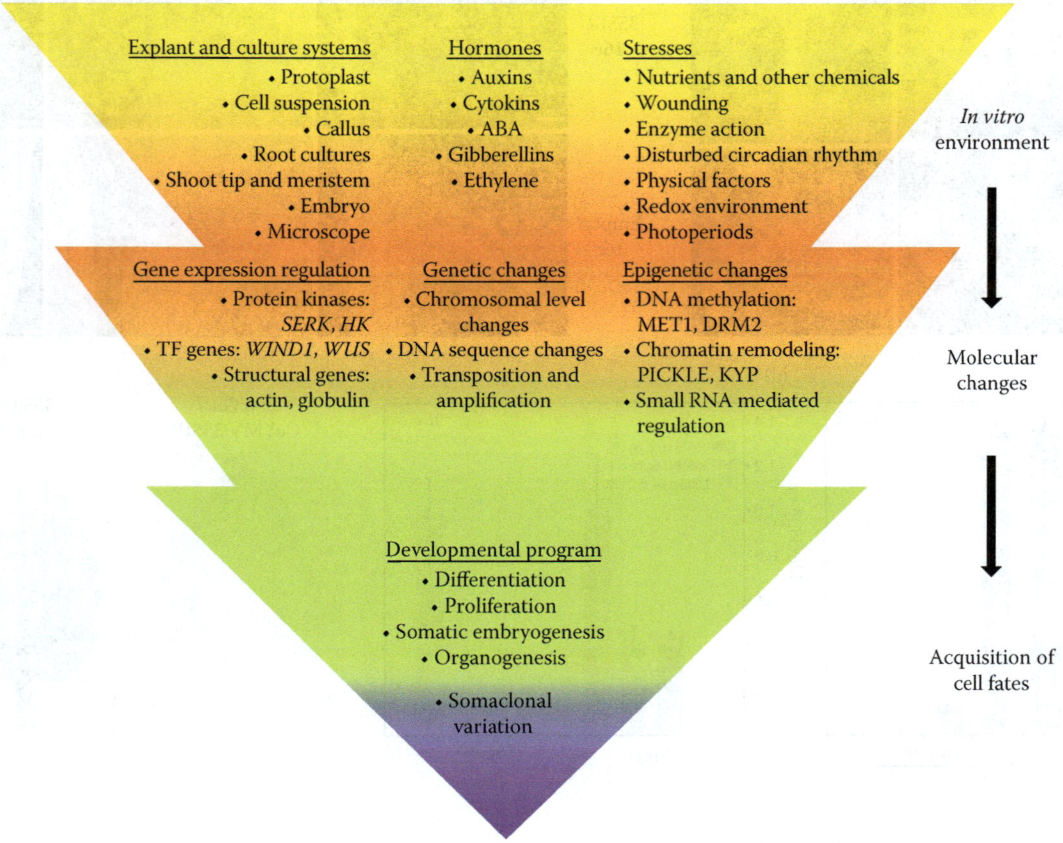

FIGURE 40.3 A word diagram of possible changes during *in vitro* culture. Genomic stress could result from wounding, physical–chemical factors, and the presence of hormones and/or enzymes, abscisic acid (ABA), coupled with the developmental events of dedifferentiation and regeneration. These changes are manifested at the gene expression level, for example, in protein kinases, transcription factors (TF), and structural genes, and can contribute to explant adaptation to stress and the reorientation of its developmental program. *In vitro* culture can also be associated with genetic changes, including chromosomal changes, DNA sequence alterations, amplifications, and transpositions. More recent discoveries point to epigenetic changes at the level of DNA methylation, chromatin modification, and sRNA-mediated regulation taking place in cultured tissues. The spectrum of genetic and epigenetic changes can potentially give rise to phenotypic changes among the regenerants that are termed here as "somaclonal variations." (From Neelakandan, A. K. and K. Wang, *Plant Cell Rep* 31:597–620, 2012.)

and physical environments. Significant advances in plant epigenetics also provide clues to the understanding of the molecular basis of somaclonal variation and the role of DNA methyltransferases, thus altering DNA methylation patterns (Phillips *et al.*, 1994), histones (alkaline proteins found in eukaryotic cell nuclei that envelope and package DNA in nucleosomes), modification enzymes, and other regulatory proteins that have essential roles in plant development (Miguel and Marum, 2011). It is increasingly accepted that the *in vitro* culture process as discussed earlier triggers the activation of transposons and retrotransposons that exist naturally in plant genomes and cause such epigenetic or somaclonal variations (Figure 40.3).

Agrobacterium rhizogenes

Since the beginning of the 20th century, *Agrobacterium* species have been known as the causal plant pathogen of tumors or galls on trees because of its ability to transfer DNA to plant cells. *Agrobacterium* can transfer DNA to a broad group of organisms, including numerous dicot and monocot angiosperm species, gymnosperms, yeasts, and basidiomycetes fungi (Gelvin, 2003). For plant propagation, our interest is more often focused on root induction, and for this, the species *A. rhizogenes* may be more valuable. It induces proliferative multi-branched adventitious "hairy roots" at the site of infection, and so, it has the potential for rooting of difficult-to-root (recalcitrant) cuttings. For example, *A. rhizogenes* induced rooting in the recalcitrant microcuttings of English walnut, almond, apple, pine, larch, and eucalyptus in a hormone-free medium and resulted in abundant callus formation in the presence of indolebutyric acid but did not induce rooting when auxin or indoleacetic acid (IAA) was present (Caboni *et al.*, 1996).

Agrobacterium species carry plasmids in their cells. These are small DNA molecules that are physically separate from chromosomal DNA within a cell, and they replicate independently. The *rol* (root loci) genes *rolA*, *rolB*, *rolC*, and *rolD* are root-inducing plasmids of *A. rhizogenes* (Cassanova *et al.*, 2005) (Figure 40.4). Transformation of plant species with *A. rhizogenes* or with its *rol* genes produces hairy roots from which adventitious shoots can be regenerated but, typically, with the characteristic hairy-root phenotype. This phenotype may include changed characteristics (traits), such as altered apical dominance, dwarfing, shortened internodes, changes in leaf shape (wrinkled, wider, or smaller), adventitious root production, altered flowering and morphology, and reduced pollen and seed production (Welander and Zhu, 2006). *Agrobacterium rhizogenes* cultures can have many other uses besides plant propagation, for example, in the production of specialized biochemicals (Figure 40.5) in hairy root cultures (Ono and Tian, 2011) or as a model system for studying physiological processes and metabolic pathways (Figure 40.6).

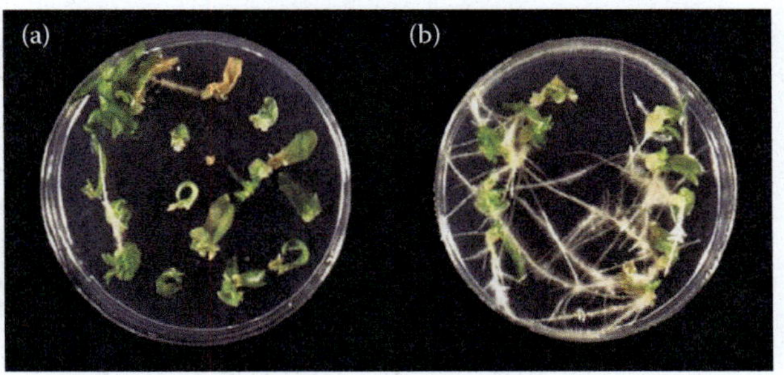

FIGURE 40.4 Comparative effect on adventitious root regeneration from detached leaflets of *in vitro*–grown carnations as affected by the presence of *rolC* transgene. Explants were cultured in Murashige and Skoog medium supplemented with 0.5 M of the auxin 1-naphthaleneacetic acid. (a) Control (*uidA*-transgenic) leaves. (b) *rolC*-transgenic leaves. (From Cassanova, E. *et al.*, *Biotech Adv* 23:3–39, 2005.)

GENES RESPONSIBLE FOR ROOTING AND RECALCITRANCE

Many environmental and endogenous factors that can influence lateral and adventitious root formation include nearly everything that can affect plant growth, for example, hormones, light quantity, light quality, oxygen, carbon dioxide, nitric oxide, free radicals, relative humidity, the pH of the growth media, the physical structure of the growth media, antioxidants, wounding, polyamines, and the concentration and types of nutrients in the media (Pijut *et al.*, 2011). Not surprisingly, auxin regulatory genes are the most frequently identified as those important for the growth and development of roots and root hairs, but complex traits, such as the growth habit of a tomato or a root system of a pine tree, are not regulated by just one, two, or a few genes. Similarly, the rooting of a cutting or its recalcitrance to rooting are complex traits that are often regulated by many genes. Such traits are usually under complex genetic control. Our understanding of such traits has been improved by the identification of quantitative trait loci (QTL) associated with the genes of rooting. These are markers closely linked to genes of interest and, therefore, almost always co-segregate with genes controlling a trait or thought to control a trait. The actual QTL does not have any direct influence on the gene(s) of interest. The genetics of such genes that regulate rooting has been best understood through controlled genetic crosses made between difficult-to-root species (such as *Populus deltoides*) and easy-to-root species (*P. euroamericana*). Subsequent progeny analysis of these crosses has revealed that the mean root length and the total number of roots initiated were under strong genetic control of five QTL. Conventional genetic studies like this have provided greater insight into the genes or genomic regions critical to root formation observed

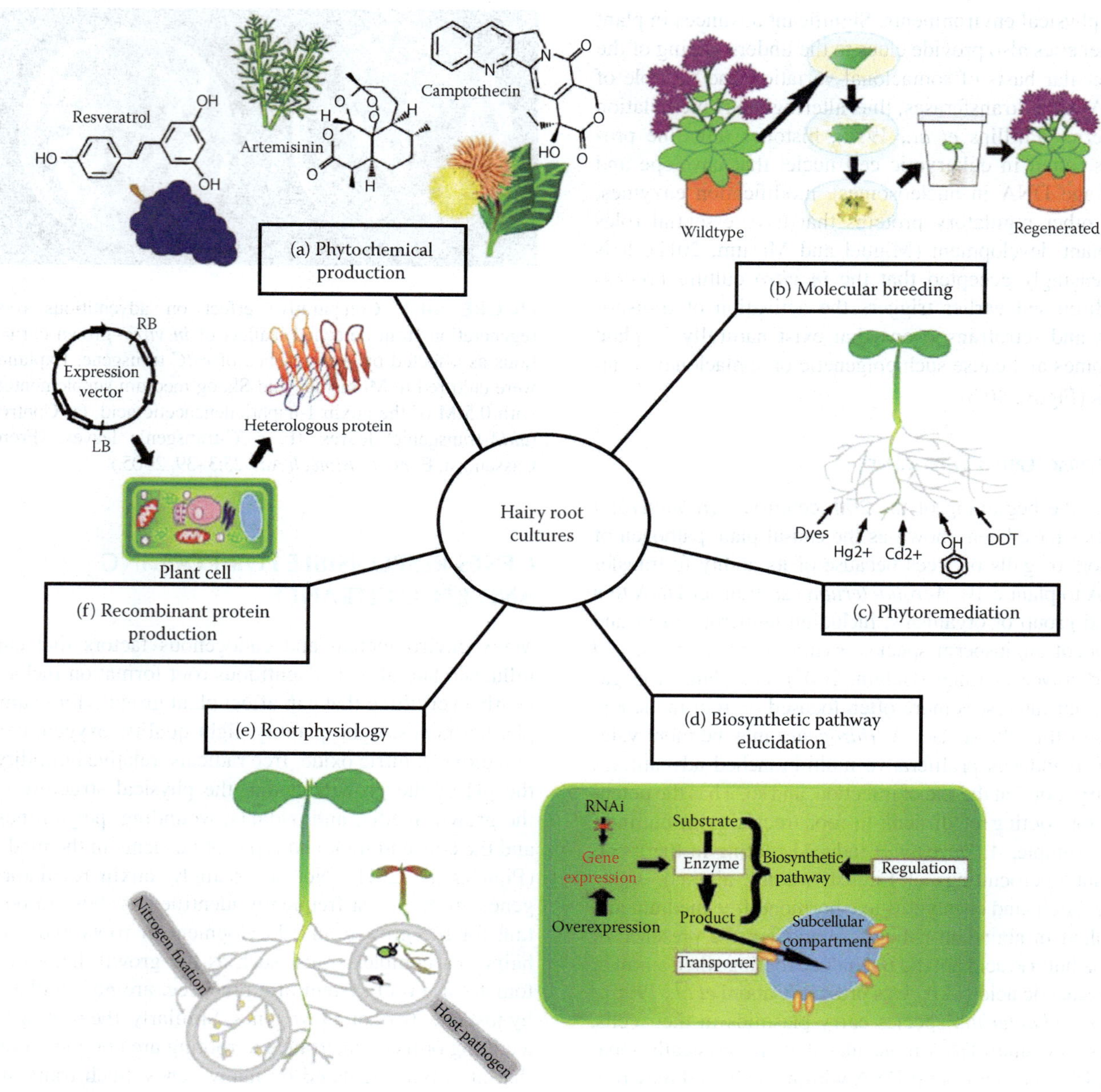

FIGURE 40.5 Various possible applications of hairy-root cultures. (a) Production of phytochemicals, including alkaloids, terpenoids, and phenolics. (b) Molecular breeding by infection of ornamental plants with *Agrobacterium rhizogenes* and regeneration of whole plants from hairy roots can yield plants with desirable phenotypes, such as compact size. (c) Hairy-root culture can serve as a model system for the phytoremediation of toxic substances and reactive dyes, dichloro-diphenyl-trichloroethane (DDT). (d) Biosynthetic pathway elucidation facilitating metabolic engineering. (e) Root physiology studies ranging from nitrogen fixation, iron deficiency, and aluminum toxicity, to host–pathogen interactions. (f) Recombinant protein production in this system as a rapid, contained, low-cost, genetically stable means of producing human antibodies, cytokines, and other protein therapeutics. (From Ono, N. and L. Tian. 2011. *Plant Sci* 180:439–446.)

in several species or conserved in nature (Zhang *et al.*, 2009). Among these QTL observed here and in other studies, genes that regulate auxin and its pathways are most prevalent, for example, nearly 40% of known *Arabidopsis thaliana* lateral root mutants affect specific components of auxin pathways (Hodge *et al.*, 2009). In roots and rooting, auxin transport and signaling are critical to lateral root formation, the maintenance of apical dominance, and adventitious root formation. These genes and their roles are summarized in Table 40.1.

Success in plant propagation often depends on the activities of existing or newly formed (*de novo*) meristems or non-zygotic (somatic) embryos. Some of the genes identified with somatic embryogenesis include Wuschel (*WUS*) associated with a transcription factor for CLAVATA (*CLV*), which maintains the undifferentiated

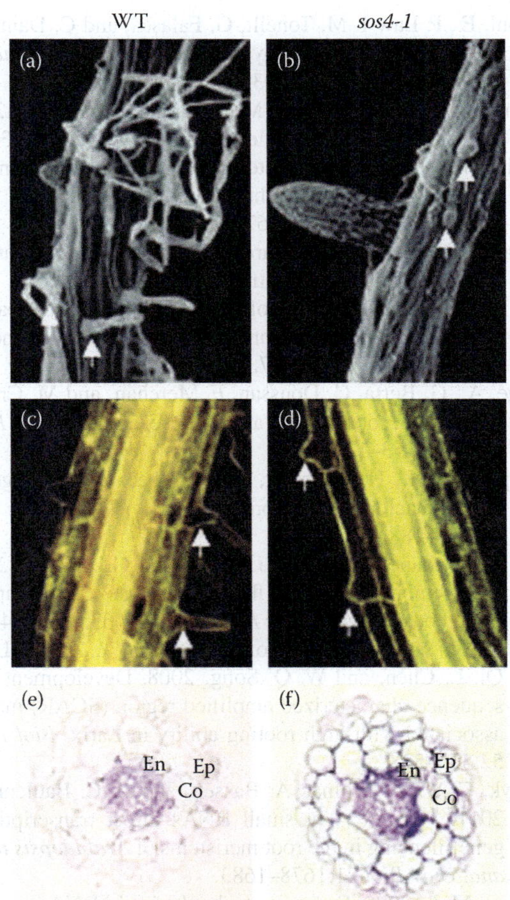

FIGURE 40.6 Root hair development is arrested at the stage of initiation and tip growth in the *Arabidopsis sos4-1* mutant. To the left (a, c, and e) is the *sos4-1* mutant and to the right (b, d, and f) are the wild type. (a and b) Primary root visualized by scanning electron microscopy. (c and d) Primary root visualized by confocal microscopy after fluorescent brightener staining. (e and f) Thin section of a primary root. Arrows point to root hairs or hair initials. Ep = epidermis; Co = cortex; En = endodermis. (From Shi, H. and J. K. Zhu. 2002. *Plant Physiol* 129: 585–593.)

(stem) cells of shoot meristems. *WUS* expresses in the heart-shaped embryo stage and, later, in the stem cell/meristematic zone in the shoot. The other gene with an interesting name—the Baby Boom (*BBM*) gene—encodes a transcription factor that promotes cell proliferation and morphogenesis during embryogenesis. The Somatic Embryogenesis Receptor Kinase (*SERK*) gene is active in the early stages of embryogenesis. It belongs to the family of Leucine-Rich Repeat Receptor-Like Kinases (*LRR-RLK*s), which influence the key processes of plant growth. The *SERK* gene was first identified in carrots, and since then, it has been found in *Arabidopsis*, maize, cacao, citrus, and potato, hence, underscoring the importance of kinases in plant propagation (Sharma *et al.*, 2008; Solis-Ramos *et al.*, 2012). In summary, the complex process of rooting and embryo formation is being better understood because of our improved understanding of auxin and other regulatory genes (Table 40.1).

FUTURE DIRECTIONS

Plant propagation is critical to human endeavors ranging from improving diet and nutrition to aesthetics and ecosystem balance. The science of propagation is rooted in the human discovery of plant functionality, and each generation of scientists claim that they have seen the most dynamic changes in the field of plant propagation compared to their predecessors. Do the past few decades represent considerable advancement? The answer is an unequivocal "yes" when we consider this era to have seen hormonal discoveries, a refinement of controlled environments, and an advancement of tissue and cell culture methods. Of course, all of these advancements have been enhanced by molecular genetic techniques, which have grown increasingly widespread in all plant sciences. So, what will the next two decades look like? Below is one perspective:

1. We will further develop the ability to sequence genes and analyze entire genomes by bioinformatics at breakneck speed and we will make greater progress in deciphering the functions of many more genes (functional genomics) whose functions have been unknown up to now.
2. The physical environment and its regulation will see changes, mainly, the stark realization that our planet does not have unlimited resources. Climate change will certainly call for more efficient water use by crops grown on our "drying planet." We will also require flood- and saline-tolerant plants in coastal areas.
3. Efficiency in lighting in a controlled environment is a case in point. We will continue to aspire and advance the utilization of energy-efficient lighting systems for plant propagation in the next few years just as we have seen the popularization of illumination sources such as fluorescent, light-emitting diodes, and beyond.
4. In controlled greenhouses, growth chambers, and tissue culture environments, the major costs are for climate control. Green and renewable energy usage (e.g., wind, solar, and geothermal) will result in more cost-effective solutions for the plant propagation industry.
5. Mycorrhiza and other competitively exclusive microorganisms that reduce potential phytopathogens (endophytic, epiphytic microorganisms) and enhance rooting and water and nutrient acquisition will be better understood and will be utilized more effectively both in the

field and in controlled environmental systems as we integrate their usage in plant production/propagation systems.
6. Educational and research resources will also reduce the cost for imparting and sharing plant biology/propagation research, especially for high economic value plant species. More specialized programs will evolve, and smarter, targeted sharing of information and knowledge will occur, capitalizing on information distributions through social media made more accessible to general and academic search engines by cloud storage to easily access publicly available databases. The trend to publish in Open Journals will aid in freer access to information generated by public resources.
7. Inter-kingdom gene transfers—We will understand the benefits accrued by plants when genes between widely unrelated organisms are exchanged. The recent discovery is that many persistent plant viruses that co-exist with plants are known to confer heat and drought resistance. So, in the future, we may be inoculating plants with naturally attenuated viruses that confer a beneficial quality attribute to plants. The other example, *Galdieria sulphuraria*, a red alga that has borrowed at least 5% of its genes from bacteria and other members of the domain *Archaea*, is adapted to extreme conditions (extremophile). These genetic adaptations now confer specific stress responses to this eukaryote, which allows it to adapt to hot volcanic pools laden with salt, acid, sulfur, and toxic metals.

The future of plant propagation is exciting, and if even a small portion of these predictions come to fulfillment, there is not much to limit how they can be applied to improve plants and how we replicate them.

REFERENCES

Abu-Abied, M., D. Szwerdszarf, I. Mordehaev, A. Levy, O. R. Stelmakh, E. Belausov, Y. Yaniv, S. Uliel *et al.* 2012. Microarray analysis revealed upregulation of nitrate reductase in juvenile cuttings of *Eucalyptus grandis*, which correlated with increased nitric oxide production and adventitious root formation. *Plant J* 71:787–799.

Bai, S., A. Kasai, K. Yamada, T. Li, and T. Harada. 2011. A mobile signal transported over long distance induces systemic transcriptional gene silencing in a grafted partner. *J Exp Botany* 62:4561–4570.

Barton, M. K. 2010. Twenty years on: The Inner Workings of the Shoot Apical Meristem, a Developmental Dynamo. *Developmental Biol* 341(1):95–113

Caboni, E., P. Lauri, B. Watillon, and C. Damiano. 1997. Isolation of mRNA species related to the rooting induction in almond and apple through differential display technique. *Biol Plant* 39:99–104.

Caboni, E., P. Law-I, M. Tonelli, G. Falasca, and C. Damiano. 1996. Root induction by *Agrobacterium rhizogenes* in walnut. *Plant Sci* 118:203–208.

Cassanova, E., M. I. Trillas, L. Moysset, and A. Vainstein. 2005. Influence of *rol* genes in floriculture. *Biotech Adv* 23:3–39.

Gelvin, S. B. 2003. Agrobacterium-mediated plant transformation: The biology behind the "gene-jockeying" tool. *Microbiol Mol Biol Rev* 67:16–37

Hilbricht, T., S. Varotto, V. Sgaramella, D. Bartels, F. Salamini, and A. Furini. 2008. Retrotransposons and siRNA have a role in the evolution of desiccation tolerance leading to ressurection of the plant Craterostigma plantagineum. *New Phytol* 179:877–887.

Hodge, A., G. Berta, C. Doussan, F. Merchan, and M. Crespi. 2009. Plant root growth, architecture, and function. *Plant Soil* 321:153–187.

Kaeppler, S. M., H. Kaeppler, and Y. Rhee. 2000. Epigentic and somaclonal variation in plants. *Plant Molec Biol* 43:179–188.

Lau, S., D. Slane, O. Herud, J. Kong, and G. Jurgens. 2012. Early embryogenesis in flowering plants: Setting up the basic body pattern. *Annu Rev Plant Biol* 63:6.1–6.24.

Li, H., S. G. Zhang, J. M. Gao, C. G. Wang, Y. Zhang, L. W. Qi, L. Chen, and W. Q. Song. 2008. Development of a sequence-characterized amplified region (SCAR) marker associated with high rooting ability in Larix. *Biol Plant* 52:525–528.

Melnyk, C. W., A. Molnar, A. Bassett, and D. C. Baulcombe. 2011. Mobile 24 nt small RNAs direct transcriptional gene silencing in the root meristems of *Arabidopsis thaliana*. *Curr Biol* 21:1678–1683.

Metzker, M. L. 2005. Emerging technologies in DNA sequencing. *Genome Res* 15:1767.

Miguel C. and L. Marum. 2011. An epigenetic view of plant cells cultured in vitro: somaclonal variation and beyond. *J Exp Bot* 62(11):3713–3725

Neelakandan, A. K. and K. Wang. 2012. Recent progress in the understanding of culture-induced genome level changes in plants and potential applications. *Plant Cell Rep* 31:597–620.

Niedringhaus, T. P., D. Milanova, M. B. Kerby, M. P. Snyder, and A. E. Barron. 2011. Landscape of next-generation sequencing technologies. *Anal Chem* 83:4327–4341.

Ono, N. and L. Tian. 2011. The multiplicity of hairy-root cultures: Prolific possibilities. *Plant Sci* 180:439–446.

Overvoorde, P., F. Hidehiro, and T. Beeckman. 2010. Auxin control of root development. *Cold Spring Harb Perspect Biol* 2:a001537.

Phillips, R. L., S. M. Kaeppler, and P. M. Olhoft. 1994. Genetic instability of plant tissue cultures: Breakdown of normal controls. *Proc Natl Acad Sci USA* 91:5222–5226.

Pijut, P. M., K. E. Woeste, and C. H. Michler. 2011. Promotion of adventitious root formation of difficult-to-root hardwood tree species. In Janick, J., ed. *Hort Rev* 38: 213–251.

Ramirez J., N. Bolduc, D. Lisch, and S. Hake 2009. Distal expression of knotted in maize leaves leads to reestablishment pf proximal/distal patterning and leaf dissection. *Plant Physiol* 151:1878–1888.

Sanger, F., S. Nicklen, and A. R. Coulson. 1977. DNA sequencing with chain-terminating inhibitors. *Proc Natl Acad Sci USA* 74:5463–5467.

Sharma, S. K., S. Millam, I. Hein, and G. J. Bryan. 2008. Cloning and molecular characterization of potato *SERK* gene transcriptionally induced during initiation of somatic embryogenesis. *Planta* 228:319–330.

Shi, H. and J. K. Zhu. 2002. *SOS4*, a pyridoxal kinase gene, is required for root hair development in *Arabidopsis*. *Plant Physiol* 129:585–593.

Solis-Ramos, L. Y., A. Andrade-Torres, L. A. S. Carbonell, C. M. O. Salin, and E. C. Serna. 2012. Somatic embryogenesis in recalcitrant plants. In Sato, K.-I., ed. *Embryogenesis*. http://www.intechopen.com/books/embryogenesis, pp. 597–618. Accessed April 8, 2014.

Wang, R., M. Okamoto, X. Xing, and N. M. Crawford. 2003. Microarray analysis of the nitrate response in *Arabidopsis* roots and shoots reveals over 1000 rapidly responding genes and new linkages to glucose, trehalose-6-phosphate, iron, and sulfate metabolism. *Plant Physiol* 132:557–567.

Welander, M. and L. H. Zhu. 2006. Rol genes: Molecular biology, physiology, morphology, breeding uses. In J. Janick, ed. *Plant Breed Rev* 26:79–97.

Weigel, D., and E. M. Meyerowitz. 1993. Activation of floral homeotic genes in Arabidopsis. *Science* 261:1723–1726.

Yadav, J. S., E. Ogwok, H. Wagaba, B. L. Patil, B. Bagewadi, T. Alicai, E. Gaitan-Solis, N. J. Taylor *et al*. 2011. RNAi-mediated resistance to cassava brown streak Uganda virus in transgenic cassava. *Mol Plant Path* 12:677–687.

Zhang, B., C. Tong, T. Yin, X. Zhang, Q. Zhuge, M. Huang, M. Wang, and R. Wu. 2009. Detection of QTL influencing growth trajectories of adventitious roots in *Populus* using functional mapping. *Tree Genet Genomes* 5:539–552.

Index

Page numbers followed by f and t indicate figures and tables, respectively.

C

D

E

H

I

J

K

L

M

N

O

S

T

U

V

W